Edited by
João F. Mano

**Biomimetic Approaches for
Biomaterials Development**

Related Titles

Pompe, W., Rödel, G., Weiss, H.-J., Mertig, M.

Bio-Nanomaterials

Designing Materials Inspired by Nature

2013
ISBN: 978-3-527-41015-6

Santin, M., Phillips, G. J. (eds.)

Biomimetic, Bioresponsive, and Bioactive Materials

An Introduction to Integrating Materials with Tissues

2012
ISBN: 978-0-470-05671-4

Li, J., He, Q., Yan, X.

Molecular Assembly of Biomimetic Systems

2011
ISBN: 978-3-527-32542-9

Kumar, C. S. S. R. (ed.)

Biomimetic and Bioinspired Nanomaterials

2010
ISBN: 978-3-527-32167-4

Behrens, P., Bäuerlein, E. (eds.)

Handbook of Biomineralization

Biomimetic and Bioinspired Chemistry

2007
ISBN: 978-3-527-31805-6

Poupon, E., Nay, B. (eds.)

Biomimetic Organic Synthesis

2011
ISBN: 978-3-527-32580-1

Edited by João F. Mano

Biomimetic Approaches for Biomaterials Development

WILEY-VCH

WILEY-VCH Verlag GmbH & Co. KGaA

The Editor

Prof. João F. Mano
University of Minho
3B's Research Group
Ave Park
4806-909 Caldas das Taipas
Guimarães
Portugal

All books published by **Wiley-VCH** are carefully produced. Nevertheless, authors, editors, and publisher do not warrant the information contained in these books, including this book, to be free of errors. Readers are advised to keep in mind that statements, data, illustrations, procedural details or other items may inadvertently be inaccurate.

Library of Congress Card No.: applied for

British Library Cataloguing-in-Publication Data
A catalogue record for this book is available from the British Library.

Bibliographic information published by the Deutsche Nationalbibliothek
The Deutsche Nationalbibliothek lists this publication in the Deutsche Nationalbibliografie; detailed bibliographic data are available on the Internet at <http://dnb.d-nb.de>.

© 2012 Wiley-VCH Verlag & Co. KGaA, Boschstr. 12, 69469 Weinheim, Germany

All rights reserved (including those of translation into other languages). No part of this book may be reproduced in any form—by photoprinting, microfilm, or any other means—nor transmitted or translated into a machine language without written permission from the publishers. Registered names, trademarks, etc. used in this book, even when not specifically marked as such, are not to be considered unprotected by law.

Composition Laserwords Private Ltd., Chennai
Printing and Binding Markono Print Media Pte Ltd, Singapore
Cover Design Formgeber, Eppelheim

Print ISBN: 978-3-527-32916-8
ePDF ISBN: 978-3-527-65230-3
ePub ISBN: 978-3-527-65229-7
mobi ISBN: 978-3-527-65228-0
oBook ISBN: 978-3-527-65227-3

Printed in Singapore
Printed on acid-free paper

Contents

Preface *XVII*
List of Contributors *XXI*

Part I Examples of Natural and Nature-Inspired Materials *1*

1   Biomaterials from Marine-Origin Biopolymers *3*
        Tiago H. Silva, Ana R.C. Duarte, Joana Moreira-Silva, João F. Mano, and Rui L. Reis
1.1   Taking Inspiration from the Sea *3*
1.2   Marine-Origin Biopolymers *6*
1.2.1   Chitosan *6*
1.2.2   Alginate *8*
1.2.3   Carrageenan *9*
1.2.4   Collagen *9*
1.2.5   Hyaluronic Acid *10*
1.2.6   Others *11*
1.3   Marine-Based Tissue Engineering Approaches *12*
1.3.1   Membranes *12*
1.3.2   Hydrogels *13*
1.3.3   Tridimensional Porous Structures *15*
1.3.4   Particles *17*
1.4   Conclusions *18*
        References *18*

2   Hydrogels from Protein Engineering *25*
        Midori Greenwood-Goodwin and Sarah C. Heilshorn
2.1   Introduction *25*
2.2   Principles of Protein Engineering *26*
2.2.1   Protein Structure and Folding *26*
2.2.2   Design of Protein-Engineered Hydrogels *28*
2.2.3   Production of Protein-Engineered Hydrogels *30*

2.3	Structural Diversity and Applications of Protein-Engineered Hydrogels	32
2.3.1	Self-Assembled Protein Hydrogels	32
2.3.2	Covalently Cross-Linked Protein Hydrogels	38
2.4	Development of Biomimetic Protein-Engineered Hydrogels for Tissue Engineering Applications	39
2.4.1	Mechanical Properties Mediate Cellular Response	40
2.4.2	Biodegradable Hydrogels for Cell Invasion	41
2.4.3	Diverse Biochemical Cues Regulate Complex Cell Behaviors	43
2.4.3.1	Cell–Extracellular Matrix Binding Domains	43
2.4.3.2	Nanoscale Patterning of Cell–Extracellular Matrix Binding Domains	44
2.4.3.3	Cell–Cell Binding Domains	45
2.4.3.4	Delivery of Soluble Cell Signaling Molecules	46
2.5	Conclusions and Future Perspective	48
	References	49
3	**Collagen-Based Biomaterials for Regenerative Medicine**	**55**
	Christophe Helary and Abhay Pandit	
3.1	Introduction	55
3.2	Collagens *In Vivo*	56
3.2.1	Collagen Structure	56
3.2.2	Collagen Fibrillogenesis	56
3.2.3	Three-Dimensional Networks of Collagen in Connective Tissues	57
3.2.4	Interactions of Cells with Collagen	57
3.3	Collagen *In Vitro*	59
3.4	Collagen Hydrogels	59
3.4.1	Collagen I Hydrogels	59
3.4.1.1	Classical Hydrogels	59
3.4.1.2	Concentrated Collagen Hydrogels	61
3.4.1.3	Dense Collagen Hydrogels Obtained by Plastic Compression	61
3.4.1.4	Dense Collagen Matrices	61
3.4.2	Cross-Linked Collagen I Hydrogels	62
3.4.2.1	Chemical Cross-Linking	62
3.4.2.2	Enzymatic Cross-Linking	62
3.4.3	Collagen II Hydrogels	63
3.4.4	Aligned Hydrogels and Extruded Fibers	64
3.4.4.1	Aligned Hydrogels	64
3.4.4.2	Extruded Collagen Fibers	65
3.5	Collagen Sponges	65
3.6	Multichannel Collagen Scaffolds	66
3.6.1	Multichannel Collagen Conduits	66
3.6.2	Multi-Channeled Collagen–Calcium Phosphate Scaffolds	66
3.7	What Tissues Do Collagen Biomaterials Mimic? (see Table 3.1)	66
3.7.1	Skin	66

3.7.2	Nerves	68
3.7.3	Tendons	68
3.7.4	Bone	69
3.7.5	Intervertebral Disk	69
3.7.6	Cartilage	70
3.8	Concluding Remarks	70
	Acknowledgments	71
	References	71

4 Silk-Based Biomaterials 75

Sílvia Gomes, Isabel B. Leonor, João F. Mano, Rui L. Reis, and David L. Kaplan

4.1	Introduction	75
4.2	Silk Proteins	76
4.2.1	*Bombyx mori* Silk	76
4.2.2	Spider Silk	77
4.2.3	Recombinant Silk	79
4.3	Mechanical Properties	82
4.4	Biomedical Applications of Silk	84
4.5	Final Remarks	87
	References	88

5 Elastin-like Macromolecules 93

Rui R. Costa, Laura Martín, João F. Mano, and José C. Rodríguez-Cabello

5.1	General Introduction	93
5.2	Materials Engineering – an Overview on Synthetic and Natural Biomaterials	94
5.3	Elastin as a Source of Inspiration for Nature-Inspired Polymers	94
5.3.1	Genetic Coding	94
5.3.2	Characteristics of Elastin	95
5.3.3	Elastin Disorders	97
5.3.4	Current Applications of Elastin as Biomaterials	97
5.3.4.1	Skin	97
5.3.4.2	Vascular Constructs	98
5.4	Nature-Inspired Biosynthetic Elastins	99
5.4.1	General Properties of Elastin-like Recombinamers	99
5.4.2	The Principle of Genetic Engineering – a Powerful Tool for Engineering Materials	100
5.4.3	From Genetic Construction to Molecules with Tailored Biofunctionality	102
5.4.4	Biocompatibility of ELRs	103
5.5	ELRs as Advanced Materials for Biomedical Applications	103
5.5.1	Tissue Engineering	104
5.5.2	Drug and Gene Delivery	106
5.5.3	Surface Engineering	108

5.6	Conclusions	*110*
	Acknowledgements	*110*
	References	*111*

6 Biomimetic Molecular Recognition Elements for Chemical Sensing *117*

Justyn Jaworski

6.1	Introduction	*117*
6.1.1	Overview	*117*
6.1.2	Biological Chemoreception	*118*
6.1.3	Host–Guest Interactions	*119*
6.1.3.1	Lock and Key	*119*
6.1.3.2	Induced Fit	*120*
6.1.3.3	Preexisting Equilibrium Model	*121*
6.1.4	Biomimetic Surfaces for Molecular Recognition	*121*
6.2	Theory of Molecular Recognition	*123*
6.2.1	Foundation of Molecular Recognition	*123*
6.2.2	Noncovalent Interactions	*123*
6.2.3	Thermodynamics of the Molecular Recognition Event	*125*
6.2.4	Putting a Figure of Merit on Molecular Recognition	*127*
6.2.5	Multiple Interactions: Avidity and Cooperativity	*128*
6.3	Molecularly Imprinted Polymers	*129*
6.3.1	A Brief History of Molecular Imprinting	*129*
6.3.2	Strategies for the Formation of Molecularly Imprinted Polymers	*129*
6.3.3	Polymer Matrix Design	*130*
6.3.4	Cross-Linking and Polymerization Approaches	*131*
6.3.5	Template Extraction	*132*
6.3.6	Limitations and Areas for Improvement	*133*
6.4	Supramolecular Chemistry	*134*
6.4.1	Introduction	*134*
6.4.2	Macrocyclic Effect	*134*
6.4.3	Chelate Effect	*135*
6.4.4	Preorganization, Rational Design, and Modeling	*135*
6.4.5	Templating Effect	*136*
6.4.6	Effective Supramolecular Receptors for Biomimetic Sensing	*137*
6.4.6.1	Calixarenes	*137*
6.4.6.2	Metalloporphyrins	*138*
6.4.7	Recent Improvement	*139*
6.5	Biomolecular Materials	*140*
6.5.1	Introduction	*140*
6.5.2	Native Biomolecules	*141*
6.5.2.1	Polypeptides	*141*
6.5.2.2	Carbohydrates	*142*
6.5.2.3	Oligonucleotides	*143*
6.5.3	Engineered Biomolecules	*144*

6.5.3.1	*In vitro* Selection of RNA/DNA Aptamers	*144*
6.5.3.2	Evolutionary Screened Peptides	*146*
6.5.3.3	Computational and Rational Design of Biomimetic Receptors	*150*
6.6	Summary and Future of Biomimetic-Sensor-Coating Materials	*151*
	References	*152*

Part II Surface Aspects *157*

7 Biology Lessons for Engineering Surfaces for Controlling Cell–Material Adhesion *159*
Ted T. Lee and Andrés J. García

7.1	Introduction	*159*
7.2	The Extracellular Matrix	*159*
7.3	Protein Structure	*160*
7.4	Basics of Protein Adsorption	*161*
7.5	Kinetics of Protein Adsorption	*162*
7.6	Cell Communication	*164*
7.6.1	Intracellular Communication	*164*
7.6.2	Intercellular Communication	*165*
7.7	Cell Adhesion Background	*166*
7.8	Integrins and Adhesive Force Generation Overview	*167*
7.9	Adhesive Interactions in Cell, and Host Responses to Biomaterials	*170*
7.10	Model Systems for Controlling Integrin-Mediated Cell Adhesion	*170*
7.11	Self-Assembling Monolayers (SAMs)	*171*
7.12	Real-World Materials for Medical Applications	*172*
7.12.1	Polymer Brush Systems	*172*
7.12.2	Hydrogels	*173*
7.13	Bio-Inspired, Adhesive Materials: New Routes to Promote Tissue Repair and Regeneration	*174*
7.14	Dynamic Biomaterials	*176*
7.14.1	Nonspecific "On" Switches	*176*
7.14.1.1	Electrochemical Desorption	*176*
7.14.1.2	Oxidative Release	*177*
7.14.2	Photobased Desorption	*178*
7.14.3	Integrin Specific "On" Switching	*178*
7.14.3.1	Photoactivation	*178*
7.14.4	Adhesion "Off" Switching	*179*
7.14.4.1	Electrochemical Off Switching	*179*
7.14.5	Reversible Adhesion Switches	*181*
7.14.5.1	Reversible Photoactive Switching	*181*
7.14.5.2	Reversible Temperature-Based Switching	*182*
7.14.6	Conclusions and Future Prospects	*184*
	References	*185*

8	**Fibronectin Fibrillogenesis at the Cell–Material Interface** *189*	
	Marco Cantini, Patricia Rico, and Manuel Salmerón-Sánchez	
8.1	Introduction *189*	
8.2	Cell-Driven Fibronectin Fibrillogenesis *189*	
8.2.1	Fibronectin Structure *190*	
8.2.2	Essential Domains for FN Assembly *192*	
8.2.3	FN Fibrillogenesis and Regulation of Matrix Assembly *194*	
8.3	Cell-Free Assembly of Fibronectin Fibrils *195*	
8.4	Material-Driven Fibronectin Fibrillogenesis *202*	
8.4.1	Physiological Organization of Fibronectin at the Material Interface *203*	
8.4.2	Biological Activity of the Material-Driven Fibronectin Fibrillogenesis *206*	
	References *210*	
9	**Nanoscale Control of Cell Behavior on Biointerfaces** *213*	
	E. Ada Cavalcanti-Adam and Dimitris Missirlis	
9.1	Nanoscale Cues in Cell Environment *213*	
9.2	Biomimetics of Cell Environment Using Interfaces *216*	
9.2.1	Surface Patterning Techniques at the Nanoscale *216*	
9.2.1.1	Surface Patterning by Nonconventional Nanolithography *216*	
9.2.1.2	Block Copolymer Micelle Lithography *217*	
9.2.2	Variation of Surface Physical Parameters at the Nanoscale *219*	
9.2.2.1	Surface Nanotopography *220*	
9.2.2.2	Interligand Spacing *221*	
9.2.2.3	Ligand Density *222*	
9.2.2.4	Substrate Mechanical Properties *223*	
9.2.2.5	Dimensionality *223*	
9.2.3	Surface Functionalization for Controlled Presentation of ECM Molecules to Cells *224*	
9.2.3.1	Proteins, Protein Fragments, and Peptides *224*	
9.2.3.2	Linking Systems *226*	
9.2.3.3	Modulation of Substrate Background *227*	
9.3	Cell Responses to Nanostructured Materials *227*	
9.3.1	Cell Adhesion and Migration *228*	
9.3.2	Cell–Cell Interactions *230*	
9.3.3	Cell Membrane Receptor Signaling *231*	
9.3.4	Applications of Nanostructures in Stem Cell Biology *232*	
9.4	The Road Ahead *233*	
	References *234*	
10	**Surfaces with Extreme Wettability Ranges for Biomedical Applications** *237*	
	Wenlong Song, Natália M. Alves, and João F. Mano	
10.1	Superhydrophobic Surfaces in Nature *237*	

10.2	Theory of Surface Wettability	*239*
10.2.1	Young's Model	*239*
10.2.2	Wenzel's Model	*240*
10.2.3	The Cassie–Baxter Model	*240*
10.2.4	Transition between the Cassie–Baxter and Wenzel Models	*240*
10.3	Fabrication of Extreme Water-Repellent Surfaces Inspired by Nature	*241*
10.3.1	Superhydrophobic Surfaces Inspired by the Lotus Leaf	*241*
10.3.2	Superhydrophobic Surfaces Inspired by the Legs of the Water Strider	*243*
10.3.3	Superhydrophobic Surfaces Inspired by the Anisotropic Superhydrophobic Surfaces in Nature	*244*
10.3.4	Other Superhydrophobic Surfaces	*245*
10.4	Applications of Surfaces with Extreme Wettability Ranges in the Biomedical Field	*245*
10.4.1	Cell Interactions with Surfaces with Extreme Wettability Ranges	*246*
10.4.2	Protein Interactions with Surfaces with Extreme Wettability Ranges	*249*
10.4.3	Blood Interactions with Surfaces with Extreme Wettability Ranges	*251*
10.4.4	High-Throughput Chips Based on Surfaces with Extreme Wettability Ranges	*252*
10.4.5	Substrates for Preparing Hydrogel and Polymeric Particles	*254*
10.5	Conclusions	*254*
	References	*255*
11	**Bio-Inspired Reversible Adhesives for Dry and Wet Conditions**	**259**
	Aránzazu del Campo and Juan Pedro Fernández-Blázquez	
11.1	Introduction	*259*
11.2	Gecko-Like Dry Adhesives	*260*
11.2.1	Fibrils with 3D Contact Shapes	*262*
11.2.2	Tilted Structures	*263*
11.2.3	Hierarchical Structures	*265*
11.2.4	Responsive Adhesion Patterns	*265*
11.3	Bioinspired Adhesives for Wet Conditions	*268*
11.4	The Future of Bio-Inspired Reversible Adhesives	*270*
	Acknowledgments	*270*
	References	*270*
12	**Lessons from Sea Organisms to Produce New Biomedical Adhesives**	**273**
	Elise Hennebert, Pierre Becker, and Patrick Flammang	
12.1	Introduction	*273*
12.2	Composition of Natural Adhesives	*274*
12.2.1	Mussels	*274*

12.2.2	Tube Worms	278
12.2.3	Barnacles	279
12.2.4	Brown Algae	280
12.3	Recombinant Adhesive Proteins	281
12.3.1	Production	281
12.3.2	Applications	283
12.4	Production of Bio-Inspired Synthetic Adhesive Polymers	284
12.4.1	Adhesives Based on Synthetic Peptides	285
12.4.2	Adhesives Based on Polysaccharides	285
12.4.3	Adhesives Based on Other Polymers	286
12.5	Perspectives	288
	Acknowledgments	288
	References	288

Part III   Hard and Mineralized Systems   *293*

13   Interfacial Forces and Interfaces in Hard Biomaterial Mechanics   *295*
Devendra K. Dubey and Vikas Tomar

13.1	Introduction	295
13.2	Hard Biological Materials	298
13.2.1	Role of Interfaces in Hard Biomaterial Mechanics	299
13.2.2	Modeling of TC–HAP and Generic Polymer–Ceramic-Type Nanocomposites at Fundamental Length Scales	301
13.2.2.1	Analytical Modeling	302
13.2.2.2	Atomistic Modeling	304
13.3	Bioengineering and Biomimetics	306
13.4	Summary	308
	References	309

14   Nacre-Inspired Biomaterials   *313*
Gisela M. Luz and João F. Mano

14.1	Introduction	313
14.2	Structure of Nacre	316
14.3	Why Is Nacre So Strong?	318
14.4	Strategies to Produce Nacre-Inspired Biomaterials	320
14.4.1	Covalent Self-Assembly or Bottom-Up Approach	320
14.4.2	Electrophoretic Deposition	322
14.4.3	Layer-by-Layer and Spin-Coating Methodologies	323
14.4.4	Template Inhibition	325
14.4.5	Freeze-Casting	326
14.4.6	Other Methodologies	326
14.5	Conclusions	328
	Acknowledgements	329
	References	329

15	**Surfaces Inducing Biomineralization** *333*
	Natália M. Alves, Isabel B. Leonor, Helena S. Azevedo, Rui. L. Reis, and João. F. Mano
15.1	Mineralized Structures in Nature: the Example of Bone *333*
15.2	Learning from Nature to the Research Laboratory *336*
15.2.1	Bioactive Ceramics and Their Bone-Bonding Mechanism *337*
15.2.2	Is a Functional Group Enough to Render Biomaterials Self-Mineralizable? *338*
15.2.2.1	How the Surface Charge of Functional Group Can Be Correlated to Apatite Formation? *338*
15.2.2.2	Designing a Properly Functionalized Surface *339*
15.3	Smart Mineralizing Surfaces *343*
15.4	*In Situ* Self-Assembly on Implant Surfaces to Direct Mineralization *345*
15.5	Conclusions *348*
	Acknowledgments *348*
	References *348*

16	**Bioactive Nanocomposites Containing Silicate Phases for Bone Replacement and Regeneration** *353*
	Melek Erol, Jasmin Hum, and Aldo R. Boccaccini
16.1	Introduction *353*
16.2	Nanostructure and Nanofeatures of the Bone *354*
16.2.1	The Structure of Bone as a Nanocomposite *354*
16.2.2	Cell Behavior at the Nanoscale *356*
16.3	Nanocomposites-Containing Silicate Nanophases *356*
16.3.1	Nanoscale Bioactive Glasses *356*
16.3.1.1	Synthetic Polymer/Nanoparticulate Bioactive Glass Composites *357*
16.3.1.2	Natural Polymer/Bioactive Glass Nanocomposites *360*
16.3.2	Nanoscaled Silica *363*
16.3.2.1	Composites Containing Silica Nanoparticles *364*
16.3.3	Nanoclays *365*
16.3.3.1	Composites Containing Clay Nanoparticles *366*
16.4	Final Considerations *372*
	References *375*

Part IV	**Systems for the Delivery of Bioactive Agents** *381*

17	**Biomimetic Nanostructured Apatitic Matrices for Drug Delivery** *383*
	Norberto Roveri and Michele Iafisco
17.1	Introduction *383*
17.2	Biomimetic Apatite Nanocrystals *384*
17.2.1	Properties *384*
17.2.2	Synthesis *386*

17.3	Biomedical Applications of Biomimetic Nanostructured Apatites	*390*
17.4	Biomimetic Nanostructured Apatite as Drug Delivery System	*394*
17.4.1	Adsorption and Release of Drugs	*397*
17.5	Adsorption and Release of Proteins	*402*
17.5.1	Adsorption and Release of Bisphosphonates	*406*
17.6	Conclusions and Perspectives	*409*
	Acknowledgments	*411*
	References	*411*

18 Nanostructures and Nanostructured Networks for Smart Drug Delivery *417*

Carmen Alvarez-Lorenzo, Ana M. Puga, and Angel Concheiro

18.1	Introduction	*417*
18.2	Stimuli-Sensitive Materials	*419*
18.2.1	pH	*419*
18.2.2	Glutathione	*420*
18.2.3	Molecule-Responsive and Imprinted Systems	*420*
18.2.4	Temperature	*422*
18.2.5	Light	*423*
18.2.6	Electrical Field	*425*
18.2.7	Magnetic Field	*426*
18.2.8	Ultrasounds	*427*
18.2.9	Autonomous Responsiveness	*428*
18.3	Stimuli-Responsive Nanostructures and Nanostructured Networks	*428*
18.3.1	Self-Assembled Polymers: Micelles and Polymersomes	*429*
18.3.2	Treelike Polymers: Dendrimers	*433*
18.3.3	Layer-by-Layer Assembly of Preformed Polymers	*436*
18.3.4	Polymeric Particles from Preformed Polymers	*438*
18.3.5	Polymeric Particles from Monomers	*439*
18.3.6	Chemically Cross-Linked Hydrogels	*444*
18.3.7	Grafting onto Medical Devices	*447*
18.4	Concluding Remarks	*449*
	Acknowledgments	*449*
	References	*450*

19 Progress in Dendrimer-Based Nanocarriers *459*

Joaquim M. Oliveira, João F. Mano, and Rui L. Reis

19.1	Fundamentals	*459*
19.2	Applications of Dendrimer-Based Polymers	*460*
19.2.1	Biomimetic/Bioinspired Materials	*460*
19.2.2	Drug Delivery Systems	*461*
19.2.3	Gene Delivery Systems	*463*
19.2.4	Biosensors	*465*

19.2.5	Theranostics *466*	
19.3	Final Remarks *467*	
	References *467*	

Part V Lessons from Nature in Regenerative Medicine *471*

20 Tissue Analogs by the Assembly of Engineered Hydrogel Blocks *473*
Shilpa Sant, Daniela F. Coutinho, Nasser Sadr, Rui L. Reis, and Ali Khademhosseini

20.1	Introduction *473*
20.2	Tissue/Organ Heterogeneity *In Vivo* *474*
20.3	Hydrogel Engineering for Obtaining Biologically Inspired Structures *477*
20.3.1	Structural Cues *477*
20.3.2	Mechanical Cues *478*
20.3.3	Biochemical Cues *480*
20.3.4	Cell–Cell Contact *482*
20.3.5	Combination of Multiple Cues *483*
20.4	Assembly of Engineered Hydrogel Blocks *485*
20.5	Conclusions *488*
	Acknowledgments *489*
	References *489*

21 Injectable *In-Situ*-Forming Scaffolds for Tissue Engineering *495*
Da Yeon Kim, Jae Ho Kim, Byoung Hyun Min, and Moon Suk Kim

21.1	Introduction *495*
21.2	Injectable *In-Situ*-Forming Scaffolds Formed by Electrostatic Interactions *496*
21.3	Injectable *In-Situ*-Forming Scaffolds Formed by Hydrophobic Interactions *497*
21.4	Immune Response of Injectable *In-Situ*-Forming Scaffolds *500*
21.5	Injectable *In-Situ*-Forming Scaffolds for Preclinical Regenerative Medicine *500*
21.6	Conclusions and Outlook *501*
	References *502*

22 Biomimetic Hydrogels for Regenerative Medicine *503*
Iris Mironi-Harpaz, Olga Kossover, Eran Ivanir, and Dror Seliktar

22.1	Introduction *503*
22.2	Natural and Synthetic Hydrogels *503*
22.3	Hydrogel Properties *505*
22.4	Engineering Strategies for Hydrogel Development *506*
22.5	Applications in Biomedicine *508*
	References *511*

23	**Bio-inspired 3D Environments for Cartilage Engineering** *515*	
	José Luis Gómez Ribelles	
23.1	Articular Cartilage Histology *515*	
23.2	Spontaneous and Forced Regeneration in Articular Cartilage *517*	
23.3	What Can Tissue Engineering Do for Articular Cartilage Regeneration? *517*	
23.4	Cell Sources for Cartilage Engineering *519*	
23.4.1	Bone Marrow Mesenchymal Cells Reaching the Cartilage Defect from Subchondral Bone *519*	
23.4.2	Autologous Mesenchymal Stem Cells from Different Sources *520*	
23.4.3	Mature Autologous Chondrocytes *521*	
23.5	The Role and Requirements of the Scaffolding Material *524*	
23.5.1	Gels Encapsulating Cells as Vehicles for Cell Transplant *524*	
23.5.2	Macroporous Scaffolds: Pore Architecture *524*	
23.5.3	Cell Adhesion Properties of the Scaffold Surfaces *525*	
23.5.4	Mechanical Properties *525*	
23.5.5	Can Scaffold Architecture Direct Tissue Organization? *526*	
23.5.6	Scaffold Biodegradation Rate *527*	
23.6	Growth Factor Delivery *In Vivo* *528*	
23.7	Conclusions *528*	
	Acknowledgment *529*	
	References *529*	
24	**Soft Constructs for Skin Tissue Engineering** *537*	
	Simone S. Silva, João F. Mano, and Rui L. Reis	
24.1	Introduction *537*	
24.2	Structure of Skin *537*	
24.2.1	Wound Healing *538*	
24.3	Current Biomaterials in Wound Healing *539*	
24.3.1	Alginate *539*	
24.3.2	Cellulose *540*	
24.3.3	Chitin/Chitosan *541*	
24.3.4	Hyaluronic Acid *543*	
24.3.5	Collagen and Other Proteins *544*	
24.3.6	Synthetic Polymers *545*	
24.4	Wound Dressings and Their Properties *545*	
24.5	Biomimetic Approaches in Skin Tissue Engineering *546*	
24.5.1	Commercially Available Skin Products *549*	
24.6	Final Remarks *549*	
	Acknowledgments *552*	
	List of Abbreviations *552*	
	References *553*	

Index *559*

Preface

Biomimetics is a rather recent multidisciplinary field that uses Nature as a model of how to conceive materials, structures, processes, systems, and strategies to solve real problems. A strong motivation for employing such paradigm is that in over 3.8 billion years of evolution Nature has introduced highly effective and power-efficient biological mechanisms, offering astonishing examples for innovation inspiration. The successful translation of these lessons into the technical world has been facilitated in the past years by the latest developments in engineering, basic sciences, nanotechnology, and biology, which have allowed to reach levels of development for effectively copying and adapting biological methods, processes, and designs.

Biomimetic approaches have been particularly attractive in the development of new materials, as natural materials exhibit unique and useful characteristics, including self-assembly and structural hierarchical organization, multifunctionality, functional and environmental adaptability, exclusive and elevated properties, and self-healing capability. Consulting Nature and applying such primary design principles can effectively cause a shift from the traditional strategies that have been used so far toward a completely new mind-set of approaching modern materials science and technology.

For example, the introduction of materials in the biomedical field started a few decades ago with the use of conventional metals, ceramics, and synthetic polymers that were adapted for several clinical needs. Although many established implantable medical devices are still immensely important, the improvement of their performance or the use of biomaterials to address other medical needs require more sophisticated approaches. The modern concepts of regenerative medicine, for instance, require a completely new perspective on how to work with materials, in order to develop devices that could re-create, to some extent, the natural process of tissue and organ formation and healing. Such enormous endeavor should be only possible if one realizes the complexity of the processes taking place during such regenerative processes and adopts an open-minded attitude to use more radical and nonevident solutions. The aim of this book is to demonstrate that nature-inspired ideas may provide nonconventional and innovative methodologies that could be used in the development of biomaterials and medical devices, not only to be implanted in the human body but also to be used *ex vivo* (for example, in diagnostic platforms or in substrates for processing biomaterials and cells). This book contains four main sections, combining general biomimetic-based strategies that could be useful in the biomedical field with some case studies of applications.

The first part of the book highlights some examples of nature-based or nature-inspired materials, focusing on macromolecules, with potential biomedical applicability. First, the sea is addressed as a tremendous potential source of biomaterials. Then, several chapters cover protein-based systems, starting with the general topic of protein-based hydrogels, followed by collagen-based biomaterials, as these proteins are the most important constituent of the extracellular matrix. Thereafter, the potential of using silk is explored, including the use of biotechnology to produce modified silk-based macromolecules with other functionalities. Biomimetic elastin-based macromolecules constitute a landmark example of how recombinant technologies permit the synthesis of high-performance stimuli-responsive biocompatible polymers, also explored in this book. In order to link this section to the next one, biomimetic molecules with the ability to be used as substrates for chemical sensing are covered as well.

The success of implantable medical devices is largely determined by the response they elicit to the surrounding biological environment. Therefore, surfaces aspects related to biomaterials have received a great deal of attention from scientists and engineers. The first chapter of the section linked to surfaces explores some fundamental aspects on cell–material interactions. To explore in more detail this, and to take into account the importance of protein adsorption in the behavior of biomaterials surfaces, the process of fibronectin fibrillogenesis is discussed. Besides the relevance of both chemical and biochemical elements that are exposed on the surface of biomaterials, it is also important to consider the influence of topography, especially at the nanoscale, on cell behavior and on other surface properties. Special topographies that are found in many examples in Nature may lead to particular peculiar behaviors, explored in two distinct chapters: biomimetic surfaces exhibiting superhydrophobic properties, which could have relevance in several biomedical applications, and bio-inspired surfaces exhibiting adhesives properties as a result of the surface topographic organization. Adhesiveness may also be the result of particular chemical characteristics of the material – again, the sea may be used as a source of inspiration to develop adhesive surfaces capable of sticking to virtually all kinds of substrates in wet (saline) conditions, thus being highly relevant to manufacture medical devices with adhesive properties toward tissues.

Most biological (natural) structural materials are composites with sophisticated microstructure and remarkable properties, many of them reinforced with a mineralized fraction having components with nanometric sizes. The chapter covering the issue of interfaces in hard biomaterials will make the liaison to the next section of the book, dealing with mineralized systems. Natural composites have been stimulating the advance of biomimetic composites with improved mechanical and osteoconductive properties, adequate to be used in orthopedic and maxillofacial applications. A case study explored in this section is related to the production of nacre-based composites, especially focusing on their biomedical potential. Strategies of obtaining biomaterials exhibiting the ability of depositing apatite are also discussed in an independent chapter, followed by another one dealing specifically on bioactive nanocomposites containing silicate phases.

The third section covers the field of systems for the delivery of bioactive agents, which applies to many biomedical applications. The first chapter of this section is still related to the previous section and covers the use of nanostructured apatite-based matrices for drug delivery. Then, biomimetic nanostructured systems and nanoparticles are discussed in the next two chapters for two distinct applications: smart systems mainly devoted to pharmaceutical applications and dendrimer-based nanocarriers especially to be used in cell and tissue engineering (the topic of the last section).

The ultimate example of employing biomimetic principles in the biomedical field is to develop methodologies that could enable the regeneration of tissues and organs. The last section of this book starts by exploring the concept of hierarchical organization of living tissues and the use of microfabricated elementary building blocks combining biomaterials and cells that could be assembled into more complex structures. The next chapter addresses the important aspect of developing injectable systems that may be used to fix cells or therapeutic molecules in specific sites in the body using minimally invasive procedures. Remarkable lessons from Nature can be used to develop biomaterials that can be degraded by the action of the cells – biomimetic hydrogels exhibiting such capability will be presented in an independent chapter. The last two chapters present two specific case studies of employing biomaterials and cells in tissue engineering strategies for the regeneration of cartilage and skin, respectively.

The collection of this set of contributions was only possible due to the superb work of all authors of this book, who have so generously shared their knowledge with us and devoted their valuable time to this project. The active and professional support from Wiley-VCH during the production of this book is also most appreciated.

João F. Mano

List of Contributors

Carmen Alvarez-Lorenzo
Universidad de Santiago de Compostela
Facultad de Farmacia
Dept. Farmacia y Tecnologia Farmaceutica
15782 Santiago de Compostela
Spain

Natália M. Alves
University of Minho
3B's Research Group—Biomaterials Biodegradables and Biomimetics
Department of Polymer Engineering
Headquarters of the European Institute of Excellence on Tissue Engineering and Regenerative Medicine
AvePark
4806-909 Caldas das Taipas
Guimarães
Portugal

and

ICVS/3B's
PT Government Associate Laboratory
Braga/Guimarães
Portugal

Helena S. Azevedo
University of Minho
3B's Research Group—Biomaterials Biodegradables and Biomimetics
Department of Polymer Engineering
Headquarters of the European Institute of Excellence on Tissue Engineering and Regenerative Medicine
AvePark
4806-909 Caldas das Taipas
Guimarães
Portugal

and

ICVS/3B's
PT Government Associate Laboratory
Braga/Guimarães
Portugal

Pierre Becker
University of Mons—UMONS
Biology of Marine Organisms and Biomimetics
20 Place du Parc
7000 Mons
Belgium

Aldo R. Boccaccini
University of
Erlangen-Nuremberg
Department of Materials Science
and Engineering
Institute of Biomaterials
Cauerstraße 6
91058 Erlangen
Germany

Marco Cantini
Universitat Politècnica de
València
Center for Biomaterials and
Tissue Engineering
Camino de Vera s/n
46022 Valencia
Spain

E. Ada Cavalcanti-Adam
University of Heidelberg
Department of Biophysical
Chemistry
Institute for Physical Chemistry
Im Neuenheimer Feld 253
69120 Heidelberg
Germany

and

Max Planck Institute for
Intelligent Systems
Department of New Materials and
Biosystems
Heisenbergstr. 3
70569 Stuttgart
Germany

Angel Concheiro
Universidad de Santiago de
Compostela
Facultad de Farmacia
Dept. Farmacia y Tecnologia
Farmaceutica
15782 Santiago de Compostela
Spain

Rui R. Costa
University of Minho
3B's Research
Group—Biomaterials
Biodegradables and Biomimetics
Headquarters of the European
Institute of Excellence on
Tissue Engineering and
Regenerative Medicine
AvePark
4806-909 Caldas das Taipas
Guimarães
Portugal

and

ICVS/3B's
PT Government Associate
Laboratory
Braga/Guimarães
Portugal

Daniela F. Coutinho
Center for Biomedical
Engineering
Department of Medicine
Brigham and Women's Hospital
Harvard Medical School
65, Landsdowne street
Cambridge, MA 02139
USA

and

Harvard-MIT dicision of Health
Science and Technology
Massachussetts Institute of
Technology
65 Landsdowne Street
Cambridge, MA 02139
USA

and

University of Minho
3B's Research
Group—Biomaterials
Biodegradables and Biomimetics
Department of Polymer
Engineering
Headquarters of the European
Institute of Excellence on
Tissue Engineering and
Regenerative Medicine
AvePark
4806-909 Caldas das Taipas
Guimarães
Portugal

and

ICVS/3B's
PT Government Associate
Laboratory
Braga/Guimarães
Portugal

Aránzazu del Campo
Max-Planck-Institut für
Polymerforschung
Minerva Group
Ackermannweg 10
55128 Mainz
Germany

Ana R.C. Duarte
University of Minho
3B's Research
Group—Biomaterials
Biodegradables and Biomimetics
Department of Polymer
Engineering
Headquarters of the European
Institute of Excellence on
Tissue Engineering and
Regenerative Medicine
AvePark
4806-909 Caldas das Taipas
Guimarães
Portugal

and

ICVS/3B's
PT Government Associate
Laboratory
Braga/Guimarães
Portugal

Devendra K. Dubey
Purdue University
School of Aeronautics and
Astronautics
West Lafayette, IN 47907
USA

Melek Erol
Istanbul Technical University
Department of Chemical
Engineering
Istanbul Technical University
Maslak
34469 Istanbul
Turkey

Juan Pedro Fernández-Blázquez
Max-Planck-Institut für Polymerforschung
Minerva Group
Ackermannweg 10
55128 Mainz
Germany

Patrick Flammang
University of Mons—UMONS
Biology of Marine Organisms and Biomimetics
20 Place du Parc
7000 Mons
Belgium

Andrés J. García
Woodruff School of Mechanical Engineering
Petit Institute for Bioengineering and Bioscience
Georgia Institute of Technology
315 Ferst Drive
Atlanta, GA 30332-0363
USA

Sílvia Gomes
University of Minho
3B's Research Group—Biomaterials Biodegradables and Biomimetics
Department of Polymer Engineering
Headquarters of the European Institute of Excellence on Tissue Engineering and Regenerative Medicine
AvePark
4806-909 Caldas das Taipas
Guimarães
Portugal

and

ICVS/3B's
PT Government Associate Laboratory
Braga/Guimarães
Portugal

and

Tufts University
Department of Biomedical Engineering
4 Colby Street
Medford, MA 02155
USA

Midori Greenwood-Goodwin
Stanford University
Bioengineering
318 Campus Drive
Stanford, CA 94305-5444
USA

Sarah C. Heilshorn
Stanford University
Materials Science and Engineering
476 Lomita Mall
Stanford, CA 94305-4045
USA

Christophe Helary
National University of Ireland
Network of Excellence for Functional Biomaterials (NFB)
IDA Business Park
Newcastle Road
Galway
Ireland

Elise Hennebert
University of Mons—UMONS
Biology of Marine Organisms and
Biomimetics
20 Place du Parc
7000 Mons
Belgium

Jasmin Hum
University of
Erlangen-Nuremberg
Department of Materials Science
and Engineering
Institute of Biomaterials
Cauerstraße 6
91058 Erlangen
Germany

Michele Iafisco
Alma Mater Studiorum
Università di Bologna
Dipartimento di Chimica
"G. Ciamician"
Via Selmi 2
40126 Bologna
Italy

Eran Ivanir
Department of Biomedical
Engineering
Technion—Israel Institute of
Technology
Technion City
Haifa 32000
Israel

Justyn Jaworski
Hanyang University
Department of Chemical
Engineering
222 Wangsimni-ro
Seongdong-gu
Seoul 133-791
South Korea

David L. Kaplan
Tufts University
Department of Biomedical
Engineering
4 Colby Street
Medford, MA 02155
USA

Ali Khademhosseini
Center for Biomedical
Engineering
Department of Medicine
Brigham and Women's Hospital
Harvard Medical School
65, Landsdowne street
Cambridge, MA 02139
USA

and

Harvard-MIT dicision of Health
Science and Technology
Massachussetts Institute of
Technology
65 Landsdowne Street
Cambridge, MA 02139
USA

and

Wyss Institute for Biologically
Inspired Engineering
Harvard University
3 Blackfan circle
Boston, MA 02116
USA

Da Yeon Kim
Ajou University
Department of Molecular Science
and Technology
Suwon 443-749
South Korea

Jae Ho Kim
Ajou University
Department of Molecular Science
and Technology
Suwon 443-749
South Korea

Moon Suk Kim
Ajou University
Department of Molecular Science
and Technology
Suwon 443-749
South Korea

Olga Kossover
Department of Biomedical
Engineering
Technion—Israel Institute of
Technology
Technion City
Haifa 32000
Israel

Ted T. Lee
Woodruff School of Mechanical
Engineering
Petit Institute for Bioengineering
and Bioscience
Georgia Institute of Technology
315 Ferst Drive
Atlanta, GA 30332-0363
USA

Isabel B. Leonor
University of Minho
3B's Research
Group—Biomaterials
Biodegradables and Biomimetics
Department of Polymer
Engineering
Headquarters of the European
Institute of Excellence on
Tissue Engineering and
Regenerative Medicine
AvePark
4806-909 Caldas das Taipas
Guimarães
Portugal

and

ICVS/3B's
PT Government Associate
Laboratory
Braga/Guimarães
Portugal

Gisela M. Luz
University of Minho
3B's Research
Group—Biomaterials
Biodegradables and Biomimetics
Department of Polymer
Engineering
Headquarters of the European
Institute of Excellence on
Tissue Engineering and
Regenerative Medicine
AvePark
4806-909 Caldas das Taipas
Guimarães
Portugal

and

ICVS/3B's
PT Government Associate
Laboratory
Braga/Guimarães
Portugal

Laura Martín
University of Valladolid
G.I.R. Bioforge
Edificio I + D
Paseo de Belén, 1
47011 Valladolid
Spain

and

Biomaterials and Nanomedicine
(CIBER BBN)
Networking Research Center on
Bioengineering
Valladolid
Spain

João F. Mano
University of Minho
3B's Research
Group—Biomaterials
Biodegradables and Biomimetics
Department of Polymer
Engineering
Headquarters of the European
Institute of Excellence on
Tissue Engineering and
Regenerative Medicine
AvePark
4806-909 Caldas das Taipas
Guimarães
Portugal

and

ICVS/3B's
PT Government Associate
Laboratory
Braga/Guimarães
Portugal

Byoung Hyun Min
Ajou University
Department of Molecular Science
and Technology
Suwon 443-749
South Korea

Iris Mironi-Harpaz
Department of Biomedical
Engineering
Technion—Israel Institute of
Technology
Technion City
Haifa 32000
Israel

Dimitris Missirlis
University of Heidelberg
Department of Biophysical
Chemistry
Institute for Physical Chemistry
Im Neuenheimer Feld 253
69120 Heidelberg
Germany

and

Max Planck Institute for
Intelligent Systems
Department of New Materials and
Biosystems
Heisenbergstr. 3
70569 Stuttgart
Germany

Joana Moreira-Silva
University of Minho
3B's Research
Group—Biomaterials
Biodegradables and Biomimetics
Department of Polymer
Engineering
Headquarters of the European
Institute of Excellence on
Tissue Engineering and
Regenerative Medicine
AvePark
4806-909 Caldas das Taipas
Guimarães
Portugal

and

ICVS/3B's
PT Government Associate
Laboratory
Braga/Guimarães
Portugal

Joaquim M. Oliveira
University of Minho
3B's Research
Group—Biomaterials
Biodegradables and Biomimetics
Department of Polymer
Engineering
Headquarters of the European
Institute of Excellence on
Tissue Engineering and
Regenerative Medicine
AvePark
4806-909 Caldas das Taipas
Guimarães
Portugal

and

ICVS/3B's
PT Government Associate
Laboratory
Braga/Guimarães
Portugal

Abhay Pandit
National University of Ireland
Network of Excellence for
Functional Biomaterials (NFB)
IDA Business Park
Newcastle Road
Galway
Ireland

Ana M. Puga
Universidad de Santiago de
Compostela
Facultad de Farmacia
Dept. Farmacia y Tecnologia
Farmaceutica
15782 Santiago de Compostela
Spain

Rui L. Reis
University of Minho
3B's Research
Group—Biomaterials
Biodegradables and Biomimetics
Department of Polymer
Engineering
Headquarters of the European
Institute of Excellence on
Tissue Engineering and
Regenerative Medicine
AvePark
4806-909 Caldas das Taipas
Guimarães
Portugal

and

ICVS/3B's
PT Government Associate Laboratory
Braga/Guimarães
Portugal

José Luis Gómez Ribelles
Universitat Politècnica de València
Center for Biomaterials and Tissue Engineering
Camino de Vera s/n
46022 Valencia
Spain

and

Networking Research Center on Bioengineering
Biomaterials and Nanomedicine (CIBER-BBN)
Valencia
Spain

Patricia Rico
Universitat Politècnica de València
Center for Biomaterials and Tissue Engineering
Camino de Vera s/n
46022 Valencia
Spain

and

CIBER de Bioingeniería
Biomateriales y Nanomedicina
c/ Eduardo Primo Yúfera 3, 46012
Valencia
Spain

José C. Rodríguez-Cabello
University of Valladolid
G.I.R. Bioforge
Edificio I + D
Paseo de Belén, 1
47011 Valladolid
Spain

and

Biomaterials and Nanomedicine (CIBER BBN)
Networking Research Center on Bioengineering
Valladolid
Spain

Norberto Roveri
Alma Mater Studiorum
Università di Bologna
Dipartimento di Chimica
"G. Ciamician"
Via Selmi 2
40126 Bologna
Italy

Nasser Sadr
Harvard-MIT dicision of Health Science and Technology
Massachussetts Institute of Technology
65 Landsdowne Street
Cambridge, MA 02139
USA

and

Wyss Institute for Biologically Inspired Engineering
Harvard University
3 Blackfan circle
Boston, MA 02116
USA

and

Bioengineering Department
Politecnico Di Milano
Piazza Leonardo Da Vinci 32
20133 Milan
Italy

Manuel Salmerón-Sánchez
Universitat Politècnica de
València
Center for Biomaterials and
Tissue Engineering
Camino de Vera s/n
46022 Valencia
Spain

and

CIBER de Bioingeniería
Biomateriales y Nanomedicina
c/ Eduardo Primo Yúfera 3, 46012
Valencia
Spain

Shilpa Sant
Center for Biomedical
Engineering
Department of Medicine
Brigham and Women's Hospital
Harvard Medical School
65, Landsdowne street
Cambridge, MA 02139
USA

and

Harvard-MIT dicision of Health
Science and Technology
Massachussetts Institute of
Technology
65 Landsdowne Street
Cambridge, MA 02139
USA

and

Wyss Institute for Biologically
Inspired Engineering
Harvard University
3 Blackfan circle
Boston, MA 02116
USA

Dror Seliktar
Department of Biomedical
Engineering
Technion—Israel Institute of
Technology
Technion City
Haifa 32000
Israel

Simone S. Silva
University of Minho
3B's Research
Group—Biomaterials
Biodegradables and Biomimetics
Department of Polymer
Engineering
Headquarters of the European
Institute of Excellence on
Tissue Engineering and
Regenerative Medicine
AvePark
4806-909 Caldas das Taipas
Guimarães
Portugal

and

ICVS/3B's
PT Government Associate
Laboratory
Braga/Guimarães
Portugal

Tiago H. Silva
University of Minho
3B's Research
Group—Biomaterials
Biodegradables and Biomimetics
Department of Polymer
Engineering
Headquarters of the European
Institute of Excellence on
Tissue Engineering and
Regenerative Medicine
AvePark
4806-909 Caldas das Taipas
Guimarães
Portugal

and

ICVS/3B's
PT Government Associate
Laboratory
Braga/Guimarães
Portugal

Wenlong Song
University of Minho
3B's Research
Group—Biomaterials
Biodegradables and Biomimetics
Department of Polymer
Engineering
Headquarters of the European
Institute of Excellence on
Tissue Engineering and
Regenerative Medicine
AvePark
4806-909 Caldas das Taipas
Guimarães
Portugal

and

ICVS/3B's
PT Government Associate
Laboratory
Braga/Guimarães
Portugal

and

Jilin University
Key Lab of Supramolecular
Structure and Materials
College of Chemistry
Qianjin Street N 2699
Changchun 130023
China

Vikas Tomar
Purdue University
School of Aeronautics and
Astronautics
West Lafayette, IN 47907
USA

Part I
Examples of Natural and Nature-Inspired Materials

1
Biomaterials from Marine-Origin Biopolymers

Tiago H. Silva, Ana R.C. Duarte, Joana Moreira-Silva, João F. Mano, and Rui L. Reis

1.1
Taking Inspiration from the Sea

Nature has a chemical diversity much broader than chemical synthesis can ever approach. In fact, on the words of Marcel Jaspars, *"Some chemists, having synthesised a few compounds believe themselves to be better chemists than nature, which, in addition to synthesising compounds too numerous to mention, synthesised those chemists as well."* Marine environment is no exception and is being increasingly chosen for the extraction of several compounds, from bioactive molecules to polymers and ceramics. Together with this great potential, one can also find such interesting structures and functions exhibited by diverse marine organisms that biomimetics appears as an extremely attractive approach. Without aiming to be exhaustive, this section presents some examples of those structures and functions and the respective biomimetic approaches.

Biomimetics has been a very attractive route for human scientists and engineers, since the solutions presented by nature to the arising challenges are real engineering wonders, being examples of maximizing functionality with reduced energy and materials. Notoriously, those are precisely the problems faced by the actual engineering challenges to which nature has already given a solution, with the additional advantage of being nonpolluting, in contrast to the majority of the human-engineered solutions [1–3].

For instance, several organisms possess complex and hierarchical structures as a result of their natural growing process, based on self-assembly principles and using relatively few constituent elements [4]. An example can be found in the tree trunk, in which fiber orientation changes with tree growth to optimize the structure and shape of the material as a response to prevailing winds or other environmental constrains, revealing an adaptive mechanical design that may be explored in new engineering solutions. Another example can be found in bone, also revealing adaptive changes to combat external loads or other environmental stimuli. In fact, bone is also an example of a hierarchical biological structure that has been the object of intense research aimed at their mimetization, namely, by tissue engineering approaches [5].

Biomimetic Approaches for Biomaterials Development, First Edition. Edited by João F. Mano.
© 2012 Wiley-VCH Verlag GmbH & Co. KGaA. Published 2012 by Wiley-VCH Verlag GmbH & Co. KGaA.

Nowadays, one of the main purposes of tissue engineering is to produce artificial tissue constructs that possess similar mechanical properties and the capability to trigger specific cellular responses adequate for the tissue to be replaced, mimicking its growth and degradation [6]. In this way, developments in tissue engineering are going toward biomaterials that are competent in biomolecular recognition of tissues adjacent to where they are implanted [7]. Also, the ability of stimulating cellular responses, mimicking extracellular matrix, and guiding new tissue formation are important characteristics that should be considered when creating new biomimetic materials [8]. Such biomaterials and hierarchical structuring, with promising tissue engineering applications, can be found in marine organisms, for instance, in mollusk shells. The mollusk shells are very interesting structures, secreted by the mantle, and composed of different layers. The middle layer is formed by calcium carbonate (calcite or aragonite) and an organic matrix. Also, in some mollusks, this layer is followed by an inner smooth and iridescent surface, the nacre. The nacre is a platelet-tough brick and mortar structure formed by aragonite embedded in a protein matrix [9–11]. Owing to their strong, lightweight hierarchical structure, seashells and, more particularly, nacre have been greatly studied. An independent chapter of this book focuses on the use of nacre-based inspiration to produce high-performance biomaterials. Kamat *et al.* [12] studied shell resistance of the mollusk *Strombus gigas* to catastrophic fracture, concluding that its high resistance is due to the shell lamellar microarchitecture. Asvanund and coauthors studied the osteogenic activity of nacre from the oyster *Pinctada maxima*. In this *in vitro* study, it was shown that nacre induced an increase in the gene expression of osteogenic markers ALP (alkaline phosphatase), BSP (bone sialoprotein), and OC (osteocalcin), demonstrating that nacre is a biomaterial with the ability to stimulate human bone regeneration [13].

Processes of mimicking nacre structure for further use in tissue engineering and regeneration have been disclosed. In 2006, Deville and coauthors unveiled a simple method for the production of a material with lamellar architecture similar to that of nacre. This method is based on the lamellar structure that seawater forms when freezing. In seawater, as freezing temperature is achieved (around $-2\,^\circ$C) salts and other particles are excreted and pure water freezes in a lamellar way. In this study, a suspension of hydroxyapatite was frozen, and after being freeze-dried, a layered nacrelike structure was formed. This type of material can be used for bone tissue regeneration [14].

More recently, another method using the layer-by-layer (LbL) technique for producing nacrelike structures has been developed [15]. In this work, the selected polycation was chitosan and the anion was bioactive glass nanoparticles. By this adjustable technique, the authors were able to obtain robust coatings with architecture similar to that of nacre. These coatings can be used in different tissue engineering constructs with applications in orthopedics.

A different marine biomimetic approach can be also envisaged, not based on structural features, but on specific functions evidenced by marine organisms, such as the extremely strong and multisurface adhesion properties of mussels and the variable stiffness exhibited by sea cucumbers and other echinoderms.

Mussels are able to adhere strongly to different wet surfaces by means of their adhesive plaques. In the mussel adhesive proteins (MAPs), the amino acid 3,4-dihydroxyphenylalanine (DOPA) is found in large quantity. It has been hypothesized that DOPA contributes to bioadhesion in sea water [16]. Monahan and Wilker [17] with their study on formation of mussel's adhesives showed that Fe^{3+} enhances the cross-linking ability of mussel adhesives. In brief, iron is extracted from sea water and used in the process of connecting the MAPs together, forming the robust byssus threads. Owing to its strong structure and ability to adhere to wet surfaces, processes of mimicking MAPs for further use in tissue engineering have been developed.

Lee and coauthors [18] synthesized linear and branched DOPA-modified poly(ethylene glycol)s (PEGs-DOPAs) containing one to four DOPA end groups that were able to cross-link into hydrogels when oxidizing reagents were used.

More recently, Burke and coauthors [19] induced the formation of gel of mimicking MAP. An oxidizing agent present in the lipid vesicles with a physiological melting transition of 37 °C is released, and when used in combination with DOPA, rapid cross-linking of the hydrogel was achieved. These stimuli-responsive gels might have the potential to be used for repair of soft tissues.

Podsiadlo *et al.* [20] were able to formulate a nanostructured composite with nacrelike architecture, using DOPA to enhance adhesion and cross-linking. This type of composite might be used for bone tissue regeneration.

Sea cucumbers and sea urchins have the ability to alter stiffness. Their skin is made of collagen fibers embedded in a matrix that can provide low to high stiffness. These collagen fibers are denominated mutable collagenous tissue (MCT). MCTs are similar to mammalian connective tissues in their composition of collagen, proteoglycan, and microfibrils [21].

Muscle is an example of a variable stiffness structure, and the understanding of the mechanism of variable stiffness might be used in "artificial muscles" production. Other potential applications have been studied as pharmacological strategies and for composite materials and might be applied when knowledge of the mechanism of MCT mechanical adaptability will be completely elucidated.

A pharmacological strategy that might be used as a therapy for fibrotic lesions in the cervix is the use of holothurian glycosaminoglycans (GAGs). As these GAGs are constituents of a very mutable connective tissue, they might be able to induce the relaxation of the fibrotic tissue when delivered to the lesion [21].

Regarding composite materials, using the knowledge of MCT, different applications for connective tissue replacement and regeneration can be employed. Wilkie [21] gives some examples of some potential applications. One example is the replacement of a complete connective tissue as the Achilles tendon; although a MCT xenograft is not (yet) possible nowadays, some developments have already been done by the study of the echinoid compass depressor ligament and peristomial membrane [22, 23]. Another example is the construction of a composite with constituents extracted from MCT, such as the collagen fibrils. Trotter and coauthors [24] proposed a composite biomaterial that brought together a synthetic interfibrillar matrix with collagen fibrils isolated from holothurians. Also, composite materials might be created through a biomimetic approach, taking inspiration from sea cucumber's MCT [21].

1.2
Marine-Origin Biopolymers

Marine environment is a source of untold diversity of materials with specific biological and chemical features, some of which are not known in terrestrial organisms. For instance, macroalgae synthesize a great diversity of polysaccharides bearing sulfate groups that find no equivalent in land plants [25] and resemble GAGs found in human extracellular matrix. In spite of this extraordinary potential, the high costs and risks and the lack of technology have hindered a deeper exploration of the marine environment [26]. Nevertheless, in the past decades, new tools and technological developments have allowed to unlock some marine knowledge and in that way, to discover new marine biomaterials to join others already known for many years [26], such as agar [27, 28].

In the present section, the more representative marine biopolymers that find application in the biomedical field are discussed, in particular their chemical nature and the process of isolation from selected marine organisms. Having normally a support function in those organisms, together with other properties, these biopolymers can be considered to be further used in the development of support systems, following a biomimetic approach, for application in tissue engineering scaffolding, discussed later. Moreover, one can see in Figure 1.1 the increasing attention that these biopolymers are receiving from the scientific community, with an increasing number of papers being published in the past 10 years, and these biopolymers are also studied for their correlation with tissue engineering.

1.2.1
Chitosan

Chitosan is composed of D-glucosamine (70–90%) and N-acetyl-D-glucosamine (10–30%) units, linked by β(1-4) glycosidic bonds, corresponding to the deacetylated form of chitin, the second most abundant natural polymer, just after cellulose [29]. The difference between chitin and chitosan is determined by the deacetylation degree, that is, the ratio of deacetylated units in the polymer chain, being higher for chitosan, which renders it soluble in dilute acid solutions because of protonation of primary amine groups.

Deacetylation degree is one of the most important characteristics of chitosan, which, together with molecular weight, and also the sequence of repeating units, is responsible for such interesting physicochemical and biological properties possessed by chitosan [29, 30].

Amine groups present in the glucosamine units can be protonated in acidic solutions, as mentioned above, and a way to make it soluble in acid aqueous solutions, which allows its further processability into membranes or gels [31, 32], or

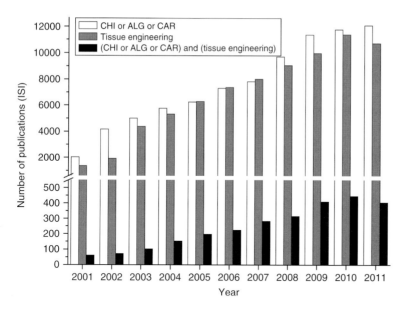

Figure 1.1 Number of published papers in the indicated year, according to the database ISI Web of Knowledge, using the terms "*chitosan* or *alginate* or *carrageenan*" as representation of marine biopolymers, "tissue engineering," and a conjugation of both.

even into particles or fibers, using adequate coagulation solutions (alkali or organic solvents) [33, 34]. Moreover, more complex structures, such as 3D porous structures, can also be obtained by using other techniques, such as freeze-drying [35, 36].

In this charged form, chitosan exhibits a polycation behavior, thus interacting with negatively charged compounds, such as GAGs on the formation of polyelectrolyte complexes [36, 37], or metal complex ions in wastewater treatment systems [38].

Basically, chitin can be isolated from raw materials by the consecutive removal of minerals (by acid treatment, such as 2.5% HCl solution [39]), proteins (by alkaline treatment, such as 2% NaOH solutions [39]), and pigments (by a solid–liquid extraction with acetone or other solvents or a mild oxidizing treatment [40–42]). Acetyl groups are then removed from chitin by treatment with concentrated alkali, thus obtaining chitosan. Two main procedures are commonly referred to in the literature [40, 43, 44] to accomplish this deacetylation: the Broussignac process, in which chitin is treated with a mixture of solid potassium hydroxide (50% w/w) in 1 : 1 96% ethanol and monoethylene glycol, and the Kurita process [45], according to which chitin is treated with hot aqueous sodium hydroxide solution (50% w/v). The reaction time and, in Kurita-based processes, the concentration of alkali and temperature of reaction are parameters that affect the deacetylation degree and concomitantly, degradation of the polysaccharide chain [40].

Commercially, chitosan is mostly produced from chitin derived from marine crustaceans, such as crabs and shrimps [46]. However, it can also be obtained from chitin isolated from other sources, in particular from cephalopods, which has a different crystallographic form, designated beta, and is characterized by a parallel chain arrangement. This parallel arrangement results in weaker intermolecular hydrogen bonds and consequently more reactive polymers [42]. In this perspective, using β-chitin as a raw material, a more reactive chitosan is produced.

1.2.2
Alginate

Alginate is an unbranched anionic polysaccharide composed of β-D-mannuronic acid (M) and α-L-guluronic acid (G) linked by 1-4 glycosidic bonds [47]. M and G are stereoisomers, differing in the configuration of the carboxyl group, and the position of each unit can also vary, so they can occur in blocks of separate (M or G) or mixed (MG) sequences [48, 49]. The variability in the ratio and sequence of M and G units significantly influences the physicochemical properties of alginate [47, 50, 51].

Alginate is well known for its gelling capacity [47], which is strongly dependent on experimental conditions such as solution viscosity and gelation agent (commonly calcium or other divalent ions) concentration, as well as on molecular characteristics such as molecular weight and structure (M/G ratio and sequence) [52, 53]. In fact, G monomers play a crucial role in the mechanism of ionic gelation, by forming intermolecular ionic bridges induced by the presence of divalent ions, such as calcium [48, 52, 54]. In this way, G-rich alginate will give transparent, stiffer, and more brittle gels, while alginate with higher M content will form more flexible gels [47, 51, 53].

Alginate is the main component of the cell wall of brown algae, thus having not only a structural function but also an important participation in ionic exchange mechanisms [47, 55]. Alginate can be responsible for up to 45% of dry weight of algae, occurring together with other polysaccharides (cellulose, fucoidan, and others) and proteins, with the amount, as well as molecular structure, being dependent on the algal species, environmental conditions, and life cycle [56–59]. For instance, alginates bearing higher G content were extracted mainly from older algae, which thus allow the preparation of stronger gels when compared with materials that are found in younger specimens [59].

The industrial extraction of alginates is mainly done from *Laminaria, Macrocystis, Ascophyllum, Eclonia, Lessonia, Durvillea,* and *Sargassum* species [60] and starts with treatment with acid solution to convert the magnesium, calcium, and sodium alginate salts found in the cell wall of algae into alginic acid, followed by extraction with sodium hydroxide solution, resulting in soluble sodium alginate. In this way, it is possible to also extract alginates originally complexed with magnesium and calcium ions, which are insoluble, and at the same time to eliminate undesired polysaccharides [61, 62]. The alginate-rich aqueous solution is then separated by filtration, and the alginate can be obtained by precipitation with ethanol, resulting in alginate salt, or by acidification of alginate solution, rendering gelatinous alginic acid [60, 63, 64].

The economically and industrially viable extraction of alginates has been extensively studied, not only to obtain a polymer with controlled and desired properties for a broad range of applications [60], such as gelling agent in food, pharmaceutical, and biomedical industries [65, 66] or as hypocholesterolemic and hypolipidemic agent [67], but also to explore the valorization of by-products, for instance, as source of dietary fibers [62].

1.2.3
Carrageenan

Carrageenans are a family of polysaccharides constituted by 3-β-D-galactopyranose and 4-α-D-galactopyranose repeating units. The structures displayed by this family of polymers allow their organization into three main types according to the number of sulfate groups per disaccharide unit: κ (kappa), ι (iota), and λ (lambda) [68], bearing one, two, and three sulfate groups per disaccharide unit, respectively. In fact, these three types correspond to the commercial names, and the respective polymers are thus obtained after alkaline modification, which involves the conversion of precursor molecules into their final form.

Carrageenans can be found in red algae (Rodophyta), being commonly extracted from the genera *Chondrus* and *Gigartina*. Depending on the life cycle stage, algae can bear different carrageenan types and amounts (between 60 and 80% of dry weight), together with proteins, Floridean starch, and several smaller compounds, some of which exhibit interesting biological activity [69]. Carrageenans can be extracted from red algae by soaking in alkaline bath to perform a chemical modification that enhances the gelling properties, followed by extraction with hot water. Carrageenan can be then recovered by precipitation with salts or organic solvents, such as ethanol [70, 71]. Depending on their use, additional purification steps can be added, in particular, dialysis and reprecipitation, being also wise to firstly remove some contaminants if an extrapure product is desired [72–74].

All types of carrageenans are soluble in water, but at low temperatures, only the λ form is soluble. Of these types, κ and ι carrageenans can be gelified, while the λ form hardly forms gels. Less sulfate content allows the formation of harder gels, but the properties can be further tuned by complexation with other alkali metal ions or by mixture with other polymers [75]. This gelling property makes carrageenans applicable in a broad range of fields, such as in fermentation processes at industrial level [76], as pharmacological excipients [77], or in food products [78, 79] (being designated in the European Union as E407) with stabilizing, thickening, and emulsifying roles [80].

1.2.4
Collagen

Collagen is the most abundant protein in mammals and is formed by three proteic chains that wrap around one another forming a triple helix, with each chain being composed of a specific set of amino acids, namely, repeating triplets of glycine and two other amino acids, of which proline and hydroxyproline are the most common [81–83].

Collagen has found a wide range of health-related applications, namely, in cosmetics, pharmaceutical capsules, dental composites, skin regeneration, ophthalmology, cardiac surgery, plastic surgery, and orthopedics [84]. Besides, collagen is also used in other sectors, mainly as gelatin, its denatured form, which is used, for instance, in not only food industry but also photography [85]. Collagen, or gelatin, is the oldest known glue, being used for about 8000 years ago near the Dead Sea [86].

Bovine and porcine bones and skins are the main industrial sources of collagen. However, owing to religious constrains mainly related to Muslim and Jewish customs avoiding porcine products and to active discussion on the risk of bovine spongiform encephalopathy (BSE) and other diseases posed to humans by the use of infected bovine-derived products, other sources of collagen and gelatin are being pursued [87, 88]. Besides recombinant technology using, for instance, the yeast *Pichia pastoris* [89], fish collagen is receiving growing attention. In marine environment, collagen can be found in several marine organisms, with the main sources being marine sponges (e.g., *Chondrosia reniformis*) [90], jellyfish [91–93], and fish bones and skins [81, 94, 95]. Collagen can be obtained from fish skins by treating them with acetic acid solution, and sometimes with the concomitant use of 10% pepsin, thus obtaining acid-soluble collagen and pepsin-soluble collagen, respectively [85]. If jellyfish is used as raw material, it should be washed to desalt and freeze-dried. The material is then treated with 0.5 M acetic acid, and the extracts are dialyzed against 0.02 M Na_2HPO_4. Solid NaCl is then added, and the precipitated fraction is redissolved in 0.5 M acetic acid, dialyzed against 0.1 M acetic acid, and finally freeze-dried [92]. As for collagen derived from fish skins, digestion with pepsin may also be done. When considering marine sponges, the available methodologies are different since sponge collagen is not soluble in acetic acid solution. Thus, Swatschek *et al.* [90] proposed an extraction methodology aimed at scale-up, based on treatment with 100 mM Tris–HCl buffer (pH 9, 10 mM EDTA, 8 M urea, 100 mM 2-mercaptoethanol) for 24 h, with stirring, at room temperature, after which the extract is centrifuged and collagen is precipitated from the supernatant by adjusting the pH to 4 with acetic acid.

1.2.5
Hyaluronic Acid

Hyaluronic acid is a nonsulfated GAG composed of alternating disaccharide units of α-1,4-D-glucuronic acid and β-1,3-*N*-acetyl-D-glucosamine linked by β(1 → 3) bonds [96], with a wide range of molecular weights, which is actually associated with its biological functions. Polymers with molecular weights as high as 10^7 Da have space-filling, antiangiogenic, and immunosuppressive properties, while smaller polymers have anti-inflammatory, immunostimulatory, angiogenic, and antiapoptotic properties [97].

Hyaluronic acid can be found in most connective tissues, such as cartilage, as well as in vitreous humor and synovial fluid [98]. In cartilage, it plays a structural role by interaction with proteoglycans and proteins [99]. In synovial fluid, hyaluronic acid acts as lubricant and shock absorber, because of its enhanced viscoelastic properties [100]. In fact, difficulty in joint movement and pain as a result of arthritic diseases are due to degradation of hyaluronic acid, which leads to reduction of its viscosity and related properties [101]. Hyaluronic acid also has an important role in the skin, protecting the cells from free radicals that may be generated by UV radiation [102].

Hyaluronic acid can be also found in marine species, in particular in cartilaginous fishes, which can be an alternative source for the production by recombinant technology or by extraction from umbilical cords and rooster combs, first described by Balazs [103] and consists of freezing the materials, thus destroying cell membranes, followed by extraction with water, and then finally precipitating the hyaluronic acid with ethanol, chloroform, or other organic solvents.

1.2.6
Others

Besides the above-mentioned polymers, that together with agar constitute the most representative, studied, and explored marine polymers, other marine polymers are highly promising for biomedical applications, in particular for tissue engineering, and are thus starting to receive growing attention from the scientific community. They are the sulfated polymers chondroitin sulfate (already known, but mainly explored in bovine and porcine cartilages), ulvan, and fucoidan.

Chondroitin sulfate is a polysaccharide that consists of a disaccharide repeating unit of D-galactosamine and D-glucuronic acid, which can exhibit sulfate ester substituents at different positions (the most common are positions C4 or C6 of the galactosamine, corresponding to chondroitin-4-sulfate and chondroitin-6-sulfate, respectively) [104]. Besides an important structural function, chondroitin sulfate also has a role in the development of central nervous system, wound repair, infection assessment, morphogenesis, and cell division [105].

Chondroitin sulfate can be found in the cartilage of several terrestrial species, being explored from bovine, porcine, and chicken cartilages [106, 107], and also in marine species, such as whale [108], shark [109, 110], squid [111], and others [112], being extracted by proteolytic digestion and further purified by precipitation with organic solvents, chromatography, or enzymatic degradation of contaminants [104]. Finally, electrophoretic techniques can be employed for the qualitative and quantitative analyses of chondroitin sulfate [113].

Ulvan represents a family of branched polysaccharides obtained from green algae *Ulva*, with a broad distribution of charge density and molecular weight [114]. It is mainly composed of rhamnose, xylose, glucuronic acid, and sulfate and also contains iduronic acid as a carbohydrate unit [114]. Ulvan is attracting attention as a source of rare sugar precursors for chemical synthesis, but important biological properties are due to its oligomers and polymers, such as antitumor, immune modulation, antiinfluenza, anticoagulant, and antioxidant activities [114].

Ulvan can be extracted from green algae by water extraction, but the purification details are quite relevant for the final use of the polymer. In this perspective,

Alves et al. [115] suggested an extraction process starting with extraction with dichloromethane and acetone to remove lipids and pigments, followed by successive extractions with hot water. The liquid resulting from filtration and further centrifugation is submitted to proteinase digestion to eliminate proteins and treated with activated charcoal to remove remaining pigments. Finally, ulvan is precipitated with ethanol and freeze-dried, resulting in a white powder.

Fucoidan is a sulfated polysaccharide existing in two different forms: the major form F-fucoidan mainly composed of L-fucose units, and U-fucoidan, with a significant amount of glucuronic acid units [85]. Several relevant biologic properties have been associated with this polysaccharide, such as antiinflammatory, anticoagulant, antitumoral, and antiviral activities [85, 116]. In particular, fucoidan has been suggested as an anticoagulant as a substituent of heparin [85] and also as an inhibitor of replication of human immunodeficiency virus [116].

Fucoidan can be extracted from some species of brown algae using a hot acid solution and can be further purified by hydrophobic chromatography and dialysis [85]. Ponce et al. [116] described a different method of extracting fucoidan from *Adenocystis utricularis* by extraction with 80% ethanol, first at room temperature and then at 70 °C, with the residue being recovered by centrifugation; by treatment with water, 2% $CaCl_2$, or 0.01 M HCl; and finally by dialysis. The use of these different solvents resulted in materials with different properties (sulfate content and monosaccharide proportion) [116].

1.3
Marine-Based Tissue Engineering Approaches

Marine-origin biopolymers have long been proposed for tissue engineering approaches. The natural base character of these materials confers on them adequate properties for applications in tissue engineering and regenerative medicine, mainly because of their low immunogenic potential and chemical/biological versatility and degradability.

Several literature reviews focus on the applications of different polymers, for bone or cartilage tissue engineering. This section specially focuses on the application of marine-based materials. There are several processing techniques available, and the choice of the appropriate one will rely not only on the characteristics of the material itself but also on the final product shape. Depending on each particular application, the material can be prepared in the form of membranes, hydrogels, 3D porous scaffolds, or particles.

1.3.1
Membranes

Membranes have been used in life-saving treatments, such as drug delivery, in diagnostic devices, and as wound dressings, because of its ease in manufacturing and self-application [117]. Solvent casting is a common technique for the preparation of polymeric membranes. This technique also allows the impregnation of active compounds such as proteins, drugs, or ceramics.

Regarding wound healing, chitosan-based membranes have been shown to be promising candidates [118]. Santos *et al.* [119] have demonstrated, for instance, that chitosan-based membranes do not elicit any inflammatory response, which confirmed the biocompatibility of these types of structures. On the other hand, surface modification of chitosan membranes, particularly by plasma treatment, is another strategy to improve cellular adhesion in matrices for wound healing that has been reported in the literature [120, 121]. An example of a composite porous network for wound healing is the use of bilayered gelatin/chondroitin-6-sulfate/hyaluronic acid membranes [122].

The reinforcement of membranes with ceramic compounds, which can be used for orthopedic applications including in hard tissue regeneration, has been reported by Caridade *et al.* [123] and Li [124].

Recently, LbL processing has gained attention in tissue engineering for the preparation of thin films. The membranes are formed by deposition of alternate layers of a polyanion and a polycation with washing steps in between. Chitosan, owing to its positive charge, is one of the most commonly used natural polycation in LbL processing of nanostructured thin films and was recently reviewed by Pavinatto *et al.* [125]. Marine-origin polyanions include alginates, carrageenans, hyaluronic acid, and chondroitin sulfate.

1.3.2
Hydrogels

Hydrogels are usually defined as water-soluble, cross-linked polymeric matrices containing covalent bonds, hydrogen bonds, physical cross-linkers, strong van der Waals interactions, or crystallite associations [126]. The suitability of hydrogels as biomedical devices is mostly due to their high swelling ability and their soft and rubbery consistency, which resembles most living tissues, and thereafter, natural-based polymers are strong candidates for the development of novel devices. van Vlierberghe and coworkers [127] reviewed the use of natural polymers for the preparation of hydrogels for tissue engineering and regenerative medicine applications. Another review by Oliveira *et al.* [128] gives an overview of the applications of polysaccharides in the development of matrices for tissue engineering. Use in blood vessels, cartilages, corneal stroma, intervertebral disks, meniscus, skin, and tendons is an application of the matrixes developed and reported in the literature. In this section, we focus on hydrogels from polysaccharides and proteins of marine origin.

The feasibility of the use of chitosan hydrogels in these applications has been described in several papers. Its use in different applications implies, however, its modification and/or blending with other polymers (either natural or synthetic) in order to achieve a material with the appropriate characteristics of the particular tissue to be regenerated. Table 1.1 gives an overview of different proposed polymer combinations and their applications.

Table 1.1 Overview of hydrogels from marine-based polymers in different tissue engineering applications.

Polymers	Application	References
Chitosan/hyaluronic acid	Cartilage TE	[129, 130]
Chitosan/alginate	Cartilage TE	[131–133]
Chitosan/gelatin	Cartilage TE	[134–136]
Chitin/nano-hydroxyapatite	Bone TE	[137, 138]
Collagen/chitosan	Blood vessel reconstruction	[139]
Chitosan/gelatin	Blood vessel reconstruction	[140]
Collagen/chitosan	Corneal regeneration	[141]
Alginate	Soft tissues	[142]
Alginate	3D neural cell culture	[143]
Carrageenan	Bone TE	[144, 145]
Carrageenan/fibrin/hyaluronic acid	Cartilage TE	[146]
Hyaluronic acid/fibronectin	Wound healing	[147]
Hyaluronic acid/marine exopolysaccharides	Cartilage TE	[148]
Hyaluronic acid/chitosan or hyaluronic acid/gelatin	Nucleus pulposus regeneration	[149]
Chondroitin sulfate	Cartilage TE	[150]

TE, tissue engineering.

Hydrogels have large applicability, for instance, in cartilage regeneration. Concerning the use of marine-origin-polymers, different hydrogels have been produced based on chitosan/hyaluronic acid [129, 130], chitosan/alginate [131–133], and chitosan/gelatin mixtures [134–136]. Chitin was also processed for the preparation of hydrogels with nanohydroxyapatite particles for bone tissue engineering [137, 138].

In terms of blood vessel reconstruction, some of the most recent works reported include combinations of chitosan/gelatin [139] or collagen/chitosan [140] prepared by freeze-drying.

Collagen/chitosan composite hydrogels have been described as good candidates for corneal treatment. Its preparation as corneal implants involved the stabilization of the matrices with different cross-linking agents [141].

Alginate-based hydrogels have also been prepared for different applications. Park and coauthors [142] evaluated the feasibility of using a novel rapid prototyping technique for the preparation of alginate hydrogels with applications in soft tissues. Alginate hydrogels were also found to be good supporting matrices for 3D neural cell culture [143]. Alginate being a hydrosoluble polymer, cross-linking reactions are required to promote stability of the matrices. Photo-cross-linking methods have now been proposed to create stable systems able to deliver cells without compromising their viability [151].

Carrageenan hydrogels have also been described as potential vehicles for drug delivery in bone tissue engineering [144], and as an injectable system combined with fibrin and hyaluronic acid for cartilage tissue engineering [146]. A comparison between different carrageenan types and alginate hydrogels as tissue engineering scaffolds was described by Mehrban and coauthors [145].

Hyaluronic acid is another marine-derived polymer that has received great attention, particularly because of its multiple roles in the angiogenic process in the body. Fibronectin—hyaluronic acid composite hydrogels were developed by Seidlits *et al.* [147] in the form of 3D hydrogel networks for wound healing. The continuous growth of the field is demonstrated by the novelty of the materials used in the work of Redderstorff *et al.* [148] in which two new GAG-like marine exopolysaccharides were dispersed in a hyaluronic acid matrix for cartilage tissue engineering. Other applications of hyaluronic-acid-based systems include injectable systems in which hyaluronic acid is combined with chitosan or gelatin for the regeneration of nucleus pulposus [149].

Regarding chondroitin sulfate, which is one of the major components of cartilage extracellular matrix, cross-linked hydrogels were proposed by Wang *et al.* [150] for cartilage tissue regeneration.

1.3.3
Tridimensional Porous Structures

The concept of tissue engineering has long passed the development of an inert matrix and is nowadays based on the development of loaded scaffolds containing bioactive molecules in order to control the cellular function or to act on the surrounding tissues. Hereafter, one of the most important stages of tissue engineering is the design and processing of a porous 3D structure, with high porosity, high interconnectivity between the pores, and uniform distribution so that the cells are able to penetrate the inner part of the network, nutrients and oxygen are accessible to the cells, and cell wastes are eliminated [152]. Conventionally, three-dimensional structures can be obtained by processes such as solvent casting–particle leaching, freeze-drying–particle leaching, thermally induced phase separation, compression molding, injection molding, extrusion, foaming, wet spinning, electrospinning, and supercritical fluid technology, among others [153].

Chitosan scaffolds have been successfully prepared by some of the above-mentioned techniques [154–158]. Figure 1.2 shows as an example a scanning electron microscopic image of a chitosan scaffold produced by supercritical assisted phase inversion.

Together with chitin, these have been by far the most widely studied polymers of marine origin [159, 160]. Chitin scaffolds produced from Verongida sponges were successfully described as matrixes for cartilage tissue regeneration [161, 162]. The difficulties in processing chitin arise due to its high crystallinity and insolubility in most solvents, which prevents its use in a broader range of applications. Alternative green solvents as ionic liquids have been proposed as good candidates for the development of 3D structures with appropriate features for biomedical applications [163, 164].

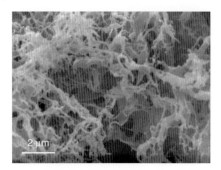

Figure 1.2 Scanning electron microscopic image of chitosan scaffolds prepared by supercritical assisted phase inversion.

Collagen extracted from jellyfish was described by Song *et al.* [92] as a promising novel material for the development of structures with a low immune response for tissue engineering applications.

The use of polysaccharides and/or proteins of marine origin as scaffolds for tissue engineering, namely, for bone tissue engineering is, however, limited because of the poor mechanical properties of these materials and the lack of inherent bioactivity. Nonetheless, this can be easily overcome by the combination of these polymers with other polymers or with bioceramics.

Three-dimensional scaffolds are generally produced to treat bone defects; therefore, special emphasis is given to this application. Polymeric mixtures of chitosan with other marine-derived polymers include combinations of gelatin and chondroitin sulfate for the preparation of 3D matrixes with enhanced properties for bone tissue engineering [165]. To overcome the lack of bioactivity of the polymers and the poor mechanical properties of the ceramics, the preparation of composite matrixes of these two materials is regarded as an interesting approach. Biodegradable composites containing hydroxyapatite-based calcium phosphates (CaP) and Bioglass® can be used to produce promising scaffolds for bone tissue engineering. Inorganic compounds, such as zirconium dioxide and titanium dioxide, are described to enhance osteogenesis, and in this sense, chitin—chitosan scaffolds have been prepared in combination with these compounds for bone tissue engineering purposes [166, 167].

Biomimetic approaches using marine-derived compounds have been reported in different research works, particularly, combining chitosan and calcium phosphates [124, 168–172]. Strategies involving chitosan and hydroxyapatite matrixes have also been described for the treatment of bone defects [173] and osteochondral defects. Oliveira and coworkers [174] describe the preparation of a bilayered scaffold by freeze-drying, while another work by Malafaya *et al.* [33] presents a processing route based on chitosan particle aggregation methodology for the production of cartilage and osteochondral tissue engineering scaffolds.

Ge *et al.* [175] suggest a hydrothermal process to convert calcium carbonate from crab shells in hydroxyapatite and developed a novel composite system of chitosan—hydroxyapatite for bone tissue engineering.

Kusmanto and coworkers [176] describe the conversion of *Phymatolithon calcareum*, a marine alga with a natural interconnected structure of calcium carbonate and hydroxyapatite, while maintaining its structure. Another work reports the production of a 3D matrix from the marine hydrocoral *Millepora dichotoma*, a natural bioactive material that has demonstrated the potential to promote mineralization and differentiation of the mesenchymal stem cells into the osteogenic lineage [177]. The development of hydroxyapatite bone structures, depicted in the work of Cunningham, by replication of three different species of sponges, has also proved to be an interesting approach to the design of novel bioactive structures [178].

1.3.4
Particles

Although 3D porous structures have been recognized as the most appropriate to sustain cell adhesion, several applications in tissue engineering may take advantage of other designs. Injectable systems that can simultaneously act as scaffolds and delivery devices are becoming more and more attractive in this field, especially because of their noninvasive approach [179].

Materials in particulate form for tissue engineering have been reviewed by Silva and coauthors in two papers where the basic concepts and the applications in bone regeneration are described [180, 181]. There is no doubt that tissue engineering strategies are moving toward systems that are able to combine materials, cells, and growth factors. Materials in particulate form can play a role in this strategy as carriers for biologically active molecules. Furthermore, a better control in parameters such as porosity, pore size, surface area, and mechanical properties can be attained in the case of materials in particulate form.

The use of particulate systems for biomedical applications is nonetheless in its embryonic stage of development, and few papers are published in the literature, especially concerning polymers of marine origin.

The applicability of chitosan particles in drug delivery systems has been reviewed by Prabaharan *et al.* [182]. Nonetheless, these particles may also be used for tissue engineering applications, namely, as cell transplantation systems, as described by Cruz *et al.* [183]. These types of particulate systems may also promote growth of bonelike apatite, after a calcium silicate treatment, according to Leonor *et al.* [184]. A technique employed to enhance the mechanical properties of microspheres is to develop multilayer particles, using the LbL technique [185].

Munarin and coworkers [186] described the preparation of calcium alginate, calcium alginate/chitosan, calcium alginate/gelatin, and pectin/chitosan microcapsules for cartilage regeneration.

A combination of chitosan/carrageenan was proposed by Grenha and coworkers [187] as a novel drug delivery device, with potential applications in tissue engineering applications as well.

The development of growth factor delivery systems is one of the most attractive fields of research in particulate engineering. The development of nano- and microscale particles of chondroitin sulfate reported by Lim *et al.* [188] represents a versatile system for the controlled delivery of positively charged growth factors for a variety of stem-cell-based applications in tissue engineering.

Particulate systems of hyaluronic acid have been developed by Sahiner et al. [189] for other types of applications such as vocal fold regeneration.

Marine sponge collagen particles have been developed and were described in two papers on transdermal drug delivery [190, 191]. This is the only work related to the use of sponge collagen in biomedical applications; however, its use can be envisaged in tissue engineering and regenerative medicine.

1.4
Conclusions

Marine life has proved to utilize an immense diversity of ingenious solutions for designing materials. Most of them are true engineering marvels that arose under the same sorts of limitations facing human engineers, such as the need to maximize functionality while minimizing costs in energy and materials. Marine systems are created using nonpolluting processes that occur at biological temperatures and in wet, often salted, environments. Therefore, many researchers have realized that the sea could inspire multiple biomimetic strategies that could be directly applied in the biomedical field. Moreover, the sea has proved to be a huge reservoir of many distinct materials, even though the available knowledge of marine materials and mechanism is still in its infancy. It is envisaged that marine-derived materials will be increasingly explored in biomedicine, in particular in tissue repair and regeneration. Biomedical devices for such applications should be processed into different shapes (fibers, membranes, hydrogels, cellular structures, particles) and sizes. In this chapter, different representative examples were given on how marine-derived biopolymers could be combined and processed into structures for well-defined applications in regenerative medicine.

References

1. Vincent, J.F.V. et al. (2006) *J. R. Soc. Interface*, **3**(9), 471–482.
2. Antonietti, M. and Fratzl, P. (2010) *Macromol. Chem. Phys.*, **211**(2), 166–170.
3. MarineBiotech (2012) Marine Biomimetics, http://www.marinebiotech.org/biomimetics.html (accessed 27 February 2012).
4. Fratzl, P. (2007) *J. R. Soc. Interface*, **4**(15), 637–642.
5. Mattheck, C. (1998) *Design in Nature: Learning from Trees*, Springer-Verlag, Berlin, Heidelberg, p. 276.
6. Langer, R. and Vacanti, J.P. (1993) *Science*, **260**(5110), 920–926.
7. Sakiyama-Elbert, S.E. and Hubbell, J.A. (2001) *Ann. Rev. Mater. Res.*, **31**, 183–201.
8. Shin, H., Jo, S., and Mikos, A.G. (2003) *Biomaterials*, **24**(24), 4353–4364.
9. Currey, J.D. (1977) *Proc. R. Soc. Lond. B, Biol. Sci.*, **196**(1125), 443–463.
10. Jackson, A.P., Vincent, J.F.V., and Turner, R.M. (1988) *Proc. R. Soc. Lond. B, Biol. Sci.*, **234**(1277), 415–440.
11. Barthelat, F., Rim, J., and Espinosa, H.D. (2009) in *Applied Scanning Probe*

Methods XIII 2009 (eds B. Bhushan and H. Fuchs), Springer, pp. 1059–1100.
12. Kamat, S. *et al.* (2000) *Nature*, **405**(6790), 1036–1040.
13. Asvanund, P., Chunhabundit, P., and Suddhasthira, T. (2011) *Implant Dent.*, **20**(1), 32–39.
14. Deville, S. *et al.* (2006) *Science*, **311**(5760), 515–518.
15. Couto, D.S., Alves, N.M., and Mano, J.F. (2009) *J. Nanosci. Nanotechnol.*, **9**(3), 1741–1748.
16. Hu, B.H. and Messersmith, P.B. (2000) *Tetrahedron Lett.*, **41**(31), 5795–5798.
17. Monahan, J. and Wilker, J.J. (2003) *Chem. Commun.*, (14), 1672–1673.
18. Lee, B.P., Dalsin, J.L., and Messersmith, P.B. (2002) *Biomacromolecules*, **3**(5), 1038–1047.
19. Burke, S.A. *et al.* (2007) *Biomed. Mater.*, **2**(4), 203–210.
20. Podsiadlo, P. *et al.* (2007) *Adv. Mater.*, **19**(7), 949–955.
21. Wilkie, I.C. (2005) *Prog. Mol. Subcell. Biol.*, **39**, 221–250.
22. Wilkie, I.C., Carnevali, M.D.C., and Bonasoro, F. (1992) *Zoomorphology*, **112**(3), 143–153.
23. Wilkie, I.C., Carnevali, M.D.C., and Andrietti, F. (1993) *Comp. Biochem. Physiol. A-Physiol.*, **105**(3), 493–501.
24. Trotter, J.A. *et al.* (2000) *Biochem. Soc. Trans.*, **28**, 357–362.
25. Querellou, J. *et al.* (2010) Marine Biotechnology: A New Vision and Strategy for Europe. European Science Foundation - Marine Board, pp. 59–63.
26. Ehrlich, H. (2010) *Biological Materials of Marine Origin*, Springer.
27. Guiseley, K.B. (1989) *Enzyme Microb. Technol.*, **11**(11), 706–716.
28. Renn, D.W. (1984) *Ind. Eng. Chem. Prod. Res. Dev.*, **23**(1), 17–21.
29. Kumar, M.N.V.R. (2000) *React. Funct. Polym.*, **46**(1), 1–27.
30. Pillai, C.K.S., Paul, W., and Sharma, C.P. (2009) *Prog. Polym. Sci.*, **34**(7), 641–678.
31. Silva, R.M. *et al.* (2004) *J. Mater. Sci. Mater. Med.*, **15**(10), 1105–1112.
32. Silva, S.S. *et al.* (2005) *J. Mater. Sci. Mater. Med.*, **16**(6), 575–579.
33. Malafaya, P.B. *et al.* (2005) *J. Mater. Sci. Mater. Med.*, **16**(12), 1077–1085.
34. Tuzlakoglu, K. *et al.* (2004) *Macromol. Biosci.*, **4**(8), 811–819.
35. Madihally, S.V. and Matthew, H.W.T. (1999) *Biomaterials*, **20**(12), 1133–1142.
36. Suh, J.K.F. and Matthew, H.W.T. (2000) *Biomaterials*, **21**(24), 2589–2598.
37. Denuziere, A. *et al.* (1998) *Biomaterials*, **19**, 1275–1285.
38. Miretzky, P. and Cirelli, A.F. (2009) *J. Hazard. Mater.*, **167**(1–3), 10–23.
39. Shahidi, F. and Synowiecki, J. (1991) *J. Agric. Food Chem.*, **39**(8), 1527–1532.
40. Tolaimate, A. *et al.* (2003) *Polymer*, **44**(26), 7939–7952.
41. Hayes, M. *et al.* (2008) *J. Biotechnol.*, **3**, 871–877.
42. Peniche, C., Argüelles-Monal, W., and Goycoolea, F. (2008) in *Monomers, Polymers and Composites from Renewable Resources* (eds M. Belgacem and A. Gandini), Elsevier, pp. 517–542.
43. Rhazi, M. *et al.* (2000) *Polym. Int.*, **49**(4), 337–344.
44. Chandumpai, A. *et al.* (2004) *Carbohydr. Polym.*, **58**(4), 467–474.
45. Kurita, K. *et al.* (1993) *J. Polym. Sci., Part A: Polym. Chem.*, **31**(2), 485–491.
46. Rinaudo, M. (2006) *Prog. Polym. Sci.*, **31**(7), 603–632.
47. Rinaudo, M. (2008) *Polym. Int.*, **57**(3), 397–430.
48. Siew, C.K., Williams, P.A., and Young, N.W.G. (2005) *Biomacromolecules*, **6**(2), 963–969.
49. Ertesvag, H. and Valla, S. (1998) *Polym. Degrad. Stab.*, **59**(1–3), 85–91.
50. Rehm, B.H.A. and Valla, S. (1997) *Appl. Microbiol. Biotechnol.*, **48**(3), 281–288.
51. Nyvall, P. *et al.* (2003) *Plant Physiol.*, **133**(2), 726–735.
52. d'Ayala, G., Malinconico, M., and Laurienzo, P. (2008) *Molecules*, **13**(9), 2069–2106.
53. Drury, J.L., Dennis, R.G., and Mooney, D.J. (2004) *Biomaterials*, **25**(16), 3187–3199.
54. Li, L.B. *et al.* (2007) *Biomacromolecules*, **8**(2), 464–468.
55. Percival, E. (1979) *Br. Phycol. J.*, **14**(2), 103–117.

56. Mizuno, H. et al. (1983) *Bull. Jpn. Soc. Sci. Fish.*, **49**(10), 1591–1593.
57. Jothisaraswathi, S., Babu, B., and Rengasamy, R. (2006) *J. Appl. Phycol.*, **18**(2), 161–166.
58. Kelly, B.J. and Brown, M.T. (2000) *J. Appl. Phycol.*, **12**(3–5), 317–324.
59. McKee, J.W.A. et al. (1992) *J. Appl. Phycol.*, **4**(4), 357–369.
60. Gomez, C.G. et al. (2009) *Int. J. Biol. Macromol.*, **44**(4), 365–371.
61. Hernandez-Carmona, G. et al. (1998) *J. Appl. Phycol.*, **10**(6), 507–513.
62. Fleury, N. and Lahaye, M. (1993) *J. Appl. Phycol.*, **5**(1), 63–69.
63. Hernandez-Carmona, G., McHugh, D.J., and Lopez-Gutierrez, F. (1999) *J. Appl. Phycol.*, **11**(6), 493–502.
64. Vauchel, P. et al. (2009) *Bioresour. Technol.*, **100**(20), 4918–4918.
65. Brownlee, I.A. et al. (2005) *Crit. Rev. Food Sci. Nutr.*, **45**(6), 497–510.
66. Tonnesen, H.H. and Karlsen, J. (2002) *Drug Dev. Ind. Pharm.*, **28**(6), 621–630.
67. Smit, A.J. (2004) *J. Appl. Phycol.*, **16**(4), 245–262.
68. Falshaw, R., Bixler, H.J., and Johndro, K. (2001) *Food Hydrocolloids*, **15**(4–6), 441–452.
69. Fleurence, J. (1999) *Trends Food Sci. Technol.*, **10**(1), 25–28.
70. Hilliou, L. et al. (2006) *Biomol. Eng.*, **23**(4), 201–208.
71. Ciancia, M. et al. (1993) *Carbohydr. Polym.*, **20**(2), 95–98.
72. Goff, D. (2004) *Dairy Ind. Int.*, **69**(2), 31–33.
73. Mangione, M.R. et al. (2005) *Biophys. Chem.*, **113**(2), 129–135.
74. Campo, V.L. et al. (2009) *Carbohydr. Polym.*, **77**(2), 167–180.
75. Yuguchi, Y., Urakawa, H., and Kajiwara, K. (2003) *Food Hydrocolloids*, **17**(4), 481–485.
76. Tye, R.J. (1989) *Carbohydr. Polym.*, **10**(4), 259–280.
77. Bornhöft, M., Thommes, M., and Kleinebudde, P. (2005) *Eur. J. Pharm. Biopharm.*, **59**(1), 127–131.
78. Baeza, R.I. et al. (2002) *Lebensm. Wiss. Technol. Food Sci. Technol.*, **35**(8), 741–747.
79. Aliste, A.J., Vieira, F.F., and Del Mastro, N.L. (2000) *Radiat. Phys. Chem.*, **57**(3–6), 305–308.
80. Rasmussen, R.S. and Morrissey, M.T. (2007) in *Advances in Food and Nutrition Research* (ed. L.T. Steve), Academic Press, pp. 237–292.
81. Gomez-Guillen, M.C. et al. (2002) *Food Hydrocolloids*, **16**(1), 25–34.
82. Kolodziejska, I., Sikorski, Z.E., and Niecikowska, C. (1999) *Food Chem.*, **66**(2), 153–157.
83. Mendis, E. et al. (2005) *Life Sci.*, **77**(17), 2166–2178.
84. Meena, C., Mengi, S., and Deshpande, S. (1999) *J. Chem. Sci.*, **111**(2), 319–329.
85. Venugopal, V. (2009) in *Marine Products for Healthcare: Functional and Bioactive Nutraceutical Compounds from the Ocean*, Functional Foods And Nutraceuticals Series (ed. G. Mazza), CRC Press, Boca Raton, FL.
86. Walker, A.A. (1998) *Archaeology*. www.archaeology.org/online/news/glue.html
87. Arvanitoyannis, I.S. and Kassaveti, A. (2008) *Int. J. Food Sci. Technol.*, **43**(4), 726–745.
88. Leary, D. et al. (2009) *Mar. Policy*, **33**(2), 183–194.
89. Olsen, D. et al. (2003) *Adv. Drug Delivery Rev.*, **55**(12), 1547–1567.
90. Swatschek, D. et al. (2002) *Eur. J. Pharm. Biopharm.*, **53**(1), 107–113.
91. Nagai, T. et al. (1999) *J. Sci. Food Agricult.*, **79**(6), 855–858.
92. Song, E. et al. (2006) *Biomaterials*, **27**(15), 2951–2961.
93. Miura, S. and Kimura, S. (1985) *J. Biol. Chem.*, **260**(28), 5352–5356.
94. Montero, P. and Gomez-Guillen, M.C. (2000) *J. Food Sci.*, **65**(3), 434–438.
95. Ramshaw, J.A.M., Bateman, J.F., and Cole, W.G. (1984) *Anal. Biochem.*, **141**(2), 361–365.
96. Laurent, T.C., Laurent, U.B.G., and Fraser, J.R.E. (1995) *Ann. Rheum. Dis.*, **54**(5), 429–432.
97. Xu, H.P. et al. (2002) *J. Biol. Chem.*, **277**(19), 17308–17314.
98. Liao, Y.H. et al. (2005) *Drug Deliv.*, **12**(6), 327–342.

99. Prehm, P. (2002) in *Biopolymers, Polysaccharides I. Polysaccharides from Prokaryotes*, vol. 5 (eds E.J. Vandamme, S. De Baets, and A. Steinbüchel), Wiley-VCH Verlag GmbH, Weinheim, pp. 379–404.
100. Nishinari, K. and Takahashi, R. (2003) *Curr. Opin. Colloid Interface Sci.*, **8**(4–5), 396–400.
101. Soltes, L. et al. (2006) *Biomacromolecules*, **7**(3), 659–668.
102. Juhlin, L. (1997) *J. Intern. Med.*, **242**(1), 61–66.
103. Balazs, E.A. *Ultra Pure, High Molecular wt. Hyaluronic Acid - is Non-antigenic and Used as Synthetic aq. Humour and Synovial Fluid, as Nerve Tissue and Wound Protector and Drug Carrier*, Biotrics Inc. (BIOT-Non-standard). US4141973-D; US4141973-A; US4141973-B, 27 Feb 1979.
104. Silva, L.F. (2006) in *Advances in Pharmacology* (ed. V. Nicola), Academic Press, pp. 21–31.
105. Sugahara, K. et al. (2003) *Curr. Opin. Struct. Biol.*, **13**(5), 612–620.
106. Lamari, F.N. et al. (2006) *Biomed. Chromatogr.*, **20**(6–7), 539–550.
107. Luo, X., Fosmire, G., and Leach, R. Jr. (2002) *Poult. Sci.*, **81**(7), 1086–1089.
108. Michelacci, Y.M. and Dietrich, C.P. (1986) *Int. J. Biol. Macromol.*, **8**(2), 108–113.
109. Seno, N. and Meyer, K. (1963) *Biochim. Biophys. Acta*, **78**(2), 258–264.
110. Nandini, C.D., Itoh, N., and Sugahara, K. (2005) *J. Biol. Chem.*, **280**(6), 4058–4069.
111. Srinivasan, S.R. et al. (1969) *Comp. Biochem. Physiol.*, **28**(1), 169–172, IN5, 173–176.
112. Lignot, B., Lahogue, V., and Bourseau, P. (2003) *J. Biotechnol.*, **103**(3), 281–284.
113. Cavari, S. and Vannucchi, S. (1996) *Clin. Chim. Acta*, **252**(2), 159–170.
114. Lahaye, M. and Robic, A. (2007) *Biomacromolecules*, **8**(6), 1765–1774.
115. Alves, A. et al. (2010) *Carbohydr. Res.*, **345**(15), 2194–2200.
116. Stortz, C.A. et al. (2003) *Carbohydr. Res.*, **338**(2), 153–165.
117. Stamatialis, D.F. et al. (2008) *J. Memb. Sci.*, **308**(1–2), 1–34.
118. Ueno, H. et al. (2001) *Adv. Drug Delivery Rev.*, **52**(2), 105–115.
119. Santos, T.C. et al. (2007) *J. Biotechnol.*, **132**(2), 218–226.
120. Luna, S.M. et al. (2011) *J. Biomater. Appl.*, **26**(1), 101–116.
121. Silva, S.S. et al. (2008) *Macromol. Biosci.*, **8**(6), 568–576.
122. Wang, T.W. et al. (2007) *J. Biomed. Mater. Res. B Appl. Biomater.*, **82B**(2), 390–399.
123. Caridade, S.G. et al. (2010) *Materials Science Forum*, **636-637**, 26–30.
124. Li, J.J. et al. (2011) *Carbohydr. Polym.*, **85**(4), 885–894.
125. Pavinatto, F.J., Caseli, L., and Oliveira, O.N. (2010) *Biomacromolecules*, **11**(8), 1897–1908.
126. Peppas, N.A. et al. (2000) *Eur. J. Pharm. Biopharm.*, **50**(1), 27–46.
127. Van Vlierberghe, S., Dubruel, P., and Schacht, E. (2011) *Biomacromolecules*, **12**(5), 1387–1408.
128. Oliveira, J.T. and Reis, R.L. (2011) *J. Tissue Eng. Regen. Med.*, **5**(6), 421–436.
129. Correia, C.R. et al. (2011) *Tissue Eng. Part C-Methods*, **17**(7), 717–730.
130. Nair, S. et al. (2011) *Carbohydr. Polym.*, **85**(4), 838–844.
131. Tigli, R.S. and Gumusdereliogu, M. (2009) *J. Mater. Sci. Mater. Med.*, **20**(3), 699–709.
132. Marsich, E. et al. (2008) *J. Biomed. Mater. Res. A*, **84A**(2), 364–376.
133. Jeong, S.I. et al. (2011) *Tissue Eng. Part A*, **17**(1–2), 59–70.
134. Tan, H.P. et al. (2007) *J. Mater. Sci. Mater. Med.*, **18**(10), 1961–1968.
135. Guo, T. et al. (2006) *Biomaterials*, **27**(7), 1095–1103.
136. Breyner, N.M. et al. (2010) *Cells Tissues Organs*, **191**(2), 119–128.
137. Kumar, P.T. et al. (2011) *Int. J. Biol. Macromol.*, **49**(1), 20–31.
138. Kumar, P.T. et al. (2011) *Carbohydr. Polym.*, **85**(3), 584–591.
139. Zhang, L. et al. (2006) *J. Biomed. Mater. Res. A*, **77A**(2), 277–284.
140. Zhu, C.H. et al. (2009) *J. Biomed. Mater. Res. A*, **89A**(2), 829–840.
141. Rafat, M. et al. (2008) *Biomaterials*, **29**(29), 3960–3972.

142. Park, S.A., Lee, S.H., and Kim, W. (2011) *Macromol. Res.*, **19**(7), 694–698.
143. Frampton, J.P. et al. (2011) *Biomed. Mater.*, **6**(1), 18.
144. Santo, V.E. et al. (2009) *Biomacromolecules*, **10**(6), 1392–1401.
145. Mehrban, N. et al. (2010) in *Comparative Study of Iota Carrageenan, Kappa Carrageenan and Alginate Hydrogels as Tissue Engineering Scaffolds*, Gums and Stabilisers for the Food Industry 15 (eds P.A. Williams and G.O. Phillips), Royal Society of Chemistry, Cambridge, pp. 407–413.
146. Pereira, R.C. et al. (2009) *J. Tissue Eng. Regen. Med.*, **3**(2), 97–106.
147. Seidlits, S.K. et al. (2011) *Acta Biomater.*, **7**(6), 2401–2409.
148. Rederstorff, E. et al. (2011) *Acta Biomater.*, **7**(5), 2119–2130.
149. Cloyd, J.M. et al. (2007) *Eur. Spine J.*, **16**(11), 1892–1898.
150. Wang, D.A. et al. (2007) *Nat. Mater.*, **6**(5), 385–392.
151. Rouillard, A.D. et al. (2011) *Tissue Eng. Part C-Methods*, **17**(2), 173–179.
152. Hutmacher, D.W. et al. (2007) *J. Tissue Eng. Regen. Med.*, **1**(4), 245–260.
153. Hutmacher, D.W. (2001) *J. Biomater. Sci. Polym. Ed.*, **12**(1), 107–124.
154. Cruz, D.M. et al. (2010) *J. Biomed. Mater. Res. A*, **95A**(4), 1182–1193.
155. Duarte, A.R.C., Mano, J.F., and Reis, R.L. (2010) in *Advanced Materials Forum V, Pt 1 and 2* (eds L.G. Rosa and F. Margarido), Trans Tech Publications Ltd, Stafa-Zurich, pp. 22–25.
156. Temtem, M. et al. (2009) *J. Supercrit. Fluids*, **48**(3), 269–277.
157. Kim, M.Y. and J. Lee, *Carbohydr. Polym.*, **84**(4), 1329–1336.
158. Ji, C.D. et al. (2011) *Acta Biomater.*, **7**(4), 1653–1664.
159. Jayakumar, R. et al. (2011) *Int. J. Mol. Sci.*, **12**(3), 1876–1887.
160. Yang, T.L. (2011) *Int. J. Mol. Sci.*, **12**(3), 1936–1963.
161. Ehrlich, H. et al. (2010) *Int. J. Biol. Macromol.*, **47**(2), 132–140.
162. Ehrlich, H. et al. (2010) *Int. J. Biol. Macromol.*, **47**(2), 141–145.
163. Silva, S.S. et al. (2011) *Acta Biomater.*, **7**(3), 1166–1172.
164. Lee, S.H. et al. (2010) in *Ionic Liquid Applications: Pharmaceuticals, Therapeutics, and Biotechnology* (ed. S.V. Malhotra), American Chemical Society, Washington, DC, pp. 115–134.
165. Emami, S.H. et al. (2010) *Int. J. Mater. Res.*, **101**(10), 1281–1285.
166. Jayakumar, R. et al. (2011) *Int. J. Biol. Macromol.*, **49**(3), 274–280.
167. Jayakumar, R. et al. (2011) *Int. J. Biol. Macromol.*, **48**(2), 336–344.
168. Tanase, C.E., Popa, M.I., and Verestiuc, L. (2011) *Mater. Lett.*, **65**(11), 1681–1683.
169. Wang, G.C. et al. (2011) *Tissue Eng. Part A*, **17**(9–10), 1341–1349.
170. Maganti, N. et al. (2011) *Adv. Eng. Mater.*, **13**(3), B108–B122.
171. Martins, A.M. et al. (2009) *Acta Biomater.*, **5**(9), 3328–3336.
172. Malafaya, P.B. and Reis, R.L. (2003) in *Bioceramics 15* (eds B. BenNissan, D. Sher, and W. Walsh), Trans Tech Publications Ltd, Zurich-Uetikon, pp. 39–42.
173. Danilchenko, S.N. et al. (2011) *J. Biomed. Mater. Res. A*, **96A**(4), 639–647.
174. Oliveira, J.M. et al. (2006) *Biomaterials*, **27**(36), 6123–6137.
175. Ge, H.R. et al. (2010) *J. Mater. Sci. Mater. Med.*, **21**(6), 1781–1787.
176. Kusmanto, F. et al. (2008) *Chem. Eng. J.*, **139**(2), 398–407.
177. Abramovitch-Gottlib, L., Geresh, S., and Vago, R. (2006) *Tissue Eng.*, **12**(4), 729–739.
178. Cunningham, E. et al. (2010) *J. Mater. Sci. Mater. Med.*, **21**(8), 2255–2261.
179. Kretlow, J.D., Klouda, L., and Mikos, A.G. (2007) *Adv. Drug Delivery Rev.*, **59**(4–5), 263–273.
180. Silva, G.A., Ducheyne, P., and Reis, R.L. (2007) *J. Tissue Eng. Regen. Med.*, **1**(1), 4–24.
181. Silva, G.A. et al. (2007) *J. Tissue Eng. Regen. Med.*, **1**(2), 97–109.
182. Prabaharan, M. and Mano, J.F. (2005) *Drug Deliv.*, **12**(1), 41–57.
183. Cruz, D.M.G. et al. (2008) *J. Tissue Eng. Regen. Med.*, **2**(6), 378–380.
184. Leonor, I.B. et al. (2008) *Acta Biomater.*, **4**(5), 1349–1359.

185. Grech, J.M.R., Mano, J.F., and Reis, R.L. (2010) *J. Mater. Sci. Mater. Med.*, **21**(6), 1855–1865.
186. Munarin, F. *et al.* (2010) *J. Mater. Sci. Mater. Med.*, **21**(1), 365–375.
187. Grenha, A. *et al.* (2010) *J. Biomed. Mater. Res. A*, **92A**(4), 1265–1272.
188. Lim, J.J. *et al.* (2011) *Acta Biomater.*, **7**(3), 986–995.
189. Sahiner, N. *et al.* (2008) *J. Biomater. Sci. Polym. Ed.*, **19**(2), 223–243.
190. Nicklas, M. *et al.* (2009) *Drug Dev. Ind. Pharm.*, **35**(9), 1035–1042.
191. Swatschek, D. *et al.* (2002) *Eur. J. Pharm. Biopharm.*, **54**(2), 125–133.

2
Hydrogels from Protein Engineering

Midori Greenwood-Goodwin and Sarah C. Heilshorn

2.1
Introduction

Protein-engineered hydrogels build upon the natural structure and function of proteins to create well-defined, multifunctional materials by cross-linking polypeptide chains together. Individual polypeptide chains are composed of independent peptide domains built from amino acid monomers. The engineered amino acid sequence is designed to form specific and reproducible structures of the selected peptide domains. The exact amino acid sequence of the polypeptide chain can be encoded into a DNA sequence, which is transfected into a host microorganism in order to produce a monodisperse batch of polypeptide chains. Interactions between individual polypeptide chains can be engineered to form hydrated macromolecular structures with tunable mechanical strength. These interchain interactions can include both noncovalent cross-linking between neighboring peptide domains as well as covalent cross-linking between amino acid residues to form stable hydrogels.

In this chapter, we discuss the development of protein-engineered hydrogels with tunable mechanical and biochemical properties for biomaterial applications. First, we discuss the fundamentals of protein structure and recombinant protein synthesis. Next, we review the diversity of peptide domains that have been utilized in protein-engineered hydrogels to date, including cell-binding domains, structural domains, and molecular recognition domains. We continue to describe the various enzymatic and synthetic chemistry strategies that have been utilized to create covalently cross-linked protein-engineered hydrogels. We conclude with an in-depth discussion of cellular interactions with protein-engineered hydrogels and of the ability to tune the biomechanics and biochemistry of the materials to direct cellular phenotype.

Development of novel protein-engineered hydrogels for therapeutic applications requires regulation of specific cellular behavior through tuning of the structural, mechanical, and biochemical properties of the hydrogel. The cellular response and activation of intracellular signaling pathways for cell adhesion and migration is different between traditional two-dimensional (2D) cell culture environments and native three-dimensional (3D) cell environments, thus the use of 3D materials for

Biomimetic Approaches for Biomaterials Development, First Edition. Edited by João F. Mano.
© 2012 Wiley-VCH Verlag GmbH & Co. KGaA. Published 2012 by Wiley-VCH Verlag GmbH & Co. KGaA.

in vitro studies provides an improved platform for understanding *in vivo* cellular behavior [1]. *In vivo*, the composition of the extracellular matrix (ECM) regulates how mechanical forces and biochemical cues are transmitted to cells, thus affecting the ability of cells to adhere, migrate, proliferate, and differentiate [2]. The ability to mimic complex 3D tissue physiology for *in vitro* studies of complex cell behaviors is a significant challenge in the biomaterials field. Protein-engineered hydrogels can be tailored to mimic the ECM and the native tissue mechanics, providing specified and reproducible 3D environments in which cells interact with each other and the surrounding material.

2.2
Principles of Protein Engineering

2.2.1
Protein Structure and Folding

The structure and chemistry of each natural protein has evolved over billions of years from a canonical set of only 20 amino acids, creating a hierarchy of structures that dictates the final 3D conformation. The unique structures of individual proteins correspond to their specialized functions, including catalytic activity, molecular recognition, and extracellular support, among many others. The *primary structure of proteins* is defined as the sequence of amino acid residues in the protein or polypeptide chain (Figure 2.1a) [3]. Amino acid monomers contain an amino group, a carboxyl group, and a side group. Each of the 20 amino acids has a unique side group and is classified by the chemical composition of the side group, which can be negatively charged, positively charged, uncharged and polar, or uncharged and nonpolar.

The *secondary structure of proteins* is defined as the localized folding patterns of the polypeptide chain. The flexibility and folding of any given polypeptide chain is restricted by steric interactions, which limit the possible bond angles between adjacent amino acids. The folding of a polypeptide chain is further constrained by several weak noncovalent bonding interactions such as hydrogen bonding, electrostatic interactions, and van der Waals attractions. Common secondary structures or local folding patterns within the protein structure include the α-helix and β-sheet (Figure 2.1b). The α-helix is formed by hydrogen bonding between the primary amine of one peptide bond with the carbonyl group of a second peptide bond, located four amino acids farther along the polypeptide chain, resulting in a regular helix formation. The β-sheet is formed by hydrogen bonding between the peptide bonds in polypeptide chains that run in the same or opposite orientation, that is, parallel or antiparallel. In addition, turns of four amino acid residues stabilized by hydrogen bonding allow the polypeptide chain to fold toward the interior and back toward the exterior [3].

The *tertiary structure of proteins* is defined as the global 3D conformation of the polypeptide chain. Proteins typically range from 60 to 2000 amino acids long

2.2 Principles of Protein Engineering

(a) Primary structure of VEGF₁₆₅

APMAEGGGQNHHEVVKFMDVYQRSYCHPIETLVDIFQEYPDEIEYIFKPSCVPLMR
CGGCCNDEGLECVPTEESNITMQIMRIKPHQGQHIGEMSFLQHNKCECRPKKDRA
RQENPCGPCSERRKHLFVQDPQTCKCSCKNTDSRCKARQLELNERTCRCDKPRR

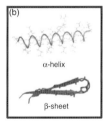

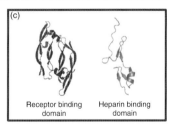

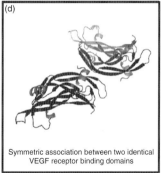

Figure 2.1 Hierarchical protein structure. (a) The primary structure, i.e. amino acid sequence, for human vascular endothelial growth factor (VEGF) spliced to 165 residues is shown [4]. (b) The most common secondary structure motifs, the α-helix and β-sheet, are shown as ribbon cartoons superimposed onto a model of the molecular structure showing individual amino acid residues. (Images generated *de novo* using PyMol; both motifs were generated based on hydrogen bonding and the α-helix motif was built based on the known leucine-zipper motif, but was not taken from a known protein structure or sequence. Any similariy to other structures is purely coincidence.) (c) The tertiary structures of the VEGF receptor binding and heparin binding domains are shown as ribbon cartoons [5, 6]. (d) The putative quaternary structure of a VEGF receptor binding complex is shown as a ribbon cartoon [7].

and adopt a limited number of protein folds. X-ray crystallization and nuclear magnetic resonance (NMR) methods have been used to identify over 1000 protein folds [8]. The final protein conformation for a given polypeptide chain generally yields a structure in which the free energy is minimized. Thus, many proteins form a hydrophobic core, in which the hydrophobic side chains of nonpolar amino acids, for example, V, L, G, and C, group toward the interior of the protein (IUPAC-IUB single letter abbreviations for amino acids are used throughout the text). The tertiary structure can be further stabilized by covalent bonding, primarily by disulfide bridges between cysteine residues.

Within a single protein there are several peptide domains. Peptide domains are established when localized secondary structure is found in a variety of proteins, and forms a stable structure typically having a similar and predictable function. Small proteins may contain only a few peptide domains, while large proteins may contain dozens of peptide domains. Peptide domains often include α-helices, β-sheets, or both in a single domain. Furthermore, peptide domains are often defined in functional terms on the basis of their observed bioactivity; for example, the protein *vascular endothelial growth factor (VEGF)* has both specialized receptor binding domains and heparin binding domains (Figure 2.1c) [5, 6]. Many structural features of peptide domains result from noncovalent bonding and form by spontaneous self-assembly; these structures can be disrupted in harsh environments and then reform in an amiable environment.

The functional domains of proteins are primarily derived from their primary, secondary, and tertiary structures, as described above. However, many proteins require an additional level of hierarchical arrangement referred to as *quaternary structure* to achieve functionality. *The quaternary structure of proteins* is defined as the arrangement of two or more distinct polypeptide chains into large protein complexes or multimeric proteins. For example, the receptor binding domain of VEGF is homodimeric, wherein two identical binding domains comprise the biologically active complex (Figure 2.1d) [7]. The quaternary structure of proteins is often globular (e.g., *hemoglobin, alcohol dehydrogenase*); however, proteins also can form long fibrous proteins (e.g., *collagen*) and helical filaments (e.g., *actin*) that span the lengths of entire cells. Such multimeric proteins can be formed by self-assembly, generating complex macromolecular structures under specific environmental conditions.

2.2.2
Design of Protein-Engineered Hydrogels

Protein-engineered materials are designed to mimic both the physical structure and biofunctionality, or bioactivity, of proteins in a reproducible and predictable manner. Multiple peptide domains, each with a discrete function, can be designed into a single polypeptide chain to create a protein-engineered hydrogel with specific structural, mechanical, and biochemical properties (Table 2.1). Because the function

Table 2.1 Representative peptide domains used in protein-engineered hydrogels.

Domain type	Function	Example peptide sequences
Cell-binding	Promotes cell adhesion to the hydrogel and/or mimics cell–cell interactions	Cell-adhesion ligands: RGDS [10], REDV [11], IKVAV [12], YIGSR, and GRILARGEINF [13]
Structure	Dictates mechanical stiffness and 3D structure of the hydrogel	Coiled-coil domain: SGDLENEVAQLEREVRS LEDEAAELEQKVSRLKNEIEDLKAE [14]
		Elastinlike domain: (VPGXG)n [15]
Degradation	Regulates local and/or bulk degradation of the hydrogel as a result of protease activity	Tissue-plasminogen-activator-sensitive domains: GTAR, TSHR, and DRIR [9]
		Matrix-metalloproteinase-sensitive domains: GCRDGPQGIWGQDRCG [13] and PVGLIG [16]
Cross-linking	Defines the type of linkage made between individual polypeptide chains, contributes to the mechanical stiffness of the hydrogel	Tissue transglutaminase cross-linking domain: VPGQG [17]
Molecular-recognition	Binds other biomacromolecules, e.g., growth factors, DNA, ligands, for therapeutic and regulatory activities	Calmodulin binding domain: PGD [18]
		Heparin binding domain: FAKLAARLYRKA [19]

of these domains is specified by the amino acid sequence, protein-engineered hydrogels allow for independent parameterization and tuning of each domain for various applications.

The incorporation of multiple selected peptide domains into a single polypeptide chain is further dependent on the ability of these domains to fold well and to form a stable overall conformation. The addition of flexible linker regions between adjacent peptide domains and the inclusion of endogenous amino acid sequences that are longer than the minimal functional sequence can promote the folding of individual peptide domains into their native conformations, conferring the desired biological and structural functionalities [14, 20]. In addition, the stability and hydrophilicity of the overall amino acid sequence of the final polypeptide chain must be analyzed. This is primarily done by using the amino acid sequence to predict the final pK_a (acid disassociation in water), solubility, secondary structure, and rate of aggregation of the polypeptide chain for a given environment [21, 22].

The cross-linking and structural domains directly contribute to the overall mechanical properties of hydrogels. Cross-linking domains link two or more individual polypeptide chains together through covalent and/or noncovalent binding. Engineering the extent and strength of cross-linking affects the mechanical stiffness of the resulting protein-engineered hydrogel. Moreover, the choice of structural domain contributes to the mechanical stiffness of the resulting protein-engineered hydrogel. For example, β-sheet structural domains are very rigid compared to α-helix structural domains and thus form stiffer hydrogels. Many structural domains used in protein-engineered hydrogels are designed to mimic or have been evolved from naturally occurring sequences. The selection of the structural domain is further dependent on the intended application, for example, elastinlike polypeptides (ELPs) and silklike polypeptides are used for their thermal responsiveness and material strength, respectively [23, 24]. One or more substitutions to the primary structure of the polypeptide chain have a direct effect on the physical and chemical properties of the protein-engineered hydrogel [25]. As such, novel amino acid sequences can be rationally designed to modulate structural behavior and function. Examples of rationally designed materials include self-assembling hydrogels produced by mixing of α-helical structures [26], macroscopic fibers produced from spider silk polypeptides [27], and volume-swelling gels formed in response to conformational changes in the polypeptide domains [28, 29].

The bioactive domains (cell-binding, molecular-recognition, and degradation domains) contribute to the ability of cells to interact with the material in a manner that mimics cellular interactions in native tissue. Cell-binding domains are incorporated to direct cellular attachment, thereby mediating cell spreading, differentiation, and migration. Molecular-recognition domains are incorporated to present DNA, RNA, hormones, proteins, and other biologically active molecules to cells, directing cellular activity by activating specific signaling pathways and regulating gene expression. Degradation domains are incorporated to control localized material degradation in response to cellular activity, promoting cell invasion. Furthermore, *de novo* peptide domains can be obtained by high throughput screening and computational modeling methods [30]. These methods allow for the development of novel peptide domains for use in protein-engineered hydrogels.

2.2.3
Production of Protein-Engineered Hydrogels

After peptide domain selection, the desired recombinant protein is synthesized by encoding the designed amino acid sequence into a recombinant DNA sequence that will be transcribed and translated in a host organism [31]. In cases of smaller polypeptide chains, about 100 amino acids or less, the polypeptide chain is produced by solid-phase synthesis [32]; however, this process is expensive, prone to sequence errors, and less adaptable for large-scale material production. Following protein expression by the host organism, the designed polypeptide chain must be purified from all other macromolecules of the organism. Finally, the designed polypeptide chain is processed to form a protein-engineered hydrogel, which may require protein concentration, hydration, refolding, self-assembly, and covalent cross-linking processes (Figure 2.2). A variety of host organisms, processing methods, and analysis techniques are used depending on the peptide domain selection, the designed polypeptide chain, and the intended application.

The choice of host organism for recombinant protein synthesis is primarily dependent on the complexity of the desired final protein product and the efficiency of protein production and purification. For recombinant proteins that require complex folding, glycosylation, or other posttranslational modifications to become functional, a mammalian or insect cell line is typically selected for expression over bacterial and yeast host organisms [33]. For less complex polypeptide chains, bacteria and yeast are often selected as host organisms because of their relatively fast production cycles, minimal medium requirements, and low overall costs. Regardless of the cell type, each amino acid is encoded as a three-nucleotide DNA sequence called a *codon*. Because several codons exist for a single amino acid, there are many possible DNA sequences that encode for the same polypeptide chain. Depending on the organism selected, the frequency of use of each codon will vary, and thus the DNA sequence designed for a recombinant gene must be optimized for expression in a particular host organism depending on how frequently the

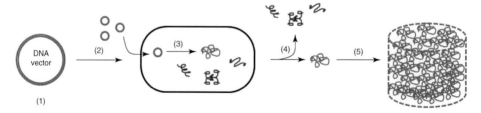

Figure 2.2 Overview of protein production and hydrogel formation process. (1) Encode amino acid sequence into a DNA vector, (2) transfect host cells with DNA, (3) express proteins as cells grow and divide, (4) purify engineered protein from host cell proteins and cell debris, and (5) concentrate protein and assemble into a hydrogel.

codon appears within the host genome. More recent efforts have focused on codon optimization Additional algorithms for codon optimization are also based on the availability of transfer RNAs, ribosome binding sites, and messenger RNA secondary structures, as well as on empirical correlations derived from multiple expression experiments [34–36]. Optimization programs have been designed to predict DNA sequences that translate efficiently based on both the host organism and the designed primary amino acid sequence [37]. Further progress in these optimization programs will continue to improve the expression, as well as the total yield of recombinant proteins. Additional methods of improving protein expression yield are to use optimized growth medium [38], to limit the expression of endogenous genes in the host that may compete for amino acids, and to use enhanced promoter sequences that improve the initiation of transcription [39].

In addition to the desired recombinant polypeptide chain, the encoded amino acid sequence may include purification and secretion tags to assist in the recovery of the engineered polypeptide chain. If the desired protein is secreted, then the media or supernatant is collected before purification. If the desired protein is not secreted, cells must be lysed, that is, broken apart, to access the targeted protein. While the choice of the host does not dictate a particular purification strategy, the required stringency of the purification is dependent on the choice. For example, expression of recombinant proteins in *Escherichia coli* for therapeutic applications requires testing for possible contamination by endotoxins, which are lipopolysaccharides found in the bacterial cell wall that can cause an immune response in mammals [40]. The methods for protein purification are selected based on the size, charge, hydrophobicity, thermal stability, and localization of the targeted protein. In many cases, the target protein can be isolated by ion-exchange, gel-filtration, or affinity-chromatography in sequential purification steps [3]. Proteins can also be purified based on their solubility in water and other solvents. For example, ELPs and ELP-fusion proteins can be purified without chromatography by inverse transition cycling. At low temperature, ELPs are soluble in aqueous solution, but at higher temperatures, the ELPs aggregate to form a protein-rich coacervate that can be easily purified by centrifugation [41]. The extent of purification of proteins can be analyzed by protein sequencing (e.g., mass spectrometry, amino acid analysis), by size analysis (e.g., gel electrophoresis, size exclusion chromatography), and by biological assays (e.g., western blots, ELISA) [42]. Overall, there are a variety of methods developed for protein purification that can be used to isolate and recover the target protein based on its physical and chemical properties.

After purification, hydrogel formation can be achieved either by self-assembly or by covalent cross-linking of the polypeptide chains under specified conditions. Protein-engineered hydrogels are water-swollen polypeptide materials that maintain a distinct 3D structure. The structure of a hydrogel can be analyzed by solid-state NMR spectroscopy [43], small-angle X-ray scattering, neutron scattering [44], mercury intrusion porosimetry, nitrogen adsorption, and scanning electron microscopy [45, 46]. Since the mechanical behavior of protein-engineered hydrogels changes based on the structural and cross-linking domains selected, the material stiffness and elasticity are further characterized and analyzed by rheological and microrheological methods [47, 48]. Additional domain features, such as the density of cell-binding, molecular-recognition, and degradation domains can be further predicted and analyzed by multi-scale modeling [49], amino acid analysis [50], elemental analysis [51], and analysis of cell–material or protein–protein interactions within the material using fluorescent resonance energy transfer (FRET) [52–54]. Further, analysis of the protein-engineered structures can be achieved by circular dichroism and isothermal titration calorimetry, which can be used to evaluate the final protein conformation [55] and protein binding interactions [56], respectively.

2.3
Structural Diversity and Applications of Protein-Engineered Hydrogels

The folding patterns of amino acid sequences can be used to create viscoelastic materials with tunable macromolecular structure. Structural elements derived from nature, including the common secondary structures in proteins, are typically used as the foundation for protein-engineered hydrogels. Because these materials are produced by recombinant technology, multiple structural and cross-linking peptide domains are easily incorporated into the bulk 3D structure. In this section, we review the functionality of several peptide domains used to form hydrogels, the protein-engineering strategies for the *de novo* development and rational design of peptide domains, as well as the current applications of protein-engineered hydrogels.

2.3.1
Self-Assembled Protein Hydrogels

Polypeptide chains can self-assemble into hydrogels by physical cross-linking, which is noncovalent and reversible. Physical cross-links result from ionic or electrostatic interactions, hydrogen bonding, and/or hydrophobic interactions. These interactions can be highly specific and tuned at the molecular level by altering the amino acid sequence or by altering the surrounding environment [57]. Several polypeptide chains have been designed and have been produced by recombinant technology to form stimuli-responsive, physically cross-linked hydrogels under varying environmental conditions, including changes to the external temperature, pH, solvent, or solutes present (Figure 2.3). Depending on the specific peptide domain(s) selected, gelation can be controlled in response to any number of environmental cues. Physically cross-linked materials are easy to process, do not contain unreacted chemical intermediates or by-products, and can be used to encapsulate biologically active molecules and cells that are sensitive to chemical cross-linkers or organic solvents.

Coiled-coil domains separated by a flexible, water-soluble peptide domain designed into a single polypeptide chain form physically cross-linked hydrogels when the coiled-coil domains aggregate. The rate and strength of coiled-coil aggregation

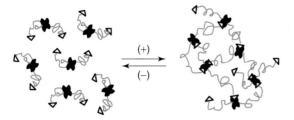

Figure 2.3 Schematic representation of a self-assembling protein-engineered hydrogel. On the left, isolated polypeptide chains are in solution. The addition (+) of an environmental stimulus results in noncovalent cross-linking between polypeptide chains and forms a solid hydrogel. This process is reversible by removal (−) of the environmental trigger.

can be engineered such that the coiled-coil domains dissociate at high pH and temperature and collapse at near-neutral pH and ambient temperature [14]. Rational design at the level of the amino acid sequence alters the electrostatic interactions between coiled-coil domains and can be used to enhance the thermal and pH stabilities of the coiled-coil aggregates, reducing the critical concentration required for coiled-coiled aggregation [58]. The thermostability of coiled-coil domains for protein-engineered hydrogels can also be enhanced through covalent disulfide bond formations [59].

Hybrid hydrogels combine peptide domains with synthetic polymer domains for engineering materials with desired structural, chemical, and biological properties. Incorporation of the coiled-coil domains into a synthetic copolymer was used to produce a temperature-sensitive hydrogel capable of physical cross-linking and with enhanced hydrogel swelling dynamics [60]. Changes to the coiled-coil domain as a result of its thermostability correlated with the hydrogel swelling, thereby enabling tuning of stimuli-responsive properties by engineering the strength of the physical cross-links [61]. For example, engineering of two different coiled-coil domains coupled to a hydrophilic synthetic polymer backbone was used to form antiparallel heterodimeric coiled-coil aggregates. The formation of coiled-coil aggregates allowed for *in situ* formation of stable hydrogels at very low concentrations that were readily degraded in the presence of a protein denaturing agent [62].

Novel protein-engineered hydrogels derived from coiled-coil domains can be developed for tissue engineering, drug delivery, and biosensor applications. The rate of erosion of hydrogels can be controlled by engineering the network topology in which a single polypeptide chain could adopt either a bridged or a looped configuration. Hydrogels formed from polypeptide chains containing a heterogeneous solution of coiled-coil domains reduce the presence of the looped configuration and are eroded 100 times more slowly than those formed from a homogenous solution of coiled-coil domains. Further development and combinations of coiled-coil domains could be used to fine-tune the erosion rate for controlled cell invasion into or drug release from a hydrogel [63]. In another example, a chimeric protein-engineered hydrogel containing coiled-coil domains was fused to the dentin matrix protein 1, which includes a hydroxyapatite nucleating domain and cell-binding domains, for bone tissue engineering applications. The resulting hydrogel supported hydroxyapatite nucleation and promoted osteoblast cell adhesion, which are both essential for bone regeneration [64]. In a third example, the incorporation of a proteolytic target peptide between the first and second coiled-coil domains created an autoinhibited coiled-coil aggregate that disassembled in response to protease activity. Following disassembly, the first coiled-coil domain was free to aggregate with a third coiled-coil domain. Optimization of both the proteolytic target peptide sequence between the first two coiled-coil domains and the strength of interaction between the first coiled-coil domain and the third coiled-coil domain yielded enhanced sensitivity to protease activity by improving the formation of a stable and functional luciferase, fused to the first and third coiled-coil domains [65]. Although this design has yet to be developed into a composite polypeptide chain, it is an excellent example of the functional diversity of the coiled-coil domain that can be further developed into a novel protein-engineered hydrogel.

Protein-engineered hydrogels with independent biological and structural functions can be designed by flanking coiled-coil domains with hydrophilic peptide domains and biologically active domains. For example, different polypeptide chains were designed to incorporate large proteins such as green fluorescent protein (GFP), enhanced cyan fluorescent protein (CFP), or Discosoma red fluorescent protein (dsRed) into the soluble domain. The addition of these large proteins did not impact the ability of the coiled-coil domains to aggregate and form a hydrogel. Furthermore, the incorporation of various fluorescence-containing polypeptide chains enabled the independent tuning of fluorescence loading from the structural stability of the hydrogel [66]. An analogous design scheme was used to form antibody binding, bioelectrocatalytic, and enzymatically active protein-engineered hydrogels [67–69]. Moreover, an enzymatic hydrogel was designed to become active upon gelation. The polypeptide chains contained inactive organophosphate hydrolase (OPH) monomers that formed active OPH dimers following gelation by coiled-coil domain aggregation. These hydrogels exhibited OPH activity and retained their hydrogel structure after five months of storage in cold buffer [70]. These results display the robustness of coiled-coil aggregation in the presence of large proteins, which can be utilized for biosensor and biomedical applications.

Elastinlike polypeptide (ELP) chains form physically cross-lined protein-engineered hydrogels with varying mechanical properties simply by altering ELP chain-solvent interactions and processing of ELP chains. ELPs are made up of oligomeric repeats of the pentapeptide sequence VPGXG, where X is any amino acid except for proline. The pentapeptide sequence is derived from the natural protein elastin, found in soft tissues such as cartilage [71]. ELPs exhibit lower critical solution temperature (LCST) behavior. Below the LCST, the polypeptide chain is soluble in aqueous solution, whereas above the LCST, the polypeptide chain undergoes a reversible phase transition to form a polymer-rich coacervate that can form a weak hydrogel under ideal and good solvent conditions. Alterations to the amino acid sequence of ELPs impact the observed LCST and can be engineered such that ELPs behave as a liquid at room temperature and form hydrogels at physiological temperature [72]. Shorter polypeptide chains based on elastin mimetic sequences also showed solution–gel phase behavior but formed ELP spherical nanoparticles instead of hydrogels [73, 74]. Thus, the tuning of the amino acid block lengths has a direct effect on the ELP gel behavior and macrostructure.

ELP-based hydrogels can be developed for tissue engineering, drug delivery, and environmental applications. For example, ELPs were engineered to form nanoparticles for use as soluble drug carriers in the treatment of solid tumors. In response to hyperthermia treatment by local heating of the whole tissue to 42 °C, the ELPs aggregated and were selectively taken up by the targeted cancer cells [75]. Furthermore, ELPs were used to encapsulate adipose-derived stem cells (ASCs) during gelation. In the presence of dexamethasone and transforming growth factor-beta (TGF-β), the encapsulated cells were cultured within the ELP-based hydrogels. These same cells showed markers for chondrocytic differentiation and increased deposition of collagen within the hydrogels compared to traditional 2D cell cultures. Thus, ELP-based hydrogels may be useful for cartilage and bone regeneration applications [41]. For environmental applications, metal binding ELP-based hydrogels were designed by incorporating a hexamer repeat of histidine into the hydrophilic segment of the ELP chain. The resulting hydrogel was able to bind 1.31 ± 0.05 nmol cadmium per nanomole of polypeptide chain. The metal could then be removed from the hydrogel, and the binding sites regenerated through a thermocycling process [76].

Physically cross-linked, protein-engineered hydrogels containing very rigid and stable β-sheet structures form by extensive hydrogen bonding between amino acid residues within a single polypeptide chain. The structural stability and self-assembly of β-sheet structures were analyzed for 63 distinct recombinant proteins sharing an identical binary pattern of polar and nonpolar residues. Despite differences in the amino acid identity at each site, all proteins formed a stable monolayer at the air–water interface and each monolayer consisted of amphiphilic β-sheets, suggesting that the β-sheet assembly is encoded into the alteration between polar and nonpolar residues [77]. The incorporation of protein hen egg white lysozyme (HEWL) into a polypeptide chain was used to form an elastic, fibril-containing hydrogel that exhibited both β-sheet and α-helix domains. The observed fibrils appeared to self-assemble along their long axes to form larger fibers that then became physically entangled and extended in the presence of water to form a hydrogel [78]. The assembly of β-sheet structures is also widely used in peptide- and recombinant-silk-based hydrogels to form fibrillar, semicrystalline polymer networks [79, 80].

Self-assembly through the heteroassembly of two or more peptide domains can be used to form a physically cross-linked hydrogel network. For example, polypeptide chains that contained repeats of either a WW domain or a proline-rich domain connected by random coil hydrophilic spacers were used to form a hydrogel upon mixing of the two polypeptide chains [81]. Gelation occurred as a result of the molecular recognition and specific binding between WW and proline-rich domains and did not require external cues such as shifts in pH or temperature. Tuning of the bulk material properties of two-component hydrogels can be achieved by altering the spacer length between binding sites, by varying the number of binding sites in one polypeptide chain, and/or by modifying the binding affinity between partners. For example, two different WW-containing polypeptide chains were designed: one derived from a native WW domain and the other derived from a computational model that resulted in a 10-fold increase in binding affinity for the proline-rich domain [82]. The use of the computationally derived high-affinity WW domain in place of the native WW domain sequence resulted in stronger cross-linking, increasing the viscoelastic properties of the hydrogel. Moreover, there is a large library of identified WW and proline-rich domains, which can be used to incorporate orthogonal binding sites within the hydrogel. The orthogonal binding sites could be used to immobilize and deliver bioactive factors or otherwise incorporate additional bifunctionality into the hydrogel. Finally, as a result of the transient physical cross-links between the WW and proline-rich domains, the resulting hydrogel has shear thinning and self-healing properties under constant environmental conditions, which is ideal for medical applications requiring injection of the hydrogel [81].

The development of novel biomaterials that assemble in the presence of a specific analyte or ligand has been achieved with a variety of peptide domains. In these materials, sensing actuation of the peptide domain is initiated by the recognition of the ligand and is then translated into a mechanical action. In one example, the covalent incorporation of a protein antibody and specific antigen domain into a synthetic polymer hydrogel allowed for reversible and tunable swelling of the hydrogel. The physical cross-linking resulted from antibody–antigen interactions and could be disrupted by the addition of soluble antigen to the surrounding media [83]. Similarly, incorporation of a lectin protein domain was used to generate glucose-sensitive hydrogel. Specifically, immobilization of the lectin domain concanavalin A (Con-A) into a covalently cross-linked hydrogel created a gel that underwent a volume phase transition at physiologically relevant temperatures in response to the presence and concentration of specific saccharides [84]. The selection of synthetic polymer, density of Con-A, and glucose delivery in the media further impacted the mechanical properties of the resulting hydrogel and enabled tuning of swelling and protein release properties [85, 86].

The EF-hand domain is a calcium-binding domain common to many calcium sensor proteins involved in intracellular signal transduction. The domain contains a helix-loop-helix structure and binds the calcium ion, Ca^{2+}, in a cooperative and ligand dependent manner [87]. Calcium-dependent binding occurs between the calmodulin protein (CaM) and calmodulin binding domains (CBDs) [88]. In response to the presence or absence of Ca^{2+} and the ligand phenothiazine, CaM forms one of three structures [89]. A genetically engineered CaM protein and the phenothiazine peptide ligand were covalently immobilized into a synthetic hydrogel. In the presence of Ca^{2+}, CaM binds the phenothiazine ligand, increasing the cross-link density of the hydrogel. Both the chelation of Ca^{2+} and the addition of soluble ligands with stronger binding affinities for CaM than phenothiazine disrupted the cross-linked network and led to increased hydrogel swelling. Modifications to the amino acid sequence of CaM were used to tune the hydrogel swelling properties in the presence of Ca^{2+}. In all cases, the observed swelling states were reversible, with minimal loss of the mechanical integrity of the hydrogel after multiple swelling and deswelling cycles [89]. The gelation and elastic modulus of protein-engineered hydrogels can also be tuned by selecting CBDs with varying binding affinities for CaM and/or varying Ca^{2+} dependency [18].

Ligand-responsive hydrogels have the potential for biosensor and drug release applications. Chemically responsive microlenses with predictable changes to the swelling of the 3D bulk network were engineered through the incorporation of CaM as a model hinge-motion binding protein. By varying the presence of competing high-affinity and low-affinity ligands for CaM, the swelling and deswelling of the microlenses were controlled. This platform can be further developed using a diversity of protein–ligand binding interactions, such that the microlenses swell only in the presence of a specific ligand and thus can be developed for biosensor applications [90]. A protein-engineered hydrogel designed from CaM fused to a thermoresponsive ELP chain was used as a biological actuator, wherein binding of Ca^{2+} to CaM induced a phase transition in the ELP chain. Because the induced transition was triggered by a specific binding interaction, the hydrogel system can be adapted for use in large-scale protein pull-down assays and biosensor applications [91]. The first demonstration of drug delivery triggered by CaM conformational changes used a polyethylene glycol (PEG)-CaM hybrid hydrogel to release VEGF. Addition of the biochemical ligand trifluperazine (TFP) resulted in the rapid release of VEGF that had been initially absorbed into the hydrogel (Figure 2.4) [92]. The ability to release biological molecules in response to specific binding interactions is advantageous compared to nonspecific triggers (e.g., temperature, light, pH) for many therapeutic applications.

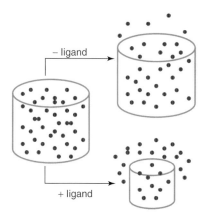

Figure 2.4 Schematic representation of a ligand-responsive protein-engineered hydrogel for drug delivery. On the left, the hydrogel is loaded with a soluble molecule (●) by mixing the molecule into the solution as the gel forms. The addition (+) of an exogenous ligand reduces the conformational flexibility of the cross-linking domains, further reducing the extension length of each polypeptide chain and resulting in gel deswelling and quick release of molecules. In the absence (−) of the exogenous ligand, the molecules are released slowly from the hydrogel.

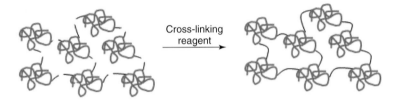

Figure 2.5 Schematic of a covalently cross-linked protein-engineered hydrogel. On the left, isolated polypeptide chains are in solution. The addition of a (bio)chemical cross-linking reagent results in covalent cross-linking between polypeptide chains at chemically reactive amino acid residues.

2.3.2
Covalently Cross-Linked Protein Hydrogels

Polypeptide chains can form hydrogels by enzymatic and/or synthetic cross-linking processes, which form covalent bonds between polypeptide chains resulting in fixed positions of individual chains within the hydrogel network (Figure 2.5). Covalently cross-linked materials provide additional control over the mechanical properties of protein-engineered hydrogels and can be used to form stable hydrogels from polypeptide chains that cannot form physical cross-links or to improve mechanical stability of a physically cross-linked hydrogel. Covalently cross-linked materials are generally stronger and stiffer than physically cross-linked hydrogels, maintain structural and functional integrity under applied stress, and are reversibly elastic.

Covalent cross-linking can be achieved by using synthetic chemicals to link specific amino acid residues or nonspecific chemical groups at various amino acid residues. Proresilin, a resilinlike protein derived from elastic proteins elastin and silk, can be recombinantly expressed and purified from *E. coli* and cast into an elastomeric hydrogel by rapid photochemical cross-linking between two tyrosine residues. The ductility and resilience of the cross-linked recombinant resilin exceeded those of the commercially available rubber polybutadiene [93]. ELPs can also form hydrogels at low protein concentrations without undergoing a phase transition by covalent cross-linking between lysine residues using the chemical tris-succinimidyl amino triacetate (TSAT). The resulting hydrogels were highly elastic; suggesting that the force applied was transmitted through the chemical cross-links, as well as through fully extended ELP chains. The TSAT cross-linker targets specific amino acid residues in contrast to γ-radiation and aldehyde cross-linkers used for photochemical cross-linking, such that the density of cross-linking can be controlled by varying the extent of the targeted amino acid residue within each polypeptide chain [94]. Furthermore, the selection of cross-linker affects both the observed gelation time and the purification steps required to remove organic solvent before swelling the hydrogel in an aqueous solution. For example, the chemical cross-linker β-[tris(hydroxymethly)phosphino-propionic acid] (THPP) or hydroxymethylphosphine can be used instead of TSAT to rapidly cross-link ELP-based hydrogels under physiological conditions [95]. Moreover, fibroblasts encapsulated within hydrogels during the cross-linking process remained viable for several days following gelation, suggesting that materials chemically cross-linked using THPP can be utilized in tissue engineering and drug delivery applications [96].

Covalent cross-linking can be achieved by using enzymes to link specific amino acid residues. For example, extensive lysyl-oxidase-mediated covalent cross-linking between lysine residues in a collagen gel was achieved by seeding cells expressing the enzyme lysyl oxidase within the gel. Compared to the cross-links produced by synthetic chemical reactions, these cross-links were less susceptible to degradation and resulted in a mechanically stronger collagen gel [97]. Another cross-linking enzyme is tissue transglutaminase (tTG) derived from animals, which catalyzes covalent bond formation between lysine and glutamine residues present in the ECM [98]. Optimization of the various lysine repeats, glutamine repeats, and spacing in a PEG–peptide hybrid hydrogel resulted in rapid tTG-mediated cross-linking, forming stable hydrogels within 2 minutes of tTG addition [99]. This method was also applied to an ELP-based hydrogel, creating a cell-responsive protein-engineered hydrogel. As a result of cell secretion of tTG, cross-linking increased over the course of four weeks, increasing the elastic modulus of the ELP-based hydrogel from 0.28 to 1.7 kPa [17]. In an independent study, combinations of tTG-susceptible and human transglutaminase (hTG)-susceptible polypeptides were used to produce protein-engineered hydrogels with varying mechanical strengths, microstructures, and swelling behaviors. Moreover, the designed hydrogels showed good cell viability and supported cell spreading [100]. These results support the use of enzymatically covalently cross-linked hydrogels for tissue engineering applications.

2.4
Development of Biomimetic Protein-Engineered Hydrogels for Tissue Engineering Applications

Cells undergo dynamic interactions between neighboring cells and the ECM in their native 3D environment (Figure 2.6). Building on the bulk assembly of 3D protein-engineered hydrogels, additional peptide domains can be included in the polypeptide chain to mimic specific interactions between the cell and the native extracellular environment. Common biofunctional features designed into protein-engineered hydrogels are cell–ECM adhesion mimics, cell–cell adhesion mimics, delivery of soluble bioactive factors, and native tissue elasticity. In addition, protein-engineered hydrogels can be engineered to respond to cell activity and environmental changes, allowing for an additional level of spatial and temporal control over cell–hydrogel interactions, hydrogel remodeling, and hydrogel degradation. The ability to mediate specific cell–hydrogel interactions is achieved by the combinatorial incorporation of independent, or modular, peptide domains, as well as nanoscale patterning of these domains. In the following sections, we detail how to control the structural, mechanical, and biochemical properties of biomimetic hybrid and protein-engineered hydrogels, as well as the impact of these properties on cell behavior.

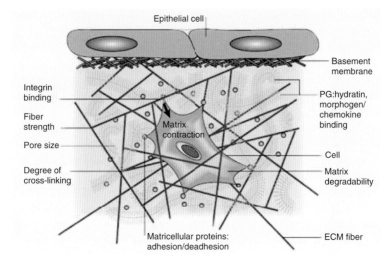

Figure 2.6 Important 3D environmental factors for tissue engineering applications. The fiber strength, degree of cross-linking, matrix degradability, and pore size regulate the ability of cells to migrate through the matrix and mediate cell response to mechanical forces. In addition, the basement membrane, produced by epithelial and endothelial cells, enhances mechanical stiffness of the 3D environment and reduces protein transport both in and out of the tissue. Integrin binding sites and matricellular proteins initiate biochemical processes that further mediate cell adhesion and migration. These biochemical cues, along with a multitude of soluble factors (e.g., growth factors, chemokines, morphogens, nutrients) also regulate cell proliferation, differentiation, and apoptosis. (Source: Reprinted with permission, Ref. [2].)

2.4.1
Mechanical Properties Mediate Cellular Response

Cells exert physical forces on their surroundings, which can be visualized as elastic distortion and wrinkling of an *in vitro* growth substrate [101]. Cells respond to differences in substrate stiffness by altering their contractility, morphology, and motility [102, 103]. For example, mouse fibroblasts cultured on 2D synthetic substrates containing both a compliant region (14 kPa) and a stiff region (30 kPa) with intermediate stiffness at the boundary exhibited higher motility rates and less spreading on the compliant region, even though the surface density of cell-adhesion domains was held constant. These results suggest the importance of the mechanical microenvironment in controlling cell locomotion. Control over substrate stiffness in 2D cultures further mediates cell proliferation, rate of apoptosis, and differentiation [104]. For example, naïve mesenchymal stem cells (MSCs) cultured on substrates with varying elasticity showed that preferential expression of differentiation markers correlated with the endogenous tissue mechanical properties. For more compliant substrates (0.1–1 kPa), MSCs expressed neurogenic cell markers; on moderately stiff substrates (8–17 kPa), MSCs expressed myogenic cell markers; and on stiff substrates (25–40 kPa), MSCs expressed osteogenic cell markers [105].

2.4 Development of Biomimetic Protein-Engineered Hydrogels for Tissue Engineering Applications

These results provide evidence that cells actively sense their mechanical environment and that such surveillance plays an integral role in regulating complex cellular behaviors such as differentiation. While similar studies on the cellular responses to mechanical properties of protein-engineered hydrogels have not yet been reported, the tunable nature of these materials and the ability to culture cells both in two and three dimensions with protein-engineered hydrogels make them ideal substrates for future studies.

In 3D cultures, cell behavior depends on overall hydrogel stiffness and the porosity of the hydrogel. When MSCs were encapsulated in a 3D synthetic hydrogel, the matrix stiffness correlated with cell fate, similar to that shown on 2D substrates [106, 107]. However, cell migration and spreading of MSCs observed in a 3D synthetic hydrogel could not be disrupted by blocking actin formation as on a 2D hydrogel. Therefore, the mechanical sensing mechanisms in 3D cultures may be fundamentally different from those observed for 2D cultures [108]. Similarly, the observed formation of focal adhesions and changes in protein expression of focal adhesion proteins analyzed for 2D and 3D cell cultures suggest that the signaling pathways for cell migration are different between 2D and 3D environments [109].

Densely cross-linked materials affect cell behavior by reducing the ability of cells to effectively contract and migrate through the material. Specifically, increasing the cross-linking density decreases the average pore size of the material, entrapping cells [110, 111]. Comparably, MSCs encapsulated in a densely cross-linked hydrogel were unable to spread, but photodegradation of the cross-linked sites allowed cells to spread out within the hydrogel [112]. Furthermore, reducing the cross-linking density and increasing pore size of a hydrogel resulted in increased cell proliferation and ECM production by encapsulated chondrocytes [113]. Control over the protein structure, protein density, cross-linking density, and cross-linking strength results in control of the bulk mechanical properties and network morphology (e.g., pore size) of protein-engineered hydrogels. Changes to these properties influence cell adhesion, migration, and differentiation and are integral to the development of protein-engineered hydrogels for tissue engineering applications.

2.4.2 Biodegradable Hydrogels for Cell Invasion

Cells not only respond to mechanical and chemical cues from the ECM but also synthesize new ECM and remodel and degrade the existing ECM during tissue regeneration. Cells control their migration *in vivo* by localized, enzymatic degradation of the ECM. Matrix metalloproteinases (MMPs), such as collagenase, have been identified as key regulators involved in 3D cell migration and matrix remodeling by cells [114]. The ability of protein-engineered hydrogels to support and promote cellular processes for tissue regeneration is facilitated by the design of cell-responsive hydrogels that can be remodeled and degraded.

Protein-engineered hydrogels can be degraded by incorporating hydrolytic or proteolytic peptide degradation sites into the material. However, the use of hydrolytic degradation is not specifically cell responsive and results in bulk degradation of the hydrogel. In contrast, the incorporation of proteolytic degradation domains results in local, cell-mediated degradation of the hydrogel. A collagenase target peptide sequence and a plasmin target peptide sequence have been previously incorporated into two discrete synthetic hydrogels made from cross-linked PEG [115]. Both the collagenase target–PEG hydrogel and the plasmin target–PEG hydrogel were selectively degraded in a dose-dependent manner by the targeted proteases. Control over the degradation rate of protease sensitive hydrogels was achieved by altering the density of proteolytic degradation domains within the hydrogel; as the density of these sites increased, the degradation rate also increased [116]. Building on these results, a protein-graft-PEG hydrogel was created containing two plasmin degradation sites within a single recombinant fibronectin peptide sequence. Fast degradation of these hydrogels occurred in the presence of human plasmin. As an assay for cell responsiveness, human foreskin fibroblasts (hFFs) were embedded within these hydrogels. Cellular outgrowth was observed as early as a 2 days and continued up to 2 weeks, when hydrogel degradation was either completed or when the hydrogel collapsed as a result of outgrowth. Furthermore, this behavior was not observed in the presence of protease inhibitors, providing evidence that the cellular outgrowth depended on cell-mediated protease degradation of the hydrogel [117]. Similar results were observed by quantitative time-lapse microscopy of the migration process in a a protein-graft-PEG hydrogel containing an MMP-sensitive peptide sequence. When hFFs were seeded within this hydrogel, the observed cellular migration *in vitro* appeared to depend on cell seeding density. Observation of hFFs up to 30 days showed localized cellular matrix remodeling such that single cells interconnected to form larger multicellular networks [118]. In an *in vivo* study, the implantation of an MMP-sensitive-PEG hydrogel into a rat cranial defect resulted in native cell invasion and bony tissue formation. Hydrogels that were sensitive to MMP degradation showed enhanced cell invasion and improved healing [119]. These results provide evidence for the importance of including proteolytic degradation domains in engineered hydrogels to enable cell migration and matrix remodeling for tissue engineering applications.

The ability to tailor the proteolytic degradation rates of protein-engineered materials was achieved by the incorporation of peptide domains targeted by tissue plasminogen activator (tPA) and urokinase plasminogen activator (uPA), which are proteases secreted by endothelial and neuronal cells [120]. By changing no more than 3% of the amino acid sequence, the kinetics of the proteolytic degradation reaction was tuned across 2 orders of magnitude, resulting in a range of slow and fast-degrading hydrogels. Moreover, by engineering the different peptide sequences into a single composite material, control of temporal and spatial degradation patterns was achieved in response to protease activity [121]. As a proof of concept, a composite protein-engineered hydrogel containing a fast-degrading region surrounded by a slow-degrading hydrogel was created. Following the addition of soluble uPA, the fast-degrading region was completely removed within 3 days, leaving a long, open channel (i.e., an internal void) within the slow-degrading hydrogel. These results show the potential for biomimetic hydrogels to provide specified cell invasion pathways, an important step in the development of biomaterials for tissue engineering applications.

2.4.3
Diverse Biochemical Cues Regulate Complex Cell Behaviors

In addition to mechanical support, the ECM also provides a multivariate network of glycoproteins and soluble bioactive factors that present biochemical signals to cells. Depending on the tissue and stage of development, cells may respond to changes in both the spatial and temporal profiles of these biochemical cues. The presentation of biochemical signals within the ECM to cells has been shown to mediate cell adhesion, viability, migration, and differentiation [122]. In addition, cell–cell receptor interactions further mediate cell adhesion, viability, migration, and differentiation [123]. Finally, the ability to regulate cell behavior can be further regulated through the efficient capture, immobilization, and delivery of soluble biochemical cues from the ECM. In this section, strategies are described to incorporate these three types of biochemical cues: cell–ECM binding domains, cell–cell binding domains, and soluble factors into protein-engineered hydrogels.

2.4.3.1 Cell–Extracellular Matrix Binding Domains
The survival, growth, proliferation, and migration of many cell types are anchorage dependent, that is, complex cell behavior requires adhesion to a substrate. Cell adhesion *in vivo* and *in vitro* is primarily mediated by integrin–ligand binding. Integrins are transmembrane proteins expressed by cells, and the ligands recognized by these integrins are located within the ECM. The resulting binding of ligands by integrins results in intracellular signaling cascades that mediate cell behavior. In a seminal study, cells showed enhanced adherence to glass coverslips coated with increasing concentrations of the naturally derived ECM glycoprotein, fibronectin [124]. Short peptide sequences, such as RGD and IKVAV, were identified by fragment analysis of large ECM proteins including fibronectin and laminin in order to determine the minimal amino acid sequences required to promote cell adhesion [125]. Protein-engineered hydrogels can incorporate these bioactive ligands into the amino acid sequence of the polypeptide chain to promote cell adhesion and regulate cell behavior.

One of the most commonly used cell-binding domains is the ligand sequence, RGD, which is present in ECM proteins fibronectin, laminin, and vitronectin. The RGD ligand binds several integrins including the $\alpha_5\beta_3$ and $\alpha_5\beta_1$ integrins [126]. Additional minimal functional sequences that have been identified as ligands involved in regulating cell adhesion in 2D and/or 3D *in vitro* applications include PHSRN, REDV, LRE, IKLLI, IKVAV, DGEA, IDAPS, GFOGER, KQAGD, and YIGSR [127]. Each ligand may mediate different integrin binding and thus can have different effects on cell phenotype depending on the cell type and integrin specificity. Incorporation of endogenous amino acid sequences that are longer than the minimal functional sequence can promote the folding of individual peptide domains into their native conformations, further modifying integrin specificity

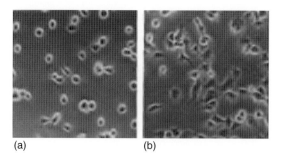

Figure 2.7 Phase contrast microscopy image of HUVECs plated on glass (a) compared to plated on elastinlike hydrogel containing CS5 cell-adhesion domains (b). Scale bar = 150 μm. (Source: Reprinted with permission, Ref. [11].)

and binding strength. For example, the incorporation of the small REDV peptide did not support the adhesion or spreading of human umbilical vein endothelial cells (HUVECs). However, the incorporation of the larger CS5 domain, which includes the REDV sequence, into a protein-engineered hydrogel did support the adhesion and spreading of HUVECs (Figure 2.7) [11]. Systematic peptide inhibition of the putative CS5 integrin showed that REDV was responsible for the observed attachment and spreading of HUVECs. In addition, optimization of the density and location of the CS5 domain relative to the cross-linking domains within the polypeptide chain, led to hydrogels that supported and sustained attachment of HUVECs placed under dynamic stresses, similar to those experienced *in vivo* [20]. Still, the incorporation of an extended peptide containing the RGD minimal-binding peptide in place of the CS5 domain, increased the rate and extent of HUVEC spreading on otherwise similar protein-engineered hydrogels. Furthermore, the cells showed formation of normal stress fibers within the cell and focal adhesions, which link the cytoskeleton of the cell to the substrate [128]. Moreover, the extent of attachment and spreading is often cell specific. For example, the conjugation of the ligand YIGSR into a synthetic hydrogel promoted endothelial cell adhesion, but not platelet adhesion [129]. These results demonstrate that cellular adhesion can be mediated by careful selection of ECM-derived ligands and that optimization of ligand selection and peptide domain sequence is essential for cell-specific adhesion to a substrate.

2.4.3.2 Nanoscale Patterning of Cell–Extracellular Matrix Binding Domains

Further control of cell adhesion and differentiation can be designed by engineering the total ligand density, integrin–ligand binding affinities, and clustering of ligands. Tuning of these ligand parameters demonstrated that an optimal level of integrin–ligand interactions promotes migration of Chinese hamster ovarian (CHO) cells [130]. Migration is thus an adhesion-mediated process, such that both low-adhesive and high-adhesive materials diminish the cell's ability to migrate on or through hydrogels. Cell–ECM binding interactions have also been used to influence cell differentiation. For example, the extent of osteogenic differentiation of MSCs was enhanced by increasing ligand specificity to the $\alpha_5\beta_1$ integrin in both 2D and 3D hydrogel cultures [131].

Tuning of ligand presentation within protein-engineered materials can be achieved by altering the spacing, clustering, and/or frequency of ligand peptide sequences within the polypeptide chains. In one example, the ligand RGDS sequence was inserted into the middle block of a polypeptide chain, and by increasing the relative ratio of RGDS-containing and unmodified polypeptide chains, the total ligand density within the self-assembling protein hydrogel was increased without altering the biomechanical properties of the hydrogel. The observed cell spreading, focal adhesion formation, production of stress fibers, and metabolic activity significantly increased at an optimal, but variable RGDS density depending on the cell type [132]. In the native ECM, ligands may be presented in a variety of nanoscale patterns; therefore, the ability to control the clustering of ligands is of interest in the development of biomimetic hydrogels. For example, the ligand YGRGD was patterned onto a synthetic polymer as single ligands, clusters of five ligands, or clusters of nine ligands, and spaced such that the average ligand density was identical across one square micrometer. All three ligand presentations supported cell attachment; however, the number of cells exhibiting stress fibers dramatically increased with ligand clustering at both low and high ligand densities [133]. Similarly, increased clustering of RGD ligands increased the presence of stable focal adhesions and persistent cell spreading [134]. While ligand nanoscale patterning has not yet been demonstrated within protein-engineered hydrogels, the exact control over amino acid sequence makes these materials ideal substrates for future studies. Taken together, control over the type, density, and presentation of ECM ligands within a biomaterial is important for tissue engineering applications, because these properties confer enhanced regulation of cell behaviors including proliferation, migration, and differentiation.

2.4.3.3 Cell–Cell Binding Domains

Cell–cell interactions include direct cell–cell contact as well as the secretion of soluble proteins and small molecules; these interactions have a direct influence on cell proliferation, migration, and differentiation. In this section, we discuss the incorporation of cell–cell adhesion proteins, including selectins, cadherins, and a subset of the immunoglobulin (Ig) superfamily, that mediate cell–cell binding, into biomimetic hydrogels to mediate cell behavior.

Cell–cell adhesion by selectins results from calcium-dependent heterophilic binding of selectins to sialylated glycans. The selectin family contains L-, E-, and P-selectins that are lectinlike adhesion receptors with an epidermal growth factor (EGF) domain [135]. Selectins are expressed by endothelial cells and bind to carbohydrate ligands expressed by leukocytes. Selectin–ligand binding results in leukocyte rolling behavior, in which cells weakly attach to endothelial cells along blood vessel walls [136]. Surface modification of a synthetic substrate with P-selectin supported rolling behavior of leukocyte cell mimics displaying the selectin ligand sialyl Lewis X (SLe^x) under physiological flow conditions [137]. Building on these results, the firm adhesion and rolling behavior of leukocytes was examined on

PEG hydrogels modified with SLe^x or nonspecific integrin binding sequences. The resulting leukocyte cell adhesion was specific, reversible, and sensitive to the ligand density and affinity, mimicking the adhesion of leukocytes on the vascular wall under physiological mechanical and flow conditions [138].

Cell–cell adhesion by cadherins results from calcium-dependent homophilic binding, where cadherins on one cell bind to the same cadherins on a neighboring cell. The cadherin family includes the N-, P-, R-, B-, and E-cadherins, among others. These proteins are essential to morphogenetic events, during which tissues segregate and organize during development and differentiation [139]. A classical example of the role of cadherins is the dynamic shift of N-cadherin expression and presentation that occurs as neural crest cells migrate from the neural tube during embryonic development [140]. Similarly, E-cadherin is expressed during development to support cell compaction, widely expressed during differentiation, and found in mature epithelial cells, where it connects the cortical actin cytoskeleton of neighboring cells. To explore the role of E-cadherin in differentiation maintenance, the extracellular domain of E-cadherin was fused to the IgG Fc region. The E-cad-Fc fusion was then adsorbed to a plastic surface through noncovalent hydrophobic interactions. Compared to unmodified surfaces, cultured hepatocytes (liver cells) maintained a differentiated state and showed decreased proliferation [141]. However, E-cadherin is not sufficient for initiating cell differentiation; isolated embryonic stem cells (ESCs) cultured on the same plates retained pluripotency and showed increased proliferation [142].

Cell–cell adhesion by Ig superfamily proteins results from recognition of the common Ig fold, a compact structure with two cysteine residues separated by two antiparallel β-sheets. Ig cell-adhesion molecules (Ig-CAMs) connect a variety of cell types and are involved in many different biological processes [143]. A well-studied system for Ig-CAMs is the developing and highly plastic adult nervous system, in which N-CAM is present on the surface of cells and mediates cell–cell adhesion between neurons, astrocytes, and glial cells. One identified functional sequence domain from N-CAM is the FGL domain. The presentation of FGL peptides promoted synapse formation in primary neuron cell cultures and was incorporated into a peptide-based hydrogel that promoted neuron cell adhesion and neurite sprouting *in vitro* [144, 145]. The ability to mimic cell–cell adhesion provides an additional level of regulation over cell-specific behaviors.

2.4.3.4 Delivery of Soluble Cell Signaling Molecules

During development, wound healing, and physiological cellular processes, concentration gradients of basic nutrients and effector molecules exist in both soluble and immobilized forms. The growth, differentiation, and regeneration of many tissues are coordinated through the spatially and temporally controlled release of bioactive factors. For example, angiogenesis, the formation of new blood vessels from existing blood vessels, requires precise and sequential release of multiple bioactive factors to promote normal cell patterning and growth [146]. Therefore, biomaterials intended for tissue engineering applications should have the capacity to efficiently capture, immobilize, and deliver bioactive factors in a manner that mimics the spatial and temporal regulation observed in native tissue. In traditional 2D cell cultures, growth media is used to support normal cell development and often includes small molecules (hormones) and proteins (growth factors) [147]. However, the delivery of these soluble components is substantially different in 3D cultures and tissue. In three dimensions, soluble gradients arise from both simple diffusion and convection, during which the components are either consumed or produced by cells. Thus, cells near the center of an *in vitro* 3D environment may behave differently than cells near the surface as a result of varying local concentrations [2].

The immobilization of bioactive factors into and onto hydrogels allows for prolonged delivery of factors *in vitro* and *in vivo*, which is not achievable in soluble systems. Hydrogels can be engineered for dual growth factor delivery; a synthetic hydrogel was designed for temporal control over the release of VEGF with platelet-derived growth factor (PDGF) in order to mimic profiles observed *in vivo* during angiogenesis. The rapid release of VEGF from the bulk material initiated angiogenesis, and the sequential slow release of PDGF from slow-degrading microspheres imbedded within the bulk material further supported vessel growth and maturation [148]. Temporal control over the release of VEGF alone has been achieved for protein-engineered hydrogels. A recombinant VEGF fused with a protease-sensitive cleavage site was covalently attached to a fibrinlike hydrogel, such that the release of VEGF could be triggered in response to local cellular activity. The release of VEGF during active proteolytic remodeling significantly increased the formation of vascular beds compared to fibrin gels, where VEGF was released by bulk degradation and passive diffusion mechanisms [149].

Further examples of immobilized growth factors include nerve growth factor (NGF), bone morphogenetic protein-2 (BMP-2), and thioredoxin (Trx) protein. An immobilized surface gradient of NGF on a laminin substrate resulted in more dense neurite outgrowth compared to soluble NGF, suggesting some variation in the intracellular signaling response as a result of the presentation of the NGF gradient [150]. BMP-2 was covalently coupled to a silk fibrin film by carbodiimide chemistry and used to culture bone marrow stromal cells. The presence of immobilized BMP-2 induced differentiation of stromal cells into bone more effectively than the addition of soluble BMP-2 with cells cultured on the same silk fibrin film [151]. Using analogous design strategies, protein-engineered hydrogels can be used for the delivery of drug and gene regulating factors to further impact cell viability, growth, and differentiation. For example, the Trx protein, an essential antioxidant in all mammals, was fused to an elastinlike domain forming Trx-ELPs. Above the LCST, Trx-ELPs aggregated and assembled on a surface modified with ELPs. In this phase, a Trx-specific antibody was able to effectively bind the Trx protein presented at the surface. Below the LCST, the (antibody-Trx)–(Trx-ELP) complex could be reversibly dissociated from the ELP-modified surface [152]. In addition, chemically cross-linked ELPs were shown to successfully trap various hydrophilic antibiotics with long-term-release profiles of four or more weeks [153]. These results provide evidence for the efficacy and efficiency of protein-engineered hydrogels in the delivery of proteins and drugs.

Similar to antibiotics, DNA and RNA complexes and nanoparticles can be encapsulated directly into protein-engineered hydrogels by designing positively charged polypeptide chains. Spatial control over the presentation of DNA and RNA complexes was achieved through optimization of DNA and RNA complex formation and encapsulation methods, which increased the loading capacity to 5 µg of DNA per 1 µl of a synthetic hydrogel, enhancing gene expression over two weeks, both *in vitro* and *in vivo* [154]. Temporal control over the release of these bioactive signals was achieved by covalently binding peptide sequences with varying protease sensitivities to DNA-modified nanoparticles and then immobilizing the particles within synthetic hydrogels. Release rates of the nanoparticles increased as a function of increased protease sensitivity, decreased number of immobilization binding sites, and increased concentration of proteases in solution. Moreover, analysis of cellular internalization showed increased uptake of nanoparticles with increased protease sensitivity, decreased number of immobilization binding sites, and increased cell expression of proteases. The development of cell-mediated release rates is ideal for tissue engineering applications, allowing for both spatial and temporal control over gene transfection based on cellular demand [155]. In addition, alterations to the polymer structure, ionic strength of media during encapsulation, and gelation time, as demonstrated using silk–elastinlike polypeptide (SELPs), influence the loading capacity and total release of naked plasmid DNA from hydrogels [156]. These results show that the development of dynamic and robust delivery platforms depends on being able to control the total loading capacity as well as the release rate of bioactive factors.

2.5
Conclusions and Future Perspective

Protein-engineered hydrogels provide a platform for the development of biomaterials containing both defined macromolecular structure and biofunctional peptide domains defined at the molecular level – the amino acid sequence. These sequences can be designed to incorporate multiple independent biomimetic domains into a single polypeptide sequence. Advances in protein-engineering methods, as well as discoveries in molecular and cellular biology, are expanding the range of possible dynamic interactions between cells and protein-engineered hydrogels. Two areas of future opportunity for this field include (i) increasing the diversity of peptide domains that can be successfully incorporated into protein-engineered hydrogels and (ii) increasing the study of protein-engineered hydrogels using *in vivo* models.

The diversity of peptide domains can be expanded by rational design of amino acid substitutions into well-characterized domains, by the computation design of *de novo* amino acid sequences, by the repurposing of evolved amino acid sequences, and by *in vitro* evolution and high throughput screening of novel binding domains and functions. First, the rational design of a fusion protein containing two different α-helical self-aggregating domains and a rigid linker domain predictably assembled into various protein nanostructures depending on the number of protein monomers present [157]. Second, underexplored but well-characterized peptide domains can be incorporated into protein-engineered hydrogels to perform novel functions. For example, the calcium-binding β-roll domain and heme-binding helix bundles have been proposed for use in ligand-responsive hydrogels, providing enhanced biosensing functionality or control over swelling properties as a result of increased binding sites per domain [158]. Third, engineered carbon-monoxide-binding heme proteins could be incorporated into a protein-engineered hydrogel for sensitive biodetection in both medical and environmental applications [159]. Another peptide domain of interest includes silicateins, which form large filaments and direct biosilica polymerization *in vitro* and *in vivo* [160]. Finally, protein-engineered hydrogels can be further developed for incorporation of novel peptide domains that directly and specifically bind DNA or metal molecules [161, 162].

Protein-engineered hydrogels have only recently been explored for *in vivo* studies and clinical applications. For therapeutic applications, protein-engineered hydrogels are being developed for delivery of drugs and other bioactive factors, for regenerative medicine through delivery of transplanted cells, and for tissue regeneration through promotion of cell invasion and growth [163]. An example of bioactive factor delivery is the noncovalent encapsulation of plasmid DNA into a silk-elastinlike polypeptide (SELP)-based hydrogel. The slow release of plasmid DNA from the SELP-based hydrogel retained bioactivity for a month, and increased the transfection rate of cells compared to naked DNA injection [164]. A regenerative medicine example is an *in situ* cross-linkable ELP-based hydrogel applied to a critical defect in a goat knee model. The ELP-treated defects showed not only cell infiltration but also partial degradation at three months. However, the untreated defects showed substantially better cartilage matrix formation compared to ELP-treated defects at six months; thus long-term stability of the ELP-based hydrogel must be further developed [165]. In addition, protein-engineered materials have now entered clinical trials. NuCore® Injectable Nucleus, which is a chemically cross-linked silk–elastin copolymer, is being explored as a treatment for intervertebral disk tissue replacement that minimizes cellular degeneration following disk surgery [166]. Given the considerable progress being made in biomimetic hydrogel development, protein-engineered hydrogels offer a substantial opportunity to impact the field of tissue engineering and clinical practice.

References

1. Cukierman, E., Pankov, R., Steven, D.R., and Yamada, K.M. (2001) *Science*, 294(5547), 1708–1712.
2. Griffith, L.G. and Swartz, M.A. (2006) *Nat. Rev. Mol. Cell Biol.*, 7(3), 211–224.
3. Lodish, H. *et al.* (2000) Hierarchical structure of proteins, *Molecular Cell Biology*, 4th edn, W.H. Freeman & Co., New York.
4. Tischer, E., Mitchell, R., Hartman, T., Silva, M., Gospodarowicz, D. *et al.* (1991) *J. Biol. Chem.*, 266(18), 11947–11954.
5. Muller, Y.A., Li, B., Christinger, H.W., Wells, J.A., Cunningham, B.C. *et al.*

(1997) *Proc. Natl. Acad. Sci. U.S.A.*, **94**(14), 7192–7197.
6. Stauffer, M.E., Skelton, N.J., and Fairbrother, W.J. (2002) *J. Biomol. NMR*, **23**(1), 57–61.
7. Muller, Y.A. and De Vos, A.M. (1997) *Structure*, **5**(10), 1325–1338.
8. Alberts, B. *et al.* (2002) The shape and structure of proteins, *Molecular Biology of the Cell*, 4th edn, Garland Science, New York, London.
9. Straley, K.S. and Heilshorn, S.C. (2009) *Soft Matter*, **5**(1), 114–124.
10. Nicol, A., Gowda, D.C., and Urry, D.W. (1992) *J. Biomed. Mater. Res. A*, **26A**(3), 393–413.
11. Panitch, A., Yamaoka, T., Fournier, M.J., Mason, T.L., and Tirrell, D.A. (1999) *Macromolecules*, **32**(5), 1701–1703.
12. Park, J., Lim, E., Back, S., Na, H., Park, Y. *et al.* (2010) *J. Biomed. Mater. Res. A*, **93A**(3), 1091–1099.
13. Straley, K.S. and Heilshorn, S.C. (2009) *Front. Neuroeng.*, **2**(9), 1–10.
14. Petka, W.A., Harden, J.L., McGrath, K.P., Writz, D., and Tirrell, D.A. (1998) *Science*, **281**(5375), 389–392.
15. Urry, D.W., Long, M.M., Cox, B.A., Ohnishi, T., Mitchell, L.W. *et al.* (1974) *Biochim. Biophys. Acta*, **371**(2), 597–602.
16. Chau, Y., Luo, Y., Cheung, A.C.Y., Nagai, Y., Zhang, S. *et al.* (2008) *Biomaterials*, **29**(11), 1713–1719.
17. McHale, M.K., Setton, L.A., and Chilkoti, A. (2005) *Tissue Eng.*, **11**(11–12), 1768–1779.
18. Topp, S., Prasad, V., Cianci, G.C., Weeks, E.R., and Gallivan, J.P. (2006) *J. Am. Chem. Soc.*, **128**(43), 13994–13995.
19. Sakiyama, S.E., Schense, J.C., and Hubbell, J.A. (1999) *FASEB J.*, **13**(15), 2214–2224.
20. Heilshorn, S.C., Di Zio, K.A., Welsh, E.R., and Tirrell, D.A. (2003) *Biomaterials*, **24**(23), 4245–4252.
21. Tynan-Connolly, B.M. and Nielsen, J.E. (2007) *Protein Sci.*, **16**(2), 239–249.
22. Xu, Y. and Xu, D. (2000) *Proteins*, **40**(3), 343–354.
23. Chilkoti, A., Christensen, T., and MacKay, J.A. (2006) *Curr. Opin. Chem. Biol.*, **10**(6), 652–657.
24. Linke, W.A. (2010) *Nat. Chem. Biol.*, **6**(10), 702–703.
25. Dandu, R., Von Cresce, A., Briber, R., Dowell, P., Cappello, J. *et al.* (2009) *Polymer*, **50**(2), 366–374.
26. Banwell, E.F., Abelardo, E.S., Adams, D.J., Birchall, M.A., Corrigan, A. *et al.* (2009) *Nat. Mater.*, **8**(7), 596–600.
27. Stark, M., Grip, S., Rising, A., Hedhammar, M., Engstrom, W. *et al.* (2007) *Biomacromolecules*, **8**(5), 1695–1701.
28. Sui, Z., King, W.J., and Murphy, W.L. (2007) *Adv. Mater.*, **19**(20), 3377–3380.
29. Sui, Z.J., King, W.J., and Murphy, W.L. (2008) *Adv. Funct. Mater.*, **18**(12), 1824–1831.
30. Petsalaki, E. and Russell, R.B. (2008) *Curr. Opin. Biotechnol.*, **19**(4), 344–350.
31. Alberts, B. *et al.* (2002) Manipulating proteins, DNA, and RNA, *Molecular Biology of the Cell*, 4th edn, Garland Science, New York, London.
32. Scott Yokum, T. and Barany, G. (2000) Strategy in solid-phase peptide synthesis, in *Solid-phase Synthesis: A Practical Guide* (eds S.A. Kates and F. Albericio), Marcel Decker Inc., New York.
33. Verma, R., Boleti, E., and George, A.J.T. (1998) *J. Immunol. Methods*, **216**(1–2), 165–181.
34. Rocha, E.P. (2004) *Genome Res.*, **14**(11), 2279–2286.
35. Gustaffsson, C., Govindarajan, S., and Minshull, J. (2004) *Trends Biotechnol.*, **22**(7), 346–353.
36. Burgess-Brown, B.A., Sharma, S., Sobott, F., Loenarz, C., Oppermann, U. *et al.* (2008) *Protein Expr. Purif.*, **59**(1), 94–102.
37. Puigbo, P., Guzman, E., Romeu, A., and Garcia-Vallve, S. (2007) *Nucleic Acids Res.*, **35**(2), W126–W131.
38. Chow, D.C., Dreher, M.R., Trabbic-Carlson, K., and Chilkoti, A. (2006) *Biotechnol. Prog.*, **22**(3), 638–646.
39. Primrose, S.B. and Twyman, R.M. (2006) Advanced transgenic technology, *Principles of Gene Manipulation and*

Genomics, 7th edn, Blackwell Publishing, Maden MA.
40. Magalhães, P.O., Lopes, A.M., Mazzola, P.G., Rangel-Yagui, C., Penna, T.C.V. et al. (2007) *J. Pharm. Pharm. Sci.*, **10**(3), 388–404.
41. Betre, H., Ong, S.R., Guilak, F., Chilkoti, A., Fermor, B. et al. (2006) *Biomaterials*, **27**(1), 91–99.
42. Scopes, R.K. (1994) Analysis for purity, *Protein Purification: Principles and Practice*, 3rd edn, Springer Science & Business Media, New York.
43. Kennedy, S.B., deAzevedo, E.R., Petka, W.A., Russell, T.P., Tirrell, D.A. et al. (2001) *Macromolecules*, **34**(25), 8675–8685.
44. Bosio, L., Johari, G.P., Oumezzine, M., and Teixeira, J. (1992) *Chem. Phys. Lett.*, **188**(1–2), 113–118.
45. Kim, S.H. and Chu, C.C. (2000) *J. Biomed. Mater. Res. A*, **53A**(3), 258–266.
46. Ferreira, L., Figueiredo, M.M., Gil, M.H., and Ramos, M.A. (2006) *J. Biomed. Res. B Appl. Biomater.*, **77B**(1), 55–64.
47. Levine, A.J. and Lubensky, T.C. (2000) *Phys. Rev. Lett.*, **85**(8), 1774–1777.
48. Mason, T.G. and Weitz, D.A. (1995) *Phys. Rev. Lett.*, **74**(7), 1250–1253.
49. Comisar, W.A., Hsion, S.X., Kong, H.J., Mooney, D.J., and Linderman, J.J. (2006) *Biomaterials*, **27**(10), 2322–2329.
50. Cooper, C., Packer, N., and Williams, K. (2000) *Amino Acid Analysis Protocols, Methods in Molecular Biology 159*, Humana Press, Totawa, New Jersey.
51. Baranov, V.I., Quinn, Z.A., Bandura, D.R., and Tanner, S.D. (2002) *J. Anal. At. Spectrom.*, **17**(9), 1148–1152.
52. Shen, W., Kornfield, J.A., and Tirrell, D.A. (2007) *Soft Matter*, **3**(1), 99–107.
53. Moorthy, J., Burgess, R., Yethiraj, A., and Beebe, D. (2007) *Anal. Chem.*, **79**(14), 5322–5327.
54. Huebsch, U.N.D. and Mooney, D.J. (2007) *Biomaterials*, **28**(15), 2424–2437.
55. Greenfield, N. and Fasman, G.D. (1969) *Biochemistry*, **8**(10), 4108–4116.
56. Bains, G. and Freire, E. (1991) *Anal. Biochem.*, **192**(1), 203–206.
57. Hennink, W.E. and van Nostrum, C.F. (2002) *Adv. Drug Delivery Rev.*, **54**(1), 13–36.
58. Xu, C.Y., Breedveld, V., and Kopecek, J. (2005) *Biomacromolecules*, **6**(3), 1739–1749.
59. Shen, W., Lammertink, R.G.H., Sakata, J.K., Kornfield, J.A., and Tirrell, D.A. (2005) *Macromolecules*, **38**(9), 3909–3916.
60. Wang, C., Stewart, R.J., and Kopecek, J. (1999) *Nature*, **397**(6718), 417–420.
61. Wang, C., Kopecek, J., and Stewart, R.J. (2001) *Biomacromolecules*, **2**(3), 912–920.
62. Yang, J., Xu, C., Wang, C., and Kopecek, J. (2006) *Biomacromolecules*, **7**(4), 1187–1195.
63. Shen, W., Zhang, K.C., Kornfield, J.A., and Tirrell, D.A. (2006) *Nat. Mater.*, **5**(2), 153–158.
64. Gajjeraman, S., He, G., Narayanan, K., and George, A. (2008) *Adv. Funct. Mater.*, **18**(24), 3972–3980.
65. Shekhawat, S.S., Porter, J.R., Sriprasad, A., and Ghosh, I. (2009) *J. Am. Chem. Soc.*, **131**(42), 15284–15290.
66. Wheeldon, I.R., Barton, S.C., and Banta, S. (2007) *Biomacromolecules*, **8**(10), 2990–2994.
67. Cao, Y. and Li, H. (2008) *Chem. Commun. (Camb.)*, **35**, 4144–4146.
68. Wheeldon, I.R., Gallaway, J.W., Barton, S.C., and Banta, S. (2008) *Proc. Natl. Acad. Sci. U.S.A.*, **105**(40), 15275–15280.
69. Wheeldon, I.R., Campbell, E., and Banta, S. (2009) *J. Mol. Biol.*, **392**(1), 129–142.
70. Lu, H.D., Wheeldon, I.R., and Banta, S. (2010) *Protein Eng. Des. Sel.*, **23**(7), 559–566.
71. Foster, J.A., Bruenger, E., Gray, W.R., and Sandberg, L.B. (1973) *J. Biol. Chem.*, **248**(8), 2876–2879.
72. Urry, D.W. (1992) *Prog. Biophys. Mol. Biol.*, **57**(1), 23–57.
73. Wright, E.R., McMillan, R.A., Cooper, A., Apkarian, R.P., and Conticello, V.P. (2002) *Adv. Funct. Mater.*, **12**(2), 149–154.
74. Wright, E.R. and Conticello, V.P. (2002) *Adv. Drug Delivery Rev.*, **54**(8), 1057–1073.

75. Raucher, D. and Chilkoti, A. (2001) *Cancer Res.*, **61**(19), 7163–7170.
76. Lao, U.L., Sun, M., Matsumoto, M., Mulchandani, A., and Chen, W. (2007) *Biomacromolecules*, **8**(12), 3736–3739.
77. Xu, G., Wang, W., Groves, J.T., and Hecht, M.H. (2001) *Proc. Natl. Acad. Sci. U.S.A.*, **98**(7), 3652–3657.
78. Yan, H., Saiani, A., Gough, J.E., and Miller, A.F. (2006) *Biomacromolecules*, **7**(10), 2776–2782.
79. Rapaport, H., Grisaru, H., and Silberstein, T. (2008) *Adv. Funct. Mater.*, **18**(19), 2889–2896.
80. Altman, G.H., Diaz, F., Jakuba, C., Calabro, T., Horan, R.L. et al. (2003) *Biomaterials*, **24**(3), 401–416.
81. Wong Po Foo, C., Lee, J.S., Mulyasasmita, W., Parisi-Amon, A., and Heilshorn, S.C. (2009) *Proc. Natl. Acad. Sci. U.S.A.*, **106**(52), 22067–22072.
82. Russ, W.P., Lowery, D.M., Mishra, P., Yaffe, M.B., and Ranganathan, R. (2005) *Nature*, **437**(7058), 579–583.
83. Miyata, T., Asami, N., and Uragami, T. (1999) *Nature*, **399**(6738), 766–769.
84. Kokufata, E., Zhang, Y., and Tanaka, T. (1991) *Nature*, **351**(6324), 302–304.
85. Lee, S.J. and Park, K. (1996) *J. Mol. Recognit.*, **9**(5–6), 549–557.
86. Obaidat, A.A. and Park, K. (1997) *Biomaterials*, **18**(11), 801–806.
87. Ikura, M. (1996) *Trends Biochem. Sci.*, **21**(1), 14–17.
88. Zhang, M. and Yuan, T. (1998) *Biochem. Cell Biol.*, **76**(2–3), 313–323.
89. Murphy, W.L., Dilmore, W.S., Modica, J., and Mrksich, M. (2007) *Angew. Chem. Int. Ed.*, **46**(17), 3066–3069.
90. Ehrick, J.D., Stokes, S., Bachas-Daunert, S., Moschou, E.A., Deo, S.K. et al. (2007) *Adv. Mater.*, **19**(22), 4024–4027.
91. Kim, B. and Chilkoti, A. (2008) *J. Am. Chem. Soc.*, **130**(52), 17867–17873.
92. King, W.J., Mohammed, J.S., and Murphy, W.L. (2009) *Soft Matter*, **5**(12), 2399–2406.
93. Elvin, C.M., Carr, A.G., Huson, M.G., Maxwell, J.M., Pearson, R.D. et al. (2005) *Nature*, **437**(7061), 999–1002.
94. Trabbic-Carlson, K., Setton, L.A., and Chilkoti, A. (2003) *Biomacromolecules*, **4**(3), 572–580.
95. Lim, D.W., Nettles, D., Setton, L.A., and Chilkoti, A. (2007) *Biomacromolecules*, **8**(5), 1463–1470.
96. Lim, D.W., Nettles, D.L., Setton, L.A., and Chilkoti, A. (2008) *Biomacromolecules*, **9**(1), 222–230.
97. Elbjeirami, W.M., Yonter, E.O., Starcher, B.C., and West, J.L. (2003) *J. Biomed. Mater. Res. A*, **66A**(3), 513–521.
98. Greenberg, C.S., Birckbichler, P.J., and Rice, R.H. (1991) *FASEB J.*, **5**(15), 3071–3077.
99. Hu, B.H. and Messersmith, P.B. (2003) *J. Am. Chem. Soc.*, **125**(47), 14298–14299.
100. Davis, N.E., Ding, S., Forster, R.E., Pinkas, D.M., and Barron, A.E. (2010) *Biomaterials*, **31**(28), 7288–7297.
101. Harris, A.K., Wild, P., and Stopak, D. (1980) *Science*, **208**(4440), 177–179.
102. Discher, D.E., Janmey, P., and Wang, Y. (2005) *Science*, **310**(5751), 1139–1143.
103. Pelham, R.J. Jr. and Wang, Y. (1997) *Proc. Natl. Acad. Sci. U.S.A.*, **94**(25), 13661–13665.
104. Lo, C.M., Wang, H.B., Dembo, M., and Wang, Y.L. (2000) *Biophys. J.*, **79**(1), 144–152.
105. Engler, A.J., Sen, S., Sweeney, H.L., and Discher, D.E. (2006) *Cell*, **126**(4), 677–689.
106. Pek, Y.S., Wan, A.C., and Ying, J.Y. (2010) *Biomaterials*, **31**(3), 385–391.
107. Huebsch, N., Arany, P.R., Mao, A.S., Shvartsman, D., Ali, O.A. et al. (2010) *Nat. Mater.*, **9**(6), 518–526.
108. Parekh, S.H., Chatterjee, K., Lin-Gibson, S., Moore, N.M., Cicerone, M.T. et al. (2011) *Biomaterials*, **32**(9), 2256–2264.
109. Fraley, S.I., Feng, Y., Krishnamurthy, R., Kim, D.H., Celedon, A. et al. (2010) *Nat. Cell Biol.*, **12**(6), 598–604.
110. Bryant, S.J. and Anseth, K.S. (2001) *J. Biomed. Mater. Res. A*, **59A**(1), 63–72.
111. Brandl, F., Sommer, F., and Goepferich, A. (2007) *Biomaterials*, **28**(2), 134–146.

112. Kloxin, A.M., Kasko, A.M., Salinas, C.N., and Anseth, K.S. (2009) *Science*, **324**(5923), 59–63.
113. Bryant, S.J., Chowdhury, T.T., Lee, D.A., Bader, D.L., and Anseth, K.S. (2004) *Ann. Biomed. Eng.*, **32**(3), 407–417.
114. Friedl, P. and Brocker, E.B. (2000) *Cell. Mol. Life Sci.*, **57**(1), 41–64.
115. West, J.L. and Hubbell, J.A. (1999) *Macromolecules*, **32**(1), 241–244.
116. Dikovsky, D., Bianco-Peled, H., and Seliktar, D. (2008) *Biophys. J.*, **94**(7), 2914–2925.
117. Halstenberg, S., Panitch, A., Rizzi, S., Hall, H., and Hubbell, J.A. (2002) *Biomacromolecules*, **3**(4), 710–723.
118. Raeber, G.P., Lutolf, M.P., and Hubbell, J.A. (2007) *Acta Biomater.*, **3**(5), 615–629.
119. Lutolf, M.P., Lauer-Fields, J.L., Schmoekel, H.G., Metters, A.T., Weber, F.E. et al. (2003) *Proc. Natl. Acad. Sci. U.S.A.*, **100**(9), 5413–5418.
120. Vassalli, J.D., Sappino, A.P., and Belin, D. (1991) *J. Clin. Invest.*, **88**(4), 1067–1072.
121. Straley, K.S. and Heilshorn, S.C. (2009) *Adv. Mater.*, **21**(41), 4148–4152.
122. Adams, J.C. and Watt, F.M. (1993) *Development*, **117**(4), 1183–1198.
123. Khetani, S.R. and Bhatia, S.N. (2006) *Curr. Opin. Biotechnol.*, **17**(5), 524–531.
124. Carter, S.B. (1965) *Nature*, **208**(16), 1183–1187.
125. Tashiro, K., Sephel, G.C., Weeks, B., Sasaki, M., Martin, G.R. et al. (1989) *J Biol Chem.*, **264**(27), 16174–16182.
126. Alberts, B. et al. (2002) Cell junctions, cell adhesion, and the extracellular matrix, *Molecular Biology of the Cell*, 4th edn, Garland Science, New York, London.
127. Petrie, T.A. and Garcia, A.J. (2009) *Biological Interactions on Materials Surfaces*, Chapter 7, Springer, New York, pp. 133–156.
128. Liu, J.C., Heilshorn, S.C., and Tirrell, D.A. (2004) *Biomacromolecules*, **5**(2), 497–504.
129. Jun, H.W. and West, J.L. (2005) *J. Biomed. Mater. Res. B Appl. Biomater.*, **72B**(1), 131–139.
130. Palecek, S.P., Loftus, J.C., Ginsberg, M.H., Lauffenburger, D.A., and Horwitz, A.F. (1997) *Nature*, **385**(6616), 537–540.
131. Martino, M.M., Mochizuki, M., Rothenfluh, D.A., Rempel, S.A., Hubbell, J.A. et al. (2009) *Biomaterials*, **30**(6), 1089–1097.
132. Fischer, S.E., Liu, X.Y., Mao, H.Q., and Harden, J.L. (2007) *Biomaterials*, **28**(22), 3325–3337.
133. Maheshwari, G., Brown, G., Lauffenburger, D.A., Wells, A., and Griffith, L.G. (2000) *J. Cell Sci.*, **113**(10), 1677–1686.
134. Cavalcanti-Adam, E.A., Volberg, T., Micoulet, A., Kessler, H., Geiger, B. et al. (2007) *Biophys. J.*, **92**(8), 2964–2974.
135. Lasky, L.A. (1992) *Science*, **258**(5084), 964–969.
136. Buttrum, S.M., Hatton, R., and Nash, G.B. (1993) *Blood*, **82**(4), 1165–1174.
137. Eniola, A.O., Willcox, P.J., and Hammer, D.A. (2003) *Biophys. J.*, **85**(4), 2720–2731.
138. Taite, L.J., Rowland, M.L., Ruffino, K.A., Smith, B.R., Lawrence, M.B. et al. (2006) *Ann. Biomed. Eng.*, **34**(11), 1705–1711.
139. Takeichi, M. (1995) *Curr. Opin. Cell Biol.*, **7**(5), 619–627.
140. Nakagawa, S. and Takeichi, M. (1998) *Development*, **125**(15), 2963–2971.
141. Nagaoka, M., Ise, H., and Akaike, T. (2002) *Biotechnol. Lett.*, **24**(22), 1857–1862.
142. Nagaoka, M., Koshimizu, U., Yuasa, S., Hattori, F., Chen, H. et al. (2006) *PLoS ONE*, **1**(1), e15.
143. Aplin, A.E., Howe, A., Alahari, S.K., and Juliano, R.L. (1998) *Pharma Rev.*, **50**(2), 197–264.
144. Cambon, K., Hansen, S.M., Venero, C., Herrero, A.I., Skibo, G. et al. (2004) *J. Neurosci.*, **24**(17), 4197–4204.
145. Zou, Z., Zheng, Q., Wu, Y., Guo, X., Yang, S. et al. (2010) *J. Biomed. Mater. Res. A*, **95A**(4), 1125–1131.
146. Alberts, B. et al. (2002) Histology: the lives and deaths of cells in tissues, *Molecular Biology of the Cell*, 4th edn, Garland Science, New York, London.

147. Lodish, H. et al. (2000) Manipulating cells and viruses in culture, *Molecular Cell Biology*, 4th edn, W.H. Freeman & Co., New York.
148. Richardson, T.P., Peters, M.C., Ennett, A.B., and Mooney, D.J. (2001) *Nat. Biotechnol.*, **19**(11), 1029–1034.
149. Ehrbar, M., Djonov, V.G., Schnell, C., Tschanz, S.A., Martiny-Baron, G. et al. (2004) *Circ. Res.*, **94**(8), 1124–1132.
150. Kapur, T.A. and Shoichet, M.S. (2004) *J. Biomed. Mater. Res. A*, **68A**(2), 235–243.
151. Karagenorgiou, V., Meinel, L., Hofmann, S., Malhotra, A., Volloch, V. et al. (2004) *J. Biomed. Mater. Res. A*, **71A**(3), 528–537.
152. Hyun, J., Lee, W.K., Nath, N., Chilkoti, A., and Zauscher, S. (2004) *J. Am. Chem. Soc.*, **126**(23), 7330–7335.
153. Adams, S.B. Jr., Shamji, M.F., Nettles, D.L., Hwang, P., and Setton, L.A. (2009) *J. Biomed. Mater. Res. Part B Appl. Biomater.*, **90B**(1), 67–74.
154. Lei, Y., Huan, S., Sharif-Kashani, P., Chen, Y., Kavehpour, P. et al. (2010) *Biomaterials*, **31**(34), 9106–9116.
155. Tokatlian, T., Shrum, C.T., Kadoya, W.M., and Segura, T. (2010) *Biomaterials*, **31**(31), 8072–8080.
156. Hwang, D., Moolchandani, V., Dandu, R., Haider, M., Cappello, J. et al. (2009) *Int. J. Pharm.*, **368**(1–2), 215–219.
157. Padilla, J.E., Colovos, C., and Yeates, T.O. (2001) *Proc. Natl. Acad. Sci. U.S.A.*, **98**(5), 2217–2221.
158. Banta, S., Wheeldon, I.R., and Blenner, M. (2010) *Annu. Rev. Biomed. Eng.*, **12**, 167–186.
159. Moffet, D.A., Case, M.A., House, J.C., Vogel, K., Williams, R.D. et al. (2001) *J. Am. Chem. Soc.*, **123**(10), 2109–2115.
160. Cha, J.N., Shimizu, K., Zhou, Y., Christiansen, S.C., Chmelka, B.F. et al. (1999) *Proc. Natl. Acad. Sci. U.S.A.*, **96**(2), 361–365.
161. Dai, H., Choe, W.S., Thai, C.K., Sarikaya, M., Traxler, B.A. et al. (2005) *J. Am. Chem. Soc.*, **127**(44), 15637–15643.
162. Wong Po Foo, C., Patwardhan, S.V., Belton, D.J., Kitchel, B., Anastasiades, D. et al. (2006) *Proc. Natl. Acad. Sci. U.S.A.*, **103**(25), 9428–9433.
163. Lutolf, M.P. and Blau, H.M. (2009) *Adv. Mater.*, **21**(32–33), 3255–3268.
164. Megeed, Z., Haider, M. Jr., Li, D., O'Malley, B.W., Cappello, J. et al. (2004) *J. Controlled Release*, **94**(2–3), 433–445.
165. Nettles, D.L., Kitaoka, K., Hanson, N.A., Flahiff, C.M., Mata, B.A. et al. (2008) *Tissue Eng. Part A*, **14**(7), 1133–1140.
166. Kitchel, S.H., Boyd, L.M., and Carter, A.J. (2006) *NuCore Injectable Disk Nucleus, Dynamic Reconstruction of the Spine*, Chapter 19, Thieme Medical Publishers, New York, pp. 142–148.

3
Collagen-Based Biomaterials for Regenerative Medicine
Christophe Helary and Abhay Pandit

3.1
Introduction

Tissue engineered materials have to mimic the physical properties of the tissue they replace and provide biological signals to promote tissue regeneration. In addition, they need to be remodeled by host cells and gradually replaced by a neotissue [1].

Two strategies can be used to replace injured tissue. The first one is the use of complex composite biomaterials obtained by decellularization of natural tissues (SIS®, Alloderm®). These biomaterials have properties resembling those of the living tissue, but they can also be a source of inflammation because of their antigenicity [2, 3]. The second strategy relies on a "bottom-up" approach that uses simple building blocks to fabricate a scaffold for tissue regeneration. At present, collagens are used as molecules to fabricate scaffolds by the bottom-up strategy. Compared with synthetic polymers, collagen is a natural molecule that is well tolerated in the body, allows for cell adhesion and proliferation, and can be degraded with cellular control [4]. As this molecule gives mechanical resistance to tissues *in vivo*, biomimetic collagen scaffolds can possess good mechanical properties. Collagen scaffolds can be designed with the aim of mimicking the structure of extracellular matrix (ECM) observed in tissue. In this case, the final construct has to attain a high degree of complexity. Another possibility is to design a precursor biomaterial favoring cell infiltration that results in ECM deposition.

This chapter is structured to first present the framework of the collagenous structures found in the body. This is followed by an overview of mimicking structures. At last, the target clinical applications are presented.

3.2
Collagens *In Vivo*

3.2.1
Collagen Structure

Collagen is a protein with a triple helical structure observed in at least one part of the protein. To date, 43 distinct polypeptide chains (termed *alpha chains*) have been assembled into 28 different collagens [5]. In the field of biomaterials, fibril-forming collagens are molecules of interest from which to fabricate scaffolds because *in vivo*, they provide strength to connective tissues. As a predominant component in the ECM, they are considered as the elementary "brick" in connective tissue generation. Five different fibril-forming collagens have been identified: collagens I, II, III, V, and XI [5]. Collagen I, the most abundant fibril-forming collagen, is found throughout the body, more specifically in tendon, bone, and skin. Collagen I is also associated *in vivo* with collagen III and V to form heterofibrils. In contrast, collagen II is specifically found in cartilage, cornea, and intervertebral disk and is associated with collagen XI to form fibrils [6].

Fibril-forming collagens are composed of three α polypeptide chains forming a right-handed supercoil without any interruption in the triple helical structure [7, 8]. Each α chain has a repeating Gly-X-Y triplet in which glycine occupies every third position to stabilize the helix. Position X is frequently proline and Y is 4-hydroxyproline [9]. The chains are linked by interchain hydrogen bonds that are strengthened by disulfide bonds [8]. Collagen I is a heterodimer that contains two identical α1 (I) chains and an α2 (I) chain. In contrast, collagen II is a homodimer composed of three identical chains $[\alpha 1(II)]_3$.

3.2.2
Collagen Fibrillogenesis

In vivo, collagen triple helices oligomerize into specific suprastructures: the fibrils. This phenomenon is called *fibrillogenesis*. The fibril formation is a self-assembly process driven by the loss of solvent from the protein surface because of the insolubilization of collagen molecule after cleavage of procollagen extremities [6, 10, 11]. After secretion of soluble procollagen in the interstitial environment, propeptides are cleaved by specific enzymes, the propeptidases [12]. The cleavage of propeptides exposes telopeptide sequences, short extensions that are not triple helical, which contain some binding sites for fibril formation, which allows for fibrillogenesis [13, 100]. So, fibrillogenesis is a self-assembly process determined by intrinsic properties of the collagen molecule [6].

Fibrils occur as 67 nm D-periodic structures because of the regular staggering of triple helical collagen molecules. Fibrils are stabilized by nonreducible covalent cross-links involving amino acids within the triple helix and telopeptides [14]. These modifications are essential for the good mechanical properties of connective tissues. Collagen fibrils can have an indeterminate length and a diameter ranging between 12 and 500 nm [6]. Fibrils are also the principal source of tensile strength in connective tissues. These suprastructures have functions in cell adhesion, cell differentiation, and structural integrity of organs. In addition, they serve as scaffold for attachment of other macromolecules [5].

3.2.3
Three-Dimensional Networks of Collagen in Connective Tissues

In specific tissues, collagen fibrils are organized in orientated networks, which impart the mechanical properties [15, 16]. For example, the alignment of fibrils in tendons can account for their capacity to stretch. Nevertheless, in some connective tissues such as skin or blood vessels, fibrils have a random orientation.

The most organized tissue encountered in vertebrate organisms is bone. This tissue is composed of an external dense cortical compact part and a porous part in the tissue core [17]. The porous part has no particular arrangement but a high porosity. In contrast, the dense cortical part is composed of cylindrical osteons formed by successive concentric lamellae around a blood vessel [18]. Each lamella is composed of fibrils similarly orientated [19]. In dark and bright osteons, fibril disposition follows an orthogonal plywood model with two principal orientations. From one lamella to the next one, fibril orientation switches at 90°. In intermediate osteons, fibril orientation follows the cholesteric structure model with a small and regular change of fibril orientation [19]. In tendons, all fibrils are longitudinally aligned along the tendon axis (Figure 3.1) [1].

3.2.4
Interactions of Cells with Collagen

The ECM acts as physiological support for cells of the connective tissues. Adhesion of cells to collagen triggers biological cues (inside-outside and outside-inside signals) that control cell behavior and ECM composition. To emulate this behavior, collagen-based biomaterials have to mimic a tridimensional environment that allows for cell adhesion, cell viability, proliferation, and remodeling [1]. In addition, biomaterial structure has to maintain the cell phenotype and promote tissue regeneration.

Integrins are the most important collagen receptors that transmit bidirectional signals between ECM and cytoplasm [20]. Four integrins specifically link fibril-forming collagen: $\alpha 1 \beta 1$, $\alpha 2 \beta 1$, $\alpha 10 \beta 1$, and $\alpha 11 \beta 1$ [21, 22]. Integrins $\alpha 1 \beta 1$ and $\alpha 2 \beta 1$ are widely distributed throughout the body. When linked to fibrillar collagen I, integrin $\alpha 1 \beta 1$ transmits signals promoting proliferation and inhibiting collagen synthesis [23]. Integrin $\alpha 2 \beta 1$ plays a role in the regulation of matrix remodeling by upregulating metalloproteinase expression [24]. Integrin $\alpha 10 \beta 1$ is more restricted to chondrocytes [25]. Finally integrin $\alpha 11 \beta 1$ is restricted to mesenchymal cells, where collagen is well organized [21, 26]. Another family of receptors is able to specifically link fibrillar collagens and influence cell behavior: discoidin domain receptor

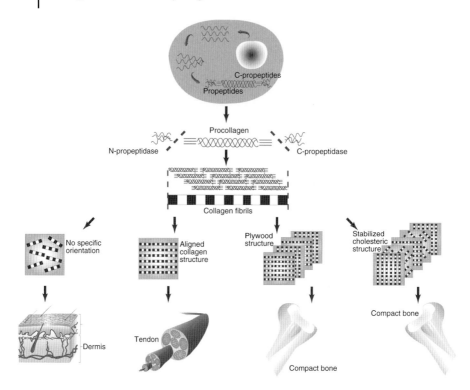

Figure 3.1 Collagen biosynthesis, fibrillogenesis, and ultrastructural arrangements of fibrils. In the cell, collagen is synthesized in a proform by association of three α chains that form a triple helix. After secretion in the interstitial space, propeptidases cleave collagen propeptides. Then, collagen triple helices self-assemble to form collagen fibrils. Subsequently, collagen fibrils can adopt specific orientation to form the ECM structure of tissues. In the skin, collagen fibrils do not have specific orientation. In tendons, fibrils are longitudinally aligned along the tendons axis. In the compact bone, collagen fibrils within lamellas are aligned but do not follow the bone vertical axis. They can adopt a plywood-like structure in which fibril orientation in each lamella is at 90° to its neighbor. The second possibility is to adopt the structure of stabilized cholesteric crystal liquid. In this case, the fibril orientation rotates a small angle between two lamellas.

(DDR). DDR1 and DDR2 are involved in control of cell adhesion, migration, and remodeling [27, 28].

The physical properties of ECM have an important influence on cell behavior. For example, fibroblasts embedded in a contracted collagen hydrogel do not withstand any tension. In this case, they are quiescent and synthesize a low amount of collagen as observed in physiological situations [29–31]. When hydrogel contraction is restrained, fibroblasts are under tension. Then, they differentiate into myofibroblasts and adopt a synthetic phenotype [32, 33]. Another example is the influence of the biomaterial stiffness on chondrocyte phenotype. Absence of appropriate mechanical properties in a hydrogel triggers their dedifferentiation into fibroblasts with the downregulation of collagen II, the marker of chondrocyte differentiation [3].

3.3
Collagen *In Vitro*

Fibrillar collagen I can be easily extracted from rat tail tendons, bovine Achilles tendons, or bovine skin to produce collagen solutions. Fibrils can be partially solubilized in acidic conditions to produce a homogeneous solution containing collagen triple helices [6]. Subsequently, telopeptides can be removed by the action of pepsin to decrease the immunogenicity of collagen [34]. Atelocollagen solutions (pepsinized collagen) are less viscous and have a better yield of extraction, but they show greater difficulty in forming D-periodic fibrils because of the absence of their telopeptides. Indeed, these nonhelicoidal parts play a crucial role in fibrillogenesis [35, 36].

Collagen II can be extracted from articular cartilages (knee or trachea), but the yield is very poor and the process of extraction is complicated. Because of the poor ability of cartilage to repair, collagen II extraction from living tissues is very difficult and this limits its utilization in tissue engineering.

Collagen molecules in acidic solution are able to self-assemble into fibrils by neutralization and warming [6]. Fibrillogenesis *in vitro* requires conditions of pH, ionic strength, and temperature to form native banded fibrils, but it does not require the presence of cells [6, 37]. Fibrils formed by neutralization of collagen acid solution have a D-periodic structure observed *in vivo*, but they have a limited length (6 μm) and a maximal cross section of 160 molecules [6]. In addition, the cross-linkage that operates by lysyl oxidase is missing, hence collagen fibrils lack sufficient strength [38]. Macroscopically, the neutralization of collagen acidic solution triggers gelling of the solution. A collagen hydrogel is then obtained. The partial removal of telopeptide by pepsin has a major effect on fibril formation. Loss of N or C telopeptides disturbs the formation of D-periodic symmetrical fibrils [6]. Hence, when atelocollagen is used, the risk of immune response is less than that with acid collagen, although the supramolecular structures characterized by fibrils are disturbed. The kinetics of fibril formation lasts longer and is not complete. Biomaterials fabricated from pepsinized collagen can be considered as less "biomimetic."

3.4
Collagen Hydrogels

3.4.1
Collagen I Hydrogels

3.4.1.1 Classical Hydrogels
Collagen I hydrogels populated by dermal fibroblasts were first used as dermal substitutes [39]. They were obtained by neutralizing and warming at 37 °C an acidic solution of weakly concentrated (0.66 mg ml^{-1}) type I collagen. Collagen I was self-assembled into D-periodic fibrils without specific orientation (Figure 3.2) [1]. As the gelling time was around 10 min, dermal fibroblasts could be encapsulated

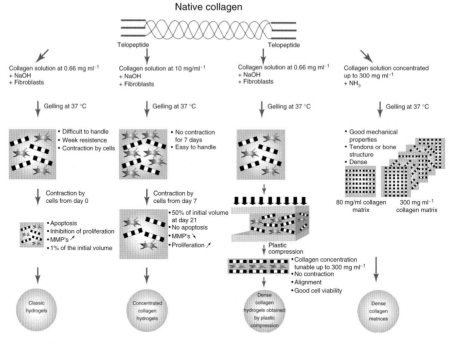

Figure 3.2 Collagen biomaterials made from native collagen molecules. They are obtained by neutralization of acidic collagen solutions that trigger the self-assembly and fibrillogenesis. Different biomaterials can be formed: (i) Classical hydrogels are obtained after gelling of a neutralized collagen solution at 0.66 mg ml^{-1} in which fibroblasts have been encapsulated. Classical hydrogels are contracted by cells and represent 1% of their initial volume after one week. (ii) Concentrated collagen hydrogels are fabricated using the same process as with classical hydrogels but from a 10 mg/ml collagen solution. Hydrogels do not shrink during the first week and have good mechanical properties. (iii) Dense collagen hydrogels are obtained by plastic compression of a classical hydrogel. Compression leads to increase of collagen concentration, decrease of volume, and alignment of fibrils when the concentration reaches more than 300 mg ml^{-1}. (iv) Dense collagen matrices are fabricated by neutralization under ammonia vapors of concentrated collagen solutions (up to 300 mg ml^{-1}). The collagen concentration within the hydrogels allows the formation of aligned or cholesteric structure observed *in vivo*.

into the classical collagen hydrogels after neutralization of the collagen solution. When cultured over three weeks, fibroblasts contracted the collagen scaffold to an area of 1% of the initial one [39]. At the end of the culture, classical hydrogels exhibited a compact structure similar to that observed in dermis *in vivo*. The study of cell phenotype revealed inhibition of proliferation after a few days [31, 40] and the presence of apoptotic cells [41, 42], which explains the decrease of cell number during the culture. In addition, the production of matrix metalloproteinases by fibroblasts was detected and collagen I synthesis was inhibited [24, 29, 43, 44]. Classical collagen hydrogels were used as the dermal part of Apligraf® [45]. Keratinocytes were cultivated and differentiated at the hydrogel surface to obtain a bilayer device used as skin substitute. Unfortunately, because of their extensive shrinkage, hydrogels exhibited a poor porosity and a weak persistence of fibroblasts inside the collagen network. These drawbacks limited the utilization of Apligraf as a permanent implant for treatment of burns [45].

3.4.1.2 Concentrated Collagen Hydrogels

Several attempts were made to improve collagen hydrogel properties; one of these was the attachment of hydrogels to their support or to a peripheral ring to prevent hydrogel contraction [46]. Unfortunately, hydrogels contracted quickly when they were released from their anchorage [42]. Because of mechanical tensions induced by the attachment, a myofibroblast phenotype was detected [30]. Different authors have shown that increasing collagen content can inhibit hydrogel contraction and improve mechanical properties [47]. By modification of the fabrication process in terms of temperature conditions, it is possible to produce concentrated collagen hydrogels up to initial concentrations of 3 and 5 mg ml^{-1} (Figure 3.2) [48, 49]. High initial collagen concentration improves the handling and mechanical properties of hydrogel. In addition, collagen concentration increases hydrogel stiffness, favors cell growth, and inhibits hydrogel contraction [50–52]. *In vivo*, these hydrogels have good integration capacities, as they are colonized completely after 15 days.

3.4.1.3 Dense Collagen Hydrogels Obtained by Plastic Compression

An important improvement in the collagen hydrogel structure has been achieved by Robert Brown and collaborators using plastic compression. This technique, relying on unconfined compression of cellularized collagen hydrogels, generates dense collagen scaffolds with a concentration up to 300 mg ml^{-1} without causing cell damage (Figure 3.2) [53].

Compressive load is applied on hyperhydrated hydrogels (99.5% of water, 1.5 mg ml^{-1} of collagen) to rapidly expel a large amount of fluid through the basal surface. As the water loss can be weighed after compression, it is possible to tune the collagen concentration within the hydrogel and to obtain a collagen scaffold with specific properties. In addition, collagen concentration is directly related to stiffness, which controls cell behavior (proliferation, macromolecule synthesis, migration). Consequently, it is possible to tune the cell phenotype within hydrogels to suit the needs of their application [53, 54]. The effect of filtration results in alignment of collagen fibrils that are arranged in the multiple layers of compacted lamella. Dense collagen hydrogels generated by plastic compression have good mechanical properties and biomimetic functions. Their structures are therefore close to the structure of connective tissues such as skin and bone [55, 56].

3.4.1.4 Dense Collagen Matrices

Dense collagen matrices have been developed to considerably improve the mechanical properties of collagen hydrogels and mimic the structures observed *in vivo*. The technique relies on the concentration by evaporation of acidic collagen solutions up to concentrations of between 40 and 300 mg ml^{-1}. The viscous solution is then neutralized by ammonia vapors to trigger self-assembly into cross-striated fibrils (D-period) and gelling (Figure 3.2) [57]. The use of ammonia vapors permits the retention of the initial collagen concentration measured in the solution and the tuning of collagen concentration within the gel. Dense collagen matrices are acellular biomaterials because their process of fabrication is incompatible with cell encapsulation (NH_3 vapors). The first studies showed that dense collagen matrices concentrated at 40 mg ml^{-1} can be colonized by dermal fibroblasts up to a density similar to that observed in human dermis [58, 59]. More recently, these matrices have been used to study the behavior of osteoblasts. Osteoblasts proliferate on dense matrices and mineralize them, and this shows their suitability for bone regeneration [60]. Collagen solution can be concentrated by evaporation or dialysis to attain a high concentration. From 80 mg ml^{-1}, collagen molecules line up and adopt a liquid crystal order [15, 18]. Neutralized by NH_3, these solutions form hydrogels with a suprastructure mimicking bone ECM. More precisely, they are stabilized cholesteric liquid crystals. Last, these structures can be mineralized by hydroxyapatite crystals that occupy the gap in the fibrils [61].

3.4.2 Cross-Linked Collagen I Hydrogels

3.4.2.1 Chemical Cross-Linking

The main drawback of classical collagen hydrogels is their lack of mechanical strength when they are not contracted by cells. To overcome this weakness, the cross-linking of collagen molecules by chemical reagents may be of great value. The first attempts of cross-linking were carried out with glutaraldehyde on decellularized tissues. Glutaraldehyde is able to form an amide bond between two collagen molecules. Different studies have shown that even though this cross-linker improves the stability, mechanical properties, and biodegradation of biomaterials, it has proved to be toxic [62, 63]. Because of their incompatibility with cell viability, cross-linkers are rarely used in cell therapy hydrogels. Carbodiimide is currently used in tissue engineering to cross-link acellular materials. Nevertheless, the molecule has to be removed by several rinses to eliminate any toxicity.

3.4.2.2 Enzymatic Cross-Linking

More recently, different studies have shown that enzymatic cross-linking of collagen hydrogels is compatible with cell viability. Microbial transglutaminase (mTGA) is an enzyme that introduces a covalent bond between glutamine and lysine residues (Figure 3.3). This reaction allows for collagen reticulation [64]. Since the bond is protease resistant, this cross-linking improves the tensile strength of collagen hydrogels and protects them from proteolytic degradation. In addition, the porosity is also increased because of collagen reticulation. Fibroblasts encapsulated within enzymatically cross-linked collagen scaffolds exhibit good viability and high proliferation. However, hydrogel contraction is not affected by this type of stabilization. Interestingly, the level of cross-linking is comparable to glutaraldehyde without any toxicity [64]. To improve the mechanical properties of transglutaminase cross-linked hydrogels, tests were carried out by introducing elastic elements such

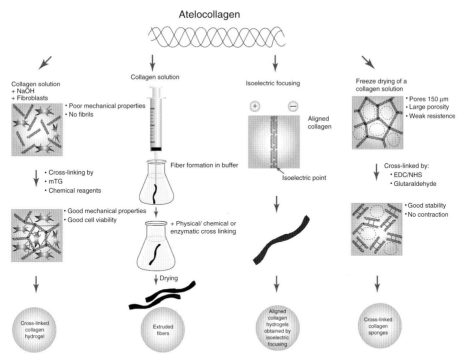

Figure 3.3 Collagen biomaterials made from atelocollagen. Different processes can be used to obtain a variety of materials. (i) Cross-linked hydrogels are made by gelling and reticulation of collagen solutions. Depending on the reagent used, cells are added before or after the cross-linking. Reticulation improves the mechanical properties of the materials. (ii) Extruded fibers are fabricated by extrusion of a collagen solution in a buffer permitting gelling. Then, extruded fibers can be cross-linked and/or dried. (iii) Aligned collagen hydrogels can be obtained by isoelectric focusing. In an electric field, collagen molecules migrate to their isoelectric point. After incubation in a buffer having pH = 7, the bundles gel and can be collected. (iv) Collagen sponges are produced by freeze-drying of a collagen solution. Collagen sponges are cross-linked by chemical reagent to avoid their contraction in physiological conditions and to improve their mechanical properties.

as an elasticlike polymer. Cross-linked by the mTGA, such hydrogels present significantly improved mechanical properties [65].

3.4.3
Collagen II Hydrogels

Collagen II is not widely used in tissue engineering because of its high cost and the difficulties of extraction. Nevertheless, its potential application remains promising because of the development of recombinant collagen technology, which could provide large amounts of monomers in the near future. Collagen II hydrogels have been developed mainly to mimic the structure of the nucleus pulposus (NP), hydrogel constituting the central part of intervertebral disks. Hydrogels are fabricated from type II atelocollagen and can be used as carrier of NP cells or mesenchymal stem cells (MSCs). Moreover, hybrid hydrogels composed of collagen II associated with hyaluronan and/or aggrecan can be obtained to mimic the natural environment of NP cells. The 3D environment leads to the differentiation of MSC to an NP cell-like phenotype or maintains the NP cell's phenotype characterized by the production of proteoglycans [66, 67]. Unfortunately, collagen II hydrogels have poor mechanical properties and weak resistance to proteases and are contracted by cells. Therefore, they need to be stabilized by cross-linkers such as mTGA. This enzyme prevents the hydrogel from contracting while favoring cell proliferation [68]. The use of carbodiimide to cross-link type II collagen hydrogels has also been investigated. Because of carbodiimide cytotoxicity, cells have to be injected inside the collagen II network after the hydrogel formation [69]. A major improvement of this system has been achieved by the stabilization of the hydrogel using a nontoxic chemical cross-linker allowing cell survival, the four-arm polyethylene glycol-succinimidyl glutarate pentaerythritol (4S Starpeg). Adipose-tissue-derived stem cells and NP cells encapsulated within the cross-linked type II collagen/Hyaluronic Acid hydrogels exhibit good viability. In addition, NP cells maintain their phenotype after culture within these hydrogels. Because of the absence of toxicity, the capability for *in situ* gelling within a short time (8 min), the good mechanical properties, and the cell phenotype maintenance, this type of hydrogel shows promise as an injectable carrier for disk regeneration [70].

3.4.4
Aligned Hydrogels and Extruded Fibers

3.4.4.1 Aligned Hydrogels

The ECM of tendons is composed of longitudinally aligned collagen fibrils. This alignment results in structural anisotropy, which provides resistance to stretching and gives tendons their functional characteristic. With the aim of mimicking tendon structure, it is possible to fabricate orientated dense collagen bundles by isoelectric focusing [71]. Collagen I molecules exposed to a pH gradient generated by water electrolysis can migrate up toward their isoelectric point and congregate there to form aligned bundles (Figure 3.3). Immediately after their electrochemical alignment, collagen bundles are incubated at 37 °C in a pH 7.4 buffer to trigger self-assembly and obtain D-periodic fibrils similar to those observed *in vivo*. In order to improve mechanical properties, collagen bundles can be chemically cross-linked, by, for example, use of genipin. The alignment dramatically improves the mechanical properties as compared to those of nonaligned bundles. Indeed, the bundles can reach half the strength of tendons and can be colonized by tendon-derived fibroblasts. This technique has been recently adapted to promote nerve regeneration. Non-cross-linked orientated collagen bundles can provide the adequate topography and biochemical cues to promote unidirectional neurite outgrowth. In addition, this system has been shown to overcome the inhibitory effect of myelin-associated glycoprotein [72].

3.4.4.2 Extruded Collagen Fibers

Another technology has been developed to mimic the structure and tensile capacities of tendons. This is the extrusion of a collagen solution from a syringe in a neutralizing buffer, to trigger fiber formation (Figure 3.3). The extruded fibers have a diameter of 300–600 μm and can subsequently be cross-linked by enzymes or physical or chemical reagents to improve their mechanical properties [38]. A dried form can also be obtained after reticulation [34]. Fibers can be fabricated from acid-soluble collagen I or atelocollagen; but in this case, few D-banded fibrils are obtained [73]. Fibers made from atelocollagen solutions and cross-linked with chemical reagents exhibit properties close to those of tendons in terms of tensile capacities. All extruded fibers are characterized by ridges and crevices running along the longitudinal axis, which favor cell attachment and migration.

3.5 Collagen Sponges

Collagen sponges are produced by the freeze-drying of atelocollagen I or II solutions to generate dried collagen scaffolds with the largest pores (Figure 3.3) [74]. In addition, it is possible to tune the porosity of these scaffolds by modifying freezing conditions (diameter ranges between 50 and 200 μm) [74]. Unfortunately, collagen sponges have a weak resistance at physiological temperature as they denaturate and contract. For this reason, sponges need to be cross-linked by chemical reagents such as glutaraldehyde, carbodiimides, or diphenyphosphorylazide [75, 76]. Because of their process of fabrication, collagen sponges are acellular biomaterials. Therefore, cells have to migrate and colonize the sponge. After three weeks in culture, fibroblasts seeded at the surface migrate into the sponge, proliferate, and fill the pores by deposition of new ECM components [76]. To fabricate sponges, atelocollagen collagen I can be associated with other molecules (collagen III, fibronectin, glycosaminoglycans) to form a hydrid scaffold [77]. At first, collagen sponges were used as dermal substitutes and commercialized under the brand name Integra®. These scaffolds are generated from a solution of collagen I and chondroitine-6 sulfate by freeze-drying and cross-linking with glutaraldehyde [78, 79]. Then, a silicon film is added to the surface to protect the skin injury from contamination and dehydration. An improvement has been achieved by the culture of fibroblasts within the sponge and keratinocytes at its surface to make a cellularized scaffold [80].

With the aim of using sponges as carrier of chondrocytes for cartilage repair, the behavior of cells seeded onto sponges made from atelocollagen I or II has been investigated [81]. Chondrocytes are able to migrate into collagen. No differences are noticed in terms of phenotype when cells are in contact with atelocollagen I or II. The phenotype of chondrocytes is characterized by low production of proteases, high Glycosamiglycans (GAG) deposition, and synthesis of ECM. The major advantage of acellular collagen sponges is their dry form, which allows for storage at room temperature. In addition, because of their cross-linking, sponges do not contract and this is crucial for a musculoskeletal application [82].

3.6 Multichannel Collagen Scaffolds

3.6.1 Multichannel Collagen Conduits

The first conduit made of collagen was commercialized under the brand name NeuraGen™. It is a single conduit clinically used to repair gaps after peripheral nerve injury [83]. Unlike other conduits, which are mainly silicone-based materials, collagen conduits are biocompatible and degradable [84]. With the aim of improving the reinervation of injuries, multichannel collagen nerve conduits have been developed. The presence of submillimeter-diameter channels (up to seven) inside the materials limits the dispersion of neurons without decreasing their regeneration [85, 86]. The conduit fabrication relies on freeze-drying of a collagen solution set in a mold containing central wires to form the channels. Molds and wires are removed after freeze-drying, and a collagen conduit is obtained. The collagen conduits are then cross-linked with carbodiimide (EDC/NHS) to provide good mechanical properties to the biomaterials.

3.6.2 Multi-Channeled Collagen–Calcium Phosphate Scaffolds

Multi-channeled collagen–calcium scaffolds are used to promote bone regeneration. They encourage tissue ingrowth by favoring cell infiltration and provide an important resistance to loading. These scaffolds are elaborated by mixing calcium phosphate with an atelocollagen solution. After a freeze-drying stage, macropores (0.6 mm in diameter) are produced using a forging technique [87, 88]. Cells attach to the scaffold, proliferate, and differentiate into osteoblasts as all phenotypic markers are detected [88]. Moreover, this scaffold can be associated with a gene delivery system to produce VEGF, a molecule promoting neovascularization [87].

3.7 What Tissues Do Collagen Biomaterials Mimic? (see Table 3.1)

3.7.1 Skin

The degree of sophistication and the structural properties of biomaterials used for skin repair depend on the application. Massive burns are characterized by the destruction of a large surface of skin. Despite the wound healing process, closure

Table 3.1 Fields of application of collagen-based biomaterials.

Tissue	Pathology	Collagen solution used	Biomaterials of interest	References
Skin	Burns	Collagen I	Cross-linked sponges	[78, 79]
	Chronic wounds	Collagen I	Classical hydrogels	[39, 45]
			Concentrated hydrogels	[48, 49]
			Enzymatically cross-linked hydrogels	[64, 65]
			Hydrogels obtained by plastic compression	[53, 89]
Cartilage	Osteoarthritis	Collagen I Collagen II	Cross-linked sponges	[81, 90]
Intervertebral disk	Degeneration	Collagen II	Cross-linked hydrogels	[66–70]
Nerve	Spinal cord injury	Collagen I	Aligned hydrogels	[71, 72]
	Peripheral nerve section	Collagen I	Multichannel conduits	[85, 86]
Tendons	Section	Collagen I	Aligned hydrogels	[71, 72]
			Extruded fibers	[34, 38, 73]
			Hydrogels obtained by plastic compression	[53]
Bone	Defect	Collagen I	Dense matrices	[61, 91]
			Multichannelled collagen–calcium phosphate scaffolds	[87, 88]

of the wound is not fast enough to prevent dehydration and contamination of the wound bed. As a consequence, the application of a permanent implant mimicking skin is required. Acellular collagen sponges such as Integra or OrCel® are used as dermal substitutes [80, 92]. Cross-linked collagen sponges have good mechanical properties, are resistant to degradation, and exhibit great porosity. They can be colonized quickly by host cells, which promotes the formation of neotissue and material integration. Collagen sponges are associated either with a silicone film or with sheets of autologous keratinocytes in order to re-create the protective barrier of the skin (epidermal substitute). Skin substitutes used for the treatment of chronic wounds need a lesser degree of complexity. They are used as wound dressing to promote wound healing, thus they do not integrate the host organism [80]. They need to be cellularized, as they act as a source of growth factors secreted by cells. In addition, they have to hydrate the wound bed and protect it against contamination. They have bilayer structure with an epidermal part made of sheets of keratinocytes and a dermal part that is a classical collagen hydrogel (Apligraf) [45]. As classical hydrogels have some drawbacks, biomaterials such as concentrated collagen hydrogels [48, 49] or hydrogels obtained by plastic compression offer promise in the improvement of materials used in chronic wound treatment [89].

3.7.2
Nerves

Despite the capacity of peripheral nerves to self-regenerate, several factors can prevent their regeneration after a traumatic injury. When nerves are sectioned, axons regrow but sometimes do not connect properly to their targets. Moreover, processes of inflammation and scarring can inhibit neuronal regeneration. When the nerve gap is less than 5 mm, nerves can be bridged by surgical coaptation of nerve stumps [84, 93]. For larger gaps, direct approximation of axonal stumps is not possible because of the excessive tension induced by coaptation. In this case, bridging the defect with autologous nerve is required. This technique is considered as the gold standard in peripheral nerve repair, even though several drawbacks exist. One such drawback is the removal of healthy sensory nerves, which can lead to donor site morbidity and ensuing sensory defects [93]. Tissue engineered autograft substitutes, termed *nerve guidance conduits*, have been developed with the aim of directing axon growth and promoting functional nerve regeneration. They guide axon sprouting and prevent excessive scarring [84]. The first conduits developed were materials made of silicon because these materials are flexible and biocompatible. They are not biodegradable; however, and do not allow for cell adhesion. So, temporary biodegradable conduits made from natural polymers have been developed in order to overcome these drawbacks. In contrast to single-channel conduits (such as Neuragen®), multichannel collagen conduits show promise for nerve regeneration because they can guide axon regeneration without dispersion, are degradable, and prevent scar formation [85, 86]. The scaffolds that could potentially be used in nerve repair are aligned hydrogels. Indeed, embryonic nerve tissue matrix is composed of lamina tubes in which fibrils are longitudinally aligned. The nerve growth operates on this matrix. Hence, using aligned collagen hydrogels serves as an advantage to guide axon regeneration [1].

3.7.3
Tendons

Tendons are composed of collagen fibrils aligned along the longitudinal axis to withstand forces of traction. The common treatment is the use of autograft from healthy donor sites. To repair sectioned tendons, the engrafted tissue has to be mechanically functional and match with the injured tendons [1]. As tendon repair is difficult to perform by grafting, tissue engineered devices offer a new hope. To mimic the properties of tendons, collagen biomaterials have to mimic the tendon structure, that is, the fibril alignment. Different materials could be used for this application, such as aligned collagen hydrogels obtained by isoelectric focusing [71, 72], which are aligned fibrils. Dense collagen matrices as well as hydrogels obtained by plastic compression also possess this aligned structure [53, 61]. Extruded collagen fibers could also be considered, as they are resistant and their diameter can be easily monitored.

3.7.4
Bone

Bone has properties of self-healing that permit its restoration in terms of shape, structure, and mechanical properties. Nevertheless, bone repair is impossible in pathological situations such as when bone losses have been caused by a trauma or a tumor resection [94]. The current treatment is the autograft, which relies on harvesting autologous bone material from iliac crest and implanting it in the bone defect. This technique, while painful and leading to morbidity of donor site, explains the development of bone substitute [95]. Bone substitutes need to have good reliance on loading and a porous network to favor cell colonization. Calcium phosphate ceramics have been developed to mimic the mechanical properties of bone and allow for its regeneration. They do not mimic the bone structure, as they do not possess the organic part made of collagen. Nevertheless, they are osteoconductive, as they permit their colonization and synthesis of ECM by osteoblasts [80]. Ceramics are made of calcium hydroxyapatite or tricalcium phosphate and possess high porosity (>500 μm) promoting their colonization. It has been evidenced that collagen is osteoinductive, that is, this molecule promotes the differentiation of osteoblastic precursors [94]. So, the association of collagen and calcium phosphate can improve the substitute structure. Moreover, this material more closely resembles the bone structure. Multichannel collagen calcium phosphate scaffolds appear promising for bone regeneration because they encourage cellular colonization without weakening the mechanical properties of the material [88]. Another material that could significantly improve the materials for bone repair is the self-assembled collagen-apatite matrix. This material is a dense collagen matrix with an aligned fibril structure. Calcium phosphate crystals have been grown inside fibril gap to obtain a bonelike structure [61, 91].

3.7.5
Intervertebral Disk

Intervertebral disk degeneration is a common cause of neck and low-back pains [96]. The current treatments are drug therapy and physiotherapy. However, when the pain is too great, an invasive surgical intervention is required to remove the disk or to carry out a vertebral fusion [97]. The intervertebral disk is composed of three parts: NP located at the center, annulus fibrosus (AF) surrounding the NP, and an outer cartilage end plate at the vertebral body. NP has a highly hydrated gel-like structure made of collagen II, proteoglycans, and hyaluronic acid and is populated by NP cells, chondrocytes, and notochondral cells. In contrast, AF is a fibrous tissue made of collagen I and is populated by fibroblastlike cells [98]. NP degeneration is characterized by a weaker disk hydration because of the decrease of proteoglycan content and cell number [69]. Tissue engineered scaffolds developed for disk regeneration offer the advantage of being degradable and can be implanted by less-invasive processes if they are injectable. Therefore, hybrid atelocollagen II/GAGs hydrogels can be materials of interest, as collagen II is the main constituent of the ECM and is able to entrap sulfated proteoglycans, thus maintaining NP hydration. When the aim is to mimic NP properties, injectable type II atelocollagen–hyaluronan hydrogels offer the most promise [70]. Their cross-linking by 4S-StarPEG provides good mechanical properties, and NP cells encapsulated inside the scaffold exhibit a good viability. In addition, atelocollagen II and hyaluronic acid can provide the conditions for an acceptable hydration after implantation.

3.7.6
Cartilage

Cartilage has a poor ability to repair. Usually, prostheses are used to substitute the defective cartilage when it is impaired. Alternative approaches can be used such as the stimulation of subchondral bone by microfractures or the autologous chondrocyte transplantation (ACT) [99]. An improvement of ACT could rely on the encapsulation of chondrocytes in a scaffold to maintain the chondrocyte phenotype. Indeed, chondrocytes have the tendency to dedifferentiate when they are not in their three-dimensional environment. In this application, collagen sponges made of atelocollagen type I or II seem to be the most appropriate, as they are dried forms and maintain the chondrocyte phenotype. Compared to hydrogels, they exhibit better mechanical properties owing to their cross-linking and are also resistant to cell degradation and contraction [81, 90].

3.8
Concluding Remarks

Collagen is the protein of choice for tissue engineering applications. This molecule has low immunogenicity, is available in large amounts, and is specifically recognized by human connective cells. Scaffolds fabricated from collagen molecules can mimic the connective tissue they have to replace in two different ways. In the first, they try to reproduce the tissue structure. In this case, native collagens are currently used because their ability to self-assemble is high. In addition, the ultrastructural arrangements of collagen fibrils observed in specific connective tissues such as tendons, skin, or bone can be achieved by modulating the collagen concentration. Scaffolds made from native collagen are mostly hydrogels and are considered as biomimetic. Nevertheless, they can still trigger an immune response when they are implanted in a host organism. The second way to mimic connective tissues is to use simpler biomaterials that provide adequate mechanical properties and a good topography to cells. Scaffolds made from atelocollagen solutions and cross-linked with chemical or physical reagents are currently used to mimic the mechanical environment of cells. In addition, a variety of biomaterials can be obtained from atelocollagen, such as aligned scaffolds, sponges, and extruded fibers.

Acknowledgments

We thank Estelle Collin and Modamad Abu-Rub for their interesting comments and discussion on this article. In addition, we thank Anthony Sloan for reviewing the language of this article. Last, we would like to thank Health Researcher Board Ireland (RP/2008/188).

References

1. Brown, R.A. and Phillips, J.B. (2007) *Int. Rev. Cytol.*, **262**, 75–150.
2. Zheng, M.H., Chen, J., Kirilak, Y., Willers, C., Xu, J., and Wood, D. (2005) *J. Biomed. Mater. Res. B: Appl. Biomater.*, **73**(1), 61–67.
3. Tibbitt, M.W. and Anseth, K.S. (2009) *Biotechnol. Bioeng.*, **103**(4), 655–663.
4. Place, E.S., Evans, N.D., and Stevens, M.M. (2009) *Nat. Mater.*, **8**(6), 457–470.
5. Kadler, K.E., Holmes, D.F., Trotter, J.A., and Chapman, J.A. (1996) *Biochem. J.*, **316**(Pt. 1), 1–11.
6. Brodsky, B. and Persikov, A.V. (2005) *Adv. Protein Chem.*, **70**, 301–339.
7. Kadler, K.E., Baldock, C., Bella, J., and Boot-Handford, R.P. (2007) *J. Cell Sci.*, **120**(Pt. 12), 1955–1958.
8. Van der Rest, M. and Garrone, R. (1991) *FASEB J.*, **5**(13), 2814–2823.
9. Kadler, K.E., Lightfoot, S.J., and Watson, R.B. (1995) *Methods Enzymol.*, **248**, 756–771.
10. Kadler, K.E. and Watson, R.B. (1995) *Methods Enzymol.*, **248**, 771–781.
11. Colige, A., Ruggiero, F., Vandenberghe, I., Dubail, J., Kesteloot, F., Van Beeumen, J. et al. (2005) *J. Biol. Chem.*, **280**(41), 34397–34408.
12. Prockop, D.J., Sieron, A.L., and Li, S.W. (1998) *Matrix Biol.*, **16**(7), 399–408.
13. Eyre, D.R., Paz, M.A., and Gallop, P.M. (1984) *Annu. Rev. Biochem.*, **53**, 717–748.
14. Giraud-Guille, M.M. and Besseau, L. (1998) *Connect. Tissue Res.*, **37**(3–4), 183–193.
15. Giraud-Guille, M.M., Besseau, L., and Martin, R. (2003) *J. Biomech.*, **36**(10), 1571–1579.
16. Rho, J.Y., Kuhn-Spearing, L., and Zioupos, P. (1998) *Med. Eng. Phys.*, **20**(2), 92–102.
17. Giraud-Guille, M.M., Mosser, G., Helary, C., and Eglin, D. (2005) *Micron*, **36**(7–8), 602–608.
18. Giraud-Guille, M.M. (1994) *Microsc. Res. Tech.*, **27**(5), 420–428.
19. Hynes, R.O. (2002) *Cell*, **110**(6), 673–687.
20. Popova, S.N., Lundgren-Akerlund, E., Wiig, H., and Gullberg, D. (2007) *Acta Physiol. (Oxf)*, **190**(3), 179–187.
21. White, D.J., Puranen, S., Johnson, M.S., and Heino, J. (2004) *Int. J. Biochem. Cell. Biol.*, **36**(8), 1405–1410.
22. Pozzi, A., Wary, K.K., Giancotti, F.G., and Gardner, H.A. (1998) *J. Cell Biol.*, **142**(2), 587–594.
23. Zigrino, P., Drescher, C., and Mauch, C. (2001) *Eur. J. Cell Biol.*, **80**(1), 68–77.
24. Bengtsson, T., Aszodi, A., Nicolae, C., Hunziker, E.B., Lundgren-Akerlund, E., and Fassler, R. (2005) *J. Cell Sci.*, **118**(Pt. 5), 929–936.
25. Tiger, C.F., Fougerousse, F., Grundstrom, G., Velling, T., and Gullberg, D. (2001) *Dev. Biol.*, **237**(1), 116–129.
26. Vogel, W.F., Abdulhussein, R., and Ford, C.E. (2006) *Cell Signal.*, **18**(8), 1108–1116.
27. Leitinger, B. and Hohenester, E. (2007) *Matrix Biol.*, **26**(3), 146–155.
28. Mauch, C., Hatamochi, A., Scharffetter, K., and Krieg, T. (1988) *Exp. Cell Res.*, **178**(2), 493–503.
29. Grinnell, F. (2003) *Trends Cell Biol.*, **13**(5), 264–269.
30. Hinz, B. and Gabbiani, G. (2003) *Thromb. Haemost.*, **90**(6), 993–1002.

31. Kessler, D., Dethlefsen, S., Haase, I., Plomann, M., Hirche, F., Krieg, T. et al. (2001) *J. Biol. Chem.*, **276**(39), 36575–36585.
32. Nicodemus, G.D. and Bryant, S.J. (2008) *Tissue Eng. Part B: Rev.*, **14**(2), 149–165.
33. Zeugolis, D.I., Paul, R.G., and Attenburrow, G. (2008) *Acta Biomater.*, **4**(6), 1646–1656.
34. Malone, J.P., George, A., and Veis, A. (2004) *Proteins*, **54**(2), 206–215.
35. Malone, J.P. and Veis, A. (2004) *Biochemistry*, **43**(49), 15358–15366.
36. Kadler, K.E., Hill, A., and Canty-Laird, E.G. (2008) *Curr. Opin. Cell Biol.*, **20**(5), 495–501.
37. Zeugolis, D.I., Paul, G.R., and Attenburrow, G. (2009) *J. Biomed. Mater. Res. A*, **89**(4), 895–908.
38. Bell, E., Ivarsson, B., and Merrill, C. (1979) *Proc. Natl. Acad. Sci. U.S.A*, **76**(3), 1274–1278.
39. Fringer, J. and Grinnell, F. (2003) *J. Biol. Chem.*, **278**(23), 20612–20617.
40. Sarber, R., Hull, B., Merrill, C., Soranno, T., and Bell, E. (1981) *Mech. Ageing Dev.*, **17**(2), 107–117.
41. Fluck, J., Querfeld, C., Cremer, A., Niland, S., Krieg, T., and Sollberg, S. (1998) *J. Investig. Dermatol.*, **110**(2), 153–157.
42. Grinnell, F., Zhu, M., Carlson, M.A., and Abrams, J.M. (1999) *Exp. Cell Res.*, **248**(2), 608–619.
43. Seltzer, J.L., Lee, A.Y., Akers, K.T., Sudbeck, B., Southon, E.A., Wayner, E.A. et al. (1994) *Exp. Cell Res.*, **213**(2), 365–374.
44. Ruangpanit, N., Chan, D., Holmbeck, K., Birkedal-Hansen, H., Polarek, J., Yang, C. et al. (2001) *Matrix Biol.*, **20**(3), 193–203.
45. Wilkins, L.M., Watson, S.R., Prosky, S.J., Meunier, S.F., and Parenteau, N.L. (1994) *Biotechnol. Bioeng.*, **43**(8), 747–756.
46. Lopez Valle, C.A., Auger, F.A., Rompre, P., Bouvard, V., and Germain, L. (1992) *Br. J. Dermatol.*, **127**(4), 365–371.
47. Rompre, P., Auger, F.A., Germain, L., Bouvard, V., Lopez Valle, C.A., Thibault, J. et al. (1990) *In Vitro Cell Dev. Biol.*, **26**(10), 983–990.
48. Helary, C., Abed, A., Mosser, G., Louedec, L., Meddahi-Pelle, A., and Giraud-Guille, M.M. (2011) *J. Tissue Eng. Regen. Med.*, **5**(3), 248–252.
49. Helary, C., Bataille, I., Abed, A., Illoul, C., Anglo, A., Louedec, L. et al. (2010) *Biomaterials*, **31**(3), 481–490.
50. Hadjipanayi, E., Mudera, V., and Brown, R.A. (2009) *J. Tissue Eng. Regen. Med.*, **3**(2), 77–84.
51. Karamichos, D., Brown, R.A., and Mudera, V. (2007) *J. Biomed. Mater. Res. A*, **83**(3), 887–894.
52. Karamichos, D., Skinner, J., Brown, R., and Mudera, V. (2008) *J. Tissue Eng. Regen. Med.*, **2**(2–3), 97–105.
53. Brown, R.A., Wiseman, M., Chuo, C.B., Cheema, U., and Nazhat, S.N. (2005) *Adv. Funct. Mater.*, **15**(11), 1762–1770.
54. Hadjipanayi, E., Brown, R.A., and Mudera, V. (2009) *J. Tissue Eng. Regen. Med.*, **3**(3), 230–241.
55. Hu, K., Shi, H., Zhu, J., Deng, D., Zhou, G., Zhang, W. et al. (2010) *Biomed. Microdevices*, **12**(4), 627–635.
56. Buxton, P.G., Bitar, M., Gellynck, K., Parkar, M., Brown, R.A., Young, A.M. et al. (2008) *Bone*, **43**(2), 377–385.
57. Besseau, L., Coulomb, B., Lebreton-Decoster, C., and Giraud-Guille, M.M. (2002) *Biomaterials*, **23**(1), 27–36.
58. Helary, C., Foucault-Bertaud, A., Godeau, G., Coulomb, B., and Giraud-Guille, M.M. (2005) *Biomaterials*, **26**(13), 1533–1543.
59. Helary, C., Ovtracht, L., Coulomb, B., Godeau, G., and Giraud-Guille, M.M. (2006) *Biomaterials*, **27**(25), 4443–4452.
60. Vigier, S., Helary, C., Fromigue, O., Marie, P., and Giraud-Guille, M.M. (2010) *J. Biomed. Mater. Res. A*, **94**(2), 556–567.
61. Nassif, N., Gobeaux, F., Seto, J., Belamie, E., Davidson, P., Panine, P. et al. (2010) *Chem. Mater.*, **22**(11), 3307–3309.
62. Speer, D.P., Chvapil, M., Eskelson, C.D., and Ulreich, J. (1980) *J. Biomed. Mater. Res.*, **14**(6), 753–764.

63. Huang-Lee, L.L., Cheung, D.T., and Nimni, M.E. (1990) *J. Biomed. Mater. Res.*, **24**(9), 1185–1201.
64. Garcia, Y., Collighan, R., Griffin, M., and Pandit, A. (2007) *J. Mater. Sci. Mater. Med.*, **18**(10), 1991–2001.
65. Garcia, Y., Hemantkumar, N., Collighan, R., Griffin, M., Rodriguez-Cabello, J.C., and Pandit, A. (2009) *Tissue Eng. Part A*, **15**(4), 887–899.
66. Sakai, D., Mochida, J., Iwashina, T., Watanabe, T., Suyama, K., Ando, K. et al. (2006) *Biomaterials*, **27**(3), 346–353.
67. Sakai, D., Mochida, J., Yamamoto, Y., Nomura, T., Okuma, M., Nishimura, K. et al. (2003) *Biomaterials*, **24**(20), 3531–3541.
68. Halloran, D.O., Grad, S., Stoddart, M., Dockery, P., Alini, M., and Pandit, A.S. (2008) *Biomaterials*, **29**(4), 438–447.
69. Calderon, L., Collin, E., Velasco-Bayon, D., Murphy, M., O'Halloran, D., and Pandit, A. (2010) *Eur. Cell Mater.*, **20**, 134–148.
70. Collin, E.C., Grad, S., Zeugolis, D.I., Vinatier, C.S., Clouet, J.R., Guicheux, J.J. et al. (2011) *Biomaterials*, **32**(11), 2862–2870.
71. Cheng, X., Gurkan, U.A., Dehen, C.J., Tate, M.P., Hillhouse, H.W., Simpson, G.J. et al. (2008) *Biomaterials*, **29**(22), 3278–3288.
72. Abu-Rub, M., Billiar, K.L., van Hes, M.H., Knight, A., Rodriguez, B.J., Zeugolis, D.I., McMahon, S., Windebank, A., and Pandit, A. (2011) *Soft Matter*, **7**(6), 2770–2781.
73. Zeugolis, D.I., Paul, R.G., and Attenburrow, G. (2008) *J. Biomed. Mater. Res. A*, **86**(4), 892–904.
74. Doillon, C.J., Whyne, C.F., Brandwein, S., and Silver, F.H. (1986) *J. Biomed. Mater. Res.*, **20**(8), 1219–1228.
75. Chevallay, B., Abdul-Malak, N., and Herbage, D. (2000) *J. Biomed. Mater. Res.*, **49**(4), 448–459.
76. Vaissiere, G., Chevallay, B., Herbage, D., and Damour, O. (2000) *Med. Biol. Eng. Comput.*, **38**(2), 205–210.
77. Doillon, C.J. and Silver, F.H. (1986) *Biomaterials*, **7**(1), 3–8.
78. Yannas, I.V., Burke, J.F., Gordon, P.L., Huang, C., and Rubenstein, R.H. (1980) *J. Biomed. Mater. Res.*, **14**(2), 107–132.
79. Yannas, I.V. and Burke, J.F. (1980) *J. Biomed. Mater. Res.*, **14**(1), 65–81.
80. Ehrenreich, M. and Ruszczak, Z. (2006) *Tissue Eng.*, **12**(9), 2407–2424.
81. Freyria, A.M., Ronziere, M.C., Cortial, D., Galois, L., Hartmann, D., Herbage, D. et al. (2009) *Tissue Eng. Part A*, **15**(6), 1233–1245.
82. Galois, L., Freyria, A.M., Grossin, L., Hubert, P., Mainard, D., Herbage, D. et al. (2004) *Biorheology*, **41**(3–4), 433–443.
83. Wangensteen, K.J. and Kalliainen, L.K. (2009) *Hand (NY)*, **5**(3), 273–277.
84. Pfister, L.A., Papaloizos, M., Merkle, H.P., and Gander, B. (2007) *J. Peripher. Nerv. Syst.*, **12**(2), 65–82.
85. Yao, L., Billiar, K.L., Windebank, A.J., and Pandit, A. (2010) *Tissue Eng. Part C: Methods*, **16**(6), 1585–1596.
86. Yao, L., de Ruiter, G.C., Wang, H., Knight, A.M., Spinner, R.J., Yaszemski, M.J. et al. (2010) *Biomaterials*, **31**(22), 5789–5797.
87. Keeney, M., van den Beucken, J.J., van der Kraan, P.M., Jansen, J.A., and Pandit, A. (2010) *Biomaterials*, **31**(10), 2893–2902.
88. Keeney, M., Collin, E., and Pandit, A. (2009) *Tissue Eng. Part C: Methods*, **15**(2), 265–273.
89. Hu, K.K., Shi, H., Zhu, J., Deng, D., Zhou, G.D., Zhang, W.J. et al. (2010) *Biomed. Microdevices*, **12**(4), 627–635.
90. Galois, L., Hutasse, S., Cortial, D., Rousseau, C.F., Grossin, L., Ronziere, M.C. et al. (2006) *Biomaterials*, **27**(1), 79–90.
91. Nassif, N., Martineau, F., Syzgantseva, O., Gobeaux, F., Willinger, M., Coradin, T. et al. (2010) *Chem. Mater.*, **22**(12), 3653–3663.
92. Ehrenreich, M. and Ruszczak, Z. (2006) *Acta Dermatovenerol. Alp. Panonica. Adriat.*, **15**(1), 5–13.
93. Johnson, E.O. and Soucacos, P.N. (2008) *Injury*, **39**(Suppl. 3), S30–S36.
94. Giannoudis, P.V., Dinopoulos, H., and Tsiridis, E. (2005) *Injury*, **36**(Suppl. 3), S20–S27.

95. Banwart, J.C., Asher, M.A., and Hassanein, R.S. (1995) *Spine (Phila PA 1976)*, **20**(9), 1055–1060.
96. Adams, M.A. and Roughley, P.J. (2006) *Spine (Phila PA 1976)*, **31**(18), 2151–2161.
97. Kishen, T.J. and Diwan, A.D. (2010) *Orthop. Clin. North. Am.*, **41**(2), 167–181.
98. O'Halloran, D.M. and Pandit, A.S. (2007) *Tissue Eng.*, **13**(8), 1927–1954.
99. Chajra, H., Rousseau, C.F., Cortial, D., Ronziere, M.C., Herbage, D., Mallein-Gerin, F. *et al.* (2008) *Biomed. Mater. Eng.*, **18**(Suppl. 1), S33–S45.
100. Khoshnoodi, J., Cartailler, J.P., Alvares, K., Veis, A., and Hudson, B.G. (2006) *J. Biol. Chem.*, **281**(50), 38117–38121.

4
Silk-Based Biomaterials
Sílvia Gomes, Isabel B. Leonor, João F. Mano, Rui L. Reis, and David L. Kaplan

4.1
Introduction

Silk is a hierarchically structured fibrous protein that has been used for thousands of years for different applications such as textile production and wound dressings [1]. Although silk proteins are produced by a large number of arthropod species such as wasps, bees, and crickets, during the past few decades, the scientific community has focused its attention in the silk produced by the mulberry silkworm *Bombyx mori* and by different spider species. Farming of the silkworm *B. mori* started around 5000 years ago in China, and since then, the cocoon case produced by this animal has been used as a source of silk for the textile industry. More recently, *B. mori* silk has been used commercially in the development of sutures for medical applications [1, 2]. In contrast to *B. mori* silk, spider silk has not been commercialized for biomedical proposes mainly because of the territorial cannibalistic behavior of spiders, which makes them difficult to farm, along with the low levels of silk production. However, in the past, spider webs have been used as bandages for wound dressing, because of its antimicrobial properties, and in other human activities such as fishing and hunting [3]. The results from recent studies demonstrating the biocompatibility of silk and the exquisite mechanical properties of silk fibers, especially in the case of spider silk, are the main reasons why *B. mori* and spider silk proteins have been the target of intense research efforts, when compared with the silk produced by other arthropods, aimed at developing new silk-based biomaterials with superior mechanical and biological properties.

On the basis of the background given above and since *B. mori* and spider silk are the most extensively characterized silk, this chapter focuses on the silk proteins produced by these animals and on the perspectives of using these proteins for future biomedical applications.

4.2
Silk Proteins

4.2.1
Bombyx mori Silk

The silk produced by B. mori is the most widely used, since this silkworm is easy to farm, allowing for steady and large-scale production [1]. The B. mori silk fiber is formed by two microfilaments embedded in a gluelike glycoprotein named *sericin*, which works as a coating and is used by the animal for cocoon assembly. Each microfilament results from the assembly of a hydrophobic ~370 kDa heavy-chain fibroin protein, with a relatively hydrophilic ~25 kDa light-chain fibroin and a 30 kDa P25 protein [4]. The heavy-chain fibroin is a large insoluble macromolecule formed by ~5263 amino acids and is rich in glycine (G, 45.9%), alanine (A, 30.3%), serine (S, 12.2%), tyrosine (Y, 5.3%), and valine (V, 1.8%) residues [1, 5]. This heavy-chain fibroin is formed by a highly repetitive crystalline fraction of 2377 repeats of GX. The X position is occupied by the A residue in 64% of the repeats, by S in 22%, by Y in 10%, by V in 3%, and by threonine (T) in 1.3% of the repeats [5, 6]. These repetitive cores form 12 domains that alternate with 11 nonrepetitive amorphous domains formed by 25 amino acids residues, which connect the crystalline domains. These amorphous regions contain proline (P) and tryptophan (W) and charged amino acids residues that are absent in the crystalline motifs [5, 7].

Marsh and coworkers were the first to propose a structural model for silk fibroin organization based on an X-ray diffraction analyses. The authors proposed a structural organization consisting of extended protein chains linked by hydrogen bonds between N–H and C=O groups forming antiparallel-chain pleated β-sheets packed on the top of each other [8]. GX repeats with X being mostly A predominate throughout these β-sheet plates with the side chains of nonglycine residues on the same side of one of the β-sheets [8].

Recently, Takahashi and coworkers [9] reevaluated the crystalline structure of B. mori silk using X-ray diffraction. The molecular structure proposed by these researchers is similar to the pleated β-sheet arrangement proposed by Marsh and colleagues. However, in Takahashi's model, the β-sheet conformation, resulting from hydrogen bonding, assumes an antipolar-antiparallel structure, with the methyl side groups from A residues pointing alternatively toward both sides of the sheet structure [9].

Heavy-chain fibroin is linked to light-chain fibroin by a disulfide bond at the C-terminus of both chains [10]. Light-chain fibroin is formed by 262 amino acids containing 14% A, 10% S, and 9% G and with an N-acetylated terminal serine residue [11]. The biological importance of the disulfide linkage between light- and heavy-fibroin chains may be related to the intracellular transport of fibroin from the endoplasmic reticulum to the Golgi apparatus [12]. Furthermore, the disulfide bond between both protein chains may control the formation of β-sheet conformation [10].

The P25 protein is bonded to the heavy- and light-chain fibroin complex by noncovalent hydrophobic interactions, in a ratio of 6 : 6 : 1 (heavy chain : light chain : P25) [13, 14]. P25 is formed by 220 amino acid residues [13, 14], and recent findings suggest that it plays a central role in the maintenance of the structural integrity of the heavy-chain/light-chain complex. The formation of the P25/heavy-chain/light-chain complex may facilitate intracellular transport of the final protein and help to avoid the formation of insoluble aggregates [15].

Finally, silk fibers are coated by a layer of sericins, which act as a glue, binding together the silk fibers. Sericin is formed by a mixture of polypeptides with molecular weights ranging from 20 to 220 kDa [16]. More recently, three polypeptides were isolated from the sericin of *B. mori* cocoons, namely, sericins M, A, and P with molecular masses of 400, 250, and 150 kDa, respectively [17]. Amino acid analysis shows that all three sericin polypeptides have high contents of serine, glycine, and alanine and differ from fibroin in having large amounts of aspartic acid and glutamic acid [16].

4.2.2
Spider Silk

Contrary to *B. mori* and other insects, spiders can produce up to seven different types of silk, mainly major ampullate, flagelliform, aggregate, minor ampullate, pyriform, aciniform, and cylindriform [1]. Spiders use silk for different purposes. *Major ampullate silk* or *dragline silk* is composed of two different proteins named major ampullate spidroin protein 1 (MaSp1) and MaSp2 in a ratio of 3 : 2 and is used as a safety line for survival and in the construction of the web anchors. *Minor ampullate silk* is formed by minor ampullate spidroin 1 (MiSp1) and 2 (MiSp2), has a structure similar to dragline silk, and is used in structural reinforcement of the spider web [18, 19]. The *flagelliform silk* fiber is believed to be composed of a single protein and forms part of the web capture spiral where it is dotted with a sticky aggregate silk, which is important for trapping prey [18, 19]. Besides dragline, minor ampullate, and flagelliform silk proteins, spiders also produce *pyriform silk*, used to attach the web to substrates [20, 21] and *aciniform* and *cylindriform silk* that are not directly related with the construction of the orb web. Aciniform silk is used by the spider to wrap and immobilize the prey, to build sperm webs, and in web decorations [22], and cylindriform silk is used to build the egg case [23].

Spider silk protein expression takes place in the columnar endothelial cells located in the upper part of the spider silk gland [24, 25]. Spider silk fibers are formed by one or more proteins designated spidroins with a molecular weight of around 350 kDa and composed of large amounts of nonpolar hydrophobic amino acids, mainly G and A, assembled into a highly repetitive core region [26]. This core region is formed by tandem repeats of certain consensus domains, which generally account for more than 90% of the protein [27]. These modular units are about 40–200 amino acids long and are repeated approximately 100 times in the core region. In major and minor ampullate silks, the core regions are composed of the consensus units A_n, $(GA)_n$, and GGX (X usually stands for alanine, leucine, glutamine, and tyrosine). In addition, in the case of the major ampullate silks, mainly MaSp2 from *Nephila clavipes* and fibroins 3 and 4 (ADF-3 and ADF-4) from

Araneus diadematus, there are a large number of GPG motifs; while MaSp1 is formed mostly by GGX and poly-A domains [28, 29].

The hydrophobic poly-A and GA blocks contribute to the formation of crystalline regions of antiparallel β-sheets, and the hydrophilic GGX regions are responsible for α-helix secondary conformations [18]. The crystalline domains are responsible for maintaining protein–protein interactions and forming silk fibers. This cohesion is possible due to the poly-A regions, where the successive alanine residues are disposed in alternate sites along the protein backbone. This arrangement allows antiparallel chains to interlock with each other because of the formation of hydrophobic interactions and hydrogen bonds [18]. In the case of GA blocks, this cohesion is not as strong since the glycine residues are unable to establish the same hydrophobic interactions as in the case of poly-A motifs, resulting in a lower binding energy [18]. The formation of helical structures in the case of GGX motifs is supported by Fourier transform infrared spectroscopy (FTIR) [30] and nuclear magnetic resonance (NMR) studies [31]. This helical conformation could act as a transition or link between the crystalline regions, allowing for less rigid protein structures [18]. The proline-containing GPG domains of MaSp2 are probably involved in the formation of type II β-turns [18]. According to the model suggested by Hayashi and colleagues [18] these β-turns are similar to springs. In the particular case of flagelliform silk, with a core region formed by the glycine-rich motif GPGXX (X often represents alanine, serine, valine, or tyrosine) [32], the β-turn springs can act as an elastic element of the silk fiber. During extension, the proline residues are forced to torque, accumulating energy that can be used for retraction [18]. Also, the presence of serine and tyrosine residues, with hydroxyls in their side chains, can contribute to stabilization of the β-springs and the different layers of coils, respectively [18, 32].

In spider silk proteins, the repetitive core is flanked by nonrepetitive carboxy- and amino-terminal domains with a globular structure formed by about 100–140 amino acid residues [33–35]. The amino-terminal sequence contains signal sequences and is considered essential for the storage of spidroins in silk glands and for secretion into the gland lumen [34]. In addition, the results obtained by Askarieh and colleagues [36] indicate that the amino-terminal domain controls spidroin assembly by preventing aggregation during storage, regulating the assembly process as the pH decreases along the spinning duct of the spider's silk gland. The carboxy-terminal region is involved in the formation of silk fibers at the spinning duct [33, 37]. Recent studies show that the carboxy-terminal domain plays an important role in the storage and assembly of spider silk fibers. During storage, the carboxy-terminal domain is responsible for stabilizing the proteins solution through the formation of higher supramolecular assemblages. The carboxy-terminal region also controls the transition between storage and assembly of silk fibers by triggering the fast and controlled salting-out of proteins and the proper alignment of β-sheets present in the repetitive regions, during shear-induced assembly [38].

Recombinant DNA technology has played a crucial role in unraveling the composition, molecular organization, and secondary structure of silk proteins described above. In this way, genetic engineering can be used to control protein structure and chemistry, allowing detailed and systematic studies of the protein hierarchical structure [39]. Also, as mentioned earlier, the main source of B. mori silk has been the cocoons produced by this worm. However, some questions have been raised in the past few years about its biocompatibility. Recent studies have shown that the variability of the source materials and, in the particular case of B. mori silk, the presence of sericin proteins are the main cause of the adverse problems with biocompatibility and hypersensitivity to B. mori silk. In order to overcome this and other problems, recombinant DNA technology has been used to produce silklike proteins. These new alternatives to silk of natural origin silk are addressed in the next section.

4.2.3
Recombinant Silk

With new technologies in the fields of polymer synthesis and processing, silk continues to be an important topic of research for biomaterial and biomedical research. In the case of B. mori silk, sericulture provides the product used by the textile industry and in medical sutures [40]. This silk is also being studied for tissue engineering in the form of scaffolds for a range of tissue needs, such as corneal regeneration [41, 42], cartilage repair [43, 44], vascular grafts [45, 46], bone regeneration [47, 48], and drug delivery [49, 50]. Since B. mori silk is available in large supplies from sericulture, it is the most commonly used silk for the above-mentioned studies. In the case of spiders, it is difficult to breed spider species because of their cannibalistic behavior. However, with the advance of recombinant DNA tools, it is now possible to bioengineer spider silk genes to produce spider silklike proteins [51] for tissue engineering and regenerative medicine applications [52–56]. Recombinant DNA technology provides well-established protocols for cloning, mutation, and gene fusion in different host cells for the expression of peptides and proteins with a broad range of sizes [57]. In addition, by using recombinant DNA protein methodologies, it is possible to overcome other problems such as immunogenic responses in *in vivo* studies and batch-to-batch variability; which are frequently associated with the polymers extracted from natural sources [58, 59]. Furthermore, the increased efficiency in making synthetic oligonucleotides and the standardization of kits and protocols for cloning and protein expression make the transgenic production approach more cost-effective for large-scale protein production [57].

Major ampullate silk was successfully expressed by bovine mammary epithelial cells, hamster kidney cells, insect cells, and in the milk of transgenic goats, generally with low yields [60]. However, bacteria can be grown at large scales and have the advantage of being easier to handle and more cost effective. Therefore, *Escherichia coli* has been actively pursued as an expression host for spider silk (Table 4.1). Since bacterial hosts have distinct codon usages, silk sequences from different spider species were reverse transcribed into cDNA, using the *E. coli* codon preferences, and double-stranded oligonucleotides coding for different domains of silk proteins were prepared [61]. These double-strand oligonucleotides were then

Table 4.1 Silk proteins expressed in E. coli and yeast recombinant systems and their potential uses.

Protein	Expression system	Advantages/ applications	References
Spider silk major ampullate from *Nephila clavipes*	*E. coli* RY-3041	Structural studies/ biomedical applications	[65, 66]
	E. coli SG 13009pREP4	Structural studies/ biomedical applications	[62]
	E. coli BL21	Structural studies/ biomedical applications	[64, 67, 68]
	E. coli M109 strain	Structural studies/ biomedical applications	[69]
	Yeast *Pichia pastoris*	Structural studies/ biomedical applications	[70]
Spider silk major dragline proteins ADF-3 and ADF-4 from *Araneus diadematus*	*E. coli* BLR strain	Structural studies/ biomedical applications	[63]
Spider silklike proteins – eADF4(C16) from *A. diadematus*	*E. coli* BLR strain	Drug delivery	[71]
Spider silk flagelliform type from *Nephila clavipes*	*E. coli* BL21 strain	Structural studies/ biomedical applications	[72]
Spider silklike proteins – NcDS, (SpI)7, and [(SpI)4/(SpII)1]4	*E. coli* BL21 strain	Structural studies/ biomedical applications	[73]

assembled into synthetic genes coding for silk proteins [60]. This cloning strategy was employed with success for the expression of *N. clavipes* consensus sequence for MaSp1 and MaSp2 [62] and the flagelliform silk protein [5] from the same species. Cloning and expression in *E. coli*, of both major ampullate silks ADF-3 and ADF-4 from *A. diadematus*, was also reported, with yields between 140 and 360 mg ml^{-1} [63]. Besides *E. coli*, other hosts for the cloning and expression of spider silks have also been explored. The yeast *Pichia pastoris* is considered an attractive host for the expression of recombinant proteins since this expression system is well developed for industrial fermentation, reaching high cell densities using low-cost media. For these reasons, it was successfully used for the expression of spider silk dragline using genes of up to 3000 codons with no evidence of truncated synthesis, a common occurrence in *E. coli* host [64]. Plants such as tobacco and *Arabidopsis thaliana* are also being explored as transgenic host systems for silk proteins [60].

Genetic engineering also offers the possibility of enriching the sequences of proteins to improve their biological activity by fusing them with other protein motifs with specific bioactivities. These new peptides or proteins are endowed with tunable properties and are considered to be promising biomaterials for medical applications. Table 4.2 gives an overview of the studies published during the past

Table 4.2 New silk chimeric proteins with potential application in the biomedical field.

Fusion protein	Expression system	Applications	References
Silk + elastin (SELP-47 K)	E. coli	Promote cell attachment and growth/tissue engineering	[80]
Spider silk + dentin matrix protein	E. coli RY-3041 strain	Biomedical applications/tissue engineering	[79]
Spider silk + bone sialoprotein	E. coli RY-3041 strain	Biomedical applications/tissue engineering	[78]
Spider silk + antimicrobial domain (HNP-2, HNP-4, and hepcidin)	E. coli RY-3041 strain	Biomedical applications/tissue engineering	[81]
Spider silk + silaffin R5 peptide from *Cylindrotheca fusiformis*	E. coli RY-3041 strain	Biomedical applications/tissue engineering	[82]
Bombyx mori silk + polyglutamic acid site [(AGSGAG)4E8AS]4	Transgenic silkworms	Biomedical applications/tissue engineering	[83]
Bombyx mori silk + RGD + elastin (FES8)	E. coli BL21 strain	Biomedical applications	[84]
RGDS + silk fibroin (RGDSx2 fibroin)	Silkworm	Facilitates chondrogenesis	[85]

few years using this approach. Silk-based block copolymers were fused with an RGD cell binding domain for intracellular gene delivery, and confocal laser scanning microscopy showed the presence of labeled DNA inside cells, demonstrating the potential of these silk bioengineered block copolymers as highly tailored gene delivery systems [74]. A recognition site for an enzyme with proteolytic activity can also be incorporated into the sequences, favoring biomaterial degradation [75]. Furthermore, the fusion of the *N. clavipes* consensus sequence for MaSp1 with proteins such as dentin matrix protein and bone sialoprotein, involved in calcium phosphate deposition in teeth and bone [76, 77], respectively, also had positive results from a biomaterials perspective [78, 79]. In both fusion proteins, the silk domain retained its self-assembly properties and the dentin matrix protein and bone sialoprotein domains maintained their ability to induce the deposition of calcium phosphates [78, 79].

Promising results were also obtained when the *N. clavipes* consensus sequence for MaSp1 was fused with antimicrobial peptides, namely, neutrophil defensins 2 HNP-2 and 4 HNP-4 and hepcidin, using a step-by-step cloning methodology [86]. The cloning and expression of these new fusion proteins expanded these chimera or fusion approaches to include antimicrobial-functionalized protein-based biomaterials [86], offering a path forward in reducing the use of antibiotics to prevent infection in implants and in the design of a new generation of protein-based materials bioengineered to prevent the onset of infections.

These studies demonstrate the potential of genetically engineered silklike proteins for application in tissue engineering and regenerative medicine through the design of new protein-based scaffolds.

4.3
Mechanical Properties

Silk is characterized by its outstanding mechanical properties in comparison to high-performance man-made fibers such as Kevlar, nylon, and high-tensile steel, and by its self-assembly, leading to fibers with a complex hierarchical arrangement [87]. These outstanding mechanical features along with its biocompatibility are the reasons why silk has been used through the millennia in such diverse applications as hunting, fabrication of paper, wound dressing, textiles, and sutures [88].

The remarkable mechanical features of the different types of silk are in part due to the presence of α-helix and β-turns, responsible for its elastic properties, alternating with β-sheets, which confer toughness to silk fibers. The strong molecular cohesion occurring with amide–amide interactions in the β-sheet crystalline regions is thought to be responsible for the stiffness of silk fibers [89]. As described earlier, in *B. mori* silk, the hexapeptide repeat GAGAGS is involved in the formation of the β-sheets. In spider silk, besides GA sequences there are also poly-A blocks and both motifs contribute for the formation of antiparallel β-sheets [89]. These poly-A and GA motifs are embedded in amorphous regions formed by either GGX or GPGXX motifs, which are believed to be responsible for the elastic features [90].

Dragline spider silk from *N. clavipes* has been extensively studied because of its extraordinary strength, 4 GPa, considerable elasticity, 35%, and the resulting remarkable toughness, 160 MJ m^{-3} [91]. These values are in line with those obtained for the major ampullate silk produced by the species *A. diadematus*, which also shows remarkable strength, 10 GPa, elasticity, 27%, and toughness, 160 MJ m^{-3} [92]. These toughness values are superior to those of commercial synthetic fibers such as synthetic rubber and nylon with the advantage that dragline spider silk is produced under benign conditions, excluding the need for harsh chemical reagents and processing conditions used in the production of these synthetic polymers [21]. Furthermore, flagelliform silk has an extensibility of 200–270%, lower than synthetic rubber materials; however, its strength of 0.5 GPa makes it exceptionally tough in comparison to its synthetic counterparts [92].

The study of silk properties by Gosline and colleagues indicates that spider silk fibers and web designs have been optimized during the course of evolution. The high toughness of major ampullate silk allows the web to absorb the large amount of kinetic energy of a flying prey, indicating that both silks were selected during the evolution process to reach an equilibrium between strength, extensibility, and viscoelasticity, resulting in a material with incredible toughness [92]. These unique physical properties are in part related to the repetitive structure of silk proteins [93].

Besides the importance of the primary structure of silk in its mechanical properties, the spinning process has been shown to affect the mechanical characteristics of spider silk [94]. The movement of the spider silk protein through the spinning duct, the accelerating elongation flow, and the shear forces lead to a transition of silk proteins from random-coil and polyproline-II helixlike conformation to a β-sheet-rich protein [26]. The stress forces generated during this stage are probably responsible for the alignment of silk proteins, resulting in a more extended conformation favoring the bonding between molecules through the formation of hydrogen bonds and giving the antiparallel conformation of the final silk fiber [21]. In the final stage, the fiber is drawn and stretched from the spigot by the spider, which leads to a reduction in fiber diameter, resulting in the improvement of the mechanical properties [26].

These remarkable mechanical properties of silk have long attracted the attention of scientists aiming to design new biomaterials with improved mechanical features. However, so far, either recombinant spidroin analogous or artificially spun regenerated natural spider silk have given rise to silk fibers that are weaker than their native counterparts, both in terms of stress and initial modulus [95]. Early studies report the use of regenerated natural silk for the production of fibers. Seidel and colleagues describe the regeneration of silk fibers from a 2.5% (w/w) solution of *N. clavipes* major ampullate silk in hexafluoro-2-propanol. The as-spun fibbers were very weak, but after post-spinning drawing, both strength and stiffness increased by 2 orders of magnitude, reaching a tensile strength of 320 MPa and a modulus of 8 GPa. Although these values are lower than the values of 875 MPa for strength and 10.9 GPa of stiffness measured for native silk, they highlight the importance of the spinning process in the mechanical properties of silk [96]. Later, Shao and colleagues described the spinning of silk fibers from an aqueous solution of raw silk from *Nephila edulis*. However, the initial modulus, 6 GPa, and strength, 0.11–0.14 GPa, for these fibers are even lower than those of silk fibers regenerated from hexafluoro-2-propanol [97]. Recently, the synthesis of meter-long fibers through the self-assembly of a miniature spidroin analog from *Euprosthenops australis* dragline was described. The structure of these fibers was similar to that of native dragline silk, and mechanical tests showed that they have features similar to those of fibers spun from redissolved silk, reaching values of ~200 MPa for tensile strength, elastic modulus of 7 GPa, and an average yield stress of 150 MPa [98].

The results described earlier in this section indicate that a deeper understanding of the complex hierarchical organization of silk proteins and of the relationships between structure and function would prompt the development of a new generation of protein-based biomaterials with tunable mechanical properties. Also, a better understanding of the control over silk conformation during the spinning process is crucial for the development of new biomaterials for load-bearing applications [94]. Even in non-load-bearing applications such as drug delivery, wound dressing, and nerve regeneration, the local stiffness of the substrate is very important. Different studies demonstrate the importance of the mechanical stiffness of the substrate in the proliferation of cells, suggesting the existence of a stiffness-dependent behavior [94]. Soft substrates mimicking brain tissue stimulate the proliferation of cells such as neurons and fibroblasts, while stiffer and rigid matrices with mechanical characteristics similar to bone tissue have osteogenic properties and sustain the proliferation of osteoblasts [99–101].

4.4
Biomedical Applications of Silk

Over the past few decades, different research studies have highlighted the potential of silk and silklike proteins for applications in the biomedical field. This growing interest is inspired by the unique characteristics of silk, mainly its impressive mechanical properties, relative environmental stability, biocompatibility, and more recently, the possibility to tailor its amino acid sequence through genetic engineering [2]. In addition, silk can be assembled into foams, films, fibers, and meshes. These are the main reasons why silk proteins have been explored for a broad range of applications in tissue engineering and regenerative medicine such as bone and cartilage regeneration, ligament and tendon replacement, cell culture, and angiogenesis [2]. Recently, silk scaffolds were approved by the Food and Drug Administration (FDA) for soft tissue repair [43]. Table 4.3 gives an overview of recent studies addressing potential applications of silk in the biomedical field.

Different studies show the ability of silk matrices to support and promote the attachment, spreading, proliferation, and differentiation of different cell types [117] such as osteoblasts, osteoclasts [118], fibroblasts [119], and mesenchymal stem cells [105]. Recently, fibroin was used as a substrate for the transplantation of tissue constructs for endothelial keratoplasty, and the results show that silk fibroin can be prepared as a transparent film capable of supporting human cornel endothelial cells [108]. Furthermore, the recombinant spider silk 4RepCT was used to produce films, foams, fibers, and meshes. The suitability of these matrices for cell culture

Table 4.3 Recent publications addressing the potential uses of silk-based materials.

Origin	System	Applications	References
Bombyx mori	Porous scaffold	Hepatic tissue engineering	[102]
		Gene delivery	[103, 104]
		Ligament tissue engineering	[105, 106]
	Fiber mesh	Bone tissue engineering	[107]
	Films	Corneal repair	[108, 109]
		Tissue engineering	[110]
		Bone tissue engineering	[111]
	Tubes	Vascular grafts	[112]
		Bladder reconstruction	[113]
	Hydrogel	Drug delivery	[114]
		Cartilage tissue engineering	[43]
	Sponges	Wound dressings	[115]
Nephila spp.	Fiber	Microsutures for nerve repair	[116]

was assessed with human primary fibroblasts, and results show high cell adhesion and proliferation [56].

To improve biological responses, silk matrices can be decorated with bioactive molecules to improve cell responses. RGD integrin recognition sequence and parathyroid hormone (PTH) have been covalently coupled to silk fibroin films. The new functionalized silk matrices were seeded with osteoblastlike cells (SaOs-2) and tested for its suitability for bone-inducing matrices [48]. Results showed a significant increase in SaOs-2 adhesion, in both RGD- and PTH-modified matrices. Furthermore, after two weeks of cell culture, the mRNA levels of alkaline phosphatase and α1 procollagen were elevated, and the osteocalcin levels and calcification were significantly higher on silk matrices functionalized with the RGD motifs [48]. A similar strategy was used to covalently bind the RGD motif to a silk wire-rope matrix to be used in ligament repair. Results demonstrated that the modified silk matrices supported human bone marrow stem cells (BMSCs) and improved anterior cruciate ligament fibroblasts (ACLFs) adhesion, with higher cell density after 14 days of culture [120]. These results allied to the unique mechanical properties of silk highlight the potential of silk-based constructs in application in bone tissue engineering and for ligament/tendon regeneration.

In the field of cartilage repair, silk-based materials have also been used as hydrogels with promising results. Silk hydrogels are biocompatible and biodegradable and can be fabricated using a water-based technology, excluding the use of harmful organic solvents. This methodology allows for a precise control over the structural and mechanical properties of the silk scaffold and fosters the development of constructs that meet the mechanical demands of the tissue to be replaced [43]. Chao and colleagues developed a new silk fibroin hydrogel for cartilage tissue engineering and used it to encapsulate primary calf chondrocytes. Results showed that both silk matrices supported chondrocyte proliferation and the expression of collagen II and GAG synthesis [43]. Besides hydrogels, silk has been used to produce porous scaffolds for cartilage repair. Scaffolds prepared with *B. mori*, *A. diadematus* egg sac, and dragline silk fibers were used for chondrocyte attachment, and results show good cell adhesion and proliferation while producing extracellular matrix [121].

Under the scope of soft tissue repair, different studies address the possibility of using silk-based materials for wound dressing. *In vitro* studies indicate that cells such as fibroblasts are able to proliferate and express extracellular matrix [122], and the *in vivo* results show revascularization of the silk graft [122], good cell colonization [115], and no signs of acute dermal toxicity [123]. The biological actions of these dressings and the good biocompatibility results indicate that new silk biomaterials are promising constructs for medical applications such as for patients suffering from chronic diseases and burn injuries.

In the field of vascular tissue engineering, silk has been used in the development of new vascular grafts for vessel replacement. Liu and colleagues [124] developed a silk fibroin nanofibrous construct with improved anticoagulant activity and the ability to support the adhesion and proliferation of endothelial and smooth muscle cells with high expression levels of phenotype-related marker genes and proteins.

In line with this approach, tubular silk scaffolds were produced using an aqueous gel spinning technique and assessed *in vitro*, in terms of thrombogenicity and cell responses, and *in vivo*. After four weeks of implantation, a layer of endothelial cells lining the tube lumen was observed [112]. Similar results were obtained with small-diameter vascular grafts implanted into the rat abdominal aorta. Results showed that endothelial cells and smooth muscle cells migrated into the silk graft early after implantation, forming an endothelium and a layer of smooth muscle cells [125]. In this way, silk matrices can be used as an alternative to synthetic polymers for the design of new vascular grafts, overcoming the thrombogenic problems sometimes attributed to synthetic materials [112].

Further applications of silk include the use of silk constructs for nerve regeneration. The most commonly used treatments for peripheral nerve injuries are end-to-end suturing and autologous grafting, with clinical outcomes often unsatisfactory. In this way, different types of materials have been tested for the repair of nerve gaps. The slow degradability, mechanical endurance, and neurobiocompatibility of silk makes it suitable for nerve regeneration [126]. Previous studies indicate that silk has good biocompatibility with dorsal root ganglia, supports the survival and migration of Schwann cells without any cytotoxic effect, and enhances axonal regrowth and remyelination [127, 128]. Furthermore, Ghaznavi and coworkers conducted an *in vivo* study to assess cell inflammatory responses and the functional recovery of a sciatic nerve defect in a rat model. The results indicated that silk possesses favorable immunogenicity and remyelination capacity for nerve repair [129].

In the drug delivery field, silk constructs have been explored as carriers for different types of bioactive molecules. The main applications of silk delivery systems have been in tissue engineering purposes, and previous studies demonstrated the ability of silk matrices to successfully deliver protein drugs and to maintain their biological activity. In this way, different growth factors and bioactive molecules have been incorporated into silk constructs for tissue engineering of bone and cartilage as well as for vascular and nerve repair and wound healing [130]. Incorporation of the osteoinductive factor bone morphogenetic protein 2 (BMP-2) has gained attention. Different studies show that silk matrices loaded with BMP-2 resulted in higher calcium deposition and upregulation of BMP-2 [131], collagen I, bone sialoprotein osteopontin, osteocalcin, and cbfa1 transcript levels [132]. *In vivo* studies with porous silk fibroin scaffolds loaded with BMP-2 and seeded with human mesenchymal stem cells implanted in critical-sized cranial defects in mice resulted in significant bone ingrowth [133].

Silk matrices have also been used as carriers for plasmids and viruses encoding a specific growth factor into the scaffold. Zhang and colleagues incorporated adenoviruses encoding BMP-7 into silk scaffolds and assessed the osteoinductive and new bone formation properties of these constructs through the creation of a critical-sized defect in mice. *In vivo* tests revealed that silk constructs were capable of delivering the adenovirus encoding BMP-7, resulting in a significant enhancement of new bone formation [103].

The examples outlined above give an overview of the range of applications for silk proteins and highlight the potential of using these proteins for the design of

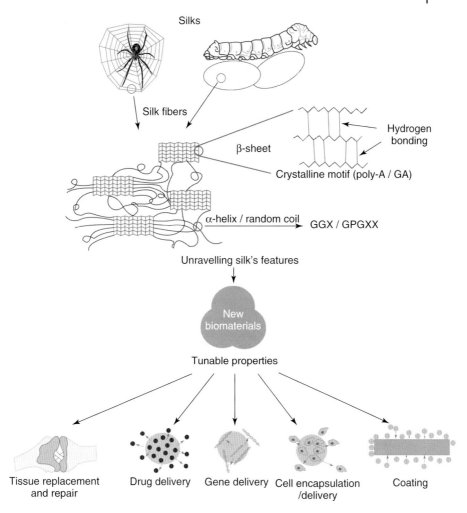

Figure 4.1 Scheme highlighting the features of silk and the potential of new silk-based biomaterials for biomedical applications.

new protein-based materials for future tissue engineering applications. Figure 4.1 highlights the importance of a better understanding of silk structure for the design of new biomaterials with tunable properties for different biomedical applications.

4.5 Final Remarks

During the past two decades, tremendous increase in the information about different aspects of silk proteins has been attained. This increase in knowledge includes protein structure and composition, spinning process, mechanical properties, and biological responses after implantation. However, despite these advances, there are still some key issues that need further enlightenment. This is the case of the relationship between protein structure, composition, and the mechanical behavior of silk fibers and also the role of the spinning process in mechanical properties of silk. A better understanding of these processes would allow for the design of new silk proteins with tunable properties that can be used as promising biomaterials for medical applications. Furthermore, the possibility of using biotechnology allows the design of new functionalized silk proteins with improved features capable of various biological, chemical, and physical responses.

References

1. Eisoldt, L., Smith, A., and Scheibel, T. (2011) *Mater. Today*, **14**(3), 81–86.
2. Altman, G.H., Diaz, F., Jakuba, C., Calabro, T., Horan, R.L., Chen, J. et al. (2003) *Biomaterials*, **24**(3), 401–416.
3. Gerritsen, V.B. (2002) *Protein Spotlight*, 24, 1–2.
4. Zhang, L., Rodriguez, J., Raez, J., Myles, A.J., Fenniri, H., and Webster, T. (2009) *Nanotechnology*, **20**, 175101–175113.
5. Zhou, C.Z., Confalonieri, F., Jacquet, M., Perasso, R., Li, Z.G., and Janin, J. (2001) *Proteins*, **44**(2), 119–122.
6. Gage, L.P. and Manning, R.F. (1980) *J. Biol. Chem.*, **255**(19), 9444–9450.
7. Zhou, C.-Z., Confalonieri, F., Medina, N., Zivanovic, Y., Esnault, C., Yang, T. et al. (2000) *Nucleic Acids Res.*, **28**(12), 2413–2419.
8. Marsh, R.E., Corey, R.B., and Pauling, L. (1955) *Biochim. Biophys. Acta*, **16**(1), 1–34.
9. Takahashi, Y., Gehoh, M., and Yuzuriha, K. (1999) *Int. J. Biol. Macromol.*, **24**(2–3), 127–138.
10. Tanaka, K., Kajiyama, N., Ishikura, K., Waga, S., Kikuchi, A., Ohtomo, K. et al. (1999) *Biochim. Biophys. Acta*, **1432**(1), 92–103.
11. Yamaguchi, K., Kikuchi, Y., Takagi, T., Kikuchi, A., Oyama, F., Shimura, K. et al. (1989) *J. Mol. Biol.*, **210**(1), 127–139.
12. Mori, K., Tanaka, K., Kikuchi, Y., Waga, M., Waga, S., and Mizuno, S. (1995) *J. Mol. Biol.*, **251**(2), 217–228.
13. Chevillard, M., Couble, P., and Prudhomme, J.C. (1986) *Nucleic Acids Res.*, **14**(15), 6341–6342.
14. Chevillard, M., Deleage, G., and Couble, P. (1986) *Sericologia*, **26**, 435–449.
15. Inoue, S., Tanaka, K., Arisaka, F., Kimura, S., Ohtomo, K., and Mizuno, S. (2000) *J. Biol. Chem.*, **275**(51), 40517–40528.
16. Sprague, K.U. (1975) *Biochemistry*, **14**(5), 925–931.
17. Takasu, Y., Yamada, H., and Tsubouchi, K. (2002) *Biosci. Biotechnol. Biochem.*, **66**(12), 2715–2718.
18. Hayashi, C.Y., Shipley, N.H., and Lewis, R.V. (1999) *Int. J. Biol. Macromol.*, **24**(2-3), 271–275.
19. Hu, X., Vasanthavada, K., Kohler, K., McNary, S., Moore, A.M.F., and Vierra, C.A. (2006) *Cell. Mol. Life Sci.*, **63**(17), 1986–1999.
20. Townley, M.A., Bernstein, D.T., Gallagher, K.S., and Tillinghast, E.K. (1991) *J. Exp. Zool.*, **259**(2), 154–165.
21. Vollrath, F. and Knight, D.P. (2001) *Nature*, **410**(6828), 541–548.
22. Hayashi, C.Y., Blackledge, T.A., and Lewis, R.V. (2004) *Mol. Biol. Evol.*, **21**(10), 1950–1959.
23. Hu, X., Kohler, K., Falick, A.M., Moore, A.M.F., Jones, P.R., and Vierra, C. (2006) *Biochemistry*, **45**(11), 3506–3516.
24. Ayoub, N.A., Garb, J.E., Tinghitella, R.M., Collin, M.A., and Hayashi, C.Y. (2007) *PLoS ONE*, **2**(6), e514.

25. Guerette, P.A., Ginzinger, D.G., Weber, B.H.F., and Gosline, J.M. (1996) *Science*, **272**(5258), 112–115.
26. Heim, M., Keerl, D., and Scheibel, T. (2009) *Angew. Chem. Int. Ed.*, **48**(20), 3584–3596.
27. Römer, L. and Scheibel, T. (2008) *Prion*, **2**(4), 154–161.
28. Hinman, M.B. and Lewis, R.V. (1992) *J. Biol. Chem.*, **267**(27), 19320–19324.
29. Xu, M. and Lewis, R.V. (1990) *Proc. Natl. Acad. Sci. U.S.A.*, **87**(18), 7120–7124.
30. Dong, Z., Lewis, R.V., and Middaugh, C.R. (1991) *Arch. Biochem. Biophys.*, **284**(1), 53–57.
31. Kümmerlen, J., van Beek, J.D., Vollrath, F., and Meier, B.H. (1996) *Macromolecules*, **29**(8), 2920–2928.
32. Hayashi, C.Y. and Lewis, R.V. (2001) *Bioessays*, **23**(8), 750–756.
33. Beckwitt, R. and Arcidiacono, S. (1994) *J. Biol. Chem.*, **269**(9), 6661–6663.
34. Rising, A., Hjälm, G., Engström, W., and Johansson, J. (2006) *Biomacromolecules*, **7**(11), 3120–3124.
35. Sponner, A., Unger, E., Grosse, F., and Weisshart, K. (2004) *Biomacromolecules*, **5**(3), 840–845.
36. Askarieh, G., Hedhammar, M., Nordling, K., Saenz, A., Casals, C., Rising, A. et al. (2010) *Nature*, **465**(7295), 236–238.
37. Beckwitt, R., Arcidiacono, S., and Stote, R. (1998) *Insect Biochem. Mol. Biol.*, **28**(3), 121–130.
38. Hagn, F., Eisoldt, L., Hardy, J.G., Vendrely, C., Coles, M., Scheibel, T. et al. (2010) *Nature*, **465**(7295), 239–242.
39. Valluzzi, R., Winkler, S., Wilson, D., and Kaplan, D.L. (2002) *Philos. Trans. R. Soc. Lond.Ser. B*, **357**(1418), 165–167.
40. Omenetto, F.G. and Kaplan, D.L. (2010) *Science*, **329**(5991), 528–531.
41. Gil, E.S., Mandal, B.B., Park, S.-H., Marchant, J.K., Omenetto, F.G., and Kaplan, D.L. (2010) *Biomaterials*, **31**(34), 8953–8963.
42. Lawrence, B.D., Marchant, J.K., Pindrus, M.A., Omenetto, F.G., and Kaplan, D.L. (2009) *Biomaterials*, **30**(7), 1299–1308.
43. Chao, P.-H.G., Yodmuang, S., Wang, X., Sun, L., Kaplan, D.L., and Vunjak-Novakovic, G. (2010) *J. Biomed. Mater. Res. Part B*, **95B**(1), 84–90.
44. Wang, Y., Bella, E., Lee, C.S.D., Migliaresi, C., Pelcastre, L., Schwartz, Z. et al. (2010) *Biomaterials*, **31**(17), 4672–4681.
45. Soffer, L., Wang, X., Zhang, X., Kluge, J., Dorfmann, L., Kaplan, D.L. et al. (2008) *J. Biomater. Sci. Polym. Ed.*, **19**(15), 653–664.
46. Zhou, J., Cao, C., and Xilan Ma, J.L. (2010) *Int. J. Biol. Macromol.*, **47**, 514–519.
47. Kim, H.J., Kim, U.-J., Kim, H.S., Li, C., Wada, M., Leisk, G.G. et al. (2008) *Bone*, **42**(6), 1226–1234.
48. Sofia, S., McCarthy, M.B., Gronowicz, G., and Kaplan, D.L. (2001) *J. Biomed. Mater. Res.*, **54**(1), 139–148.
49. Lammel, A.S., Hu, X., Park, S.-H., Kaplan, D.L., and Scheibel, T.R. (2010) *Biomaterials*, **31**(16), 4583–4591.
50. Uebersax, L., Merkle, H.P., and Meinel, L. (2008) *J. Controlled Release*, **127**(1), 12–21.
51. Spiess, K., Lammel, A., and Scheibel, T. (2010) *Macromol. Biosci.*, **10**(9), 998–1007.
52. Allmeling, C., Jokuszies, A.K.K.R., Kall, S., and Vogt, P.M. (2006) *J. Cell. Mol. Med.*, **10**(3), 770–777.
53. Allmeling, C., Jokuszies, A., Reimers, K., Kall, S., Choi, C.Y., Brandes, G. et al. (2008) *Cell Prolif.*, **41**(3), 408–420.
54. Baoyong, L., Jian, Z., Denglong, C., and Min, L. (2010) *Burns*, **36**(6), 891–896.
55. Kluge, J.A., Rabotyagova, O., Leisk, G.G., and Kaplan, D.L. (2008) *Trends Biotechnol.*, **26**(5), 244–251.
56. Widhe, M., Bysell, H., Nystedt, S., Schenning, I., Malmsten, M., Johansson, J. et al. (2010) *Biomaterials*, doi: 10.1016/j.biomaterials.2010.08.061
57. Kyle, S., Aggeli, A., Ingham, E., and McPherson, M.J. (2009) *Trends Biotechnol.*, **27**(7), 423–433.
58. Langer, R. and Tirrell, D.A. (2004) *Nature*, **428**, 487–492.
59. Romano, N.H., Sengupta, D., Chung, C., and Heilshorn, S.C.

(2010) *Biochim. Biophys. Acta.*, doi: 10.1016/j.bbagen.2010.07.005
60. Vendrely, C. and Scheibel, T. (2007) *Macromol. Biosci.*, **7**(4), 401–409.
61. Scheibel, T. (2004) *Microb. Cell Fact.*, **3**(1), 14.
62. Prince, J.T., McGrath, K.P., DiGirolamo, C.M., and Kaplan, D.L. (1995) *Biochemistry*, **34**(34), 10879–10885.
63. Huemmerich, D., Helsen, C.W., Quedzuweit, S., Oschmann, J., Rudolph, R., and Scheibel, T. (2004) *Biochemistry*, **43**(42), 13604–13612.
64. Fahnestock, S.R. and Irwin, S.L. (1997) *Appl. Microbiol. Biotechnol.*, **47**(1), 23–32.
65. Rabotyagova, O., Cebe, P., and Kaplan, D.L. (2009) *Biomacromolecules*, **10**, 229–236.
66. Rabotyagova, O.S., Cebe, P., and Kaplan, D.L. (2010) *Macromol. Biosci.*, **10**, 49–59.
67. Arcidiacono, S., Mello, C., Kaplan, D.L., Cheley, S., and Bayley, H. (1998) *Appl. Microbiol. Biotechnol.*, **49**(1), 31–38.
68. Lewis, R.V., Hinman, M., Kothakota, S., and Fournier, M.J. (1996) *Protein Expression Purif.*, **7**(4), 400–406.
69. Fukushima, Y. (1998) *Biopolymers*, **45**(4), 269–279.
70. Fahnestock, S.R. and Bedzyk, L.A. (1997) *Appl. Microbiol. Biotechnol.*, **47**(1), 33–39.
71. Lammel, A., Schwab, M., Hofer, M., Winter, G., and Scheibel, T. (2011) *Biomaterials*, **32**(8), 2233–2240.
72. Zhou, Y., Wu, S., and Conticello, V.P. (2001) *Biomacromolecules*, **2**(1), 111–125.
73. Mello, C.M., Soares, J.W., Arcidiacono, S., and Butle, M.M. (2004) *Biomacromolecules*, **5**(5), 1849–1852.
74. Numata, K.J.J.H., Subramanian, B., and Kaplan, D.L. (2010) *J. Controlled Release*, **146**(1), 136–143.
75. Arias, F.J., Reboto, V., Martín, S., López, I., and Rodríguez-Cabello, J.C. (2006) *Biotechnol. Lett.*, **28**(10), 687–695.
76. Benesch, J., Mano, J.F., and Reis, R.L. (2008) *Tissue Eng. Part B*, **14**(4), 433–445.
77. Ganss, B., Kim, R.H., and Sodek, J. (1999) *Crit. Rev. Oral Biol. Med.*, **10**(1), 79–98.
78. Gomes, S., Leonor, I.B., Mano, J.F., Reis, R.L., and Kaplan, D.L. (2010) *Soft Matter.*, Accepted. **7**, 4964–4973.
79. Huang, J., Wong, C., George, A., and Kaplan, D.L. (2007) *Biomaterials*, **28**, 2358–2367.
80. Qiu, W., Huang, Y., Teng, W., Cohn, C.M., Cappello, J., and Wu, X. (2010) *Biomacromolecules*, **11**(12), 3219–3227.
81. Gomes, S., Leonor, I.B., Mano, J.F., Reis, R.L., and Kaplan, D.L. (2010) *Biomaterials*, Accepted. **32**, 4255–4266.
82. Mieszawska, A.J., Nadkarni, L.D., Perry, C.C., and Kaplan, D.L. (2010) *Chem. Mater.*, **22**(20), 5780–5785.
83. Nagano, A., Tanioka, Y., Sakurai, N., Sezutsu, H., Kuboyama, N., Kiba, H. et al. (2011) *Acta Biomater.*, **7**(3), 1192–1201.
84. Yang, M., Tanaka, C., Yamauchi, K., Ohgo, K., Kurokawa, M., and Asakura, T. (2008) *J. Biomed. Mater. Res. A*, **84**(2), 353–363.
85. Kambe, Y., Yamamoto, K., Kojima, K., Tamada, Y., and Tomita, N. (2010) *Biomaterials*, **31**(29), 7503–7511.
86. Gomes, S., Leonor, I.B., Mano, J.F., Reis, R.L., and Kaplan, D.L. (2010) *Biomaterials*, doi: 10.1016/j.biomaterials.2011.02.040
87. Heim, M., Römer, L., and Scheibel, T. (2010) *Chem. Soc. Rev.*, **39**(1), 156–164.
88. Hardy, J.G. and Scheibel, T.R. (2009) *Biochem. Soc. Trans.*, **37**, 677–681.
89. Sponner, A., Vater, W., Monajembashi, S., Unger, E., Grosse, F., and Weisshart, K. (2007) *PLoS ONE*, **2**(10), e998.
90. Beek, J.Dv., Hess, S., Vollrath, F., and Meier, B.H. (2002) *Proc. Natl. Acad. Sci. U.S.A.*, **99**(16), 10266–10271.
91. Teulé, F., Furin, W.A., Cooper, A.R., Duncan, J.R., and Lewis, R.V. (2007) *J. Mater. Sci.*, **42**(21), 8974–8985.
92. Gosline, J.M., Guerette, P.A., Ortlepp, C.S., and Savage, K.N. (1999) *J. Exp. Biol.*, **2002** (Pt 23), 3295–3303.
93. Hinman, M.B., Jones, J.A., and Lewis, R.V. (2000) *Trends Biotechnol.*, **18**(9), 374–379.

94. Brown, C.P., Rosei, F., Traversa, E., and Licoccia, S. (2011) *Nanoscale*, **3**(3), 870–876.
95. Grip, S., Johansson, J., and Hedhammar, M. (2009) *Protein Sci.*, **18**(5), 1012–1022.
96. Seidel, A., Liivak, O., Calve, S., Adaska, J., Ji, G., Yang, Z. et al. (2000) *Macromolecules*, **33**(3), 775–780.
97. Shao, Z., Vollrath, F., Yang, Y., and Thøgersen, H.C. (2003) *Macromolecules*, **36**(4), 1157–1161.
98. Stark, M., Grip, S., Rising, A., Hedhammar, M., Engström, W., Hjälm, G. et al. (2007) *Biomacromolecules*, **8**(5), 1695–1701.
99. Engler, A.J., Sen, S., Sweeney, H.L., and Discher, D.E. (2006) *Cell*, **126**(4), 677–689.
100. Georges, P.C. and Janmey, P.A. (2004) *J. Appl. Physiol.*, **98**(4), 1547–1553.
101. Khatiwala, C.B., Peyton, S.R., and Putnam, A.J. (2006) *Am. J. Physiol. Cell. Physiol.*, **290**(6), C1640–C1650.
102. Gotoh, Y., Ishizuka, Y., Matsuura, T., and Niimi, S. (2011) *Biomacromolecules.*, doi: 10.1021/bm101495c
103. Zhang, Y., Fan, W., Nothdurft, L., Wu, C., Zhou, Y., Crawford, R. et al. (2011) *Tissue Eng. Part C*, doi: 10.1089/ten.TEA.2010.0453
104. Lü, K., Xu, L., Xia, L., Zhang, Y., Zhang, X., Kaplan, D.L. et al. (2011) *J. Biomater. Sci. Polym. Ed.*, doi: 10.1163/092050610X552861
105. Teh, T.K.H., Toh, S.L., and Goh, J.C.H. (2011) *Tissue Eng. Part C*, doi: 10.1089/ten.TEA.2010.0513
106. Mandal, B.B., Park, S.-H., Gil, E.S., and Kaplan, D.L. (2011) *Biomaterials*, **32**(2), 639–651.
107. Wei, K., Li, Y., Kim, K.-O., Nakagawa, Y., Kim, B.-S., Abe, K. et al. (2011) *J. Biomed. Mater. Res. A*, doi: 10.1002/jbm.a.33054
108. Madden, P.W., Lai, J.N.X., George, K.A., Giovenco, T., Harkin, D.G., and Chirila, T.V. (2011) *Biomaterials*, doi: 10.1016/j.biomaterials.2010.12.034
109. Higa, K., Takeshima, N., Moro, F., Kawakita, T., Kawashima, M., Demura, M. et al. (2010) *J. Biomater. Sci. Polym. Ed.*, doi: 10.1163/092050610X538218
110. Kharlampieva, E., Kozlovskaya, V., Wallet, B., Shevchenko, V.V., Naik, R.R., Vaia, R. et al. (2010) *ACS Nano*, **4**(12), 7053–7063.
111. Mieszawska, A.J., Fourligas, N., Georgakoudi, I., Ouhib, N.M., Beltonc, D.J., Perry, C.C. et al. (2010) *Biomaterials*, **31**(34), 8902–8910.
112. Lovett, M., Eng, G., Kluge, J.A., Cannizzaro, C., Vunjak-Novakovic, G., and Kaplan, D.L. (2010) *Organogenesis*, **6**(4), 217–224.
113. Mauney, J.R., Cannon, G.M., Lovett, M.L., Gong, E.M., Vizio, D.D., Gomez, P. III et al. (2011) *Biomaterials*, **32**(3), 808–818.
114. Guziewicz, N., Best, A., Perez-Ramirez, B., and Kaplan, D.L. (2011) *Biomaterials*, **32**(10), 2642–2650.
115. Min, S., Gao, X., Han, C., Chen, Y., Yang, M., Zhu, L. et al. (2010) *J. Biomater. Sci. Polym. Ed.*, doi: 10.1163/092050610X543609
116. Kuhbier, J.W., Reimers, K., Kasper, C., Allmeling, C., Hillmer, A., Menger, B. et al. (2011) *J. Biomed. Mater. Res. B*, **97**(2), 381–387.
117. Unger, R.E., Wolf, M., Peters, K., Motta, A., Migliaresi, C., and Kirkpatrick, C.J. (2004) *Biomaterials*, **25**(6), 1069–1075.
118. Jones, G.L., Motta, A., Marshall, M.J., Haj, A.J.E., and Cartmell, S.H. (2009) *Biomaterials*, **30**(29), 5376–5384.
119. Mai-ngam, K., Boonkitpattarakul, K., Jaipaew, J., and Mai-ngam, B. (2010) *J. Biomater. Sci. Polym. Ed.*, doi: 10.1163/092050610X530964
120. Chen, J., Altman, G.H., Karageorgiou, V., Horan, R., Collette, A., Volloch, V. et al. (2003) *J. Biomed. Mater. Res. A*, **67**(2), 559–570.
121. Gellynck, K., Verdonk, P.C.M., Nimmen, E.V., Almqvist, K.F., Gheysens, T., Schoukens, G. et al. (2008) *J. Mater. Sci. Mater. Med.*, **19**(11), 3399–3409.
122. Etienne, O., Schneider, A., Kluge, J.A., Bellemin-Laponnaz, C., Polidori, C., Leisk, G.G. et al. (2009) *J. Periodontol.*, **80**(11), 1852–1858.
123. Padol, A.R., Jayakumar, K., Shridhar, N.B., Swamy, H.D.N., Swamy, M.N.,

and Mohan, K. (2011) *Toxicol. Int.*, **18**(1), 17–21.
124. Liu, H., Li, X., Zhou, G., Fan, H., and Fan, Y. (2011) *Biomaterials*, **32**(15), 3784–3793.
125. Nakazawa, Y., Sato, M., Takahashi, R., Aytemiz, D., Takabayashi, C., Tamura, T. et al. (2010) *J. Biomater. Sci. Polym. Ed.*, **22**(1–3), 195–206.
126. Madduri, S., Papaloïzos, M., and Gander, B. (2010) *Biomaterials*, **31**(8), 2323–2334.
127. Radtke, C., Allmeling, C., Waldmann, K.-H., Reimers, K., Thies, K., Schenk, H.C. et al. (2011) *PLoS ONE*, **6**(2), e16990.
128. Yang, Y., Chen, X., Ding, F., Zhang, P., Liu, J., and Gu, X. (2007) *Biomaterials*, **28**(9), 1643–1652.
129. Ghaznavi, A.M., Kokai, L.E., Lovett, M.L., Kaplan, D.L., and Marra, K.G. (2011) *Ann. Plast. Surg.*, **66**(3), 273–279.
130. Wenk, E., Merkle, H.P., and Meinel, L. (2011) *J. Controlled Release*, **150**(2), 128–141.
131. Li, C., Vepari, C., Jin, H.-J., Kim, H.J., and Kaplan, D.L. (2006) *Biomaterials*, **27**(16), 3115–3124.
132. Karageorgiou, V., Meinel, L., Hofmann, S., Malhotra, A., Volloch, V., and Kaplan, D.L. (2004) *J. Biomed. Mater. Res. A*, **71**(3), 528–537.
133. Karageorgiou, V., Tomkins, M., Fajardo, R., Meinel, L., Snyder, B., Wade, K. et al. (2006) *J. Biomed. Mater. Res. A*, **78**(2), 324–334.

5
Elastin-like Macromolecules
Rui R. Costa, Laura Martín, João F. Mano, and José C. Rodríguez-Cabello

5.1
General Introduction

Scientists have long discovered that macromolecules are the best option for obtaining highly functional and complex materials. Hierarchical organization, self-assembly, or "intelligence" are common to many natural macromolecules and are desirable for several applications in biology and biotechnology. Proteins, nucleic acids, and polysaccharides are fine examples of structures that have been employed in many biomimetic concepts displaying such properties.

In biomedicine, several advances have been made in the past decades regarding the design and conception of new state-of-the-art devices with a great potential for future in-depth studies and, ultimately, use in human health care services. Some of them emerged from the increasing use of elastin-derived materials, consisting in viable and functional protein-based molecules as alternatives to the conventional polymers used in biomedical applications, including in tissue engineering and regenerative medicine (TERM).

Elastin-like macromolecules are genetically engineered materials rooted in the repeating sequence of natural elastin, a vital protein component of the extracellular matrix (ECM) found in various mammalian tissues [1]. As biomaterials, they became popular for exhibiting excellent mechanical properties [2–6] and biocompatibility [7–10]. They have also shown the potential to self-assemble and respond to changes in the environment or physiological conditions [11–14].

Because they are genetically engineered, they are often called elastin-like polymers (ELPs) or, more recently, elastin-like recombinamers (ELRs) [15]. This class of materials can be modified to virtually include any desired peptide sequence, especially functional sequences with relevant biological role [16–18], and has already been studied extensively for applications in drug delivery [19–21] and cell and tissue engineering [22–24].

In the following sections, a comprehensive discussion of the importance of elastin-like macromolecules in the conception of TERM devices is presented. Here, not only their properties but also their origin and natural inspiration are presented.

Biomimetic Approaches for Biomaterials Development, First Edition. Edited by João F. Mano.
© 2012 Wiley-VCH Verlag GmbH & Co. KGaA. Published 2012 by Wiley-VCH Verlag GmbH & Co. KGaA.

5.2
Materials Engineering – an Overview on Synthetic and Natural Biomaterials

Biomaterials have been applied in medicine since the early 1950s and have expanded in four traditional directions: metals, polymers, ceramics, and combinations thereof – composites [25, 26]. More recently, biological or natural materials, found in animals and plants, were added as the fifth direction. Natural materials are intricate structures that have risen from hundreds of millions of years of evolution. Compared to our current technology, Nature was able to conceive structures more complex than man-made materials, forming complex arrays, hierarchical structures, and multifunctionality [25, 27]. According to Wegst and Ashby [28], natural materials can be categorized into four groups: (i) ceramics and ceramic composites, including shells, bones, or any material in which the mineral component is prevalent; (ii) polymer and polymer composites, for example, ligaments and silk; (iii) elastomers, such as elastic soft tissues in the body that undergo large stretches or strains; and (iv) cellular materials – typically light-weight materials prevalent, for example, in feathers, wood, and cancellous bone.

On implantation, natural biomaterials can be recognized by the body and processed through established metabolic pathways [29, 30]. The metabolic products are simple sugars, amino acids, or minerals. This eliminates the possibility of cytotoxic and inflammatory response from the host, a situation occurring with wear particles of long-term orthopedic joint prostheses made of metal or synthetic polymers [31]. In this regard, inspiration was found in Nature to develop new materials and new methods of processing. In particular, the scientific community has been exploiting biological systems to produce tailored materials. Such synergy between the idealized material and its final construction is the basis of biomimetism [32, 33]. This principle has already started to be applied through genetic engineering, through the construction of specific DNA sequences in order to obtain a desired peptide [34]. Elastin-like macromolecules are successful examples of this methodology, mimicking natural elastin, which is discussed in the following sections.

5.3
Elastin as a Source of Inspiration for Nature-Inspired Polymers

Elastin is a vital biopolymer with elastic properties. It can be found in structures that require elasticity as part of their function, stretching and relaxing more than a billion times during life [1]. For instance, the content of elastin in skin and ligament is 5–10% and 10–15%, respectively, and in the aorta, it is as high as 40–50% [35].

5.3.1
Genetic Coding

The elastin gene is a single-copy gene found in chromosome 7 in humans [36]. Tropoelastin is the product of this gene and the precursor of human elastin. It is a 72 kDa soluble precursor with 760 residues [37, 38] and is expressed by various cell types during the pre- and neonatal stages of development, including smooth muscle [39] and endothelial cells [40], fibroblasts [41], and chondrocytes [42]. At present, the full amino acid composition of tropoelastin is known for several species besides humans, such as rats and bovines. Although some variations exist, there is still a similarity of 70% among the different tropoelastin species [36, 43–46].

The elastin gene possesses 36 exons, encoding two major alternating domains: hydrophobic (responsible for elasticity) and hydrophilic cross-linking domains. The composition of the hydrophobic domain is a combination of glycine (G), valine (V), proline (P), alanine (A), leucine (L), and isoleucine (I) residues – the first two being the most abundant – often repeated in sequences of (GVGVP), (GVPGV), and (GVGVAP). The hydrophilic domains contain several lysine (K) and A residues, which shows an important role in the cross-linking of tropoelastin [1, 36, 47].

In order to synthesize mature elastin, the precursor must undergo a process called *elastogenesis* (Figure 5.1), a highly complex phenomenon composed of all the events that lead to the synthesis of a fully functional elastic fiber [48]. The process starts with the synthesis of tropoelastin itself, in the rough endoplasmic reticulum. As soon as it is formed, galactolectin – an elastin-binding protein – acts as a chaperone that prevents premature intracellular aggregation [48]. This association lasts until the complex is excreted into the extracellular medium. The binding protein has a higher affinity for galactosugars of the microfibrilar component of the extracellular medium. Outside the cells, the microfibrils act as scaffolds for the deposition and alignment of tropoelastin molecules, which end up cross-linked to each other in K-containing sites [49, 50]. Cross-linking is the culmination of elastogenesis, resulting in mature elastin, which is insoluble but extremely stable. In fact, it is one of the most stable proteins known and has a half-life ranging from 40 to 70 years [51, 52].

5.3.2
Characteristics of Elastin

Mature elastin is an insoluble polymer composed of several or more tropoelastin molecules covalently bound to each other by cross-links. Despite its essentially hydrophobic nature, elastin can be highly hydrated by water molecules from both hydration and solvent water. In between the very rigid cross-linking domains, the hydrophobic segments exhibit a considerable mobility [36, 53].

Despite several decades of research, the nature of elastin itself has hindered the study of its properties and structure, mostly because of the insolubility in water and backbone mobility. Nevertheless, scientists were able to obtain important knowledge regarding this polymer. For example, calorimetric studies have shown that entropy is the primary source of its remarkable elasticity and swelling capability *in vivo* [54–56].

There are two main models that attempt to explain elastin's behavior: the single-phase model (also known as the *random-chain model*) and the two-phase model. The latter can be further divided into the liquid-drop and the oiled-coil

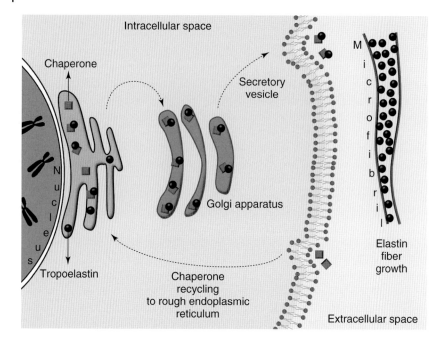

Figure 5.1 The elastogenesis phenomenon, starting with the translation of the elastin gene and proceeding to its alignment by microfibrils.

models. The single-phase model considers elastin to be like a typical rubber with randomly distributed polymeric chains in any solvent system [55, 57]. The two-phase model states that water-swollen elastin is composed of hydrophobic domains between cross-linked proteins and that the aqueous solvent is driven to the spaces between its domains. Specifically, the liquid-drop model [58] assumes that the peptide segments between the cross-links are globular domains. The oiled-coil model [59] is similar to the liquid-drop model, but assumes that the so called "oiled-coils" are broader in comparison and are composed of a series of β-turns between the cross-linking domains.

Unfortunately, none of the models described provide sufficient insight into the molecular basis of its elastomeric properties at the atomic level. The groups of Urry and Tamburro carried out a variety of physicochemical studies using synthetic elastin-based pentapeptides to investigate the relationship between its structure and function [4, 56, 60, 61]. Their studies led to the conception of yet another two-phase perspective: the β-spiral or fibrillar model, stating that there is one type II β-turn per pentameric unit, stabilized by intraspiral, interturn, and interspiral hydrophobic contacts. The repetition of this conformational unit results in an organized helical arrangement called the *β-spiral*, hence the name of the model. The properties of these biosynthesized elastin-like macromolecules are discussed in detail in Sections 5.4 and 5.5.

5.3.3
Elastin Disorders

Mutations involving elastic fibers are important for understanding the function of elastin and the process of elastogenesis and associated disorders. Congenital supravalvular aortic stenosis (SVAS) is a pathological condition associated with a mutation in the elastin gene. Its symptom is obstruction of the outflow from the left ventricle narrowing of the arterial lumen, because of the failure to regulate cellular proliferation and matrix deposition of elastin [62–64]. At the molecular level, the cause is a lack of some cross-linking domains or deficient coacervation of tropoelastin molecules [62, 65]. SVAS can be inherited as an autosomal dominant trait or as part of the Williams syndrome, a developmental disorder involving the central nervous system and the connective tissues [66]. Another pathology linked to the elastin gene is cutis laxa, a rare autosomal dominant disease caused by abnormally branched and fragmented elastic fibers with reduced tropoelastin deposition in the elastic fibers and few microfibrils in the dermis. The symptoms are inelastic loose-hanging skin, hernias, and emphysema [67, 68]. Changes in other molecules found in elastic fibers may also result in deficient elastogenesis or abnormal structures. Marfan's syndrome, for example, is an autosomal dominant disorder associated with mutations in the gene encoding fibrillin-1, characterized by pleiotropic manifestations in the ocular, skeletal, and cardiovascular systems [69].

5.3.4
Current Applications of Elastin as Biomaterials

Elastin can be used as a biomaterial in different forms. It can be found in its insoluble form in purified elastin preparations or can be hydrolyzed to obtain its soluble forms. One problem associated with elastin use in biomedical devices is calcification because it serves as a nucleation site for mineralization, a problem to be considered, especially in cardiovascular implants. Fortunately, it can be controlled by using molecules that prevent calcification, such as ethanol/EDTA treatments or the addition of glycosaminoglycans or basic fibroblast growth factor [70–72].

Naturally, elastin can also be found also in autografts, allografts, xenografts, and decellularized ECMs. Well-known examples are split skin autografts for burn wounds, autologous saphenous veins and umbilical vein allografts for coronary artery bypass graft surgery, and aortic heart valve xenografts. Incorporation of elastin in biomaterials is especially significant when its elasticity or biological effects can be exploited.

5.3.4.1 Skin
Skin is the largest organ of the human body. It is composed of two layers: the epidermis and dermis, which lie on subcutaneous fat. The epidermis mainly consists of layers of keratinocytes scattered with other cell types, among them melanocytes and Langerhans cells. It is separated from the dermis by the basement membrane.

The dermis is composed of papillary and reticular compartments that contain an ECM made of collagen, reticulum fibers, elastin, and glycosaminoglycans. The cellular constituents are mainly fibroblasts [73, 74].

Skin substitutes are usually applied to treat either burn or chronic wounds. There are several commercial solutions currently available, such as Apligraf® [75]. Apligraf is a type I collagen gel cultured with human neonatal foreskin dermal fibroblasts and keratinocytes and has been used for the treatment of venous leg [76] and diabetic foot ulcers [77]. Like others, it contains little elastin, so autologous skin grafting is still the gold standard for skin substitutes. Other solutions have also been investigated. Haffeman et al. [78] developed a membrane of collagen and elastin of porcine origin to serve as a matrix for neodermis growth with acceptable morphology and function. Using rats as animal models, these membranes showed complete vascularization and colonization by fibroblasts and cells from the immune systems within three weeks.

More recently, Nillesen et al. [79] designed an acellular double-layered skin construct, both cross-linked, to improve wound regeneration. The epidermal layer was essentially made of collagen containing heparin and fibroblast growth factor7 (FGF7) to stimulate the proliferation of keratinocytes. The dermal layer was porous and consisted of type I collagen fibrils, solubilized elastin from equine ligamentum nuchae, dermatan sulfate, heparin, FGF2, and vascular endothelial growth factor (VEGF). Using a rat model, and in comparison to another commercial skin substitutes, Integra®DRT, the double-layered construct showed more cell influx, significantly less contraction, and increased blood vessel formation at early time points. As late as 112 days, it also contained the most elastic fibers, and hair-follicle-like and sebaceous-gland-like structures could be identified. This work addressed the current problem of wound closure, which requires minimal contraction, vascularization, and elastogenesis in the early stages.

5.3.4.2 Vascular Constructs

Type I collagen and elastin are the main components present in the arterial wall matrix, supplying the artery with strength and elasticity, respectively. Blood vessels are structured in three layers: intima, media, and adventitia. In the typical three-layered architecture, elastin is found in the intimal and medial layers. Collagen is also found in the medial and adventitial sites [80, 81].

Coronary artery and peripheral vascular diseases cause high mortality in Western societies. The conventional treatment is surgery to restore blood flow using autologous vessels or valves [80, 82]. Although autologous vessels remain the standard for small-diameter grafts, many patients do not have a vessel suitable for grafting because of vascular disease, amputation, or previous harvest. Artificial grafts made from expanded poly(tetrafluoroethylene) (ePTFE) or Invista Dacron® are an alternative option [83, 84], but for smaller vessels (diameter less than 5 mm), the possible formation of clots from thrombotic events may rapidly close them. In the past two decades, tissue-engineered solutions have emerged, using elastin as a crucial ingredient. Stitzel et al. [85] have developed vascular graft scaffolds fabricated by electrospinning using polymer blends of type I collagen from calf skin, elastin from bovine neck ligament, and poly (D,L-lactide-co-glycolide) (PLGA). The scaffolds were tested both *in vitro* and *in vivo*, using a mouse model, and an absence of elicit local and systemic toxic effects was observed. They also possessed a tissue composition and mechanical properties similar to that of native vessels. Koens *et al.* [80] developed a triple-layered construct consisting of an inner layer of elastin fibers, a middle (porous) film layer of collagen fibrils, and an outer scaffold layer, also made of collagen. The substitute did not evoke platelet aggregation *in vitro*. The structure could also be sutured and was considered adequate for *in vivo* application.

5.4
Nature-Inspired Biosynthetic Elastins

Typically, aggressive methods of extraction are used to isolate elastin from tissues. These involve the use of 0.1 M NaOH at 98 °C, or autoclaving [86, 87]. More recent approaches follow stepwise protocols under mild conditions by peptide bond cleavage [88]. However, the yield of these approaches is usually low. Based on exploitation of biological systems, an approach that helps addressing this problem by the production of tailored molecules mimicking natural elastin through genetic engineering was developed. The pentapeptide poly(VPGVG) was the first to be extensively studied, for being the most abundant sequence in natural elastin, but variations of this model soon followed. It was possible to produce recombinant peptides with precision in the primary amino acid content, being able to introduce relevant functional sequences, for example, to improve cell adhesion and to trigger biomineralization and vascularization. Owing to their characteristics and peptide sequence, these molecules are known as *elastin-like recombinamers*, a class of recombinant ELPs [15].

5.4.1
General Properties of Elastin-like Recombinamers

ELRs are a promising class of biocompatible protein-based polymers inspired by the mammalian elastin protein sequences, or modifications thereof [4, 34, 89]. In aqueous solution and below a certain transition temperature (T_t), the free polymer chains are disordered, consisting of fully hydrated random coils, mainly by hydrophobic hydration. This kind of hydration is characterized by ordered clathratelike water structures surrounding the nonpolar moieties of the polymer with a structure somewhat similar to that described for crystalline gas hydrates, although a more heterogeneous one that varies in terms of perfection and stability [4, 13]. When the temperature is increased above T_t, the structure loses the water molecules by hydrophobic hydration and the chains fold and assemble. A phase-separated state is formed, consisting of 63% water and 37% polymer in weight. In this self-assembled separated state, the chains adopt a regular nonrandom structure known as β-*spiral conformation* [60, 61, 90].

Urry studied this phenomenon and stated that a polypeptide with a correct balance of polar and nonpolar amino acids can be soluble in water at low temperatures, and yet undergo a phase separation from the aqueous medium on raising the temperature [91]. In the latter case, the structures assemble into anisotropic fibers because of hydrophobic associations between the β-spirals. This folding is completely reversible if the temperature of the sample is decreased below T_t [4, 34]. This transition phenomenon is called inverse temperature transition (ITT) and has become the key issue in the development of new peptide-based polymers as molecular machines and materials [92]. Although the phenomenology of ELRs is similar to that of amphiphilic polymers with a lower critical solution temperature (LCST), the presence of an ordered state above the T_t led to the usage of the term *"inverse temperature transition"* as a descriptive term for these polymers' behavior [92, 93].

The self-assembly of ELRs is also dependent on others factors and external stimuli besides temperature. There is a strong dependence of the T_t on concentration in the range of 0.01 to 5–10 mg ml^{-1}, showing a decrease in T_t with increasing macromolecule concentration, because of the facilitated aggregation of molecules in higher number [13, 34]. Above this range, the T_t does not show further significant changes with increasing concentrations up to 150–200 mg ml^{-1} [13, 94]. Increasing the ionic strength of an ELR solution also influences the T_t and is particularly useful to trigger the ITT phenomenon when its concentration is low. Reguera *et al.* [95] suggested that an increase in the salt concentration increases the polarity of the solvent. This creates a higher difference in polarity with respect to the hydrophobic moieties of the polymer, recruiting more and more ordered water molecules to surround the polymer chains. Increasing the molecular weight of the ELR has also been demonstrated to decrease the T_t [96, 97]. ELRs can be modified in order to exhibit smart behavior toward other stimuli. By modifying the composition of the polymer, it is possible to conceive materials that react to stimuli such as pH, redox reactions, and even light [90, 98, 99].

Some ELR systems that are successful in maintaining some of the main characteristics of natural elastin have been developed. For example, the cross-linked matrices of these polymers retain most of the mechanical properties of elastin [100, 101], which becomes important when this behavior is accompanied by other interesting properties, such as biocompatibility, stimuli-responsive behavior, and the ability to self-assemble. Elastin-like systems and possible applications are further discussed in Section 5.5.

5.4.2
The Principle of Genetic Engineering – a Powerful Tool for Engineering Materials

In the past decade, protein biosynthesis showed the ability to directly produce high-molecular-weight polypeptides of exact amino acid sequence with high fidelity. Its usefulness to biosynthesize ELRs proceeds with near-absolute control of their macromolecular architecture in size, composition, sequence, topology, and stereochemistry [102, 103]. There are a few limitations, however, with respect

5.4 Nature-Inspired Biosynthetic Elastins

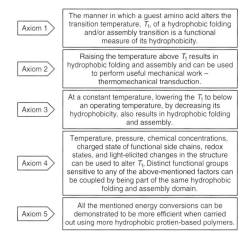

Figure 5.2 The five axioms of elastin-like recombinamers, according to Urry [10].

to the sequence and its final properties. In the general model (VPGXG), X cannot be proline [15, 104]. Substitution of amino acids in other positions is also limited. For example, poly(APGVG) becomes an irreversible granular precipitate instead of a reversible viscoelastic coacervate on increasing the temperature [105].

Recombinant DNA technology and bacterial protein expression have been employed for the biosynthesis of repetitive polypeptides based on the 20 naturally occurring amino acids, with no direct parallels in nature [91, 102, 106]. It is a technology that clearly opens infinite possibilities, with far more potential than most common synthetic ways of production, limited only by the living molecular machinery itself. Urry has presented five essential axioms to conceive protein-based macromolecules capable of undergoing hydrophobic folding and assembly because of ITTs, presented in Figure 5.2. In short, the axioms describe that the ITT depends on the amino acid composition of an ELR and that external factors can also influence its conformation and T_t.

Typically, ELRs are produced within cytoplasmatic inclusion bodies by *Escherichia coli* strains through bacterial fermentation, accounting for nearly 80% of the cell's volume by the end of the process. Low production costs can be achieved by producing an elastin sequence via the machinery of living cells, independent of its complexity. The production cost of these materials is not related to their complexity since the most costly task in terms of both time and money is the gene construction. However, once the plasmid is inserted into the microorganism's genetic information, the fast and cheap production of the polymer rapidly compensates for the costs associated with the molecular biology steps. This intrinsic advantage also has environmental benefits since recombinant protein-based materials are obtained by an easily scalable fermentation technology that uses only moderate amounts of energy and temperatures, with water as the only solvent [15, 91, 107, 108].

5.4.3
From Genetic Construction to Molecules with Tailored Biofunctionality

The design of an appropriate strategy at the gene level is critical for the efficient synthesis of the protein encoding sequence and for the production of a uniform protein product with an optimal quality and yield. The biosynthesis of any artificial protein generally includes (i) design and construction of a coding synthetic gene for the protein of interest in a plasmid with close transcription control; (ii) cloning of a recombinant gene with the necessary transcriptional regulatory elements into competent host cells; (iii) screening of plasmids containing the desired clones and verification of their DNA sequence; (iv) transformation of the chosen plasmids into expression-competent host cells; (v) growth of appropriate volumes of host cells and induction of protein expression; and (vi) purification of the protein of interest from cell lysates [109].

Molecular biology techniques are typically employed to self-ligate monomer DNA fragments in an oligomerization process that relies on restriction-enzyme-based approaches when designing genes that encode repetitive recombinamers. In this case, the monomeric fragments must be oligomerized in a "head-to-tail" orientation and can either be seamless in sequence or contain intervening linkers between the desired repeats. Approaches to oligomerization can be classified as follows. (i) Iterative method, in which a DNA segment is oligomerized in a series of single, uniform steps. Each step extends the oligomer by one unit of length of the monomer gene. (ii) Random method or "concatemerization," in which an uncontrolled number of monomer DNA segments are oligomerized in a single step to create a population of oligomerized clones with different lengths. This method creates a library of genes of different lengths that encode oligomeric polypeptides with the same repeat sequences but sacrifices the precision of the oligomerization process. (iii) Recursive directional ligation (RDL), an alternative method in which DNA segments with two different restriction sites flanking the insert are joined in sequential steps, with the length of the ligated segments growing geometrically in each step. This approach is suitable for the synthesis of repetitive polypeptides with a specific and predetermined chain length, as it seamlessly joins the two monomeric inserts and also eliminates the restriction sites placed at either ends of the dimerized gene [110].

An obvious disadvantage of oligomerization would seem to be the incapability to generate biopolymers with unnatural amino acids, such as those not encoded by the conventional codons or ones containing chemical modifications, cyclic amino acids, aromatic amino acids, or β-amino acids, among others. Fortunately, this problem can be resolved by native chemical ligation [111, 112]. Recently, Liu *et al.* [113] have developed a new approach that allows unnatural amino acids to be genetically encoded in mammalian cells.

One of the main goals in the materials field is to develop materials with an increasing and programmable degree of complexity, which is closely related to the application. More complex materials are capable of performing more complex and tailored functions (Figure 5.3). The primary sequence of ELRs is responsible

Figure 5.3 Schematic diagram of increasing complexity of ELRs with different functionality for use in various applications. Recombination techniques allow obtaining several degrees of complexity: the scheme represents (a) a simple model of recombinant elastin, (b) the introduction of specific peptide sequences and blocks with distinct properties, (c) the inclusion of a bioactive motif, and (d) the inclusion of more than one in the same peptidic formulation.

for the formation of well-defined secondary structures, such as α-helices, β-sheets, or β-turns, assembling into supramolecular structures. Furthermore, they can be switched by several factors, such as temperature, pH, and light in aqueous solution, and can include cross-linking or self-assembling domains and diverse biofunctionalities along the polypeptide chain, including targeting ligands, cell adhesion, or specific biodegradation sequences, and fluorescent or contrast agents. Combined with their good expected biocompatibility and favorable degradation rate and products, they constitute promising materials for multiple applications, with particular emphasis on biomedical and nanotechnological applications [114, 115].

5.4.4
Biocompatibility of ELRs

The biocompatibility feature of ELRs, which is relevant for their use in advanced biomedical applications, was evaluated by the complete series of the American Society for Testing Materials (ASTM) generic biological tests for materials and devices in contact with tissues, tissue fluids, and blood [101, 116, 117]. Regarding these biopolymers, it has not been possible to obtain monoclonal antibodies against them; it would seem that the immune system cannot distinguish them from the natural elastin. In a more detailed interpretation of this effect, Urry [118] has suggested that the β-spiral strongly contributes to prevent the identification of these foreign materials by the immune system. In addition, the secondary products of its biodegradation are just simple and natural amino acids that can be metabolized by the body to produce nontoxic degradation products. More detailed examples of their biocompatibility and multifunctional possibilities is discussed in the next section.

5.5
ELRs as Advanced Materials for Biomedical Applications

ELRs have found widespread use because of their good biocompatibility and non-immunogenicity in the fields of drug delivery and tissue engineering. As they can be genetically synthesized, the precisely defined molecular weight is ideal for drug delivery, because the molecular weight is a key parameter in the route of clearance from the body and *in vivo* half-life of the polymer [104]. The ability to display stimuli-driven variations in their structure as a consequence of their self-assembling behavior in aqueous solution has been exploited in useful functional structures and devices combining these properties, with an adequate arrangement of specific building blocks and physicochemical and mechanical properties.

5.5.1
Tissue Engineering

New classes of materials with additional tissue-specific properties or ones that could be tailored to several tissue systems are required for TERM. The ECM is an important model for the design of biomaterials. The goal of mimicking the ECM structure and biological functions requires the design of artificial scaffolds that reproduce one or, preferably, more properties and functionalities of the natural tissue [119, 120]. The scaffold must be biocompatible and biodegradable, and it should possess properties that support tissue morphogenesis. This generally requires a multifunctional artificial ECM that can supply temporary mechanical support until the engineered tissue has sufficient mechanical integrity to support itself. Future advances in tissue engineering will depend on the development of biomimetic materials that actively participate in the formation of new functional tissues [27].

Recombinant DNA technologies allow designing and expressing artificial genes to prepare artificial analogs of ECM proteins with controlled mechanical properties that incorporate domains to modulate cell behavior. Cross-linked ELRs of poly(VPGVG) resulted in a matrix showing a mechanical response similar to that of natural elastin [116], which is an important factor to provide a structural basis that properly transmits the forces from the environment to the growing tissue. However, these cross-linked matrices showed antifouling properties, limiting cell adhesion. Although lacking biofunctionality that would promote unspecific cell adhesion, the poly(VPGVG) concept matrix was used as the starting material with good mechanical and biocompatible properties, and short peptides having specific bioactivities were inserted into the polymer chain, even though by chemical synthesis. Once genetic engineering techniques became the production method of choice, the molecular design started to increase in complexity. To increase the complexity of ELRs, a simple substitution of the amino acid X in the elastin repeated unit (VPGXG) results in the addition of cross-linking domains to obtain more stable substrates, which are usually based on lysine residues [2, 121–123].

Several modular ELRs have been designed to include bioactive peptides into the polymer chain such as the well-known, general-purpose cell adhesion sequence RGD [124], found in several ECM proteins [18, 125–127], and REDV [128], included in the CS5 domain of fibronectin and specific for endothelial cells [17, 129]. An extra feature – namely, degradation responsiveness – has been included into more advanced versions of ELRs for tissue repair to allow the renovation and replacement of natural ECM by living tissue [17, 130]. In this regard, ELR hydrogels

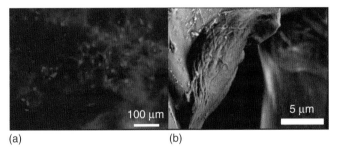

(a) (b)

Figure 5.4 Microscopic assessment of HUVECs seeded in macroporous ELP hydrogels after 48 h of incubation in (a) fluorescence microscopy with Phalloidin Alexa Fluor488 and DAPI staining and (b) SEM magnified view. (Source: Reprinted (adapted) with permission from Ref. [133]. Copyright 2009 American Chemical Society.)

have been obtained by different cross-linking methods such as photoinitiation [131], irradiation [132], or enzymatic cross-linking [122]. The physical properties can be tuned by the cross-linking conditions and degree, while retaining the stimuli-responsive behavior. This way, appropriate bioactive substrates can be obtained for a wide range of applications. Highly porous hydrogels have been conceived by chemical cross-linking of ELRs containing REDV sequence: salt leaching/gas foaming was performed to obtain suitable 3D scaffolds promoting infiltration of HUVEC cells inside the porous network (Figure 5.4) [133].

Another potential application of these bioactive ELRs is the preparation of films to be used for ocular surface tissue engineering [134, 135]. Recently, Srivastava *et al.* [135] have successfully investigated the use of ELRs as a substrate to maintain the growth, phenotype, and functional characteristics of retinal pigment epithelial cells *in vitro* to obtain a suitable carrier for transplantation in the treatment of age-related macular degeneration.

Hybrid scaffolds composed of collagen and different proportions of this ELR-REDV have been obtained by enzymatic cross-linking by tissue transglutaminase, affecting physical properties such as porosity, thermal behavior, and mechanical strength. Such approach led to a differential colonization of the scaffolds with diverse cell lines, which is very useful for wound healing in vascular tissue and skin [122]. Other complex hybrid scaffolds were created to obtain an acellular arterial substitute as artificial vessel-like platforms: they were produced by the combination of several layers of an ELR matrix reinforced with collagen microfibers, resulting in constructs with appropriate mechanical properties, such as compliance and suture retention strength [136]. Other types of smart bilayer scaffolds of an ELR containing RGD and collagen have also been investigated, namely, foams, fibers, and foam-fiber bilayer scaffolds [137]. A study of the structural and mechanical properties indicated that the incorporation of ELRs into the scaffold improved the uniformity of the pore network and decreased the fiber diameter. The culture of human fibroblasts and epithelial cells in these scaffolds showed the positive contribution of the bioactive ELR to the proliferation of both cell types, compared to collagen foam. These results were promising for the reconstruction of full-thickness skin and oral mucosa equivalents.

Other strategies consist in the incorporation of ELR coatings onto different materials to improve cell attachment or stimuli-responsiveness. Barbosa et al. [138] developed chemically cross-linked chitosan hydrogels with an ELR coating containing an osteoconductive sequence, inducing precipitation of calcium phosphate when the gels were soaked in simulated body fluid. More recently, an ELR incorporating octaglutamic acid with well-defined charged backbone has been used to design multifunctional bone cements, resulting in materials with hydroxyapatite binding ability and improved mechanical strength [139]. Prieto et al. [140] have designed a set of ELRs containing the SN_A15 statherin analog domain. In vitro mineralization showed interesting results with regard to the ELR composition being a key parameter in controlling the calcium phosphate nucleation. These studies show the high potential of ELRs for biomedical hybrid materials development.

Another attractive alternative is based on the thermal behavior of modular amphiphilic ELRs to form in situ stable gels under physiological conditions [141, 142]. They constitute low-viscosity solutions below their T_t and form by self-assembling physically cross linked hydrogels that are suitable for fixing materials and biological elements in specific sites in the body through minimally invasive procedures. When greater mechanical integrity is desired, the mechanical properties can be enhanced by including additional chemical cross-linking domains in the ELR chain [117, 143]. Sallach et al. [144] have also highlighted the good biocompatibility of these ELR physical gels, showing a minimal inflammatory response and robust in vivo stability for periods exceeding one year. These self-gelating materials offer the possibility of obtaining multifunctional bioactive versions, which can also be used as carriers for controlled drug release or as biocompatible surface coatings. Oliveira et al. [145] used an ELR containing the RGD cell adhesion domain to fabricate cell-induced microparticle aggregates as scaffolds for TERM. Using an osteoblastlike cell line (SaOs-2) and microparticles with varying degrees of cross-linking, they have shown that for higher degrees of cross-linking, the cell proliferation was more favorable and formed cell-induced aggregation scaffold.

5.5.2
Drug and Gene Delivery

Drug delivery carriers have several advantages over free drug delivery systems such as specific site targeting, sustained release, minimized systemic exposure, and reduced toxicity. In order to be effective, systemic drug delivery carriers must be biocompatible and must have controllable composition and molecular weight. Although a wide range of materials have been employed for controlled and targeted drug release, the most common carriers are polymer-based systems [146, 147].

ELRs exhibit several properties that can be useful for drug delivery purposes, such as precise control over molecular weight and composition, self-assembly, and stimuli-responsive behavior, as well as biocompatibility, nontoxic properties, and efficient pharmacokinetics. In addition, their genetic nature allows incorporating targeting ligands, cell membrane fusion sequences, or reactive sites for drugs conjugation [21, 148, 149]. Their self-assembly can also be designed in response to an extrinsic temperature stimulus, such as the local application of heat, ultrasound, or light, or a stimulus intrinsic to the site environment, such as low extracellular pH or upregulated protease expression for cancer drug delivery [150].

The ability of ELRs to form stable micro- and nanoparticles that are able to trap active substances has facilitated the development of smart systems for therapeutic release. Poly(VPAVG) was shown to form stable particles in water solution, with a size below 3 µm above its T_t. They exhibited hysteresis behavior and were capable of encapsulating significant amounts dexamethasone phosphate with a sustained-release profile for about 30 days [151]. Later, two bone morphogenetic proteins (BMP-2 and BMP-14) were encapsulated for combined release in a sustained way for 14 days, promising for bone tissue engineering [19]. Electrospraying has been used by Wu et al. [152] to generate stimuli-responsive ELR nanoparticles (300–400 nm in diameter) that can encapsulate drugs. The molecular weight and solution concentration of the ELRs were found to have a significant influence on the morphology and size distribution of the nanoparticles. Doxorubicin, a chemotherapeutic agent for cancer, was released by the ITT-driven solubility of the particles at 37 °C. In another recent report, Dash et al. [153] have fabricated tunable monodispersed cross-linked ELR hollow spheres ranging from 100 to 1000 nm by using a template-based method. Plasmid DNA and polyplexes were efficiently loaded inside hollow spheres by diffusion and charge interactions. These polyplex-loaded spheres displayed controlled release and transfection ability in their use as gene transfection agents, with the possibility to include future functionalities to the spheres for targeted gene delivery applications.

Some of the recent studies have focused more on controlling the size and monodispersity of ELR particles since these are two key factors in improving therapeutic efficacy. Genetic engineering of protein-based block copolymers is a suitable way to program defined polypeptide block sequences to study the influence of macromolecular architecture in the hierarchical structural organization and morphology. ELR-based amphiphilic block copolymers undergo ordered nanoscale self-assembly forming proteinlike micellar systems with controlled structure and function. Strategies that describe stable monodispersed ELR particles include those of Chilkoti and coworkers [150, 154, 155], who fabricated sub-100 nm sized micelles conjugated to diverse hydrophobic molecules and chemotherapeutics. The goal of these studies was to investigate the design rules for tuning the self-assembly of different ELRs forming multivalent spherical micelles within the clinically relevant hyperthermic temperature range 37–42 °C and target cancer cells. Similarly, Kim et al. [156] improved the stability of micelles from ELR-diblock copolymers incorporating cysteine residues at the core–shell interface, which undergo covalent cross-linking through disulfide bond formation. Most of these studies involve diblocks, although there are some examples based on triblocks, such as the one that relies on a reversible switch of secondary structures [157]. The triblock–ELR copolymer can form monodispersed temperature-responsive micelles in dilute solutions below its T_t, and an abrupt increase in the micelle compactness with a reduction in particle size occurs on raising the temperature.

5.5.3
Surface Engineering

In the field of biomaterials science, the control and modification of surfaces and interfaces has opened novel frontiers for developing advanced materials and devices with distinct and specific features for biomedical applications [158, 159]. Surfaces can be functionalized by multiple approaches involving physical or chemical modifications, such as coatings and grafts. They allow introducing small biological ligands, such as peptides or proteins with fouling/antifouling features, specific groups for cell–material interactions, smart behavior (stimuli responsive or environmentally sensitive), or micro- and nanopatterns [160–162].

Surfaces modified with stimuli-responsive polymers, called *"smart surfaces,"* undergo physical and chemical changes in response to external variations [163–165]. In this regard, ELRs exhibit some additional advantages that make them excellent candidates for the development of smart surfaces: their stimuli-responsive behavior can depend on other factors such as pH, light, or ionic strength. Genetic engineering also allows for a precise control of the reactive sites on the polypeptide chain for use in surface grafting, leading to an extensive potential of self-assembly exhibited by these polymers. Biosensing surfaces displaying topologically modified self-assembly with ELRs have been developed by Chilkoti's group. The "thermodynamically reversible addressing of proteins" (TRAPs) technology obtained by covalent micropatterning of ELRs onto glass surfaces against an inert background allows a reversible, spatiotemporal modulation of ligand binding triggered by the phase transition of the ELR at a solid–liquid interface. Such technology can be applied in different systems for bioanalytical sensors in order to detect single biomolecules [158, 164]. Fluorophore-labeled glutamine binding proteins (QBPs) and derivatives – which can be coupled to a designed hydrophobic polypeptide – can be constructed and adhered onto solid surfaces [166]. Another novel fusion protein for controlling cellular functions combines the RGD sequence, epidermal growth factor (EGF), and a hydrophobic sequence into a single molecule: it was shown to exhibit both cell-adhesive and growth factor activity. In addition, the hydrophobic sequence contributed to self-assembly of the RGD, retaining its activity on a solid-phase surface and proving its use for wound healing [167].

For tissue engineering, novel temperature-responsive culture surfaces coated with ELR have been developed for cell harvesting. By reducing the temperature using a polyvinylidene difluoride (PVDF) membrane, the cell sheet was detached from the coated surface and subsequently transferred to new surfaces, suggesting its potential for the fabrication of multilayer cell sheets for transplantation [24]. For bone tissue engineering, an ELR-RGD has been adsorbed onto micropatterned poly(N-isopropyl acrylamide) films displaying a positive maintenance of the cell attachment in the scaffolds under dynamic culture conditions [168]. Hyun *et al.* [169] have also developed a lysine-containing ELR conjugated onto an aldehyde–glass surface by micropatterning. The sharp and controlled phase transition of ELR enabled reversible cell adhesion on the surface by changing the temperature or salt concentration, demonstrating potential applications for cell-based microdevices.

Other techniques using ELRs include the alternating adsorption of polyelectrolytes. This is a simple technique for generating bioactive surfaces, driven by complementary forces present between two distinct materials [170]. These ultrathin nanoscale coatings promote cell adhesion and proliferation, and the results show that the thickness and mechanical integrity of the multilayer assembly modulates the cell response. Swierczewska *et al.* [170] have reported the use of ELRs – modified either with polyethyleneimine (PEI) or with polyacrylic acid (PAA) to provide positive or negative charge, respectively – to conceive films in a sequential manner. Another similar work described a similar approach but with ELRs containing negatively charged residues [171]. Costa *et al.* also reported the construction of thin coatings [172] and nanostructured multilayers [173] made of chitosan and RGD-containing ELRs with smart properties toward temperature (Figure 5.5a) and pH (Figure 5.5b) and also improved cell adhesion. These examples opened up a field where polymeric coatings include specific bio-functional responses.

Progress in the past few years combining surface chemistry with microfabrication techniques has provided new tools to study cell–material interactions with their environment. Using lithography and patterning techniques, peptides and proteins can be deposited with absolute spatial control on specific regions of a surface [174, 175]. The ability to obtain nano- and microstructured surfaces of patterned biomacromolecules is of great importance for several applications, including biological assays, miniaturized biosensors, and biomedical diagnostics. In this regard, ELRs have been employed in the design and development of biosensors and microfluidic bioanalytical devices, as reported by Nath and Chilkoti [176]. Nanostructured surfaces taking advantage of the self-assembly properties of ELRs

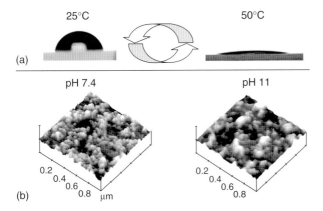

Figure 5.5 Physical changes induced in ELR-modified surfaces. (a) Wettability variations with different temperatures. (b) Topography at distinct pHs. (Source: (a) Reprinted (adapted) with permission from Ref. [172]. Copyright 2009 Wiley-VCH. (b) Reprinted (adapted) with permission from Ref. [173]. Copyright 2011 Wiley-VCH.)

have been obtained for capturing and releasing proteins by combining ELRs and dip-pen nanolithography.

Recently, the replica molding method has been adapted to obtain 3D microstructured thermo-responsive ELR hydrogels by Martín et al. [121]. In this study, the thermally responsive behavior of macroscopic and micropatterned features in response to different dimensions and spacing was studied, showing dimensional changes with no alteration of the topography. This feature could be used during cell culture, adjusting the T_t in the correct range to study cell behavior by changing the dimensions of the micropattern and subsequent mechanical properties.

5.6
Conclusions

Elastin is the root of inspiration for a class of versatile materials with physical and biological properties of great interest for TERM applications. The cumulative studies of elastin and elastin-like macromolecules now provide the required background for understanding in more detail the structure-related aspects of such biomaterials from a functional point of view.

Taking advantage of genetic recombination approaches, one can easily "engineer" a peptide-based macromolecule exhibiting characteristics that can be tuned for specific fields such as skin substitutes, vascular constructs, drug delivery devices, and functional interfaces. Their functionality can be further expanded by adding sequences with biological relevance to the common elastin structure, such as cell adhesion motif, tissue-specific motif, and degradation motif. Complex, well-defined and tailored polymers can be obtained with a wide range of properties and possibilities for new devices. One could say that such degree of complexity is only limited by the imagination of the scientist; however, it still needs to be faithful to the original properties of elastin-like macromolecules, since an unfavorable modification can lead to a nonfunctional polypeptide.

The study and use of elastin-like macromolecules is leading to the design of new cutting-edge macromolecules, and more complex ones are emerging as promising candidates for future treatments. Their smart behavior and self-assembly properties may be exploited for further tuning and ultimate control of the performance of an elastin-based system, such as how a specific cell phenotype will bind to a substrate or how an active agent will be released under stimuli variations and biological conditions.

Acknowledgements

We acknowledge financial support from the Portuguese "Fundação para a Ciência e Tecnologia" – FCT (Grant SFRH/BD/61126/2009); "Fundo Social Europeu" – FSE; "Programa Diferencial de Potencial Humano" – POPH; the European Commission, through the 7th Framework Programme (Projects

HEALTH-F4-2011-278557 and NMP3-LA-2011-263363); from the Spanish Minister of Economy and Competitiveness – MEC (Projects MAT2009-14195-C03-03, MAT-2010-15982, MAT2010-15310, IT2009-0089, ACI2009-0890, and PRI-PIBAR-2011-1403); the Junta de Castilla y León-JCyL (Projects VA034A09 and VA0049A11-2); the CIBER-BBN (Project CB06-01-1038) and the JCyL; and the Instituto de Salud Carlos III, under the "Network Center of Regenerative Medicine and Cellular Therapy of Castilla and León."

References

1. Almine, J.F., Bax, D.V., Mithieux, S.M., Nivison-Smith, L., Rnjak, J., Waterhouse, A., Wise, S.G., and Weiss, A.S. (2010) *Chem. Soc. Rev.*, **39**, 3371–3379.
2. Nowatzki, P.J. and Tirrell, D.A. (2004) *Biomaterials*, **25**, 1261–1267.
3. Trabbic-Carlson, K., Setton, L.A., and Chilkoti, A. (2003) *Biomacromolecules*, **4**, 572–580.
4. Urry, D.W. (1993) *Angew. Chem., Int. Ed. Engl.*, **32**, 819–841.
5. Urry, D.W. and Parker, T.M. (2002) *J. Muscle Res. Cell Motil.*, **23**, 543–559.
6. Welsh, E.R. and Tirrell, D.A. (2000) *Biomacromolecules*, **1**, 23–30.
7. Gigante, A., Chillemi, C., Bevilacqua, C., Greco, F., Bisaccia, F., and Tamburro, A.M. (2003) *J. Mater. Sci. Mater. Med.*, **14**, 717–720.
8. Ito, S., Ishimaru, S., and Wilson, S.E. (1998) *Angiology*, **49**, 289–297.
9. Leach, J.B., Wolinsky, J.B., Stone, P.J., and Wong, J.Y. (2005) *Acta Biomater.*, **1**, 155–164.
10. Urry, D.W., Asima, P., Xu, J., Woods, T.C., McPherson, D.T., and Partker, T.M. (1998) *J. Biomater. Sci. Polym. Ed.*, **9**, 1015–1048.
11. Meyer, D.E. and Chilkoti, A. (2004) *Biomacromolecules*, **5**, 846–851.
12. Mithieux, S.M., Rasko, J.E.J., and Weiss, A.S.A.S. (2004) *Biomaterials*, **25**, 4921–4927.
13. Rodríguez-Cabello, J.C., Alonso, M., Pérez, T., and Herguedas, M.M. (2000) *Biopolymers*, **54**, 282–288.
14. Sciortino, F., Urry, D.W., Palma, M.U., and Prasad, K.U. (1990) *Biopolymers*, **29**, 1401–1407.
15. Rodríguez-Cabello, J.C., Martín, L., Alonso, M., Arias, F.J., and Testera, A.M. (2009) *Polymer*, **50**, 5159–5169.
16. Arias, F., Reboto, V., Martín, S., López, I., and Rodríguez-Cabello, J. (2006) *Biotechnol. Lett.*, **28**, 687–695.
17. Girotti, A., Reguera, J., Rodríguez-Cabello, J., Arias, F., Alonso, M., and Testera, A. (2004) *J. Mater. Sci. Mater. Med.*, **15**, 479–484.
18. Nicol, A., Channe Gowda, D., and Urry, D.W. (1992) *J. Biomed. Mater. Res.*, **26**, 393–413.
19. Bessa, P.C., Machado, R., Nürnberger, S., Dopler, D., Banerjee, A., Cunha, A.M., Rodríguez-Cabello, J.C., Redl, H., van Griensven, M., Reis, R.L., and Casal, M. (2010) *J. Controlled Release*, **142**, 312–318.
20. Dreher, M.R., Raucher, D., Balu, N., Colvin, O.M., Ludeman, S.M., and Chilkoti, A. (2003) *J. Controlled Release*, **91**, 31–43.
21. Furgeson, D.Y., Dreher, M.R., and Chilkoti, A. (2006) *J. Controlled Release*, **110**, 362–369.
22. Betre, H., Setton, L.A., Meyer, D.E., and Chilkoti, A. (2002) *Biomacromolecules*, **3**, 910–916.
23. McHale, M.K., Setton, L.A., and Chilkoti, A. (2005) *Tissue Eng.*, **11**, 1768–1779.
24. Zhang, H., Iwama, M., Akaike, T., Urry, D.W., Pattanaik, A., Parker, T.M., Konishi, I., and Nikaido, T. (2006) *Tissue Eng.*, **12**, 391–401.
25. Meyers, M.A., Chen, P.-Y., Lin, A.Y.-M., and Seki, Y. (2008) *Prog. Mater. Sci.*, **53**, 1–206.
26. Ashby, M.F. and Jones, D.R.H. (2006) *Engineering Materials 2: An Introduction*

to Microstructures, Processing and Design, 3rd edn, Elsevier, Oxford.
27. Langer, R. and Tirrell, D.A. (2004) *Nature*, **428**, 487–492.
28. Wegst, U.G.K. and Ashby, M.F. (2004) *Philos. Mag.*, **84**, 2167–2186.
29. Mano, J.F., Silva, G.A., Azevedo, H.S., Malafaya, P.B., Sousa, R.A., Silva, S.S., Boesel, L.F., Oliveira, J.M., Santos, T.C., Marques, A.P., Neves, N.M., and Reis, R.L. (2007) *J. R. Soc. Interface*, **4**, 999–1030.
30. Nair, L.S. and Laurencin, C.T. (2007) *Prog. Polym. Sci.*, **32**, 762–798.
31. Anderson, J.M. (2001) *Annu. Rev. Mater. Res.*, **31**, 81–110.
32. Hubbell, J.A. (1995) *Nat. Biotechnol.*, **13**, 565–576.
33. Sanchez, C., Arribart, H., and Giraud Guille, M.M. (2005) *Nat. Mater.*, **4**, 277–288.
34. Rodríguez-Cabello, J., Reguera, J., Girotti, A., Arias, F., and Alonso, M. (2006) in *Ordered Polymeric Nanostructures at Surfaces*, vol. 200 (eds G.J. Vancso and G. Reiter), Springer, Berlin, pp. 119–167.
35. Holzapfel, G.A. (2001) in *Handbook of Materials Behavior Models* (ed. J. Lemaitre), Academic Press, New York, pp. 1057–1071.
36. Debelle, L. and Tamburro, A.M. (1999) *Int. J. Biochem. Cell Biol.*, **31**, 261–272.
37. Bashir, M.M., Indik, Z., Yeh, H., Ornstein-Goldstein, N., Rosenbloom, J.C., Abrams, W., Fazio, M., Uitto, J., and Rosenbloom, J. (1989) *J. Biol. Chem.*, **264**, 8887–8891.
38. Indik, Z., Yeh, H., Ornstein-Goldstein, N., Sheppard, P., Anderson, N., Rosenbloom, J.C., Peltonen, L., and Rosenbloom, J. (1987) *Proc. Natl. Acad. Sci. U.S.A.*, **84**, 5680–5684.
39. Jones, P.A., Scott-Burden, T., and Gevers, W. (1979) *Proc. Natl. Acad. Sci. U.S.A.*, **76**, 353–357.
40. Mecham, R.P., Madaras, J., McDonald, J.A., and Ryan, U. (1983) *J. Cell. Physiol.*, **116**, 282–288.
41. Sephel, G.C. and Davidson, J.M. (1986) *J. Invest. Dermatol.*, **86**, 279–285.
42. Moskalewski, S. (1976) *Am. J. Anat.*, **146**, 443–448.
43. Hsiao, H., Stone, P.J., Toselli, P., Rosenbloom, J., Franzblau, C., and Schreiber, B.M. (1999) *Connect. Tissue Res.*, **40**, 83–95.
44. Mauch, J.C., Sandberg, L.B., Roos, P.J., Jimenez, F., Christiano, A.M., Deak, S.B., and Boyd, C.D. (1995) *Matrix Biol.*, **14**, 635–641.
45. Pierce, R.A., Deak, S.B., Stolle, C.A., and Boyd, C.D. (1990) *Biochemistry*, **29**, 9677–9683.
46. Raju, K. and Anwar, R.A. (1987) *J. Biol. Chem.*, **262**, 5755–5762.
47. Daamen, W.F., Veerkamp, J.H., van Hest, J.C.M., and van Kuppevelt, T.H. (2007) *Biomaterials*, **28**, 4378–4398.
48. Hinek, A. and Rabinovitch, M. (1994) *J. Cell Biol.*, **126**, 563–574.
49. Brown-Augsburger, P., Broekelmann, T., Rosenbloom, J., and Mecham, R.P. (1996) *Biochem. J.*, **318**, 149–155.
50. Rosenbloom, J., Abrams, W., and Mecham, R. (1993) *FASEB J.*, **7**, 1208–1218.
51. Martyn, C.N. and Greenwald, S.E. (1997) *Lancet*, **350**, 953–955.
52. Powell, J.T., Vine, N., and Crossman, M. (1992) *Atherosclerosis*, **97**, 201–208.
53. Li, B. and Daggett, V. (2002) *J. Muscle Res. Cell Motil.*, **23**, 561–573.
54. Andrady, A.L. and Mark, J.E. (1980) *Biopolymers*, **19**, 849–855.
55. Hoeve, C.A.J. and Flory, P.J. (1958) *J. Am. Chem. Soc.*, **80**, 6523–6526.
56. Tamburro, A.M., Guantieri, V., Daga-Gordini, D., and Abatangelo, G. (1978) *J. Biol. Chem.*, **253**, 2893–2894.
57. Dorrington, K.L. and McCrum, N.G. (1977) *Biopolymers*, **16**, 1201–1222.
58. Weis-Fogh, T. and Andersen, S.O. (1970) *Nature*, **227**, 718–721.
59. Gray, W.R., Sandberg, L.B., and Foster, J.A. (1973) *Nature*, **246**, 461–466.
60. Tamburro, A.M., Guantieri, V., and Gordini, D.D. (1992) *J. Biomol. Struct. Dyn.*, **10**, 441–454.
61. Urry, D.W., Shaw, R.G., and Prasad, K.U. (1985) *Biochem. Biophys. Res. Commun.*, **130**, 50–57.
62. Ewart, A.K., Jin, W., Atkinson, D., Morris, C.A., and Keating, M.T. (1994) *J. Clin. Invest.*, **93**, 1071–1077.
63. Kumar, A., Olson, T.M., Thibodeau, S.N., Michels, V.V., Schaid, D.J., and

Wallace, M.R. (1994) *Am. J. Cardiol.*, **74**, 1281–1283.

64. Li, D.Y., Toland, A.E., Boak, B.B., Atkinson, D.L., Ensing, G.J., Morris, C.A., and Keating, M.T. (1997) *Hum. Mol. Genet.*, **6**, 1021–1028.

65. Wu, W.J. and Weiss, A.S. (1999) *Eur. J. Biochem.*, **266**, 308–314.

66. Ewart, A.K., Morris, C.A., Atkinson, D., Jin, W., Sternes, K., Spallone, P., Stock, A.D., Leppert, M., and Keating, M.T. (1993) *Nat. Genet.*, **5**, 11–16.

67. Rodriguez-Revenga, L., Iranzo, P., Badenas, C., Puig, S., Carrio, A., and Mila, M. (2004) *Arch. Dermatol.*, **140**, 1135–1139.

68. Zhang, M.-C., He, L., Giro, M., Yong, S.L., Tiller, G.E., and Davidson, J.M. (1999) *J. Biol. Chem.*, **274**, 981–986.

69. Robinson, P.N. and Godfrey, M. (2000) *J. Med. Genet.*, **37**, 9–25.

70. Kurane, A., Simionescu, D.T., and Vyavahare, N.R. (2007) *Biomaterials*, **28**, 2830–2838.

71. Schinke, T. and Karsenty, G. (2000) *Nephrol. Dial. Transplant.*, **15**, 1272–1274.

72. Singla, A. and Lee, C.H. (2003) *J. Biomed. Mater. Res. A*, **64A**, 706–713.

73. Phillips, T.J. and Gilchrest, B.A. (1992) *Epithelial Cell Biol.*, **1**, 39–46.

74. Pomahaè, B., Svensjö, T., Yao, F., Brown, H., and Eriksson, E. (1998) *Crit. Rev. Oral. Biol. Med.*, **9**, 333–344.

75. Ehrenreich, M. and Ruszczak, Z. (2006) *Tissue Eng.*, **12**, 2407–2424.

76. Falanga, V. and Sabolinski, M. (1999) *Wound Repair Regen.*, **7**, 201–207.

77. Ramsey, S.D., Newton, K., Blough, D., McCulloch, D.K., Sandhu, N., Reiber, G.E., and Wagner, E.H. (1999) *Diabetes Care*, **22**, 382–387.

78. Hafemann, B., Ensslen, S., Erdmann, C., Niedballa, R., Zuhlke, A., Ghofrani, K., and Kirkpatrick, C.J. (1999) *Burns*, **25**, 373–384.

79. Nillesen, S.T.M., Lammers, G., Wismans, R.G., Ulrich, M.M., Middelkoop, E., Spauwen, P.H., Faraj, K.A., Schalkwijk, J., Daamen, W.F., and van Kuppevelt, T.H. (2011) *Acta Biomater.*, **7**, 1063–1071.

80. Koens, M.J.W., Faraj, K.A., Wismans, R.G., van der Vliet, J.A., Krasznai, A.G., Cuijpers, V.M.J.I., Jansen, J.A., Daamen, W.F., and vanKuppevelt, T.H. (2010) *Acta Biomater.*, **6**, 4666–4674.

81. Patel, A., Fine, B., Sandig, M., and Mequanint, K. (2006) *Cardiovasc. Res.*, **71**, 40–49.

82. Gotlieb, A.I. (2005) *Cardiovasc. Pathol.*, **14**, 181–184.

83. Isenberg, B.C., Williams, C., and Tranquillo, R.T. (2006) *Circ. Res.*, **98**, 25–35.

84. Nerem, R.M. and Seliktar, D. (2001) *Annu. Rev. Biomed. Eng.*, **3**, 225–243.

85. Stitzel, J., Liu, J., Lee, S.J., Komura, M., Berry, J., Soker, S., Lim, G., Van Dyke, M., Czerw, R., Yoo, J.J., and Atala, A. (2006) *Biomaterials*, **27**, 1088–1094.

86. Neuman, R.E. and Logan, M.A. (1950) *J. Biol. Chem.*, **186**, 549–556.

87. Rasmussen, B.L., Bruenger, E., and Sandberg, L.B. (1975) *Anal. Biochem.*, **64**, 255–259.

88. Daamen, W.F., Hafmans, T., Veerkamp, J.H., and Kuppevelt, T.H.V. (2005) *Tissue Eng.*, **11**, 1168–1176.

89. Tamburro, A.M., Pepe, A., Bochicchio, B., Quaglino, D., and Ronchetti, I.P. (2005) *J. Biol. Chem.*, **280**, 2682–2690.

90. Rodríguez-Cabello, J.C., Reguera, J., Alonso, M., Parker, T.M., McPherson, D.T., and Urry, D.W. (2004) *Chem. Phys. Lett.*, **388**, 127–131.

91. Urry, D.W. (1999) *Trends Biotechnol.*, **17**, 249–257.

92. Lee, J., MacOsko, C.W., and Urry, D.W. (2001) *J. Biomater. Sci. Polym. Ed.*, **12**, 229–242.

93. Urry, D.W. (1997) *J. Phys. Chem. B*, **101**, 11007–11028.

94. Rodríguez-Cabello, J.C., Reguera, J., Girotti, A., Alonso, M., and Testera, A.M. (2005) *Prog. Polym. Sci.*, **30**, 1119–1145.

95. Reguera, J., Urry, D.W., Parker, T.M., McPherson, D.T., and Rodríguez-Cabello, J.C. (2007) *Biomacromolecules*, **8**, 354–358.

96. Meyer, D.E. and Chilkoti, A. (2002) *Biomacromolecules*, **3**, 357–367.

97. Urry, D.W., Hayes, L.C., Gowda, D.C., Harris, C.M., and Harris, R.D. (1992) *Biochem. Biophys. Res. Commun.*, **188**, 611–617.

98. Strzegowski, L.A., Martinez, M.B., Gowda, D.C., Urry, D.W., and Tirrell, D.A. (1994) *J. Am. Chem. Soc.*, **116**, 813–814.
99. Urry, D.W., Hayes, L.C., Gowda, D.C., and Parker, T.M. (1991) *Chem. Phys. Lett.*, **182**, 101–106.
100. Di Zio, K. and Tirrell, D.A. (2003) *Macromolecules*, **36**, 1553–1558.
101. Urry, D.W., Parker, T.M., Reid, M.C., and Gowda, D.C. (1991) *J. Bioact. Compat. Polym.*, **6**, 263–282.
102. McMillan, R.A., Lee, T.A.T., and Conticello, V.P. (1999) *Macromolecules*, **32**, 3643–3648.
103. Padgett, K.A. and Sorge, J.A. (1996) *Gene*, **168**, 31–35.
104. Simnick, A.J., Lim, D.W., Chow, D., and Chilkoti, A. (2007) *Polym. Rev.*, **47**, 121–154.
105. Urry, D.W., Luan, C.H., Parker, T.M., Gowda, D.C., Prasad, K.U., Reid, M.C., and Safavy, A. (1991) *J. Am. Chem. Soc.*, **113**, 4346–4348.
106. McPherson, D.T., Xu, J., and Urry, D.W. (1996) *Protein Expr. Purif.*, **7**, 51–57.
107. Fothergill-Gilmore, L.A. (1993) in *Protein Biotechnology: Isolation, Characterization, and Stabilization* (ed. F. Franks), Humana Press, Totowa, NJ, pp. 254–256.
108. Guda, C., Zhang, X., McPherson, D.T., Xu, J., Cherry, J.H., Urry, D.W., and Daniell, H. (1995) *Biotechnol. Lett.*, **17**, 745–750.
109. Mi, L.X. (2006) *Biomacromolecules*, **7**, 2099–2107.
110. Chilkoti, A., Dreher, M.R., and Meyer, D.E. (2002) *Adv. Drug Delivery Rev.*, **54**, 1093–1111.
111. Paramonov, S.E., Gauba, V., and Hartgerink, J.D. (2005) *Macromolecules*, **38**, 7555–7561.
112. Dawson, P.E., Muir, T.W., Clarklewis, I., and Kent, S.B.H. (1994) *Science*, **266**, 776–779.
113. Liu, W.S., Brock, A., Chen, S., Chen, S.B., and Schultz, P.G. (2007) *Nat. Methods*, **4**, 239–244.
114. Rodriguez-Cabello, J.C., Martin, L., Girotti, A., Garcia-Arevalo, C., Arias, F.J., and Alonso, M. (2011) *Nanomedicine*, **6**, 111–122.
115. Rabotyagova, O.S., Cebe, P., and Kaplan, D.L. (2011) *Biomacromolecules*, **12**, 269–289.
116. Li, B., Alonso, D.O.V., Bennion, B.J., and Daggett, V. (2001) *J. Am. Chem. Soc.*, **123**, 11991–11998.
117. Sallach, R.E., Cui, W., Wen, J., Martinez, A., Conticello, V.P., and Chaikof, E.L. (2009) *Biomaterials*, **30**, 409–422.
118. Urry, D.W. (2006) *What Sustains Life? Consilient Mechanisms for Protein-based Machines and Materials*, Springer-Verlag, New York.
119. Furth, M.E., Atala, A., and Van Dyke, M.E. (2007) *Biomaterials*, **28**, 5068–5073.
120. Badylak, S.F., Freytes, D.O., and Gilbert, T.W. (2009) *Acta Biomater.*, **5**, 1–13.
121. Martín, L., Alonso, M., Moller, M., Rodríguez-Cabello, J.C., and Mela, P. (2009) *Soft Matter*, **5**, 1591–1593.
122. Garcia, Y., Hemantkumar, N., Collighan, R., Griffin, M., Rodriguez-Cabello, J.C., and Pandit, A. (2009) *Tissue Eng. Part A*, **15**, 887–899.
123. Lim, D.W., Nettles, D.L., Setton, L.A., and Chilkoti, A. (2007) *Biomacromolecules*, **8**, 1463–1470.
124. Hersel, U., Dahmen, C., and Kessler, H. (2003) *Biomaterials*, **24**, 4385–4415.
125. Straley, K.S. and Heilshorn, S.C. (2009) *Front. Neuroeng.*, **2**, 9.
126. Panitch, A., Yamaoka, T., Fournier, M.J., Mason, T.L., and Tirrell, D.A. (1999) *Macromolecules*, **32**, 1701–1703.
127. Liu, J.C., Heilshorn, S.C., and Tirrell, D.A. (2004) *Biomacromolecules*, **5**, 497–504.
128. Heilshorn, S.C., DiZio, K.A., Welsh, E.R., and Tirrell, D.A. (2003) *Biomaterials*, **24**, 4245–4252.
129. Plouffe, B.D., Njoka, D.N., Harris, J., Liao, J., Horick, N.K., Radisic, M., and Murthy, S.K. (2007) *Langmuir*, **23**, 5050–5055.
130. Straley, K.S. and Heilshorn, S.C. (2009) *Soft Matter*, **5**, 114–124.
131. Nagapudi, K., Brinkman, W.T., Leisen, J.E., Huang, L., McMillan, R.A., Apkarian, R.P., Conticello, V.P., and Chaikof, E.L. (2002) *Macromolecules*, **35**, 1730–1737.

132. Lee, J., Macosko, C.W., and Urry, D.W. (2001) *Macromolecules*, **34**, 4114–4123.
133. Martín, L., Alonso, M., Girotti, A., Arias, F.J., and Rodríguez-Cabello, J.C. (2009) *Biomacromolecules*, **10**, 3015–3022.
134. Martínez-Osorio, H., Juárez-Campo, M., Diebold, Y., Girotti, A., Alonso, M., Arias, F.J., Rodríguez-Cabello, J.C., García-Vázquez, C., and Calonge, M. (2009) *Curr. Eye Res.*, **34**, 48–56.
135. Srivastava, G.K., Martín, L., Singh, A.K., Fernandez-Bueno, I., Gayoso, M.J., Garcia-Gutierrez, M.T., Girotti, A., Alonso, M., Rodríguez-Cabello, J.C., and Pastor, J.C. (2011) *J. Biomed. Mater. Res. A*, **97A**, 243–250.
136. Caves, J.M., Kumar, V.A., Martinez, A.W., Kim, J., Ripberger, C.M., Haller, C.A., and Chaikof, E.L. (2010) *Biomaterials*, **31**, 7175–7182.
137. Kinikoglu, B., Rodríguez-Cabello, J., Damour, O., and Hasirci, V. (2011) *J. Mater. Sci. Mater. Med.*, **22**, 1541–1554.
138. Barbosa, J.S., Ribeiro, A., Testera, A.M., Alonso, M., Arias, F.J., Rodriguez-Cabello, J.C., and Mano, J.F. (2010) *Adv. Eng. Mater.*, **12**, B37–B44.
139. Wang, E.D., Lee, S.H., and Lee, S.W. (2011) *Biomacromolecules*, **12**, 672–680.
140. Prieto, S., Shkilnyy, A., Rumplasch, C., Ribeiro, A., Arias, F.J., Rodríguez-Cabello, J.C., and Taubert, A. (2011) *Biomacromolecules*, **12**, 1480–1486.
141. Wu, X., Sallach, R.E., Caves, J.M., Conticello, V.P., and Chaikof, E.L. (2008) *Biomacromolecules*, **9**, 1787–1794.
142. Martín, L., Arias, F.J., Alonso, M., García-Arévalo, C., and Rodríguez-Cabello, J.C. (2010) *Soft Matter*, **6**, 1121–1124.
143. Wu, X.Y., Sallach, R., Haller, C.A., Caves, J.A., Nagapudi, K., Conticello, V.P., Levenston, M.E., and Chaikof, E.L. (2005) *Biomacromolecules*, **6**, 3037–3044.
144. Sallach, R.E., Cui, W.X., Balderrama, F., Martinez, A.W., Wen, J., Haller, C.A., Taylor, J.V., Wright, E.R., Long, R.C., and Chaiko, E.L. (2010) *Biomaterials*, **31**, 779–791.
145. Oliveira, M.B., Song, W., Martin, L., Oliveira, S.M., Caridade, S.G., Alonso, M., Rodriguez-Cabello, J.C., and Mano, J.F. (2011) *Soft Matter*, **7**, 6426–6434.
146. Kost, J. and Langer, R. (1991) *Adv. Drug Delivery Rev.*, **6**, 19–50.
147. Chilkoti, A., Dreher, M.R., Meyer, D.E., and Raucher, D. (2002) *Adv. Drug Delivery Rev.*, **54**, 613–630.
148. Bidwell, G.L., Davis, A.N., and Raucher, D. (2009) *J. Controlled Release*, **135**, 2–10.
149. Massodi, I., Bidwell, G.L., and Raucher, D. (2005) *J. Controlled Release*, **108**, 396–408.
150. MacEwan, S.R., Callahan, D.J., and Chilkoti, A. (2010) *Nanomedicine*, **5**, 793–806.
151. Herrero-Vanrell, R., Rincón, A.C., Alonso, M., Reboto, V., Molina-Martínez, I.T., and Rodríguez-Cabello, J.C. (2005) *J. Controlled Release*, **102**, 113–122.
152. Wu, Y.Q., MacKay, J.A., McDaniel, J.R., Chilkoti, A., and Clark, R.L. (2009) *Biomacromolecules*, **10**, 19–24.
153. Dash, B.C., Mahor, S., Carroll, O., Mathew, A., Wang, W., Woodhouse, K.A., and Pandit, A. (2011) *J. Controlled Release*, **152**, 382–392.
154. MacKay, J.A., Chen, M.N., McDaniel, J.R., Liu, W.G., Simnick, A.J., and Chilkoti, A. (2009) *Nat. Mater.*, **8**, 993–999.
155. Dreher, M.R., Simnick, A.J., Fischer, K., Smith, R.J., Patel, A., Schmidt, M., and Chilkoti, A. (2008) *J. Am. Chem. Soc.*, **130**, 687–694.
156. Kim, W., Thevenot, J., Ibarboure, E., Lecommandoux, S., and Chaikof, E.L. (2010) *Angew. Chem. Int. Ed.*, **49**, 4257–4260.
157. Sallach, R.E., Wei, M., Biswas, N., Conticello, V.P., Lecommandoux, S., Dluhy, R.A., and Chaikof, E.L. (2006) *J. Am. Chem. Soc.*, **128**, 12014–12019.
158. Nath, N., Hyun, J., Ma, H., and Chilkoti, A. (2004) *Surf. Sci.*, **570**, 98–110.
159. Akiyama, Y., Kikuchi, A., Yamato, M., and Okano, T. (2004) *Langmuir*, **20**, 5506–5511.

160. Kato, R., Kaga, C., Kunimatsu, M., Kobayashi, T., and Honda, H. (2006) *J. Biosci. Bioeng.*, **101**, 485–495.
161. Luzinov, I., Minko, S., and Tsukruk, V.V. (2004) *Prog. Polym. Sci.*, **29**, 635–698.
162. Hyun, J., Lee, W.K., Nath, N., Chilkoti, A., and Zauscher, S. (2004) *J. Am. Chem. Soc.*, **126**, 7330–7335.
163. Idota, N., Tsukahara, T., Sato, K., Okano, T., and Kitamori, T. (2009) *Biomaterials*, **30**, 2095–2101.
164. Nath, N. and Chilkoti, A. (2002) *Adv. Mater.*, **14**, 1243–1247.
165. Shi, J., Alves, N.M., and Mano, J.F. (2007) *Adv. Funct. Mater.*, **17**, 3312–3318.
166. Wada, A., Mie, M., Aizawa, M., Lahoud, P., Cass, A.E.G., and Kobatake, E. (2003) *J. Am. Chem. Soc.*, **125**, 16228–16234.
167. Elloumi, I., Kobayashi, R., Funabashi, H., Mie, M., and Kobatake, E. (2006) *Biomaterials*, **27**, 3451–3458.
168. Ozturk, N., Girotti, A., Kose, G.T., Rodríguez-Cabello, J.C., and Hasirci, V. (2009) *Biomaterials*, **30**, 5417–5426.
169. Na, K., Jung, J., Kim, O., Lee, J., Lee, T.G., Park, Y.H., and Hyun, J. (2008) *Langmuir*, **24**, 4917–4923.
170. Swierczewska, M., Hajicharalambous, C.S., Janorkar, A.V., Megeed, Z., Yarmush, M.L., and Rajagopalan, P. (2008) *Acta Biomater.*, **4**, 827–837.
171. Barbosa, J., Costa, R., Testera, A., Alonso, M., Rodríguez-Cabello, J., and Mano, J. (2009) *Nanoscale Res. Lett.*, **4**, 1247–1253.
172. Costa, R.R., Custódio, C.A., Testera, A.M., Arias, F.J., Rodríguez-Cabello, J.C., Alves, N.M., and Mano, J.F. (2009) *Adv. Funct. Mater.*, **19**, 3210–3218.
173. Costa, R.R., Custódio, C.A., Arias, F.J., Rodríguez-Cabello, J.C., and Mano, J.F. (2011) *Small*, **7**, 2640–2649.
174. Offenhausser, A., Bocker-Meffert, S., Decker, T., Helpenstein, R., Gasteier, P., Groll, J., Moller, M., Reska, A., Schafer, S., Schulte, P., and Vogt-Eisele, A. (2007) *Soft Matter*, **3**, 290–298.
175. Bernard, A., Renault, J.P., Michel, B., Bosshard, H.R., and Delamarche, E. (2000) *Adv. Mater.*, **12**, 1067–1070.
176. Nath, N. and Chilkoti, A. (2003) *Anal. Chem.*, **75**, 709–715.

6
Biomimetic Molecular Recognition Elements for Chemical Sensing
Justyn Jaworski

6.1
Introduction

6.1.1
Overview

In this chapter, we focus on chemically receptive materials that are capable of selectively binding to specific molecular targets. These materials are biomimetic in nature by way of their adherence to the laws that govern analogous chemical receptors found in biological systems. In this chapter, we discuss these laws and how they apply to the invention of selective biomimetic surface coatings for chemical-sensing systems. Throughout this discussion, specific attention is made to the theory of molecular recognition as well as categorical spotlights of significant recent developments in molecular imprinting, supramolecular analytical chemistry, and bioreceptor coating research. In this survey, the achievements, trends, and prospects for improved biomimetic molecular recognition are highlighted.

Within the context of chemical sensing, a material that mimics biological function is highly desired, as biological selectivity is unsurpassed. The main concepts in achieving selectivity are explored in this chapter by first providing the foundational background theory for noncovalent molecular interactions. Subsequent case examples in this chapter draw attention to biomimetic materials for chemical sensor coatings. By learning from the principles provided by nature's chemical receptors, researchers have imitated those ideas to design selective molecular recognition elements to allow binding of specific target analytes to the surface of chemical sensors. This use of biomimetic receptive materials has led to the rapid increase in the development of genomics and proteomics screening tools and selective analytical systems. In addition, the last decade has seen dramatic improvements in our ability to selectively detect small-molecule compounds including pesticides, explosives, and other chemicals that may harmfully impact our health through water and food stocks. While chemical sensing remains a major application of chemically receptive biomimetic materials, it is important to keep in mind the broad impact that biomimetic molecular recognition has on the world around us. Research in

Biomimetic Approaches for Biomaterials Development, First Edition. Edited by João F. Mano.
© 2012 Wiley-VCH Verlag GmbH & Co. KGaA. Published 2012 by Wiley-VCH Verlag GmbH & Co. KGaA.

catalysis, separation, and even therapeutics relies heavily on the same principles, and work in these fields is impacted greatly by increasing our understanding of the mechanisms of biomimetic recognition. It is the goal of this chapter to provide the core understanding of these principles and to present the current status of the field in terms of recent advances, perceived limitations, and future prospects for improving biomimetic surfaces for selective molecular recognition.

6.1.2
Biological Chemoreception

Biological molecular recognition is ubiquitous in our daily lives and is a vital part of maintaining the functions necessary for sustaining life. Among the various aspects of molecular recognition, two well-studied biological systems of particular interest to selective biomimetic materials research are the immune response system and the olfactory/gustatory system.

In the immune system, antibodies provide exceptionally selective binding capabilities in order to counteract the presence of foreign objects, also known as *antigens*, in the body. The antibodies, produced by plasma cells, are either secreted in order to search for antigens or bound to the membrane of B cells in order to act as cell-based infection sensors. While the structure of most antibodies is generally similar, the light chain ends of the large "Y"-shaped protein complex contain a variable domain that is responsible for molecular recognition of the antigen and is termed the *paratope*. The paratope is highly variable allowing for millions of different amino acid sequences at this region. Slight changes in the sequence of the paratope will produce a unique structure that can serve as a receptor (analogous to a lock) for a specific antigen (analogous to a key). The diversity of antibodies for specific antigens can therefore be extremely vast, thereby allowing the immune system to respond effectively by binding to an unknown antigen entering the body. This concept of a biological lock and key proposed by Emil Fischer over 100 years ago to describe the binding of natural receptors [1] has been a fundamental starting point for chemical sensor coating technologies, as the ability to mimic the extraordinary selectivity achievable by antibodies is a highly sought after goal to eliminate the occurrence of false-positive signals, which plague most engineered sensing platforms.

Another biomimetic approach to achieving surfaces capable of molecular recognition is borrowed from our body's own chemical-sensing systems of smell (olfactory system) and taste (gustatory system). The key components of these systems are G-protein-coupled receptors that are bound to the membrane of cells and activate the opening of ion channels to a certain degree upon analyte binding, depending on the extent of cyclic AMP formation induced by the extent of analyte binding. In this respect, the olfactory and gustatory system are presently believed to achieve recognition of molecules in a different way than that of the immune system. While the immune system has a vast number of potential receptors for different antigens, the olfactory systems possess approximately 1000 different receptors (of which not all are functional). In a set of paradigm-shifting experiments by Buck and Axel, each odorant cell was found to possess unique G-protein-coupled odorant receptors that demonstrated the ability to be activated to different degrees by interaction with different target odorants. As opposed to the idea that one unique receptor exists for a unique odorant molecule, findings have demonstrated that a single receptor can bind to multiple different odorants, thereby creating the concept of a complex cross-reactive array. This combinatorial interaction code offers an explanation for our ability to sense thousands of diverse smells despite having a far smaller number of different receptors. Currently, there is no theory that completely explains olfaction; however, it is widely believed that different chemical features of odorant chemicals are perceived by different receptors according to the characteristic shapes of the molecules, termed *shape theory*, or due to the molecular vibrational frequency of the odorant as outlined in the vibration theory [2]. The resulting interactions between an odorant and several different receptors occur to varying degrees, which create a unique odor "fingerprint" that is processed as a specified smell attributed to the chemical. Importantly, this cross-reactive nature of the olfactory system has inspired researchers to implement pattern-recognition-based approaches to biomimetic chemical-sensing system with great success.

6.1.3
Host–Guest Interactions

In order to make use of the evolutionary advantage of nature, researchers have tried to identify and imitate the underlying mechanisms of biological molecular recognition. In the field of host–guest chemistry, particular importance is given to creating complementarity between the host (receptor) and the guest (analyte) in terms of size, shape, and functionality in order to maintain selectivity in molecular recognition. The pioneering works by Pedersen, Lehn, and Cram in the development and understanding of synthetic receptors led to their receipt of the 1987 Nobel Prize for chemistry. The impact of their works led to the use of specific control of receptor structures to limit their association with different target analytes via noncovalent interactions. This form of molecular complementarity is analogous to the original mechanism proposed for biological recognition akin to lock and key binding, which has become widely accepted as an effective means of molecular recognition. Exceptions and distinctions to this theory have resulted in dividing the mechanism into several model systems. The following sections highlight the different aspects of three such model systems: lock and key, induced fit, and preexisting equilibrium (Figure 6.1). The divergence among these models arises when the receptor is entirely rigid, conformable between two states, or considerably flexible, respectively.

6.1.3.1 Lock and Key
To understand molecular recognition in the case of lock and key binding as proposed by Emil Fisher in 1894, we must take the case of a purely rigid receptor and target. In terms of selectivity, the predefined binding pocket may accommodate only specific shapes and functionalities of guest molecules. In addition to shape

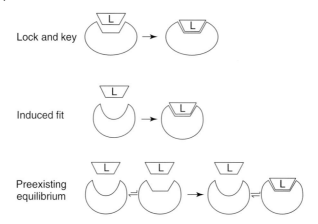

Figure 6.1 Model mechanisms for receptor-based recognition of ligands, L.

and functionality, the size of the receptor site essentially restricts the mismatch of larger competing molecules. However, competing analytes possessing molecular similarity to the guest but with smaller sizes than the intended target may still exhibit binding to the predefined receptor site. In practice, this biomimetic mechanism is widely used in the design of macrocycle systems for molecular sensing and separation. Moreover, the constrained nature of the macrocycles has proven an enhancement in binding stability due to the preorganized rigidity of the receptor framework. This is attributed to both the lack of entropic loss that would be exhibited if using flexible or unconstrained receptors and the avidity effect requiring multiple binding sites to simultaneously break in order to release the guest.

6.1.3.2 Induced Fit

Formerly proposed by Daniel Koshland in 1958 as applicable to certain enzyme–substrate interactions, induced fit proposes a mechanism for increasing the stability of correct host–guest interactions in conformable systems. Illustratively, consider a system in which the binding site is slightly larger than the size of the intended target. Initially, weak binding interactions occur between the host site and the guest molecule that stimulate the host to undergo a small conformational change resulting in the formation of stronger molecular recognition events with the correct target guest. The energy cost required for the host site to conform increases quickly for greater bending, thereby requiring only small mismatch difference to effectively enhance the selectivity [3]. If a smaller competing molecule enters the host cavity, then the amount of binding required to fully interact is larger than that required for the intended target molecule and hence more energetically unfavorable. Stabilization of intramolecular and intermolecular bonds helps to ensure the specificity of the host–guest system. The selectivity enhancement of this approach is used widely throughout biological systems in which induced fit deformations have been observed to be as large as tens of angstroms. Several biomimetic molecular recognition systems have made use of this biologically inspired mechanism by utilizing elements that induce conformational changes for enhanced binding to achieve molecular tweezers, switches, and receptor systems.

6.1.3.3 Preexisting Equilibrium Model

The work of Foote and Milstein challenged the assumption that the antibody's binding site has a single structure. In their pivotal experiments, they supported the existence of equilibrium between several antibody conformations, in which the guest molecule bound preferentially to a single structural isomer [4]. The preexisting equilibrium model (PEEM), ultimately based on the energy landscape and molecular dynamics theory of protein folding and binding, proposes that the natural state of a receptor exhibits a variety of different binding site conformations [5–7]. Throughout these conformational fluctuations, a guest molecule will bind to an active conformation of the host thereby biasing the equilibrium toward the bound conformation [8]. Work by James *et al.* [9, 10] demonstrated that different conformations of unbound antibody can bind to structurally different guest molecules, in which one binding mode exhibits a shallow groove, while another form displays a deep hole. This type of conformational diversity results in cross-reactivity, which attributes to multispecificity or, in other cases, nonspecific interactions. In another example, researchers examined an antibody capable of binding various sets of DNA [11] to find that the binding site is fundamentally unstructured in the absence of the guest but after binding the host site becomes structurally ordered. It is believed that the conformational flexibility allows the host to dramatically modify the size and shape of the binding site to enhance complementarity to other guest molecules, thereby accounting for multispecificity [12]. Schultz and coworkers [13] have found that mutations in such flexible domains can result in a rigid binding site with optimal guest complementarity capable of exhibiting much higher-affinity binding (30 000 times) than their flexible counterparts. In addition, they found that locking the antibody into the fixed confirmation resulted in the introduction of structural features that interfered with nonguest interactions thereby rendering increased specificity.

6.1.4
Biomimetic Surfaces for Molecular Recognition

It may not be surprising that in living systems, most molecular recognition events take place at interfacial environments [14]. On the basis of the biological reaction mechanisms described above, researchers have explored a number of biomimetic material designs in order to achieve effective sensor coatings. Perhaps, the oldest engineering of molecular recognition is based on the lock and key mechanism through the use of molecular imprinting. As discussed later, this field utilizes polymeric surfaces with complementary features to target guests in order to achieve surfaces capable of binding targets with selectivity among enantiomers. Supramolecular chemistry has also provided foundational understanding of the lock and key mechanism of binding through the design and analysis of structurally constrained macrocyclic receptors for target-specific binding.

Recently, supramolecular chemists have also applied preexisting equilibrium to analogous binders with promiscuous cross-reactivity, in which a complex can still be formed even if the guest molecule does not perfectly match the host site. While the quality of molecular recognition in terms of selectivity is compromised, researchers take advantage of the multispecific binding in order to create chemical sensor coatings, which are capable of recognition of distinct binding patterns in a biomimetic manner, as described previously. In such cases, a flexible recognition element, which does not exhibit perfect structural matching, may provide advantages over rigid host receptors in order to increase the diversity of specificity. Ultimately, these biomimetic systems depend on the flexibility and molecular similarity of the host and guest, respectively.

As for the induced fit model of binding, researchers are actively pursuing the design of receptive materials that exploit the free energy barrier imposed by the conformational change to create a threshold for discriminating target binders from nontarget binders. This biomimetic concept utilizes structural mismatches between the host and guest to enhance recognition specificity among a noisy background of competing chemicals [15]. As this is highly applicable to sensor coating design, researchers have made use of biopolymer receptors engineered to contain key structural mismatches. These mismatches can reduce the initial interaction of both the correct and incorrect targets; however, the inhibitory effect on the rate of incorrect target binding is much more significant. As a result, these biomimetic receptor systems have demonstrated astonishing selectivity as sensor coatings.

In addition to designing selectivity of the receptive elements using the strategies mentioned above, other aspects must be engineered into biomimetic surfaces for sensing purposes. During the past several decades, extensive research on molecular recognition at interfaces has been performed using monolayer and lipid assemblies to identify the critical surface effects to be considered in order to maintain efficiency of the selectivity elements [14]. Biological systems have optimized their molecular recognition formats over eons of selection and evolution. To accommodate sufficient space in biological systems, for instance, B cells are known to clear the presence of surface groups in the area of thousands of angstroms surrounding attached antibodies [16]. Biomimetic sensor coatings must also be designed to minimize the occurrence of competition or steric hindrance associated with surface saturation of receptive motifs. Surface attachment itself is an important consideration that can often become problematic, as it depends highly on the material of the sensing platform. Depending on the transduction mechanism, spacer sequences may be placed between the sensing surface and the receptive motif to deter surface fouling effects or to increase hydration [17]. Sufficient spacing between the active site and the surface can also ensure effective presentation of the receptive motif. The functional site of active receptors must be properly displayed in order to achieve useful molecular recognition. Antibodies, if randomly attached to a sensing surface, will have a fraction of the recognition sites inaccessible. In order to accommodate exposure of the active binding site, attachment approaches to control the directionality of surface receptors generally utilize site-specific chemical modifications (i.e., biotin) or functionalization by affinity tag binding.

As an example, researchers have also utilized the natural Fc binder, protein A, to anchor antibodies to a sensing surface in an orientation that provides full exposure of the antigen binding site [18]. Maximizing the accessibility of the recognition site by controlling host site orientation is often overlooked, although it is a major consideration for the effective design of biomimetic sensing surfaces.

6.2 Theory of Molecular Recognition

6.2.1 Foundation of Molecular Recognition

As we now know, molecular recognition is an essential component of biological systems. While often referred to as a *binding event*, this is a bit of an oversimplification. One of the pioneers of host–guest chemistry, Jean-Marie Lehn, has defined molecular recognition as involving both binding and selection through structurally well-defined intermolecular interactions [19]. Thus, complementarity in size, shape, and functionality akin to that described in the lock and key model is fundamental to achieving molecular recognition [20]. Throughout biological and biomimetic systems, receptive sites often present themselves as a concave surface laden with acids, bases, amino acids, metal ions, and so on to offer noncovalent interactions in a structurally controlled space to receive the complementary molecular guest [21]. From a steric viewpoint, a complementary guest implies that the van der Waals radii of the guest molecule's atoms cannot go beyond that of the host and correspondingly should serve to fill the host cavity as densely as possible [22]. Although complete structural complementarity may not be required for molecular recognition [23], this section considers such molecular recognition system in order to begin to explore the elements and thermodynamics of the binding event as well as the means of analyzing and quantifying the host–guest molecular interaction.

6.2.2 Noncovalent Interactions

Effective functional complementarity between a recognition site and a guest molecule relies on the occurrence of favorable noncovalent interactions. Forms of noncovalent interactions include hydrogen bonding, hydrophobic forces, metal coordination, van der Waals forces, aromatic interactions, and electrostatic forces. Given a target molecule, there exist a large number of possible configurations of host receptors that can facilitate recognition through these interactions. Spatial optimization, as mentioned, can be designed to accommodate maximization of attractive dispersion forces. However, even if surface complementarity is low, effective molecular recognition elements can still be designed through electrostatic complementarity and hydrophobic complementarity. The presence of complementarity of the molecular electrostatic potential relies in essence on maximization of the ionic and polar interactions, such as hydrogen bonds, for a given target molecule. Long-range electrostatic forces, such as coulomb forces, may lead to strong accumulative binding and can contribute extensively to the overall binding energy [24]. It is noteworthy that the neighboring charges of individual contact atoms in a host–guest complex need not necessarily show complementarity because from a coulombic aspect, it is the property of the entire independent host and guest systems that is most dominant [25]. As such, the potential in a host cavity will benefit from being opposite in sign to that attributed to a guest molecule. While significant, these coulomb forces can diminish other important interactions responsible for precise molecular recognition [24]. Such interactions may include metal coordination, aromatic interactions, and hydrogen bonding. Metal binding can subtly change or constrain the confirmation of a host site and even confer metal ion specificity and affinity. The existence of aromatic interactions is ever present in biologic recognition systems, as pi–pi stacking is a major stabilizing factor among binding events, particularly in those involving DNA and RNA. As for hydrogen bonding, it is essentially electrostatic in nature and is a highly important interaction necessary for tuning the selectivity of the recognition site.

The hydrogen bond is effectively an attractive interaction between a hydrogen acceptor and a hydrogen atom from a functional motif that is more electronegative than hydrogen, such as fluorine, oxygen, and nitrogen. Atoms, groups of atoms, anions, pi bonds, metal complexes, and even pi electrons from aromatic rings may act as hydrogen bond acceptors [26]. The strength of the interaction depends on the nature of the bonding components, for example, weak polar interactions may involve pi-system acceptor groups and CH donor groups [27]. In addition, the interaction has directional preference and usually becomes stronger as the bond angle approaches linear. As a result, researchers can manipulate the location and charge of the hydrogen bonding sites to change the capabilities of synthetic receptors to function in different solvents [28].

Aromatic components can offer a variety of bonding interactions, including the formation of hydrogen bonds via their pi system as described above, pi–pi interactions, and cation–pi interactions, which act as major forces to stabilize host–guest associations. Aromatic groups are capable of forming T-shaped arrangements allowing edge-to-face interactions as well as parallel-displaced arrangements allowing aromatic pi–pi stacking interactions [29]. High edge-to-face attraction is found when partial positively charged H atoms interact with aromatic rings enhanced by pi-electron density. Aromatic partners may also form charge-transfer complexes under parallel alignments. These pi–pi stacking interactions are affected by the electrostatics of component atoms in addition to the geometric alignment of molecular dipoles. These contributing factors determining how the pi systems interact and thus affect the strength of the pi–pi interaction based on the stacking geometry. Perhaps the strongest interaction arising from aromatics occurs between a cation in conjunction with the aromatic face of the pi system [30]. Cation–pi interactions, which have been show to contribute to ion selectivity in potassium channels [31], depend critically on the aromatic system and charge of the cation. A more positively charged cation will interact more strongly with the electron-rich pi system resulting in a stronger molecular interaction. As such, the nature of the aromatic components is important in determining the strength of the interaction, with stronger cation–pi binding occurring when electron-donating groups are present. In a related manner, one may consider the formation of anion–pi interactions in aromatic systems if strong electron-withdrawing components are placed within the pi system [32]. While biological substrates and cofactors have predominantly been anionic, researchers have recently explored this area to implement the recognition of specific anions for applications in chemical-sensing materials [33]. In contrast to the true aryl pi system interaction seen with cations, anion binding is considered to be primarily due to local dipoles induced by the added substituents and not due to the attraction with the aromatic pi system [34].

Finally, hydrophobic complementarity, which relates to the tendency of nonpolar groups to associate in an aqueous medium, also plays a critical role in molecular recognition. In general, this phenomenon results in the minimization of contact between nonpolar and polar regions. The burial of charged groups in a nonpolar recognition site, for instance, is highly unfavorable. This manifestation of the "similis simili gaudet" principle offers that association of similar species (i.e., ionic, polar, or nonpolar components) contributes to a more stable system as compared to association of dissimilar components [35]. Hydrophobic interactions by themselves may be viewed as amorphous and nonselective; however, they can be used in conjunction with spatial restriction to confer stability and limited specificity. With the support of polar interactions, which have optimal distance and directional requirements, highly selective molecular recognition is achieved [36] (Table 6.1).

6.2.3
Thermodynamics of the Molecular Recognition Event

Molecular recognition obeys the same rules as any spontaneous process and therefore will occur given a negative change in free binding energy (ΔG). One may hope to engineer a host receptor for a given target molecule such that the free

Table 6.1 Strengths of representative bonds and interactions in molecular recognition processes.

Magnitude of interaction energies

Type of noncovalent interaction	Strength of interaction (kJ mol^{-1})
Hydrogen bonding	10–65
Coulomb interaction	250
$\pi-\pi$	<50
Cation–π	5–80
Ion–dipole	50–200
Dipole–dipole	5–50
van der Waals	<5

Source: Adapted from Hoeben et al. [37].

energy of binding is favorable; however, this remains a difficult task. To attempt such an achievement, it is important to first demystify the free energy of binding by looking at the components of enthalpy (ΔH) and entropy ($-T\Delta S$). These two components are inherently related to the structure of the host and guest as well as the solvation effects of the system, thereby making the design of a molecular recognition element a daunting task [38]. One design factor that is widely accepted is the burial of a nonpolar site within the recognition complex, as this favorable free energy contribution of desolvation is conventionally attributable to a gain in the entropy due to the randomized release of solvent and/or ions [36]. While this exists, nonpolar groups also directly provide an enthalpic driving force for binding by a nonclassical hydrophobic effect [29]. The favorable enthalpy term arises during complexation from at least two factors. The first enthalpic contribution due to nonpolar burial is caused by an increase in solvent cohesive interactions (hydrogen bonding) between water molecules on release of the surface-solvating water molecules into the bulk. This favorable outcome is directly a factor of nonpolar surfaces participating in fewer strong hydrogen bonds. The second aspect of this replacement is that nonpolar components will have less favorable interactions with water as compared to after complexation in which attractive dispersion forces between nonpolar components at the guest–host interface will promote stability due to similar polarizabilities [29]. When we consider polar binding sites at the surface of the host and guest molecule, there may exist a negative contribution to the binding enthalpy attributable to desolvation of these polar binding sites. In such circumstances, it may be advantageous to design hydrogen bond donor sites with close proximity to the host recognition site in order to spatially limit its solvation capabilities [28]. Yet another alternative strategy may be through the design of CH/pi hydrogen binding interactions in the host–guest system, as this is expected to work well in a variety of solvent systems [39].

Solvation effects as well as structural effects are often interconnected phenomena when considering the thermodynamic aspects of molecular recognition. For instance, a host–guest complex akin to the lock and key model provides significant structural complementarity, which is more effective than a loose complex in terms of maximizing the strongly distance-dependent attractive van der Waals interactions [40]. Conversely, a cavity larger than the guest molecule will provide significant desolvation entropy on binding. In both circumstances, an additional entropic term must be accounted for that is associated with the initial state of the host and guest relative to the freezing of bond rotations as well as loss in translational and rotational freedom of the interaction component on complexation. This inherent entropic cost is hard to avoid but may be minimized by the use of constrained recognition sites. Conversely, positively cooperative interactions exist throughout biology; for instance, in the streptavidin–biotin complex, where structural tightening, which unfavorably decreases conformational entropy, results in the formation of additional interactions with shorter distances, which is enthalpically favorable, thereby providing a net benefit to the molecular recognition event [41]. Alternatively, a partially bound state model also exists to explain such net losses in free energy on complexation, in which the balance shifts from an entropically favorable state to a less mobile enthalpically favorable state. This model is proposed to account for mechanisms of enthalpic chelate effect and enthalpy–entropy compensation found in certain biomolecular recognition events [42].

6.2.4
Putting a Figure of Merit on Molecular Recognition

Quantification of molecular recognition relies on identifying the preferential affinity of a host receptor for a particular guest molecule as compared to nonintended competing guest molecules. This affinity, generally quantified as the binding constant, relates the ratio of unbound and bound molecules at a state of equilibrium between binding and dissociation. Particularly for bimolecular association, the ratio of host–guest complex concentration to the product of host and guest concentrations provides the association constant, K_a, while the inverse represents the dissociation constant, K_d, which is commonly used to represent the affinity. Because the change in Gibbs free energy is related to the equilibrium association constant, $\Delta G = RT \ln K_a$, the thermodynamic parameters of the association event may be calculated. In order to measure the binding constant for such a bimolecular recognition event, one must find a way to measure the amount of complex formed over a range of initial guest concentrations. Experimentally, it is possible to identify the host–guest complex concentration as a function of the initial concentration by various quantitative techniques including UV–vis and fluorescence spectroscopy, nuclear magnetic resonance spectroscopy, surface plasmon resonance (SPR), Raman spectroscopy, and isothermal titration calorimetry. Using approximations and measuring the equilibrium concentration of host–guest complex formed over a range of starting guest concentrations, a binding isotherm plot can be obtained in order to identify K_a of the molecular recognition event. Commonly considered the gold standard in accurate measurement of the true equilibrium data, isothermal titration calorimetry (ITC) directly determines the heat flow on host–guest complexation with which we can measure the enthalpy, binding affinity, and the binding stoichiometry of the interacting system. This data is subsequently used to calculate the entropy and Gibbs free energy of the system. Generally, this approach works well for recognition events with binding constants in the range of millimolar to nanomolar. Other techniques, including SPR detection systems, measure the kinetic rates of association and dissociation from which the binding constant can be determined; however, these techniques are arguably less reliable. In general, though, thermodynamic values are inherently sensitive to preparation conditions such as salt concentrations and buffer. For instance, weak solvation of hydrogen bonding sites in certain nonpolar organic solvent systems can cause strong enthalpic driving force for binding, which may not be realistic to a biomimetic chemical-sensing surface. As such, subtle experimental variations can make obtaining consistent thermodynamic values of enthalpy and entropy a challenge. In contrast, observing the binding constant and hence the net overall free energy change remains a more dependable strategy to quantify the molecular recognition event without deconvoluting the associated structural changes, long-range interactions, and desolvation–solvation considerations.

6.2.5
Multiple Interactions: Avidity and Cooperativity

The above-mentioned case of a simple biomolecular recognition event provides a relatively easy scenario of a single host interacting with a single guest. In many biological systems, particularly protein–protein interactions, a host–guest interaction may occur via more than one recognition event. For instance, if both the host and guest sites exist as a dimer, the initial binding event brings the host and guest in close proximity to facilitate the subsequent incidence of binding at the other site. This occurrence of multivalent binding provides an additional stabilizing component to increasing the host–guest interaction by lowering the off rate of the complete system. This effect, known as *avidity*, is often used to enhance the performance of biomimetic receptive surfaces for chemical sensing and is a natural component of antibody-based recognition. As compared to the case of dissociation from a monovalent interaction, multivalent antibodies are less likely to diffuse away from an antigen when a single site dissociates, thereby making it more likely for the bond to reassociate. The multiple simultaneous interactions provided by avidity thereby provide a strengthening mechanism for molecular interactions that affect the affinity of the entire system [43]. Importantly, the affinity of the entire system cannot quantitatively be determined as just a collection of the individual monovalent interactions, as there must exist a linker between multimeric binding sites. As such, extensive design considerations must be taken into account to optimize the linker in terms of rigidity, length, orientation, and proximity between binding sites, depending on the characteristics of the system [44].

Another phenomenon that may occur when more than one binding site exists on a receptor is cooperativity, which can arise from conformational changes or long-range interactions as a result of an initial binding event [15]. Specifically, if there is an increase in the binding affinity of subsequent guest molecules as the result of prior guest binding, then the receptor exhibits positive cooperativity. Conversely, if the binding of the first guest molecules causes a decrease in affinity for the incoming guests, then it results in negative cooperativity. Assessing the effectiveness of these systems requires a more complex formulation than that previously described for the case of simple bimolecular interactions; however, receptor–target models accounting for components of multivalency and cooperativity have been proposed [15, 45]. The use of positive cooperative recognition in chemical sensing has been explored as a viable molecular recognition design strategy, and it is appearing with increased prevalence in the biomimetic receptor community for the creation of high-affinity interactions via binding-induced strain and folding [46, 47].

6.3
Molecularly Imprinted Polymers

6.3.1
A Brief History of Molecular Imprinting

From our prior discussion, we see that biology is replete with molecular recognition events based on host–guest complementarity in size, shape, and chemical functionality. Biomimetic recognition systems may utilize these same principles in the creation of synthetic chemical receptors, which may have an increased stability than their biomolecular counterparts. The use of molecular-imprinting-based biomimetic sensors thereby provides an attractive alternative. The concept of molecular imprinting has existed for many years as a technique exploited through silica matrices, as identified by the Soviet chemist M.V. Polyakov in 1931 [48]. Four decades later, the molecular imprinting technique was implemented in organic polymers to create surfaces with predetermined target selectivity [49, 50]. In general, predetermined target template molecules are present during the casting stage and removed to provide an imprinted cavity in the molecularly imprinted polymer (MIP). The MIP then contains a recognition site available for subsequent interaction with the target molecule [51]. The use of molecular imprinting received increased attention in the area of biomimetic receptive surfaces after the creation of theophylline-recognizing MIPs that showed strong binding and cross-reactivity profiles similar to those of antibodies [52]. The ability to use molecular imprinting for selective molecular recognitions on a synthetic surface continues to be of interest in chemical sensor coating research as selectivity remains the key component in attaining effective detection systems. In addition, the rigidity of the polymer matrix and tolerance to extreme environments allow these surfaces to remain stable in practical scenarios.

6.3.2
Strategies for the Formation of Molecularly Imprinted Polymers

MIPs can be created with high-affinity binding sites for a target molecule by two different strategies: self-assembly and preorganization (Figure 6.2). Generally, these strategies utilize covalent/noncovalent preassociation complexes formed between a template molecule and the proper monomers to provide complementary size, shape, and chemical functionality [53]. In the self-assembly approach of molecular imprinting, intermolecular noncovalent interactions are formed between monomer precursors and the template molecule. The choice of monomer is important, as this dictates the complex host–guest binding interactions that will be present in the final MIP. On polymerization, the shape and distance of these interactions are structurally constrained depending on the extent of cross-linking. The template is then extracted and the shape of the complementary site is maintained in the polymer matrix, with the arrangement of the functional groups optimized to provide a recognition site for the intended target. In the preorganized molecular

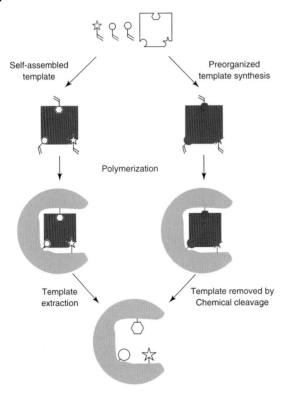

Figure 6.2 General strategy for the formation of molecular imprinted polymers by either using a template capable of self-assembling with the polymerizable monomers or synthesizing a polymerizable template.

imprinting approach, monomers are linked to the template molecule via cleavable covalent bonds before polymerizing the polymer matrix. After polymerization, the template molecule is rigidly locked into the MIP. Cleavage of the bonds holding the template allows its extraction from the polymer matrix, resulting in a polymeric cavity with recognition site complementarity. Combinations of these strategies utilize a covalent template–monomer complex for imprinting, but on cleavage and removal, an entirely noncovalent interaction is used for binding [54]. This approach provides a more homogeneous recognition site with faster binding kinetics than the conventional covalent approach [55].

6.3.3
Polymer Matrix Design

After deciding on a self-assembly or preorganization approach, the polymeric system must be selected. The choice of monomers and solvent are critical design parameters in achieving an effective MIP. It has become a general rule that the molecular recognition capability of an MIP is best when operating in the same solvent as that used during polymerization [56]. The use of acrylic and vinylic monomers offers a variety of functional motifs for noncovalent template interaction, and the monomers are thus widely implemented when designing a polymer matrix based on the self-assembly strategy [57]. In the preorganization approach, there exists an intrinsic limitation in the type of template target molecule that can be imprinted. Since the process requires covalent attachment, the template must be prefabricated into a polymerizable derivative of the intended target molecule. Classically, this preorganization approach has utilized template–monomer complexes formed by carboxylic ester linkages, boronate esters, ether linkages, imines, and ketal bonds, among others [53, 58, 59]. Despite the limitation in the functional groups that can be imprinted using the preorganization approach, effective MIP designs have demonstrated highly selective molecular recognition capabilities [60]. This has been attributed to the distinct bonding between the template and monomer during polymerization, which results in a higher uniformity of the recognition sites as compared to the self-assembly approach [61]. The self-assembly approach, however, is more mimetic of antibody-based recognition, in that noncovalent bonds dominate the recognition event.

In addition to having an appropriately functional monomer, it has become a general rule that the molecular recognition capability of an MIP is best when operating in the same solvent used during polymerization [56]. The solvent's purpose is to both dissolve the components and generate a highly porous structure to allow transport of the template and later provide access to the recognition site. From the Flory–Huggins solution theory, a thermodynamically good solvent will have increased polymer–solvent interactions, which may create more porous polymeric structures with higher specific surface areas as compared to a thermodynamically poor solvent [62]. In addition, the solvent must not hinder the formation of the necessary template–monomer interactions. Nonpolar solvents are generally used in noncovalent imprinting to promote the formation of template–monomer complexes. Water, often believed to prohibit the formation of template–monomer interactions, is usually replaced with perfluorocarbons [63]. Typical solvents include acetonitrile, chloroform, and toluene with a monomer/solvent ratio of 3 : 4 by volume [64]. Aside from the monomer and solvent, researchers have used other components to improve the design of the polymer matrix. Particularly, researchers have created cleavable sacrificial spacers, which are covalently linked between the monomers and template molecule, to reduce steric hindrance in the recognition site. With these elements in place, the formation of a structurally defined stable polymeric framework can be achieved through effective cross-linking and template removal to create a biomimetic receptive MIP or as it is often referred to a *"plastic antibody."*

6.3.4
Cross-Linking and Polymerization Approaches

The most commonly used polymerization method for the formation of MIPs is the free radical vinyl polymerization [65]. As mentioned above, there are a wide variety of monomers available for this polymerization scheme with diverse chemical structures and functionality. In addition, these reactions have the added benefit of occurring under mild reaction conditions and are not easily susceptible to impurities. One important caveat is that different vinyl groups may be incorporated at different rates in the polymeric system depending on the functionality. An alternative, yet still successful, approach of step-growth polymerization has been used by Dickert *et al.* to make polyurethane-based MIPs with biomimetic recognition capabilities toward sensing a variety of templates including vapor targets [66], oils [67], and even whole cells [68, 69]. Among the polymerization systems, one parameter that seems most pertinent is the ratios of monomer, cross-linker, and template, which is typically 8 : 40 : 1 respectively [64]. Often in noncovalent MIPs, the types of monomer implemented and their ratios are varied using automation in order to optimize the recognition sites formed in the MIP [70, 71]. Importantly there also remains an optimum cross-linking density for achieving selectivity, which will depend implicitly on the template as well as the matrix design strategy [72]. For example, attempts to make MIPs for larger biomolecules such as proteins and whole cells have been found to be more successful if using a lower cross-linking density as compared to MIPs for small molecules [64]. In principle, the type and density of cross-linker used is important in determining the stability of the binding site as well as the mechanical properties and porosity of the entire polymer matrix.

Another key point in the formation of MIPs is the use of initiators for free radical polymerization. In such systems using noncovalent MIP formation, polymerization strategies have been developed that allow the initiator concentration and temperature to be extremely low since hydrogen bonds forming the template–monomer complex are thermally less stable [72]. In such instances, photochemically active initiators are favored since they operate well at low temperatures. In contrast, if the template molecule is photosensitive, then a thermally triggered initiator may be preferred.

6.3.5
Template Extraction

After the formation of a constrained recognition domain by polymerization around the template molecule, removal of the template is necessary for use in ensuing molecular recognition. The molecular recognition sites are created in the polymer cavity by properly located functional groups that provide complementary interactions with the intended target molecule [73]. For removal of template from MIPs formed by preorganization, the process may require harsh conditions for template desorption to break the reversible covalent bonds. Covalent-based imprinting is therefore generally more successful if the bonds used between the functional group and template are cleavable under relatively mild conditions [54]. By chemical cleavage of the supporting covalent bonds, the template is released from the MIP and the corresponding functional groups located in the recognition site cavity are free for future interactions [54]. In some circumstances, the embedded cavity is formed by a combination of the covalent and noncovalent approaches. In this approach, sacrificial spacer groups, used to improve subsequent noncovalent binding geometries, are cleaved and eliminated during template extraction [65]. In the case of MIPs formed by purely noncovalent interactions, the template is removed from the embedded recognition site in a much easier manner. Generally, this is accomplished by several extraction cycles using a simple solvent system such as methanol/acetic acid followed by methanol and afterward drying under vacuum [64].

In some instances, it has been found that target molecules unintentionally remain in the MIP after the initial extraction step despite thorough washing. Larger templates are usually problematic, as they can easily become trapped in the polymeric matrices. A biomimetic endeavor to overcome this problem employs a fragment or the key epitope of the target analyte as the template structure [74, 75]. This approach is parallel to the immune system in which biomolecular recognition events focus on recognizing key features of the antigen rather than the entire surface of a protein. Another difficulty related to the template extraction concern may take place during the analytical use of the MIP. After target exposure, removal of the target from the recognition site may be problematic regardless of continuous washing. To compensate for this predicament, it has proven effective to use template molecules that are analogous to the intended target analyte; however, this enhancement in sensitivity may be at the cost of selectivity [76].

6.3.6
Limitations and Areas for Improvement

While MIPs provide the advantages of being made from chemically and physically stable materials and allowing tailor-made recognition sites with high specificity, there are still areas for improvement. The practical use of MIPs as biomimetic receptive surfaces may require exposure to aqueous buffer. As these polymers are originally imprinted in nonpolar solvents, aqueous environments can alter the selectivity of the recognition sites, hence limiting the use to nonpolar organic solvents [64]. To alleviate this problem, researchers have proposed the use of metal complexes and covalent imprinting techniques in polar solvents to promote the formation of strong bonds between the polymer and template target molecule [72, 77]. Another inherent problem with MIP is the randomized placement of the template molecule through the system during the polymerization process. After extraction of arbitrarily oriented templates, a heterogeneous population of binding sites exists with varying affinities for the intended target molecule. Unfortunately, the percentage of high-affinity sites in a noncovalently imprinted polymer is usually less than 1% of the binding sites [64]. There is evidence that covalent imprinting can provide a more homogeneous population of binding sites [61]. Improvements in these areas are still thriving areas of research, so are the biomimetic recognition capabilities of MIP-sensing surfaces to detect large biomolecules, which will prove valuable in advancing medical diagnostic systems.

6.4
Supramolecular Chemistry

6.4.1
Introduction

Supramolecular chemistry has been fundamental in increasing our understanding of molecular recognition. This section focuses on biomimetic supramolecular chemistry approaches to the creation of receptive analytical systems. Ever since the works of Pedersen, Lehn, and Cram opened the arena for biomimetic host–guest systems, receptive materials have taken on a number of functional forms including clathrates, cryptophanes, and cyclotriveratrylenes As there is immense space for discussion on this topic, we limit the focus of this section to select examples of metalloporphyrins, crown ethers, and calixarenes in chemical-sensing surfaces. In this area, a large focus has been on anion recognition. The first such system was shown to operate via macrobicyclic amines through the combination of electrostatic interactions and directional hydrogen bonding [78]. Before that, the use of macrocyclic systems had seen a recent development as Charles Pederson found that crown ethers were able to spontaneously complex with alkali metals [79]. The cation solvation properties of crown ethers were then elucidated by a variety of polyether rings in terms of size and functionality. Asymmetric host molecules were developed soon after by Cram and Cram [80] as a new biomimetic class of crown ethers, which offered the capability of distinguishing between enantiomers by charge and shape complementarity. The work of Lehn and coworkers [81–83] also reflected on the importance of multivalent binding and constrained shapes in achieving molecular recognition. Particularly, their synthesis and analysis of cagelike cryptands demonstrated improved binding and selectivity over conventional crown ethers, thereby helping to open the area of supramolecular chemistry for the creation of concave receptor cavities that may be lined with binding sites intended for a specific guest [81, 84]. More recent work incorporating the use of hydrogen bonding components for chiral recognition proved to expand these previous approaches [85, 86]. Since its inception, supramolecular-chemistry-based recognition systems have grown in their selectivity and complexity. As an example, synthetic systems have been demonstrated to be complementary in topology and dimensions to hydrogen sulfate, with the cavity operating through an interlocked receptor mechanism [87]. These elaborate systems continue to lead the field of analytical supramolecular chemistry to improved recognition materials for biomimetic sensing surface.

6.4.2
Macrocyclic Effect

An important component of the recognition capabilities of supramolecular chemical systems, in addition to the ability to control shape, size, and chemical functionality, is exploitation of the macrocyclic effect. Macrocyclic host systems, cyclic molecules capable of multiple guest interaction, have enhanced complex formation stability over a linear multivalent counterpart. Because the interaction sites are prefixed into a preferred conformation, there is less configurational entropy for macrocyclic receptors to lose on binding. Also, the preconstrained macrocycle will not exhibit an enthalpy loss due to intramolecular repulsion that can occur in the linear case. Given the size of the guest and host coinciding, an additional benefit to macrocyclic constraint can be realized through the use of a proper solvent. Macrocycles are not as heavily solvated as compared to free linear analogs; hence, on complexation, they will have a more favorable solvation enthalpy contribution. This can also be applied to multimacrocycles, which explains the stronger complex formation for cryptates over conventional crown ether formations, as mentioned earlier.

6.4.3
Chelate Effect

The use of chelation has obvious advantages in supramolecular systems. In chelation, formation of the first bond between a host and guest capable of multiple interactions increases the local effective concentration of the nearby association sites. As a result, subsequent bond formation is more likely to occur in contrast to a monovalent binding event. Dissociation of a host–guest complex requires the dissociation of several bonds in a multivalent recognition event. On dissociation of the first bond, there remain points of attachment that prevent the immediate release of the guest, which is the source of the chelate effect, resulting in very fast bond reformation. Correspondingly, a macrocycle is still a chelate so it can benefit from this effect, though, to an even greater extent than a linear receptor. The dissociation of a complex is generally initiated by breaking intermolecular bonds one by one beginning from a particular end group. In a macrocyclic host–guest complex, the structure does not have an equivalent end group, resulting in very stable complexes with extremely low dissociation rates. Creating such systems by optimizing the flexibility and topology of the host–guest interaction allows for the creation of highly stable molecular recognition events.

6.4.4
Preorganization, Rational Design, and Modeling

The ability of synthetic receptor hosts to be tailor-made with well-defined shapes, sizes, and chemical functionalities is the key link between supramolecular chemistry and molecular recognition. Initially, it was the coordination features of many supramolecular chemistry building blocks that facilitated their use in the selective recognition of metal ions [88]. Preorganization of host structures around a coordination sphere provided effectively stable receptors with well-defined recognition site geometries for select target ions; however, too much preorganization may prove to make it difficult to initially form the complex [89, 90]. Since the addition of metal ions can cause different binding modes to occur, which in turn can affect the selectivity, rational design by controlling structural properties alone could be problematic. Although it may be appealing to design a host system solely based on rational size, shape, and chemical complementarity to a guest analyte, using a rational design approach in combination with computational modeling is a more effective means of creating new host systems [91]. Using modeling methods, researchers aim to estimate the binding capability as well as analyte selectivity of their designed receptor. One of the first quantitative studies between a rigid neutral cavity of a host receptor and neutral guest molecules revealed the expected outcome, as predicted by molecular dynamic simulations, to be different from that obtained by experimental comparison. After improving the van der Waals parameters from their standard force field model, the experimental binding behavior was predictable by the computational model. This reflects the importance of empirical evidence for proper improvement in parameterization of host–guest systems [92]. Depending on the level of detail desired, host–guest complex stabilization energies may be predicted via molecular modeling by force field methods such as MM3 [93]. While not definitive, modeling allows researchers to assess the feasibility of analyte incorporation or at least obtain a better understanding of host and guest geometries in relation to complex formation [94]. While the geometries and individual interactions of the binding event are important, predicting molecular recognition has added complexity due to cooperative effects, in which the strength of interactions is changed by the formation of other contacts [92]. In addition, solvent effects are often the key driving forces in complex formation; however, it has been noted that their contribution to binding may be accounted for by using a polarizable model. Innovative processes are continually arising for receptor–target modeling, such as the use of the shared-electron number (SEN) method for calculation of hydrogen bond energies [95]. Some agree that the complexity of host–guest molecular recognition is best to be broken down into basic components, which can be assigned enthalpies and entropies through a thermochemical model that can be confirmed empirically. Overall, the various modeling methods available are refining our understanding of natural and synthetic receptors and are widely being used in the design and modification of beta-cyclodextrins and supramolecular recognition elements for chemical-sensing applications.

6.4.5
Templating Effect

While supramolecular chemistry is diverse in its repertoire of host–guest systems, macrocycles represent a major synthetic component of biomimetic recognition systems. Researchers often utilize smaller, linear molecules to synthesize macrocyclic host systems. Creating a closed ring structure classically involves similar mechanisms as polymerization and requires extremely high dilutions to favor the intramolecular reaction. Cyclization was later improved by the template effect in which transition metals could be used to position reactive sites into a specific geometry. The choice of an appropriate linear molecule and transition metal template can allow for steric control and formation of a preorganized macrocyclic product. The properties of the resulting host system will depend greatly on the coordination and should therefore be carefully considered when designing a receptive host system. Crown ethers represent a class of host that has been templated by alkali metals, while porphyrins do not require a metal templating ion. Calixarenes, another class of oligomeric macrocycle, are typically formed via a catalyzed phenol-formaldehyde condensation. The reaction conditions and starting materials in each of these cases must be optimized to produce the desired cyclic structures.

6.4.6
Effective Supramolecular Receptors for Biomimetic Sensing

6.4.6.1 Calixarenes

Owing to their ease of modification with respect to shape and size, calixarenes are an important scaffold for designing a host system to selectively recognize specific guest molecules. They are typically shaped in a cuplike configuration characterized by a well-defined upper rim, lower rim, and central annulus (Figure 6.3) [96]. This shape enables calixarenes to act as effective receptors as a result of their natural cavities. Favorable host cavities are essentially formed by cyclic intramolecular hydrogen bonding of the hydroxyl groups to create the cup or cone shape of the calixarene [97]. Researchers have found it possible to control the host selectivity to ion or small-molecule binding by functional modification of the upper and lower rims [96]. This can effectively be used to lock the calixarene into a specific confirmation. Alternatively, it has been observed that the cavity can expand for large guests or shrink for smaller ones depending on the amount of plasticity in the upper-rim or lower-rim components [98]. As such, an optimal shape may conform on binding to certain guest molecules, representative of an induced fit binding mechanism [99].

Using force field calculations to model the recognition event, researchers have been able to design calixarene receptors with strong interactions to specific targets, including halogenated or aromatic hydrocarbons, based on host/guest inclusion principles [99]. Such polyaromatic macrocycles are also highly effective receptors for organometallic anions by charge pairing interaction via hydrogen bonding residues or by preorganization of the upper-rim charge [100]. Changing the upper- and lower-rim charges is known to effectively change the ion selectivity of the host

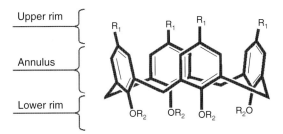

Figure 6.3 Basic structure of calix[4]arene in which the upper and lower rims may be functionalized.

system. For instance, calixarenes with oxygen donor atoms on the upper rim turned outward have demonstrated selectivity for binding of alkali ions [96]. In some cases, the upper-rim components may shield particular analytes from forming an endo complex. The appearance of aromatic groups on the periphery of the calixarene can also result in CH-pi stacking interactions that lead to exo-complex formation in which the guest binds outside the cavity [101], which may be the case seen in the binding exhibited by certain saccharide targets [102]. Another variable parameter used to control selectivity in the calixarene system is deciding the number of aromatic units in the cyclic oligomer. As an example, calix[4]arenes, tetramers having four phenolic units, show selectivity for sodium ions, while larger cavities, such as calix[6]arene, have higher-affinity binding for larger metal cations such as potassium and rubidium [96].

While aromatic monomers may be designed as identical units, interesting functional modifications of individual monomers and synthesis of low-symmetry calixarenes have opened the way to a range of selective receptor materials. In an example of induced fit binding, calix[6]arene cores have been functionalized with alternating amino arms that project inward after selective target binding, thus closing the calixarene [98]. This particular case reflects the important contribution of having a flexible and polarized hydrophobic scaffold for selective molecular recognition. Depending on the choice of functional group, modifications can instill chromogenic properties in the recognition material as has been demonstrated by Shinkai and others [103, 104]. Using calixarenes with multiple different functional groups incorporated on the lower rim has shown that the photophysical properties of these systems can be easily influenced by different target binding modes. Aside from chromogenic sensors, these biomimetic receptor materials have shown successful incorporation into a variety of sensor platforms, including quartz crystal microbalance and surface acoustic wave oscillator base sensors, as the selective coatings [97]. The shape and chiral selective capabilities of calixarene-based coatings have shown to offer a solution to the selectivity problem of several sensing platforms. As another recent example, researchers have demonstrated that chiral calix[4]arenes functionalized with aminonaphthol moieties were capable of selective chiral recognition of mandelic acid, a useful precursor to various drugs [105]. As future biomimetic calixarene systems emerge and are refined, we will continue to witness improvements in our chemical-sensing capability as well as our analytical systems for assessing enantiomeric purity of pharmaceuticals.

6.4.6.2 Metalloporphyrins

Researchers have explored an enormous variety of metalloporphyrin systems over the years in order to better understand and mimic the biological efficiency of a variety of biochemical functions found in natural systems [106]. Porphyrin is a form of heterocyclic macrocycle with open coordination sites available for axial ligation. Owing to the highly conjugated nature of porphyrin, unique spectral properties of these systems have often been exploited in biomimetic chemical recognition systems [107]. Researchers have found the arrangement and combination of hydrogen acceptor (N) sites, hydrogen donor (OH), and Lewis acid (Zn) sites in metalloporphyrins to be the influential factors in determining their target preference [108]. The selectivity of these systems is not as high as other molecular recognition approaches; however, Suslick *et al.* have utilized the shape-selective features of metalloporphyrins in their colorimetric sensor arrays. Their arrays are arranged such that the targets interact with the various host sites on the surface in a cross-reactive manner; that is, the host receptor molecules are not highly selective for a single analyte, rather the analyte binds with various hosts depending on their corresponding electronic and shape characteristics. Specifically, cross-reactive arrays have used various metal-incorporated tetraphenylporphyrins to provide discrimination based on coordination [109]. The steric selectivity in these systems arise from the incorporation of large-, medium-, and small-sized silylether groups on both faces of porphyrin to differentiate analytes based on size and shape [109]. Binding and coordination via the porphyrin's metal center result in large spectral shifts, which can be detected by alterations in the colorimetric properties of the array. In contrast to a quantized on-off response, these systems operate to distinguish analytes based on their polarizability, ligation affinities, and associated photophysical color changes that occur on binding [110].

As mentioned previously, molecular recognition in these cross-reactive systems relies on weak interactions and does not occur through a highly selective host–guest recognition event. Instead, the recognition occurs by the analyte creating a unique distributed response pattern across the various hosts in the array, which can be assessed as the analyte's signature [111]. This biomimetic approach is akin to that of the olfactory and gustatory systems, which we utilize for chemical sensing of the environments around us. Interestingly, the importance of Lewis acids in natural chemical sensing has also been supported, as there are indications that the olfactory receptors are metalloproteins [112]. In essence, the Lewis acidity of the metals in metalloporphyrins makes these hosts useful for the recognition of analytes with Lewis acid/base capabilities, while the presence of bulky side groups and hydrogen bonding components can restrict, to some extent, the selectivity of the interaction event [113]. This inherent lack of complete selectivity is the fundamental aspect of utilizing these synthetic receptors in biomimetic arrays. Through the use of pattern recognition and the ability of the individual receptors to interact differently depending on the analyte, these systems have proven useful in chemical sensing at discriminating a large variety of complex analytes arising from food [114], beverages [115, 116], bacteria [117], and even explosives [118]. In the future, cross-reactive metalloporphyrins may be used in a variety of diagnostic settings due to the wide chemical diversity capable of being displayed by these biomimetic receptor systems.

6.4.7
Recent Improvement

While supramolecular receptive materials work well for small-molecule systems, synthetic receptors are claimed to be no match for the selective recognition of medium- and large-complex analytes achievable with antibodies and aptamers

[111]. Particularly, the use of cross-reactive arrays based on the pattern recognition/differential receptor paradigm has difficulty in discriminating complex protein targets. Biological systems utilize binding models similar to the induced fit mechanism, which we have previously discussed, and as such, supramolecular chemistry must mimic this process in order to achieve the same level of molecular recognition. Recently, a new generation of macrocyclic host has emerged that can mimic the induced fit mechanisms of biology. In the recently developed systems, known as *heterocalixaromatics*, the aromatic rings are linked by different heteroatoms including oxygen and nitrogen, whose bond angles and lengths can vary resulting in finely tuned cavities, rather than the conventional methylene bridge [119]. For example, azacalix[4]pyridine can adopt a different electronic configuration to alter its cavity conformation for providing the maximum and most efficient interaction with its guest species [119]. Impressive structural features such as this continue to improve the molecular recognition capabilities of supramolecular chemical systems, although constructing biologically analogous synthetic receptors still remains difficult due to the complex atomic networks needed to form large binding sites with precisely positioned binding functionality [120].

6.5
Biomolecular Materials

6.5.1
Introduction

The third class of biomimetic receptive material is derived directly from the natural building blocks of the nature of biological systems. In terms of chemical sensing, biological selectivity remains unrivaled. Accordingly, biomolecular materials offer a unique advantage in terms of molecular selectivity due to their inherent diversity arising from sequence heterogeneity. Conceivably, the most well-known example of molecular recognition in biology was discovered by James Watson and Francis Crick in 1953. Specifically, they found the hydrogen bonding patterns between purines and pyrimidines of certain types of DNA, referred to as *base pairing*, that provided a framework for controlling the hybridization between complementary strands of nucleotides. Because the molecular recognition events are spatially constrained to the limits imposed for stable helix formation, effective binding can occur more reliably when a high fidelity of sequence complementarity is maintained. If the facile base pairing requirements are not met, the resulting complex exists with a certain degree of instability, which may likely cause dissociation or inhibit the initial strand hybridization. Using this inherent sequence specificity, researchers can create custom host–guest systems with precise molecular recognition capabilities. Already, the biomimetics community has created an abundance of biomolecular materials exploiting the canonical Watson–Crick base pairing for use as receptive coatings. As discussed later, DNA chip technologies have made particularly effective use of these principles for genomic screening and analysis. Also highlighted are the recent achievements using engineered nucleic acid systems that have opened the way for a new class of biomimetic materials capable of selective chemical and biomolecular detection.

In addition to nucleic acids, the heterogeneity of amino acid sequences has allowed a large repertoire of diverse receptors and enzymes to evolve with exquisite specificity for unique target ligands and substrates. One of the strongest protein recognition events, which is widely used throughout the biotechnology industry, is the complex formation between biotin and streptavidin, having a dissociation constant of approximately 10^{-14} M [121]. The protein exists as a homotetramer, each with binding domains for the small-molecule biotin. Multiple hydrogen bonding and van der Waals interactions facilitate high-affinity binding in addition to structural closure of surface loops to bury the biotin in the interior of the protein through quaternary changes [122]. This system is often used to capture and immobilize proteins, DNA, or other molecules onto beads and surfaces for various biological assays, selections, or purifications. Biomimetic attempts to harness the active site constituent of streptavidin have yielded several peptide variants, which function as stand-alone recognition elements for biotin, with similar selectivity.

A key strategy of the biomolecular materials approach to molecular recognition makes use of sequence rearrangements of biopolymers composed of natural building blocks, such as nucleic acids and amino acids, to create a diverse collection of candidate receptors and further imitate the evolutionary process. Using cleverly designed combinatorial libraries and directed evolution strategies, researchers have aimed to recapitulate the highly specific molecular recognition events obtained through native biological systems in order to obtain biomimetic receptors for their own targets of interest. This section discusses the *in vivo*, *in vitro*, and even *in silico* selection strategies used to obtain biomimetic receptors derived from biomolecular materials with target selectivity comparable to that achieved by nature.

6.5.2
Native Biomolecules

Over 10 million polypeptide sequences have been identified (as of January 2011), and a small percentage of these have their three-dimensional structures solved [123]. Among this collection of known proteins, several forms of natural receptors have been discovered, which offer unparalleled selectivity for their target ligand. It is therefore attractive to consider the use of native biomolecules (i.e., polypeptides, polynucleotides, and polysaccharides) as true biomimetic molecular recognition elements for sensor coating applications. The following sections highlight some of the key natural recognition elements that have been utilized for sensor coating applications.

6.5.2.1 **Polypeptides**
Ranging from short chains of six amino acid residues to nearly 37 000 amino acid containing proteins, polypeptides offer diverse functional and structural components that are used to carry out the biological processes of life through

selective molecular recognition. Research on sensing applications has focused on two broad uses of natural polypeptides known for their substrate specificity (catalytic recognition and affinity-based recognition). In the context of this chapter, we focus on biomimetic molecular recognition through affinity-based means, with the most prominent examples being antibodies and receptor proteins. These proteins offer high affinity and target specificity improved over a long natural evolutionary process. In addition, the diversity of existing receptors is enormous, with a number of useful targets including toxins, metabolites, and other relevant chemicals [124]. Aside from existing receptor–ligand pairs, antibodies for a specific target can be obtained from the immune system of healthy animals by introducing the antigen (target chemical or protein) into the blood stream. The healthy immune response will activate B lymphocytes to produce antibodies against the target antigen. These cells are harvested from the animal and fused to an immortal cell line to allow for continual production of the antibody of interest. Researchers have utilized this approach to produce thousands of different antibodies with specificity for a range of biological and chemical targets, including dyes, explosives, carcinogens, and pesticides for sensor applications [125–127]. Effective immobilization of the molecular recognition elements is critical for ensuring the efficacy of the sensing surface. For example, burial of the recognition domain because of randomly oriented immobilization may render the selectivity element inaccessible to target binding. In addition, excessive linkages between the surface and the recognition element may result in conformational changes of the polypeptide that will alter the shape of the recognition domain. A caveat of many natural receptors is that they are membrane proteins requiring a membrane framework, for example, Langmuir–Blodgett monolayer films, in order to create a recognition surface for sensing purposes; however, antibodies do not have such requirements and hence are more widely used for sensing applications [128]. These membrane receptors can host a variety of different targets of biological importance and are generally classified into ion channel receptors, G-protein-linked receptors, and single transmembrane receptors, among others. As future receptor protein structures become resolved, we may gain a better understanding of the recognition site selectivity imposed by these biopolymers through directly observing the amino acids responsible for binding with the target.

6.5.2.2 Carbohydrates

Complex carbohydrates exist on the surface of eukaryotic cells and play important roles in molecular recognition for cell–cell recognition, adhesion, and activation. Researchers have identified that polysaccharides can even specifically interact with polynucleotides [129]. Understanding polysaccharides and how their structures confer selectivity is still an eventful topic of study. For instance, microbes utilize glycoconjugates (carbohydrate-functionalized biomolecules) on the surface of host cells and tissues for infecting them, as in the case for influenza virus binding to sialic acid, but they also may utilize similar carbohydrate structures in their own microbial surface to mimic the host in order to escape immune defense [130]. Unlike the phosphate or peptide bond of other biopolymers, polysaccharides utilize several different glycosidic linkages between various different monosaccharides to allow a significant diversity in their structures. As an example, gangliosides are a sialic-acid-containing glycoconjugate with great heterogeneity and diversity in the structures of their carbohydrate chains for participating in cell–cell recognition, adhesion, and signal transduction [131]. Because of their capability to be used in the molecular recognition of certain protein targets, glycoconjugates have been incorporated into chemical sensor platforms as selective sensing surfaces for cholera toxin and influenza virus [132, 133]. By incorporating these glycoconjugates into colorimetric sensors based on conjugated liposomes and lipid membrane film, the work by Charych and coworkers effectively demonstrated the ability to probe selective molecular interactions by mimicking the receptive strategy used by cellular surfaces. Because of the ease of incorporation into such systems, natural carbohydrate-based receptors offer significant benefits to the *de novo* design of artificial receptors. The actively explored research area of glycobiology will undoubtedly provide a future insight into the binding capability, affinity, and selectivity of these native receptors, as we continue to develop a better understanding of their critical molecular recognition roles in signal transduction and other vital life processes.

6.5.2.3 Oligonucleotides

The most widely utilized native biopolymers for selective molecular recognition have been the oligonucleotides, DNA and RNA. Single-stranded DNA (ssDNA) and RNA offer the unique advantage of selective molecular recognition achievable through linear sequence complementarity. As such, researchers have been able to take advantage of this simplistic binding paradigm in order to create oligonucleotide-embedded materials and surfaces capable of selective biomimetic recognition of target strands with some of the highest specificities of any molecular recognition system. The high selectivity and affinity of these natural components has become further popularized by the creation of DNA microarray technologies, most notably those developed by Affymetrix, which encode a significant repertoire of genes from a given species' genome. Because of the high specificity of complementary base pairing, a copy of ssDNA can provide quantitative detection of a single complementary strand in the background of an extremely complex mixture [134, 135]. In short, DNA microarrays offer a molecular recognition surface composed of oligonucleotide segments, representing all known genetic elements of an organism, which can be exposed to a sample of cDNA to estimate genome-wide transcription profiles of the sample's host.

While these high-throughput systems offer the advantage of collecting a large amount of information about gene expression levels, other oligonucleotide-based recognition elements composed of RNA have been utilized by researchers for a different but just as interesting purpose. Natural noncoding RNA, which serves to control biochemical pathways, has been found to have exquisite molecular recognition capabilities for various metabolites. These so-called riboswitches can undergo a conformational change to bind with high affinity to ligands including vitamin B_{12}, thiamine pyrophosphate, and flavin mononucleotide [136]. In addition, the selectivity of these molecular recognition elements are inherently superb; for example, thiamine pyrophosphate sensor riboswitches are found to have 1000-fold preferential binding to target over the unphosphorylated precursor metabolite [137]. Owing to these advantageous properties, natural oligonucleotide-based receptors are often exploited in biomimetic sensing materials. As discussed in the next section, biopolymers including RNA can also be engineered computationally and by *in vitro* screening to achieve high-affinity receptive motifs with designed selectivity for a desired target of interest.

6.5.3
Engineered Biomolecules

While native biopolymers, such as G-protein-coupled receptors, antibodies, and RNA riboswitches, offer high-affinity molecular recognition elements, the ability to tailor-make a biomimetic molecular recognition element with predetermined specificity is an attractive prospect for chemical-sensing applications. The following section discusses the strategies for engineering biomolecular materials for use as selective chemical-sensing components. Particularly, selection technologies for DNA and RNA aptamers, directed evolution of recombinant antibody mimics, and the computational design of receptive biomolecules are focused on.

6.5.3.1 *In vitro* Selection of RNA/DNA Aptamers

Before riboswitches were discovered to play a critical cellular role via metabolite interaction, research found that DNA or RNA could be engineered to form aptamers, a selective molecular recognition motif capable of high-affinity binding of targets including small chemical targets as well as proteins [138, 139]. In 1990, the laboratories of Szostak and Gold found that large libraries of approximately 10^{10} random sequence RNA molecules could be screened through *in vitro* selection, often referred to as *SELEX*, to identify selective molecular recognition elements with unexpectedly high affinity for predestined target ligands. Since then, DNA and RNA aptamers have been identified by systematic evolution of ligands by exponential enrichment for targets ranging from simple ions, small-molecule metabolites, and peptides to large proteins, organelles, and viruses with library sizes up to 10^{15} randomized oligonucleotides [140]. For the most part, published binding constants for such aptamers are generally in the low nanomolar to picomolar range [141]. This is accomplished by nucleic acid aptamers offering numerous hydrogen bond acceptors and donors to facilitate the molecular recognition of their target. The recognition process encompasses shape complementarity, stacking and electrostatic interactions, and other aspects of intermolecular interaction, which were discussed thoroughly in previous sections. In addition, the flexibility of polynucleotides allows for the formation of recognition cavities for the selective enclosure of small-molecule targets. While DNA is intrinsically more chemically stable than RNA, there is inherently no difference in the capability of forming highly selective aptamers with either polynucleotide. It should be noted though that an ssDNA aptamer sequence if converted into the RNA equivalent typically

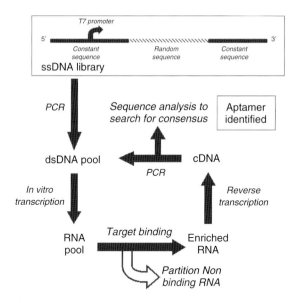

Figure 6.4 Outline of a typical scheme for *in vitro* selection of aptamer RNA by cyclic screening.

will not have the same molecular recognition capabilities due to the role of the 2'-OH contributing to interactions that stabilize the aptamer and complex [140]. Thus, deciding between either RNA or DNA screening strategies is the first step in selecting high-affinity nucleic-acid-based receptors. Subsequently, a cyclic procedure is performed in which RNA- or DNA-based aptamers, capable of binding to a given target, are selected from pools of variant sequences followed by amplification of those sequences that bind well to the target (Figure 6.4).

As outlined, the SELEX strategy can utilize randomized RNA or DNA sequences of 20–80 nucleotides long, which can be flanked by a constant region of known sequence for the purposes of amplification, transcription, or reverse transcription. After a few cycles of selection and amplification, dominant selective binding sequences tend to populate the observed sequence space given that the initial diversity of the library was sufficiently large to contain at least one such binding sequence. A variety of selection strategies can be applied to remove unwanted low-affinity binding sequences, such as washing strength, time, and buffer conditions. Other selection pressures can include a short target exposure time for binding, which will preferentially enhance the proportion of sequences with fast on-rates. The results of this approach yield a consensus sequence motif, generally purine rich, with dominant binding capabilities for the intended target facilitated by a properly formed H-bonding surface, sometimes via irregular oligonucleotide topologies and noncanonical base pairing [140]. Hence, the self-annealing properties of the aptamer allow for three-dimensional structures that specifically recognize the intended target molecule with high affinity. By providing a means for identifying a molecular recognition element for a target of interest, this screening strategy offers several benefits to the diagnostic and sensor community [142]. In addition to their widespread use as highly selective recognition elements, these systems have also demonstrated that binding-induced conformational changes in the structure of these nucleic-acid-based receptors can be utilized for transduction of the binding signal. Such engineered RNA aptamers have been modularly designed to recognize small-molecule targets and consequently undergo strand displacement to act as a reporter on binding to a number of chemicals of interest [143, 144]. This offers a smart approach, as the precise target binding sites and corresponding conformational changes are generally unknown beforehand.

In contrast to the use of antibodies as biomimetic molecular recognition elements for sensing chemicals, aptamers can offer several advantages. First of all, aptamers are considerably smaller that makes them easier to synthesize and enables them to access areas that might otherwise be inaccessible to the bulkier structure of the native antibody [145]. This qualification can also facilitate a sensor surface to achieve higher functional coating densities when using aptamers as compared to antibodies. In addition, SELEX screening can be performed using nonphysiological conditions and be used to screen otherwise toxic targets. Despite these advantages, nucleic-acid-based aptamers have drawbacks as well. While aptamers have a long shelf life, they do remain highly susceptible to enzymatic degradation and are quickly destroyed in biological fluids [141]. In addition, not every target may be effectively screened by SELEX, particularly those with hydrophobic or negatively charged surfaces, thereby leaving room for improvement in this strategy's future [146].

In summary, oligonucleotide library sequence screening allows for key aptamers to be selectively bound to a target, amplified, and enriched by way of elimination of nonbinding sequences to yield useful recognition elements. Their ease of chemical modification and immobilization without loss of function is expected to bring about the replacement of antibody-based assays with aptamer-based sensing materials [141]. The value of these nucleic-acid-based aptamers lies in their flexible nature, simple synthesis, high selectivity, and high affinity, thereby allowing their implementation in a variety of biosensor platforms with successes rivaling those of antibody-based recognition elements [128, 147, 148].

6.5.3.2 Evolutionary Screened Peptides

In an approach akin to the previously mentioned SELEX strategy, amino acid libraries may also be screened for selective binding sequences. One may consider using a similar combinatorial chemistry approach to synthesize large peptide libraries, which may prove useful in some limited applications. Unfortunately, combinatorial chemical synthesis of peptide libraries does not provide the same link between genotype and phenotype, as was seen in the previous section using oligonucleotide libraries, thus making it difficult or impossible to identify single peptide molecules capable of target binding and subsequent amplification. As such, peptides are best screened using an evolutionary selection method. Herein, we briefly discuss a few approaches to identifying biomimetic molecular recognition element through the evolutionary screening of peptides, generally referred to as *display techniques*.

The potential of rapid evolutionary screening of peptide libraries was first realized through the seminal work of George Smith in 1985. Therein, he revealed that peptide sequences could be displayed as fusions to protein 3 (p3) on the coat of filamentous bacteriophages and thereafter be specifically enriched 1000-fold (over wild-type phage) by affinity selection against an antibody target for the displayed peptide [149]. The resulting process, referred to as *phage display*, is now widely used as a high-throughput screening procedure for identifying high-affinity molecular recognition elements for target molecules, proteins, and crystals, among other uses [150]. Since its inception, the concept of displaying peptides on the protein coat of phage had sparked immediate interest in producing combinatorial libraries or collections of phage (typically having a diversity of 10^9) with different displayed peptide sequences capable of serving as recognition elements [151]. The efforts of Sir Gregory Winter and coworkers have been pivotal in improving the functionality and capabilities of the evolutionary screening of phage for creating highly selective receptors for predetermined targets. One of their most important contributions revealed that large libraries, or collections of phages, could be created in which the p3 phage coat was fused with a short variable antibody fragment, which may act as a receptor for a given target. By introducing the library of different receptive antibody fragments to a target antigen, they could selectively remove the nonbinding phage and isolate the phage that displayed antibody fragments with high affinity for the target antigen [152]. In doing so, they could analyze the peptide sequence to effectively identify an antibody engineered for a given target by phage display selection to that target, thereby providing receptive peptides with the high affinity and selectivity for antibodies but in a much smaller size. The use of such antibody fragments is less expensive and less time consuming compared to polyclonal or monoclonal antibody production, and they have been successfully used to detect toxins, proteins, and a number of other targets of interest [128]. In addition to discovering new receptors for target molecules, the phage display system is also an effective format for directed evolution to improve the molecular recognition capabilities of existing low-affinity receptors [153]. By incorporating an existing low-affinity antibody into the phage protein coat and varying key amino acids in the binding site, a receptive antibody with better selectivity and higher affinity may be identified through the high-throughput screening process.

The general strategy for using phage display to identify high-affinity molecular recognition elements first requires an appropriate library of candidate receptor to be created. The library may be designed with varying diversity, with varying lengths of amino acid sequences, and by varying the placement of amino acids within the scaffold, and may even be designed to have constrained loops to avoid unfavorable entropic folding on target binding. Once the phage library is cloned and expressed in *Escherichia coli*, the phage may be collected for use in screening against a target molecule of interest. The target molecule may be immobilized onto a surface, allowing physical separation of the target from the solution containing the phage library. The target is incubated with the library to allow the phages bearing specific receptors to bind to the target, while the nonspecific binders are consequently washed away. The bound phages, as compared to the original library, on average contain better receptor sequences for the target of interest. After isolating and capturing the bound phage, they may be replicated by infecting *E. coli* to produce many copies of themselves. Generally, the phages are subjected to four or more cycles of the screening procedure (in some cases, with more stringent binding conditions) to enrich primarily the target binding sequences [154]. Since the protein sequence displayed corresponds to the genetic sequence inside the phage, the DNA sequence can be read to identify the amino acids comprising the target-specific recognition motif. A number of improvements and variations can be employed as useful strategies for enhancing the capabilities and applications of the systems. Before screening, for instance, the phages may be chemically modified to produce constrained peptide loops with an organic core. Such strategies have been effective at enhancing the binding functionality as well as the stability of the receptors to protease degradation [155]. These methodologies are critical in obtaining a recognition element with properties matching the necessary application. Moreover, the facile nature of chemically modifying these engineered peptides has allowed them to be rapidly incorporated into a variety of chemical-sensing platforms with great success [156–158].

An influential factor for finding a suitable receptor to a target using display technologies is the diversity of the library. Larger libraries will inherently have more potential for containing sequences that bind with high affinity to a target of interest. As such, researchers strive to extend the diversity of their combinatorial phage libraries. It is generally accepted that the bacterial transformation requirement in the production of phage display libraries is the limiting factor, thus restricting starting phage libraries to a diversity range of $10^9 - 10^{10}$ varying peptide sequences [159]. Another screening approach, known as *ribosome display*, offers a fully *in vitro* alternative to phage display screening and has been effective in screening random peptide libraries to achieve target-selective binding with nanomolar affinities [160]. Before its development, there were no effective methods to link phenotype and genotype for peptide screening [160, 161]. By avoiding the low efficiency of DNA uptake required for *in vivo* peptide screening systems, ribosome display (in addition to the related mRNA display discussed later) can offer better diversity. As such, initial studies showed decapeptide libraries with 10^{12} diversity can be effectively screened for enrichment of binders [160].

While the basics of ribosome display are similar to other screening procedures in that alternating steps of affinity selection and amplification of binders are performed, there are significant differences in the details and complexity of the processes. In the particular case of screening a peptide–antibody fragment library by ribosome display, an antibody fragment DNA library must be amplified by PCR to contain a T7 promoter for transcription, a ribosome binding site for translation, a spacer segment to allow sufficient space between the antibody fragment and the ribosome, and stem loops at the 5′ and 3′ ends to prevent destruction by exonucleases. The sequence is then transcribed into mRNA that is translated by the *E. coli* S-30 *in vitro* system. During translation, the activity of the ribosome is stopped by cooling on ice, with further stabilization provided by increasing the level of magnesium [160]. These conditions must be preserved in order to maintain the complex formed between mRNA, ribosome, and the translating antibody fragment. The antibody fragment complexes are then subjected to affinity selection and removal of nonspecific binders. The complexes are then eluted directly, or by dissociation of the mRNA from the complex, by EDTA exposure, at which point the isolated mRNA is amplified by RT-PCR and is ready for the next round of screening. Using this concept, researchers have successfully demonstrated that single-chain fragment antibodies can be enriched 10^8-fold over five cycles of the above-mentioned transcription, translation, affinity selection, and amplification [160].

While the above ribosome display system has proven successful in the discovery of many receptors capable of molecular recognition, researchers sought out a means of improvement. In a very similar approach known as *mRNA display*, researchers were able to directly link the mRNA to the nascent peptide by use of an adapter molecule, thereby making the system more robust by eliminating the need for stabilization of the ribosome complex [162]. The elegant adapter molecule is linked to the 3′ end of the oligonucleotide by way of a DNA spacer directly attached to the mRNA. As the translating ribosome reaches the junction between the mRNA and DNA, the ribosome becomes stalled. At this point, the adapter molecule, puromycin, mimics the necessary aminoacyl end of tRNA by entering the A site of the ribosome where it forms a stable amide linkage with the nascent peptide. The resulting fusion is an mRNA–peptide hybrid that effectively forms a direct link between genotype and phenotype, with enhanced stability over the complex used in ribosome display. Since the capabilities of the mRNA display system indicate that more diverse libraries (10^{13}) can be created by the *in vitro* screening systems over conventional phage display (10^9), there is an inherent implication that mRNA display is a preferable method for creating longer peptide libraries in order to possess a more complete repertoire of the randomized sequence space [159]. Indeed this is also seen in practice, as typical phage display libraries tend to utilize randomized domains in the size range of 6–14 amino acids, while mRNA display has shown to be effective for peptides that are 27 amino acids long [162]. In some cases, this increased length and diversity achievable by mRNA display has resulted in selected peptides with higher affinity than those obtained from shorter and less complete phage display libraries [163]. However, despite the apparent advantages, phage display remains a more commonly used screening platform, perhaps due to its ease of use and inexpensive commercially available kits [150].

Regardless of the approach, the molecular recognition capabilities achievable through evolutionary screening can yield peptide-based receptors, with direct applications as biomolecular coating materials for chemical sensors. Over the years, evolutionary selection of peptides has been fruitful in achieving highly selective binders with superb affinities for targets including crystal surfaces [164], small molecules [17], and proteins [165] by an array of innovative screening strategies. As we continue to explore the use of more constrained peptide morphologies in these systems for enhancement of binding affinities as well as the use of organic chemical core hybrid systems for expanding the functionality and immobilization capabilities of these systems, we will undoubtedly find the incorporation into analytical devices and sensor platforms more frequent as biomimetic molecular recognition elements.

6.5.3.3 Computational and Rational Design of Biomimetic Receptors

Given the large amount of sequence and structural data available for receptor proteins, it may not be surprising to know that researchers are able to create new biological receptors for different target molecules by *de novo* design and *in silico* testing. While the capabilities of tailor designing an effective biomolecule-based receptor for any given target of interest may be still under development, the foundation of such work has been laid by pioneers using computational protein engineering to convert the specificity of existing biomolecules [166, 167]. Progress in the area of receptor design had generally been hindered due to the large number of structures that must be considered [168]. Using existing protein scaffolds, such as proteins from the periplasmic binding protein superfamily, research could limit the range of the potential structure space to examine specific classes of receptor–target interactions [169]. In doing so, the complexity of identifying the desired electrostatic and structural complementarity between a target and receptor becomes less overwhelming. The design of a receptor in this fashion may often utilize an iterative process of sequence modification and assessment of internal strain, desolvation, and van der Waals and coulombic interactions within the primary coordination sphere of the receptor–target complex to maximize estimated binding constants and selectivity.

Several members of the periplasmic binding protein superfamily, such as the phosphate-binding protein, have proven to be effective biomolecular materials for detection of their natural targets [170]. Using computational approaches to reengineer the maltose-binding protein, Hellinga and coworkers [171] have shown the capability of altering the specificity of the protein to a number of nonnative substrates including the explosive trinitrotoluene. Throughout their efforts, they have redesigned the protein-based biomimetic sensors for targets including zinc, pinacolyl methylphosphonic acid, lactate, and serotonin with enhanced affinities; however, a recent reassessment by crystallographic structural data revealed that the binding mechanisms were not the same as those predicted [172, 173]. Nonetheless, experimental evaluation showed as low as nanomolar affinities for some of the designed receptor–target interactions, which signifies the potential for these systems as molecular recognition elements in biomimetic chemical-sensing systems [171]. Moreover, this rational design approach allows researchers to expand the repertoire of biomimetic sensor targets outside those for which receptors naturally evolved [174]. The strategies utilized in these design approaches recapitulate the advantage of using biomimetic materials as sensor coating in that we need not revamp an entire sensing platform if we wish to alter the selectivity, but rather the problem of tailor-made chemical sensing is best achieved from a modular recognition element approach [175].

6.6
Summary and Future of Biomimetic-Sensor-Coating Materials

By exploring the productive history behind the use of biomimetic materials for surface-based molecular recognition, we hope to have provided a broad knowledge base for better understanding the current field of chemically receptive coatings. In addition, this provided a perspective for appreciating the recent improvements, trends, and needs in developing molecular recognition elements as we highlighted in the areas of molecular imprinting, supramolecular chemistry, and biomolecular-material-based recognition elements. While these three areas represent distinct strategies with respective benefits and drawbacks as selective molecular recognition materials, a comprehensive understanding allows future researchers to avoid pitfalls, which result in unproductive development. As an example, purely computational strategies for the design of receptor–ligand pairs may be ineffective, as an algorithm may not be able to capture all the characteristics necessary for accurately predicting complex formation. As a result, researchers may utilize a combination of structural studies and experimental validations to improve the computational design strategies. In addition, the wealth of data obtainable from *in vitro* screening of RNA, DNA, or peptide libraries may provide a starting point for the combination of theoretical design strategies with the observable evolution and adaptability of target binding sites. The remaining goal in chemical sensor coatings is to find a strategy for the rapid and effective creation of a surface coating material capable of high affinity and selectivity for a predetermined target of interest. While the traditional concepts of lock and key mechanistic fitting may still be alluring due to their simplicity, recent findings from the case of biological molecular recognition show that kinetic proofreading and multistep dynamic binding events may provide a better strategy to selective recognition in complex environments by overcoming noise. In addition, these receptive materials that undergo structural changes on binding to their target of interest may have other immediate benefits as sensor surface coatings, as small perturbations have the potential for being transduced into detectable signals.

With increased research in the area of biomolecular materials development, RNA, DNA, or peptides may likely become the dominant form of sensor coating material in the near future. Aside from the commercial availability of easy-to-use, step-by-step kits for finding peptide-based receptors via phage display, the continuing progress in modifying biomolecules to be more stable for sensor platform technologies will likely result in their widespread use as highly robust and effective biomaterial sensor coatings. Interestingly, the fields of catalysis, separation, and therapeutics utilize many of the same underlying principles of molecular recognition and will thus be impacted significantly by the continued understanding of biomimetic surfaces for selective molecular recognition. The ability to create on-demand selectivity for a target of interest will also have a profound impact on not only the research community but also, more importantly, to the public in the areas of medical diagnostics and analytical equipment for assessing chemical exposure risk. We expect that gaining a better understanding of these principles, selection strategies, design concepts, and computational models will aid researchers in overcoming some of the current limitations outlined earlier on biomimetic materials for selective molecular recognition in sensing.

References

1. Fischer, E. (1894) *Ber. Dtsch. Chem. Ges.*, **27**(3), 2985–2993.
2. Franco, M.I., Turin, L., Mershin, A., and Skoulakis, E.M.C. (2011) *Proc. Natl. Acad. Sci.*, **108**, 3797–3802.
3. Savir, Y. and Tlusty, T. (2007) *PLoS ONE*, **2**(5), e468.
4. Foote, J. and Milstein, C. (1994) *Proc. Natl. Acad. Sci.*, **91**(22), 10370–10374.
5. Ma, B., Kumar, S., Tsai, C.J., and Nussinov, R. (1999) *Protein Eng.*, **12**(9), 713–720.
6. Frauenfelder, H., Sligar, S.G., and Wolynes, P.G. (1991) *Science*, **254**(5038), 1598–1603.
7. Dill, K.A. and Chan, H.S. (1997) *Nat. Struct. Mol. Biol.*, **4**(1), 10–19.
8. Goh, C.S., Milburn, D., and Gerstein, M. (2004) *Curr. Opin. Struct. Biol.*, **14**(1), 104–109.
9. James, L.C. and Tawfik, D.S. (2005) *Proc. Natl. Acad. Sci. U.S.A.*, **102**(36), 12730–12735.
10. James, L.C., Roversi, P., and Tawfik, D.S. (2003) *Science*, **299**(5611), 1362–1367.
11. Schuermann, J.P., Prewitt, S.P., Davies, C., Deutscher, S.L., and Tanner, J.J. (2005) *J. Mol. Biol.*, **347**(5), 965–978.
12. Yin, J., Beuscher, A.E. IV, Andryski, S.E., Stevens, R.C., and Schultz, P.G. (2003) *J. Mol. Biol.*, **330**(4), 651–656.
13. Wedemayer, G.J., Patten, P.A., Wang, L.H., Schultz, P.G., and Stevens, R.C. (1997) *Science*, **276**(5319), 1665–1669.
14. Ariga, K., Hill, J., and Endo, H. (2007) *Int. J. Mol. Sci.*, **8**(8), 864–883.
15. Schamel, W.W.A., Risue, R.M., Minguet, S., Ort, A.R., and Alarc, B. (2006) *Trends Immunol.*, **27**(4), 176–182.
16. Greer, J.P., Foerster, J., Lukens, J. Rodgers, G.M., and Paraskevas, F. (2004) *Wintrobe's Clinical Hematology*, 11th edn, Lippincott Williams & Wilkins, pp. 453–456.
17. Jaworski, J.W., Raorane, D., Huh, J.H., Majumdar, A., and Lee, S.W. (2008) *Langmuir*, **24**(9), 4938–4943.
18. Park, J.S., Cho, M.K., Lee, E.J., Ahn, K.Y., Lee, K.E., Jung, J.H., Cho, Y., Han, S.S., Kim, Y.K., and Lee, J. (2009) *Nat. Nano*, **4**(4), 259–264.
19. Lehn, J.M. (1988) *Angew. Chem. Int. Ed. Engl.*, **27**(1), 89–112.
20. Lehn, J.M. (1985) *Science*, **227**(4689), 849–856.
21. Rebek, J. (1987) *Science*, **235**(4795), 1478–1484.
22. Náray-Szabó, G. (1993) *J. Mol. Recognit.*, **6**(4), 205–210.
23. Apaya, R.P., Lucchese, B., Price, S.L., and Vinter, J.G. (1995) *J. Comput.-Aided Mol. Des.*, **9**(1), 33–43.
24. Schneider, H.J., Blatter, T., Eliseev, A., Rudiger, V., and Raevsky, O.A. (1993) *Electrostatics in molecular recognition: from ion pairs and inophores to nucleotides and DNA*, Pure and Applied Chemistry, **65**, Research Triangle Park, NC, ETATS-UNIS, p. 56.
25. Dean, P.M., Chau, P.L., and Barakat, M.T. (1992) *J. Mol. Struct. THEOCHE*, **256**, 75–89.
26. Perutz, M.F. (1993) *Philos. Trans. R. Soc. London, Ser. A*, **345**(1674), 105–112.
27. Plevin, M.J., Bryce, D.L., and Boisbouvier, J. (2010) *Nat. Chem.*, **2**(6), 466–471.
28. Fan, E., Van Arman, S.A., Kincaid, S., and Hamilton, A.D. (1993) *J. Am. Chem. Soc.*, **115**(1), 369–370.
29. Meyer, E.A., Castellano, R.K., and Diederich, F. (2003) *Angew. Chem. Int. Ed.*, **42**(11), 1210–1250.
30. Vijay, D. and Sastry, G.N. (2010) *Chem. Phys. Lett.*, **485**(1–3), 235–242.
31. Kumpf, R.A. and Dougherty, D.A. (1993) *Science*, **261**(5129), 1708–1710.
32. Quiñonero, D., Garau, C., Rotger, C., Frontera, A., Ballester, P., Costa, A.,

and Deyà, P.M. (2002) *Angew. Chem. Int. Ed.*, **41**(18), 3389–3392.
33. de Hoog, P., Gamez, P., Mutikainen, I., Turpeinen, U., and Reedijk, J. (2004) *Angew. Chem.*, **116**(43), 5939–5941.
34. Houk, K.N. and Wheeler, S.E. (2010) *J. Phys. Chem. A*, **114**(33), 8658–8664.
35. Nàray-Szabó, G. (1989) *J. Mol. Graphics*, **7**(2), 76–81.
36. Olsson, T.S.G., Williams, M.A., Pitt, W.R., and Ladbury, J.E. (2008) *J. Mol. Biol.*, **384**(4), 1002–1017.
37. Hoeben, F.J.M., Jonkheijm, P., Meijer, E.W., and Schenning, A.P.H.J. (2005) *Chem. Rev.*, **105**(4), 1491–1546.
38. Bissantz, C., Kuhn, B., and Stahl, M. (2010) *J. Med. Chem.*, **53**(14), 5061–5084.
39. Nishio, M. (2011) *Phys. Chem. Chem. Phys.*, **13**(31), 13873–13900.
40. Mecozzi, S. and Rebek, J.J. (1998) *Chem. Eur. J.*, **4**(6), 1016–1022.
41. Frederick, K.K., Marlow, M.S., Valentine, K.G., and Wand, A.J. (2007) *Nature*, **448**(7151), 325–329.
42. Hunter, C.A. and Tomas, S. (2003) *Chem. Biol.*, **10**(11), 1023–1032.
43. Badjic, J.D., Nelson, A., Cantrill, S.J., Turnbull, W.B., and Stoddart, J.F. (2005) *Acc. Chem. Res.*, **38**(9), 723–732.
44. Handl, H.L., Vagner, J., Han, H., Mash, E., Hruby, V.J., and Gillies, R.J. (2004) *Expert Opin. Ther. Targets*, **8**(6), 565–586.
45. Rao, J., Lahiri, J., Weis, R.M., and Whitesides, G.M. (2000) *J. Am. Chem. Soc.*, **122**(12), 2698–2710.
46. Ercolani, G. (2005) *Org. Lett.*, **7**(5), 803–805.
47. Glass, T.E. (2000) *J. Am. Chem. Soc.*, **122**(18), 4522–4523.
48. Polyakov, M. (1931) *Zhur. Fiz. Khim.*, **2**, 799.
49. Wulff, G. and Sarhan, A. (1972) *Angew. Chem.*, **84**(8), 364–364.
50. Klotz, I.M., Royer, G.P., and Scarpa, I.S. (1971) *Proc. Natl. Acad. Sci.*, **68**(2), 263–264.
51. Andersson, H.S., Nicholls, I.A., and Sellergren, B. (2000) *Techniques and Instrumentation in Analytical Chemistry*, vol. **23**, Chapter 1, Elsevier, pp. 1–19.
52. Vlatakis, G., Andersson, L.I., Muller, R., and Mosbach, K. (1993) *Nature*, **361**(6413), 645–647.
53. Kriz, D., Ramstrom, O., and Mosbach, K. (1997) *Anal. Chem.*, **69**(11), 345A–349A.
54. Maier, N. and Lindner, W. (2007) *Anal. Bioanal. Chem.*, **389**(2), 377–397.
55. Umpleby, R.J., Bode, M., and Shimizu, K.D. (2000) *Analyst*, **125**(7), 1261–1265.
56. Theodoridis, G. and Manesiotis, P. (2002) *J. Chromatogr. A*, **948**(1–2), 163–169.
57. Haupt, K. and Mosbach, K. (2000) *Chem. Rev.*, **100**(7), 2495–2504.
58. Shea, K.J., Sasaki, D.Y., and Stoddard, G.J. (1989) *Macromolecules*, **22**(4), 1722–1730.
59. Shea, K.J. and Dougherty, T.K. (1986) *J. Am. Chem. Soc.*, **108**(5), 1091–1093.
60. Damen, J. and Neckers, D.C. (1980) *J. Am. Chem. Soc.*, **102**(9), 3265–3267.
61. Whitcombe, M.J., Rodriguez, M.E., Villar, P., and Vulfson, E.N. (1995) *J. Am. Chem. Soc.*, **117**(27), 7105–7111.
62. Fried, J.R. (2003) *Polymer Science and Technology*, Chapter 3, 2nd edn, Prentice Hall.
63. Mayes, A.G. and Mosbach, K. (1996) *Anal. Chem.*, **68**(21), 3769–3774.
64. Haupt, K. and Mosbach, K. (1998) *Trends Biotechnol.*, **16**(11), 468–475.
65. Alexander, C., Andersson, H.S., Andersson, L.I., Ansell, R.J., Kirsch, N., Nicholls, I.A., O'Mahony, J., and Whitcombe, M.J. (2006) *J. Mol. Recognit.*, **19**(2), 106–180.
66. Dickert, F.L. and Thierer, S. (1996) *Adv. Mater.*, **8**(12), 987–990.
67. Dickert, F.L., Forth, P., Lieberzeit, P.A., and Voigt, G. (2000) *Fresen. J. Anal. Chem.*, **366**(8), 802–806.
68. Hayden, O., Bindeus, R., Haderspöck, C., Mann, K.J., Wirl, B., and Dickert, F.L. (2003) *Sens. Actuators, B*, **91**(1–3), 316–319.
69. Dickert, F.L., Hayden, O., and Halikias, K.P. (2001) *Analyst*, **126**(6), 766–771.
70. Takeuchi, T., Fukuma, D., and Matsui, J. (1998) *Anal. Chem.*, **71**(2), 285–290.
71. Lanza, F. and Sellergren, B. (1999) *Anal. Chem.*, **71**(11), 2092–2096.

72. Hart, B.R. and Shea, K.J. (2002) *Macromolecules*, **35**(16), 6192–6201.
73. Schwarz, L.J., Danylec, B., Harris, S.J., Boysen, R.I., and Hearn, M.T.W. (2011) *J. Chromatogr. A*, **1218**(16), 2189–2195.
74. Minouraa, N., Rachkov, A., Higuchi, M., and Shimizu, T. (2001) *Bioseparation*, **10**(6), 399–407.
75. Rachkov, A. and Minoura, N. (2001) *Biochim. Biophys. Acta, Protein Struct. Mol. Enzymol.*, **1544**(1–2), 255–266.
76. Andersson, L., Paprica, A., and Arvidsson, T. (1997) *Chromatographia*, **46**(1), 57–62.
77. Rao, R.N., Maurya, P.K., and Khalid, S. (2011) *Talanta*, **85**(2), 950–957.
78. Park, C.H. and Simmons, H.E. (1968) *J. Am. Chem. Soc.*, **90**(9), 2431–2432.
79. Pedersen, C.J. (1967) *J. Am. Chem. Soc.*, **89**(26), 7017–7036.
80. Cram, D.J. and Cram, J.M. (1974) *Science*, **183**(4127), 803–809.
81. Graf, E. and Lehn, J.M. (1975) *J. Am. Chem. Soc.*, **97**(17), 5022–5024.
82. Graf, E. and Lehn, J.M. (1976) *J. Am. Chem. Soc.*, **98**(20), 6403–6405.
83. Lehn, J.M., Pine, S.H., Watanabe, E.I., and Willard, A.K. (1977) *J. Am. Chem. Soc.*, **99**(20), 6766–6768.
84. Lehn, J.M. (1988) *J. Inclusion Phenom.Macrocyclic Chem.*, **6**(4), 351–396.
85. Jeong, K.S., Muehldorf, A.V., and Rebek, J.J.R. (1990) *Molecular Recognition. Asymmetric Complexation of Diketopiperazines*, vol. **112**, American Chemical Society, Washington, DC, p. 2, ETATS-UNIS.
86. Jeong, K.S., Muehldorf, A., Deslongchamps, G., Famulok, M., and Rebek, J.J.R. (1991) *Convergent Functional Groups. X, Molecular Recognition of Neutral Substrates*, vol. **113**, American Chemical Society, Washington, DC, p. 9, TATS-UNIS.
87. Curiel, D. and Beer, P.D. (2005) *Chem. Commun.*, (14), 1909–1911.
88. Albrecht, M., Fiege, M., and Osetska, O. (2008) *Coord. Chem. Rev.*, **252**(8–9), 812–824.
89. Hancock, R.D., Melton, D.L., Harrington, J.M., McDonald, F.C., Gephart, R.T., Boone, L.L., Jones, S.B., Dean, N.E., Whitehead, J.R., and Cockrell, G.M. (2007) *Coord. Chem. Rev.*, **251**(13–14), 1678–1689.
90. Halime, Z., Lachkar, M., Furet, E., Halet, J.F., and Boitrel, B. (2006) *Inorg. Chem.*, **45**(26), 10661–10669.
91. Vidyadhar, H., Chi-Ying, H., Puttannachetty, M., Cunningham, R., Fpner, T., and Thummel, R.P. (1993) *Design of Receptors for Urea Derivatives Based on the Pyrido[3,2-g]indole Subunit*, vol. **115**, American Chemical Society, Washington, DC, ETATS-UNIS.
92. Schalley, C.A. (2007) *Analytical Methods in Supramolecular Chemistry*, Chapter 12, Wiley-VCH Verlag GmbH.
93. Harada, T., Rudzinski, J.M., Osawa, E., and Shinkai, S. (1993) *Tetrahedron*, **49**(27), 5941–5954.
94. Dickert, F., Reif, H., and Stathopulos, H. (1996) *J. Mol. Model.*, **2**(10), 410–416.
95. Kretschmer, R., Kinzel, D., and González, L. (2011) *Int. J. Quantum Chem.*, **112**, 1786–1795.
96. McHahon, G., O'Malley, S., and Nolan, K. (2004) *ChemInform*, **35**(23), 23–31.
97. Dickert, F.L. and Schuster, O. (1995) *Microchim. Acta*, **119**(1), 55–62.
98. Coquiere, D., Le Gac, S., Darbost, U., Seneque, O., Jabin, I., and Reinaud, O. (2009) *Org. Biomol. Chem.*, **7**(12), 2485–2500.
99. Dickert, F.L. and Schuster, O. (1993) *Adv. Mater.*, **5**(11), 826–829.
100. Travis Holman, K., William Orr, G.L., Atwood, J.W., and Steed, J. (1998) *Chem. Commun.*, (19), 2109–2110.
101. Perrin, M., Gharnati, F., Oehler, D., Perrin, R., and Lecocq, S. (1992) *J. Inclusion Phenom. Macrocyclic Chem.*, **14**(3), 257–270.
102. Walker, D., Joshi, G., and Davis, A. (2009) *Cell. Mol. Life Sci.*, **66**(19), 3177–3191.
103. Gravett, D.M. and Guillet, J.E. (1996) *Macromolecules*, **29**(2), 617–624.
104. Kim, J.S. and Quang, D.T. (2007) *Chem. Rev.*, **107**(9), 3780–3799.
105. Durmaz, M., Yilmaz, M., and Sirit, A. (2010) *Org. Biomol. Chem.*, **9**(2), 571–580.
106. Tatsumi, K. and Hoffmann, R. (1981) *J. Am. Chem. Soc.*, **103**(12), 3328–3341.

107. Zhao, S. and Luong, J.H.T. (1995) *J. Chem. Soc., Chem. Commun.*, (6), 663–664.
108. Mizutani, T., Kurahashi, T., Murakami, T., Matsumi, N., and Ogoshi, H. (1997) *J. Am. Chem. Soc.*, **119**(38), 8991–9001.
109. Suslick, K.S., Rakow, N.A., and Sen, A. (2004) *Tetrahedron*, **60**(49), 11133–11138.
110. Rakow, N.A. and Suslick, K.S. (2000) *Nature*, **406**(6797), 710–713.
111. Anslyn, E.V. (2007) *J. Org. Chem.*, **72**(3), 687–699.
112. Wang, J., Luthey-Schulten, Z.A., and Suslick, K.S. (2003) *Proc. Natl. Acad. Sci.*, **100**(6), 3035–3039.
113. Sen, A. and Suslick, K.S. (2000) *J. Am. Chem. Soc.*, **122**(46), 11565–11566.
114. Musto, C.J. and Suslick, K.S. (2010) *Curr. Opin. Chem. Biol.*, **14**(6), 758–766.
115. Suslick, B.A., Feng, L., and Suslick, K.S. (2010) *Anal. Chem.*, **82**(5), 2067–2073.
116. Zhang, C., Bailey, D.P., and Suslick, K.S. (2006) *J. Agric. Food. Chem.*, **54**(14), 4925–4931.
117. Carey, J.R., Suslick, K.S., Hulkower, K.I., Imlay, J.A., Imlay, K.R.C., Ingison, C.K., Ponder, J.B., Sen, A., and Wittrig, A.E. (2011) *J. Am. Chem. Soc.*, **133**(19), 7571–7576.
118. Lin, H. and Suslick, K.S. (2010) *J. Am. Chem. Soc.*, **132**(44), 15519–15521.
119. Wang, M.X. (2008) *Chem. Commun.*, (38), 4541–4551.
120. Yoon, S.S. and Still, W.C. (1993) *J. Am. Chem. Soc.*, **115**(2), 823–824.
121. Michael, N. and Avidin, G. (1975) in *Advances in Protein Chemistry*, vol. **29** (eds C.B. Anfinsen, J.T. Edsall, and F.M. Richards), Academic Press, pp. 85–133.
122. Weber, P., Ohlendorf, D., Wendoloski, J., and Salemme, F. (1989) *Science*, **243**(4887), 85–88.
123. Márcio, D., Luciana, S.B., and Luis, C.L. (2011) Combining machine learning and optimization techniques to determine 3-D structure of polypeptides. Proceedings of the 22nd International Joint Conference on Artificial Intelligence, 2011, pp. 2794–2795.
124. Subrahmanyam, S., Piletsky, S.A., and Turner, A.P.F. (2002) *Anal. Chem.*, **74**(16), 3942–3951.
125. Matsumoto, K., Torimaru, A., Ishitobi, S., Sakai, T., Ishikawa, H., Toko, K., Miura, N., and Imato, T. (2005) *Talanta*, **68**(2), 305–311.
126. Ju, C., Tang, Y., Fan, H., and Chen, J. (2008) *Anal. Chim. Acta*, **621**(2), 200–206.
127. Alcocer, M.J.C., Dillon, P.P., Manning, B.M., Doyen, C., Lee, H.A., Daly, S.J., O'Kennedy, R., and Morgan, M.R.A. (2000) *J. Agric. Food. Chem.*, **48**(6), 2228–2233.
128. Chambers, J.P., Arulanandam, B.P., Matta, L.L., Weis, A., and Valdes, J.J. (2008) *Curr. Issues Mol. Biol.*, **10**(1), 1–12.
129. Sakurai, K. and Shinkai, S. (2000) *J. Am. Chem. Soc.*, **122**(18), 4520–4521.
130. Wang, D., Liu, S., Trummer, B.J., Deng, C., and Wang, A. (2002) *Nat. Biotech.*, **20**(3), 275–281.
131. Yu, R.K., Tsai, Y.T., Ariga, T., and Yanagisawa, M. (2011) *J. Oleo Sci.*, **60**(10), 537–544.
132. Charych, D., Cheng, Q., Reichert, A., Kuziemko, G., Stroh, M., Nagy, J.O., Spevak, W., and Stevens, R.C. (1996) *Chem. Biol.*, **3**(2), 113–120.
133. Pan, J.J. and Charych, D.H. (1997) Molecular Recognition and Optical Detection of Biological Pathogens at Biomimetic Membrane Interfaces, 1997, pp. 211–217.
134. Brown, P.O. and Botstein, D. (1999) *Nat. Genet.*, **21**, 33–37.
135. Pease, A.C., Solas, D., Sullivan, E.J., Cronin, M.T., Holmes, C.P., and Fodor, S.P. (1994) *Proc. Natl. Acad. Sci.*, **91**(11), 5022–5026.
136. Kaempfer, R. (2003) *EMBO Rep.*, **4**(11), 1043–1047.
137. Mandal, M., Boese, B., Barrick, J.E., Winkler, W.C., and Breaker, R.R. (2003) *Cell*, **113**(5), 577–586.
138. Tuerk, C. and Gold, L. (1990) *Science*, **249**(4968), 505–510.
139. Ellington, A.D. and Szostak, J.W. (1990) *Nature*, **346**(6287), 818–822.

140. Wilson, D.S. and Szostak, J.W. (1999) *Annu. Rev. Biochem.*, **68**, 611–647.
141. Stoltenburg, R., Reinemann, C., and Strehlitz, B. (2007) *Biomol. Eng.*, **24**(4), 381–403.
142. Jenison, R., Gill, S., Pardi, A., and Polisky, B. (1994) *Science*, **263**(5152), 1425–1429.
143. Nutiu, R. and Li, Y. (2005) *Angew. Chem. Int. Ed.*, **44**(7), 1061–1065.
144. Win, M.N. and Smolke, C.D. (2007) *Proc. Natl. Acad. Sci.*, **104**(36), 14283–14288.
145. Lee, J.F., Stovall, G.M., and Ellington, A.D. (2006) *Curr. Opin. Chem. Biol.*, **10**(3), 282–289.
146. Hermann, T. and Patel, D.J. (2000) *Science*, **287**(5454), 820–825.
147. Savran, C.A., Knudsen, S.M., Ellington, A.D., and Manalis, S.R. (2004) *Anal. Chem.*, **76**(11), 3194–3198.
148. Stadtherr, K., Wolf, H., and Lindner, P. (2005) *Anal. Chem.*, **77**(11), 3437–3443.
149. Smith, G. (1985) *Science*, **228**(4705), 1315–1317.
150. Bratkoviè, T. (2010) *Cell. Mol. Life Sci.*, **67**(5), 749–767.
151. Huse, W., Sastry, L., Iverson, S., Kang, A., Alting-Mees, M., Burton, D., Benkovic, S., and Lerner, R. (1989) *Science*, **246**(4935), 1275–1281.
152. Clackson, T., Hoogenboom, H.R., Griffiths, A.D., and Winter, G. (1991) *Nature*, **352**(6336), 624–628.
153. Rader, C. and Barbas, C.F. III (1997) *Curr. Opin. Biotechnol.*, **8**(4), 503–508.
154. Pande, J., Szewczyk, M.M., and Grover, A.K. (2010) *Biotechnol. Adv.*, **28**(6), 849–858.
155. Heinis, C., Rutherford, T., Freund, S., and Winter, G. (2009) *Nat. Chem. Biol.*, **5**(7), 502–507.
156. Cerruti, M., Jaworski, J., Raorane, D., Zueger, C., Varadarajan, J., Carraro, C., Lee, S.W., Maboudian, R., and Majumdar, A. (2009) *Anal. Chem.*, **81**(11), 4192–4199.
157. Jaworski, J., Yokoyama, K., Zueger, C., Chung, W.J., Lee, S.W., and Majumdar, A. (2011) *Langmuir*, **27**(6), 3180–3187.
158. Kim, T.H., Lee, B.Y., Jaworski, J., Yokoyama, K., Chung, W.J., Wang, E., Hong, S., Majumdar, A., and Lee, S.W. (2011) *ACS Nano*, **5**(4), 2824–2830.
159. Gold, L. (2001) *Proc. Natl. Acad. Sci.*, **98**(9), 4825–4826.
160. Hanes, J. and Plückthun, A. (1997) *Proc. Natl. Acad. Sci.*, **94**(10), 4937–4942.
161. Mattheakis, L.C., Bhatt, R.R., and Dower, W.J. (1994) *Proc. Natl. Acad. Sci.*, **91**(19), 9022–9026.
162. Roberts, R.W. and Szostak, J.W. (1997) *Proc. Natl. Acad. Sci.*, **94**(23), 12297–12302.
163. Wilson, D.S., Keefe, A.D., and Szostak, J.W. (2001) *Proc. Natl. Acad. Sci. U.S.A.*, **98**(7), 3750–3755.
164. Whaley, S.R., English, D.S., Hu, E.L., Barbara, P.F., and Belcher, A.M. (2000) *Nature*, **405**(6787), 665–668.
165. Matsumura, N., Tsuji, T., Sumida, T., Kokubo, M., Onimaru, M., Doi, N., Takashima, H., Miyamoto-Sato, E., and Yanagawa, H. (2010) *FASEB J.*, **24**(7), 2201–2210.
166. Bolon, D.N. and Mayo, S.L. (2001) *Proc. Natl. Acad. Sci.*, **98**(25), 14274–14279.
167. Dwyer, M.A., Looger, L.L., and Hellinga, H.W. (2004) *Science*, **304**(5679), 1967–1971.
168. Moon, J.B. and Howe, W.J. (1991) *Proteins Struct. Funct. Bioinf.*, **11**(4), 314–328.
169. Dwyer, M.A. and Hellinga, H.W. (2004) *Curr. Opin. Struct. Biol.*, **14**(4), 495–504.
170. Salins, L.L.E., Deo, S.K., and Daunert, S. (2004) *Sens. Actuators, B*, **97**(1), 81–89.
171. Looger, L.L., Dwyer, M.A., Smith, J.J., and Hellinga, H.W. (2003) *Nature*, **423**(6936), 185–190.
172. Schreier, B., Stumpp, C., Wiesner, S., and Höcker, B. (2009) *Proc. Natl. Acad. Sci.*, **106**(44), 18491–18496.
173. Vallée-Bélisle, A. and Plaxco, K.W. (2010) *Curr. Opin. Struct. Biol.*, **20**(4), 518–526.
174. Marvin, J.S. and Hellinga, H.W. (2001) *Proc. Natl. Acad. Sci.*, **98**(9), 4955–4960.
175. Hellinga, H.W. and Marvin, J.S. (1998) *Trends Biotechnol.*, **16**(4), 183–189.

Part II
Surface Aspects

7
Biology Lessons for Engineering Surfaces for Controlling Cell–Material Adhesion

Ted T. Lee and Andrés J. García

7.1
Introduction

In surface aspects, we discuss important biology lessons that are required for understanding cell–material adhesion. This chapter discusses the basics of the extracellular matrix (ECM), protein structure and adsorption to surfaces, and cellular communication and signaling proteins. It finally gives a review of current engineered cell adhesion technologies. These lessons are critical to understanding how to engineer biomaterial surfaces for use in both *in vitro* and *in vivo* applications.

7.2
The Extracellular Matrix

This section provides a brief overview of the ECM and its functions. This overview provides knowledge important to understand the composition of the ECM and engineer synthetic ECMs to tune cell-fate-based processes.

ECMs are composed of a complex, insoluble, three-dimensional mixture of secreted macromolecules, including collagens and noncollagenous proteins, such as elastin and fibronectin (FN), glycosaminoglycans, and proteoglycans, which are present between cells [1]. In addition, provisional fibrin-based networks constitute specialized matrices for wound healing and tissue repair. The ECM's function is to provide structure and order to the extracellular space and regulate multiple functions associated with the establishment, maintenance, and remodeling of differentiated tissues [2]. Matrix components such as FN and laminin mediate adhesive interactions that support cell anchorage, migration, and tissue organization. These all serve to activate signaling pathways that direct cell survival, proliferation, and differentiation.

ECM components can interact with growth and differentiation factors, chemokines, and other soluble factors that regulate cell cycle progression and differentiation to control their availability and activity. By using various strategies to immobilize and order these proteins (ligands), ECMs control the spatial and temporal profiles of these signals and generate gradients necessary for vectorial responses. Moreover, structural elements within ECMs, namely, collagens, elastin, and proteoglycans, contribute to the mechanical integrity, rigidity, and the viscoelasticity of skin, cartilage, vasculature, and other tissues. Finally, the composition and structure of ECMs are dynamically modulated by the cells within them, reflecting the highly regulated and bidirectional communication between cells and ECMs.

Therefore, cell–material interactions are highly dependent on the ECM that is present on surfaces. By designing simplified protein sequences, we can take an engineered approach to decoding, and possibly controlling, these complex interactions with cells.

7.3
Protein Structure

Proteins are structurally complex and functionally sophisticated biomolecules. They serve a variety of functions in all living organisms and have been tuned by natural selection over millions of years. In the following section, we explore the basic chemistry of proteins in order to understand more about protein adsorption onto biomaterial surfaces.

Proteins are polymer chains that consist of an exact sequence of amino acids [3]. There are 20 known basic amino acids in nature and a number of amino acid derivatives with modified structure. All amino acids contain conserved structural regions, including an alpha carbon (α-carbon), an amino group, a carboxyl group, and a variable side chain (R group). It is the side chain group that determines the chemical properties of each amino acid. There are nonpolar, polar, acidic, and basic side chains. As a result, the combined action of all side chains largely dictates a protein's overall three-dimensional structure, biochemical activity, and adsorption kinetics.

Proteins are made of individual amino acids that are linked via a peptide bond (Figure 7.1). This bond has a resonance-stabilized structure that keeps the peptide bond in a planar conformation. A chain of many amino acids creates a *polypeptide*. All polypeptides have two free ends, one a carboxy terminus (C-terminus) and the other an amino terminus (N-terminus).

Protein structure can be broken into four discrete levels of organization: primary, secondary, tertiary, and quaternary structure. The physical arrangement of amino acids in a chain is known as the *primary structure*. Hydrogen bonding interactions between the functional groups of an amino acid chain cause the formation of local ordered or repeating units known as the *secondary structure*. Examples of secondary structures include α-helices, β-sheets, and turns. The *tertiary* structure of a protein consists of the spatial orientation of various structural units within a single protein chain. These interactions are largely dictated by weaker interactions such as van der Waals, ionic, and hydrophobic interactions. The complex of various individual

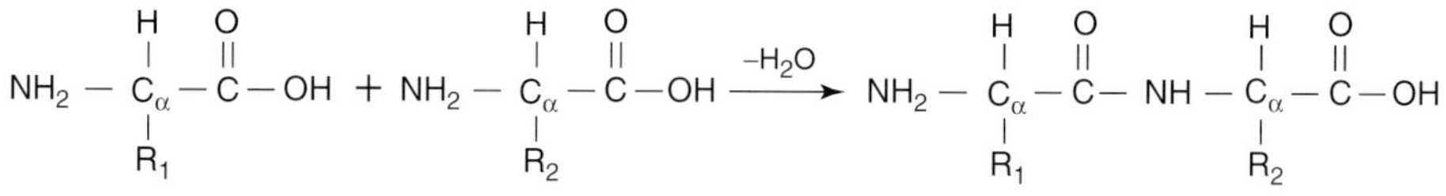

$$NH_2 - \underset{\underset{R_1}{|}}{\overset{\overset{H}{|}}{C_\alpha}} - \overset{\overset{O}{\|}}{C} - OH + NH_2 - \underset{\underset{R_2}{|}}{\overset{\overset{H}{|}}{C_\alpha}} - \overset{\overset{O}{\|}}{C} - OH \xrightarrow{-H_2O} NH_2 - \underset{\underset{R_1}{|}}{\overset{\overset{H}{|}}{C_\alpha}} - \overset{\overset{O}{\|}}{C} - NH - \underset{\underset{R_2}{|}}{\overset{\overset{H}{|}}{C_\alpha}} - \overset{\overset{O}{\|}}{C} - OH$$

Figure 7.1 Structural diagram of amino acid structure and formation of peptide bond. C_α the alpha carbon and R_1/R_2 are the functional groups. There are 20 basic functional groups.

polypeptide chains forms a protein's *quaternary* structure. Ultimately, variations in protein structure modulate both its function and activity in biological systems.

7.4
Basics of Protein Adsorption

The first event in the cell adhesion cascade to material surfaces is protein adsorption to a surface, and it is one of the major mechanisms regulating cell–material interactions [4]. As a consequence, control of protein adsorption becomes critical. Protein adsorption is a phenomenon driven by entropy that is influenced by the chemical and physical properties of both the protein and the receiving surface. In general, protein adsorption to surfaces is controlled by *noncovalent interactions* that are mediated by the amino acid side chains present in different parts of the protein. These interactions include hydrophobic interactions, van der Waals, hydrogen bonding, and electrostatic forces.

When looking at any protein, we can make a couple of generalizations about its interactions with surfaces and surrounding fluids. Under physiological conditions, protein folding is dictated by hydrophobic interactions to exclude water from highly hydrophilic regions. There are two main types of proteins, those that are membrane bound and those that largely exist as soluble globular proteins. Soluble proteins often have hydrophobic domains that are buried in the protein core, in order to minimize the entropy of the protein–solution system. The domains of membrane-bound or transmembrane proteins often have outward-facing hydrophobic domains that interact with the lipid bilayer. These hydrophobic interactions are all driven by the minimization of free energy of the system and are the primary mediators of protein tertiary structure.

The most common surface properties that biomaterial engineers are interested in are *surface hydrophobicity* and *surface charge*. Hydrophobicity is the manner in which a material responds to the presence of water, and it is generally quantified by contact angle measurement (Figure 7.2).

Generally, with increasing hydrophobicity of the surface, there will be greater protein adsorption; however, this does not mean that the protein activity is maintained. Protein charge may also influence the conformation of a protein when it adsorbs to a surface. A negatively charged domain interacting with a negatively charged surface will not be energetically favorable; therefore, one would expect counter ions from solution to shield the protein from these unfavorable events.

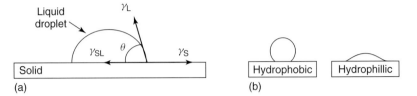

Figure 7.2 Contact angle. (a) Measurement of contact angle on a surface. θ denotes contact angle and γ denotes respective surface tensions. (b) Basic schematic representation of water droplets on hydrophobic and hydrophilic surfaces.

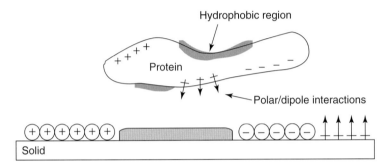

Figure 7.3 Protein on a heterogeneous surface. + denotes positively charged region, − denotes negatively charged region, gray area denotes hydrophobic regions, and arrows indicate polar regions with dipole.

Other surface properties that can influence protein interaction include topography, composition, heterogeneity, and electrical potential (Figure 7.3). The topography of a surface can influence the available surface area for protein adsorption. The *surface composition* of a material, or its chemical moieties, will influence interactions with amino acid side chains and determine the intermolecular forces that will be exerted on a protein. The *heterogeneity* of a material will increase the types of protein "domains" that can interact with a surface. *Electrical potential* will also affect the conformation of a protein on a surface and perhaps affect the distribution of ions within a protein.

7.5
Kinetics of Protein Adsorption

Before adsorption can occur, we must look at the transport of proteins to surfaces. Transport occurs through two main modes: diffusion and convection. The mode of transport that is most relevant to cell–material interactions is near the boundary layer between the surface and protein solution, where the protein adsorption rate is primarily dictated by diffusion alone. However, in certain microfluidic setups and blood-contacting surfaces, convection-mediated adsorption may also be important.

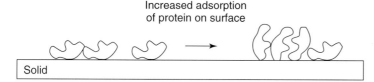

Figure 7.4 Packing density affects protein adsorption. As the packing density of a protein increases on a surface it can affect its overall orientation, conformation, and activity.

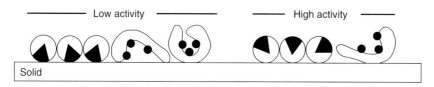

Figure 7.5 Protein orientation versus activity. Proteins with active domains oriented toward cells will exhibit higher overall activity compared to proteins with active domains facing the substrate surface. Black region denotes active binding domain.

In the adsorption process, proteins are constantly in motion, arranging themselves in the most entropically favorable orientation and packing density. After a protein has begun binding to a surface, as protein density increases, one can expect the protein to adopt different conformations for more efficient packing (Figure 7.4). Interactions between the protein and the surface, as well as between other proteins, will influence protein conformation and its interactions with surfaces. Usually, this is accompanied by limited protein unfolding and spreading. After a period of time, protein adsorption is thought to be irreversible since it would be unlikely that all the modalities for interaction would fail simultaneously. After adsorption, the final conformations of the protein with respect to the surface dictate its activity and function. Depending on the orientation of the protein, its activity can be severely hindered (Figure 7.5).

Even though protein adsorption is classically thought as irreversible, in the context of heterogeneous protein solution, a protein can be displaced from a surface by another protein of higher surface affinity. As mentioned before, proteins on a surface are highly dynamic. At the instant when a protein changes its orientation or breaks a series of noncovalent interactions with the surface, it is possible for a second protein to compete for that physical space on the surface. As a consequence, if this second protein has higher affinity for the surface, it could displace the original protein. Therefore, the final protein concentration on a surface is influenced by both its concentration and protein affinity. In a mixed population of proteins, those with higher concentration will be adsorbed first; however, they will eventually be replaced by proteins with higher surface affinity. This is known as the *Vroman effect* (Figure 7.6) [4, 5].

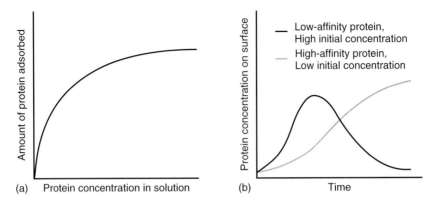

Figure 7.6 Vroman effect. (a) Protein adsorption isotherm of single protein species. (b) Vroman effect with two proteins of different initial concentration and affinities for surface.

7.6
Cell Communication

When considering strategies for controlling cell–material interactions, it is important to understand the mechanisms dictating cell communication, whether it is to another cell, from an ECM, or from a surface. We consider first the different types of signals that cells can respond to, then the receptors to which signals bind, the intracellular signaling pathways, and finally, the terminal effects of these signals. Cellular communication is a critical regulator of all cell behavior and can alter metabolism, gene expression, or cell shape/movement, with both spatial and temporal control.

7.6.1
Intracellular Communication

The main regulator of almost all cell-binding events is protein-based cell receptors. A typical cell membrane is about 50% proteins by mass. In this signaling cascade, first, a signaling molecule, such as a growth factor or an ECM protein, binds to its target cell through a specific protein called a *receptor*. This binding event initiates downstream signal transduction and amplification, ultimately leading to a change in cell behavior (Figure 7.7).

There are two main classes of receptors; transmembrane receptors that span cell plasma membranes and bind to extracellular ligands and intracellular receptors that reside in the cytoplasm or nucleus of the cell. In the case of transmembrane receptors, after binding to the target receptor occurs, the receptor becomes activated and initiates a cascade of downstream signaling events. Alternatively, intracellular membrane receptors can be targeted via small hydrophobic molecules that can freely diffuse across the cell membrane. In either case, these binding events activate *secondary messengers* that are generated in large numbers and activate downstream signaling proteins. There are two types of secondary messengers, those that are

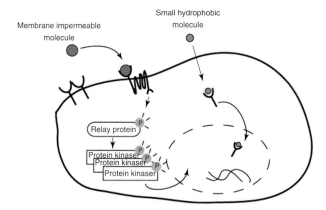

Figure 7.7 Cell signaling. General schematic representation of intracellular signal transduction of membrane-impermeable and membrane-permeable signaling molecules.

soluble in the cytosol and those that are membrane bound. Either way, they serve to amplify the original signal from the receptor–signaling molecule ligation.

There is also a class of larger signaling molecules known as *intracellular signaling proteins*. These proteins, when activated by secondary messengers, activate their own downstream small and large protein signaling molecules. There are 10 main modalities for downstream signal propagation, also known as *signal transduction*. Some of the proteins in this cascade include relay, messenger, adaptor, amplifier, transducer, bifurcation, integrator, latent gene regulatory, modulator, and scaffold proteins. For the purpose of our discussion, the exact action of these protein classes is not important, but more information is found in other sources [1]. Ultimately, signal transduction from cell–material interactions leads to complex symphonies of signals, each having their own concentrations and temporal activation states.

7.6.2
Intercellular Communication

There are five main cell–cell communication modalities: contact-dependent, paracrine, synaptic, endocrine, and autocrine (Figure 7.8). Contact-dependent and, to a lesser extent, paracrine modalities are the most important when designing cell–material interactions. Oftentimes, when designing materials to alter cell behavior, we immobilize proteins/signaling molecules, also known as *ligands*, to the surface. Direct cell interaction with these immobilized ligands ultimately leads to downstream signaling. Small populations of cells can also regulate their behavior using autocrine/paracrine signaling.

Clearly, there is a large interplay between various signaling modalities, and this should be taken into consideration when attempting to engineer a surface for a specific cell–material interaction.

A cell, in general, is exposed to hundreds, if not thousands, of different signals in its own local microenvironment. However, cells have adapted to respond to only

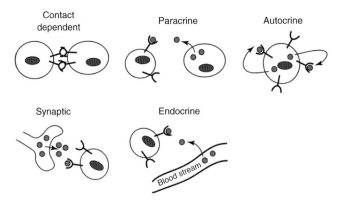

Figure 7.8 Five basic modes of cell communication.

very specific combinations of signals for various behaviors. One combination of signals may result in cell survival, while another set of signals result in migration and proliferation.

The spatiotemporal coordination of soluble, ECM-bound, and cellular signals create virtually unlimited signal combinations. Also, cell behavior to a particular signal is not conserved across cell types. For example, a cardiac myocyte on exposure to acetylcholine, a common neurotransmitter, decreases the rate of contraction, whereas a salivary gland cell responds to acetylcholine via secretion of amylase. It is clear that cellular communication is a complex symphony of many signals. In order to effectively discover exactly which combinations of signals is required for a particular cellular behavior, we must have an empirical and experimentally sound platform for discovery, where we can expose a cell to a single or multiple signals in a highly defined manner. In the following sections, we cover basic concepts required for designing cell–material interactions and also review the current methodologies used to study cell–material signaling events.

7.7
Cell Adhesion Background

Most mammalian cells adhere to ECMs via specific cellular receptors. In the previous section, we discussed how the ECM functions as a scaffold facilitating the transfer of signals to adhering cells via specific proteins such as those containing adhesive domains such as fibrinogen, vitronectin, collagen, and FN. Cells bind these adhesive ligands through a special class of receptors known as *integrins* [6, 7]. Through integrin-mediated signaling, these ligands regulate a range of specific cellular processes such as adhesion, migration, proliferation, secretion, gene expression, and apoptosis. In the following section, we highlight the importance of integrin-mediated cell adhesion before we begin our discussion on controlling cell–matrix interactions.

Integrin-mediated cell adhesion to the ECM is central to the organization, maintenance, and repair of tissues by providing mechanical anchorage and signals that direct cell survival, migration, cell cycle progression, and the expression of differentiated phenotypes [8, 9]. In fact, integrin-mediated adhesion is required for mammalian development. Animal models containing deletions for integrin receptors, integrin ligands, and focal adhesion (FA) components result in absolute lethality at early embryonic stages [10–13]. Abnormalities in adhesive interactions are often involved in pathological states, including blood clotting and wound healing defects, as well as malignant tumor formation [14, 15]. Furthermore, adhesive interactions are responsible for regulating cellular and host responses to implanted biomedical devices, tissue-engineered constructs, and biotechnological systems [7].

Whereas significant progress has been achieved in identifying key adhesion components and how these participate in cell spreading, migration, and signaling, the mechanical interactions between a cell and its environment remain poorly understood. It is increasingly evident that mechanotransduction, or signaling triggered by mechanical stimulation, between cells and their environment, regulates gene expression, cell fate, and even malignant transformation [16]. Therefore, engineering and controlling specific cell–matrix interactions are essential to unraveling these complex interactions [17–22].

7.8
Integrins and Adhesive Force Generation Overview

Cell adhesion to native ECM components, such as FN and laminin, is primarily mediated by the integrin family of heterodimeric ($\alpha\beta$) receptors [6]. The integrin receptor has a large extracellular domain formed by both α and β subunits, a single transmembrane pass, and two short cytoplasmic tails that do not contain catalytic motifs. Integrin-mediated adhesion is a highly regulated process involving receptor activation and mechanical coupling to extracellular ligands [23–26]. Bound integrins rapidly associate with the actin cytoskeleton and cluster together to form FAs – discrete supramolecular complexes (Figure 7.9).

FAs are supramolecular complexes that contain both structural proteins, such as vinculin, talin, and α-actinin, and signaling molecules, including FAK, SRC, and paxillin [27]. FAs function as structural links, allowing for strong cell-adhesive forces, and as signal transduction elements between the cell and its extracellular environment. These adhesive complexes are dynamic structures that are actively remodeled during cell migration [28, 29]. Assembly and disassembly of FAs is regulated by numerous pathways in response to external stimuli, including growth factors and mechanical force [30–32]. In particular, the Rho GTPase effector Rho-kinase (also designated ROCK and ROKαII) plays a central role in serum-induced formation of FAs and stress fibers. Rho-kinase controls FA and stress fiber formation by regulating actomyosin contractility through direct phosphorylation of the myosin light chain and inactivation of myosin phosphatase

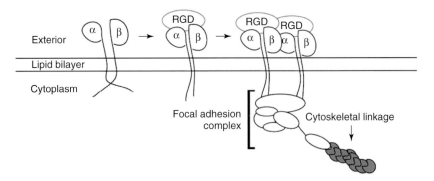

Figure 7.9 Basic diagram of integrin cascade. Inactive integrin becomes activated, creating a conformation change in the molecule, unfolding the cytoplasmic domains. This allows interactions with focal adhesion proteins, ultimately resulting in linkage with the cell cytoskeleton.

[33, 34]. Actomyosin contractility then drives the formation of FAs and stress fibers by an unknown mechanism [35, 36].

Mechanical interactions between a cell and its ECM are composed of different spatiotemporal force components (Figure 7.10). For example, migrating cells utilize complex spatiotemporal patterns of traction forces via FA assembly dynamics that ultimately generate directional cell movement [28, 37]. Also, a cell attached to a surface exhibits an equilibrium balance between its own internal contractile forces and the anchorage forces to the underlying surface. This equilibrium is dictated by the size and distribution of cell–material adhesive structures, cytoskeletal architecture, and actomyosin contractile forces. What further confounds the analysis is that all these adhesive interactions are interrelated and exhibit complex, often nonlinear relationships. For example, migration speed exhibits a biphasic dependence on adhesive strength and ligand density [38], while epithelial cell scattering correlates with adhesion strength and actomyosin contractility but not with migration speed [39].

The accepted model for focal complex-generated adhesive forces, proposed by McClay and Erickson, postulates a two-step process consisting of initial integrin–ligand binding followed by a rapid strengthening [40]. The strengthening response is a consequence of (i) increases in cell–substrate contact area (spreading), (ii) receptor recruitment to anchoring sites (recruitment and clustering), and (iii) interactions with cytoskeletal elements that lead to enhanced force distribution among bound receptors via local membrane stiffening (FA assembly) (Figure 7.11).

Integrin clustering was one of the first observed and most intensively studied events in the adhesive process. Using chimeric receptors, LaFlamme et al., showed that ligand binding targets integrin receptors to sites of integrin–FN adhesion [41]. Integrin clustering is a critical step in the adhesive process by promoting the recruitment of cytoskeletal components and initiating activation of signaling molecules. On the basis of immunostaining analyses, Yamada and colleagues

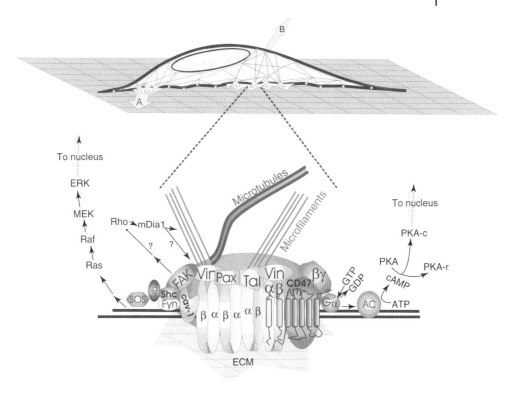

Figure 7.10 Detailed cartoon of mechanotransduction pathway. (A denotes attachment force, B denotes cytoskeletal contractile force) (Source: Copyright Ingber et al.)

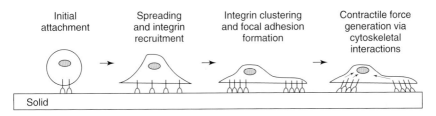

Figure 7.11 Cell adhesion cascade on a surface. Initial attachment leads to integrin recruitment, spreading, and integrin clustering; ultimately resulting in the generation of contractile forces by the cell on the surface.

demonstrated that integrin binding and clustering have synergistic effects on integrin function [42, 43]. Integrin occupancy by a single monovalent ligand induced receptor redistribution but no recruitment of cytoskeletal or signaling elements. In contrast, antibody-mediated integrin aggregation induced recruitment of FAK and tensin but no other FA components. Both integrin occupancy and aggregation are required for the robust assembly of FAs and activation of signaling pathways.

7.9
Adhesive Interactions in Cell, and Host Responses to Biomaterials

Integrins are critically involved in host and cellular responses to biomaterials because of their essential roles in cell adhesion to ECM components. In platelet aggregation, the platelet integrin $\alpha_{IIb}\beta_3$ (GPIIb/IIa) binds to several ligands such as fibrinogen, von Willebrand factor, and FN [6]. This receptor is the essential mediator of initial events in the blood-activation cascade after blood has come into contact with a synthetic material [44, 45]. Leukocyte-specific β_2 integrins, such as $\alpha_M\beta_2$ (Mac-1), mediate monocyte and macrophage adhesion to various ligands, including fibrinogen, FN, IgG, and complement fragment iC3b. These receptors all play central roles in inflammatory responses *in vivo* [46, 47]. Binding of $\alpha_M\beta_2$ integrin to fibrinogen P1 and P2 domains exposed on adsorption to biomaterial surfaces controls the recruitment and accumulation of inflammatory cells on implanted devices [48]. This integrin also plays a critical role in macrophage adhesion and fusion into giant foreign body cells [47, 48]. For many types of tissues such as connective, muscular, neural, and epithelial tissues, the β_1-family of integrins is largely responsible for adhesion to ECM ligands [7].

Integrins mediate cell–materials interactions by binding to adhesive extracellular ligands. These can be adsorbed from solution, blood, plasma, or serum; secreted and deposited onto the biomaterial surface by cells; or specifically engineered onto a biomaterial interface. All these interactions are highly dynamic, and the dominant adhesive mechanism could vary over time and is often different among cell types. For example, in blood plasma, the primary adhesive ligand is fibrinogen, while in serum, vitronectin plays a dominant role [49, 50]. As mentioned in previous sections, adhesive ligands can be displaced and replaced by other proteins in the surrounding medium via the Vroman effect. Therefore, a common surface engineering strategy is to coat a surface with FN to promote cell adhesion and survival, and then allow cells to secrete their own adhesive proteins and ECM onto the material surface. It is important to note that integrin expression and activity profiles on a particular cell change over time. Most cells exhibit several integrins specific for the same ligand, and the binding activity of these receptors can be rapidly regulated via changes in integrin conformation. However, integrin expression profiles do not necessarily correlate with integrin function on a particular substrate. In most cellular responses, multiple integrins are typically involved.

7.10
Model Systems for Controlling Integrin-Mediated Cell Adhesion

A key to studying and controlling cell–material interactions is starting with a material that provides a "clean-slate" background – one that prevents nonspecific protein adsorption and cell adhesion and signaling. Surfaces that are able to resist nonspecific adsorption of biomolecules are known as *"bioinert"* or *"nonfouling."* These types of nonfouling surfaces are required for the detailed study of singular bioactive peptide sequences or molecules. By tethering a single or multiple bioactive species in a highly characterized manner, detailed investigations of specific cell–material interactions can be completed.

Poly(ethylene glycol) (PEG) groups are highly resistant to protein adsorption and remain the benchmark for nonfouling surfaces [51]. The mechanism of resistance to protein adsorption is thought to be the combination of the polymer chain's ability to retain interfacial water (osmotic repulsion) and the resistance of the polymer chain to compression (entropic repulsion) [52]. Well-packed, self-assembled monolayers (SAMs) of ethylene glycol $(EG)_n$ repeats as short as $n = 3$, have excellent nonfouling properties [53, 54]. The nonfouling characteristics of these surfaces are highly dependent on the conformation of the $(EG)_n$ chain. Chains in either a helical or amorphous conformation have significantly higher resistance to protein adsorption than the all-trans conformation, probably because of interactions between the EG chains and interfacial water. While PEG groups are excellent at preventing protein adsorption, surfaces that completely abrogate protein adsorption have not been attained.

7.11
Self-Assembling Monolayers (SAMs)

SAMs are an organized layer of amphiphilic molecules that contain a head, alkane core, and functional group. In the context of cell adhesion, alkane-thiol-based SAM assembly on gold surfaces has been an extensively used model system to impart both specificity and nonfouling nature to surfaces. In this alkane-thiol-based SAM, the head group contains a –SH, or a thiol group, that has a specific, yet reversible affinity for Au (gold). In the assembly of the monolayer, the head groups are first attached to the gold surface via a near-covalent interaction ($\sim$100 kJ mol^{-1}) either in the liquid or in the gas phase. Then, the tail groups, which consist of long-chain alkanes, or single bonded carbon chains, begin to orient and form semicrystalline to crystalline structures. These tail group interactions stabilize the close packing of SAM molecules via hydrophobic interactions and van der Waals interactions (Figure 7.12).

In vitro, SAMs present a method to generate and study model substrates presenting specific ligands to which cells can adhere. An advantage of SAMs over other model substrates is the level of control over the composition of the substrate as well as the ability to characterize these biological interactions by surface plasmon resonance [55].

Self-assembling molecules terminated with oligo- or poly(ethylene glycol) units are generally used to render surfaces nonfouling to cell adhesion. Nonfouling SAMs are an effective way to prevent nonspecific protein adsorption to model surfaces. When used in conjunction with bioactive or other functional groups that support protein adsorption, nonfouling SAMs can be used to create a mixed monolayer. Mixed SAMs for the immobilization and attachment of cells only require 0.01–1% of the cell supporting SAM species [56]. Cells can then adhere

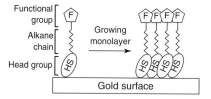

Figure 7.12 Basic self-assembling monolayer structure for controlling cell adhesion.

to these surfaces via nonspecific interactions. Nevertheless, cells will continually remodel their microenvironment over time; therefore, it might be difficult to control exactly which ECM protein and what ratio of these proteins the cell will secrete. In order to rigorously study the effects of a single ligand, the ligand can be tethered to a mixed SAM surface, to present integrin binding peptide sequences [57]. While these model surfaces are excellent systems for *in vitro* analysis, their utility is limited in *in vivo* applications because of poor stability (Figure 7.13).

7.12
Real-World Materials for Medical Applications

Although SAMs have been used extensively in *in vitro* applications, they have several limitations when used in *in vivo* medical applications. It is difficult to create robust noble metal coatings on biomedical implants. SAM layers are limited to gold and silver substrates, and SAMs still suffer from long-term instability and loss of bioresistance [58–61]. As a consequence, in the past decade, there has been a concerted effort toward developing medical-grade polymer brush systems and hydrogels.

7.12.1
Polymer Brush Systems

Polymer brush systems offer a more robust method of tethering biomolecules to biomedical device surfaces. These tethering chemistries have utilized the –OH groups on oxidized titanium and silicon substrates for silane immobilization. There are two main methodologies for functionalizing these surfaces, "grafting-to" and "grafting-from" approaches. Grafting-to involves tethering a polymer brush to the metal surface directly; however, this method suffers from low packing density because of steric hindrances from grafted groups. Grafting-from involves first immobilizing a silane layer on top of the metal to provide a high density of functional groups for direct chain-growth polymerization. Previously, adsorption of end-functionalized PEG onto titanium metal allowed the surface to be nonfouling [62]. Using surface-initiated atom-transfer radical polymerization (SI-ATRP), oligo(ethylene glycol)methacrylate (OEGMA) has been grafted from a gold surface modified with a thiol monolayer of α-bromo ester. Furthermore, poly(OEGMA) coatings have been successfully grown from titanium substrates

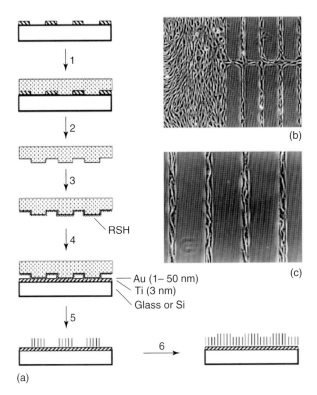

Figure 7.13 Microcontact printing. (a) Microcontact printing workflow. A polydimethylsiloxane (PDMS) stamp cast from a silicon wafer (1). The stamp is removed from the master (2), inked with SAM solution (3), and brought into contact with a gold-coated substrate (4). When the stamp is removed, a patterned monolayer remains only in the region of contact between the stamp and substrate (5). Backfilling of the substrate in a solution of a second alkane thiol results in the formation of a different monolayer in the remaining regions of gold (6). (b) An optical micrograph of capillary endothelial cells attached to a monolayer that was patterned into regions terminated by methyl groups (to which cells attached) and hexa(ethylene glycol) groups (which prevented the attachment of cells). (c) A micrograph at higher resolution shows the confinement of cell spreading to the pattern of underlying monolayer. (Source: Adapted from Ref. [56].)

and have been subsequently modified for tethering to bioactive peptide sequences [63] (Figure 7.14).

7.12.2
Hydrogels

Hydrogel technology has the ability to mimic the 3D architecture of the native ECM to control the function of cells and guide the spatially and temporally complex multicellular processes of tissue repair and regeneration. There are two classes of hydrogels, natural and synthetic – each with their own advantages and disadvantages. Natural hydrogels such as collagen and alginate are derived from

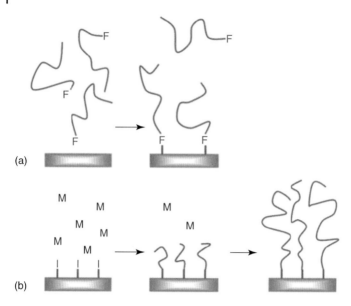

Figure 7.14 (a) The grafting-to method; the initially tethered polymer impedes further deposition, resulting in low density of brushes (F denotes coupling functional group). (b) Grafting-from method; the initiator-bearing substrate allows for high density of polymer brushes (M denotes monomer, I denotes initiator). (Source: Adapted from Ref. [64].)

nature; therefore, they already present receptor-binding ligands and are subject to cell-based enzymatic degradation and modeling. However, there are issues with purification, immunogenicity, and pathogen transmission. Synthetic hydrogels on the other hand are engineered to have specific structure–function relationships and can be fine-tuned to suit various biological applications via conjugation to bioactive molecules, adhesive moieties, and enzymatically degradable links. In terms of basic science research, synthetic hydrogels have been favored because of the ability to study single and multiple cell receptor–ligand interaction events in a highly empirical and stepwise manner. For an excellent review regarding the use of ECM-mimetic synthetic hydrogels, the readers are referred to [65].

7.13
Bio-Inspired, Adhesive Materials: New Routes to Promote Tissue Repair and Regeneration

The ECM is naturally complex and presents to cells a myriad of signals that convolute analysis of various cell processes. Using simplified synthetic ECM-mimetic peptides, model surfaces can be constructed that present only a particular signal or adhesive motif. Some of the most prevalent adhesion motifs consist of arginine-glycine-aspartic acid (RGD) tripeptide for FN [66] and the tyrosine-isoleucine-glycine-serine-arginine (YIGSR) oligopeptide for laminin

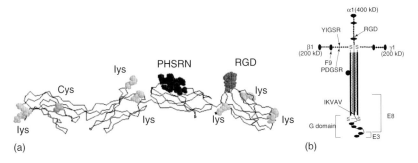

Figure 7.15 Schematic of RGD and YIGSR sites. (a) Schematic diagram of RGD and synergy site PHSRN on the fibronectin III domain. (b) Schematic diagram of YIGSR and RGD binding sites on laminin. (Source: Adapted from Ref. [86, 87].)

[67]. These short bioadhesive peptides have been tethered to both synthetic and natural surfaces, as well as three-dimensional scaffolds, to promote adhesion and migration in various cell types (as reviewed in [68–71]). In conjunction with these adhesive peptides, nonfouling, or protein-"resistant," supports, such as PEG, poly(acrylamide), and alginate, are often used to reduce nonspecific protein adhesion and background adhesion. The density of tethered peptide is an important design parameter. Cell adhesion, FA assembly, spreading and migration [72–75], neurite extension and neural differentiation [76, 77], smooth muscle cell activities [78], as well as osteoblast and myoblast differentiation [79–81] exhibit peptide-density-dependent effects. Also, tethering of bioadhesive ligands onto biomaterial surfaces and scaffolds enhances *in vivo* responses, such as bone formation and integration [63, 82], nerve regeneration [83, 84], and corneal tissue repair [85] (Figure 7.15).

These results indicate that functionalization of biomaterials with short adhesive oligopeptides significantly enhances certain cellular activities. In addition, conveying biospecificity while avoiding unwanted interactions with other regions of the native ligand, short bioadhesive peptides allow facile incorporation into synthetic backbones and enhanced stability of the tethered motif. However, these strategies still have their limitations; namely, (i) low activity of oligopeptides compared to the native ligand because of the absence of modulatory domains, (ii) limited specificity for adhesion receptors and cell types, and (iii) inability to bind certain receptors because of conformation differences compared to the native ligand [7]. Strategies to improve ligand activity include use of conformationally constrained (e.g., cyclic) RGD, oligopeptide mixtures, and recombinant protein fragments spanning binding domains of native ligands [63, 88–92]. Also, self-assembling peptides reconstituting the triple helical structure of type I collagen have been used to target collagen integrin receptors and promote enhanced osteoblastic differentiation and mineralization on biomaterial supports [93, 94]. The improved activity and selectivity of these materials result in enhanced cellular activities and *in vivo* responses [63, 95] (Figure 7.16).

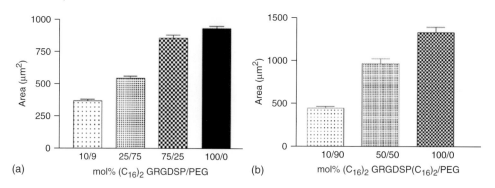

Figure 7.16 Dose-dependent spreading on linear versus cyclic RGD. Area of spread HUVEC to supported mixed monolayers of (a) (C16)2GRGDSP and PEG and (b) looped GRGDSP and PEG, after 60 min incubation at 37 °C in basal medium supplemented with 0.1% bovine serum albumin (error bars represent SEM). (Source: Adapted from Ref. [90].)

7.14
Dynamic Biomaterials

The previous sections highlighted the importance of bioadhesive ligand presentation in terms of density and specificity. However, current approaches to ligand presentation are often composed of static, constant densities of ligands. In contrast, natural cell-adhesive interactions with ECMs exhibit spatiotemporal patterns of binding and activation. Therefore, a key to future research in controlling cell–material interactions will be the development of materials that can respond to external stimuli. The following section presents both the model systems currently being employed and the emerging technologies in this field.

There are many modalities to control the presentation of ligands on a biomaterial. Some of these methods include, electrochemical desorption, oxidative release, light-controlled desorption, and enzyme-controlled activation.

7.14.1
Nonspecific "On" Switches

7.14.1.1 Electrochemical Desorption

Whitesides and coworkers created a surface that could be used to screen drug candidates via controlling cell migration. Using microcontact-printed (μCP) $HS(CH_2)_{17}CH_3$ and $HS(CH_2)_{11}(OCH_2OCH_2)_3OH(C_{11}EG_3)$ as a nonfouling barrier to protein adsorption, bovine capillary endothelial (BCE) cells were seeded and spread on the nonpatterned regions (Figure 7.17). By applying a < -1.2 V potential for 30 s, the complete desorption of these nonfouling SAMs occurred, and the BCE cells were allowed to migrate and proliferate [96].

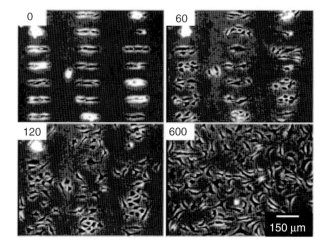

Figure 7.17 BCE cells spread on patterned surface with nonfouling barrier $C_{11}EG_3$ and C_{18}. Images show time lapse after application of cathodic voltage pulse (−1.2 V for 30 s). The numbers indicate the time elapsed (in minutes) after the voltage pulse. (Source: Adapted from Ref. [96].)

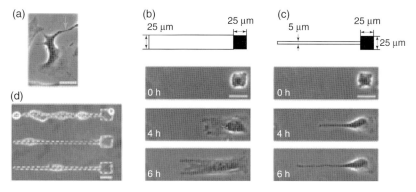

Figure 7.18 Selective initiation of two different characteristic protrusions of NIH3T3 cells. (a) Example of lamellipodia (dark arrow) and filopodium (light arrow) on unpatterned fibronectin surface. (b,c) Schematic of photo-patterned desorption along with time-lapsed protrusion. (d) Cell adhesion onto substrate with patterns illuminated simultaneously. (Source: Adapted from Ref. [101].)

7.14.1.2 Oxidative Release

Wittstock and coworkers developed an oxidative release mechanism for oligoethylene-glycol (OEG)-terminated thiol SAMs using ultramicroelectrodes [97]. By generating Br_2 *in situ* using electrochemical processes, an OEG-terminated SAM was oxidized to a cell-adhesive state. This process still had shortcomings in patterning fidelity because of limiting factors of electrode size and diffusion of bromine away from the electrode. The new method that has been employed by

Wittstock and coworkers involves the use of scanning electrochemical microscopy (SECM) to generate bromine *in situ* [98].

Nishizawa and coworkers created a pretreated surface with cell-adhesion-resistant bovine serum albumin (BSA) [99]. By the electrochemical generation of hydrobromic acid *in situ*, the BSA was removed, and other serum proteins from media were allowed to attach to the desorbed regions. Nevertheless, these methods also have the same problems as electrochemical desorption methods since they allow ECM proteins to adsorb, with little overall control of composition.

7.14.2
Photobased Desorption

Light-based control schemes allow for the high-resolution patterning of cells in culture in a highly spatially repeatable and specific manner. Strategies by Nakanishi *et al.* were based on silane-grafted SAMs containing a 2-nitrobenzyl group. A glass coverslip was modified with 1-(2-nitrophenyl)ethyl-5-trichloro silylpentanoate (NPE-TCSP) that contained nitrobenzyl ester group that could support the hydrophobic adsorption of nonadhesive BSA. Selective areas were irradiated with $\lambda = 365$ nm light, which converted exposed nitrobenzyl ester groups to carboxylic acid groups that could support FN adsorption [100]. Using a photomask, this technology was able to form subcellular adhesive islands to study FA formation and cell migration [101]. While these surfaces have the added benefit of spatiotemporal control via UV irradiation, it still has the same problems as the previously described surfaces in that nonspecific protein adsorption can still occur (Figure 7.18).

7.14.3
Integrin Specific "On" Switching

7.14.3.1 **Photoactivation**
Mrksich and colleagues [102] have generated substrates that used a Diels–Alders reaction for the immobilization of peptide-diene conjugates on quinine-terminated SAMs. The base substrate was a nitroveratryloxycarbonyl (NVOC)-protected hydroquinone-terminated SAM (1%) with a tri(ethylene glycol)-terminated alkanthiol background (99%). When selectively irradiated with $\lambda = 365$ nm for 2 min, the NVOC-protecting group is cleaved, leaving a reactive hydroquinone that can be electrochemically oxidized to yield a benzoquinone, which then reacts with a peptide diene. This system allowed for the selective immobilization of cells via the RGD peptide (Figures 7.19 and 7.20).

Del Campo and coworkers have demonstrated photocontrolled cell adhesion to SAMs via a caged-cyclic RGD-containing peptide [103]. A photolabile caging group consisting of a 3-(4,5 dimethoxy-2-nitrophenyl)-2-butyl ester (DMNPB) is tethered to the carboxylic acid side chain of the aspartic acid residue. On UV irradiation at 361 nm, the DMNPB group is cleaved, effectively removing the cage and leaving behind standard cyclic RGDfK. The caging group mechanism of action has not been fully characterized, but it is possible that additional steric hindrance, conformation

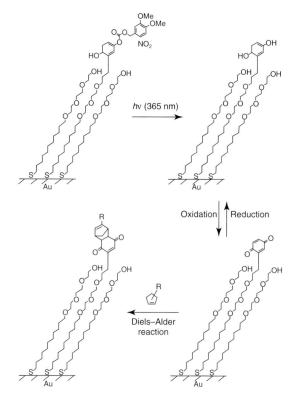

Figure 7.19 Strategy for patterning immobilized ligands to a self-assembled monolayer via NVOC-protected hydroquinone. On irradiation with UV light (365 nm), NVOC is cleaved leaving hydroquinone, which is reversibly oxidized to quinine. Quinone undergoes Diels–Alder reaction with cyclopentadiene-conjugated ligand, immobilizing the ligand on SAM. (Source: Adapted from Ref. [102].)

restriction, or changes in the charge distribution of the peptide are responsible for blocking integrin recognition (Figure 7.21).

7.14.4
Adhesion "Off" Switching

7.14.4.1 Electrochemical Off Switching

An improvement of Mrksich's NVOC system resulted in the generation of an electroactive quinone ester SAM that has a tethered RGD [104]. On electrochemical reduction (5 min at 550 mV), the quinone forms a hydroquinone, which then performs a rapid cyclization, ultimately cleaving the RGD peptide. A further improvement allowed Mrksich to generate two different redox-active tethers based on the quinone ester and O-silyl hydroquinone, which respond to −650 and 650 mV, respectively [105]. This allows for specific dynamic control of two different ligands on the same surface (Figure 7.22).

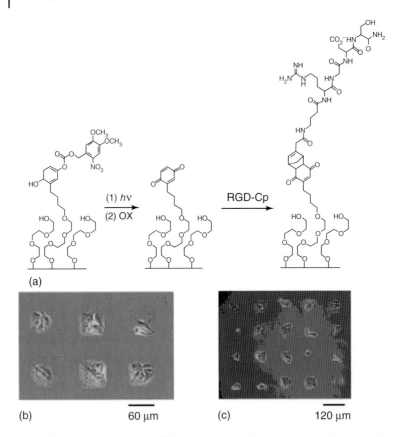

Figure 7.20 Photodeprotection/Diels–Alder immobilization strategy. (a) Schematic representation of overall process flow. (b,c) Swiss 3T3 fibroblasts attached to regions of substrate illuminated with UV light, oxidized (+400 mV, 15 s) and then treated with RGD-Cp (5 mM, 4 h). (Source: Adapted from Ref. [102].)

Del Campo and colleagues have recently developed an intercalated 4,5-dialkoxy 1-(2-nitrophenyl)ethyl photolabile group that can be attached to free amine groups at the surface of a substrate via a carbamate bond [106]. A tetraethylene glycol (TEG) spacer is included in the structure in order to provide a protein- and cell-repellent surface before attaching bioactive ligands. A bioactive ligand, such as biotin or RGD oligopeptide, can be attached to the surface by reacting with the free amine at the end of the TEG spacer. When used in conjunction with amine-terminated SAMs, this methodology allowed for the removal of human umbilical vein endothelial cells from model surfaces using irradiation at $\lambda = 350$ nm. After 8 min of irradiation, 70% cleavage of RGD was attained. This system allows for the selective immobilization of a peptide containing a free carboxylic acid and its subsequent cleavage by UV irradiation (Figure 7.23).

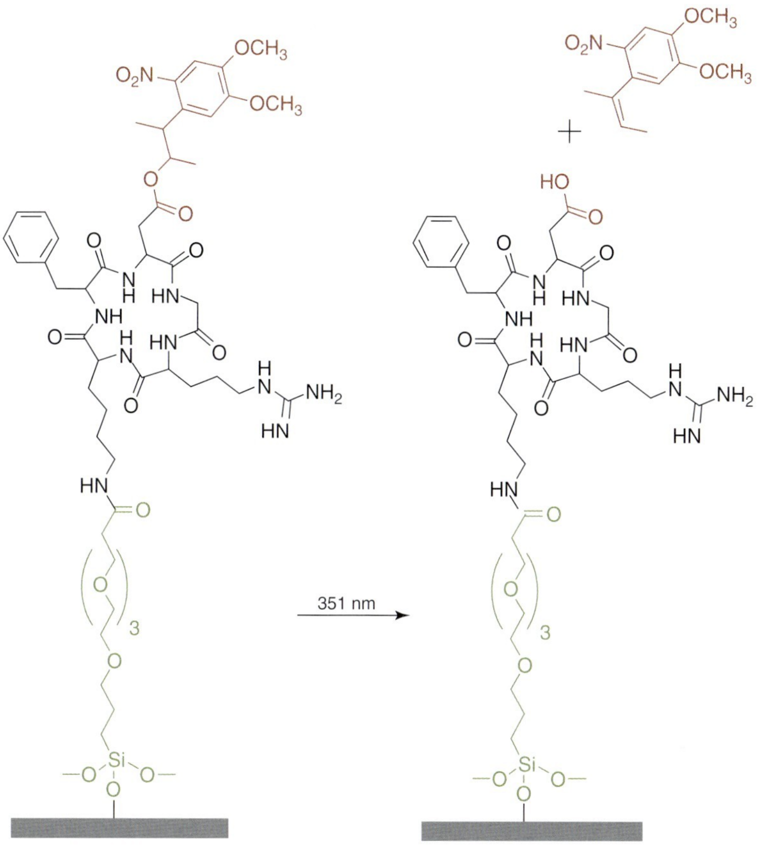

Figure 7.21 Caged RGD compound immobilized to surface. Chemical structure of cyclo[RGD(DMNPBfK)] (DMNPB in red) with TEG-based silane immobilization strategy (TEG in green). On UV irradiation at 365 nm, DMNPB group is cleaved from aspartic acid residue, restoring bioactive cyclo(RGD). (Source: Adapted from Ref. [103].)

7.14.5
Reversible Adhesion Switches

7.14.5.1 Reversible Photoactive Switching

Liu et al. [107] have demonstrated a switchable SAM system that can reversibly modulate RGD presentation. This design employed an azobenzene unit that upon UV irradiation ($\lambda = 340\text{–}380$ nm) can be switched from a thermally stable E-isomer to the thermally unstable Z-isomer. Additional irradiation at $\lambda = 450\text{–}490$ nm reverts this SAM to its original E-state. When an RGD group is tethered to the end of this molecule, in the E-isomer form, the substrate supports cell adhesion, while in the Z-isomer form, cell-binding is significantly attenuated. Using a similar method, Kessler et al. used the switching from the E- to Z-isomer directly to modulate the percentage of cells on the surface [108]. These methods, however,

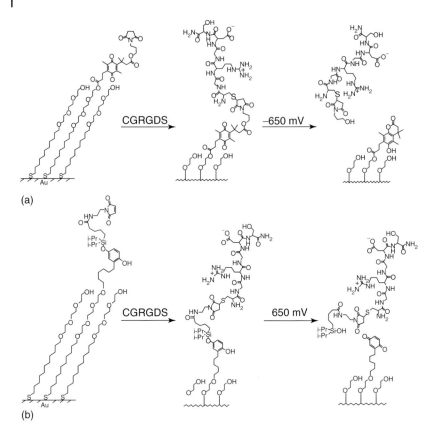

Figure 7.22 Molecular strategies to prepare dynamic substrates. (a) Monolayer presenting maleimide tethered to electroactive quinine ester reacts with cysteine-terminated RGD peptides to immobilize the ligand; electrochemical reduction of quinine releases RGD ligand. (b) Monolayer presenting a maleimide group tethered to an electroactive O-silyl hydroquinone is used to immobilize cysteine-terminated RGD, electrochemical oxidation releases RGD ligand. (Source: Adapted from Ref. [105].)

are potentially limited by damaging pre-adhered cells due to the UV exposure (Figure 7.24).

7.14.5.2 Reversible Temperature-Based Switching

One of the most promising technologies to control the adhesion of cells and tissues to a substrate has been the temperature-responsive polymer poly(N-isopropylacrylamide) (PIPAAm). This polymer, which can be covalently tethered to cell culture surface, has the unique characteristic of being able to control protein adsorption based on temperature alone. At temperatures above the lower critical solution temperature (LCST) of 32 °C, poly(NIPAMM) surfaces are hydrophobic and support the adsorption of proteins from serum and media, allowing cells to attach and proliferate. On lowering the temperature below the LCST, the surface is rendered hydrophilic and causes a hydrated layer to form

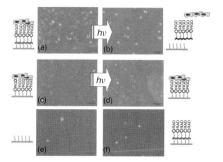

Figure 7.23 Schematic representation and microscopy images of intercalated 4,5-dialkoxy 1-(2-nitrophenyl)ethyl photolabile linker. (a,c) HUVECs attached to RGD immobilized onto photolabile linker and noncleavable linker control, respectively. (b,d) Resulting HUVEC detachment post UV irradiation. (e) HUVECs seeded 2 h on amine-terminated SAM. (f) HUVECs incubated for 2 h on cyclo(RADfk) nonadhesive control coupled to amine-terminated SAM with linker. (Source: Adapted from Ref. [106].)

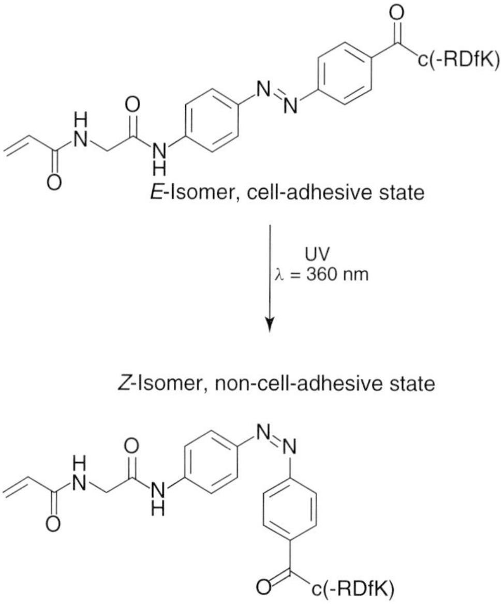

Figure 7.24 Cyclic RGD peptide presented on photoswitchable 4-[(4-aminophenyl)azo]benzocarbonyl unit]. c-(RDfK) denotes cyclic(RGD) peptide. (Source: Adapted from Ref. [109].)

between the cell sheet and the culture surface [110]. This causes all cells and ECM components to detach from the polymer surface in a sheet, allowing for *cell-sheet*-based tissue engineering. This technology has experienced great success in recent years and is currently being utilized in early-phase clinical trials [111] (Figure 7.25).

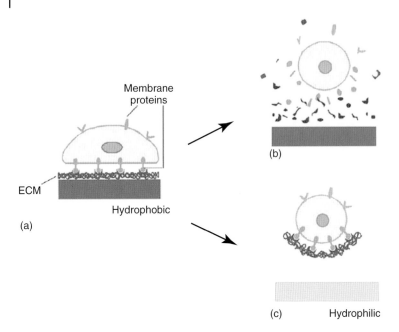

Figure 7.25 Cell harvest mechanism by using temperature-responsive culture surfaces. (a) Cells attach to hydrophobic surfaces via cell membrane proteins and adsorbed ECM proteins from serum. (b) Enzymatic digestion of membrane and ECM components result in cell detachment. (c) On poly(NIPAAm) surfaces, interconnection between ECM and cells remained intact post release. (Source: Adapted from Ref. [112].)

7.14.6
Conclusions and Future Prospects

In summary, controlling cell–adhesive interactions is paramount to the study of cell signaling and biomaterials development. Cell–matrix interactions rely on specific interactions with either the native or synthetic ECM. This ECM is comprised of various biomolecules including proteins. Proteins are biomolecules with complex structure and can modulate function and activity of biological systems. Engineering surfaces with simplified protein sequences allows for an empirical approach to decoding these complex cell–material interactions. Through the rational design of material surfaces, protein adsorption and receptor–ligand interaction can be tightly controlled both spatially and temporally. Because of this level of control, researchers have begun unraveling the complex symphony of signals for cellular events such as adhesion and cell spreading.

Cell–material interactions are primarily mediated by receptor–ligand interactions between adsorbed or covalently bound ligands to a surface and cell receptors called integrins. Integrin-mediated signaling has been implicated in many cellular processes such as adhesion, migration, growth, secretion, gene expression, and apoptosis. These adhesive interactions also regulate cellular and host responses to implanted biomedical devices, tissue-engineered constructs, and biotechnological systems. It is increasingly evident that mechanotransduction between cells and their environment regulates gene expression, cell fate, and even malignant transformation.

The cell adhesion cascade involves initial attachment, integrin recruitment and spreading, integrin clustering and FA formation, and contractile force generation via cytoskeletal interactions. Cell adhesion to native ECM components, such as FN and laminin, is mediated by the activation and mechanical coupling to integrins. Bound integrins rapidly associate with the actin cytoskeleton and cluster to form FA complexes.

Model surfaces have been created to explore in-depth specific mechanisms for cell adhesion and spreading to various peptide sequences and adhesive proteins. Simplified protein sequences such as RGD and YIGSR are tethered to both synthetic and natural surfaces in order to study the effects of a single protein motif. Dynamic biomaterials can modulate the spatiotemporal presentation of these motifs through electrochemical, enzyme, and light-controlled modalities. The development of additional dynamic biomaterials is critical to unraveling the complex adhesive interactions between cells and surfaces. Applications of adhesive control in three dimensions and in *in vivo* applications will undoubtedly provide key insights into cellular behavior in more complex ECM constructs and may also provide new avenues for clinical therapies and treatments.

References

1. Alberts, B. (2008) *Molecular Biology of the Cell*, Garland Science.
2. Reichardt, L.F., Kreis, T., and Vale, R. (1999) in *Guidebook to the Extracellular Matrix, Anchor, and Adhesion Proteins*. Oxford University Press, New York, pp. 335–344.
3. Berg, J.M., Tymoczko, J.L., and Stryer, L. (2007) *Biochemistry*, W. H. Freeman.
4. Andrade, J. *et al.* (1992) *Clin. Mater.*, **11**(1–4), 67–84.
5. Horbett, T.A. and Brash, J.L. (1987) *Proteins at Interfaces: Current Issues and Future Prospects*, ACS Publications.
6. Hynes, R.O. (2002) *Cell*, **110**(6), 673–687.
7. Garcia, A.J. (2005) *Biomaterials*, **26**(36), 7525–7529.
8. De Arcangelis, A. and Georges-Labouesse, E. (2000) *Trends Genet.*, **16**(9), 389–395.
9. Danen, E.H. and Sonnenberg, A. (2003) *J. Pathol.*, **201**(4), 632–641.
10. George, E.L. *et al.* (1993) *Development*, **119**(4), 1079–1091.
11. Stephens, L.E. *et al.* (1995) *Genes Dev.*, **9**(15), 1883–1895.
12. Furuta, Y. *et al.* (1995) *Oncogene*, **11**(10), 1989–1995.
13. Xu, W., Baribault, H., and Adamson, E.D. (1998) *Development*, **125**(2), 327–337.
14. Wehrle-Haller, B. and Imhof, B.A. (2003) *J. Pathol.*, **200**(4), 481–487.
15. Jin, H. and Varner, J. (2004) *Br. J. Cancer*, **90**(3), 561–565.
16. Chen, C.S. (2008) *J. Cell Sci.*, **121**(20), 3285.
17. Wozniak, M.A. *et al.* (2003) *J. Cell Biol.*, **163**(3), 583–595.
18. McBeath, R. *et al.* (2004) *Dev. Cell*, **6**(4), 483–495.
19. Mammoto, A. *et al.* (2004) *J. Biol. Chem.*, **279**(25), 26323–26330.
20. Polte, T.R. *et al.* (2004) *Am. J. Physiol. Cell Physiol.*, **286**(3), C518–C528.

21. Engler, A.J. et al. (2004) *J. Cell Biol.*, **166**(6), 877–887.
22. Paszek, M.J. et al. (2005) *Cancer Cell*, **8**(3), 241–254.
23. Faull, R.J. et al. (1993) *J. Cell Biol.*, **121**(1), 155–162.
24. Choquet, D., Felsenfeld, D.P., and Sheetz, M.P. (1997) *Cell*, **88**(1), 39–48.
25. Garcia, A.J., Huber, F., and Boettiger, D. (1998) *J. Biol. Chem.*, **273**(18), 10988–10993.
26. Ginsberg, M.H., Partridge, A., and Shattil, S.J. (2005) *Curr. Opin. Cell Biol.*, **17**(5), 509–516.
27. Geiger, B. and Bershadsky, A. (2001) *Curr. Opin. Cell Biol.*, **13**(5), 584–592.
28. Gupton, S.L. and Waterman-Storer, C.M. (2006) *Cell*, **125**(7), 1361–1374.
29. Ridley, A.J. et al. (2003) *Science*, **302**(5651), 1704–1709.
30. Ridley, A.J. and Hall, A. (1992) *Cell*, **70**(3), 389–399.
31. Greenwood, J.A. et al. (2000) *J. Cell Biol.*, **150**(3), 627–641.
32. Riveline, D. et al. (2001) *J. Cell Biol.*, **153**(6), 1175–1185.
33. Amano, M. et al. (1997) *Science*, **275**(5304), 1308–1311.
34. Kimura, K. et al. (1996) *Science*, **273**(5272), 245–248.
35. Chrzanowska-Wodnicka, M. and Burridge, K. (1996) *J. Cell Biol.*, **133**(6), 1403–1415.
36. Amano, M. et al. (1998) *Genes Cells*, **3**(3), 177–188.
37. Munevar, S., Wang, Y.L., and Dembo, M. (2001) *Mol. Biol. Cell*, **12**(12), 3947–3954.
38. Palecek, S.P. et al. (1997) *Nature*, **385**(6616), 537–540.
39. de Rooij, J. et al. (2005) *J. Cell Biol.*, **171**(1), 153–164.
40. Lotz, M.M. et al. (1989) *J. Cell Biol.*, **109**(4, Pt 1), 1795–1805.
41. LaFlamme, S.E., Akiyama, S.K., and Yamada, K.M. (1992) *J. Cell Biol.*, **117**(2), 437–447.
42. Miyamoto, S., Akiyama, S.K., and Yamada, K.M. (1995) *Science*, **267**(5199), 883–885.
43. Miyamoto, S. et al. (1995) *J. Cell Biol.*, **131**(3), 791–805.
44. Broberg, M., Eriksson, C., and Nygren, H. (2002) *J. Lab. Clin. Med*, **139**(3), 163–172.
45. Gorbet, M.B. and Sefton, M.V. (2003) *J. Biomed. Mater. Res. A*, **67**(3), 792–800.
46. Tang, L. et al. (1996) *J. Clin. Invest.*, **97**(5), 1329.
47. Flick, M.J. et al. (2004) *J. Clin. Invest.*, **113**(11), 1596–1606.
48. Hu, W.J. et al. (2001) *Blood*, **98**(4), 1231–1238.
49. Tang, L. and Eaton, J.W. (1993) *J. Exp. Med.*, **178**(6), 2147.
50. Howlett, C.R. et al. (1994) *Biomaterials*, **15**(3), 213–222.
51. Merrill, E.W. (1992) *Poly (Ethylene Oxide) and Blood Contact*, Plenum Press, New York.
52. Hoffman, A.S. (1999) *J. Biomater. Sci., Polym. Ed*, **10**(10), 1011.
53. Prime, K.L. and Whitesides, G.M. (1991) *Science*, **252**(5010), 1164.
54. Prime, K.L. and Whitesides, G.M. (1993) *J. Am. Chem. Soc.*, **115**(23), 10714–10721.
55. Mrksich, M., Sigal, G.B., and Whitesides, G.M. (1995) *Langmuir*, **11**(11), 4383–4385.
56. Mrksich, M. (2000) *Chem. Soc. Rev.*, **29**(4), 267–273.
57. Houseman, B.T. and Mrksich, M. (1998) *J. Org. Chem.*, **63**, 7552–7555.
58. Flynn, N.T. et al. (2003) *Langmuir*, **19**(26), 10909–10915.
59. Nelson, C.M. et al. (2003) *Langmuir*, **19**(5), 1493–1499.
60. Mrksich, M. et al. (1997) *Exp. Cell Res.*, **235**(2), 305–313.
61. Yang, Y. et al. (2006) *Adv. Funct. Mater.*, **16**(8), 1001–1014.
62. Fan, X. et al. (2005) *J. Am. Chem. Soc.*, **127**(45), 15843–15847.
63. Petrie, T.A. et al. (2008) *Biomaterials*, **29**(19), 2849–2857.
64. Raynor, J.E. et al. (2009) Polymer brushes and self-assembled monolayers: Versatile platforms to control cell adhesion to biomaterials (Review), AVS. *Biointerphases*, **4**(2), FA3–FA16.
65. Lutolf, M. and Hubbell, J. (2005) *Nat. Biotechnol.*, **23**(1), 47–55.
66. Ruoslahti, E. and Pierschbacher, M.D. (1987) *Science*, **238**, 491–497.

67. Graf, J. et al. (1987) *Biochemistry*, **26**(22), 6896–6900.
68. Hubbell, J.A. (1999) *Curr. Opin. Biotechnol.*, **10**(2), 123–129.
69. Hubbell, J.A. (2003) *Curr. Opin. Biotechnol.*, **14**(5), 551–558.
70. Shakesheff, K., Cannizzaro, S., and Langer, R. (1998) *J. Biomater. Sci., Polym. Ed.*, **9**(5), 507–518.
71. Hersel, U., Dahmen, C., and Kessler, H. (2003) *Biomaterials*, **24**(24), 4385–4415.
72. Massia, S.P. and Hubbell, J.A. (1991) *J. Cell Biol.*, **114**, 1089–1100.
73. Maheshwari, G. et al. (2000) *J. Cell Sci.*, **113** (Pt. 10), 1677–1686.
74. Shin, H., Jo, S., and Mikos, A.G. (2002) *J. Biomed. Mater. Res.*, **61**(2), 169–179.
75. Sagnella, S.M. et al. (2004) *Biomaterials*, **25**(7–8), 1249–1259.
76. Schense, J.C. and Hubbell, J.A. (2000) *J. Biol. Chem.*, **275**(10), 6813–6818.
77. Silva, G.A. et al. (2004) *Science*, **303**(5662), 1352–1355.
78. Mann, B.K. and West, J.L. (2002) *J. Biomed. Mater. Res.*, **60**(1), 86–93.
79. Petrie, T.A. et al. (2006) *Biomaterials*, **27**(31), 5459–5470.
80. Rowley, J.A. and Mooney, D.J. (2002) *J. Biomed. Mater. Res.*, **60**(2), 217–223.
81. Rezania, A. and Healy, K.E. (2000) *J. Biomed. Mater. Res.*, **52**(4), 595–600.
82. Alsberg, E. et al. (2002) *Proc. Natl. Acad. Sci. U.S.A*, **99**(19), 12025–12030.
83. Schense, J.C. et al. (2000) *Nat. Biotechnol.*, **18**(4), 415–419.
84. Yu, X. and Bellamkonda, R.V. (2003) *Tissue Eng.*, **9**(3), 421–430.
85. Li, F. et al. (2003) *Proc. Natl. Acad. Sci. U.S.A*, **100**(26), 15346–15351.
86. Cutler, S.M. and Garcia, A.J. (2003) *Biomaterials*, **24**(10), 1759–1770.
87. Nomizu, M. et al. (1995) *J. Biol. Chem.*, **270**(35), 20583.
88. Humphries, J.D. et al. (2000) *J. Biol. Chem.*, **275**(27), 20337–20345.
89. Kao, W.J. et al. (2001) *J. Biomed. Mater. Res.*, **55**(1), 79–88.
90. Ochsenhirt, S.E. et al. (2006) *Biomaterials*, **27**(20), 3863–3874.
91. Rezania, A. and Healy, K.E. (1999) *Biotechnol. Prog.*, **15**(1), 19–32.
92. Dee, K.C., Andersen, T.T., and Bizios, R. (1998) *J. Biomed. Mater. Res.*, **40**(3), 371–377.
93. Reyes, C.D. and Garcia, A.J. (2003) *J. Biomed. Mater. Res.*, **65A**, 511–523.
94. Reyes, C.D. and Garcia, A.J. (2004) *J. Biomed. Mater. Res.*, **69A**(4), 591–600.
95. Reyes, C.D. et al. (2007) *Biomaterials*, **28**(21), 3228–3235.
96. Jiang, X. et al. (2003) *J. Am. Chem. Soc.*, **125**(9), 2366–2367.
97. Zhao, C., Witte, I., and Wittstock, G. (2006) *Angew. Chem. Int. Ed.*, **45**(33), 5469–5471.
98. Wittstock, G. et al. (2007) *Angew. Chem. Int. Ed.*, **46**(10), 1584–1617.
99. Kaji, H. et al. (2004) *Langmuir*, **20**(1), 16–19.
100. Yamaguchi, K. et al. (2000) *Chem. Lett.*, **29**(3), 228–229.
101. Nakanishi, J. et al. (2007) *J. Am. Chem. Soc.*, **129**(21), 6694–6695.
102. Dillmore, W.S., Yousaf, M.N., and Mrksich, M. (2004) *Langmuir*, **20**(17), 7223–7231.
103. Petersen, S. et al. (2008) *Angew. Chem. Int. Ed.*, **47**(17), 3192–3195.
104. Yeo, W.S., Hodneland, C.D., and Mrksich, M. (2001) *ChemBioChem*, **2**(7–8), 590–593.
105. Yeo, W.S. and Mrksich, M. (2006) *Langmuir*, **22**(25), 10816–10820.
106. Wirkner, M. et al. (2011) *Adv. Mater.*, **23**, 3907–3910.
107. Liu, D. et al. (2009) *Angew. Chem. Int. Ed.*, **48**(24), 4406–4408.
108. Auernheimer, J. et al. (2005) *J. Am. Chem. Soc.*, **127**(46), 16107–16110.
109. Robertus, J., Browne, W.R., and Feringa, B.L. (2010) *Chem. Soc. Rev.*, **39**(1), 354–378.
110. Okano, T. et al. (1995) *Biomaterials*, **16**(4), 297–303.
111. Yang, J. et al. (2007) *Biomaterials*, **28**(34), 5033–5043.
112. Shimizu, T. et al. (2003) *Biomaterials*, **24**(13), 2309–2316.

8
Fibronectin Fibrillogenesis at the Cell–Material Interface*

Marco Cantini, Patricia Rico, and Manuel Salmerón-Sánchez

8.1
Introduction

Cells within tissues are surrounded by fibrillar extracellular matrices (ECMs) that support cell adhesion, migration, proliferation, and differentiation. Fibronectin (FN) is an ECM protein that is organized into fibrillar networks by cells through an integrin-mediated process that involves contractile forces. This assembly allows for the unfolding of the FN molecule, exposing cryptic domains that are not available in the native globular FN structure and activating intracellular signaling complexes. The first part of this chapter describes the so-called cell-mediated FN fibrillogenesis, an important physiological process to elaborate the ECM. However, the production of FN fibrils *in vitro* in the absence of cells has been investigated making use of several methods, seeking to artificially reproduce FN fibrils. These methods range from the use of denaturants to the use of mechanical force to stretch and assemble the molecules into fibrils and are revised in the second part of the chapter. Finally, we have recently proposed the organization of FN into a physiological fibrillar network upon adsorption onto a concrete surface chemistry; the last part of the chapter demonstrates cell-free, material-induced FN fibrillogenesis into a biological matrix with enhanced cellular activity.

8.2
Cell-Driven Fibronectin Fibrillogenesis

The ECM is a dynamic and heterogeneous meshwork of fibrillar and nonfibrillar components that provides an active microenvironment for cell adhesion, differentiation, migration, and proliferation. Indeed, it acts as a reservoir for growth factors and fluids and can be assembled into elaborate structures participating in basement membranes and providing a scaffold for cells and tissue organization. It also regulates numerous cell functions by activating multiple signaling pathways

---

*Marco Cantini and Patricia Rico contributed equally to this work.

Biomimetic Approaches for Biomaterials Development, First Edition. Edited by João F. Mano.
© 2012 Wiley-VCH Verlag GmbH & Co. KGaA. Published 2012 by Wiley-VCH Verlag GmbH & Co. KGaA.

at adhesion sites. ECM is composed of collagens (COLs), laminins, and other glycoproteins such as FN. Intriguingly, the ECM components are secreted by cells as nonfunctional protein units, which are assembled into functional supramolecular structures in a highly regulated manner [1–3]. The ECM also plays an important role in diseases: defects in assembly stops embryogenesis; deranged assembly promotes scarring, tumorigenesis, and fibrotic disease; and delayed assembly provokes birth defects, chronic wounds, and skeletal malformations [4].

FN is a ubiquitous glycoprotein and the core component of ECM. It is synthesized by adherent cells, which then assemble it into a fibrillar network in an integrin-binding-dependent mechanism [5]. FN–integrin interactions initiate a stepwise process involving conformational changes of FN structure outside cells and organization of the actin cytoskeleton inside them. The conformational changes of FN are cell-mediated and induce the exposition of cryptic FN binding sites, allowing intermolecular interactions needed for FN fibrillar matrix formation [6]. The assembly of FN matrix is the initial step that orchestrates the assembly of other ECM proteins and promotes cell adhesion, migration, and signaling. FN matrix is formed by fibrils that form linear and interconnected networks. Thin fibrils predominate in early stages of matrix assembly (5 nm diameter), and, as the matrix matures, these fibrils cluster together into thicker fibrils (25 nm diameter) [4].

The functional properties of the FN matrix are diverse. The fibrils possess binding sites for multiple ECM components needed for the assembly of several other ECM proteins. FN matrix provides support for cell adhesion, and the adhesion receptors (most notably integrins) transduce signals that trigger cell fate [1, 2]. FN matrix also controls the availability of growth factors, for example, regulating the activation of TGF-β [7]. Therefore, FN matrix has an important role in normal cell adhesion and growth and plays a critical role in early development [8].

8.2.1
Fibronectin Structure

FN is a single-gene-encoded protein. Its 8 kB mRNA can be alternatively spliced, allowing the expression of 20 monomeric isoforms in humans and up to 12 in mouse [9], which may result in an even larger variety of FN isoforms if we consider the possible combination between monomers to form FN dimers. The most common mechanism of splicing takes place at three sites of the molecule, thus generating the two major different forms of FN: cellular fibronectin (cFN) and plasma fibronectin (pFN). pFN is produced by hepatocytes and secreted into the blood, where it remains in a soluble form in order to avoid fibrillar formation that may cause severe diseases. In addition, in blood, hematopoietic cells are adapted to express their integrins in an inactive conformation; hence, both mechanisms prevent the binding of pFN to cells. In case of a vascular injury or wound, integrins shift to their active conformation by a platelet-mediated mechanism, and pFN can bind and assemble into fibrils that are required for thrombus growth and

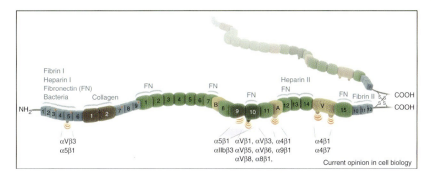

Figure 8.1 Molecular structure of FN that consists of three different modules (type I, blue; type II, brown; and type III, green). The alternatively spliced extradomains B and A and the variable region (V) are colored in ocher. The FN dimer forms via two disulfide bonds at the C-terminus. Integrin binding sites are indicated, as well as other binding domains for FN, collagen, fibrin, heparin, and bacteria. (Source: With permission from Ref. [13].)

stability [10, 11]. Interestingly, a recent study has shown that a major fraction of FN component of ECM is plasma derived [12].

cFN is secreted by cells as a dimer in a compact globular structure and is then assembled into fibrils (insoluble form) in a cell-dependent process. cFN and pFN differentiate between each other in the extra type III repeats included in their modular structure and produced by alternative splicing.

FN is a multidomain protein as illustrated in Figure 8.1. It contains domains to interact with other ECM proteins, glycosaminoglycans (GAGs), integrins, other FN molecules, and also pathogens such as bacteria [6]. This combination of domains allows the simultaneous binding of FN to cells and other molecules. Each subunit of the FN molecule ranges in size from 230 to 270 kDa, depending on alternative splicing of the mRNA, and organizes into dimers via two disulfide bonds at the C-terminus of the protein. FN contains three types of repeating modules, types I, II, and III (Figure 8.1). There are 12 type I modules, 2 type II modules, and 15–17 type III modules (15 constitutively expressed and 2 alternatively spliced). The type I and II units contain two intramolecular disulfide bonds to stabilize the folded structure, while type III units lack these kinds of bridges. Both type I and II protein modules are structured in β-sheets enclosing a hydrophobic core containing highly conserved aromatic amino acids [4, 6].

The extra type III repeats produced by alternative splicing occur by exon skipping mechanism at the EIIIA and EIIIB modules, while the variable region (V) is produced by exon subdivision mechanism at the V module [14]. Both extra type III repeats are included in cFN molecules, but they are not present in pFN. The functional roles of these extra III repeats are unclear since the *in vitro* and *in vivo* results obtained are controversial. Cultured cells lacking FN EIIIA or EIIIB regions have a slower growth rate and reduced FN matrix assembly. Nevertheless, EIIIA and EIIIB knockout mice are developmentally normal. Only few effects in abnormal skin wound healing and smaller atherosclerotic lesions have been found in EIIIA null mice. Double-null mice EIIIA/EIIIB presented defects in vascular development during embryogenesis, but matrix incorporation of mutant FN was not affected. In accordance with these results, it seems that the alternative exons are not necessary for matrix assembly but their absence may affect matrix levels [4, 6].

The V region is present in the vast majority of cFN subunits, but it is only present in one subunit of the pFN dimer. Results concerning this region suggest that it is essential for FN dimer secretion [14] and provides the binding site for $\alpha_4\beta_1$ integrin [15]. Indeed, there is a splice variant that lacks the V region and is only present in cartilaginous tissue, although it is not clear if this variant is able to assemble into fibrils [16, 17].

8.2.2
Essential Domains for FN Assembly

FN matrix assembly is a cell-dependent process and is mediated by the binding of FN dimers to integrin receptors. Integrins link FN to the actin cytoskeleton and other cytoskeletal-associated proteins, and this linkage is important for matrix assembly. Thus, FN–integrin binding promotes FN–FN association and fibril formation.

The essential domains for FN assembly are FN dimerization, the N-terminal type I repeats 1–5, the 70 kDa fragment, the conserved sequence Arg-Gly-Asp (RGD), and the synergy site (PHSRN).

FN dimerization depends on covalent association of the subunits mediated by a pair of disulfide bonds at the C-terminus of the FN molecule (Figure 8.1). Studies realized with recombinant monomeric FN lacking these cysteines ablated dimerization. The resulting monomeric FN was secreted but did not form fibrils [18]. Other results indicate that the dimer structure is involved in matrix incorporation even in the absence of cell binding [4].

The N-terminal assembly domain is composed of the first five repeats of type I units and is also part of the 70 kDa fragment. FN fibrils formation and consequently FN matrix assembly depends directly on this domain because it is the nexus point that binds FN molecules to each other by noncovalent interactions, as revealed by functional analyses using various mutant recombinant fragments [19]. Recombinant FN lacking all or part of the five type I repeats is unable to assemble into fibrils [18]. Furthermore, this part of the molecule can also interact with a number of other sites along the FN molecule, maybe providing overlap between interacting FN dimers.

The 70 kDa fragment extends from type I1 to I9 including the assembly and the COL/gelatin binding domains. This fragment binds to cells in monolayer culture and, when added in excess, it blocks FN matrix assembly. Similarly, the assembly is blocked when antibodies to this region are used [20, 21]. The inhibitory effect of FN I1-9 has been attributed to the high-affinity binding sites for FN, thus blocking FN–FN interactions required to align and cross-link FN molecules into fibrils. Within the 70 kDa fragment, the 40 kDa COL/gelatin binding modules do not appear to play a direct role in the assembly. It seems that the binding activity resides only in type I1–5 portion of the molecule (the N-terminal assembly domain).

Recently, potential integrin binding sites in FN I1-9 have been described. The Asn-Gly-Arg (NGR) sequence located in the first five repeats of type I modules is a high-affinity binding site for $\alpha_v\beta_3$ integrin [22]. The presence of this site offers an explanation for results obtained in some reports, showing that the initial binding of $\alpha_5\beta_1$ integrin to the RGD sequence is not the only way to promote fibrillogenesis. Such a site offers an alternative independent of $\alpha_5\beta_1$ binding in early embryogenesis [13].

FN assembly is a cell-mediated process that requires the binding of the primary receptor $\alpha_5\beta_1$ integrin to the RGD sequence located in repeat III10 and the synergy sequence PHSRN located in repeat III9 [23]. Blocking antibodies directed against either the cell binding domain of FN or the $\alpha_5\beta_1$ receptor inhibits fibril formation [20, 24]. The RGD sequence is required only during the initial steps of matrix assembly. Mutant recombinant FN lacking the RGD sequence is able to coassemble into fibrils if wild-type FN is added [18, 21]. Although both sites are required for fibril formation, the synergy site is not essential because matrix levels are drastically reduced but not ablated with FN lacking this sequence [25]. Some studies proposed an indirect function of the synergy region that provides a high-affinity binding of $\alpha_5\beta_1$ by optimally exposing the flexible RGD motif or by inducing appropriate electrostatic steering [26].

Experiment in homozygous mice that express an FN molecule in which the RGD motif was substituted with an integrin binding inactive RGE motif showed a lethal phenotype in early embryonic development but normal quantities of FN fibrils. The results indicate that there is an independent α_5 integrin binding site probably supplied by the α_v integrin family [13].

Not only $\alpha_5\beta_1$ integrin (Figure 8.1) but also several additional integrins can bind to the conserved RGD motif of FN, including all members of the α_v subfamily, $\alpha_8\beta_1$, $\alpha_9\beta_1$, and the platelet-specific $\alpha_{IIb}\beta_3$. Interestingly, in the absence of $\alpha_5\beta_1$ integrin expression in cells or ablation of α_5 integrin gene in mice, FN can still be assembled by the operation of other integrins, most notably the α_v integrin subfamily [27, 28]. In fact, the double knockout of α_v and α_5 integrin genes causes lack of capacity of FN fibril formation [29]. However, FN fibrils are different if assembled by α_5 or α_v integrins. The α_v-class-produced fibrils are shorter and thicker than those produced by α_5. The possible role of the coexistence of both types of fibrils is still an unanswered question: maybe diverse types of FN fibrils could provide tissues with FN matrices with different qualities, implicating different functional properties [13].

As mentioned above, the N-terminal assembly domain I1-5 is essential for FN fibrillogenesis, although other FN binding domains are also implicated in matrix assembly. III1-2 and III12-14 can bind FN. In addition, III1 can bind to III7, and III2-3 can interact with III12-14 [6]. All these domains can promote FN fibrillogenesis because of the property of binding FN, but they can also participate in intramolecular interactions that keep soluble FN in a compact form [30]. In addition, some of them are cryptic domains, for example, III1, III5, III8, and III10. The cryptic nature of these sites may indicate that perhaps they participate in the formation of stable insoluble matrix, as they are available only when FN molecule unfolds [6], but experiments performed using recombinant proteins lacking these sites showed that they are not essential for FN fibril formation. Another mechanism for the exposition of cryptic sites is the proteolysis process that occurs during matrix remodeling. An example of this is the case of $\alpha_4\beta_1$ interaction with FN matrix induced by proteolysis [4].

FN has two heparin binding domains on I1-5 and III13-15, and both participate in the binding of proteoglycan cell surface receptors. This binding promotes FN fibrillogenesis via at least two mechanisms. First, it can activate protein kinases such as FAK, which promotes integrin clustering and focal adhesion assembly. Second, they activate small GTPases such as RhoA and Rac1 via an $\alpha_5\beta_1$-bonded integrin-mediated process. The activation of such GTPases is required to generate the necessary force to unfold FN molecules for matrix formation [13].

8.2.3
FN Fibrillogenesis and Regulation of Matrix Assembly

FN in solution has a compact conformation and does not form fibrils even at extremely high concentrations. This compact form is maintained by intramolecular interactions between III2-3 and III12-14 type modules [30]. Soluble FN can bind selectively to cell surface receptors, integrins such as $\alpha_5\beta_1$, which is the primary integrin for FN matrix assembly via binding to the RGD and synergy sites. FN–integrin binding induces integrin clustering, which groups together cytoplasmic molecules such as FAK, Src kinase, paxilin, and others, promoting the formation of focal complexes. These complexes activate the polymerization of the actin cytoskeleton and kinase cascade-mediated intracellular signaling pathways [31]. Receptor clustering by dimeric FN helps to organize FN into short fibrils. After that, the contractility of the cytoskeleton contributes to FN fibril formation; the process is controlled by Rho GTPases that stimulate Rho kinases to enhance cell contractility by inducing actin–myosin interactions and actin rearrangement into stress fibers [32]. Rho activation also stimulates FN incorporation into a matrix [33]. The stretching due to cell contractility provokes a progressive extension of the FN molecule and the exposition of binding sites that would mediate lateral interactions between FN molecules. Indeed, FN has intrinsic protein–disulfide isomerase activity in the C-terminal type I module I12, and this activity may introduce intermolecular disulfide bonds within fibrils [34]. Initial thin fibrils then grow in length and thickness as matrix matures, and FN fibrils are converted to an insoluble form [4].

There is also continuous FN polymerization needed for matrix stabilization, a phenomenon that shows a tight relationship between FN polymerization and turnover, mediated by endocytosis of soluble FN [35].

Thus, the dynamic interactions between FN, integrins, and intracellular proteins are essential for FN matrix assembly and regulation. Proteoglycans also play an important role in FN fibrillogenesis. The concomitant binding of integrins and transmembrane proteoglycans stimulates Rho GTPases and FAK activities.

Deletion of actin rearrangement with inhibition of Rho GTPase causes loss of FN matrix [36].

In fact, certain pathways are critical for the initiation and maintenance of the matrix assembly. FAK plays a central role in integrin signaling, and loss of talin binding to β_1 integrin reduces FN assembly [37]. Other focal adhesion proteins participate in the assembly, but they are not essential, such as the Src kinase family, which, among other functions, phosphorylates paxilin. Inhibition of Src kinase caused a loss of matrix because of reduction in paxilin phosphorylation levels [38]. As mentioned above, a continuous FN polymerization for matrix maintenance is needed. The same mechanism would act with focal adhesion proteins involved in matrix assembly. They are needed not only in the initial steps of assembly but also to maintain matrix association with the cell surface.

Clearly, proper integration of extracellular signals with active intracellular pathways plays a crucial role in the initiation and progression of FN matrix assembly.

8.3
Cell-Free Assembly of Fibronectin Fibrils

The need for controllable and reproducible *in vitro* models of FN networks and for new synthetic materials able to serve as bio-inspired scaffolds for tissue engineering has driven the efforts in biology and regenerative medicine for the identification of cell-free routes able to induce FN fibrillogenesis. These routes are based on the assumption that unfolding of soluble FN dimers from their globular conformation is needed for FN–FN interactions to occur, leading eventually to FN polymerization and fibril formation. The methods that have been used in literature include

- addition of reducing [39, 40] or oxidizing [41] agents to the protein solution;
- use of denaturing [42–45], cationic [46, 47], or anionic [48, 49] compounds;
- use of peptidic FN fragments [50–52];
- force-based assembly, via application of mechanical tension [36, 53–56] or shear forces [57–67];
- surface-initiated assembly [68–72].

First, Williams and coworkers found that reduction of FN by dithiothreitol (DTT) induced the self-assembly of plasma FN via unfolding of disulfide-stabilized globular domains and subsequent promotion of noncovalent binding interactions among FN molecules [39]. Sakai *et al.* [40] further proved that multimerization of FN can occur under a physiologically possible condition (of pH, temperature, and ionic strength) in the presence of a low concentration of DTT, leading to FN aggregates with a fibrillar structure. A disulfide exchange mechanism from intramolecular to intermolecular bonds was demonstrated to be involved in FN multimer formation: the first reaction to occur is the reduction of the intramolecular disulfide bonds, followed by reformation of new disulfide bonds in an oxidative environment. The authors also suggested the involvement of the terminal region of the FN chains in the DTT-induced multimerization process, as well as of type I and

II repeats. Vartio also reported disulfide-bonded polymerization of FN, induced by low concentrations of strong oxidants, such as $FeCl_3$ and $CuSO_4$ [41]. Metal ions (Fe^{3+} and Cu^{2+}) can in fact participate in oxidation reactions and stimulate the formation of disulfide bonds, either via the free sulfhydryl groups (in the case of $CuSO_4$) or also via a disulfide exchange mechanism (in the presence of $FeCl_3$).

Another cell-free method to induce FN fibrillogenesis *in vitro* involves the addition of denaturants to a protein solution. Mosher and coworkers [42, 43] first used guanidine as the denaturing agent; denatured FN dimers partially expose their free sulfhydryl groups, allowing multimeric FN to form. Peters and collaborators studied the macromolecular structure of multimeric FN obtained via denaturation with guanidine and incubation in dialysis tubing and observed fibrillar structures similar to the ones formed in culture by human skin fibroblasts using cryo-high-resolution scanning electron microscopy (cryo-HRSEM) [44, 45]. Cell-free fibrillogenesis produced an extensive network of fibrils, highly intertwined with one another [44]. These fibrils had at least two distinct structural arrangements: the majority had a ropelike structure that appeared to consist of nodules (10–13 nm in diameter), while few of them adopted a straight conformation with a smooth surface. The former fibrils had one free end, while the latter had both ends attached to the substrate, suggesting that a factor favoring one conformational state over the other might have been the tension applied to the fibrils [44]. Moreover, both conformations could exist within an individual fibril, indicating that FN fibrils are capable of undergoing localized conformational changes and are highly flexible. The nodules were proposed to represent discrete domains of three to four type III repeats, as they could be labeled with the monoclonal antibody IST-2 to the III_{13-14} repeats in FN and they were observed in FN fragments that only contain type III repeats. On the other hand, smooth fibrils were never recognized by the IST-2 antibody, suggesting that certain epitopes of FN may be buried, or exposed, depending on the conformation of the fibril [45].

Also anionic molecules, specifically heparin, were found to induce FN fibrillogenesis in the absence of cells [48, 49]. Jilek and Hörmann found that FN was partially precipitated by heparin, and electron micrographs showed the presence of filamentous structures [48]; the authors suggested that heparin induces the transition of FN from a globular to an elongated form, exposing masked binding sites responsible for self-association and therefore leading to the formation of fibrillar precipitates. Later, Richter *et al.* [49] observed the formation of fibrils, visible by phase-contrast microscopy, when heparin-induced FN precipitation was performed onto hydrophobized glass cover slides. Precipitation was considerably dependent on the ionic strength, indicating that electrostatic forces played a major role in the aggregation of FN. Electrostatic interactions were also involved in FN polymerization induced by polyamines (cationic molecules) [46, 47]. Vuento and coworkers [46] found that polyamines were able to aggregate proteins from human plasma and serum, and identified fibrinogen (FNG) and FN as the major molecular components of these aggregates. Further studies by the same group revealed that soluble FN was polymerized by polyamines (at a concentration of 1–5 mM and at low ionic strength) into filamentous structures, evident via observation with electron and phase-contrast microscopy [47]. Thick filaments with diameters varying from 50 nm to several hundreds of nanometers were seen; in addition, thin filaments with diameter of 2–3 nm and nonpolymerized material at the background of the filaments were present. The thick filaments could often be resolved to be composed of side-to-side associated thin filaments with a diameter of 2–3 nm.

Interaction of FN dimers with purified recombinant fragments of FN was shown to induce fibrillogenesis in the absence of cells [50–52]. Morla and collaborators [50] showed that anastellin, a C-terminal recombinant fragment from the first type III repeat of FN (III$_1$-C) bound to FN and induced spontaneous disulfide cross-linking of the molecule (probably through disulfide exchange) into multimers of high relative molecular mass, which resembled matrix fibrils. Treatment of FN with this aggregation-inducing fragment also converted FN into a form with greatly enhanced adhesive properties (termed "*superfibronectin*"), which suppressed cell migration. Briknarová *et al.* [52] later proved that the main structural features of anastellin resemble those of amyloid fibril precursors, implying some similarity between FN and amyloid fibril formation. Hocking and coworkers [51] obtained the formation of high-molecular-mass-FN multimers using a recombinant III$_{10}$ FN module. The authors suggested a mechanism by which the interaction of the III$_{10}$ fragment with the III$_1$ domain in intact FN promoted a conformational change within the III$_1$ module, which unmasked the amino-terminal binding site and triggered the self-polymerization of FN, leading to the formation of disulfide-stabilized multimers in the absence of cells.

Zhong *et al.* [36] confirmed the importance of molecular unfolding for FN fibrillogenesis by applying mechanical tension to expose the self-assembly sites within FN, thereby enhancing the binding of soluble FN in the absence of cells. Specifically, the application of a 30–35% stretch to FN immobilized onto flexible rubber culture dishes induced binding of additional FN molecules, of the 70 kDa N-terminal FN fragment (which is critical for matrix assembly), and of the L8 monoclonal antibody (which binds to I$_9$/I$_{10}$ and appears to be specific for an epitope that is exposed by tension exerted on the FN molecule). The authors suggested that tension on FN could expose self-assembly sites either by unfolding a large region of FN that is looped back on itself, or by partially unfolding FN type III domains, therefore allowing the nucleation of FN assembly [36]. The role of mechanical tension for the formation of FN fibrils was further investigated by Baneyx and Vogel, who observed FN assembly into extended fibrillar networks after adsorption to a dipalmitoyl phosphatidylcholine (DPPC) monolayer in contact with physiological buffer [53]. FN was adsorbed to receptor-free membrane-mimetic interfaces, prepared by spreading DPPC monolayers at the air/buffer interface in a Langmuir trough, whose movable barriers were used to control the physical state and, thus, the molecular packing of the lipid monolayer itself. Specifically, FN was injected into the buffer subphase after compression of the monolayer to a state in which liquid condensed (LC) and liquid expanded (LE) domains coexisted, and was found to partially insert into the expanded phase, although it did not bind to DPPC in the condensed phase. Then, entropic and/or electrostatic forces led to the migration of FN from the LE phase to the domain boundaries, resulting in FN enrichment at these locations. At this point, expansion of the monolayer initiated the pulling of FN into extended fibrils by virtue of an increasing distance between neighboring domain boundaries; when the LC domains eventually melted away, fibrils protruding from the domain edges were pulled back together and intersected, resulting in a fibrillar protein network. The authors suggested that mechanical tension caused by domain separation pulled the proteins into an extended conformation, exposing cryptic self-assembly sites and allowing the formation of networks stabilized by disulfide cross-linking [53].

Spatz et al. [54–56] also induced fibrillogenesis by applying mechanical tension to FN. This group developed a system in which cell-free FN fibrillogenesis could be achieved through a two-stage process initiated by the shear-stress independent partitioning of globular FN molecules into the air/liquid interface, where they formed an insoluble two-dimensional sheet, followed by force-dependent fibrillogenesis along a superhydrophobic surface made of elastic micropillars (Figure 8.2a,b). Specifically, globular FN molecules with an average diameter of $\sim$32 nm were found to assemble into an insoluble sheet at the air/water interface (Figure 8.2e); the partitioning of FN into the interface apparently induced a mechanical deformation of the globular FN, allowing it to preassemble into a stable network. This is in agreement with the expectations that FN fiber formation requires proper alignment of the outer FN-I domain, enabling specific intermolecular interactions between the FN-I domains of different chains [73, 74]. Subsequent application of force to this FN sheet attached to a micropillar array, due to wetting of the microstructure, initially resulted in the formation of rough fibers with globular subdomains, whose sizes were affected by the tension applied to the FN layer (Figure 8.2f,g). Two major forms of rough fibers were noted, with globular domains measuring either 24 nm $\times$ 16 nm or 19 nm $\times$ 9 nm. On the other hand, strongly aligned fibers displayed a uniform diameter of 14 nm and smooth surfaces (Figure 8.2h); these fibers tended to split at their ends into two 7 nm subfibrils, while thicker fibers were formed through lateral association of the 14 nm wide fibers (Figure 8.2i). The average force applied to the FN particles by micropillar bending (estimated by measuring its displacement) was $\sim$8 pN per molecule, which is within the range necessary for FN unfolding [75–77]. HRSEM images of immunogold-labeled FN showed that force-induced fibrils displayed much more intense labeling than the relaxed FN particles, suggesting that mechanical stretching of FN fibers exposes cryptic antigenic sites that are not available in its globular conformation [54]. Finally, Kaiser and Spatz adapted this force-induced FN fibrillogenesis to obtain a method for the production of highly regular arrays of nanofibrils from other ECM proteins besides FN, including COL I and laminin isolated from Engelbreth-Holm-Swarm tumors (EHS-LAM) [56]. The mean nanofibril diameter (in the range of 20–160 nm) was controlled by adjusting the dewetting parameters of a protein solution on a hydrophobic silicon micropillar structure; the array geometry predetermined the nanofibril diameter, while protein concentration and dewetting speed had to be adapted to yield high regularity. The results were congruent with the model proposed by Guan et al. [78] for the fabrication of DNA nanowires, indicating a

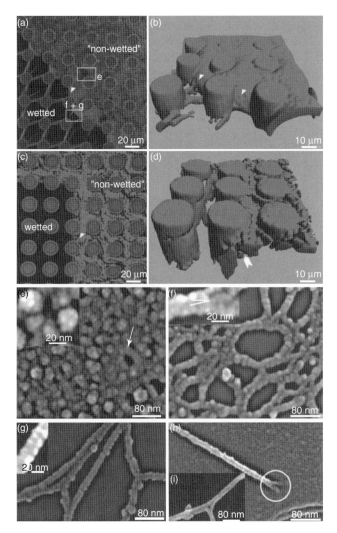

Figure 8.2 Fluorescence microscopy and high-resolution SEM images depicting FN fibrillogenesis on micropillar arrays. (a) Confocal fluorescence micrograph of a sample during the wetting process of a micropillar interface after a thin layer of FN was partitioned into the air/solution interface, showing the tendency of FN to generate fibrils. (b) Reconstruction of 3D confocal fluorescence micrographs, indicating that the pillars are bent (white arrowheads). (c,d) Fluorescent confocal images of BSA layers indicating the accumulation of a thin protein film at the air/water interface; no fiber formation was observed. Instead, the protein layer disintegrated into small fragments on wetting of the micropillars. (e–g) SEM micrographs, revealing FN structures from regions represented by the areas in (a) bordered in white. The white arrow in (e) indicates an imperfection of the FN sheet where the globular particles start aligning into a rough fibril. (h) The thinnest fibers displayed two subfibers at their end (white circle) and formed thicker fibers by lateral association (i). (Source: Reproduced with permission from Ref. [54].)

common mechanism of polymer fibrillogenesis induced by dewetting. The necessity for self-association sites was suggested by the observation that bovine serum albumin (BSA), even at high concentrations, did not yield nanofibrils, despite its ability to concentrate at the air/buffer interface (Figure 8.2c,d) [54, 56]. Moreover, the engineered ECM could be transferred from the micropillar array onto polyethylene glycol (PEG) hydrogels, reconstituting the nanofibrils in buffer and simultaneously eliminating the substrate microtopography.

Other cell-free routes for the force-induced assembly of FN were based on the application of shear forces to an FN solution in different setups, which included manual pulling of single fibrils out of a concentrated drop of FN [57, 59, 61, 63, 66, 67], stirring of an FN solution in an ultrafiltration cell [57, 58, 60, 65], and wet extrusion of a concentrated FN solution [62, 64, 65]. The first method allowed to obtain thin microscale FN fibrils by simply drawing them out from a small pool of a concentrated FN solution with a thin glass rod or a pipette tip and depositing them onto a substrate [57, 59, 61, 63, 66]; FN was allowed to self-associate under the influence of a directional shear force [57]. Manually deposited fibers pulled from concentrated solutions of soluble FN resembled *in vivo* FN fibers in diameter and composition [59]. The average fiber diameter could be moderately adjusted via the pulling procedure or by varying the concentration of the FN solution; on the other hand, the length of a fiber could be controlled by drawing it out of a solution to a desired length before bringing it in contact with the substrate [66]. Moreover, fluorescence images of manually deposited fibers showed that submicron fibrils emerged from the surface of the FN solution and bundled together to form larger cables of fibrils [66]. Mechanical characterization of these fibers revealed that they show signs of breakage only when stretched over five to six times their resting length, that high extensions involve unfolding of type III modules, and that the mechanically strained fibers clearly bind more N-terminal 70 kDa FN fragment than the relaxed ones, suggesting that mechanical tension exposes cryptic binding sites for this fragment [66]. Another study suggested that the fibers can withstand more than eightfold extension with respect to their resting length before 50% of them are broken [67]. This extraordinary extensibility requires the mechanical unfolding of secondary structure, and remarkably, release of mechanical stress permitted the recovery of the fiber's original mechanical properties. Moreover, cryptic molecular recognition sites could be switched on by force, as demonstrated for the exposure of cryptic cysteines on FN-III$_7$ and FN-III$_{15}$ [67].

Instead of fibers, the concentration of an FN solution under a continuous unidirectional stirring motion allowed to produce oriented FN mats [57, 58, 60, 65]. The yield of FN mat was drastically improved in the presence of urea, which denatures FN by unfolding it into an elongated conformation [79] that favors lateral association and fibril growth [58]. Finally, shear-dependent fibrillogenesis was achieved via wet extrusion of urea-denatured concentrated protein solutions [62, 64, 65]. This method led to the production of large FN cables (up to 1.5 cm in diameter), which, once hydrated, had a parallel fibril alignment and a porous cross-sectional structure, with pore sizes between 10 and 100 μm in diameter [65].

The last cell-free route of FN fibrillogenesis is based on the unfolding of soluble FN as a result of its adsorption onto a material surface [68–72]. For example, Nelea and Kaartinen described FN filament formation on a surface with negative potential in the absence of cells [70]. They adsorbed FN at very low ionic strength (0.5 mM NaCl) and at a concentration of 1 µg ml^{-1} on bare and differently charged silicon wafers. AFM analyses showed that FN formed compact, rounded particles on all positively charged surfaces and on bare silicon. On the other hand, in the case of a polysulfonated substrate, that has a negatively charged surface, the assembly of FN into nanofilaments was observed. Most of these filaments showed lengths in the range of 0.5–2 µm, with apparent widths of 15–20 nm. The height and width of FN filaments were close to those characteristic of the extended FN dimer, indicating that these filaments were being formed by alignment of extended FN, and as such, they could represent an initial alignment process during FN assembly. From the height profile analysis of AFM scans, the filaments were revealed to be formed of a chain with periodic arrangement of connected beads, giving a "bead-on-a-string" appearance. The average interbead distance of 58 nm (±15 nm) represented half of the length of an elongated FN dimer, that is, the length of a monomer arm. Filament heights were low, in the range of 0.2–2.6 nm, with the largest values residing at the level of the beads. These measured heights demonstrated that the beads were not derived from chains of compact FN, which have considerably larger heights (generally 4–5 nm). The beads were 18–30 nm long and 12–22 nm wide, while the segments connecting them were thinner, 8–15 nm. Since the extended FN molecule show widths that range from 7 to 13 nm, the filament was likely composed of only single FN molecules interacting in a chain [70].

The formation of FN networks after adsorption onto a polysulfonated surface was reported in a previous work by Pernodet *et al.* [68], which described FN fibrillogenesis on a negatively charged surface at relatively low resolution. In this work, the adsorption of FN onto a surface with high negative charge density was suggested to open and extend the molecule via interaction with the III$_{12-14}$ modules [68]. Interactions between I$_{1-5}$ and III$_{12-14}$ modules of two FN molecules would then create a staggered alignment, which has been proposed in previous theories [80–82], and position two type III$_{1-7}$ modules so that they interact with each other; the region spanning type III$_{1-7}$ modules has been reported to drive FN fibrillogenesis [83]. Since type III modules are larger than other modules, two type III$_{1-7}$ regions could create a thicker stretch in the filament with the appearance of a bead. The theoretical type III$_{1-7}$ module overlap would be about 24.5 nm, which would fit the bead diameter observed by Nelea and Kaartinen; moreover, this type of alignment would create an interbead distance of about 60 nm, in accordance with the experimental data [70].

Another example of surface-induced FN fibrillogenesis was presented by Pellenc and coworkers [69], who observed the formation of fibrillar structures after FN adsorption onto a model mineral matrix of hydroxyapatite; however, the pathway of FN assembly was not clear. Finally, Feinberg and Parker demonstrated that surface-initiated assembly could be used to engineer multiscale, free-standing nanofabrics using a variety of ECM proteins (FN, LAM, FNG, COL$_I$, and COL$_{IV}$)

[71]. They used protein–surface interactions to unfold ECM proteins and trigger their assembly. Specifically, the process involved adsorbing nanometer-thick layers of ECM proteins from a solution onto a hydrophobic surface at high density to partially unfold them and expose cryptic binding domains [84, 85], transferring the ECM proteins in the unfolded state to a relatively hydrophilic, dissolvable surface, and thermally triggering surface dissolution to synchronize matrix assembly and its nondestructive release. ECM proteins could be deposited in spatially defined patterns by microcontact printing (µCP) with polydimethylsiloxane (PDMS) stamps onto an anhydrous poly(N-isopropylacrylamide) (PIPAAm) film spin cast on glass coverslips. The µCP process step could be repeated multiple times with multiple proteins in multiple geometries to produce interconnected nanofabrics with variable thread counts and weaves, rip-stop properties, and both chemical and mechanical anisotropies [71]. Control experiments using BSA and immunoglobulin G (IgG) failed to form nanofabrics, proving that nanofabric formation requires biopolymers with intrinsic self-binding domains. Thickness was customized by controlling the density of protein molecules adsorbed to the surface via the concentration of the FN solution used to ink the µCP PDMS stamp. For example, dry FN nanofabrics on PIPAAm were 1–5 nm thick, comparable in size to FN dimers unfolded on hydrophobic and hydrophilic mica surfaces [84, 86], and increased to ~8 nm when hydrated. Free-standing FN nanofabrics captured on PDMS substrates had a dense network of interconnected fibrils, structurally similar to the ones observed for force-induced FN matrix assembly [54]. AFM and SEM analyses revealed nodular and fibrillar nanostructures in FN nanofabrics under strain; contracted regions had high nodular density, while stretched regions had lower nodular density. AFM helped resolving fibrillar structures connecting the nodules together and "bead-on-a-string" structures oriented in the direction of mechanical strain. Closer examination of the region between nodules revealed a fibrous meshwork interconnecting them, similar in height and diameter with unfolded FN dimers [71]. Being capable of multiscale deformation via a combination of nanoscale molecular folding, microscale buckling, and macroscale fabric architecture, FN nanofabrics have unique mechanical properties that cause them to outperform synthetic polymers, potentially leading to a new class of advanced, high-performance biomolecular materials [71].

The formation of a physiological-like FN network after adsorption onto a particular surface chemistry was reported by us [72]. Details of this process are provided in the next section.

8.4
Material-Driven Fibronectin Fibrillogenesis

Significant efforts have focused on engineering materials that recapitulate the characteristics of ECM, such as the presentation of cell-adhesive motifs or protease-degradable cross-links, in order to direct cellular responses [87, 88]. However, materials-based approaches to reconstitute the network structure and bioactivity of FN fibrillar matrices have not been established yet. As previously commented, the use of denaturing or unfolding agents and applied forces to promote FN fibril assembly indicate that changes in the structure of FN are required to expose sites within the molecule to drive assembly into fibers [42, 50, 53, 54]. In our group, we have hypothesized that adsorption of individual FN molecules onto particular surface chemistries would induce exposure of self-assembly sites to drive FN fibril assembly and identified poly(ethyl acrylate) (PEA) as a potential surface chemistry to generate FN fibrils [89, 90]. In addition, we have recently investigated the organization of FN molecules at the material (PEA) interface and the analogy with the physiological cell-induced FN fibrillogenesis [72].

8.4.1
Physiological Organization of Fibronectin at the Material Interface

We compared the organization of FN at the material interface on two similar chemistries: PEA and poly(methyl acrylate) (PMA), which differ in one single carbon in the side chain (Figure 8.3a). Both surface chemistries show similar wettability and total amount of adsorbed FN (Figure 8.3b). However, the conformation and distribution of the protein following passive adsorption onto these surfaces are completely different (Figure 8.3c). Interconnected FN fibrils are organized on adsorption from a solution of concentration $20\,\mu g\,ml^{-1}$ on PEA (Figure 8.3c), whereas only dispersed molecules are present on PMA (Figure 8.3c). FN fibril formation on PEA is dependent on the FN solution concentration, as lower concentrations result in dispersed adsorbed molecules [72]. We next examined the requirements for the 70 kDa amino-terminal domain of FN in this material-driven fibrillogenesis. The 70 kDa amino-terminal regions are essential for cell-mediated FN assembly and, within this region, the I_{1-5} repeats confer FN binding activity [18]. This domain is not accessible in the folded, compact structure of FN in solution, and a conformational change of the molecule is mandatory for physiological matrix assembly to occur [6]. Strikingly, material-driven fibrillogenesis absolutely requires the 70 kDa amino-terminal region of FN. Addition of the 70 kDa fragment completely blocks the organization of FN at the material interface, and only discrete molecular aggregates can be observed (Figure 8.3c) without any trace of the assembled FN network. The resulting adsorbed FN resembles the FN adsorbed onto the control PMA polymer. These results demonstrate that a particular polymer chemistry (PEA) drives assembly of adsorbed FN molecules into FN fibrils and that this material-driven fibrillogenesis requires the 70 kDa amino-terminal domain of FN.

The dynamics of the formation of the FN network on PEA was followed with AFM after different adsorption times. AFM images of the FN adsorbed on PEA after different adsorption times (10, 30, 60, and 180 s) from $20\,\mu g\,ml^{-1}$ protein solution allows one to grasp the intermediate stages of the organization process of FN at the material interface [91]. At the very beginning of the adsorption process (10 s), isolated FN molecules are homogeneously distributed on the material. After 30 s of adsorption, FN molecules tend to align, suggesting the initial formation of

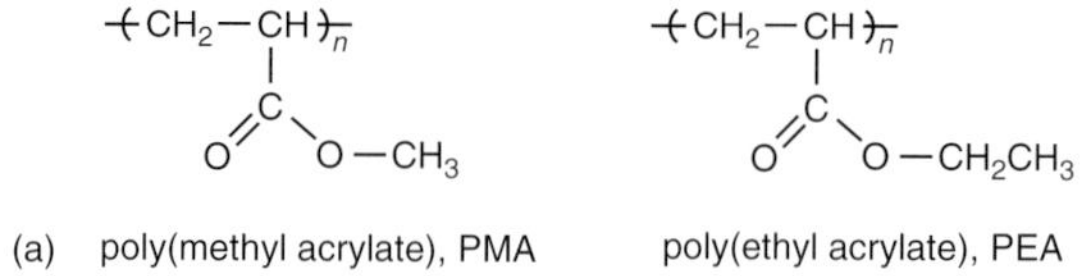

(b)

Substrate	WCA (°)	FN surface density ($ng\ cm^{-2}$)
PMA	74 ± 2	470 ± 43
PEA	80 ± 2	450 ± 45

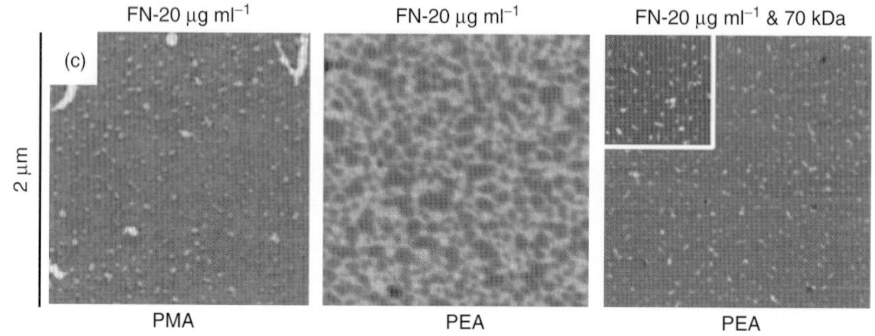

Figure 8.3 FN adsorption on material substrates. (a) Chemical structure of the polymers PEA and PMA. (b) Water contact angle on the different substrates and equilibrium surface density of adsorbed FN from a solution of concentration 20 $\mu g\ ml^{-1}$. (c) FN distribution on material substrates as obtained by AFM: globular aggregates on PMA and FN network on PEA after adsorption from a solution of concentration 20 $\mu g\ ml^{-1}$. FN fibrillogenesis is blocked on PEA in the presence of the amino-terminal 70 kDa FN fragment, leading to dispersed molecular aggregates. (Source: Adapted with permission from Ref. [72].)

intermolecular connections, which result in protein–protein contacts through the surface. After 60 s, AFM images reveal the formation of a protein network on the material surface. Increasing the adsorption time results in thickening the fibrils, which make up the protein network [91]. These results allow one to conclude that FN organization on the material interface occurs in a timescale that is adequate to be followed via AFM, whose acquisition time is in the minute range.

Seeking to follow the adsorption process in a more detailed way, a different experiment was planned. Instead of fixing the concentration of the solution from which the protein is adsorbed and vary the adsorption time, we fixed the adsorption time from solutions of increasing concentration of FN [90]. Figure 8.4 shows AFM images for FN adsorbed onto PEA substrates after immersion for 10 min in protein solutions of different concentrations: 2, 2.5, 3.3, 5, 20, and 50 $\mu g\ ml^{-1}$. The lowest concentration (Figure 8.4a) results in isolated extended FN molecules homogeneously distributed on the material. For a concentration of 2.5 $\mu g\ ml^{-1}$ (Figure 8.4b), FN molecules are observed in a higher density. Extended FN molecules tend to align, suggesting the initial formation of intermolecular connections (Figure 8.4b). FN conformation in Figure 8.4c suggests the incipient

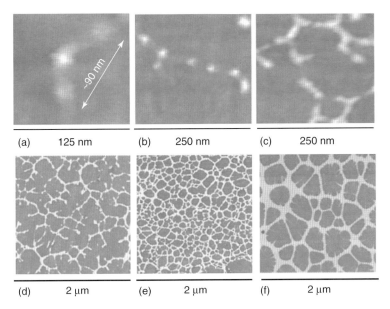

Figure 8.4 AFM phase image of single FN molecules on PEA: (a) isolated molecule, (b) two FN molecules interacting through the amino-terminal (I_{1-5}) domains, (c) assembly of FN molecules into an incipient network, (d) assembly of FN into a network that is not completely interconnected, (e) interconnected FN network, and (f) thickening of protein arms at higher concentrations. FN was adsorbed for 10 min from solutions with concentration of 2, 2.5, 3.3, 5, 20, and 50 µg ml^{-1}, respectively. (Source: Reproduced with permission from Ref. [90].)

formation of a protein network on the material when FN was adsorbed from a solution with a concentration of 3.3 µg ml^{-1}. Protein adsorption from higher solution concentrations gives rise to FN networks on the material with higher cross-link density, that is, a higher number of cross-link points and lower distance between them [90].

The development of an FN network in the absence of cells gains a distinct bioengineering interest because it is a way to improve the biocompatibility of materials. It is well documented that cells recognize faster and with higher affinity already assembled FN fibrils compared to adsorbed protein [6, 31]. The existence of cell-free routes able to induce the formation of FN fibrils from isolated molecules have been described previously in this chapter and include (i) the exposition of sulfhydryl groups (able to form disulfide-bonded FN multimers) by adding denaturants, for example, guanidine 3 M [42], (ii) the reduction of disulfide bonds that promotes unfolding of the molecule and noncovalent binding interactions among FN molecules [39], and (iii) the addition of a fragment from the first type III repeat of FN that induces spontaneous disulfide cross-linking of the molecule into multimers [50]. Apart from these biochemical routes to induce FN assembly, it has been shown that FN can be assembled into networks after adsorption on certain substrates, (iv) underneath DPPC monolayers, which resemble the major lipid fraction of cell membranes [53], and (v) onto a superhydrophobic surface made of elastic micropillars [54]. Both these situations require the existence of mechanical tension to drive the process. The formation of FN networks on PEA (Figure 8.4) must be a consequence of the following sequence of events:

1) Conformational change of FN upon adsorption onto PEA. It is known that FN has a compact folded structure in physiological buffer that is stabilized through ionic interactions between arms [30]. The FN interaction with chemical groups of the substrate (a vinyl backbone with $-COOCH_2CH_3$ side chain) gives rise to conformational changes in the molecule that lead to the extension of the protein arms (Figure 8.4a). Adsorption of FN on slightly charged surfaces (negative neat group in the $-COO^-$ group) gives rise to elongated structures of the molecule, as obtained for SiO_2 and glass [84, 85]. It is likely that FN orients at the surface, so that its hydrophobic segments interact with the methyl groups in PEA, maybe throughout the heparin binding fragment, as proposed for the FN–DPPC interaction [53], but with more efficient arm extension because of the neat negative charge of the surface.
2) Enhanced FN–FN interaction on the PEA substrate. The adequate conformation of individual FN molecules as the adsorption process continues favors FN–FN interactions involving the amino-terminal 70 kDa fragment [84], probably throughout the interaction between I_{1-5} and III_{1-2} domains located near their amino side [9]. Figure 8.4b shows the relative orientation of two FN molecules compatible with this hypothesis.
3) New FN molecules are preferentially adsorbed in close contact to FN molecules already present on the substrate (Figure 8.4c), probably as a consequence of the presence of polar-oriented FN molecules enhancing the collision rate of FN self-assembly sites [9], which finally gives rise to the initial formation of a protein network on the substrate. This process leads to a well-interconnected network of FN on the surface of the substrate (Figure 8.4) [90]. Adsorption from solutions of higher concentrations leads to the formation of a protein network with thicker arms (Figure 8.4d–f). The formation of an FN network on PEA is not a universal property of this protein. For example, a similar network was found for FNG [92] and COL IV [93], but only globular isolated molecules were observed after laminin [94] and vitronectin adsorption [95].

8.4.2
Biological Activity of the Material-Driven Fibronectin Fibrillogenesis

The biological activity of the FN network assembled on PEA was assessed by investigating cell adhesion on electrospun fibers of this polymer. Random and aligned fibers of the polymer were obtained seeking to mimic the spatial organization of the ECM. The existence of an FN network assembled on the electrospun PEA fibers was assessed by AFM [90]. However, it is more important that this FN network that assembles spontaneously on PEA fibers is biologically active, even more than FN directly adsorbed on the underlying glass (Figure 8.5), which is supporting the

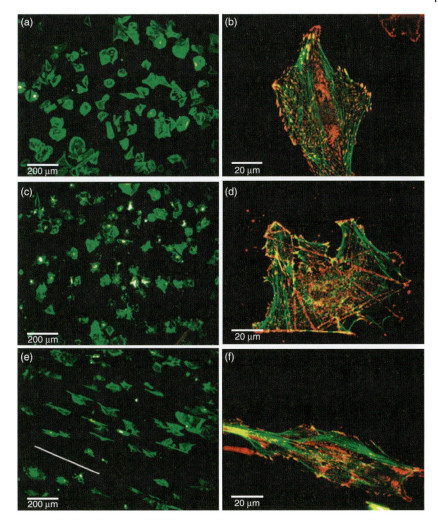

Figure 8.5 Overall morphology of fibroblasts adhering on (c,d) random and (e,f) aligned FN-coated PEA fibers compared to the (a,b) control of FN-coated glass visualized by actin (a,c,e) or double stained for vinculin (red) and actin (green) (b,d,f). The line in E represents the main direction of PEA fiber alignment. (a,c,e) Magnification = 10×: bar, 200 μm. (b,d,f) Magnification = 100×; bar, 20 μm. (Source: Reproduced with permission from Ref. [90].)

hypothesis that the material-driven FN network shares some similarities with the physiological one (both of them assembled via the 70 kDa domain as explained previously). It is evident that cells tend to interact with PEA fibers rather than glass and start to orient, modifying their characteristic spread morphology, following the fiber's direction (Figure 8.5e,f). Complementary material in Ref. [90] shows a movie of living cells that migrate throughout the surface, in a real tactile exploration, until a PEA fiber is found. Afterward, cell morphology is modified to adapt cell–substrate contact along the electrospun PEA fiber; also, the cells may "jump" from the supporting glass to the PEA fibers, preferring to move on the FN network. The well-developed focal adhesion complexes and the insertions of actin stress fibers in these complexes point to a proper transmission of signals by integrins to the cell interior. Interestingly, when adhering on random PEA fibers, cells tend to develop a rounded morphology with multiple projections, which resemble stellate morphology, characteristic for cells in the 3D environment. This also means that cells receive proper signals coming from the FN network on the PEA fibers. Moreover, the same applies to the oriented fibers as well, because fibroblasts immediately acquire extended morphology with asymmetrical organization of the adhesive complexes (Figure 8.5f), suggesting activation of their motility.

Late FN matrix formation, after three days of culture, was again influenced by the formation of the FN network on the PEA fibers, reflecting the high biological activity of the preorganized protein. Matrix formation was excellent on both the PEA fibers and control glass; however, in agreement with the initial cell adhesion, newly synthesized FN is preferentially deposited on PEA fibers [90]. Thus, one can assume that, by tailoring the fiber orientation, we can control the organization of the provisional FN matrix secreted by the cells.

We also compare cell differentiation on the substrate-assembled FN network with those substrates that do not promote FN fibrillogenesis. We hypothesized that material-driven FN fibrils have enhanced biological activity compared to adsorbed FN molecules because the fibrillar structure recapitulates the native structure of FN matrices.

We evaluated the biological activity of the material-driven FN networks by examining the myogenic differentiation process [72]. Sarcomeric myosin expression and cell bipolar alignment and fusion into myotubes, markers of myogenesis, were significantly higher on the substrate-induced FN network (Figure 8.6a, PEA-20), as compared to the same substrate coated with a lower density of FN, which does not lead to the fibrillar organization of the protein at the material interface, and PMA polymer coated with the same density of FN but lacking any fibrillar organization (Figure 8.6a). Surprisingly, myogenic differentiation was considerably more robust on the PEA-assembled FN matrix than on COL type I, which represents the standard substrate for myogenic differentiation [72]. It is important to emphasize that these differences in myogenic differentiation are not due to differences in the number of cells on the substrates [72]. To further evaluate the biological activity of the material-driven FN fibrils, we used different blocking antibodies and FN fragments to assess the importance of different parts of the FN molecule on the myogenic differentiation process (Figure 8.6b). Addition of the adhesion-blocking HFN7.1 antibody inhibited differentiation on PEA to levels found for those substrates on which FN is not organized into a network (Figure 8.6b, PEA-HFN7.1). The HFN7.1 antibody binds to the flexible linker between the ninth and tenth type III repeats of FN where the integrin binding domain is located, demonstrating that the integrin binding domain of FN is essential for myogenic differentiation on FN fibrils. Moreover, HFN7.1 is specific for human FN (adsorbed on the substrate) and does not cross-react with mouse FN (cell secreted), indicating that the human FN

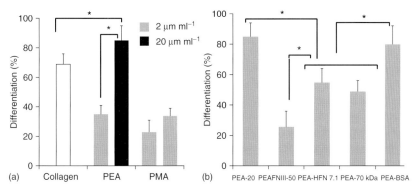

Figure 8.6 (a) Myogenic differentiation as determined by the percentage of sarcomeric myosin-positive cells on the different substrates after adsorbing FN from solutions of concentrations 2 and 20 µg ml^{-1}. (b) Both the central FN domain (FNIII$_{7-10}$) and the amino-terminal fragment (70 kDa) involved in FN fibrillogenesis enhance myogenic differentiation on the material-driven FN network on PEA. (b) Myogenic differentiation as determined by the percentage of sarcomeric myosin-positive cells on the substrate-induced FN network (PEA-20), after coating PEA with a recombinant fragment of FN (FNIII7-10), blocking the central FN domain with HFN7.1 antibody (PEA-HFN7.1), and after adsorbing FN altogether with the 70 kDa fragment, which blocked the formation of the FN network (PEA-70 kDa), as well as control experiment for the 70 kDa fragment using BSA instead (PEA-BSA). Statistically significant differences ($p < 0.05$) are indicated with ∗. (Source: Adapted with permission from Ref. [72].)

adsorbed onto the substrate before cell seeding provided the dominant signal for differentiation. We next examined whether a recombinant fragment of FN spanning the seventh to tenth type III repeats (FNIII$_{7-10}$) that encompass the integrin binding domain recapitulates the material-dependent differences in myogenic differentiation [96]. In contrast to complete FN, adsorption of FNIII$_{7-10}$ onto PEA yielded minimal levels of myogenic differentiation (Figure 8.6b, PEA-FNIII$_{7-10}$). FNIII$_{7-10}$ does not contain domains involved in FN–FN interactions (I$_{1-5}$ and III$_{1-2}$ or III$_{12-14}$ domains), demonstrating that both integrin binding domains and domains involved in FN–FN interactions are required for the enhanced myogenic differentiation on PEA-driven FN matrices. Consistent with our observations for FN network formation, addition of the 70 kDa amino-terminal fragment during the adsorption process of FN blocked the differentiation of myoblasts on PEA (Figure 8.6b, PEA-70 kDa). This result demonstrates the importance of FN–FN interactions and the fibrillar structure of the protein on the cell differentiation process: only when the FN network is assembled on PEA via interactions involving the 70 kDa amino-terminal fragment does the differentiation process occur. As a control, when albumin (a protein with similar molecular weight to the 70 kDa fragment) is added at the same concentration during the network assembly process of FN at the material interface, the formation of the network is not disturbed and myoblast differentiation levels remain as in the original material-assembled FN network (Figure 8.6b, PEA-BSA). Importantly, the addition of the 70 kDa FN fragment once the FN network is already assembled on the material surface has no effect on subsequent myoblast differentiation [72].

References

1. Larsen, M., Artym, V.V., Green, J.A., and Yamada, K.M. (2006) *Curr. Opin. Cell Biol.*, **18**, 463–471.
2. Vakonakis, I. and Campbell, I.D. (2007) *Curr. Opin. Cell Biol.*, **19**, 578–583.
3. Zhou, X., Rowe, R.G., Hiraoka, N., George, J.P., Wirtz, D., Mosher, D.F., Virtanen, I., Chernousov, M.A., and Weiss, S.J. (2008) *Genes Dev.*, **22**, 1231–1243.
4. Singh, P., Carraher, C., and Schwarzbauer, J.E. (2010) *Annu. Rev. Cell Dev. Biol.*, **26**, 397–419.
5. Hynes, R.O. (1990) *Fibronectins. Springer Series in Molecular Biology*, Springer, New York.
6. Mao, Y. and Schwarzbauer, J.E. (2005) *Matrix Biol.*, **24**, 389–399.
7. Fontana, L., Chen, Y., Prijatelj, P., Sakai, T., Fässler, R., Sakai, L.Y., and Rifkin, D.B. (2005) *FASEB J.*, **19**, 1798–1808.
8. George, E.L., Georges-Labouesse, E.N., Patel-King, R.S., Rayburn, H., and Hynes, R.O. (1993) *Development*, **119**, 1079–1091.
9. Pankov, R. and Yamada, K.M. (2002) *J. Cell Sci.*, **115**, 3861–3863.
10. Cho, J. and Mosher, D.F. (2006) *J. Thromb. Hemost.*, **4**, 1461–1469.
11. Ni, H., Yuen, P.S.T., Papalia, J.M., Trevithick, J.E., Sakai, T., Fässler, R., Hynes, R.O., and Wagner, D.D. (2003) *Proc. Natl. Acad. Sci. U.S.A.*, **100**, 2415–2419.
12. Moretti, F.A., Chauhan, A.K., Iaconcig, A., Porro, F., Baralle, F.E., and Muro, A.F. (2007) *J. Biol. Chem.*, **282**, 28057–28062.
13. Leiss, M., Beckmann, K., Girós, A., Costell, M., and Fässler, R. (2008) *Curr. Opin. Cell Biol.*, **20**, 502–507.
14. Schwarzbauer, J.E. (1991) *Curr. Opin. Cell Biol.*, **3**, 786–791.
15. Guan, J.L. and Hynes, R.O. (1990) *Cell*, **60**, 53–61.
16. Chen, H., Gu, D.N., Burton-Wurster, N., and MacLeod, J.N. (2002) *J. Biol. Chem.*, **277**, 20095–20103.
17. Kozaki, T., Matsui, Y., Gu, J., Nishiuchi, R., Sugiura, N., Kimata, K., Ozono, K., Yoshikawa, H., and Sekiguchi, K. (2003) *J. Biol. Chem.*, **278**, 50546–50553.
18. Schwarzbauer, J.E. (1991) *J. Cell Biol.*, **113**, 1463–1473.
19. Sottile, J., Schwarzbauer, J., Selegue, J., and Mosher, D.F. (1991) *J. Biol. Chem.*, **266**, 12840–12843.
20. Mcdonald, J.A., Quade, B.J., Broekelmann, T.J., LaChance, R., Forsman, K., Hasegawa, E., and Akiyama, S. (1987) *J. Biol. Chem.*, **262**, 2957–2967.
21. Sechler, J.L., Corbett, S.A., Wenk, M.B., and Schwarzbauer, J.E. (1998) *Ann. N.Y. Acad. Sci.*, **857**, 143–154.
22. Curnis, F., Longhi, R., Crippa, L., Cattaneo, A., Dondossola, E., Bachi, A., and Corti, A. (2006) *J. Biol. Chem.*, **281**, 36466–36476.
23. Ruoslahti, E. and Obrink, B. (1996) *Exp. Cell Res.*, **227**, 1–11.
24. Fogerty, F.J., Akiyama, S.K., Yamada, K.M., and Mosher, D.E. (1990) *J. Cell Biol.*, **111**, 699–708.
25. Sechler, J.L., Corbett, S.A., and Schwarzbauer, J.E. (1997) *Mol. Cell. Biol.*, **8**, 2563–2573.
26. Takagi, J., Strokovich, K., Springer, T.A., and Walz, T. (2003) *EMBO J.*, **22**, 4607–4615.
27. Wennerberg, K., Lohikangas, L., Gullberg, D., Pfaff, M., Johansson, S., and Fässler, R. (1996) *J. Cell Biol.*, **132**, 227–238.
28. Wu, C., Keivens, V.M., Otoole, T.E., Mcdonald, J.A., and Ginsberg, M.H. (1995) *Cell*, **83**, 715–724.
29. Yang, J.T., Bader, B.L., Kreidberg, J.A., Ullman-Culleré, M., Trevithick, J.E., and Hynes, R.O. (1999) *Dev. Biol.*, **215**, 264–277.

30. Johnson, K.J., Sage, H., Briscoe, G., and Erickson, H.P. (1999) *J. Biol. Chem.*, **274**, 15473–15479.
31. Geiger, B., Bershadsky, A., Pankov, R., and Yamada, K.M. (2001) *Nat. Rev. Mol. Cell Biol.*, **2**, 793–805.
32. Hall, A. (2005) *Biochem. Soc. Trans.*, **33**, 891–895.
33. Yoneda, A., Ushakov, D., Multhaupt, H.A.B., and Couchman, J.R. (2007) *Mol. Cell. Biol.*, **18**, 66–75.
34. Langenbach, K.J. and Sottile, J. (1999) *J. Biol. Chem.*, **274**, 7032–7038.
35. Shi, F. and Sottile, J. (2008) *J. Cell Sci.*, **121**, 2360–2371.
36. Zhong, C., Chrzanowska-Wodnicka, M., Brown, J., Shaub, A., Belkin, A.M., and Burridge, K. (1998) *J. Cell Biol.*, **141**, 539–551.
37. Green, J.A., Berrier, A.L., Pankov, R., and Yamada, K.M. (2009) *J. Biol. Chem.*, **284**, 8148–8159.
38. Wierzbicka-Patynowski, I., Mao, Y., and Schwarzbauer, J.E. (2007) *J. Cell. Physiol.*, **210**, 750–756.
39. Williams, E.C., Janmeyn, P.A., Johnson, R.B., and Mosher, F. (1983) *J. Biol. Chem.*, **258**, 5911–5914.
40. Sakai, K., Fujii, T., and Hayashi, T. (1994) *J. Biochem.*, **115**, 415–421.
41. Vartio, T. (1986) *J. Biol. Chem.*, **261**, 9433–9437.
42. Mosher, D.F. and Johnson, R.B. (1983) *J. Biol. Chem.*, **258**, 6595–6601.
43. Mosher, D.F. (1987) Process for Preparation of Multimeric Plasma Fibronectin.
44. Chen, Y., Zardi, L., and Peters, D.M.P. (1997) *Scanning*, **19**, 349–355.
45. Peters, D.M.P., Chen, Y., Zardi, L., and Brummel, S. (1998) *Microsc. Microanal.*, **4**, 385–396.
46. Vuento, M., Salonen, E., and Riepponen, P. (1980) *Biochimie*, **62**, 99–104.
47. Vuento, M., Vartio, T., Saraste, M., von Bonsdorff, C.H., and Vaheri, A. (1980) *Eur. J. Biochem.*, **105**, 33–42.
48. Jilek, F. and Hörmann, H. (1979) *Hoppe-Seyler's Z. Physiol. Chem.*, **360**, 597–604.
49. Richter, H., Wendt, C., and Hörmann, H. (1985) *Biol. Chem. Hoppe-Seyler*, **366**, 509–514.
50. Morla, A., Zhang, Z., and Ruoslahti, E. (1994) *Nature*, **367**, 193–196.
51. Hocking, D.C., Smith, R.K., and McKeown-Longo, P.J. (1996) *J. Cell Biol.*, **133**, 431–444.
52. Briknarová, K., Åkerman, M.E., Hoyt, D.W., Ruoslahti, E., and Ely, K.R. (2003) *J. Mol. Biol.*, **332**, 205–215.
53. Baneyx, G. and Vogel, V. (1999) *Proc. Natl. Acad. Sci. U.S.A.*, **96**, 12518–12523.
54. Ulmer, J., Geiger, B., and Spatz, J.P. (2008) *Soft Matter*, **4**, 1998–2007.
55. Volberg, T., Ulmer, J., Spatz, J., and Geiger, B. (2010) in *IUTAM Symposium on Cellular, Molecular and Tissue Mechanics* (eds K. Garikipati and E.M. Arruda), Springer, The Netherlands, pp. 69–79.
56. Kaiser, P. and Spatz, J.P. (2010) *Soft Matter*, **6**, 113–119.
57. Ejim, O.S., Blunn, G.W., and Brown, R.A. (1993) *Biomaterials*, **14**, 743–748.
58. Brown, R.A., Blunn, G.W., and Ejim, O.S. (1994) *Biomaterials*, **15**, 457–464.
59. Wòjciak-Stothard, B., Denyer, M., Mishra, M., and Brown, R.A. (1997) *In Vitro Cell. Dev. Biol. Anim.*, **33**, 110–117.
60. Ahmed, Z., Idowu, B.D., and Brown, R.A. (1999) *Biomaterials*, **20**, 201–209.
61. Ahmed, Z. and Brown, R.A. (1999) *Cell Motil. Cytoskeleton*, **42**, 331–343.
62. Underwood, S., Afoke, A., Brown, R.A., MacLeod, A.J., and Dunnill, P. (1999) *Bioprocess Eng.*, **20**, 239–248.
63. Harding, S.I., Underwood, S., Brown, R.A., and Dunnill, P. (2000) *Bioprocess Eng.*, **22**, 159–164.
64. Underwood, S., Afoke, A., Brown, R.A., MacLeod, A.J., Shamlou, P., and Dunnill, P. (2001) *Biotechnol. Bioeng.*, **73**, 295–305.
65. Ahmed, Z., Underwood, S., and Brown, R.A. (2003) *Tissue Eng.*, **9**, 219–231.
66. Little, W.C., Smith, M.L., Ebneter, U., and Vogel, V. (2008) *Matrix Biol.*, **27**, 451–461.
67. Klotzsch, E., Smith, M.L., Kubow, K.E., Muntwyler, S., Little, W.C., Beyeler, F., Gourdon, D., Nelson, B.J., and Vogel, V. (2009) *Proc. Natl. Acad. Sci. U.S.A.*, **106**, 18267–18272.
68. Pernodet, N., Rafailovich, M., Sokolov, J., Xu, D., Yang, N.-L., and McLeod, K. (2003) *J. Biomed. Mater. Res. A*, **64**, 684–692.

69. Pellenc, D., Berry, H., and Gallet, O. (2006) *J. Colloid Interface Sci.*, **298**, 132–144.
70. Nelea, V. and Kaartinen, M.T. (2010) *J. Struct. Biol.*, **170**, 50–59.
71. Feinberg, A.W. and Parker, K.K. (2010) *Nano Lett.*, **10**, 2184–2191.
72. Salmeròn-Sánchez, M., Rico, P., Moratal, D., Lee, T.T., Schwarzbauer, J.E., and García, A.J. (2011) *Biomaterials*, **32**, 2099–2105.
73. Erickson, H.P. (2002) *J. Muscle Res. Cell Motil.*, **23**, 575–580.
74. de Jongh, H.H.J., Kosters, H.A., Kudryashova, E., Meinders, M.B.J., Trofimova, D., and Wierenga, P.A. (2004) *Biopolymers*, **74**, 131–135.
75. Oberhauser, A.F., Badilla-Fernandez, C., Carrion-Vazquez, M., and Fernandez, J.M. (2002) *J. Mol. Biol.*, **319**, 433–447.
76. Rief, M., Gautel, M., and Gaub, H.E. (2000) in *Advances in Experimental Medicine and Biology: Elastic Filaments Of The Cell* (eds H.L. Granzier and G.H. Pollack), Springer, New York, pp. 129–141.
77. Krammer, A., Lu, H., Isralewitz, B., Schulten, K., and Vogel, V. (1999) *Proc. Natl. Acad. Sci. U.S.A.*, **96**, 1351–1356.
78. Guan, J., Ferrell, N., Yu, B., Hansford, D.J., and Lee, L.J. (2007) *Soft Matter*, **3**, 1369–1371.
79. Nelea, V., Nakano, Y., and Kaartinen, M.T. (2008) *Protein J.*, **27**, 223–233.
80. Peters, D.M.P. and Mosher, D.F. (1989) in *Extracellular Matrix Assembly and Structure* (eds P.D. Yurchenco, D.E. Birk, and R.P. Mecham), Academic Press, San Diego, CA, pp. 315–350.
81. Mosher, D.F., Fogerty, F.J., Chernousov, M.A., and Barry, E.L.R. (1991) *Ann. N.Y. Acad. Sci.*, **614**, 167–180.
82. Mosher, D.F. (1993) *Curr. Opin. Struct. Biol.*, **3**, 214–222.
83. Sechler, J.L., Rao, H., Cumiskey, A.M., Vega-Colín, I., Smith, M.S., Murata, T., and Schwarzbauer, J.E. (2001) *J. Cell Biol.*, **154**, 1081–1088.
84. Bergkvist, M., Carlsson, J., and Oscarsson, S. (2003) *J. Biomed. Mater. Res. A*, **64**, 349–356.
85. Baugh, L. and Vogel, V. (2004) *J. Biomed. Mater. Res. A*, **69**, 525–534.
86. Hull, J.R., Tamura, G.S., and Castner, D.G. (2007) *Biophys. J.*, **93**, 2852–2860.
87. Lutolf, M.P., Gilbert, P.M., and Blau, H.M. (2009) *Nature*, **462**, 433–441.
88. Petrie, T.A., Raynor, J.E., Dumbauld, D.W., Lee, T.T., Jagtap, S., Templeman, K.L., Collard, D.M., and García, A.J. (2010) *Sci. Transl. Med.*, **2**, 45ra60.
89. Rico, P., Rodríguez Hernández, J.C., Moratal, D., Altankov, G., Monleón Pradas, M., and Salmerón-Sánchez, M. (2009) *Tissue Eng. A*, **15**, 3271–3281.
90. Gugutkov, D., González-García, C., Rodríguez Hernández, J.C., Altankov, G., and Salmerón-Sánchez, M. (2009) *Langmuir*, **25**, 10893–10900.
91. Gugutkov, D., Altankov, G., Rodríguez Hernández, J.C., Monleón Pradas, M., and Salmerón Sánchez, M. (2010) *J. Biomed. Mater. Res. A*, **92**, 322–331.
92. Rodríguez Hernández, J.C., Rico, P., Moratal, D., Monleón Pradas, M., and Salmerón-Sánchez, M. (2009) *Macromol. Biosci.*, **9**, 766–775.
93. Coelho, N.M., González-García, C., Planell, J.A., Salmerón-Sánchez, M., and Altankov, G. (2010) *Eur. Cell Mater.*, **19**, 262–272.
94. Rodríguez Hernández, J.C., Salmerón Sánchez, M., Soria, J.M., Gómez Ribelles, J.L., and Monleón Pradas, M. (2007) *Biophys. J.*, **93**, 202–207.
95. Toromanov, G., González-García, C., Altankov, G., and Salmerón-Sánchez, M. (2010) *Polymer*, **51**, 2329–2336.
96. García, A.J., Vega, M.D., and Boettiger, D. (1999) *Mol. Cell. Biol.*, **10**, 785–798.

9
Nanoscale Control of Cell Behavior on Biointerfaces

E. Ada Cavalcanti-Adam and Dimitris Missirlis

9.1
Nanoscale Cues in Cell Environment

Cells in living tissues reside inside a complex microenvironment composed of other cells, soluble biomolecules, and a protein network, which is known as the extracellular matrix (ECM). Understanding the intricate interactions and cross talk between microenvironment and cells remains an active endeavor in biology and medicine. Over the past few decades, studies on how structural and biological properties of the ECM affect cell function have culminated in remarkable scientific and technological advances. The use of ECM proteins and ECM-derived peptides has become routine in cell biology studies that investigate the role of adhesion-mediated processes in cell function and are moving toward use in medical products aimed at restoring tissue function. In the meantime, physical properties of the ECM, such as topography and elasticity, have also emerged as key regulators of cell fate.

The sensitivity of cells to local ECM architecture and biomolecular patterns extends down to the nanometer scale. Different topologies and nanostructures formed by ECM proteins *in vivo* have been shown to be pivotal for cell–matrix interactions and to contribute to matrix–cell signaling. For example, in the myocardium, the ECM immediately underlying the aligned cell arrays is present in the form of 100 nm thick fibers, arranged parallel to the direction of cells [1]. In this tissue, it is hypothesized that the nanoscale cues provided by the ECM exercise multiscale control over tissue assembly by directing the different embedded cell types in a defined structure and arrangement according to nanoscale topographical and molecular patterns.

Bone constitutes another example of a hierarchically structured tissue with nanoscale organization (Figure 9.1). The ECM secreted by bone-lining cells is mostly composed of collagen type I fibrils, which exhibit a characteristic nanotopography. The collagen monomer is $\sim$300 nm long and 1.5 nm wide and forms extended fibrils ranging from 260 to 410 nm in diameter [3]. Interestingly, collagen normally has a 66 nm axial period, which shifts to a 240 nm periodicity in neoplastic tissues [4]. Nanoindentation of collagen fibrils showed that they are axially heterogeneous, with differences in mechanical properties in gap and overlay regions [5]. The gap

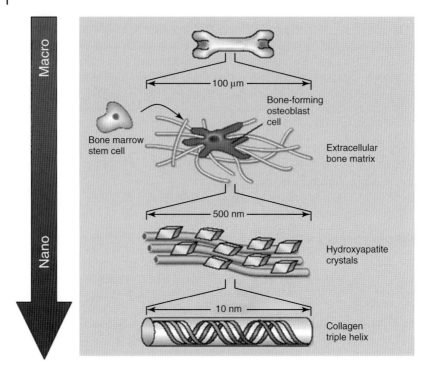

Figure 9.1 Bone is a nanostructured material. Bone-forming cells are embedded in a network mainly composed of collagen fibrils and hydroxyapatite crystals. These structures show defined spatial organization down to the nanometer length scale. (Source: Adapted from [2].)

regions are believed to serve as nucleation sites for hydroxyapatite crystals, a process that leads to matrix mineralization. While the structures at the micrometer scale in bone may vary depending on the type of bone, these mineralized collagen fibrils are highly conserved and constitute the primary building blocks of bone. Thus, the characteristic load-bearing properties of bone, which are tightly correlated to its structure, can most likely be traced to these characteristic nanostructures [6].

The impact of nanoscale ECM molecular organization is also observed in tissues in which the ECM is not predominantly in fibrillar form. The basement membrane of epithelial tissues, for example, consists primarily of collagen IV and laminin arranged in a meshwork. However, minor components embedded in the network, such as nidogen/entactin aggregates, are essential for the maintenance of tissue structure and function. They stabilize the matrix and directly interact with laminin, forming nanoscale complexes, which in turn directly influence cell adhesion and function. Similarly, in bone, minor components such as osteopontin form a meshwork that provides mechanical support and acts as an adhesive layer for mineralized fibrils [7].

The ECM further provides the physical substrate on which cells exert forces during cell adhesion and migration. These mechanical cues are regulated by

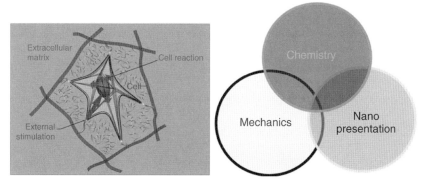

Figure 9.2 Mimicking cell–ECM interactions with biomaterials. The external stimulation of cells, triggered by chemical, mechanical and spatial cues at the nanoscale, results in cell reactions. These comprise changes in cell adhesion and architecture as well as initiation of signaling cascades and activation of gene expression.

ECM composition and structure and highlight an additional manner by which the extracellular space may control cell function. The question of how variations in the mechanical properties of a cell niche between tissues and during development affect cell behavior remains open [8]. Moreover, in different pathological situations, such as myocardial infarction and cancer progression, a local increase in ECM rigidity has been observed [9, 10]. The source of this enhanced rigidity is primarily local accumulation of a dense, cross-linked ECM [11–13].

The examples cited above, along with many analogous studies, demonstrate that tissue functions are strongly dependent on ECM components as well as on the location of these components relative to each other. Resolving ECM organization and impact *in vivo* remains a significant challenge because of the small scale of the underlying interactions and the complexity of the concerted action of physicochemical signals present. Moreover, each type of solid tissue has its own typical and distinct ECM composition and organization, which can exhibit significant internal heterogeneity. Therefore, there is a pressing need to construct *in vitro* platforms able to reproduce the specific arrangement and organization of ECM in a controlled manner. With the advent of nanotechnology and its biological and biomedical applications, researchers have the tools to accomplish this task [1, 14]. The nanoscale control achieved till date has opened the way to a more detailed picture of how cells interact with their microenvironment [15, 16]. Nevertheless, many questions and tasks lie ahead, and the ongoing development of appropriate synthetic nanostructured biomaterial surfaces will certainly be invaluable in dealing with them (Figure 9.2).

In this chapter, current strategies for surface patterning at the nanoscale and the systematic variation of relevant surface parameters are described. Then, the controlled presentation of ECM-derived biomolecules and peptides to cells is discussed. Finally, examples of cell responses to nanostructured materials are presented, with an emphasis on regulation of adhesion, migration, and intracellular signaling.

9.2
Biomimetics of Cell Environment Using Interfaces

9.2.1
Surface Patterning Techniques at the Nanoscale

The development of cell interfaces for basic science and biomedical applications is progressing rapidly, with increased emphasis on nanoscale control. A variety of traditional and novel surface fabrication technologies have been proposed to produce substrates that mimic the extracellular space and organization down to its smallest details. The aim of such nanopatterned interfaces is primarily to control where, how, and how many biomolecules are presented to the cells. In this section, the use of nonconventional nanolithographic techniques as tools for biomolecule immobilization at the nanoscale is presented.

9.2.1.1 Surface Patterning by Nonconventional Nanolithography

Lithographic methods for patterning nanostructures onto surfaces have been historically motivated by technological applications in electronics, where decreasing the components size minimizes energy loss, computational time, and cost. These lithographic methods were soon adapted for applications in biotechnology and biomedicine, in order to control biomolecule immobilization on surfaces and thus direct cell–material interactions. Nanolithography encompasses a variety of methods to modify surfaces by etching, writing, and printing at the nanoscopic level. To circumvent complexity in the procedure or unsuitability for handling a large variety of organic and biological molecules, variations in conventional lithography techniques have been developed to achieve ultrahigh resolution and capabilities. While most of them still involve several elaborate steps, a steady progress to overcome instrument limitations has allowed sufficient throughput. The latest advances in lithography for surface patterning at the nanoscale comprise a range of different techniques, referred to collectively as *nonconventional nanolithography*. These can be divided into two groups: direct *printing methods*, such as dip-pen nanolithography, and soft-lithography or *stamping methods*, such as nanoimprint lithography and nanocontact printing (Figure 9.3).

Dip-pen nanolithography allows the direct deposition of various materials onto surfaces down to nanoscale resolution using an atomic force microscopy (AFM). First described by Jaschke *et al.* [19] and further developed by Mirkin *et al.* [20], this technique is based on coating the tip of an AFM cantilever, having a radius curvature of ∼20 nm, with an "ink," containing the biomolecule to be patterned. The tip then makes contact with the surface at predefined points, depositing the biomolecules in the desired patterns. Therefore, it is a direct-write lithographic technique that allows patterning of nanostructures with precision and resolution below 50 nm. The versatility of dip-pen nanolithography regarding the type of molecules and surfaces for patterning makes it a widely used technique with extending capabilities [21]. However, its most notable limitations remain the inability to provide continuous ink flow and the difficulty in transferring high-molecular-weight inks.

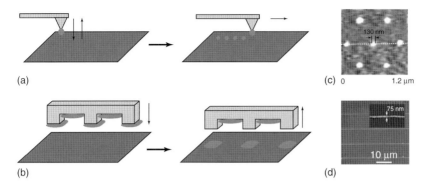

Figure 9.3 Nonconventional nanolithography approaches for nanopattern formation. Schematic representations of (a) a printing method (dip-pen nanolithography) and (b) a stamping method (nanoimprint lithography) and corresponding examples of fabricated patterns (c,d) are shown. (c) AFM topography image of nanopatterned antibodies using dip-pen nanolithography [17]. (d) SEM image of oxide nanolines formed by nanoimprinting nanolithography. (Source: Reprinted with permission from Ref. [18]. Copyright 2004 American Chemical Society.)

An alternative nanolithography technique, first proposed in 1995 by Chou *et al.* [22] is nanoimprint lithography . This process involves an initial imprint step of pressing a nanopatterned mold onto a thin polymer resist, which has been previously cast on a surface. In the second step, the desired nanopattern is generated on the surface either by a deposition or by an etching process. In the latter case, the residual resist is removed from compressed areas, exposing the desired nanopattern. A significant advantage of nanoimprinting is the possibility to pattern structures below 25 nm over large areas with low cost [23]. To achieve even higher resolution, nanoimprinting was recently combined with self-aligned pattern transfer [24]. In this process, the postimprint deposition of an angle-evaporated hard mask facilitates the liftoff step while producing high pattern uniformity and feature size reduction. In fact, surfaces displaying gold nanoparticles with diameters below 5 nm were fabricated. The nanoparticles were arranged in local clusters, with as few as four nanoparticles per cluster and interparticle spacing was controlled with nanometer resolution. The open challenge is now to create extended patterns with such unprecedented resolution on larger surface areas.

9.2.1.2 Block Copolymer Micelle Lithography

The aforementioned techniques fall in the category of top-down manufacturing approaches. In contrast, the spontaneous formation of ordered structures by molecular self-assembly provides a dynamic alternative to produce nanopatterned surfaces. The strategy behind this approach, commonly referred to as the *bottom-up approach*, lies in the ability of molecular building blocks to assemble reproducibly into defined architectures, based on energetic and entropic principles. In recent years, an intensive effort has been directed to modulate the organization of amphiphilic block copolymers in monolayers to create nanopatterned surfaces. The chemical link between the different blocks counteracts phase separation at a

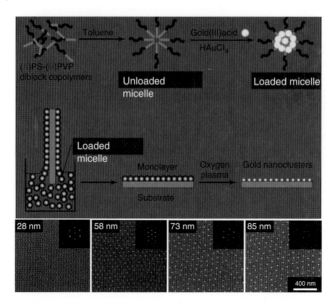

Figure 9.4 Block copolymer micelle nanolithography for decoration of surfaces with gold nanoparticles. The process of micelle self-assembly, film formation, and particle fabrication is schematically presented. On the bottom, scanning electron micrographs of the deposited gold nanoparticles arranged in patterns with variable spacings are shown. The numbers indicate the inderdot distance, and insets are fast Fourier transforms showing the hexagonal lattice of the dots.

macroscopic length scale and drives instead phase separation in the nanometer range, leading to regular structures with periodicity. The packing, and the morphology of nanometer domains are dependent on the composition of the amphiphilic copolymers and the interaction between the constituent blocks. Manipulation of these structural parameters enables the use of block copolymers for the design of self-organized assemblies in a variety of organic, inorganic, metallic, or biologically relevant materials.

An example of a bottom-up approach is *block copolymer micelle lithography (BCML)*, based on the self-assembly of polystyrene-block-poly(2-vinylpyridine) (PS-b-P2VP) into reverse micelles in toluene [25–27]. In the core of the micelle, the P2VP block is complexed with a metal precursor (HAuCl$_4$). Dipping and retracting a substrate from a solution of micelles results in deposition of uniform and extended monomicellar films on the substrate. Subsequent removal of the organic parts with oxygen or hydrogen gas plasma treatment results in the deposition of highly regular metal (gold) nanoparticles. The nanoparticles form a quasihexagonal pattern on solid interfaces such as glass or silicon wafers. The size of the nanoparticles may be varied between 1 and 20 nm by adjusting the amount of metal precursor added to the micellar solution (Figure 9.4). The spacing between nanoparticles may also be adjusted from 15 to 250 nm, by choosing the appropriate molecular weight and composition of PS-b-P2VP or by changing the retraction speed.

This lithographic technique can be used for both conductive and nonconductive substrates. For example, BCML was employed to decorate Teflon AF surfaces with gold nanoparticles, which were then functionalized with peptides that promoted endothelial cell adhesion [28]. Gold nanoparticle patterns were also successfully transferred on hydrogel substrates. Following pattern fabrication on glass by BCML, the nanoparticles were functionalized with acryloyl groups via a thiol linker. The acryloyl groups then participated in a radical, cross-linking polymerization of poly(ethylene glycol) (PEG) diacrylates, effectively linking the gold particles to the hydrogel. Removal of the glass substrate then exposed the gold particles on the hydrogel surface [29].

To further control the position of individual nanoparticles on surfaces, a unique approach, termed *scanning probe block copolymer lithography*, has been recently introduced [30]. This technique relies on dip-pen or polymer-pen lithography to deposit on a surface very small volumes of block copolymer micelles containing metal ions inside their core. The metal nanoparticles formed after plasma treatment have diameters as small as 5 nm, significantly lower than the initial micelle dimensions.

BCML is considerably less costly than electron beam lithography and allows the fabrication of structures with sizes not easily accessible by optical lithography. High throughput and simplicity in the production process are also considered advantageous over other templating methods. Moreover, the technique can be combined with traditional lithography to fabricate hierarchical structures, comprising nanoscale patterns arranged in micrometer-sized domains. Precise control over polymer self-assembly and the resulting structures, in terms of uniformity in size and density, still represents technical challenges and inspires further modifications of this technique.

The choice of method to fabricate nanoscale interfaces for cell studies is determined by a number of criteria. Some of these nanofabrication techniques, such as electron beam lithography, enable the production of features as small as 5 nm, but are still limited by time demand, high costs, and handling of difficult equipment. Novel material selection is of critical importance as cell–nanotopography interactions continue to be utilized in tissue engineering applications. Synthetic and natural materials must be selected not only on the basis of biocompatibility but also on the basis of compatibility with nanofabrication processes. For basic research in the laboratory, complexity of preparation protocols may be of secondary importance compared to the requirement for superior control over surface properties. However, if the results of these studies are to be translated to clinical products, issues such as cost, reproducibility, and ease of handling will become increasingly significant in developing appropriate methodologies.

9.2.2
Variation of Surface Physical Parameters at the Nanoscale

Control over individual parameters of a surface is essential to assess their corresponding impact on cell behavior and provide the design criteria for novel biomaterials. Several physical determinants of cell behavior at the nanoscale have been identified over the years, including (nano)topography, ligand distribution, spacing, and clustering. The influence of these parameters individually or in combination with bulk substrate properties such as elasticity and dimensionality is starting to provide a more comprehensive picture of cell–surface interactions at the nanoscale. The role of the physical parameters of nanopatterned surfaces is briefly discussed in the next section.

9.2.2.1 Surface Nanotopography

The observation that *in vivo* ECM topography exhibits significant heterogeneity has fueled interest to construct cell interfaces able to mimic ECM features at different length scales. At the micrometer scale, surface topography is already a well-established factor in modulating cell adhesion and cell functions *in vitro*. At an even smaller scale, nanotopography has also become an important and increasingly studied physical determinant of cell behavior. Fabrication of surfaces with controlled geometric features in the form of wells, pillars, or channels in an ordered manner has revealed the potential of this approach in modulating basic cell function.

For example, nanogratings in vascular grafts were able to guide alignment of smooth muscle cells and endothelial cells or enhance neurite migration during peripheral nerve regeneration [31]. In an attempt to mimic myocardial tissue, biocompatible PEG hydrogel arrays were developed with defined nanotopography [32]. The engineered substrates displayed grooves and ridges of varying widths (ridge width : groove width 150 : 50 to 800 : 800 nm) and different heights (200–500 nm). Therefore, the highly variable distance between individual fibrils in the ECM of myocardial tissue was replicated and allowed detailed biochemical studies to be performed.

Biomaterial nanotopography has been proved to additionally influence early osteointegration of bone and dental implants. A key requirement for the success of a bone or tooth implant is its ability to promote osteogenic and stem cell attachment to its surface. Substrates displaying different nanotopographies were used to determine which structures induce selective adhesion of osteoblast precursors and how cell response can be guided. These studies have shown that regularly spaced pits or pillars negatively affect adhesion, while grooves and steps positively regulate adhesion [33]. In another example, titanium dioxide nanotubes and poly(methyl methacrylate) (PMMA) nanopatterns were shown to poorly induce osteogenic differentiation of stem cells, while random nanotopographies with distinct geometries triggered differentiation even in the absence of osteogenic factors [34, 35].

Overall, it is evident that order, symmetry, and shape of nanofeatures are critical determinants of cell response with potential in improving scaffold design [36]. Advances in nanofabrication methods enable the production of nanotopographies with increasing intricacy and precision (Figure 9.5). However, even though an increasing amount of literature is available on the effects of such modifications

9.2 Biomimetics of Cell Environment Using Interfaces | 221

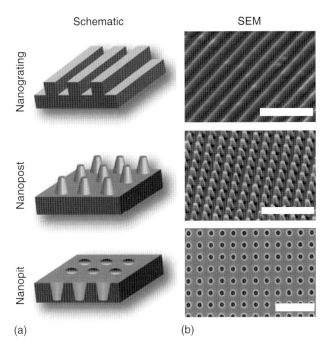

Figure 9.5 Examples of nanotopography features on surfaces. A schematic representation (a) and the corresponding SEM images (b) of common topographies fabricated on surfaces are shown, including nanogratings (top; scale bar 5 μm), nanopost arrays (middle; scale bar 5 μm), and nanopit arrays (bottom; scale bar 1 μm) [37].

on biomaterial performance, the lack of in-depth understanding of the mechanisms by which these different nanotopographies affect cell function often leads to unpredicted results. One underlying reason is the lack of control over presentation of biochemical ligands on the surface. Different nanotopographies could favor different protein adsorption patterns or limit the surface area available for cell interactions. Moreover, the same nanotopography on different substrate materials (e.g., hydrophobic vs. hydrophilic) might have an opposite effect on cells. It is therefore necessary to consider the entirety of surface parameters and carefully functionalize these and not only focus on developing more intricate nanotopographies.

9.2.2.2 Interligand Spacing

Cells possess a multitude of specific plasma membrane receptors to sense their microenvironment and interact with their surroundings. The extracellular domains of these receptors are generally a few nanometers in size and recognize soluble biochemical signals or immobilized components of the ECM. Usually, ligand binding triggers conformational changes that initiate some signaling cascade. However, receptor clustering into larger, functional complexes is often necessary for signal propagation. For example, the death signaling cascade mediated by binding of tumor necrosis factor to its receptor at the cell membrane occurs only after receptor clustering [38]. Clustering can be imposed by the architecture of the extracellular ligand or can be a result of receptor activation on ligand recognition. In both cases, the nanoscale distribution of ligands is key to transferring efficiently the information to the intracellular compartment. It is therefore crucial that ligand spacing and density be tightly controlled on synthetic, nanostructured surfaces.

The earliest and simplest approach to vary interligand spacing on a surface was to randomly disperse the ligand of interest on it. The average interligand distance is then a function of the concentration used, assuming a homogeneous coverage. For example, various homogeneous surfaces functionalized with the cell-adhesive peptide arginine-glycine-aspartic acid (RGD) were employed to study ligand density effects on cell adhesion [39]. However, these approaches did not allow control over the exact distance between ligands and their surface distribution. To overcome this problem and achieve well-defined patterns, ligands can be immobilized on previously formed stable nanostructures that exhibit the desired control over distribution. Among the various nanostructures developed, gold nanodot arrays are of particular interest, first, because gold nanoparticles are easily functionalized using thiol-gold chemistry and second, because their small size ensures that each functionalized nanoparticle can interact with only one receptor at a time. However, a challenge for functionalizing neighboring dots with different ligands remains for this system. A possible solution is to sequentially deposit different types of particles and then functionalize them with orthogonal chemistries [40].

An alternative technique with potential to create such patterns is DNA origami nanotechnology [41]. This technology uses the DNA code as a building block to assemble programable templates on two or three dimensions. Precise positioning of particles, or directly of biomolecules, is then possible using complementary RNA linkers. A few examples of this emerging approach include linearly arranging quantum dots with different periodicity [42], assembly of different-sized gold nanoparticles on a template [43], or dually functionalizing with organic and inorganic particles a DNA nanostructure [44]. However, DNA templating is still challenging to achieve on large scales, and it remains to be shown how the produced templates can be rendered stable enough as cell interfaces.

9.2.2.3 Ligand Density

Cells typically interact with a large number of surface ligands. The amount of occupied receptors regulates signaling cascades, and often, a threshold of such interactions is required for signaling activation. When interligand spacing is modified in a homogeneous surface, the overall ligand density on that surface is also modified. Ideally, to determine whether global ligand availability or just local ligand density is regulating a given cell response, surfaces should be fabricated with control over one parameter at a time. In order to keep these two parameters decoupled, micropatterning techniques can be combined with nanopatterning ones. For example, micron-sized areas were patterned on a surface and decorated with gold nanodot arrays of variable interdot distances, allowing formation of surfaces displaying the same amount of RGD ligands but different interligand spacing [45]. Alternatively, ligand density can be manipulated by ligand clustering. For example, Comisar et al. used RGD-functionalized alginate to investigate preosteoblast behavior as a function of local versus global adhesion density. By independently varying the degree of functionalization of each polymer chain and the ratio of functionalized to unmodified alginate, they were able to decouple these two parameters.

Surfaces exhibiting a gradual change in ligand spacing have also been used to study the effects of ligand density. Cells sense such gradients and move toward areas that promote higher cell adhesion in a process termed *haptotaxis* [46]. Even though gradients have been used in the past, constructing them with nanometer resolution was only recently made possible [47]. This has allowed probing length scales much smaller compared to the cell body and revealed that cells have mechanisms to sense with exquisite sensitivity variations of few nanometers in ligand presentation.

9.2.2.4 Substrate Mechanical Properties

Among the important physical features of a cell interface are the mechanical properties of the substrate. An exponentially increasing number of studies are linking phenomenological and biochemical cell alterations to physical cues from the microenvironment, a phenomenon that is generally termed *mechanotransduction*. Cells sense their mechanical microenvironment via receptor complexes that connect the ECM to the cytoskeleton and adapt to the generated forces by a multitude of different pathways that lead to dramatic shifts in adhesion, motility, and proliferation. The responses to dynamic or static changes in substrate elasticity are manifested at very different time scales. Cells adjust their spreading on sudden jumps in substrate stiffness within seconds to minutes [48], whereas commitment of stem cells to different lineages is apparent after several days of culturing on substrates, differing only in stiffness [49].

A large number of substrates with mechanical properties more relevant to physiological conditions, compared to polystyrene or glass, have been developed. Hydrogels are especially attractive for this purpose because they provide a cell repellent background, which can be coated with ECM proteins or components. A qualitative step forward with respect to homogeneous hydrogels of tunable elasticity is to include nanopatterns on their surface. As a result, the investigation of mechanical properties in combination with additional parameters, for example, ligand spacing, is possible. For example, Aydin et al. [50] developed a protocol to transfer gold nanopatterns of varying ligand spacing and density on hydrogels with tunable elasticity. Rat embryonic fibroblasts were viable on these gels and preferentially interacted with the biofunctionalized pattern. These substrates are now being employed to determine the optimal parameters for growth of different cell types, providing information for manipulating cell-specific functions.

9.2.2.5 Dimensionality

The investigation of cells using 2D nanopatterned surfaces constitutes a reductionist approach since cells in solid tissues *in vivo* reside in a 3D microenvironment. Ultimately, it would be beneficial to fabricate surfaces allowing contact formation in three rather than two dimensions, given the reported marked differences in cell adhesion and morphology [51]. The design and fabrication of biomaterials for cell encapsulation is not in the scope of this chapter, and the reader is encouraged to refer to several chapters of this book dealing with this issue. An intermediate situation is the construction of surfaces, which allow cells to form contacts in three dimensions while maintaining control over nanoscale properties. This can be achieved by fabrication of microwells of varying size and shape in which cells can be cultured [52]. Alternatively, channels and chambers can be formed via two-photon confocal patterning of light-sensitive bulk hydrogels. Using light-degradable cross-linkers and focusing the beam according to a specified pattern, micrometer guides for cell migration were constructed inside the hydrogel [53]. Combination of these technologies with a zoomed-in detail at the nanoscale presents itself as a worthwhile challenge. Characterization of cell–biomaterial interactions in such systems poses additional technical challenges and is a field evolving in parallel to material development. Fluorescence techniques such as FRET capable of providing information on distances between individual fluorophores in the order of one to a few nanometers are being developed to probe interactions between cell receptors and their matrix-immobilized ligands [54].

Counterintuitive as it may seem at first glance, reducing dimensionality can also provide a more physiological environment for studying cell behavior. Many ECM proteins assemble in fibrils with diameters much smaller than the cell body. By patterning one-dimensional adhesive lines on a surface, it was shown that cell migration and focal adhesion (FA) formation were more reminiscent of the situation in 3D than corresponding 2D surfaces [55]. The minimum line width in this study was around 1 μm, leaving open the question of how cells would behave in even smaller lines and whether there is a threshold for efficient attachment and motility. Taking these one-dimensional lines to the third dimension in form of fibers is also attracting substantial interest. Peptide-based, self-assembled biomaterials have been developed that are held together by a combination of hydrophobic, electrostatic, and hydrogen bonding interactions in the form of nanofibers [56] or nanotubes [57]. These can present high concentrations of bioactive peptides on their surface, arranged in three dimensions. Mixing of different peptides on the same fiber and control over interligand spacing are still achieved through statistical means. However, it is easy to envision phase-separated systems in which patches of ligands form via favorable energetics of their self-assembling parts, whether these are phase-separated lipids or polymers [58].

9.2.3
Surface Functionalization for Controlled Presentation of ECM Molecules to Cells

9.2.3.1 Proteins, Protein Fragments, and Peptides
Efficient presentation of the appropriate biochemical signals to cells in the context of nanoengineered biomaterials requires consideration of the pertinent conjugation chemistry. Natural or synthetic proteins, protein fragments, and short bioactive peptides can be employed to functionalize surfaces (Figure 9.6). Each

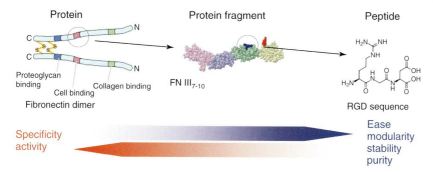

Figure 9.6 Biomolecules for surface functionalization. Proteins, recombinant protein fragments, and peptides may be used to functionalize nanopatterned surfaces according to considerations of size, bioactivity, and usability. As an example, cell adhesion may be promoted by the usage of fibronectin, the recombinant fragment from the seventh to the tenth type III repeat of the whole protein or the RGD peptide sequence contained within this fragment.

of these types of biomolecule offers distinct advantages and poses different limitations in the functionalization and handling of the biomaterial, as is discussed below.

An obvious choice for promoting cell attachment and growth on substrates is coating them with reconstituted ECM proteins, either as purified mixtures or as isolated components. ECM proteins tend to be large, complex biomolecules that usually contain several binding domains for different receptors and other ECM components, making it difficult to determine the individual contribution of each ligand and receptor. Moreover, protein mixtures are difficult to purify at levels sufficient to exclude contribution from contaminants. Purification is critical for the applicability in medical devices, where contaminations may lead to infection or inflammation. In order to overcome these limitations, a strategy often employed is to determine the minimal peptide sequence of the parental protein that retains activity and employ these peptides for functionalization.

Peptides can be obtained at high purities and offer ease of synthesis and handling. Their small size and chemical versatility facilitates their conjugation on a variety of surfaces, and site-specific modification for labeling or attachment to linkers is more accessible compared to proteins, allowing controlled orientation. Peptides can be mixed on a surface to produce multicomponent surfaces that can synergistically act on cells. Alternatively, multiepitope peptides can be designed where several bioactive sequences can be linked via covalent bonds. Finally, peptides can be presented at densities that exceed the physiological ones, enabling higher binding affinities. Nevertheless, the use of peptides entails several drawbacks. Frequently, they are an incomplete substitute of the biological signal and exhibit lower specificity and/or affinity for their receptor. One reason for this reduction in activity is that the interaction with the receptor may entail secondary attachment or synergy sites. For example, fibronectin possesses a synergistic site for integrin α subunits with the sequence PHSRN, aside from the RGD sequence [59]. The two peptides are spaced a few nanometers apart and cooperatively increase the binding affinity to fibronectin substantially compared to the RGD peptide alone. Another reason for reduced affinity of peptides is the loss of their natural configuration that occurs when peptides are isolated from their parental protein. An example for this is the higher binding affinity of cyclic RGD compared to its linear counterpart [60]. Accordingly, many strategies are being investigated with the aim of inducing the peptide native structure on the interface. For example, the collagen triple helix has been synthetically mimicked to increase peptide activity [61].

In many cases, the active site of the protein of interest may be unknown or interactions may not be mediated by short amino acid sequences but instead by multiple weak interactions. In such cases, the design and production of protein fragments can also be employed to replace the whole ECM protein. Protein fragments provide a shorter, more amenable substitute and usually conserve the three-dimensional structure of the ligand while presenting only the binding site of interest. When compared to corresponding short peptides, specificity and affinity are generally enhanced. For example, the recombinant fibronectin fragment $FNIII_{7-10}$ was shown to promote higher adhesion strength and enhanced cell proliferation compared to RGD and RGD-PHSRN peptides, while retaining its integrin specificity [62].

9.2.3.2 Linking Systems

Efficient surface functionalization is synonymous to controlled surface functionalization. The quantity and orientation of the biomolecule of interest on the substrate should be tightly controlled, while activity should not be hindered by the immobilization protocol in order to achieve consistent and reliable results. Considering the size of proteins and the scale at which we try to interrogate cells using nanostructured interfaces, it becomes obvious that inefficiencies in active ligand presentation can easily skew results. Even though physical adsorption may be used for larger ligands owing to their large interfacial areas, it generally suffers from lack of specificity in protein orientation and presentation. Covalent linking to the substrate offers superior stability and control of molecular positioning. Chemical immobilization faces its own challenges, especially for large, multifunctional proteins. Nonspecific modification using accessible lysines on the protein surface can interfere with activity by directly modifying the active site or by making it inaccessible. To overcome such obstacles, site-specific methods are preferred [63]. For example, linking biomolecules expressing histidine tags with nickel-nitrilotriacetic acid (Ni-NTA) resulted in specific and highly efficient functionalization of gold dot nanopatterns, with 75% of these containing a single ligand [64]. Other approaches include biotin–avidin linkages and introduction of unnatural amino acids, which can then be used for specific modification.

Peptides are generally easier to manipulate and link to the substrate via covalent bonding. In order for the peptides to be sufficiently accessible for interactions with cell-bound receptors, inert linkers are usually introduced between the bioactive sequence and the support. Linkers of varying lengths, rigidities, and chemistries have been tested to position the ligand at a distance from the substrate and avoid steric hindrances with the underlying surface. There is no ideal universal linker, and linker selection should be optimized for each case. Nevertheless, a large numbers of studies have established that the use of short ethylene oxide linkers between peptide and surface generally improves bioactivity.

In many cases, in order to increase avidity to a receptor or induce receptor clustering, multiple copies of the biomolecule may be presented at each functional site. Linkers presenting several reactive groups such as dendrimers or starlike polymers offer this possibility. Alternatively, recombinant proteins containing a protein fragment and a sequence for promoting self-assembly may be designed. Using this approach, monomers, dimers, trimers, and pentamers of $FNIII_{7-10}$ were synthesized and compared in respect to cell binding [65]. Even if the receptor interacts with one biomolecule at a time, the availability of additional ligands around the receptor–ligand bond increases the local receptor concentration and thus the possibility of immediate binding following dissociation.

9.2.3.3 Modulation of Substrate Background

An important consideration in ligand presentation with nanoscale organization is to select an appropriate background that prevents interactions with cells or interferes with ligand activity. ECM proteins, present in serum or excreted by cells, will rapidly adsorb nonspecifically to most surfaces providing attachment sites for the cells, distorting the effects of any elaborate nanostructure. Therefore, to ensure limited interactions, the substrate background is normally designed to be protein repellent. The most commonly used molecules to achieve this goal are flexible, hydrophilic oligomers or polymers, with PEG considered the golden standard [66]. Physical adsorption of amphiphilic PEG copolymers or covalent linking of end-functionalized PEG to the surface is used to achieve the desired passivation of the surface. An alternative approach is to use a very dense oligo(ethylene) glycol coating in the form of a self-assembled monolayer (SAM) [67]. SAMs are highly homogeneous and can be used to coat surfaces of different geometries, even with very high curvatures such as the surface of nanoparticles [68]. Self-assembled lipid layers also produce molecularly smooth surfaces and have been considered as the background on which ligands are presented. They can be formed as bilayers on hydrophilic substrates, as monolayers on hydrophobic ones, or even as tethered bilayers at a defined distance from the underlying substrate. In contrast to SAMs, lipids are free to laterally diffuse and present a more physiological background, especially when cell–cell interactions are studied.

9.3
Cell Responses to Nanostructured Materials

The use of surfaces with nanoscale control over the aforementioned physical and chemical properties is starting to reveal its potential for answering basic biological questions. At the same time, conclusions of such studies are attracting more and more attention by the biomedical industry for the design of new biomaterials and optimization of existing ones. In this section, we discuss a few examples where controlling ligand presentation in the nanoscale has been proved to be invaluable for elucidating cell function.

9.3.1
Cell Adhesion and Migration

The majority of cells in solid tissues need to adhere to their ECM in order to proliferate, migrate, and differentiate. attachment of cells to the ECM is mediated by transmembrane receptors, of which the most prominently studied ones are the integrins. This family of heterodimeric receptors can transmit signals bidirectionally and is key in mechanosensing. Integrin specificity arises from the large number of distinct types (24 to date) that recognize epitopes on ECM proteins. Accordingly, cells employ different integrins to attach to specific substrates and regulate their expression levels depending on their surrounding ECM [69, 70].

Integrins participate in the dynamic process of formation of contact points between cells and substrate. On 2D substrates, multiprotein complexes assemble and mature into focal adhesions (FAs), which provide the mechanical anchor to the substrate and through which cells sense extracellular forces. The assembly and interactions between FA proteins has unveiled high complexity at the molecular level. More than 100 different proteins have been identified as FA components [71], which come together in a spatially and temporally controlled manner, ultimately linking the ECM to the actin cytoskeleton [72]. While FAs can be a few microns in size, they are apparently composed of molecular aggregates (*FA particles*), which are merely 25 nm in size [73]. Considering further that a single integrin has a width of 9–12 nm [74], it becomes evident that micrometer-sized cells regulate attachment to their substrata at the nanometer scale.

Indeed, it is now well established that a minimum lateral distance between integrins and their ligands is necessary for cell attachment and spreading. Initially, however, studies reported values that spanned more than 1 order of magnitude for the critical spacing between cell-adhesion ligands [75]. An underlying reason for the absence of a consensus was the lack of tight control over interligand spacing. Nanostructured surfaces were thus employed to unambiguously provide the answer.

Using BCML, surfaces decorated with gold nanoparticles were fabricated, with interparticle distances ranging systematically from a few tens to more than a hundred nanometers and an accuracy of only a few nanometers. The particles were then functionalized with RGD peptides, while the area between these dots was passivated by covalently immobilizing PEG (Figure 9.7). The small size (8 nm) of each gold particle ensured that adhesion of only one integrin heterodimer was possible, allowing the direct investigation of the role of lateral spacing for integrin clustering.

Cell spreading and FA formation was compromised when interligand spacing exceeded 58 nm. (Figure 9.7) [76, 77]. This threshold for promoting cell adhesion

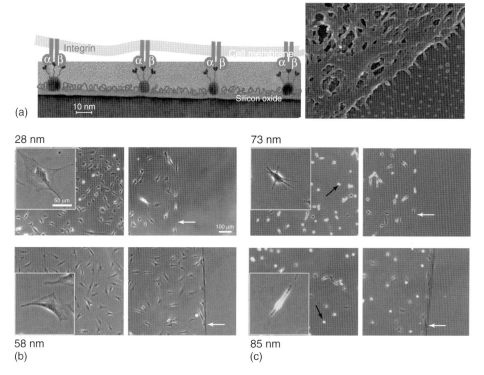

Figure 9.7 The effect of cell-adhesion ligand spacing on cell attachment and spreading. Gold nanodots fabricated on a glass substrate were decorated with integrin ligands, and the area between the dots was passivated with PEG (cartoon a). Owing to the small size of the particles, only one integrin can bind to the ligands. A scanning electron micrograph shows a rat embryonic fibroblast (REF52 cells) adhering to a gold nanopatterned surface (b). Interligand spacing affects adhesion and spreading of MC3T3 murine preosteoblastic cells (c). Cells adhere specifically to the nanopatterns and spread on surfaces when interligand distance is smaller than 73 nm. White arrows indicate the border of the nanopattern, while black arrows indicate nonadherent cells. (Source: Figure adapted from Ref. [76].)

was subsequently confirmed in several studies on integrin $\alpha_v\beta_3$ clustering in cells of mesenchymal origin. It is also critical for FA strength, with increasing ligand densities causing higher adhesion [78].

The determination of a universal spacing raised the question of the origin of the large span of values reported previously. An elegant study performed by Huang *et al.* using gold nanoparticle arrays having the same average spacing, but presenting the ligand in an ordered hexagonal packing in one case and in a disordered pattern in the other, revealed efficient cell spreading even when average interparticle distances were greater than 73 nm [79]. Therefore, simply dispersing the ligands and statistically calculating the distance can be problematic when trying to determine an exact spacing. Another reason that RGD spacing may have been overestimated is that cells can move ligands when these are presented at ends of flexible tethers or presented in a fluid bilayer, creating locally high enough concentrations for integrin clustering.

Having established the minimal distance required for integrin clustering that leads to cell spreading and FA formation, the question arose as to whether that distance merely reflects a certain overall concentration of ligands/area or if it represents an actual threshold above which the cell cannot form contacts. To answer this question, Deeg *et al.* [45] prepared surfaces having the same local density but different global distances by patterning an island covered with nanodots with a few square micrometers surface area dispersed in a nonadhesive substrate. Their results showed that, indeed, local ligand density is the dominant parameter in FA formation and strength. Schvartzman *et al.* [80] reached the same conclusion using a similar system and went one step further by suggesting that at least four integrins are necessary for the formation of the complex. The aforementioned studies provide a striking demonstration of the significance of controlling cell interfaces at the nanoscale.

The detailed experimental information gathered at such scale has also fueled interest in understanding the physical means by which FAs mature and what is actually determining the universal spacing observed. FAs transmit stresses from the ECM to the cytoskeleton, transducing force cues into biochemical signals that regulate cell functions. It is tempting to explain results on the basis of the size of some FA cross-linking proteins (Figure 9.8). Talin, with a size of ~60 nm, is a prime candidate for sensing interligand distance. When integrin spacing exceeds this size, talin would not be able to bridge these receptors, leading to unstable FAs. However, physical models purely relying on force exerted on the FA by the actin cytoskeleton [81] or on membrane energetics during deformation [82], were found to predict minimal spacing of integrin receptor for FA stabilization. The power of such models lies in that they make testable predictions. For example, [81] predict that at very high ligand density adhesion, strength should increase above physiological levels.

9.3.2
Cell–Cell Interactions

Nanopatterned surfaces can also be used to investigate cell–cell communication and responses to direct cell–cell contact. Cell–cell adhesion is mediated by transmembrane proteins, with cadherins being the most notable members. At the same time, specialized protein ligands and their receptors are utilized to transfer information between cells of the same or different types. For example, the immunological synapse between T cells and antigen-presenting complex is a multiprotein complex with defined architecture that directs T-cell activation [83]. Reconstituting cell surface ligands and ECM ligands differ in that the former are mobile on the plasma membrane. Consequently, studies of cell–cell contacts are usually performed on supported lipid bilayers, which mimic the plasma membrane and are therefore more physiologically relevant. However, lipid bilayers are fluid,

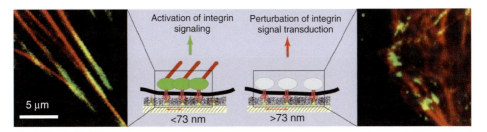

Figure 9.8 Focal adhesion assembly and integrin signaling are regulated by nanoscale spacing of integrin ligands. MC3T3 murine preosteblasts cultured on gold nanodot arrays exhibited efficient focal adhesion formation and activation of integrin signaling pathways only when RGD ligand spacing was lower than 73 nm. Fluorescence microscopy images were stained for vinculin (green) and actin (red).

and therefore, a strategy is required to confine ligands within nanoscale areas. One solution is to use immobile barriers to trap bilayers – and consequently the receptors in them – within defined areas, as small as 100 nm × 100 nm. Using this system, it was possible to answer questions regarding the importance of receptor clustering in immunological synapse signaling [84] and EphA2 receptor tyrosine kinase activity in response to its ligand ephrin-A1 [85]. These studies utilized compositionally homogeneous bilayers and concentrated on preventing lateral mobility of incorporated ligands. Combining this technology with microfluidics, it is possible to create gradients [86] or mixed compositions [87], which could tackle a wider range of questions relating to how nanoscale organization of ligands and receptors affects cell response.

9.3.3
Cell Membrane Receptor Signaling

Elucidating the dependence of cell motility, survival, and differentiation on their adhesion state to the ECM is a challenging question with implications in development and disease, as well as in design of biomimetic materials. The growth of anchorage-dependent cells is regulated by cell contacts to the ECM, which act as signaling hubs. Local accumulation of cytoplasmic signaling molecules and cytoskeletal components ensue the formation of FAs and trigger protein phosphorylation, which lead to signaling cascades to the nucleus [88]. Nanostructured surfaces offer the possibility to unravel the effect of cell attachment on signaling pathways by modulating adhesion with nanoscale precision. For example, it is likely that some signaling events simply relate to the number of bonds formed between ECM ligands and corresponding cell receptors. Hence, nanopatterns provide a tool to specify the number of active bonds and thus investigate this assumption.

Cell changes can be monitored both at the cellular level (changes in motility, morphology, proliferation, etc.) and at the molecular level (biomarker quantitation, phosphorylation events, etc.). Generally, changes in cell adhesion and morphology affect the cytoskeleton and intracellular signaling, thereby influencing cell fate [89]. For example, integrin expression levels have been correlated to different substrate nanotopographies [90]. For cells of mesenchymal origin, there seems to be a critical value of 40 nm for feature height, above which cells spread less, form less FAs, and in turn show altered shape [91]. Alterations in cell shape and mechanotransduction have been observed on nanograting platforms, because of modulation of cytoskeletal assembly and signaling [92].

To date, little is known regarding the exact role of material nanoscale features on integrin-dependent downstream signaling. Important mediators of adhesion signaling are nonreceptor tyrosine kinases, such as focal adhesion kinase (FAK). FAK regulates several cellular processes, including growth and differentiation. Surfaces promoting integrin binding have been shown to activate FAK signaling and direct osteogenic differentiation [93–95]. Differentiation into bone cell phenotype was also observed on submicron-groove arrays and linked to FA growth and FAK-mediated activation of the ERK/MAPK signaling pathway [96]. It seems that effects of nanoscale features of surfaces are directed toward cell cycle regulation, rather than cell viability [91]. For example, proliferation of several cell types has been shown on substrates with roughness below 45 nm and with large fractal dimension [97].

Uncoupling the effects of integrin–ligand interactions and the effects due to deposition of ECM proteins, which provide additional integrin binding motifs, still remains a challenge for studies on nanoscale surface modifications. Finally, an effort should be made to control adhesion and proliferation of desired cell types, while discouraging adhesion of other types, via individual surface parameters. For example, choosing different integrin binding motifs and therefore enhancing adhesion via specific integrin types, might differentially affect signaling pathways and lead to controlled differentiation.

9.3.4
Applications of Nanostructures in Stem Cell Biology

Self-renewal and differentiation of stem cell *in vivo* are governed by the unique local microenvironment in which these cells reside. This is known as the *stem cell niche*, and it encompasses the collective interactions with neighboring cells and complex mixtures of soluble and insoluble ECM signals. A key challenge in stem cell research is to develop *in vitro* culture systems that recapitulate functions of the *in vivo* microenvironment and thus provide a way to manipulate stem cell fate accordingly. On one hand, there is a need to maintain an undifferentiated, proliferating cell population, since stem cell therapies would require a large number of stem cells. On the other hand, the ability to direct differentiation down a specific cell lineage in a noninvasive manner is necessary to generate patient-specific tissues on demand.

In view of these requirements, researchers have studied the effects of nanostructured substrates on stem cell behavior. Current research shows that nanoscale features of materials indeed have the ability to elicit specific cues and promote the controlled differentiation of stem cells *in vitro*. Substrate topography influences a variety of stem cell behaviors in a way that is distinct from surface chemistry. Geometric features that impose a control over cell shape play a significant role in promoting stem cell differentiation [98]. For nanoscale cues, the regulation of adhesion and presence of intracellular stresses could result in cell structure changes and signaling cascades sufficient to trigger differentiation.

Increase in adhesion is associated with cell proliferation and preservation of cell pluripotency, as has been shown for embryonic stem cells on highly cross-linked PLL/HA nanofilms [99]. In contrast, nanopatterns that limit adhesion area have been shown to enhance differentiation via adhesion signaling regulation. For example, cells on vertically aligned TiO_2 nanotubes of 15 nm diameter exhibited higher differentiation, which was correlated with increased FAK and ERK activation [100]. Cells on such nanotube patterns show sensitivity to their diameter and not to their height [101]. Nanofibers are also able to provide extracellular scaffolding, which promotes regeneration of specific tissues [102].

A recent focus in material development has been on strategies to trigger stem cell differentiation in a highly controlled manner without the need for exogenous factors or supplements. Oh *et al.* showed that nanotubes dictated stem cell fate without aid of soluble factors [103] and also demonstrated that nanopatterned surfaces prepared by dip-pen nanolithography with various spacing could control cell growth without specific growth factors added to the media.

In principle, by carefully controlling the nanotopography and surface chemistry, it should be possible to design a material that enhances a selective population to grow and differentiate. Although many challenges lie ahead, scaffolds with defined nanoscale features could be used in the future to direct the regeneration of damaged tissues making use of stem cell migration and differentiation.

9.4
The Road Ahead

The quest to mimic natural structures down to their finest details in order to manipulate cell function has already attracted significant attention from biologists, material scientists, and engineers. Inspired by nature and using the toolbox provided by modern nanotechnology, researchers have set out to construct interfaces equipped with all the necessary information to predict and manipulate cell behavior. The initial steps toward this overarching goal constitute a basis full of promise for the future. The examples given in this chapter highlight how initial, rudimentary approaches were able to expand our comprehension of how cells regulate their behavior *in vitro*.

Most of the approaches so far have used only a few ligand types and were often limited in the number of free parameters, while leaving many others undefined. The logical evolution of these approaches is to expand the ligand types, incorporate multiple functionalities, and control more parameters at a time. The way each parameter affects cell behavior and their interplay may not always be straightforward. In order to unravel these interactions, the use and development of high-throughput technologies for screening will be necessary. At the same time, integrating all the relevant parameters into models will require the involvement of system biology tools.

Nonetheless, cells possess some intrinsic probabilistic behavior. It is interesting to explore the borders of deterministic behavior of cells using interfaces that maximize control over all relevant parameters. Ultimately, single cell studies are needed to see the origins of the remaining variability.

Perhaps a more important issue for practical applications is to validate the findings of *in vitro* studies to the dynamic biological environment *in vivo*. In vitro experimentation by its very nature reduces the complexity and interactions present *in vivo*, and this should be kept in mind. Nevertheless, the conclusions of *in vitro* studies using interfaces are calling for their implementation in the design of superior medical devices. Medical implants or grafts have been traditionally produced mainly considering macroscopic properties (e.g., mechanical properties, handling ease) of the biomaterial, neglecting its structure at the nanoscale. The growing evidence that it is exactly at this scale that cells probe their substrate and determine their functions is slowly shifting the balance.

Efforts to incorporate surface modifications on clinical biomaterials have initiated and are slowly finding their way to the production lines. In this process of translation to the clinics, it is critical to consider the actual application in question and the handling of the biomaterial. For example, if a biomaterial is to be administered at a site where bleeding occurs, it is possible that all nanostructures will be quickly masked by clotting. Mechanical shear and tissue movement might also impose additional considerations. Finally, it is important to keep in mind that the underlying principles emerging from the *in vitro* and *in vivo* studies are the ones relevant for the design of the substrates, rather than the actual methodologies used. For example, creating gold nanodot arrays with highly defined interligand spacings might be invaluable to unveil the mechanisms of cell adhesion and propagation, but the patterns can easily be substituted by some other system that enforces ligand spacing to promote or impair adhesions. The methodologies used must be robust and easily integrated to existing manufacturing process. Establishment of new manufacturing processes will ensue provided that nanostructured biomaterials fulfill their promise as critical factors of biomaterial performance.

The road ahead in biomaterial research appears to encompass a higher degree of details at nanodimensions. An interdisciplinary commitment from different research fields, such as biology, engineering, and chemistry, is certainly an extremely valuable basis for advancing the design of cell interfaces for basic research and translating the results to applications in regenerative medicine and therapeutic biomaterials.

References

1. Kim, D.H. *et al.* (2010) *Adv. Mater.*, **22**, 4551–4566.
2. Taton T.A. (2001) *Nature*, **412**(6846), 491–492.

3. Tzaphlidou, M. (2001) *Micron*, **32**(3), 337–339.
4. Eyden, B. and Tzaphlidou, M. (2001) *Micron*, **32**(3), 287–300.
5. Minary-Jolandan, M. and Yu, M.F. (2009) *Biomacromolecules*, **10**(9), 2565–2570.
6. Fratzl, P. et al. (2009) *Nanomedicine, Nanotechnology*, Vol. 5, Chapter 12, Wiley-VCH Verlag GmbH, Weinheim, pp. 345–360.
7. Fantner, G.E. et al. (2007) *Nano Lett.*, **7**(8), 2491–2498.
8. Reilly, G.C. and Engler, A.J. (2010) *J. Biomech.*, **43**(1), 55–62.
9. Dean, R.G. et al. (2005) *J. Histochem. Cytochem.*, **53**(10), 1245–1256.
10. Huang, S. and Ingber, D.E. (2005) *Cancer Cell*, **8**(3), 175–176.
11. Paszek, M.J. et al. (2005) *Cancer Cell*, **8**(3), 241–254.
12. Tilghman, R.W. et al. (2010) *PLoS ONE*, **5**(9), e12905.
13. Levental, K.R. et al. (2009) *Cell*, **139**(5), 891–906.
14. Von der Mark, K. et al. (2010) *Cell Tissue Res.*, **339**, 131–153.
15. Cavalcanti-Adam, E.A. et al. (2008) *HFSP J.*, **2**(5), 276–285.
16. Girard, P.P. et al. (2007) *Soft Matter*, **3**, 307–326.
17. Lim, J.H. et al. (2003) *Angew. Chem. Int. Ed.*, **42**, 2309–2312.
18. Hoff, J.D. et al. (2004) *Nano Lett.*, **4**(5), 853–857.
19. Jaschke, M. et al. (1995) *Langmuir*, **11**, 1061–1064.
20. Mirkin, C.A. et al. (2001) *ChemPhysChem*, **2**, 37–39.
21. Kramer, M.A. et al. (2010) *J. Am. Chem. Soc.*, **132**(13), 4532–4533.
22. Chou, S.Y. et al. (1995) *Appl. Phys. Lett.*, **67**(21), 3114–3116.
23. Chou, S.Y. et al. (1996) *Science*, **272**(5258), 85–87.
24. Schvartzman, M. and Wind, S.J. (2009) *Nano Lett.*, **9**(10), 3629–3634.
25. Spatz, J.P. et al. (1999) *Adv. Mater.*, **11**(2), 149–153.
26. Spatz, J.P. et al. (2000) *Langmuir*, **16**(2), 407–415.
27. Spatz, J.P. (2002) *Angew. Chem. Int. Ed.*, **41**(18), 3359–3362.
28. Kruss, S. et al. (2010) *Adv. Mater.*, **22**(48), 5499–5506.
29. Graeter, S.V. et al. (2007) *Nano Lett.*, **7**(5), 1413–1418.
30. Chai, J. et al. (2010) *Proc. Natl. Acad. Sci. U.S.A.*, **107**(47), 20202–20206.
31. Yim, E.K. and Leong, K.W. (2005) *Nanomedicine*, **1**(1), 10–21.
32. Kim, D.H. et al. (2010) *Proc. Natl. Acad. Sci. U.S.A.*, **107**(2), 565–570.
33. Curtis, A. and Wilkinson, C. (2001) *Trends Biotechnol.*, **19**(3), 97–101.
34. Oh, S. et al. (2009) *Proc. Natl. Acad. Sci. U.S.A.*, **106**(7), 2130–2135.
35. Dalby, M.J. et al. (2007) *Nat. Mater.*, **6**(12), 997–1003.
36. Dvir, T. et al. (2011) *Nat. Nanotechnol.*, **6**, 13–22.
37. Bettinger, C.J. et al. (2009) *Angew. Chem. Int. Ed.*, **48**(30), 5406–5415.
38. Ashkenazi, A. and Dixit, V.M. (1998) *Science*, **281**, 1305–1308.
39. Hersel, U. et al. (2003) *Biomaterials*, **24**, 4385–4415.
40. Polleux, J. et al. (2011) *ACS Nano*, **5**(8), 6355–6364.
41. Aldaye, F.A. et al. (2008) *Science*, **321**, 1795–1799.
42. Bui, H. et al. (2010) *Nano Lett.*, **10**, 3367–3372.
43. Ding, B. et al. (2010) *J. Am. Chem. Soc.*, **132**, 3248–3249.
44. Kuzuya, A. et al. (2010) *Small*, **6**(23), 2664–2667.
45. Deeg, J. et al. (2011) *Nano Lett.*, **11**(4), 1469–1476.
46. Carter, S.B. (1965) *Nature*, **208**, 1183–1187.
47. Arnold, M. et al. (2008) *Nano Lett.*, **8**(7), 2063–2069.
48. Yoshikawa, H.Y. et al. (2010) *J. Am. Chem. Soc.*, **133**, 1367–1374.
49. Engler, A.J. et al. (2006) *Cell*, **126**, 677–689.
50. Aydin, D. et al. (2010) *Langmuir*, **26**(19), 15472–15480.
51. Cukierman, E. et al. (2001) *Science*, **294**, 1708–1712.
52. Ochsner, M. et al. (2007) *Lab Chip*, **7**, 1074–1077.
53. Kloxin, A.M. et al. (2009) *Science*, **324**, 59–63.
54. Huebsch, N. et al. (2010) *Nat. Mater.*, **9**, 518–526.

55. Doyle, A.D. et al. (2009) *J. Cell Biol.*, **184**(4), 481–490.
56. Cui, H. et al. (2010) *Biopolymers*, **94**(1), 1–18.
57. Gazit, E. (2007) *Chem. Soc. Rev.*, **36**(8), 1263–1269.
58. Cui, H. et al. (2007) *Science*, **317**, 647–650.
59. Redick, S.D. et al. (2000) *J. Cell Biol.*, **149**(2), 521–527.
60. Koivunen, E. et al. (1993) *J. Biol. Chem.*, **268**(27), 20205–20210.
61. Reyes, C.D. and Garcia, A.J. (2003) *J. Biomed. Mater. Res. A*, **65A**(4), 511–523.
62. Petrie, T.A. et al. (2006) *Biomaterials*, **27**(31), 5459–5470.
63. Rusmini, F. et al. (2007) *Biomacromolecules*, **8**, 1775–1789.
64. Wolfram, T. et al. (2007) *Biointerphases*, **2**(1), 44–48.
65. Coussen, F. et al. (2002) *J. Cell Sci.*, **115**(12), 2581–2590.
66. Vermette, P. and Meagher, L. (2003) *Colloids Surf., B*, **28**, 153–198.
67. Love, J.C. et al. (2005) *Chem. Rev.*, **105**, 1103–1169.
68. Jackson, A.M. et al. (2004) *Nat. Mater.*, **3**, 330–336.
69. Barczyk, M. et al. (2010) *Cell Tissue Res.*, **339**, 269–280.
70. Luo, B.H. et al. (2007) *Annu. Rev. Immunol.*, **25**, 619–647.
71. Zaidel-Bar, R. and Geiger, B. (2010) *J. Cell Sci.*, **123**, 1385–1388.
72. Kanchanawong, P. et al. (2010) *Nature*, **468**(7323), 580–584.
73. Patla, I. et al. (2010) *Nat. Cell Biol.*, **12**(9), 909–915.
74. Xiong, J.P. et al. (2001) *Science*, **294**, 339–345.
75. Massia, S.P. and Hubbell, J.A. (1991) *J. Cell Biol.*, **114**(5), 1089–1100.
76. Arnold, M. et al. (2004) *ChemPhysChem*, **5**(3), 383–388.
77. Cavalcanti-Adam, E.A. et al. (2007) *Biophys. J.*, **92**(8), 2964–2974.
78. Selhuber-Unkel, C. et al. (2010) *Biophys. J.*, **98**, 543–551.
79. Huang, J. et al. (2009) *Nano Lett.*, **9**(3), 1111–1116.
80. Schvartzman, M. et al. (2011) *Nano Lett.*, **11**(3), 1306–1312.
81. de Beer, A.G. et al. (2010) *Phys. Rev. E Stat. Nonlin. Soft Matter Phys.*, **81**(5, Pt 1), 051914.
82. Wei, Y. (2008) *Langmuir*, **24**, 5644–5646.
83. Manz, B.N. and Groves, J.T. (2010) *Nat. Rev. Mol. Cell Biol.*, **11**, 342–352.
84. Mossman, K.D. et al. (2005) *Science*, **310**, 1191–1193.
85. Salaita, K. et al. (2010) *Science*, **327**, 1380–1385.
86. Stroumpoulis, D. et al. (2007) *Langmuir*, **23**, 3849–3856.
87. Shen, K. et al. (2009) *J. Am. Chem. Soc.*, **131**, 13204–13205.
88. Zhao, J.H. et al. (1998) *J. Cell Biol.*, **143**(7), 1997–2008.
89. Itano, N. et al. (2003) *Proc. Natl. Acad. Sci. U.S.A.*, **100**(9), 5181–5186.
90. Lim, J. Y. et al. (2007) *Biomaterials*, **28**(10), 1787–1797.
91. Pennisi, C.P. et al. (2011) *Colloids Surf., B*, **85**(2), 189–197.
92. Yim, E.K. et al. (2005) *Biomaterials*, **26**(26), 5405–5413.
93. Keselowsky, B.G. et al. (2007) *J. Biomed. Mater. Res. A*, **80**(3), 700–710.
94. Cavalcanti-Adam, E.A. et al. (2002) *J. Bone Miner. Res.*, **17**(12), 2130–2140.
95. Grigoriou, V. et al. (2005) *J. Biol. Chem.*, **280**(3), 1733.
96. Biggs, M.J. et al. (2009) *Biomaterials*, **30**(28), 5094–5103.
97. Gentile, F. et al. (2010) *Biomaterials*, **31**(28), 7205–7212.
98. Kilian, K.A. et al. (2010) *Proc. Natl. Acad. Sci.*, **107**(11), 4872–4877.
99. Blin, G. et al. (2010) *Biomaterials*, **31**(7), 1742–1750.
100. Park, J. et al. (2007) *Nano Lett.*, **7**(6), 1686–1691.
101. Bauer, S. et al. (2009) *Integr. Biol.*, **1**(8–9), 525–532.
102. Smith, L.A. et al. (2010) *Biomaterials*, **31**(21), 5526–5535.
103. Oh, S et al. (2009) *Proc Natl Acad Sci USA*, **106**(7), 2130–2135.

10
Surfaces with Extreme Wettability Ranges for Biomedical Applications
Wenlong Song, Natália M. Alves, and João F. Mano

10.1
Superhydrophobic Surfaces in Nature

Learning from nature has become a shortcut to synthesize new materials or to improve their properties for different applications [1–6]. In particular, surfaces with extreme wettability found in nature have attracted great interest because they can find applications in industry, agriculture, and daily life [7, 8]. Among them, superhydrophobic surfaces, that is, those that show a water contact angle (WCA) >150° and a sliding angle <5° have been especially investigated in recent years [9] because of the foreseeable applications in biomedicine and tissue engineering. One can find this unique property in many plants or insects, as shown in Figure 10.1, such as in lotus leaves, legs of water strider, butterfly wings, and moth or mosquito eyes [10–13]. The lotus leaf, the most well-known example of superhydrophobicity in nature, also presents the so-called self-cleaning effect, as reported by Barthlott and Ehler in 1977 for the first time [14]. Furthermore, surface analysis indicates that the lotus leaf surface is full of micropapillae and that nanoscale structures can be observed on each papilla. These hierarchical micro- and nanostructures and the covered thin layer of biowax are responsible for the superhydrophobicity of the lotus leaves. This kind of surface is also called isotropic superhydrophobic surface because of the isotropy of the surface roughness geometry. Anisotropic superhydrophobic surfaces can also be found in nature, for example, a wheat leaf or a butterfly wing, in which wettability changes with the direction [15, 16]. These surfaces could be widely used for designing new microfluidic [17] and bioanalyzing devices that require directional (2D) and spatial (3D) variations of controlled wettability, adhesive force, and friction force. Regarding the superhydrophobic cases of insects in nature, and taking the water strider as an example, the ability to "walk" on water surface has been attributed to the combined action of the covered biowax and the nano/microscale hierarchical structures of the small hairs (setae) on the legs [18, 19].

From the previous examples taken from nature, it is clear that the special hierarchical micro- and nanostructures are the determinants of the superhydrophobicity of a surface. Because the adhesion of cells or proteins on a surface is also affected by the surface wettability, these special hierarchical structures can be foreseen to

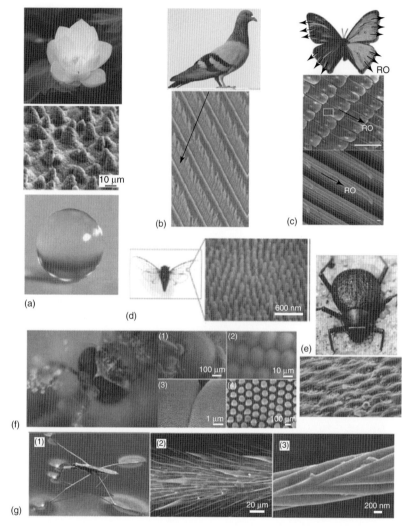

Figure 10.1 Some examples from nature. (a) Lotus effect. (b) Waterproof pigeon feathers. (c) Butterfly wings. (d) Surface nanostructure on the wing of *Cicada orni*. (e) Water-harvesting wing surface of the Namib desert beetle. (f) Antireflective moth's eye. (g) Water strider walking on water. (Source: (a) Reproduced with permission from Ref. [6] Copyright 2009, the Royal Society. (b) Reproduced with permission from Bormashenko *et al. J. Coll. Interf. Sci.* **311** (2007) 212. (c) Reproduced with permission from Ref. [15]. (d) Reproduced with permission from Lee *et al. Langmuir* **20** (2004) 7665. (e) Reproduced with permission from Scientific American, July 21 (2008). (f) Reproduced with permission from Ref. [13] Copyright 2007, Wiley-VCH Verlag GmbH & Co. KGaA. (g) Reproduced with permission from Ref. [18].)

be strongly related to the biological adhesion. In this chapter, a brief description of the theory of surface wettability is presented and, thereafter, recent developments in superhydrophobic surfaces inspired by nature and the related applications in biomedicine and tissue engineering are specially addressed.

10.2 Theory of Surface Wettability

10.2.1 Young's Model

Wettability is a fundamental property of a solid surface governed by the surface chemical composition and structure. A basic theoretical model can be demonstrated by Young's equation [20]

$$\cos \theta_{\text{Young}} = \frac{(\gamma_{\text{SV}} - \gamma_{\text{SL}})}{\gamma_{\text{LV}}} \qquad (10.1)$$

where γ_{SV}, γ_{SL}, and γ_{LV} are the surface energies of solid–vapor, solid–liquid, and liquid–vapor, respectively. In this model (Figure 10.2a), the surface is assumed to be absolutely smooth; the liquid drop contacts with the surface completely.

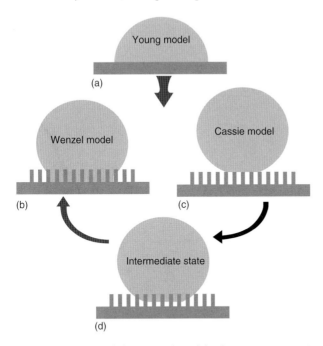

Figure 10.2 Wetting behavior on the solid substrates. (a) Young's mode, (b) Wenzel's mode, (c) Cassie's mode, and (d) intermediate state between the Wenzel and Cassie modes.

However, real surfaces will never be completely flat, that is, roughness always exists. Therefore, two other theoretical models, those by Wenzel and Cassie–Baxter, were proposed to explain the relationship between the contact angle and surface roughness.

10.2.2
Wenzel's Model

In Wenzel's model [21] (Figure 10.2b), the liquid is assumed to in complete contact with the rough surface, therefore the contact area between the liquid and the surface is larger than the situation in Young's model. A corrected roughness factor R_f is introduced in Wenzel's equation:

$$\cos\theta_{Wenzel} = R_f \cos\theta_{Young} \tag{10.2}$$

R_f is defined as the ratio between the real surface area and the apparent one and is always higher than 1. On the basis of Wenzel's equation, if the ideal surface shows hydrophobicity ($\theta_{Young} \geq 90°$), the roughness will increase the surface hydrophobic character; on the contrary, if the ideal surface shows hydrophilicity ($\theta_{Young} < 90°$), the roughness will increase the surface hydrophilic character. Herein, the roughness induces a wettable amplification of the ideal flat surface.

10.2.3
The Cassie–Baxter Model

The Cassie–Baxter model [22] (Figure 10.2c) assumes that liquid can only have contact with the top surface of solid rough structures and that air is trapped between the liquid and the solid structures. The Cassie–Baxter equation can be described by

$$\cos\theta_{CB} = f_V \cos\theta_V + f_S \cos\theta_S \tag{10.3}$$

Here, f_V and f_S correspond to the fractional areas of vapor and solid on the surface, respectively; θ_V is the contact angle of vapor (usually air) with a value of 180°; and θ_S is the contact angle of an ideal solid surface (i.e., where only the surface tensions are involved), namely, θ_{Young}. Therefore, Eq. (10.3) can be written as [23]

$$\cos\theta_{CB} = f_S \cos\theta_S + f_S - 1$$
$$\text{or } \cos\theta_{CB} = f_S R_f \cos\theta_{Young} + f_S - 1 \tag{10.4}$$

From Eq. (10.4), one can also find that even for some hydrophilic surfaces, the contact angle still increases with the roughness.

10.2.4
Transition between the Cassie–Baxter and Wenzel Models

It has been reported that the contact model of a liquid and a solid can be changed from the Cassie–Baxter model to the Wenzel model in a periodic nanopillared hydrophobic surface by applying, for example, a physical pressure to the droplet, or by vibration of the droplet [9, 24–29] (Figure 10.2d). The surface geometrical conditions such as the height of nanopillars, space between pillars, intrinsic contact angle, and impinging velocity of water nanodroplets can strongly affect this transition.

10.3
Fabrication of Extreme Water-Repellent Surfaces Inspired by Nature

Superhydrophobic surfaces are commonly produced by adopting two distinct strategies: creating micro/nanostructures on hydrophobic surfaces or chemically modifying a micro/nanostructured surface with a material of low surface energy. Various methods have been used such as crystallization control, phase separation, electrochemical deposition, chemical vapor deposition (CVD), or even lithographic imprinting using a polydimethylsiloxane (PDMS) stamp that is a negative of lotus leaf. Coating rough substrates with a fluoroalkylsilane coating or roughening of polytetrafluoroethylene (PTFE) substrates by plasma treatment are also common methods. In the case of metals, as they usually do not display hydrophobic characteristics, they are typically roughened and then coated with a hydrophobic material such as PDMS or are oxidized and further modified by CVD in order to obtain a superhydrophobic surface. A detailed description of the different preparation methods can be found elsewhere [30, 31]. Some representative examples of superhydrophobic surfaces fabricated using nature as an inspiration are given in this section.

10.3.1
Superhydrophobic Surfaces Inspired by the Lotus Leaf

Distinct classes of materials have been employed to produce superhydrophobic surfaces that mimic the structure of lotus leaves, such as natural polymers [32, 33]; synthetic polymers [34–36]; synthetic organic, inorganic, or hybridized organic–inorganic materials [37–41]; and metals [42–44].

Regarding polymeric surfaces, superhydrophobic poly(L-lactic acid) (PLLA) with WCA >150° was prepared by a simple and low-cost phase inversion method [32, 45]. It is clear that in these PLLA surfaces, besides the roughness at the micrometer level, the individual papillaelike structures exhibit roughness at the nanolevel (Figure 10.3). Such morphology is remarkably similar to the papillae nanostructure of the lotus leaf. Moreover, different durations of Ar plasma treatment were used to increase the hydrophilicity on these PLLA rough samples and to achieve controlled WCAs down to the superhydrophilic range. Such a technique permits the construction of surfaces with gradient wettability and may be used to produce patterned surfaces with spatial control of wettability. Chitosan has been proposed for a series of biomedical applications because of its biocompatibility, processability, and possibility of chemical modification [46, 47]. Chitosan films

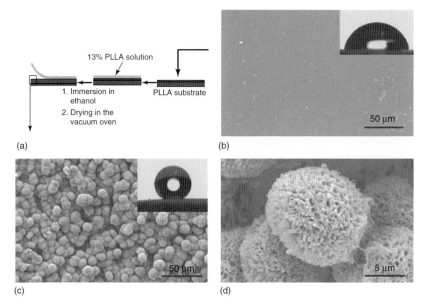

Figure 10.3 (a) Schematic representation of the experimental process to produce superhydrophobic surfaces, (b) scanning electron microphotograph of the smooth polylactic acid surfaces, (c) scanning electron microphotograph of the rough surface, and (d) magnified scanning electron microphotograph of a protrusion on the superhydrophobic surface. Insets show water drops on the corresponding surfaces. (Source: Reproduced with permission [34] Copyright 2009, Wiley-VCH Verlag GmbH & Co. KGaA.)

exhibiting superhydrophobicity over the whole pH range were also produced by a phase separation method [32]. The roughness of such films exhibited three levels of hierarchical organization, as shown in Figure 10.4.

Jiang et al. [48] prepared a superhydrophobic polystyrene (PS) film with a novel composite structure consisting of porous microspheres and nanofibers by electrospinning. The morphology of these materials could be controlled by adjusting the concentration of the starting solution. The porous microspheres contribute to the superhydrophobicity by increasing surface roughness; meanwhile, nanofibers interweave to form a three-dimensional network that reinforces the composite film.

Li et al. [33] produced an extremely superhydrophobic cellulose surface based on the reaction of hydroxyl groups and liquid trichloromethylsilane by CVD. The surface presented both nano- and microscale roughness. This approach is versatile and can be conducted on a variety of organic and inorganic substrates with hydroxyl-group-functionalized surfaces.

Regarding metallic surfaces, Hwang et al. [49] fabricated uniform superhydrophobic films with micro- and nanohierarchical structures similar to those of the lotus leaf by blasting an aluminum foil with sodium bicarbonate and anodizing the dimpled foil. This fabrication procedure is simple and cost effective compared to normal electrochemically deposited methods. Furthermore, the fabricated

10.3 Fabrication of Extreme Water-Repellent Surfaces Inspired by Nature

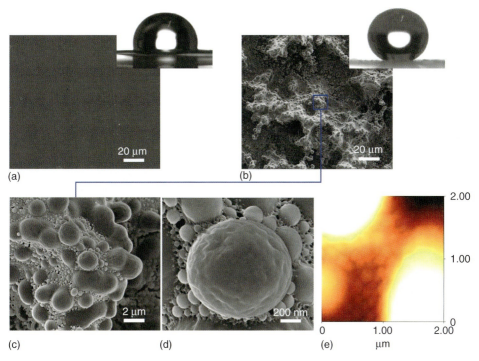

Figure 10.4 (a,b) Representative SEM images of smooth and rough chitosan film; on the top right corner of these micrographs, the profiles of the water drops deposited onto the films are shown, with contact angles of $103.6 \pm 2.7°$ and $151.5 \pm 1.9°$, respectively. (c,d) Magnified images of the superhydrophobic surface, which displayed abundant of micro- and nanoparticles. (e) AFM image on the superhydrophobic surface. (Source: Reproduced with permission [32] Copyright 2010, Elsevier.)

hierarchical structures have high contact angles, and the relatively low contact angle hysteresis decreases as nanofibers become longer.

Nakanishi *et al.* [50] described a superhydrophobic fullerene-C_{60} derivative film prepared by hierarchical self-organization. All derivatives self-assembled into well-defined three-dimensional superstructures and microparticles with nanoflaked or plate-assembled outer surfaces depending on the experimental parameters, such as molecular structure, solvent, and temperature.

10.3.2
Superhydrophobic Surfaces Inspired by the Legs of the Water Strider

Researchers have recently tried to fabricate surfaces inspired by the leg structure of the water striders, which allowed them to "walk" on water surface, using different methods such as solution immersion, electrochemical technique, or self-assembling technique. For example, Yao *et al.* [51] have prepared novel water-repellent array surfaces decorated with ribbed and conical nanoneedles consisting of copper hydroxide, which is similar to the unique structure of the setae of strider legs. They also demonstrated that stressing and relaxing drops between such surfaces would result in a fully reversible change of the solid surface by the liquid, which is distinct from other superhydrophobic surfaces. They concluded that the lateral nanogrooves could not only strongly support the enclosed liquid–air interface when a force was applied on a drop but also offered stable contact lines for the easy depinning of the deformed interface when the force was released from the drop.

Wu et al. [52] developed a simple way to fabricate a stable superhydrophobic surface by combining the electrochemical technique and surface reaction. Superhydrophobic copper wires were prepared with surface structures similar to the legs of water striders. Their method could be easily extended to any other materials surfaces, such as metal, silicon or glass wafer, and PTFE, for preparing stable superhydrophobic coatings.

Jiang et al. [53] prepared superhydrophobic surfaces composed of crystalline micro- and nanowires of an organic semiconductor. Their results showed that if a model strider leg was coated with these wires, the maximum supporting force of the "leg" would increase at least 2.4 times, which ensured that the model water striders could really "walk" on the water surface.

10.3.3
Superhydrophobic Surfaces Inspired by the Anisotropic Superhydrophobic Surfaces in Nature

The examples of anisotropic wettability in nature, such as butterfly wings and wheat leaves, have recently attracted great interest because of the advantage of controlling liquid flow into a desired direction, which has great applied potential in microfluidic devices. Many efforts have been carried out to achieve anisotropic wettabilities by chemical or physical methods, such as surface patterning or surface roughing.

Chung et al. [54] fabricated anisotropic microwrinkled surfaces by mechanical compression of ultraviolet-ozone (UVO)-treated PDMS. The water droplet on the prepared surfaces with geometrically anisotropic patterns showed a special wetting character when compared with droplets on isotropically patterned surfaces. As the height of the grooves increased, an increase in conjoint barriers was observed and attributed to the amplification of the contact angle measured perpendicular to the direction of the wrinkles. They also concluded that the change of contact angles on a rough surface was strongly affected by the nature of the three-phase contact line structure, rather than by simply increasing surface roughness. Xia and Brueck [55] reported a strongly anisotropic wetting behavior on one-dimensional nanopatterned surfaces, developed by interferometric lithography, on which hydrophobic and hydrophilic areas were regularly expressed. Wu et al. [56] used an electrospinning technique combined with specially designed nanofiber collectors to prepare surfaces containing polymer nanofibers with different patterns that could mimic lotus leaves, bamboo leaves, goose feathers, and water strider's legs. All the wetting properties, such as strong water repellence and anisotropic wetting in different directions, can be observed on the prepared nanofiber patterned surfaces. Using this novel fabrication method [56], one can prepare large-area patterned nanofibers with functional wetting surfaces without being confined to particular substrate geometries or introducing complex fabricated equipments.

Gao et al. [57] fabricated smart anisotropic surfaces by directly replicating biological surface structures including lotus and rice leaves using regular replica molding and temperature-induced phase separation micromolding. The PDMS-negative replicas of biological surface structures are durable molds for multiple replication. The positive replicas of biological surface structures are obtained by the second step of replication using phase separation micromolding of poly(N-isopropylacrylamide) (PNIPAAm) aqueous solution, which can be simply carried out by adjusting the temperature. The replicated artificial lotus leaf and rice leaf obtained using PNIPAAm showed typical micro- and nanostructures and also thermally responsive wettability. It was anticipated that this temperature-induced phase separation micromolding approach can be used to fabricate more smart biostructure-based surfaces.

10.3.4
Other Superhydrophobic Surfaces

In this section, the production of superhydrophobic surfaces inspired by rose petals and spider silk is briefly introduced.

Feng et al. [58] fabricated superhydrophobic surfaces by a two-step replication using rose petal as the mold. The unique property of this "sticked" superhydrophobic surface can be attributed to the hierarchical micro- and nanostructures of the rose petal, on which water droplet can keep spherical conformation and will not roll off even when the surface is turned upside down. This work proposes a simple methodology in the field of microfabrication.

The spider silk net is capable of efficiently collecting water from air. In order to mimic this unique property, Jiang et al. [59, 60] fabricated a series of artificial spider silks with spindle knots using different polymers including poly(vinyl acetate), poly(methyl methacrylate), PS, and poly(vinylidene fluoride). Their work showed that water drops could be driven in the desired direction ("toward" or "away from" the knot) by optimizing factors such as curvature, chemistry, and roughness gradients on the fibber surfaces.

10.4
Applications of Surfaces with Extreme Wettability Ranges in the Biomedical Field

Cell response to a surface is dictated by the interactions of the proteins in the culture medium and the surface. In fact, when a biomaterial has the first contact with the culture medium, interstitial fluids, or blood, the first event is the adsorption of protein onto the material surface. Therefore, it is very important to understand how proteins adsorb onto a surface and which mechanisms control cells adhesion and proliferation on the biomaterial surface. It is well known that cell adhesion and protein adsorption onto a substrate are highly affected by distinct surface properties such as topography, roughness, electrical charge, biochemical cues, stiffness, functional groups, and wettability [61–70]. Regarding the last mentioned property, some studies have mainly focused on surfaces ranging from hydrophilic to hydrophobic, as flat surfaces have been typically used. The effect of wettability on protein adsorption has been controversial in the literature. Some authors have reported increased protein adsorption onto hydrophilic substrates [71], whereas the majority has found that proteins tend to absorb more extensively onto hydrophobic surfaces [72]. Regarding the effect of wettability on cell response, it can be said that it is quite dependent both on the cell type and other surface properties.

There are both fundamental and practical interests in extending such studies toward the superhydrophilic and superhydrophobic limits. For example, new insights may be obtained on the influence of such extreme environments on the physiological response of cells, including their contractile characteristics and signaling activity, which may influence adhesion, morphology/anisotropy, migration, proliferation, and differentiation. Recent studies on the interactions between cells or proteins and substrates with extreme wettability ranges are presented in this section. The interactions between blood platelets and superhydrophobic substrates, with obvious relevance to the biomedical field, are analyzed. The interest of using superhydrophobic substrates in *ex vivo* biomedical applications is addressed. For examples, the possibility of using superhydrophobic surfaces for developing high-throughput chips is discussed, as well as the use of superhydrophobic substrates to produce particles for cell or drug encapsulation.

10.4.1
Cell Interactions with Surfaces with Extreme Wettability Ranges

The interaction of the superhydrophobic PLLA substrates mentioned in Section 10.3.1 with mammalian cells was investigated [34]. The influence of the exposition time to Ar plasma on cell adhesion for the developed surfaces was analyzed. Figure 10.5 shows fluorescence microscopy and scanning electron microscopy images with cells adsorbed onto the surfaces (Figure 10.5a–d), and the corresponding quantitative analysis was carried out using a DNA assay (Figure 10.5). On the smooth surfaces, no significant difference in cell attachment was observed for different times of treatment. In the absence of plasma treatment, few cells adhered onto the superhydrophobic surfaces (Figure 10.5b), compared with the number of cells adhered onto the smooth ones (Figure 10.5a). The water repellency of superhydrophobic surfaces may prevent the medium and cells from coming in contact with the entire surface, as predicted by the Cassie–Baxter model. Therefore, cells mainly adhered to some points of the asperities at the surface. The cells adopted a round shape (Figure 10.5b, inset) because of such unfavorable anchorage situation. For the rough surfaces, the Ar plasma treatment enhanced cell attachment and cell number reached the highest value after 50 s of treatment.

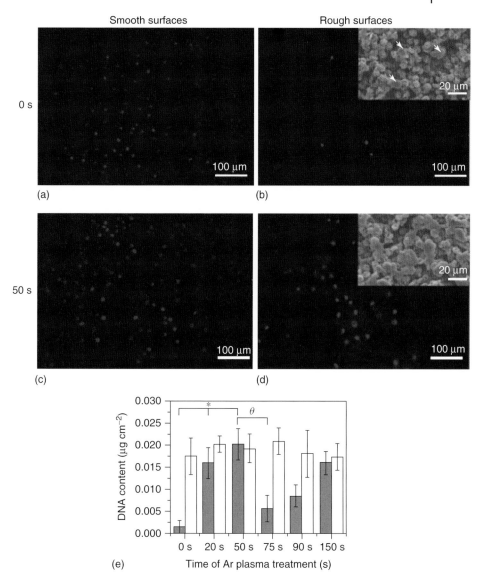

Figure 10.5 Fluorescent microscopy images of smooth and rough surfaces treated with argon plasma or untreated (blue DAPI staining for the nuclei of cells). (a) Smooth polylactic acid, untreated. (b) Rough polylactic acid, untreated. The arrows indicate the few cells that adhered to this surface. (c) Smooth surface treated with argon plasma for 50 s. (d) Rough surface treated with argon plasma for 50 s. Insets: SEM images of cells attached to the rough surfaces. Inset scale bar: 20 mm. (e) DNA content in the untreated and treated samples after 24 h of culture. White and gray bars correspond to the smooth and rough surfaces, respectively. Bars represent main statistical differences ($n \geq 6$; *: $p < 0.05$ in comparison with 0 s plasma treatment of the rough surfaces; θ: $p < 0.05$ in comparison with 50 s of plasma treatment of the rough surfaces). (Source: Reproduced with permission [34] Copyright 2009, Wiley-VCH Verlag GmbH & Co. KGaA.)

These results suggested that for the surfaces treated under this condition, the combination of roughness, surface chemistry, and wettability presented the best environment for the cells. In this case, the cells exhibited a more flattened and extended morphology (Figure 10.5d, inset).

PS is a widely used material in biology and, hence, is an obvious choice to study cell interactions with surfaces with extreme wettability ranges. Mundo et al. [73] analyzed the response of SaOs-2 cells to superhydrophobic PS surfaces that presented nanoscale dots produced by a tailored plasma-etching process. Their results indicated that SaOs-2 cells respond to surfaces with different nanoscale roughness and show a certain inhibition in cell adhesion when the nanoscale roughness is particularly small. The *in vitro* performance of different cell lines was analyzed on superhydrophobic PS surfaces produced by phase separation [74]. Compared to standard tissue culture PS, it was found that ATDC5 and SaOs-2 cell lines were not able to proliferate on such surfaces and the cell morphology was affected. Ballester-Beltrán et al. [75] also reported that mouse osteoblastic cells (MC3T3-E1) adhered much less and proliferated slower onto a superhydrophobic PS substrate when compared with the standard PS.

The attachment, morphology, and proliferation of SaOs-2 were also studied on rough and smooth PS surfaces, with wettability controlled by UVO irradiation, ranging from superhydrophobic to superhydrophilic [74]. After 4 h in culture, the attachment of SaOs-2 was higher on the surfaces treated for 18 min, namely, on rough superhydrophilic and highly hydrophilic smooth PS surfaces. It was also found that, for these PS surfaces, the proliferation after six days in culture was always higher in surfaces with WCAs ranging from 13° to 30°, irrespective of being rough or smooth. Superhydrophilic regions were also patterned on the superhydrophobic PS surfaces using hollowed masks to control the location of the UVO irradiation [74]. It was found that the cells can be preferably confined on such superhydrophilic spots, avoiding the attachment onto the superhydrophobic regions (Figure 10.6).

Piret et al. [76] analyzed the behavior of mammalian cells (Chinese Hamster Ovary K1) on patterned superhydrophilic/superhydrophobic silicon nanowire arrays and observed that the cells adhered preferentially to the superhydrophilic regions. Fibroblasts cultured on micropatterned superhydrophobic/superhydrophilic surfaces fabricated by CVD and vacuum ultraviolet (VUV) irradiation [77] also attached to the superhydrophilic regions in a selective manner. These studies evidenced that bio-inspired platforms where cell attachment/proliferation could be controlled by patterning wettable spots on superhydrophobic substrates can be produced. Applications of such platforms are discussed in more detail in Section 10.4.4.

Besides the works on cell lines or primary cells, there are also a few works that analyzed the behavior of stem cells on surfaces with extreme wettability ranges. Bauer et al. [78] reported the response of mesenchymal stem cells to superhydrophobic substrates prepared by organic modification of TiO_2 nanotubes with self-assembled monolayers of octadecylphosphonic acid. Cell attachment was considerably enhanced after 24 h on the superhydrophobic surfaces. However,

10.4 Applications of Surfaces with Extreme Wettability Ranges in the Biomedical Field | 249

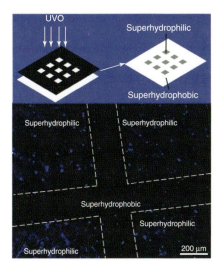

Figure 10.6 Schematic representation of the patterning of superhydrophilic regions on superhydrophobic PS surfaces by UVO irradiation and using a hollowed mask. Fluorescent staining of the SaOs-2 cell nucleus with DAPI after six days in culture. (Source: Adapted from Ref. [74].)

this effect was temporary and cell adhesion was lost after three days. Superhydrophobic PLLA surfaces were also found to prevent adhesion and proliferation of bone-marrow-derived cells of rats, when compared with the smooth ones [45]. Cha *et al.* [79] analyzed the response of adipose-derived stem cells on PS surfaces fabricated with a lotus surface structure by a hot-embossing process. They found that, when compared with the flat substrate, the superhydrophobic one induced higher cell attachment rate but did not change the cell proliferation rate significantly and the cells exhibited a narrower spreading morphology and less-organized cytoskeleton. Topological modification has been pointed as a promising tool for controlling stem cell differentiation. These authors [79] also analyzed for the first time the effect of using the peculiar topography of a lotus-inspired surface on stem cell differentiation. They found that the superhydrophobic surfaces induced higher adipogenic differentiation of the cells and a decreased chondrogenic and osteogenic differentiation than the flat surfaces. So, such surfaces could be potentially used as culture dishes, for an efficient increase of adipogenic differentiation of stem cells, which is attractive in cosmetic industry.

10.4.2
Protein Interactions with Surfaces with Extreme Wettability Ranges

The human serum albumin (HSA) and human plasma fibronectin (HFN) adsorptions on PS surfaces with wettability controlled by UVO irradiation, ranging from superhydrophobic to superhydrophilic, were analyzed [80]. These proteins were chosen because HSA is presented in high quantities in human blood plasma and HFN is known to facilitate cell attachment onto biomaterial surfaces. Thus, HSA and HFN are representative of a large number of serum proteins. A strong decrease in protein adsorption was observed for the case of the PS superhydrophilic surface [80]: the two proteins essentially do not adsorb. Molecular dynamics studies showed that highly wettable surfaces produce large repulsive forces on the proteins, leading to lower protein adsorption [81], in agreement with these results. Moreover, it was found that the adsorbed density was almost the same for the superhydrophobic sample and the control smooth sample and tended to decrease slowly for both proteins from the superhydrophobic situation to the hydrophilic state [80], this decrease being more evident for HSA. Ballester-Beltran et al. [75] also reported HFN adsorption on superhydrophobic PS. However, HFN was adsorbed on superhydrophobic PS surfaces at lower density and altered conformation as compared with the corresponding standard smooth PS. As a consequence, they found that MC3T3-E1 adhesion occurred without formation of mature focal adhesion plaques and scarce phosphorylation of focal adhesion kinases. Under these circumstances, cell contractility, reorganization of adsorbed HFN, and secretion of newly synthesized HFN fibrils did not occur on the superhydrophobic PS.

Koc et al. [82] analyzed bovine serum albumin (BSA) adsorption on superhydrophobic surfaces obtained by chemical modification with either hydrocarbon or fluorocarbon groups. BSA adsorption was not significantly decreased on the superhydrophobic ones when compared with the smooth controls, under static conditions. They also observed that under flow conditions almost all the protein film was completely removed from the superhydrophobic surface [82]. Hence, superhydrophobic surfaces may be useful in environments, such as microfluidics, where protein sticking is a problem and fluid flow is present, to reduce protein adsorption and to promote desorption.

On the other hand, Zimmermann et al. [83] fabricated superhydrophobic silicone nanofilaments and reported a low affinity toward adsorption of β-lactoglobulin, α-chymotrypsin, and lysozyme. They also performed subsequent chemical modifications onto these superhydrophobic nanofilaments to obtain either amino- or carboxyl-ended surfaces, which resulted in superhydrophilic surfaces. In this case, the combination of the presence of the amino and carboxyl groups with the high contact area resulting from the nanoroughness allowed to obtain surfaces with excellent protein retention properties of high specificity toward these charged proteins, although these surfaces were highly wettable.

All the works referred to in this section evidenced that protein adsorption onto superhydrophobic surfaces may occur, although its dependency on the rough feature sizes of the surface and the used protein might be expected. Two opposite effects are present regarding this situation. On one hand, one could expect an increase in protein adsorption on rough surfaces because of the increase in surface area. It may also be hypothesized that roughening may create surfaces with a highly heterogeneous pattern of surface free energy that might promote protein adsorption [84]. On the other hand, the topographic features of the surfaces have particular consequences in terms of wettability: the presence of air pockets in superhydrophobic surfaces, according to the Cassie–Baxter model could have an opposite effect of the increase of surface area in the rough surfaces, inhibiting protein adsorption. This is in accordance with the work of Leibner et al. [85] that investigated the adsorption of HSA onto a commercial superhydrophobic PTFE by two different analyzed methods: standard radiometry and solution depletion of unlabeled protein. They found that when the adsorption was performed with degassed protein solutions under vacuum (reduced air trapped in the rough structures), the adsorption results were similar for the two methods. These results evidenced that the air trapped within the interstices of the superhydrophobic material surface prevents intimate contact with the protein solution and may strongly inhibit protein adsorption.

10.4.3
Blood Interactions with Surfaces with Extreme Wettability Ranges

The understanding of interactions between blood platelets and material surfaces is very important for constructing artificial heart valves, vascular stents, and circulatory support devices [86–91]. Moreover, blood coagulation and thrombosis are greatly affected by blood platelet adhesion and activation on the used substrates. The surface wettability is an important parameter in the adhesion and activation of platelets.

Sun et al. [92] analyzed blood platelet adhesion on superhydrophobic carbon nanotubes and found that almost no platelets adhered and there was much less platelet activation onto these materials; when compared with the smooth carbon nanotubes. Khorasani et al. [93] prepared superhydrophobic PDMS surfaces through CO_2-pulsed laser treatment and superhydrophilic surfaces by further grafting of hydroxyethylmethacrylate phosphatidylcholine onto these superhydrophobic substrates. The *in vitro* results showed that both superhydrophobic and superhydrophilic surfaces reduced platelet adhesion and that these two extreme wettabilities exhibited better blood compatibility than the control smooth samples [93].

Recently Yang et al. [89] used a combination of electrochemical anodization and surface self-assembly technique to construct superhydrophobic TiO_2 nanotube layers on a titanium substrate. Superhydrophilic TiO_2 nanotube layers were also produced by exposing the superhydrophobic surface to UV irradiation. They found that the superhydrophobic surfaces presented remarkable blood compatibility and the ability of resisting the adhesion and activation of blood platelets. On the other hand, large quantities of platelets adhered and spread either on the polished flat titanium substrate or on the superhydrophilic surfaces.

So, the inhibition for blood platelet adhesion and blood compatibility generally observed in surfaces with extreme wettability could be advantageous in a variety of intracorporeal or extracorporeal medical devices in contact with blood, such as blood vessels or circulatory support devices.

10.4.4
High-Throughput Chips Based on Surfaces with Extreme Wettability Ranges

High-throughput screening permits us to assess the relationships between many combinations of materials, surfaces, and the corresponding biological responses, including cell adhesion, growth, and differentiation or gene expression in a single experiment. Different methods have been employed to produce such high-throughput screening, including surfaces varying in roughness, surface chemistry/energy, mechanical properties, or density of biochemical elements [94, 95]. The substrates for such kind of high-throughput screening have been fabricated by distinct methods such as robotic DNA spotter; microfabrication masking techniques, such as photolithography, soft lithography, microfluidics, templating, imprint lithography, microelectronics, and magnetic forces; and direct microfabrication techniques such as contact printing and noncontact printing, ink-jet printing, electron beam lithography, and dip-pen nanolithography [95–102]. The microarray format enables the rapid deposition of different materials and, thereafter, screening a large library of multiple biomaterials and microenvironments.

However, in most cases, all the spots employed in the chip are usually tested in the same biological environment, which means the entire device is immersed in a unique culture medium. Advances in this field should offer a possibility to screen individually different combinations of biomaterials under different conditions in the same chip, including different cells, culture media, or solutions with different proteins or other molecules. In this context, a new low-cost platform for high-throughput analysis that permitted screening of the biological performance of independent combinations of biomaterials, cells, and culture media was proposed [80]. Patterned superhydrophobic PS substrates with controlled wettable spots were used to produce microarray chips for accelerated multiplexing evaluation. As volumes may be confined in the wettable regions because of strong contrasts in surface tension, this simple methodology enables to deposit with high control materials, cells, and other substances. These patterned superhydrophobic PS substrates having preadsorbed combinations of proteins (HSA and HFN) were used for cell studies to demonstrate their applicability for high-throughput analysis (Figure 10.7). As a tendency, for the same total protein concentration, more cells are detected in the spots treated with relatively higher amounts of HFN (Figure 10.7). For the same HFN/HSA composition, the number of cells also tends to increase with increasing total protein concentration, also corresponding to an increase in the total amount of HFN. Such findings are consistent with the fact that albumin is a passivating protein and HFN has cell-adhesive properties because of the existence of integrin binding domains in its structure. This inexpensive and simple bench-top method, or simple adaptations of it, could be integrated in tests involving larger libraries of substances that could be tested under distinct biological conditions, constituting a new tool accessible to virtually anyone to be used in the field of tissue engineering/regenerative medicine, cellular biology, diagnosis, drug discovery, and drug delivery monitoring. In particular, this technology was transposed to assess

10.4 Applications of Surfaces with Extreme Wettability Ranges in the Biomedical Field | 253

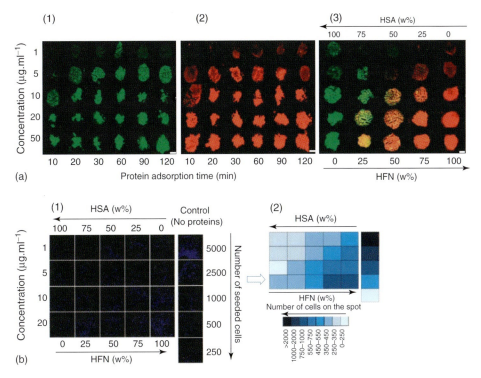

Figure 10.7 (a) Fluorescent microscope images of substrates, (a,1) HSA (green) and (a,2) HFN (red) with different concentrations (vertical axis) and during different adsorption times (horizontal axis). (a,3) HSA and HFN fluorescent fingerprints in patterned surfaces after different relative amounts and protein concentrations were deposited in the hydrophilic spots. (b,1) Confocal microscope pictures of osteoblastlike cells cultured for 4 h on the micropatterned array preadsorbed with different protein quantities (equivalent to the array of (a,3)). The column on the right represents the number of seeded cells attached on hydrophilic spots without the protein preadsorbed. (b,2) The heat map for the cell number per spot corresponding to the same array tested with different combinations of proteins. Scale bars, 500 μm. (Source: Reproduced with permission [80] Copyright 2011, the Royal Society of Chemistry.)

the biological response of 3D biomaterials, by encapsulating cells in arrays of miniaturized hydrogels dispensed onto patterned superhydrophobic substrates [103].

A high-throughput protein microarray was also proposed by Shiu *et al.* [104]. Superhydrophobic Teflon AF substrates were prepared by plasma treatment, and a switchable wetting character that could change from superhydrophobic to superhydrophilic was achieved using the electrowetting technology. Micropatterns were fabricated on these switchable superhydrophobic substrates [104]. Each spot on this microarray can be addressed individually, and different types of proteins or other biomolecules can be deposited on the microarray without losing their activity.

10.4.5
Substrates for Preparing Hydrogel and Polymeric Particles

Polymeric particles are ideal vehicles for controlled delivery applications because of their ability to encapsulate a variety of substances, namely low- and high-molecular-mass therapeutics, antigens, or DNA [105]. Moreover, polymeric microparticles have been applied in tissue engineering, as building blocks of hybrid structures and matrices for the delivery of soluble factors, either by fusion giving rise to porous scaffolds or as injectable systems for *in situ* scaffold formation [106].

It was shown that superhydrophobic substrates could be used to prepare spherical hydrogel or polymeric particles with high encapsulation efficiency, by dispensing almost spherical droplets of liquid precursors over superhydrophobic inert substrates, which then harden into solid particles [107]. In this fabrication process, spheres are produced involving basically liquid–air interfaces, which means that spheres harden in a dry environment preventing any losses of the encapsulating material. Owing to the mild conditions that can be employed, the methodology permits cell encapsulation [107] and also encapsulation of proteins in stimuli-responsive matrices [108].

10.5
Conclusions

In this chapter, a brief review of the surfaces recently produced with extreme wettabilities inspired by the superhydrophobic examples found in nature, such as lotus leaves, wheat leaves, butterfly wings, water striders, and mosquito eyes, was presented. Furthermore, biomedical applications of these bio-inspired substrates with extreme wettabilities ranges were introduced, which included the controlled adhesion of cells, blood platelet, and proteins. From the presented studies, it can be said that although the behavior is dependent on the cell type and the material used, in general, few cells adhere to superhydrophobic surfaces according to the Cassie–Baxter model and when cell attachment occurs, cells typically present a round morphology. When both superhydrophobic and superhydrophilic regions are present in the surface, cells adhere selectively to the superhydrophilic regions. Regarding protein adsorption, superhydrophilic surfaces seem to be ideal for repelling proteins and adsorption could occur on superhydrophobic surfaces depending on the chosen protein and mainly on the balance between the rough feature sizes, the low wettability, and the presence of air pockets at the surface. However, more studies are needed to understand exactly how roughness and extreme wettabilities affect protein conformation. In the case of blood interaction with surfaces with extreme wettabilities, the studies performed until now have shown that such surfaces generally inhibit blood platelet adhesion and exhibit blood compatibility, properties that could be obviously useful in a variety of intracorporeal or extracorporeal medical devices in contact with blood. The use of superhydrophobic surfaces to develop high-throughput microarray chips, which could be applied in preparing biomedical microfluidic chips and high-efficient bioanalyzing devices, was also discussed.

Another application of such highly repellent surfaces is to produce hydrogels or polymeric particles to encapsulate cells or molecules with high efficiency and under mild conditions. All these bio-inspired works will help us to search new insights on designing or improving devices for biomedicine and tissue engineering.

References

1. Sanchez, C., Arribart, H., and Guille, M.M.G. (2005) *Nat. Mater.*, **4**, 277.
2. Meyers, M.A., Chen, P.Y., Lin, A.Y.M., and Seki, Y. (2008) *Prog. Mater. Sci.*, **53**, 1.
3. Reis, R.L. and Weiner, S. (2004) *Learning from Nature How to Design New Implantable Biomaterials: From Biomineralization Fundamentals to Biomimetic Materials and Processing Routes*, Kluwer Academic Publishers, Netherlands.
4. Mano, J.F. and Reis, R.L. (2005) *Mater. Sci. Eng., C*, **25**, 93.
5. Bhushan, B. (2009) *Philos. Trans. R. Soc. A*, **367**, 1445.
6. Bhushan, B. and Jung, Y.C. (2011) *Prog. Mater. Sci.*, **56**, 1.
7. Liu, K.S., Yao, X., and Jiang, L. (2010) *Chem. Soc. Rev.*, **39**, 3240.
8. Sun, T.L., Qing, G.Y., Su, B.L., and Jiang, L. (2011) *Chem. Soc. Rev.*, doi: 10.1039/C0CS00124D
9. Lafuma, A. and Quéré, D. (2003) *Nat. Mater.*, **2**, 457.
10. Koch, K., Bhushan, B., and Barthlott, W. (2008) *Soft Matter*, **4**, 1943.
11. Blossey, R. (2003) *Nat. Mater.*, **2**, 301.
12. Koch, K. and Barthlott, W. (2009) *Philos. Trans. R. Soc. A*, **367**, 1487.
13. Gao, X.F., Yan, X., Yao, X., Xu, L., Zhang, K., Zhang, J.H., Yang, B., and Jiang, L. (2007) *Adv. Mater.*, **19**, 2213.
14. Barthlott, W. and Ehler, N. (1977) *Rasterelektronmikroskopie Der Epidermis-OberfläChen Von Spermatophyten, Tropische Und Subtropische Pflanzenwelt, Akademie Der Wissenschaften Und Literatur, Mainz*, Franz Steiner Verlag GmbH, Wiesbaden.
15. Zheng, Y.M., Gao, X.F., and Jiang, L. (2007) *Soft Matter*, **3**, 178.
16. Semprebon, C., Mistura, G., Orlandini, E., Bissacco, G., Segato, A., and Yeomans, J.M. (2009) *Langmuir*, **25**, 5619.
17. Zhao, B., Moore, J.S., and Beebe, D.J. (2001) *Science*, **291**, 1023.
18. Gao, X.F. and Jiang, L. (2004) *Nature*, **432**, 36.
19. Bush, J.W.M., Hu, D.L., and Prakash, M. (2007) *Adv. Insect. Physiol.*, **34**, 117.
20. Young, T. (1805) *Philos. Trans. R. Soc.*, **95**, 65.
21. Wenzel, R.N. (1936) *Ind. Eng. Chem.*, **28**, 988.
22. Cassie, A.B.D. and Baxter, S. (1944) *Trans. Faraday Soc.*, **40**, 546.
23. Nosonovsky, M. and Bhushan, B. (2008) *Langmuir*, **24**, 1525.
24. Bico, J., Marzolin, C., and Quéré, D. (1999) *Europhys. Lett.*, **47**, 220.
25. Ishino, C., Okumura, K., and Quéré, D. (2004) *Europhys. Lett.*, **68**, 419.
26. Jung, Y.C. and Bhushan, B. (2009) *Langmuir*, **25**, 9208.
27. Koishi, T., Yasuoka, K., Fujikawa, S., Ebisuzaki, T., and Zeng, X.C. (2009) *Proc. Natl. Acad. Sci.*, **106**, 8435.
28. Nosonovsky, M. and Bhushan, B. (2006) *Microsyst. Technol.*, **12**, 231.
29. Patankar, N.A. (2004) *Langmuir*, **20**, 7097.
30. Feng, X. and Jiang, L. (2006) *Adv. Mater.*, **18**, 3063.
31. Crick, C.R. and Parkin, I.P. (2010) *Chem. Eur.*, **16**, 3568.
32. Song, W.L., Gaware, V.S., Rúnarsson, Ö.V., Masson, M., and Mano, J.F. (2010) *Carbohydr. Polym.*, **81**, 140.
33. Li, S., Xie, H., Zhang, S., and Wang, X. (2007) *Chem. Commun.*, **46**, 4857.
34. Song, W.L., Veiga, D.D., Custódio, C.A., and Mano, J.F. (2009) *Adv. Mater.*, **21**, 1830.
35. Buck, M.E., Schwartz, S.C., and Lynn, D.M. (2010) *Chem. Mater.*, **22**, 6319.
36. Ting, W., Chen, C., Dai, S.A., Suen, S., Yang, I., Liu, Y., Chen, F.M.C., and Jeng, R. (2009) *J. Mater. Chem.*, **19**, 4819.

37. Nakata, K., Nishimoto, S., Yuda, Y., Ochiai, T., Murakami, T., and Fujishima, A. (2010) *Langmuir*, **26**, 11628.
38. Zhao, X.D., Fan, H.M., Liu, X.Y., Pan, H., and Xu, H.Y. (2011) *Langmuir*, **27**, 3224.
39. Ishizaki, T. and Saito, N. (2010) *Langmuir*, **26**, 9749.
40. Kwak, G., Lee, M., and Yong, K. (2010) *Langmuir*, **26**, 9964.
41. Zimmermann, J., Reifler, F.A., Fortunato, G., Gerhardt, L., and Seeger, S. (2008) *Adv. Funct. Mater.*, **18**, 3662.
42. Nguyen, J.G. and Cohen, S.M. (2010) *J. Am. Chem. Soc.*, **132**, 4560.
43. Bayer, I.S., Biswas, A., and Ellialtioglu, G. (2011) *Polym. Compos.*, **32**, 576.
44. Luo, Z.Z., Zhang, Z.Z., Hu, L.T., Liu, W.M., Guo, Z.G., Zhang, H.J., and Wang, W.J. (2008) *Adv. Mater.*, **20**, 970.
45. Alves, N.M., Shi, J., Oramas, E., Santos, J.L., Tomás, H., and Mano, J.F. (2009) *J. Biomed. Mater. Res. A*, **91A**, 480.
46. Alves, N.M. and Mano, J.F. (2008) *Int. J. Biol. Macromol.*, **43**, 401.
47. Suh, J.K.F. and Matthew, H.W.T. (2000) *Biomaterials*, **21**, 2589.
48. Jiang, L., Zhao, Y., and Zhai, J. (2004) *Angew. Chem. Int. Ed.*, **43**, 4338.
49. Lee, S., Kim, D., and Hwang, W. (2011) *Curr. Appl. Phys.*, **11**, 800.
50. Nakanishi, T., Shen, Y., Wang, J., Li, H., Fernandes, P., Yoshida, K., Yagai, S., Takeuchi, M., Ariga, K., Kurth, D.G., and Möhwald, H. (2010) *J. Mater. Chem.*, **20**, 1253.
51. Yao, X., Chen, Q., Xu, L., Li, Q., Song, Y., Gao, X., Quere, D., and Jiang, L. (2010) *Adv. Funct. Mater.*, **20**, 656.
52. Wu, X. and Shi, G. (2006) *J. Phys. Chem. B*, **110**, 11247.
53. Jiang, L., Yao, X., Li, H., Fu, Y., Chen, L., Meng, Q., Hu, W., and Jiang, L. (2010) *Adv. Mater.*, **22**, 376.
54. Chung, J.Y., Youngblood, J.P., and Stafford, C.M. (2007) *Soft Matter*, **3**, 1163–1169.
55. Xia, D. and Brueck, S.R.J. (2008) *Nano Lett.*, **8**, 2819.
56. Wu, H., Zhang, R., Sun, Y., Lin, D., Sun, Z., Pan, W., and Downs, P. (2008) *Soft Matter*, **4**, 2429.
57. Gao, J., Liu, Y., Xu, H., Wang, Z., and Zhang, X. (2010) *Langmuir*, **26**, 9673.
58. Feng, L., Zhang, Y., Xi, J., Zhu, Y., Wang, N., Xia, F., and Jiang, L. (2008) *Langmuir*, **24**, 4114.
59. Zheng, Y., Bai, H., Huang, Z., Tian, X., Nie, F.Q., Zhao, Y., Zhai, J., and Jiang, L. (2010) *Nature*, **463**, 640.
60. Bai, H., Tian, X., Zheng, Y., Ju, J., Zhao, Y., and Jiang, L. (2010) *Adv. Mater.*, **22**, 5521.
61. Alves, N.M., Pashkuleva, I., Reis, R.L., and Mano, J.F. (2010) *Small*, **6**, 2208.
62. Girard, P., Cavalcanti-Adam, E., Kemkemer, R., and Spatz, J. (2007) *Soft Matter*, **3**, 307.
63. Craighead, H.G., James, C.D., and Turner, A.M.P. (2001) *Curr. Opin. Solid State. Mater.*, **5**, 177.
64. Dalby, M.J., Riehle, M.O., Johnstone, H., Affrossman, S., and Curtis, A.S.G. (2002) *Biomaterials*, **23**, 2945.
65. Dalby, M.J., Childs, S., Riehle, M.O., Johnstone, H.J.H., Affrossman, S., and Curtis, A.S.G. (2003) *Biomaterials*, **24**, 927.
66. Jungbauer, S., Kemkemer, R., Gruler, H., Kaufmann, D., and Spatz, J.P. (2004) *ChemPhysChem*, **5**, 85.
67. Chen, C.S., Mrksich, M., Huang, S., Whitesides, G.M., and Ingber, D.E. (1997) *Science*, **276**, 1425.
68. Prime, K.L. and Whitesides, G.M. (1991) *Science*, **24**, 1164.
69. Wilson, C.J., Clegg, R.E., Leavesley, D.I., and Pearcy, M.J. (2005) *Tissue Eng.*, **11**, 1.
70. Ganesan, R., Kratz, K., and Lendlein, A. (2010) *J. Mater. Chem.*, **20**, 7322.
71. (a) Kobel, S. and Lutolf, M.P. (2010) *Biotechniques*, **48**, 9.; (b) Hook, A.L., Anderson, D.G., Langer, R., Williams, P., Davies, M.C., and Alexander, M.R. (2010) *Biomaterials*, **31**, 187.
72. Liu, M.J., Zheng, Y.M., Zhai, J., and Jiang, L. (2010) *Acc. Chem. Res.*, **43**, 368.
73. Mundo, R.D., Nardulli, M., Milella, A., Favia, P., D'Agostino, R., and Gristina, R. (2011) *Langmuir*, **27**, 4914.

74. Oliveira, S.M., Song, W., Alves, N.M., and Mano, J.F. (2011) *Soft Matter*, **7**, 8932.
75. Ballester-Beltran, J., Patricia Rico, P., Moratal, D., Wenlong Song, W., Mano, J.F., and Salmeron-Sanchez, M. (2011) *Soft Matter*, **7**, 4147.
76. Piret, G. *et al.* (2011) *Soft matter*, **7**, 8642.
77. IshizaKi, T., Saito, N., and Takai, O. (2010) *Langmuir*, **26**, 8147.
78. Bauer, S., Park, J., von der Mark, K., and Schmuki, P. (2008) *Acta Biomater.*, **4**, 1576.
79. Cha, K.J., Park, K.S., Kang, S.W., Cha, B.W., Lee, B.K., Han, I.B., Shin, D.A., and Lee, S.H. (2011) *Macrom. Biosci.*, **11**, 1357.
80. Neto, A.I., Custódio, C.A., Song, W.L., and Mano, J.F. (2011) *Soft Matter*, **7**, 4147.
81. Zheng, J., Li, L., Tsao, H.K., Sheng, Y.J., Chen, S., and Jiang, S. (2005) *Biophys. J.*, **89**, 158.
82. Koc, Y., de Mello, A.J., McHale, G., Newton, M.I., Roach, P., and Shirtcliffe, N.J. (2008) *Lab Chip*, **8**, 582.
83. Zimmermann, J., Rabe, M., Verdes, D., and Seeger, S. (2008) *Langmuir*, **24**, 1053.
84. Muller, R., Hiller, K.A., Schmalz, G., and Ruhl, S. (2006) *Anal. Biochem.*, **359**, 194.
85. Leibner, E.S., Barnthip, N., Chen, W., Baumrucker, C.R., Badding, J.V., Pishko, M., and Vogler, E.A. (2009) *Acta Biomater.*, **5**, 1389.
86. Zhou, J., Yun, J., Zang, X., Shen, J., and Lin, S. (2005) *Colloids Surf., B*, **41**, 55.
87. Kwok, S.C.H., Wang, J., and Chu, P.K. (2005) *Diamond Relat. Mater.*, **14**, 78.
88. Hou, X., Wang, X., Zhu, Q., Bao, J., Mao, C., Jiang, L., and Shen, J. (2010) *Colloids Surf., B*, **80**, 247.
89. Yang, Y., Lai, Y., Zhang, Q., Wu, K., Zhang, L., Lin, C., and Tang, P. (2010) *Colloids Surf., B*, **79**, 309.
90. Zhang, Z., Zhang, M., Chen, S., Horbett, T.A., Ratner, B.D., and Jiang, S. (2008) *Biomaterials*, **29**, 4285.
91. Kaladhar, K. and Sharma, C.P. (2004) *Langmuir*, **20**, 11115.
92. Sun, T., Tan, H., Han, D., Fu, Q., and Jiang, L. (2005) *Small*, **1**, 959.
93. Khorasani, M.T. and Mirzadeh, H. (2004) *J. Appl. Polym. Sci.*, **91**, 2042.
94. Peters, A., Brey, D.M., and Burdick, J.A. (2009) *Tissue Eng. B*, **15**, 225.
95. Yliperttula, M., Chung, B.G., Navaladi, A., Manbachi, A., and Urtti, A. (2008) *Eur. J. Pharm. Sci.*, **35**, 151.
96. Flaim, C.J., Chien, S., and Bhatia, S.N. (2005) *Nat. Methods*, **2**, 19.
97. Mei, Y., Saha, K., Bogatyrev, S.R., Yang, J., Hook, A.L., Kalcioglu, Z.I., Cho, S.W., Mitalipova, M., Pyzocha, N., Rojas, F., Van, K.J., Davies, M.C., Alexander, M.R., Langer, R., Jaenisch, R., and Anderson, D.G. (2010) *Nat. Mater.*, **9**, 768.
98. Simon, C.G. Jr. and Lin-Gibson, S. (2011) *Adv. Mater.*, **23**, 369.
99. Peters, A., Brey, D.M., and Burdick, J.A. (2009) *Tissue Eng. Part B Rev.*, **15**, 225.
100. Hook, A.L., Anderson, D.G., Langer, R., Williams, P., Davies, M.C., and Alexander, M.R. (2010) *Biomaterials*, **31**, 187.
101. Scopelliti, P.E., Bongiorno, G., and Milani, P. (2011) *Comb. Chem. High Throughput Screen.*, **14**, 206.
102. Davies, M.C., Alexander, M.R., Hook, A.L., Yang, J., Mei, Y., Taylor, M., Urquhart, A.J., Langer, R., and Anderson, D.G. (2010) *J. Drug Targeting*, **18**, 741.
103. Salgado, C.L., Oliveira, M.B., and Mano, J.F. *Integrative Biology*, in press.
104. Shiu, J.Y. and Chen, P.L. (2007) *Adv. Funct. Mater.*, **17**, 2680.
105. Lima, A.C., Sher, P., and Mano, J.F. (2012) *Expert Opin. Drug Delivery*, **9**, 231.
106. Oliveira, M.B. and Mano, J.F. (2011) *Biotechnol. Prog.*, **27**, 897.
107. Song, W., Lima, A.C., and Mano, J.F. (2010) *Soft Matter*, **6**, 5868.
108. Lima, A.C., Song, W., Blanco-Fernandez, B. *et al.* (2011) *Pharm. Res.*, **28**, 1294.

11
Bio-Inspired Reversible Adhesives for Dry and Wet Conditions

Aránzazu del Campo and Juan Pedro Fernández-Blázquez

11.1
Introduction

Biological adhesives allow attachment of organisms or animals to surfaces, either temporally or permanently. Microscopic bacteria, fungi, and larger marine algae as well as invertebrates, insects, frogs, and terrestrial vertebrates (i.e., gecko) use specialized adhesive organs and secretions for this purpose. The performance of these adhesives is remarkable, and their diversity suggests the potential for developing adhesion concepts and artificial adhesive materials that are markedly different from those currently available. These bioinspired adhesives will provide elegant solutions to contemporary engineering and biomedical adhesive requirements, that is, high adhesion strength and reliability, reversible attachment, and attachment to any surface surfaces including aquatic and fluid environments.

Adhesion to a surface can be mediated by strong chemical bonds or by physical interactions. Covalent and hydrogen bonds are generally used in the animal world for irreversible bonding. In most cases, surface-reactive chemical species are secreted by the animal and used as a bioglue to stick to a surface, for example, catechol units in DOPA-containing adhesive proteins secreted by mussels. When reversible adhesion is required, as in adhesive pads used for locomotion, nature utilizes physical interactions enhanced by complex topographical designs. Two main types of adhesive designs are found in the attachment pads of animals: "fibrillar" (in geckos, insects, and spiders) [1] and "soft" pads (in crickets [2] and tree and torrent frogs [3–6]). Both systems allow conformal contact to the substrate independent of the substrate roughness, either using flexible hairs or highly deformable material. Figure 11.1 shows examples of both types.

Gecko's adhesive pads display a complex surface hyperstructure consisting of long keratinous hairs ("setae") that are 30–130 μm long and contain hundreds of projections terminated by 200–500 nm spatula-shaped structures (Figure 11.1). Gecko's dry adhesion mechanism has been extensively studied in the past decade [2, 7–14], and it is now well accepted that its complex fibrillar design enables it to exploit van der Waals and capillary forces to obtain strong but reversible adherence [10, 15–17]. Attempts to engineer gecko's fibrillar pads using artificial surfaces

Biomimetic Approaches for Biomaterials Development, First Edition. Edited by João F. Mano.
© 2012 Wiley-VCH Verlag GmbH & Co. KGaA. Published 2012 by Wiley-VCH Verlag GmbH & Co. KGaA.

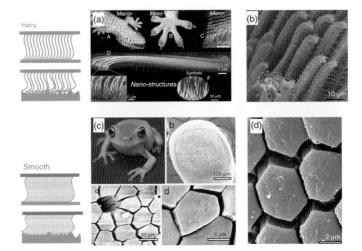

Figure 11.1 Diagram of "hairy" and "smooth" attachment systems on smooth and structured substrata showing that both systems are able to adapt to the surface profile [20]. SEM pictures of different hairy ((a) gecko [7, 8], (b) tarantula *Grammostola rosea* [21]) and soft ((c) great green bush-cricket *Tettigonia viridissima* [2], and (d) White's tree frog *Litoria caerulea* [22]) pads.

have been made in the past decade and have been recently reviewed [18, 19]. Recent advances are presented in Section 11.2.

Tree frogs' attachment pads consist of a regular hexagonal pattern with 10 μm epithelial cells separated by 1 μm wide channels, subdivided into a submicrometric array of nanopillars of ∼300–400 nm diameter (Figure 11.1a–d) [22]. Wide pores ranging from 5 to 10 μm are distributed randomly within the array, and these secrete a mucus layer with unknown properties and unclear role in the attachment process. Tree frogs offer solutions to reversible adhesion under wet conditions. Reported measurements with living frogs [5, 6, 23, 24] and a few theoretical studies [25, 26] have suggested contributions from different forces to the wet adhesion mechanism: capillary forces [27], friction forces [24, 25], and viscous forces [6, 23]. Water-draining effects assisted by the microchannels have also been predicted as plausible wet adhesion mechanism [25]. However, no clear picture about the extent to which each of these forces contribute to the attachment is available yet. Initial mimics have been reported, but both theories and bio-inspired reversible wet adhesive systems are still much less developed than reversible dry adhesives. Section 11.3 describes recent progress in this field.

11.2
Gecko-Like Dry Adhesives

In contrast to conventional adhesives, the design of gecko-like adhesives is not chemistry dominated, but geometry dominated. The diameter, length, and density of the fibers constitute the basic design criteria for adhesion enhancement, as predicted by the "contact splitting" principle [28]. The orientation and tilting of the individual fibers plays a crucial role for the reversibility of the attachment and, therefore, enables "easy" detachment of the gecko during locomotion. The hierarchical arrangement within the array seems to assist in adaptability to surface roughness of any kind as well as in mechanical stability of the array and self-cleaning properties. The spatulalike terminals also play a major role in adhesion enhancement.

The mechanical stability and surface energy of the hairs impose boundary conditions on the design. Long fibers tend to be mechanically unstable and may collapse as a consequence of their own weight, or cluster if the adhesive forces between the contact tips become stronger than the forces required for bending. Such condensation reduces the contact area, decreasing the overall adhesion of the array. This means that fibers need to be made of a material with high mechanical stability. In fact, hairs in natural structures are built of β-keratin, with a Young's modulus lying between 2 and 4 GPa. However, fibers with a too high modulus will result in a reduced capability to elastically deform and build the contact area. The optimum combination of height, thickness, elasticity, and spacing of the fibers has been displayed in "design maps," which can be used to guide the design of artificial adhesives [29].

A last material-related issue should be mentioned. The complex geometrical design of natural attachment systems is frequently accompanied by a complex and anisotropic design of the materials properties [29]. This fact may allow geometrical designs far more complex than those calculated for isotropic fibers, as it has been predicted theoretically [30].

Over the past 10 years, different structuring methods have been applied to generate gecko-inspired fibrillar substrates. Initial efforts focused on obtaining densely packed arrays of fibrils, following predicted adhesion enhancement via "contact splitting." These "first-generation" gecko-inspired adhesives have been extensively described in recent review articles [18, 19, 31]. Most examples contained microsized pillars obtained by casting elastomeric precursors (soft lithography) onto lithographic templates [32–39]. The obtained microstructured surfaces (with radii between 1 and 25 µm and lengths between 5 and 80 µm) showed moderate adhesion performance but allowed accurate quantitative adhesion studies because of their regularity and well-defined dimensions. Using highly sophisticated nanofabrication techniques (i.e., e-beam [40] and nanoimprint [41–44] lithography and carbon nanotube (CNT) deposition [45–50]), arrays of pillars with submicrometric diameters were obtained (Figure 11.2). Higher adhesion performance was achieved, especially in the case of arrays of CNTs. However, nanofibrillar structures tend to stick to their closest neighbors, forming loosely packed "bundles," which was detrimental for adhesion performance and hindered reversibility.

The main reason for adhesion failure of the first-generation gecko-like adhesives was the oversimplification of the geometrical design of the artificial mimics. No hyperstructure, tilt, or optimized tip geometry (e.g., spatula like) was implemented in the designs, mainly because of the intrinsic difficulty and scarce availability of 3D micro- and nanofabrication methods to obtain such structures over medium-size

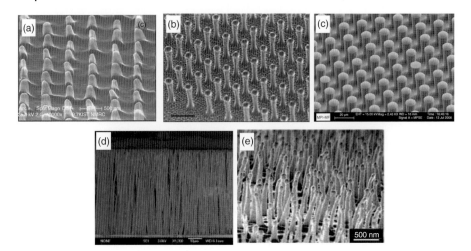

Figure 11.2 First-generation gecko-inspired adhesive surfaces containing densely packed pillars terminated with flat tips. (a) poly(methyl methacrylate) (PMMA) patterns made by molding a thin PMMA film at 120 °C [41]. (b) Arrays of Poly(imide), (PI) pillars microfabricated using e-beam lithography and pattern transfer by dry etching in oxygen plasma [40]. (c) Structured polydimethylsiloxane (PDMS) films obtained via soft lithography [35]. (Reprinted with permission from reference [35]. Copyright (2007) American Chemical Society.) (d) SEM images of aligned PS nanotubes prepared by filling Anodic Alumina (AA) templates [52]. (e) Poly(styrene), PS fibers obtained by fiber drawing method using a mold with cavities [44].

areas (at least several square centimeters) required for adhesion testing. Moreover, although some of the structured systems showed higher adhesion force than flat analogs, none of them showed the additional properties of gecko's attachment pads such as easy detachment, adaptability to any kind or roughness, or reusability. In the past five years "second-generation" gecko-like adhesives have been reported, including fibrillar surfaces with more complex geometries such as spatula-ended fibrils, tilted fibrils, or hierarchical fibrillar structures. In addition, responsive materials were used to incorporate the reversible and adaptive character of gecko adhesion to artificial systems. In the following section, the fabrication methods used and structured surfaces obtained so far are described.

11.2.1
Fibrils with 3D Contact Shapes

Spherical, conical, filamentlike, bandlike, suckerlike, spatulalike, flat, and toroidal tip shapes have been observed in the fibrillar attachment pads of different animals [53]. The relevance of the 3D shape of the fibril tip for adhesion of fibrillar structures has also been studied theoretically [53]. Experiments in artificial model systems have also demonstrated a clear impact of the contact geometry on adhesion performance. Among different tip geometries, mushroom-shaped tips, as closest

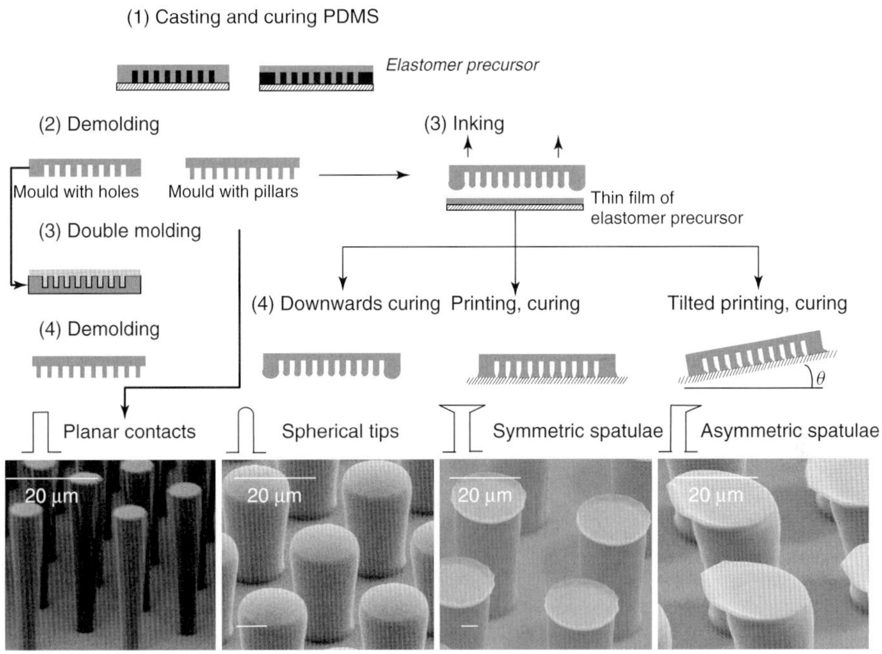

Figure 11.3 Structuring methods for different 3D tip shapes [55].

mimics of gecko spatulae, have so far given the highest adhesion strength values of all polymeric fibers [33, 54–58].

Fabrication of arrays of spatula-terminated fibers require multistep processing [33, 54–56, 58–60]. Most reported structures have been obtained by inking a fibrillar surface in a thin film of polymer precursor and pressing against a flat substrate until it solidifies (Figure 11.3). This renders fibrils with a flexible and flat roof ("mushroom shape"), with a diameter that depends on the inked precursor drop. The roofs can be made asymmetric by tilting the flat substrate before curing. This method has been also been used to obtain spatular tips onto tilted microfibres [56, 58]. This design, containing two independent tilted components (fiber and spatula), represents the most complex structure obtained to date with artificial systems. However, the inking/printing process has only been demonstrated on micrometric structures. Transfer of this fabrication approach to the nanoscale in order to achieve higher adhesion performance is not trivial, but, if achieved, this could well represent an interesting possibility for an upscaled manufacture of gecko-inspired reversible adhesives.

11.2.2
Tilted Structures

An important feature of such biological adhesion systems is the ability to switch between strong attachment and easy detachment, which is crucial for animal

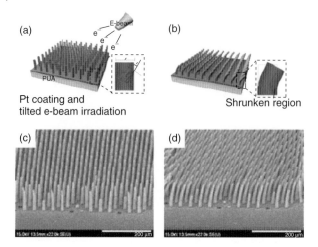

Figure 11.4 Tilted Poly(urethane acrylate), (PUA) nanofibers fabricated by local softening of the polymer using electron-beam irradiation [66, 69].

locomotion. Recent investigations have suggested that the asymmetric, naturally curved setae with a directional tilt play a crucial role in gecko's reversible adhesion mechanism [61]. This suggestion is supported by the observation that geckos curl their toes backward (digital hyperextension) while detaching from a surface. These observations have pushed research in bio-inspired systems into the fabrication of tilted fibrillar surfaces in order to confer "directionality" to the adhesive design.

Tilted fibrillar structures have been obtained by replication of photolithographic masters obtained by inclined lithography or anisotropic etching [39, 58, 62–64]. Tilted fibrils with angles between 0 and 50° with the substrate, diameter of 4–35 μm, and aspect ratios of up to 10 have been reported. Alternatively, arrays of tilted fibrils with submicrometric dimensions were obtained by mechanically processing a patterned film with vertical pillars through two heated rollers [65], or by a postmolding electron-beam [66, 67] or e-beam exposure step of vertical pillars that caused tilting at controlled angles [68] (Figure 11.4).

Angled fibers were shown to possess a large difference in shear stresses when tested parallel to the tilting direction ("gripping direction") or antiparallel ("releasing direction") [39, 56, 58, 63, 66], with the latter situation showing lower stresses. Shearing and normal contact experiments revealed higher pull-off forces for inclined structures than for vertical fibrils [58, 63, 65]. Experiments performed on tilted fibers with nontilted mushroom tips did not show a significant difference between shear forces, indicating that the two tilted components of the fibril may need to be combined to obtain reversible adhesives [58]. In fact, recent adhesion studies performed on tilted fibers with angled mushroom tips showed shear forces almost six times higher in the gripping direction than in the releasing direction [56]. Visual observation of the contact region revealed that the tips adhere and stretch when displaced in the gripping direction, whereas they flip over and slide when displaced in the releasing direction (Figure 11.5).

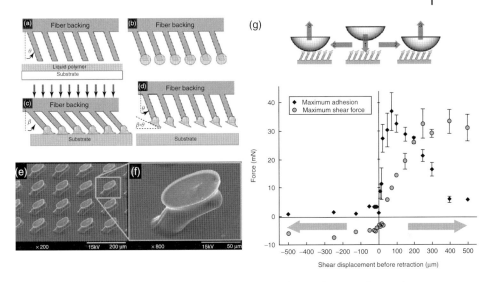

Figure 11.5 (a–d) Fabrication of angled tips to tilted fibers. (e,f) SEM images of the obtained tilted and spatula-terminated gecko mimics. (g) Demonstration of directional adhesion (in shear). The curve shows adhesion values obtained during shear displacement in the two opposite directions (gripping and releasing) [56].

11.2.3
Hierarchical Structures

Theoretical works have predicted that the hierarchical arrangement of setae of gecko provides (i) mechanical stability of the fibrillar design, (ii) the ability to conform to roughness at different length scales, and (iii) enhanced adhesion performance compared to single-level structures [30, 64, 70–77]. From a fabrication point of view, the generation of hierarchical structures at the micro- and nanoscale required either multistep processing of a hierarchical mold that can be replicated [77–82] or the combination of several molding steps that render a stacked arrangement of fibrils with decreasing size [42, 43, 64, 83]. Experimental results on gecko mimics have only proved marginal improvements (if at all) over single-level structures when attaching to planar substrates [83]. At this point, therefore, it remains questionable whether hierarchy is a relevant parameter for adhesion to planar surfaces, although it will certainly play an important role on rough surfaces (Figure 11.6).

11.2.4
Responsive Adhesion Patterns

An important difference between animal adhesive pads and the micro- and nanofabricated adhesives is the fact that gecko pads are "actuated." A mechanical stimulus, performed by the animal via gross leg movements, causes reorientation of setae shaft angle and establishes the adhesive contacts. A similar but opposite process occurs during detachment. Different strategies have been recently reported that

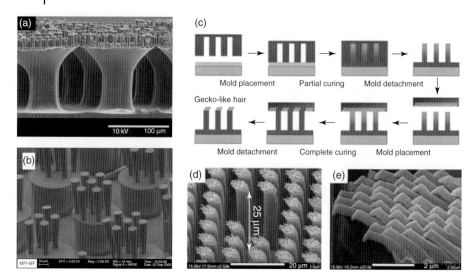

Figure 11.6 Hierarchical structures obtained with different methods. (a) PU structures with a complex tip geometry prepared by soft lithography and capillary molding [83]. (Reprinted with permission from reference [83]. Copyright (2009) American Chemical Society.) (b) Soft molding PDMS onto SU-8 templates fabricated by double exposure photolithography [78]. (c) Schematic illustration of the fabrication of hierarchical polyurethane acrylate nanohairs by sequential application of a molding process [64, 69]. (Reproduced with permission of the Journal of Experimental Biology.) (d) and (e) show the resulting structures.

use reorganization of the pattern geometry by use of external stimuli in order to switch between adhesive and nonadhesive states and are described in this section.

Activation of adhesion by means of external stimuli is feasible if responsive polymers are used to fabricate the patterned surfaces. The stimulus causes a change in the pillar orientation and consequently in the adhesion performance. This has been demonstrated by using shape-memory polymers for pillar fabrication. These materials are known for their ability to recover the predefined original shape (programmed during processing) when exposed to an external stimulus, even when deformed into a very different temporary shape [21]. In the case of thermosensitive shape-memory polymers, the temporary shape can be changed to the permanent shape by reheating the material above a specific transition temperature, T_{trans}. This strategy has been exploited to obtain micropatterned surfaces with triggerable adhesion due to a change in the tilting angle of the fibrils on heating the system to T_{trans} [84] (Figure 11.7).

In a different approach, tunable adhesive micropatterns were obtained that contained micropillars supported onto stretchable elastomeric films [85]. Mechanical stretching of the film caused reorientation of the pillars, and adhesion was activated. When the mechanical force was removed, the film became wrinkled and the adhesive pillars were no longer able to contact the countersurface. A similar strategy was used by other authors to tune adhesion of a film-terminated fibrillar surface (Figure 11.8) [86].

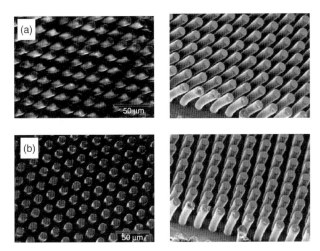

Figure 11.7 SEM images of fibrils made by soft molding a shape-memory polymer with a masters with holes of radius 10 μm, height 100 μm, and fibril spacing 20 μm [84]. (a) Top and side views of tilted fibrils after deformation above the shape-memory transition temperature and fixation in the deformed state by cooling (nonadhesive state). (b) Top and side views of recovered fibrils after reheating the sample above the transition temperature (adhesive state).

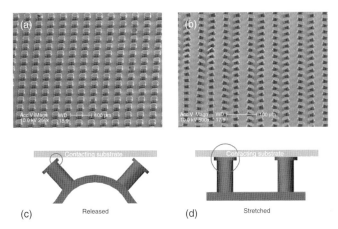

Figure 11.8 Switching between adhesive states by means of prestrain. When the film is relaxed (no prestrain), the pillars fold and the contact area is significantly reduced (nonadhesive state). Under prestrain, the pillars are vertical and are able to make contact with the surface (adhesive state) [87]. (Reprinted with permission from reference [87]. Copyright (2010) American Chemical Society.)

It is important to point out that none of these systems allows "reversible" adhesion at low energy cost, as the gecko does during detachment by curling its toes. They only provide an initial "on" or "off" adhesive state, but no peeling strategy to deactivate adhesion once the surfaces are in contact. This has only been demonstrated for nonpatterned bilayer films that contain a pressure-sensitive

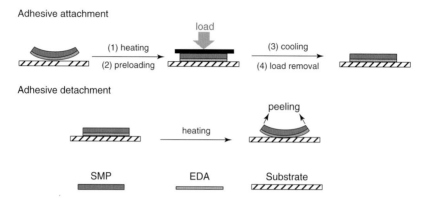

Figure 11.9 The strategy of reversible adhesion using bilayered films, including a shape-memory polymer that allows easy peeling-off the adhesive film on heating [90].

adhesive as "adhesive layer" and a shape-memory polymer as backing layer in a curved configuration (Figure 11.9) [88, 89]. For adhesive attachment, the film is heated above the glass-transition temperature of the shape-memory polymer, pressed against the countersurface, and left to cool to enable bonding. On reheating above T_g, the shape-memory polymer recovers its original curved shape, creating a peeling force that drives detachment. The combination of this idea with directional adhesive patterns (as described in Section 11.2.2) would be the closest mimic to the natural system. However, this has not been realized yet.

11.3
Bioinspired Adhesives for Wet Conditions

The interest in adhesive patterned surfaces has turned to also include adhesion under wet environments, especially because of possible applications in the medical field. Although the gecko pads are not designed for adhering in watery environments, a few groups have tested the adhesion of gecko-like fibrillar surfaces in wet conditions. Patterned elastomeric substrates showed modest adhesion performance against tissue [91], although mechanical interlocking and not gecko-like effects are expected to be involved and dominate adhesion to soft tissue. In a different approach, the adhesion of micropatterned surfaces coated with DOPA, a key component of wet adhesive proteins found in mussel holdfasts, was tested in the presence of a fluid layer [92, 93]. DOPA-coated patterns showed better adhesion performance than noncoated controls, indicating that the combination of mussel and gecko adhesive tools might extend the application of fibrillar adhesives to wet environments (Figure 11.10) [92, 93].

The adhesive structures of insects, containing fibrillar structures and an oily fluid, have also been mimicked. Arrays of flat-terminated micropillars showed up to sixfold adhesion enhancement when a thick oil film was placed at the interface [94]. Mushroom-terminated pillars showed a modest 20% increase in adhesion

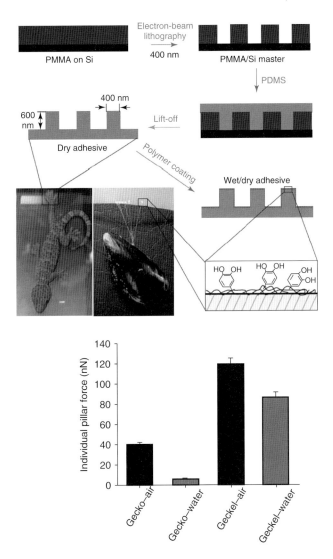

Figure 11.10 Example of a bio-inspired wet adhesive combining adhesive features from geckos and mussels ("geckel") and its adhesion performance [92].

underwater [95], and the authors claim that suction forces were responsible for that. Using a more elaborate surface design containing hexagonal micropillars with narrow channels, higher friction forces were reported in the microstructure than in the flat surface in the presence of a fluid water layer [96]. However, reported adhesion/friction values are modest and already indicate that the surface design is far from being the optimum one. Tree-frog-inspired patterns have not been reported until now, but they are certainly the ideal system to look at for enhancing adhesion under wet conditions using surface patterns.

11.4
The Future of Bio-Inspired Reversible Adhesives

Chemistry, physics, applied mechanics, and materials science have joined forces during the past years to understand and exploit bio-inspired patterned adhesives. For artificial surfaces to function, the mechanical, structural, and chemical aspects have to be optimized at the same time. The number of possible design parameters is huge, and research work so far, while having reached a very high level in recent years, has only been able to elucidate some of the first-order effects in bioadhesion. It is to be expected that the strategy for artificial adhesive surfaces will still be adapted and reformulated as more and more information of the animal systems is unraveled.

Bio-inspired adhesives would find many potential applications: temporary fastening in the construction industry, temporary labeling, optimization of surfaces for sports equipment, biomedical materials and devices, and fixation for household items are only a few examples. While adhesion "on demand" has been demonstrated for patterned chemical adhesives, the switching of the adhesive interactions is still in an exploratory stage. The intrinsic difficulty of low-cost manufacture of micro- and nanopatterned surfaces will certainly limit the application of these systems to consumer products, unless new technologies develop.

Acknowledgments

The financial support from the SPP 1420 Program "Biomimetic Materials Research: Functionality by Hierarchical Structuring of Materials" from the Deutsche Forschung Gemeinschaft is acknowledged.

References

1. Federle, W. (2006) *J. Exp. Biol.*, **209**, 2611.
2. Gorb, S., Jiao, Y.K., and Scherge, M. (2000) *J. Comp. Physiol. A Neuroethol. Sens. Neural Behav. Physiol.*, **186**, 821.
3. Barnes, J., Smith, J., and Platter, J. (2007) *Comp. Biochem. Phys. A*, **146**, S144.
4. Barnes, W.J.P. (2006) *J. Comp. Physiol. A*, **192**, 1165.
5. Barnes, W.J.P., Oines, C., and Smith, J.M. (2006) *J. Comp. Physiol. A*, **192**, 1179.
6. Hanna, G. and Barnes, W.J.P. (1991) *J. Exp. Biol.*, **155**, 103.
7. Autumn, K. (2007) *MRS Bull.*, **32**, 473.
8. Autumn, K. and Gravish, N. (2008) *Philos. Trans. R. Soc. A Math. Phys. Eng. Sci.*, **366**, 1575.
9. Hiller, U. (1968) *Z. Morphol. Tiere*, **62**, 307.
10. Autumn, K., Liang, Y.A., Hsieh, S.T., Zesch, W., Chan, W.P., Kenny, T.W., Fearing, R., and Full, R.J. (2000) *Nature*, **405**, 681.
11. Autumn, K. and Peattie, A.M. (2002) *Integr. Comp. Biol.*, **42**, 1081.
12. Jiao, Y.K., Gorb, S., and Scherge, M. (2000) *J. Exp. Biol.*, **203**, 1887.
13. Scherge, M. and Gorb, S. (2000) *Proc. R. Soc. Lond. B*, **267**, 1239.
14. Scherge, M. and Gorb, S.N. (2001) *Biological Micro- and Nanotribology: Nature's Solutions*, Springer, Berlin.
15. Autumn, K., Sitti, M., Liang, Y.A., Peattie, A.M., Hansen, W.R., Sponberg, S., Kenny, T.W., Fearing, R., Israelachvili, J.N., and Full, R.J.

(2002) *Proc. Natl. Acad. Sci. U.S.A.*, **99**, 12252.
16. Huber, G., Gorb, S.N., Spolenak, R., and Arzt, E. (2005) *Biol. Lett.*, **1**, 2.
17. Huber, G., Mantz, H., Spolenak, R., Mecke, K., Jacobs, K., Gorb, S.N., and Arzt, E. (2005) *Proc. Natl. Acad. Sci. U.S.A.*, **102**, 16293.
18. Boesel, L.F., Greiner, C., arzt, E., and del Compo, A. (2010) *Adv. Mater.*, **22**, 2125.
19. del Campo, A. and Arzt, E. (2007) *Macromol. Biosci.*, **7**, 118.
20. Creton, C. and Gorb, S. (2007) *MRS Bull.*, **32**, 466.
21. Peattie, A.M., Dirks, J.H., Henriques, S., and Federle, W. (2011) *PLoS ONE*, **6**, article number: e20485.
22. Scholz, I., Barnes, W.J.P., Smith, J.M., and Baumgartner, W. (2009) *J. Exp. Biol.*, **212**, 155.
23. Smith, J.M., Barnes, W.J.P., Downie, J.R., and Ruxton, G.D. (2006) *J. Comp. Physiol. A*, **192**, 1193.
24. Federle, W., Barnes, W.J.P., Baumgartner, W., Drechsler, P., and Smith, J.M. (2006) *J. R. Soc. Interface*, **3**, 689.
25. Persson, B.N.J. (2007) *J. Phys. Condens. Matter*, **19**, article number: 376110.
26. Barnes, W.J.P. (2007) *MRS Bull.*, **32**, 479.
27. Emerson, S.B. and Diehl, D. (1980) *Biol. J. Linn. Soc.*, **13**, 199.
28. Arzt, E., Gorb, S., and Spolenak, R. (2003) *Proc. Natl. Acad. Sci. U.S.A.*, **100**, 10603.
29. Alibardi, L. (2009) *Zoology*, **112**, 403.
30. Yao, H. and Gao, H. (2006) *J. Mech. Phys. Solids*, **54**, 1120.
31. Kamperman, M., Kroner, E., del Campo, A., McMeeking, R.M., and Arzt, E. (2010) *Adv. Eng. Mater*, **12**, 335.
32. Crosby, A.J., Hageman, M., and Duncan, A. (2005) *Langmuir*, **21**, 11738.
33. del Campo, A., Greiner, C., and Arzt, E. (2007) *Langmuir*, **23**, 10235.
34. Glassmaker, N.J., Jagota, A., Hui, C.Y., and Kim, J. (2004) *J. R. Soc. Interface*, **1**, 23.
35. Greiner, C., del Campo, A., and Arzt, E. (2007) *Langmuir*, **23**, 3495.
36. Sitti, M. and Fearing, R.S. (2003) *J. Adhes. Sci. Technol.*, **17**, 1055.
37. Peressadko, A. and Gorb, S.N. (2004) *J. Adhes.*, **80**, 247.
38. Lamblet, M., Verneuil, E., Vilmin, T., Buguin, A., Solberzan, P., and Léger, L. (2007) *Langmuir*, **23**, 6966.
39. Aksak, B., Murphy, M.P., and Sitti, M. (2007) *Langmuir*, **23**, 3322.
40. Geim, A.K., Dubonos, S.V., Grigorieva, I.V., Novoselov, K.S., Zhukov, A.A., and Shapoval, S.Y. (2003) *Nat. Mater.*, **2**, 461.
41. Yoon, E.S., Singh, R.A., Kong, H., Kim, B., Kim, D.H., Jeong, H.E., and Suh, K.Y. (2006) *Tribol. Lett.*, **21**, 31.
42. Jeong, H.E., Lee, S.H., Kim, J.K., and Suh, K.Y. (2006) *Langmuir*, **22**, 1640.
43. Jeong, H.E., Lee, S.H., Kim, P., and Suh, K.Y. (2008) *Colloids Surf. A Physicochem. Eng. Aspects*, **313–314**, 359.
44. Jeong, H.E., Lee, S.H., Kim, P., and Suh, K.Y. (2006) *Nano Lett.*, **6**, 1508.
45. Qu, L. and Dai, L. (2007) *Adv. Mater.*, **19**, 3844.
46. Yurdumakan, B., Raravikar, N.R., Ajayan, P.M., and Dhinojwala, A. (2005) *Chem. Commun.*, 3799.
47. Ge, L., Sethi, S., Ci, L., Ajayan, P.M., and Dhinojwala, A. (2007) *Proc. Natl. Acad. Sci. U.S.A.*, **104**, 10792.
48. Zhao, Y., Tong, T., Delzeit, L., Kashani, A., Meyyappan, M., and Majumdar, A. (2006) *J. Vac. Sci. Technol. B*, **24**, 331.
49. Qu, L., Dai, L., Stone, M., Xia, Z., and Wang, Z.L. (2008) *Science*, **322**, 238.
50. Maeno, Y. and Nakayama, Y. (2009) *Appl. Phys. Lett.*, **94**, 012103.
51. Majidi, C., Groff, R.E., and Fearing, R. Proceedings of IMECE, California 2004.
52. Jin, M.H., Feng, X.J., Feng, L., Sun, T.L., Zhai, J., Li, T.J., and Jiang, L. (2005) *Adv. Mater.*, **17**, 1977.
53. Spolenak, R., Gorb, S., Gao, H.J., and Arzt, E. (2005) *Proc. R. Soc. Lond. Ser. A Math. Phys. Eng. Sci.*, **461**, 305.
54. del Campo, A., Álvarez, I., Filipe, S., and Wilhelm, M. (2007) *Adv. Funct. Mater.*, **17**, 3590.
55. del Campo, A., Greiner, C., Álvarez, I., and Arzt, E. (2007) *Adv. Mater.*, **19**, 1973.
56. Murphy, M.P., Aksak, B., and Sitti, M. (2009) *Small*, **5**, 170.

57. Seok, K., Aksak, B., and Sitti, M. (2007) *Appl. Phys. Lett.*, **91**, article number: 221913-1-3.
58. Murphy, M.P., Aksak, B., and Sitti, M. (2007) *J. Adhes. Sci. Technol.*, **21**, 1281.
59. Kim, S. and Sitti, M. (2006) *Appl. Phys. Lett.*, **89**, 261911.
60. Davies, J., Haq, S., Hawke, T., and Sargent, J.P. (2009) *Int. J. Adhes. Adhes.*, **29**, 380.
61. Zhao, B.X., Pesika, N., Zeng, H.B., Wei, Z.S., Chen, Y.F., Autumn, K., Turner, K., and Israelachvili, J. (2009) *J. Phys. Chem. B*, **113**, 3615.
62. Kehagias, N., Zelsmann, M., Torres, C.M.S., Pfeiffer, K., Ahrens, G., and Gruetzner, G. (2005) *J. Vac. Sci. Technol. B*, **23**, 2954.
63. Santos, D., Spenko, M., Parness, A., Kim, S., and Cutkosky, M. (2007) *J. Adhes. Sci. Technol.*, **21**, 1317.
64. Jeong, H.E., Lee, J.K., Kim, H.N., Moon, S.H., and Suh, K.Y. (2009) *Proc. Natl. Acad. Sci. U.S.A.*, **106**, 5639.
65. Lee, J., Fearing, R.S., and Komvopoulos, K. (2008) *Appl. Phys. Lett.*, **93**, 191910.
66. Kim, T.-I., Jeong, H.E., Suh, K.Y., and Lee, H.H. (2009) *Adv. Mater.*, **21**, 2276.
67. Kim, T.-I., Pang, C., and Suh, K.Y. (2009) *Langmuir*, **25**, 8879–8882.
68. Vaziri, A., Moon, M.W., Cha, T.G., Lee, K.R., and Kim, H.Y. (2010) *Soft Matter*, **6**, 3924.
69. Kwak, M.K., Jeong, H.E., Kim, T.I., Yoon, H., and Suh, K.Y. (2010) *Soft Matter*, **6**, 1849.
70. Persson, B.N.J. (2007) *J. Adhes. Sci. Technol.*, **21**, 1145.
71. Kim, T.W. and Bhushan, B. (2007) *J Adhes. Sci. Technol.*, **21**, 1.
72. Kim, T.W. and Bhushan, B. (2007) *Ultramicroscopy*, **107**, 902.
73. Yao, H. and Gao, H. (2007) *J. Adhes. Sci. Technol.*, **21**, 1185.
74. Yao, H. and Gao, H. (2007) *Bull. Pol. Acad. Sci. Techn. Sci.*, **55**, 141.
75. Bhushan, B., Peressadko, A., and Kim, T.W. (2006) *J. Adhes. Sci. Technol.*, **20**, 1475.
76. Porwal, P.K. and Hui, C.Y. (2008) *J. R. Soc. Interface*, **5**, 441.
77. Greiner, C., Arzt, E., and del Campo, A. (2009) *Adv. Mater.*, **21**, 479.
78. del Campo, A. and Greiner, C. (2007) *J. Micromech. Microeng.*, **17**, R81.
79. Kustandi, T.S., Samper, V.D., Ng, W.S., Chong, A.S., and Gao, H. (2007) *J. Micromech. Microeng.*, **17**, N75.
80. Northen, M.T. and Turner, K.L. (2005) *Nanotechnology*, **16**, 1159.
81. Northen, M.T. and Turner, K.L. (2006) *Sens. Actuators A Phys.*, **130**, 583.
82. Rodriguez, I., Ho, A.Y.Y., Yeo, L.P., and Lam, Y.C. (2011) *ACS Nano*, **5**, 1897.
83. Murphy, M.P., Kim, S., and Sitti, M. (2009) *Appl. Mater. Interfaces*, **1**, 849.
84. Reddy, S., del Campo, A., and Arzt, E. (2007) *Adv. Mater.*, **19**, 3833.
85. Suh, K.Y., Jeong, H.E., and Kwak, M.K. (2010) *Langmuir*, **26**, 2223.
86. Jagota, A., Nadermann, N., Ning, J., and Hui, C.Y. (2010) *Langmuir*, **26**, 15464.
87. Jeong, H.E., Kwak, M.K., and Suh, K.Y. (2010) *Langmuir*, **26**, 2223.
88. Xie, T., Wang, R.M., and Xiao, X.C. (2010) *Macromol. Rapid Commun.*, **31**, 295.
89. Xie, T. and Xiao, X.C. (2008) *Chem. Mater.*, **20**, 2866.
90. Wang, R.M., Xiao, X.C., and Xie, T. (2010) *Macromol. Rapid Commun.*, **31**, 295.
91. Mahdavi, A., Ferreira, L., Sundback, C., Nichol, J.W., Chan, E.P., Carter, D.J.D., Bettinger, C.J., Patanavanich, S., Chignozha, L., Ben-Joseph, E., Galakatos, A., Pryor, H., Pomerantseva, I., Masiakos, P.T., Faquin, W., Zumbuehl, A., Hong, S., Borenstein, J., Vacanti, J., Langer, R., and Karp, J.M. (2008) *Proc. Natl. Acad. Sci. U.S.A.*, **105**, 2307.
92. Lee, H., Lee, B.P., and Messersmith, P.B. (2007) *Nature*, **448**, 338.
93. Glass, P., Chung, H., Washburn, N.R., and Sitti, M. (2009) *Langmuir*, **25**, 6607.
94. Cheung, E. and Sitti, M. (2008) *J. Adhes. Sci. Technol.*, **22**, 569.
95. Varenberg, M. and Gorb, S. (2008) *J. R. Soc. Interface*, **5**, 383.
96. Varenberg, M. and Gorb, S.N. (2009) *Adv. Mater.*, **21**, 483.

12
Lessons from Sea Organisms to Produce New Biomedical Adhesives

Elise Hennebert, Pierre Becker, and Patrick Flammang

12.1
Introduction

Of all biological phenomena that have been investigated with a view to biomimetics, adhesion in nature has perhaps received the most interest. Indeed, biological adhesives often offer impressive performance in their natural context and, therewith, the potential to inspire novel, superior industrial adhesives for an increasing variety of high-tech applications [1–3]. Yet, because of the complexity of biological adhesion and the multidisciplinarity needed to tackle it, there has been little visible progress in the development of bio-inspired adhesives and many technological challenges remain, such as the development of adhesives that can function underwater or in aqueous environments. The marine environment is a place of choice for the search of inspiration for such adhesives. Indeed, numerous invertebrates such as mollusks, worms, or sea stars produce adhesive secretions that are able to form long-lasting attachments, even when totally immersed [3–5]. Most of these adhesive secretions are made up of complex blends of proteins, often associated with other components.

Obviously, parallels exist between the marine and human physiological environments, and a strategy that works well in one context may be useful in the other [6]. It is usually admitted that adhesives suitable for medical applications should possess the following general characteristics [7–9]:

- biocompatibility (nontoxic, nonirritant, nonallergenic);
- minimal interference with the tissue, with the natural healing process, and, eventually, ultimately biodegradability;
- spreading, bonding, and curing in moist environments allowing the construction of a strong, flexible bond;
- appropriate adhesive and cohesive strength in the range of 0.01–6 MPa, depending on the particular bonding site.

Biological marine adhesives are likely to gather these different characteristics. The first attempts to exploit the potentialities of marine glues for medical applications were undertaken in the 1980s. Native barnacle cement was shown to be effective in tests of rabbit bone repair [10]. Adhesive proteins extracted from mussels appeared to be nontoxic and well tolerated by biological systems when used in transplantation experiments *in vivo* (e.g., corneal transplantation in rabbit; [11]). However, in both cases, the difficulty to collect sufficient quantities of adhesive material prevented further use of this material in concrete medical procedures. Indeed, the extraction and purification of adhesive proteins directly from the tissues of marine organisms for commercial purposes is feasible but is economically and ecologically problematic. In mussels and barnacles, tens of thousands individuals are needed to obtain 1 g of proteins [8, 10, 12–14]. A commercial mussel extract, consisting of a mixture of adhesive proteins, is commercially available (Cell-Tak™, BD Biosciences) and is sold as an adhesive for the *in vitro* immobilization of cells and tissue sections on various surfaces, including plastic, glass, metal, and biological materials [6, 15, 16]. However, for most industrial applications (including medical applications), harvesting the natural adhesive is an unrealistic solution. To bypass the problem of obtaining adhesive proteins directly from marine organisms, biomimetic adhesives have been developed. These bio-inspired molecules are produced either in the form of recombinant preparations of the adhesive proteins or in the form of chemically synthesized polymers incorporating the functional adhesive motifs of marine adhesive proteins.

Among marine organisms, a more intense research effort has been devoted to the elucidation of the mechanisms allowing permanent adhesion of sessile invertebrates and algae [13, 17]. These organisms are indeed involved in the economically important problem of fouling in the marine environment. Because biomimetic efforts are only possible when composition and key molecular components of biological adhesives are understood, so far, only a very limited number of organisms have been used for the development of bio-inspired adhesives. The best-characterized marine bioadhesive is that from the mussel, and it has inspired most of the biomimetic adhesives currently available (see [18] for review). Recently, other organisms such as tube worms, barnacles, and brown algae have also been used as models. The following paragraphs focus on these four organisms, from the features of their adhesives to the imaginative applications that have been developed for their biomimetic counterparts.

12.2
Composition of Natural Adhesives

12.2.1
Mussels

To attach themselves to the substratum, mussels produce a byssus (Figure12.1a), which consists of a bundle of proteinaceous threads connected proximally to the base of the animal's foot, within the shell, and terminating distally with a flattened plaque that mediates adhesion to the substratum [18, 19]. These plaques are formed by the autoassembly of secretory products originating from four distinct

Figure 12.1 Model organisms used as an inspiration source for the development of biomimetic adhesives. (a) Mussels of the species *Mytilus edulis* attached to each other by means of the byssus (original). The byssus consists of a bundle of byssal threads (BTs), each terminating with a flattened plaque (P), which mediates adhesion to the surface. Barnacles of the species *Balanus crenatus* are attached on some of the mussels. (b) A polychaete of the species *Sabellaria alveolata* in its tube made up of cemented sand grains (original). (c) The brown alga *Fucus serratus*. (Courtesy of H. Bianco-Peled, Technion – Israel Institute of Technology.)

glands enclosed in the mussel foot. These products are composed of a collagenous substance, a mucous material, a mixture of polyphenolic proteins (known as foot proteins 2–6, abbreviated as fps-2–6), and an accessory protein (fp-1) [19–23]. Only the different fps-1–6, also known as mussel adhesive proteins (MAPs), are considered here.

Since the characterization of fp-1 in the early 1980s, MAPs have been the subject of a very large number of studies, leading to a detailed knowledge on their structures, functions, and interactions within the byssal attachment plaque. Proteins fp-2 and fp-4 form the central core of the plaque; fp-3, fp-5, and fp-6 are located at the interface between the plaque and the substratum (primer layer); and fp-1 forms a hard cuticle protecting the core from hydrolysis, abrasion, and microbial attack [1, 18, 22].

The 6 fps constituting the attachment plaques have been mostly characterized from mussels of the genus *Mytilus*. Their characteristics are summarized in

Table 12.1 Characteristics of the proteins constituting the byssus of mussels from the genus *Mytilus*, the cement of the tube worm *Phragmatopoma californica*, and the cement of the barnacle *Megabalanus rosa*.

Protein	Mass (kDa)	Function	Repeated unit (frequency)	DOPA (mol%)	Feature	References
Mussels (genus *Mytilus*)						
fp1	110	Protection, antimicrobial	AKPSYPPTYK (80)	10–15	Hyp, DiHyp	[22, 25]
fp2	47	Structural framework	EGF-like motif (11)	2–3	Cys	[26]
fp3	6	Surface coupling	None	20–25	Arg-OH	[27]
fp4	93	Structural framework	HVHTHRVLHK (36, in N-term half) DDHVNDIAQTA (16, in C-term half)	<2	His	[28]
fp5	9.5	Surface coupling	None	27	p-Ser	[29]
fp6	12	Structural framework	None	4	p-Ser, Cys	[23]
Tube worms (*Phragmatopoma californica*)						
Pc-1	18	Nd	VGGYGYGGKK (15)	9.8	basic	[30]
Pc-2	21	Nd	HPAVHKALGGYG (8)	7.3	basic	[30]
Pc-3[a]	10–52	Nd	Poly(S)	Nd	p-Ser (about 80 mol%), highly acidic	[30]
Barnacles (*Megabalanus rosa*)						
cp-16k	16	Antimicrobial	None	None	similar to lysozyme	[31, 32]
cp-19k	19	Surface coupling	None	None	bias toward Ser, Thr, Gly, Ala, Lys, Val (these residues amount for 67% of the total amino acids)	[33]
cp-20k	20	Surface coupling	Six Cys-rich repeated sequences	None	Cys, charged amino acids	[34]
cp-52k	52	Structural framework	Repetitive sequence	None	Hydrophobic, N-glycosylated	[32, 95]
cp-68k	68	Surface coupling	None	None	Bias toward Ser, Thr, Ala, Gly (these residues amount for 60% of the total amino acids); glycosylated	[32]
Cp-100k	100	Structural framework	None	None	Hydrophobic	[35]

[a]The mass of Pc-3 varies according to the variant considered.
Ala, alanine; Arg-OH, 4-hydroxyarginine; Cys, cysteine; DiHyp, 3,4 dihydroxyproline; Gly, glycine; His, histidine; Hyp, 4-hydroxyproline; Lys, lysine; Nd, not determined; Ser, serine; p-Ser, O-phosphoserine; Thr, threonine; Val, valine.

Table 12.1. Most MAPs exhibit repeated sequence motifs, whose number and amino acid composition vary according to the species considered. Moreover, all the proteins identified in the plaque share a common distinctive feature: the presence of 3,4-dihydroxyphenylalanine (DOPA), a residue formed by the posttranslational hydroxylation of tyrosine. This modified amino acid performs two important roles in the attachment plaque (Figure 12.2): (i) it is involved in the formation of cross-links between the different fps (cohesion) and (ii) it mediates physicochemical interactions with the surface (adhesion) [1, 6, 17]. It is generally accepted that cross-linking reactions are related to the oxidation of DOPA to DOPA-quinone, a reaction catalyzed by a catecholoxidase in the byssus. DOPA-quinones may

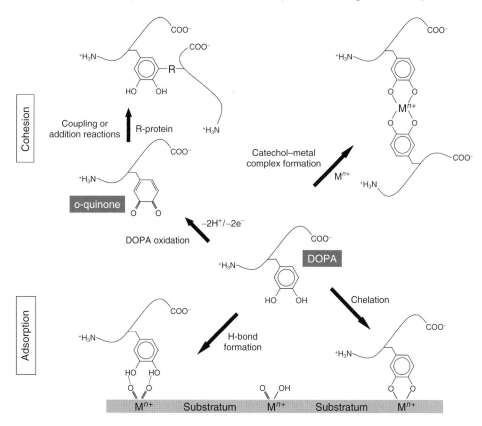

Figure 12.2 Dual functionality of peptidyl DOPA groups in mussel adhesive proteins. The catechol functions of DOPA-containing proteins contribute to adhesive adsorption through hydrogen bonding of the phenolic OH groups to the oxygen atoms of the surface (lower left) or by forming mono-bidentate complexes with metal ions at the surface of mineral, metal oxide, or metal hydroxide substrata (lower right). Peptidyl DOPA groups also contribute to adhesive cohesion by forming bis- or tris-catechol–metal complexes (upper right) or by forming intermolecular cross-links. Cross-linking follows the oxidation of DOPA to an o-quinone that reacts with amino acid side chains (R) of other proteins, in which R can belong to cysteine, histidine, lysine, and other DOPA residues (upper left).

also result from redox reactions involving transition metal ions or may form spontaneously at alkaline pH. Once formed, DOPA-quinone is capable of participating in a number of different reaction pathways, leading to intermolecular cross-link formation. DOPA also confers to proteins a capacity for intermolecular metal complexation through the formation of tris- and/or bis-catechol–metal complexes. In addition, DOPA is involved in surface coupling, either through hydrogen bonds or by forming complexes with metal ions and metal oxides present in mineral surfaces. These complexes possess some of the highest known stability constants of metal–ligand chelates [1, 17, 24]. Some of the MAPs also contain other posttranslationally modified amino acids such as O-phosphoserine (fp-5 and fp-6), which could mediate adhesion to calcareous substrata, or 4-hydroxyproline (fp-1), 3,4 dihydroxyproline (fp-1), and 4-hydroxyarginine (fp-3), which could make hydrogen bonds with surfaces [17, 18, 22].

12.2.2
Tube Worms

Some polychaete worms of the family Sabellariidae are tube dwelling and live in the intertidal zone [36]. They are commonly called *honeycomb worms* or *sandcastle worms* because they are gregarious and the tubes of all individuals are closely imbricated to form large reeflike mounds. To build the tube in which they live (Figure 12.1b), they collect particles such as sand grains or shell fragments with their tentacles from the water column and sea bottom. These particles are then conveyed to the building organ, which is a crescent-shaped structure near the mouth. There, the particles are dabbed with spots of cement secreted by two types of unicellular glands located in the worm's thorax and are added to the end of the preexisting tube by the building organ [37–39].

Cement composition has been investigated in the species *Phragmatopoma californica*. The adhesive mostly consists of three proteins (known as *Phragmatopoma* cement proteins 1–3, abbreviated as Pc-1–3) and large amounts of Mg^{2+} and Ca^{2+} ions [30, 39, 40]. The three cement proteins (cps) have highly repetitive and blocky primary structures with limited amino acid diversity (Table 12.1; [30]). As it is the case for plaque proteins in mussels, Pc-1 and 3 are present in the form of several variants. Pc-1 and 2 contain basic residues with amine side chains and DOPA residues, some of which are halogenated into 2-chloro-DOPA residues. DOPA residues presumably play the same functions as in MAPs (Figure 12.2), contributing to both adsorptive interactions with the surface (adhesion) and cross-link formation (cohesion) [38, 41]. Regarding the chloro-DOPA residues, they could be involved in the protection of the cement from microbial fouling and degradation [42]. Pc-3 is particularly rich in serine (72.9 mol%), and careful calculations have indicated that up to 90% of these serine residues are posttranslationally phosphorylated. Pc-3 is therefore a remarkably acidic protein ($pI = 1$) [30]. Phosphorylation is thought to impart a potential for both cohesive (by Ca^{2+} or Mg^{2+} bridging) and adhesive contributions to the cement [38, 41, 43].

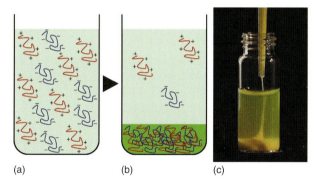

Figure 12.3 Complex coacervation is the associative phase separation in a solution of positively and negatively charged macroions. From a stable colloidal solution of polyelectrolyte complexes (a), a trigger such as a pH change leads to the formation of two immiscible aqueous phases: a dilute equilibrium phase and a denser solute-rich phase (b). The latter, called the *coacervate phase*, is an isotropic liquid containing amorphous associative particles that move freely relative to one another (c), (Courtesy of R. Stewart, University of Utah, USA).

Co-occurrence of the positively charged Pc-1 and Pc-2 proteins and the negatively charged Pc-3 protein in the tube worm cement led to the hypothesis that a phenomenon called *complex coacervation* could play a role in the condensation of the adhesive in the form of a dense water-immiscible fluid [30, 41]. Complex coacervation is the spontaneous separation of an aqueous solution of two oppositely charged polyelectrolytes into two immiscible aqueous phases, a dilute equilibrium phase and a denser solute-rich phase (Figure 12.3). Coacervation occurs when the charges of the polyelectrolytes are balanced. This phenomenon is therefore pH dependent, occurring to the maximum extent at the pH at which the solution is electrically neutral; is dependent on the ionic strength, since shielding of charges can change the charge balance of the system; and is dependent on the ratio of polyelectrolyte concentrations. Whether complex coacervation is really involved in the formation of the natural cement or not is still under debate. However, the ideal material and rheological properties of complex coacervates provide a valuable blueprint for the synthesis of bio-inspired, water-borne, underwater adhesives ([41]; see below).

12.2.3
Barnacles

Barnacles are sessile crustaceans that attach firmly and massively not only to a large variety of underwater natural substrata such as rocks (Figure 12.1a) but also to man-made substrata such as ship hulls, therefore causing major economical losses. In these organisms, attachment is mediated by the release of a permanent adhesive called *cement* [2, 44]. In acorn barnacles, this cement is produced by large isolated secretory cells (the cement cells) joined together by ducts that open onto the base of the animal [44–46]. The cement is composed of ~90% proteins, with the remainder being carbohydrates (1%), lipids (1%), and inorganic ash (4%, of which calcium accounts for 30%) [47]. More than 10 proteins have been found in the cement (cement proteins, abbreviated as cp), of which 6 have been purified and characterized, originally from the species *Megabalanus rosa* and later from other species [2, 32, 48]. Their features are summarized in Table 12.1. Among these proteins, three (cp-19k, cp-20k, and cp-68k) have a surface coupling function, two (cp-52k and cp-100k) have a bulk function, and the last one (cp-16k) is an enzyme whose possible function is the protection of the cement from microbial degradation [2, 32, 48]. Little or no posttranslational modification has been found so far in individual barnacle cps. Their strong attachment to surfaces does not therefore involve the "DOPA-system." Only two of the cps, cp-52k and cp-68k, appear to be posttranslationally modified. These proteins are indeed glycosylated, although the glycosyl moiety is limited in amount [2, 32, 48, 95]. Barnacles thus display a contrasting adhesive system compared to mussels and tube worms, which are characterized by their reliance on posttranslational modification of adhesive proteins.

12.2.4
Brown Algae

Brown algae live firmly attached on hard substrata in the subtidal and intertidal rocky shores, where they are often exposed to high gradients of turbulence. They have therefore developed adhesive strategies to attach themselves strongly and durably throughout their life cycle, from the microscopic reproductive cell stages to the large thalli of mature algae (Figure 12.1c) [49]. In contrast to invertebrate adhesive secretions, enzymatic extractions and cytochemical studies indicated that the adhesive mucilage produced by algae is mostly carbohydrate based, containing polysaccharides and glycoproteins [49, 50]. Phenolic polymers also play a role in

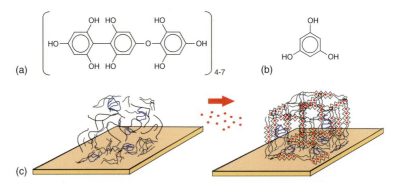

Figure 12.4 Adhesion mechanism of the brown alga *Fucus serratus*. (a) Chemical structure of a polyphenol. (b) Phloroglucinol monomer. (c) The natural adhesive is made up of polyphenols (blue) and alginate (black). In the biomimetic adhesive, the polyphenols are replaced by synthetic phloroglucinol units. In both cases, when calcium ions are added (red), a rigid network is formed. (Courtesy of H. Bianco-Peled, Technion - Israel Institute of Technology.)

brown algae adhesion. They appear to be secreted a few hours after fertilization, allowing initial substratum adhesion and, after germination, are localized at the site of adhesive mucilage formation, that is the rhizoid tip in the case of *Fucus* [51]. These polyphenols are composed of phloroglucinol units linked by carbon–carbon and ether bonds (Figure 12.4a,b) [51]. They are reminiscent of the DOPA-containing proteins of mussels and tube worms in that they can be oxidized and form cross-links. The proposed mechanism of algal adhesion indeed postulates that the polyphenols are oxidized by an extracellular vanadium bromoperoxidase to enable their cross-linking with the polysaccharide components of the adhesive, such as alginate (Figure 12.4c) [49, 51].

12.3
Recombinant Adhesive Proteins

12.3.1
Production

Recombinant DNA technology has been used in order to obtain large quantities of marine adhesive proteins (see [10, 18, 52] for review). Attempts to produce recombinant forms of MAPs started in the early 1990s with the expression, production, and purification of complete fp-1 and of synthetic fp-1 protein analogs consisting of 6–20 repeats of the consensus fp-1 decapeptide (Ala-Lys-Pro-Ser-Tyr-Pro-Pro-Thr-Tyr-Lys) in the yeast *Saccharomyces cerevisiae* or in the bacterium *Escherichia coli* [25, 53]. However, these first attempts failed to obtain large amounts of recombinant proteins (Table 12.2) for different reasons, including the highly biased amino acid composition of fp-1 and differences in

Table 12.2 Comparison of several recombinant mussel adhesive proteins.

Recombinant protein	Host	Production yield (mg l^{-1}) after purification	Solubility (g l^{-1})	References
fp-1	*Saccharomyces cerevisiae* (Y)	Nd	Nd	[18, 25]
	Escherichia coli (B)	4–10	Nd	[53–55]
fp-2	*S. cerevisiae* (Y)	Nd	Nd	[18]
fp-3	*Kluyveromyces lactis* (Y)	1	Nd	[56]
	E. coli (B)	0.8–3	1	[57]
fp-5	*E. coli* (B)	2.8	1	[58]
fp-131	*E. coli* (B)	Nd	Nd	[59]
fp-151	*E. coli* (B)	1000	330	[60]
	Sf9 cells (I)	23	Nd	[96]
fp-353	*E. coli* (B)	39	90	[61]

B, bacterium; I, insect; Y, yeast; Nd, not determined.

codon usage between mussels and the heterologous expression systems. Bacterial expression of recombinant fp-1 was enhanced by the fusion of fp-1 with an E. coli signal peptide (OmpASP), which boosts secretion by directing the expressed protein to the periplasm [54]. Another problem with the heterologous expression of fp-1 and MAPs, in general, is that the recombinant proteins lack the post-translational modifications such as the hydroxylation of proline, arginine or, most importantly, tyrosine residues. The production of functional recombinant MAPs therefore requires an additional *in vitro* modification step: the enzyme-catalyzed modification of tyrosine residues to DOPA [55]. This is usually done using a commercially available mushroom tyrosinase.

Fp-3 and fp-5 are other interesting adhesive proteins, as they have a high DOPA content and are present at the interface between the byssus attachment plaques and the substratum. Both were cloned in expression vectors containing a histidine tag sequence, which allows purification of the protein using metal affinity chromatography [57, 58]. The vectors were expressed in *E. coli*, with subsequent *in vitro* treatment by mushroom tyrosinase for DOPA formation. Recombinant fp-5 presented a low purification yield (Table 12.2) but remarkable adsorption and adhesive abilities on various surfaces. Indeed, adsorption and adhesion forces were comparable to – and sometimes exceeding – those of Cell-Tak. A similar purification yield was achieved for recombinant fp-3 (Table 12.2). The adhesion capabilities of this protein were, however, lower than those of recombinant fp-5, but still comparable to those of Cell-Tak. Recombinant fp-5 was also tested successfully as a cell-adhesion material for *in vitro* cell culture.

The successful production of recombinant MAPs and their good performance in microscale adhesion tests were encouraging. However, macroscale testing and large-scale applications were still prevented by poor production due to postinduction bacterial cell growth inhibition and by the low solubility of purified proteins in aqueous buffers (Table 12.2) [18, 52]. To overcome these limitations, recombinant hybrid adhesive proteins were designed, the so-called fp-353, fp-151, and fp-131 [59–61]. The first is a fusion protein with fp-3 at each terminus of fp-5. Because fp-353 formed inclusion bodies, host cell growth inhibition did not occur [61]. In addition, the solubility of fp-353 was better than that of fp-3 or fp-5 alone, permitting the preparation of a viscous concentrated glue solution for large-scale adhesion strength measurements. The fp-151 and fp-131 hybrids resulted from the fusion of six fp-1 decapeptide repeats at each terminus of fp-5 and fp-3, respectively [60, 61]. Recombinant fp-151 displayed the highest production yield, reaching $1\,\text{g}\,\text{l}^{-1}$ of batch-type flask culture (Table 12.2). In fed-batch-type bioreactor cultures, similarly high production levels were maintained through coexpression of fp-151 with bacterial hemoglobin, which facilitates the utilization of cellular oxygen [62]. Of all recombinant MAPs, fp-151 possesses the highest solubility (up to $330\,\text{g}\,\text{l}^{-1}$), allowing the concentration of the protein solution into a sufficiently viscous liquid for practical adhesive application [60]. In addition, fp-151 showed better adsorption and similar adhesion force, compared to recombinant fp-5, and its adhesive strength ($\sim$0.8 MPa on cowhide, $\sim$1.1 MPa on aluminum, and $\sim$1.8 MPa on poly(methyl methacrylate (PMMA) (Plexiglas)) is always largely in excess of that of the commercially available fibrin glue Tisseel™. The recombinant fusion proteins, marketed in milligram quantities (Kollodis, Inc), are less expensive than extracted MAPs (Cell-Tak) [10, 96]. Recently, fp-151 was successfully produced in insect cells (Table 12.2) [96]. Importantly, the insect-derived adhesive protein contained DOPA residues converted by *in vivo* post-translational modification, but also other post-translational modifications including phosphorylation of serine and hydroxylation of proline.

Two barnacle adhesive proteins, cp-19k and cp-20k, have been produced in bacteria at the laboratory scale. Recombinant cp-19k was demonstrated to present underwater irreversible adsorption activity to a variety of surface materials, including positively charged, negatively charged, and hydrophobic ones [33]. Recombinant cp-20k, On the other hand, was adsorbed to calcite and metal oxides in seawater, but not to glass and synthetic polymers [63]. Both proteins are functional in their recombinant forms produced in *E. coli*, and fusion with other functional proteins or biological motifs enabled their immobilization on material surfaces [48].

12.3.2
Applications

Fp-151 proved to be biocompatible for the *in vitro* culture of various cell types including both anchorage-dependent and anchorage-independent cells [60]. In this context, fp-151 was also fused to an RGD peptide, a sequence identified as a cell adhesion recognition motif by integrins, and used in culture plate coating [64]. The resulting hybrid, fp-151-RGD, not only maintained the high production yield of fp-151 but also presented superior spreading and cell-adhesion abilities compared to the commercially produced cell-adhesion materials poly-L-lysine (PLL) and Cell-Tak, and this regardless of the mammalian cell line used. The excellent adhesion and spreading abilities of fp-151-RGD might be due to the fact that it combines three types of cell-binding mechanisms: DOPA adhesion of Cell-Tak, cationic binding force of PLL, and RGD sequence-mediated adhesion of fibronectin [64]. These characteristics make the two hybrid proteins fp-151 and fp-151-RGD suitable for use as cell-adhesion material in cell culture or tissue engineering [97]. In a totally different context, fp-151 has been proposed as a potential gene delivery material in view of its similarity of amino acid composition to histone proteins, which are known as effective mediators of transfection [65]. The fusion protein displayed comparable transfection efficiency in human and mouse cells compared to the widely used transfection agent Lipofectamine™ (Invitrogen).

The tube worm cement coacervation hypothesis prompted efforts to form complex coacervates with the mussel hybrid fusion proteins fp-151 and fp-131 [59, 66]. The MAP hybrids are insoluble at physiological pH, so coacervation was done at pH 3.8 with hyaluronic acid (HA) at HA:fp-151 (or fp-131) ratios of ~1:3 [59, 66]. The coacervates showed shear-thinning viscosity, high friction coefficient, and low interfacial energy, three promising features for delivery, spreading, and adhesive properties for future wet adhesive and coating technologies [66]. Moreover, it was found that the highly condensed complex coacervates significantly increased

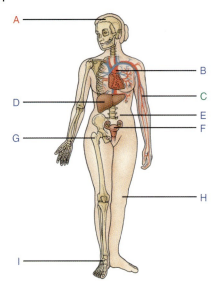

Figure 12.5 Potential medical applications of adhesive materials inspired by mussels (blue), tube worms (red), and brown algae (green). (A) Bone adhesive for the reconstruction of craniofacial features [67], (B) adhesive-coated vascular scaffold promoting endothelial cell proliferation [68], (C) vascular sealant for surgical adjunctive leakage control [69], (D) adhesive for immobilization of pancreatic transplants on liver surface [70], (E) adhesive-coated biologic meshes for hernia repair [71], (F) sealant for fetal membrane repair [72], (G) adhesive-coated titanium bone implant promoting osteoblast proliferation [66], (H) adhesive for skin incision closure [73], and (I) adhesive-coated biologic scaffold for Achilles tendon repair [74]. All these biomimetic adhesives have been successfully tested *in vitro* with cultured cells or *in vivo* in animal models.

the bulk adhesive strength of MAPs in both dry and wet environments (up to ~4 MPa on aluminum) [59]. In addition, oil droplets were successfully incorporated in the coacervate, forming a microencapsulation system that could be useful in the development of self-adhesive microencapsulated drug carriers, for use in biotechnological and biomedical applications [59]. Coacervates were also prepared using HA and fp-151-RGD. The low interfacial energy of the coacervate was exploited to coat titanium (Ti), a metal widely used in implant materials. The coacervate effectively distributed both HA and fp-151-RGD over the Ti surfaces and enhanced osteoblast proliferation (Figure 12.5) [66].

12.4
Production of Bio-Inspired Synthetic Adhesive Polymers

Another way to mimic marine adhesive proteins is the chemical synthesis of important functional parts of these adhesives. This has been done either by peptide synthesis or by the functionalization of various other polymers with adhesive protein reactive groups, mostly DOPA or catechol groups.

12.4.1
Adhesives Based on Synthetic Peptides

Various synthetic polypeptides inspired by MAPs were studied for adhesion strength (see [75] for review). This approach was used to experimentally identify the exact functions of the amino acids that are active in the chemistry of the adhesive proteins. The importance of particular amino acids overexpressed in MAP, mainly Lys and Tyr/DOPA, was therefore tested through different polytripeptides and polydipeptides such as poly(X-Tyr-Lys), poly(Gly-Lys), poly(Tyr-Lys), and poly(Gly-Tyr), each peptide being subsequently incubated with tyrosinase for the enzymatic hydroxylation of tyrosines into DOPA [76]. The adhesive strength of poly(X-Tyr-Lys) peptides, irrespective of whether X was Gly, Ala, Pro, Ser, Leu, Ile, or Phe, measured on pig skin (i.e., 0.01 MPa) was higher than that of poly(Gly-Lys) peptides but similar to the one of poly(Tyr-Lys) peptides and of a poly(fp-1 decapeptide). Tyrosine (DOPA) and lysine thus appear to be key amino acids for adhesion efficiency, with little importance of the other residues and of the primary structure of the MAPs [73]. A poly(Gly-Tyr-Lys) peptide was also used as a surgical glue to close a skin incision in a living pig (Figure 12.5). Good incision adhesion and reduced immunological response after one week were observed [73].

Peptide synthesis also allows the direct incorporation of DOPA into the polymers, thus avoiding the complications associated with the oxidation of precursor tyrosine residues [10, 14, 75]. Water-soluble copolypeptides containing DOPA and Lys residues were prepared by polymerization of α-amino acid N-carboxyanhydride monomers. Using lap shear tensile adhesive measurements, these copolymers were found to form moisture-resistant adhesive bonds to a variety of substrata after cross-linking with different oxidizing agents [77]. Adhesion was strongest to metals and polar surfaces such as glass (3–5 MPa). Adhesive strength of about 0.2 MPa was also measured on porcine skin and bone *in vitro*. Furthermore, outcomes of endothelial cell cultures demonstrated that the copolypeptide presented a good cell affinity, which would provide basic data for its application in the biomaterial field [78].

The protein cp-20k from barnacle cement was used for designing self-assembling peptides that may present a novel source of inspiration for the design of new materials [48, 79]. A single repetitive sequence from this protein has been chemically synthesized to obtain a water-soluble peptide solution. The addition of salt at a final concentration of 1 M triggered an irreversible self-assembly into a three-dimensional meshlike structure made up of interwoven nanofilaments [79].

12.4.2
Adhesives Based on Polysaccharides

The adhesive of the brown alga *Fucus* inspired a biomimetic adhesive composed of synthetic phloroglucinol units, alginate, and calcium (Figure 12.4) [80]. This synthetic adhesive was shown to be capable of adhering to a variety of both hydrophilic and hydrophobic surfaces, with strength similar to that of the alga-born natural adhesive. It also adhered well to porcine tissue and was shown to be safe for living cells [80]. The biomimetic adhesive was also tested on porcine muscle tissue and presented a tensile strength similar to that of fibrin glue [69]. It is marketed as a vascular sealant for surgical adjunctive leakage control (Figure 12.5) (SEAlantis; H. Bianco-Peled, personal communication).

12.4.3
Adhesives Based on Other Polymers

DOPA being considered as a key component of both mussel and tube worm adhesion, functionalization of synthetic polymers other than polypeptides with catechol groups has emerged as a promising strategy for the development of new biomimetic adhesives [6, 10, 81]. Following this approach, simplified polymer mimics of mussel fps have been designed, in which a polystyrene backbone takes the place of the protein polyamide chain [81]. Hardening of this poly[(3,4-dihydroxystyrene)-*co*-styrene] by treatment with various oxidizing agents including dichromate and Fe^{3+} yielded adhesive strength of up to 1.2 MPa on aluminum. Linear and branched poly(ethylene glycol) (PEG) molecules were also chemically coupled to one to four DOPA end groups [6]. PEG has a low toxicity and is nonallergenic, therefore allowing its utilization as a convenient platform in medicine. The addition of oxidizing agents resulted in the polymerization of PEG-DOPA via DOPA-quinone cross-links, forming polymer networks and rapid gelation [82]. Although these gels were not adhesive, it has been shown that unoxidized four-armed PEG polymers functionalized with a single DOPA residue at the extremity of each arm (PEG-DOPA$_4$) adsorbed strongly on mucin, this mucoadsorption being largely due to the presence of DOPA [83]. A tissue adhesive was investigated by using liposomes in order to compartmentalize an oxidizing reagent (NaIO$_4$) in a solution containing PEG-DOPA$_4$ [84]. While sodium periodate is sequestered in these vesicles at room temperature, it is released in the polymer solution when liposomes melt at wound site (37 °C). Oxidation of PEG-DOPA$_4$ resulted in hydrogel formation with interesting adhesive properties. Indeed, the lap shear strength between two porcine dermal tissues was five times higher than that of fibrin glue [84]. A similar strategy was used for efficient immobilization of transplanted pancreatic islets in mice and for fetal membrane repair (Figure 12.5), both with minimal toxic and inflammatory responses [70, 72]. Polymers made up of PEG-DOPA coupled to polycaprolactone (PCL) were also used to coat the biological meshes and scaffolds used for hernia and Achilles tendon repair, respectively (Figure 12.5) [71, 74].

On the basis of tube worm model, poly(meth)acrylate polymers were synthesized with phosphate, amine, and catechol side chains with molar ratios similar to the natural adhesive proteins Pc-1–3 [41, 85]. Analogs of Pc-1 and Pc-2 were created by copolymerization of *N*-(3-aminopropyl)methacrylamide hydrochloride (APMA) and acrylamide, while mimics of Pc-3 were synthesized by free radical copolymerization of monoacryloxyethyl phosphate (MAEP), acrylamide, and dopamine methacrylate (DMA), DMA being a DOPA analog monomer [85]. Both copolymers were mixed,

12.4 Production of Bio-Inspired Synthetic Adhesive Polymers

with divalent cations, in similar proportions as the natural proteins. Around neutral pH, they condensed into a liquid complex coacervate (Figure 12.3c), while at pH 10, they formed a hydrogel through cross-links between DOPA-quinone and amine side chains [85]. After oxidation with sodium periodate, the underwater bond strength of the biomimetic adhesive on aluminum was ~0.8 MPa, about twice the value estimated for the natural glue [41]. Moreover, a biodegradable version of the adhesive coacervate was designed by replacing the APMA/acrylamide copolymer by an amine-modified collagen hydrolysate [86]. This adhesive coacervate was used to repair rat calvarial bone defect and was capable of maintaining three-dimensional bone alignment in freely moving rats over a 12 week indwelling period [67]. Histological evaluation demonstrated that the adhesive was gradually resorbed and replaced by new bone with an inflammatory response commensurate with normal wound healing. This noncytotoxic degradable coacervate therefore appears to be suitable for use in the reconstruction of craniofacial features (Figure 12.5) [67].

The use of toxic oxidizing reagents for solidification of hydrogels represents a medical concern for *in vivo* applications [6, 82]. To avoid such reagents, DOPA was conjugated to poly(ethyleneoxide)-poly(propyleneoxide)-poly(ethyleneoxide) (PEO-PPO-PEO) block copolymers [87]. These DOPA-modified block copolymers were soluble in cold water but self-aggregated into micelles at a higher temperature, which was dependent on the block copolymer concentration. For instance, a 20 wt% solution formed gels when heated to body temperature [87]. Photopolymerization is another approach to achieve oxidation-free cross-linking, and monomers combining DOPA and a UV (or visible light)-polymerizable methacrylate group were therefore copolymerized with PEG diacrylate [88, 89]. In both heat- and light-triggered polymerization, gelation did not result from DOPA-quinone cross-links. DOPA groups were thus available for adhesion, but the adhesive strength of these hydrogels was not investigated. Self-assembly of amphiphilic triblock copolymers is also an alternative strategy to obtain oxidation-free polymerization [90]. Triblock copolymers were made up of hydrophobic endblocks consisting of PMMA and of a poly(methacrylic acid) (PMAA) water-soluble midblock. DOPA was incorporated into the hydrophilic PMAA [90]. Hydrogels were then obtained by exposing triblock copolymers solution in DMSO to water vapor. As water diffused into the solution, the hydrophobic endblocks formed aggregates that were bridged by water-soluble midblocks [90]. Underwater adhesive properties of a DOPA-modified PMMA-PMAA-PMMA membrane on titanium and pig skin were assessed and showed strong adhesion in both cases [91].

The strategy centered on the functionality of the catechol group reached its highest point with the report of a method to form multifunctional polymer coatings through simple dip coating of objects in an aqueous solution of dopamine, a DOPA analog [92]. Indeed, dopamine self-polymerization forms thin, surface-adherent polydopamine films on a wide range of inorganic and organic materials including metals, oxides, polymers, semiconductors, and ceramics. Secondary reactions can then be used to create a variety of ad-layers. These polydopamine films can promote cell adhesion on any type of material surface, including the well-known antiadhesive substrate poly(tetrafluoroethylene), and can therefore convert a variety of bioinert substrates into bioactive ones [93]. They also improve the blood compatibility of various materials [94] and promote endothelial cell growth on vascular scaffolds [68].

12.5
Perspectives

From the biologist's perspective, marine adhesives may serve one or more of the following functions: (i) the temporary or permanent attachment of an organism to a surface, including dynamic attachment during locomotion and maintenance of position; (ii) the collection or capture of food items; or (iii) the building of tubes or burrows. The evolutionary background and biology of the species on the one hand and the environmental constraints on the other hand influence the specific composition and properties of adhesive secretions in a particular organism. The diversity of marine adhesives is therefore huge and daunting for researchers but, in return, it will also allow a great deal of flexibility in applications. To date, only a very limited number of marine organisms have been used as models for the development of biomimetic adhesives, and most of this diversity remains unexploited. The study of new model organisms representative of the different types of adhesion is therefore required in order to extract essential structural, mechanical, and chemical principles from their adhesives. Simplification of natural systems while retaining the desired efficacy is, indeed, one of the foundation stones of biomimetics. As quoted by Herb Waite [3], it is no longer absurd to predict that a compilation of bio-inspired adhesive designs will someday fill a textbook with recipes for a whole range of medical applications. Indeed, the necessity to repair and seal different organs, will require tissue adhesives specifically adapted to both tissue surface chemistry and mechanical challenges during organ function.

Acknowledgments

This work was supported in part by the "Service Public de Wallonie – Programme Winnomat 2" and by the "Communauté française de Belgique – Actions de Recherche Concertées." E. Hennebert and P. Flammang are, respectively, postdoctoral researcher and senior research associate of the Fund for Scientific Research of Belgium (F.R.S.-FNRS). This study is a contribution of the "Centre Interuniversitaire de Biologie Marine" (CIBIM).

References

1. Waite, J.H., Andersen, N.H., Jewhurst, S., and Sun, C. (2005) *J. Adhes.*, **81**, 297–317.
2. Kamino, K. (2008) *Mar. Biotechnol.*, **10**, 111–121.
3. von Byern J. and Grunwald, I. (eds) (2010) *Biological Adhesive Systems-From Nature to Technical and Medical Application*, Springer, Wien.
4. Walker, G. (1987) in *Synthetic Adhesives and Sealants* (ed. W.C. Wake), John Wiley & Sons, Ltd, Chichester, pp. 112–135.

5. Smith, A.M. and Callow, J.A. (eds) (2006) *Biological Adhesives*, Springer-Verlag, Berlin.
6. Lee, B.P., Dalsin, J.L., and Messersmith, P.B. (2006a) in *Biological Adhesives* (eds A.M. Smith and J.A. Callow), Springer, Berlin, pp. 257–278.
7. Strausberg, R.L. and Link, R. (1990) *Trends Biotechnol.*, **8**, 53–57.
8. Donkerwolcke, M., Burny, F., and Muster, D. (1998) *Biomaterials*, **19**, 1461–1466.
9. Blume, J. and Schwotzer, W. (2010) in *Biological Adhesive Systems-From Nature to Technical and Medical Application* (eds J. von Byern and I. Grunwald), Springer, Wien, pp. 213–224.
10. Stewart, R.J. (2011) *Appl. Microbiol. Biotechnol.*, **89**, 27–33.
11. Robin, J.B., Picciano, P., Kusleika, R.S., Salazar, J., and Benedict, C. (1988) *Arch. Ophthalmol.*, **106**, 973–977.
12. Burzio, L.O., Bursio, V.A., Silva, T., Burzio, L.A., and Pardo, J. (1997) *Curr. Opin. Biotechnol.*, **8**, 309–312.
13. Taylor, S.W. and Waite, J.H. (1997) in *Protein-Based Materials* (eds K. McGraph and D. Kaplan), Birkhäuster, Boston, pp. 217–248.
14. Wilker, J.J. (2010) *Curr. Opin. Chem. Biol.*, **14**, 276–283.
15. Shao, J.Y., Ting-Beall, H.P., and Hochmuth, R.M. (1998) *Proc. Natl. Acad. Sci.*, **95**, 6797–6802.
16. Schedlich, L.J., Young, T.F., Firth, S.M., and Baxter, R.C. (1998) *J. Biol. Chem.*, **273**, 18347–18352.
17. Wiegemann, M. (2005) *Aquat. Sci.*, **67**, 166–176.
18. Silverman, H.G. and Roberto, F.F. (2007) *Mar. Biotechnol.*, **9**, 661–681.
19. Waite, J.H. (1983) *Biol. Rev.*, **58**, 209–231.
20. Tamarin, A., Lewis, P., and Askey, J. (1976) *J. Morphol.*, **149**, 199–222.
21. Benedict, C.V. and Waite, J.H. (1986) *J. Morphol.*, **189**, 261–270.
22. Waite, J.H. (2002) *Integr. Comp. Biol.*, **42**, 1172–1180.
23. Zhao, H. and Waite, J.H. (2006a) *J. Biol. Chem.*, **281**, 26150–26158.
24. Lee, H., Scherer, N.F., and Messersmith, P.B. (2006b) *Proc. Natl. Acad. Sci.*, **103**, 12999–13003.
25. Filpula, D.R., Lee, S.M., Link, R.P., Strausberg, S.L., and Strausberg, R.L. (1990) *Biotechnol. Prog.*, **6**, 171–177.
26. Inoue, K. *et al.* (1995) *J. Biol. Chem.*, **270**, 6698–6701.
27. Papov, V.V. *et al.* (1995) *J. Biol. Chem.*, **270**, 20183–20192.
28. Zhao, H. and Waite, J.H. (2006) *Biochemistry*, **45**, 14223–14231.
29. Waite, J.H. and Qin, X.-X. (2001) *Biochemistry*, **40**, 2887–2893.
30. Zhao, H., Sun, C., Stewart, R.J., and Waite, J.H. (2005) *J. Biol. Chem.*, **280**, 42938–42944.
31. Kamino, K. and Shizuri, Y. (1998) in *New Developments in Marine Biotechnology* (eds Y. Le Gal and H. Halvorson), Plenum Press, NewYork, pp. 77–80.
32. Kamino, K. (2006) in *Biological Adhesives* (eds A.M. Smith and J.A. Callow), Springer-Verlag, Berlin, pp. 145–166.
33. Urushida, Y., Nakano, M., Matsuda, S., Inoue, N., Kanai, S., Kitamura, N., Nishino, T., and Kamino, K. (2007) *FEBS J.*, **274**, 4336–4346.
34. Kamino, K. (2001) *Biochem. J.*, **356**, 503–507.
35. Kamino, K. *et al.* (2000) *J. Biol. Chem.*, **275**, 27360–27365.
36. Rouse, G. and Pleijel, F. (eds) (2001) *Polychaetes*, Oxford University Press, Oxford.
37. Vovelle, J. (1965) *Arch. Zool. Exp. Gén.*, **106**, 1–187.
38. Stewart, R.J., Weaver, J.C., Morse, D.E., and Waite, J.H. (2004) *J. Exp. Biol.*, **207**, 4727–4734.
39. Stevens, M.J., Steren, R.E., Vlamidir, H., and Stewart, R.J. (2007) *Langmuir*, **20**, 5045–5049.
40. Endrizzi, B.J. and Stewart, R.J. (2009) *J. Adhes.*, **85**, 546–559.
41. Stewart, R.J., Wang, C.S., and Shao, H. (2010) *Adv. Colloid Interface Sci.*, doi:
42. Sun, C., Shrivastava, A., Reifert, J.R., and Waite, J.H. (2009) *J. Adhes.*, **85**, 126–138.
43. Sun, C., Fantner, G.E., Adams, J., Hansma, P.K., and Waite, J.H. (2007) *J. Exp. Biol.*, **210**, 1481–1488.
44. Walker, G. (1981) *J. Adhes.*, **12**, 51–58.
45. Lacombe, D. (1970) *Biol. Bull.*, **139**, 164–179.

46. Power, A.M., Klepal, W., Zheden, V., Jonker, J., McEvilly, P., and von Byern, J. (2010) in *Biological Adhesive Systems-From Nature to Technical and Medical Application* (eds J. von Byern and I. Grunwald), Springer, Wien, pp. 153–168.
47. Walker, G. (1972) *J. Mar. Biol. Assoc. U.K.*, **52**, 429–435.
48. Kamino, K. (2010) *J. Adhes.*, **86**, 96–110.
49. Potin, P. and Leblanc, C. (2006) in *Biological Adhesives* (eds A.M. Smith and J.A. Callow), Springer-Verlag, Berlin, pp. 105–124.
50. Oliveira, L., Walker, D.C., and Bisalputra, T. (1980) *Protoplasma*, **104**, 1–15.
51. Vreeland, V., Waite, J.H., and Epstein, L. (1998) *J. Phycol.*, **34**, 1–8.
52. Cha, H.J., Hwang, D.S., and Lim, S. (2008) *Biotechnol. J.*, **3**, 1–8.
53. Salerno, A.J. and Goldberg, I. (1993) *Appl. Microbiol. Biotechnol.*, **39**, 221–226.
54. Lee, S.J., Han, Y.H., Nam, B.H., Kim, Y.O., and Reeves, P.R. (2008) *Mol. Cells*, **26**, 34–40.
55. Kitamura, M., Kawakami, K., Nakamura, N., Tsumoto, K., Uchiyama, H., Ueda, Y., Kumagai, I., and Nakaya, T. (1999) *J. Polym. Sci., Part A: Polym. Chem.*, **37**, 729–736.
56. Platko, J.D. et al. (2008) *Protein Expression Purif.*, **57**, 57–62.
57. Hwang, D.S., Gim, Y., and Cha, H.J. (2005) *Biotechnol. Prog.*, **21**, 965–970.
58. Hwang, D.S., Yoo, H.J., Jun, J.H., Moon, W.K., and Cha, H.J. (2004) *Appl. Environ. Microbiol.*, **70**, 3352–3359.
59. Lim, S., Choi, Y.S., Kang, D.G., Song, Y.H., and Cha, H.J. (2010) *Biomaterials*, **31**, 3715–3722.
60. Hwang, D.S., Gim, Y., Yoo, H.J., and Cha, H.J. (2007a) *Biomaterials*, **28**, 3560–3568.
61. Gim, Y., Hwang, D.S., Lim, S., Song, Y.H., and Cha, H.J. (2008) *Biotechnol. Prog.*, **24**, 1272–1277.
62. Kim, D., Hwang, D.S., Kang, D.G., Kim, J.Y.H., and Cha, H.J. (2008) *Biotechnol. Prog.*, **24**, 663–666.
63. Mori, Y., Urushida, Y., Nakano, M., Uchiyama, S., and Kamino, K. (2007) *FEBS J.*, **274**, 6436–6446.
64. Hwang, D.S., Sim, S.B., and Cha, H.J. (2007) *Biomaterials*, **28**, 4039–4046.
65. Hwang, D.S., Kim, K.R., Lim, S., Choi, Y.S., and Cha, H.J. (2009) *Biotechnol. Bioeng.*, **102**, 616–623.
66. Hwang, D.S., Waite, J.H., and Tirrell, M. (2010) *Biomaterials*, **31**, 1080–1084.
67. Winslow, B.D., Shao, H., Stewart, R.J., and Tresco, P.A. (2010) *Biomaterials*, **31**, 9373–9381.
68. Ku, S.H. and Park, C.B. (2010) *Biomaterials*, **31**, 9431–9437.
69. Bitton, R., Josef, E., Shimshelashvili, I., Shapira, K., Seliktar, D., and Bianco-Peled, H. (2009) *Acta Biomater.*, **5**, 1582–1587.
70. Bilic, G., Brubaker, C., Messersmith, P.B., Mallik, A.S., Quinn, T.M., Haller, C., Done, E., Gucciardo, L., Zeisberger, S.M., Zimmermann, R., Deprest, J., and Zisch, A.H. (2010) *Am. J. Obstet. Gynecol.*, **202**, 85.e1–85.e9.
71. Murphy, J.L., Vollenweider, L., Xu, F., and Lee, B.P. (2010) *Biomacromolecules*, **11**, 2976–2984.
72. Brubaker, C.E., Kissler, H., Wang, L.J., Kaufman, D.B., and Messersmith, P.B. (2010) *Biomaterials*, **31**, 420–427.
73. Tatehata, H., Mochizuki, A., Ohkawa, K., Yamada, M., and Yamamoto, H. (2001) *J. Adhes. Sci. Technol.*, **15**(9), 1003–1013.
74. Brodie, M., Vollenweider, L., Murphy, J.L., Xu, F., Lyman, A., Lew, W.D., and Lee, B.P. (2011) *Biomed. Mater.*, **6**, doi:
75. Deming, T.J. (2007) *Prog. Polym. Sci.*, **32**, 858–875.
76. Tatehata, H., Mochizuki, A., Kawashima, T., Yamashita, S., and Yamamoto, H. (2000) *J. Appl. Polym. Sci.*, **76**, 929–937.
77. Yu, M. and Deming, T.J. (1998) *Macromolecules*, **31**, 4739–4745.
78. Wang, J., Liu, C., Lu, X., and Yin, M. (2007) *Biomaterials*, **28**, 3456–3468.
79. Nakano, M., Shen, J.R., and Kamino, K. (2007) *Biomacromolecules*, **8**, 1830–1835.
80. Bitton, R. and Bianco-Peled, H. (2008) *Macromol. Biosci.*, **8**, 393–400.
81. Westwood, G., Horton, T.N., and Wilker, J.J. (2007) *Macromolecules*, **40**, 3960–3964.

82. Lee, B.P., Dalsin, J.L., and Messersmith, P.B. (2002) *Biomacromolecules*, **3**, 1038–1047.
83. Catron, N.D., Lee, H., and Messersmith, P.B. (2006) *Biointerphases*, **1**, 134–141.
84. Burke, S.A., Ritter-Jones, M., Lee, B.P., and Messersmith, P.B. (2007) *Biomed. Mater.*, **2**, 203–210.
85. Shao, H., Bachus, K.N., and Stewart, R.J. (2008) *Macromol. Biosci.*, **8**, 464–471.
86. Shao, H. and Stewart, R.J. (2010) *Adv. Mater.*, **22**, 729–733.
87. Huang, K., Lee, B.P., Ingram, D.R., and Messersmith, P.B. (2002) *Biomacromolecules*, **3**, 397–406.
88. Lee, B.P., Huang, K., Nunalee, F.N., Shull, K.R., and Messersmith, P.B. (2004) *J. Biomater. Sci., Polym. Ed.*, **15**, 449–464.
89. Lee, B.P., Chao, C.Y., Nunalee, F.N., Motan, E., Shull, K.R., and Messersmith, P.B. (2006) *Macromolecules*, **39**, 1740–1748.
90. Guvendiren, M., Messersmith, P.B., and Shull, K.R. (2008) *Biomacromolecules*, **9**, 122–128.
91. Guvendiren, M., Brass, D.A., Messersmith, P.B., and Shull, K.R. (2009) *J. Adhes.*, **85**, 631–645.
92. Lee, H., Dellatore, S.M., Miller, W.M., and Messersmith, P.B. (2007) *Science*, **318**, 426–430.
93. Ku, S.H., Ryu, J., Hong, S.K., Lee, H., and Park, C.B. (2010) *Biomaterials*, **31**, 2535–2541.
94. Wei, Q., Li, B., Yi, N., Su, B., Yin, Z., Zhang, F., Li, J., and Zhao, C. (2010) *J. Biomed. Mater. Res. A*, **96**, 38–45.
95. Kamino, K., Nakano, M., Kanai, (2012) *FEBS J.*, **279**, 1750–1760.
96. Lim, S., Kim, K.R., Choi, Y.S. Kim, D.K., Hwang, D. Cha, H.J., (2011) *Biotechnol. Prog.*, **27**, 1390–1396.
97. Hong, J.M., Kim, B.J., Shim, J.-H. Kang, K.S., Kim, K.-J. Rhie, J.W., Cha, H.J. Cho, D.-W., (2012) *Acta Biomater.*, **8**, 2578–2586.

Part III
Hard and Mineralized Systems

13
Interfacial Forces and Interfaces in Hard Biomaterial Mechanics
Devendra K. Dubey and Vikas Tomar

13.1
Introduction

Biological materials have evolved over millions of years and are often found as complex composites with superior properties compared to their relatively weak original constituents. The toughness of spider silk, the strength and lightweight of bamboos, self-healing of bone, high toughness of nacre, and the adhesion abilities of the gecko's feet are a few of the many examples of high-performance natural materials. Hard biomaterials such as bone, nacre, and dentin have intrigued researchers for decades for their high stiffness, toughness, multifunctionality, and self-healing capabilities. For example, nacre has 3000 times more toughness compared to its mineral constituent [1]. Tooth enamel is 1000 times stiffer than its constituent protein polymer collagen [2]. The general mechanical performance of these composites is quite remarkable. In particular, they combine two properties that are usually quite contradictory but essential for the function of these materials. Bones, for example, need to be stiff to prevent bending and buckling, but they must also be tough since they should not break catastrophically even when the load exceeds the normal range. Such hard biological materials are not only light weight but also possess high toughness and mechanical strength.

One of the defining features of such biological composites is that they are highly hierarchical with different structures at different length scales. Often, they are complex nanocomposites of soft fibrous polymeric phase and hard mineral phase. For instance, bone has up to seven levels of hierarchy [3, 4] (Figure 13.1), and nacre shows up to six levels of hierarchial structure (Figure 13.2). Materials such as bone and nacre have such multilevel hierarchical structural design that concept of stress concentration at flaws remains invalid, leading to flaw-tolerant structure [5]. In spite of complex hierarchical structures, the smallest building blocks in such biological materials are at the nanometer length scale. For example, at the lowest level in bone, nanometer-sized crystals of carbonate apatite are embedded in the fibrous protein collagen in a well-organized staggered arrangement (Figure 13.1f).

In nanocomposite materials, the volume fraction of the protein–mineral interface can be enormous as the mineral bits have nanoscale size. For example, in a

Biomimetic Approaches for Biomaterials Development, First Edition. Edited by João F. Mano.
© 2012 Wiley-VCH Verlag GmbH & Co. KGaA. Published 2012 by Wiley-VCH Verlag GmbH & Co. KGaA.

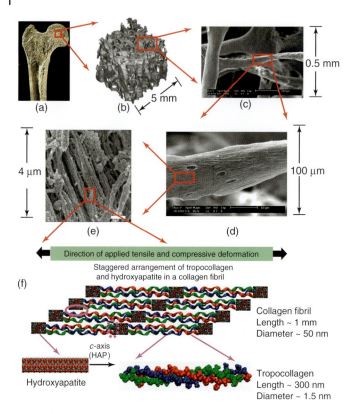

Figure 13.1 A representation of the hierarchical structure in bone, (a) longitudinal cross section of the end of human femur showing the trabecular bone inside, (b) microstructure of trabecular bone resembling the honeycomb structure, (c) high-resolution view of the strutlike structures in microporous structure, (d) showing one such strut called *trabecula*, having aligned mineralized collagen fibers and lacunae (cavities left after cell apoptosis), (e) fracture surface of human bone showing mineralized collagen fibrils, and (f) a schematic of staggered and layered assembly of TC molecules and HAP platelets to form a mineralized collagen fibril. Three different colors in TC molecules depict three polypeptide chains forming a triple helix. HAP crystal's *c*-axis is along the loading direction. Solid and dotted ellipses in pink show two possible types of interactions between TC and HAP surface, one with TC perpendicular to HAP surface and other with TC parallel to HAP surface. Images in (b–d) are borrowed from [6] and (e) from [7].

raindrop size volume of a nanocomposite, the area of interfacial region can be as large as a football field [9]. Interfaces play crucial role in regulating the overall mechanical properties of nanocomposites. In the case of hard biomaterials such as bone, dentin, and nacre, they have primarily an organic phase (e.g., tropocollagen (TC) or chitin) and a mineral phase (e.g., hydroxyapatite (HAP) or aragonite) arranged in a staggered arrangement. In bone, the crystalline mineral phase is preferentially aligned along the longitudinal axis of the polypeptide molecules permitting maximum contact area in a staggered arrangement [10–13] (Figure 13.1). The extent of interfacial interaction and the interfacial arrangement

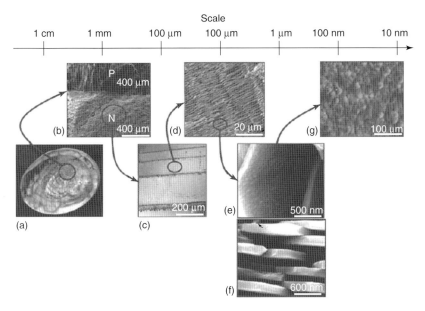

Figure 13.2 Hierarchical organization in nacre showing at least six structural levels. This image is borrowed from [8]. However, authors themselves borrowed parts of this image from other references.

are important determinants of the structure–function property relationship of biomaterials and influence the mechanical strength substantially [14–16]. Such biological materials have been reviewed in appreciable detail, in the context of their hierarchical structure, material properties, and failure mechanisms [3, 17–19].

An important aspect to focus in biomaterials engineering of hard biomaterials is the chemomechanics of the organic–inorganic interfaces and its correlation with overall mechanical behavior. This understanding is vital for selecting appropriate constituents, their size scales, and their relative arrangements, which in turn is governed by the functional requirements of the composite materials. For example, a three-dimensional (3D) explicit atomistic failure analyses of model TC–HAP interfacial biomaterial (similar to material found in bone tissues) performed at the nanoscopic length scale [10, 20]) have pointed out that maximizing the contact area between the TC and HAP phases result in higher interfacial strength as well as higher fracture strength. Analyses have also shown that high toughness and strain-hardening behavior of such biomaterial is due to reconstitution of columbic interactions between TC and HAP surfaces during interfacial sliding due to mechanical deformation. Recently, it has also been shown that changes in the residue sequences of TC molecules at the interface can affect the material mechanical strength considerably [14–16, 21]. Furthermore, it has been shown earlier that moisture can play a major role in affecting the strength of such hybrid interfaces in biological materials [22–25].

Challenges lie in identifying nature's mechanisms behind imparting such properties and its pathways in fabricating and optimizing these composites. The route frequently acquired by nature is by embedding submicrometer- or nanosized mineral particles in protein matrix in a well-organized hierarchical arrangement. The key here is the formation of large amount of precisely and carefully designed organic–inorganic interfaces and synergy of mechanisms acting over multiple scales to distribute loads and damage, dissipate energy, and resist change in properties owing to damages such as cracking.

The length scale and complexity of microstructure of hybrid interfaces in biological materials makes it difficult to study them and understand the underlying mechanical principles that are responsible for their extraordinary mechanical performance. For this reason, the governing mechanisms for the mechanical behavior for such biomaterials are not understood completely. At the same time, such building-block-level understanding is important not only for the evolution of biological materials science but also to the development of bio-inspired materials. Attractive mechanical properties of hard biological materials have given birth to biomimetism and new biomaterials, which have been reviewed in past by many [2, 19, 26–33].

This chapter presents a brief overview of the role of interfacial structural design and interfacial forces in imparting superior mechanical performance to hard biological materials. Focus is on understanding the underlying engineering principles of nature's materials for use in biomedical engineering and biomaterials development.

13.2
Hard Biological Materials

Bone is an excellent example of stiff biological nanocomposite materials. From the mechanical strength viewpoint, its main constituents are TC molecules and biological HAP mineral. Biological HAP is a poorly crystalline impure form of the compound HAP (chemical formula: $Ca_5(PO_4)_3OH$), containing constituents such as carbonate, citrate, magnesium, fluoride, and strontium [34, 35]. Despite the fact that TC is a soft phase and HAP is brittle, together they form a material of high mechanical strength and fracture toughness [36]. The structure of bone is organized over several length scales, with six to seven levels of hierarchy. Figure 13.1 shows the several hierarchical features from macroscopic to atomic scale. TC molecules assemble into collagen fibrils in a hydrated environment, which mineralize by the formation of HAP crystals in the gap regions that exist owing to the staggered arrangement. These mineralized collagen fibrils (MCF s) assemble together with extrafibrillar matrix to form next hierarchical layer of bone. While the structures at scales larger than MCFs vary for different bone types, MCFs are highly conserved, nanostructural primary building blocks of bone that are found universally [4, 37–39]. As shown in Figure 13.1, at the fundamental scale, collagen fibrils are formed by staggered self-assembly of 300 nm long TC molecules and mineral HAP occupies the gap regions along with water. Collagen fibril consists of some other constituents as well (such as noncollagenous proteins, glycans, mineral impurities, etc.). However, interfacial interactions between the TC and the HAP phases along with their hierarchical arrangement are thought to be the most important factors imparting high mechanical strength [40]. As shown using dotted and solid ellipses in Figure 13.1, a staggered arrangement is responsible for possible tension–shear type of load transfer between TC molecules and HAP crystals in bone and in similar other biomaterials [41]. There have been growth- and mineralization-based explanations behind existence of such an arrangement [11, 13, 42]. However, it may also be possible that the existence of such structural arrangement is driven by its important role in imparting high mechanical strength.

Nacre is another example of high toughness natural biomaterials found in the interior layers of the hard outer shells of mollusks. The exterior layers of the shell are typically brittle, but hard, and provide resistance to penetration from external impact [43, 44]. While nacre, present in the inner layers of shell, provides toughening by dissipating the mechanical energy owing to its capability of undergoing large inelastic deformations [45]. Main constituents of nacre are aragonite (a crystallographic form of calcium carbonate, 95% by volume) and elastic organic material (proteins such as chitin and polysaccharides). Nacre has a hierarchical brick-and-mortar structure (Figure 13.2) with polygonal aragonite tablets as bricks and organic matrix as the mortar acting as the filler and adhesive between such tablets [46]. The aragonite tiles are roughly 0.5 μm thick and 5–8 μm in diameter. In some of the cases, such as red abalone, nacre shows some level of organization across different layers, where tablets are stacked in columns with some overlap between tablets from adjacent columns. The interface between the tablets is roughly 30 nm thick and is composed of organic materials [47], nanoasperities [45], and direct mineral bridges connecting two adjacent tablets [48, 49]. In fact, the tablets themselves are formed by aragonite nanograins (Figure 13.2g), which again are separated by a very fine three-dimensional network of organic material [50, 51].

13.2.1
Role of Interfaces in Hard Biomaterial Mechanics

The high toughness in bone is attributed to the ability of its microstructure to dissipate deformation energy without propagation of crack. Different toughening mechanisms have been reported for bone [52], such as crack deflection and crack blunting at the interlamellar interfaces, formation of nonconnected microcracks ahead of the crack tip, and crack bridging in the wake zone of the crack. Important contributions to high toughness and defect tolerance of natural biomineralized composites are believed to arise from the nanometer-scale structural motifs such as collagen fibrils in bone. When under loading, both mineral nanoparticles and mineralized fibrils deform at first elastically but to different degrees [39]. The hierarchical nature of bone deformation is exemplified by a staggered model of load transfer in bone matrix, which is shown in Figure 13.3. The long and thin (100–200 nm diameter) mineralized fibrils lie parallel to each other and are separated by a thin layer (1–2 nm thick) of extrafibrillar matrix [40]. When external tensile load is applied to the tissue, it is resolved into a tensile deformation of

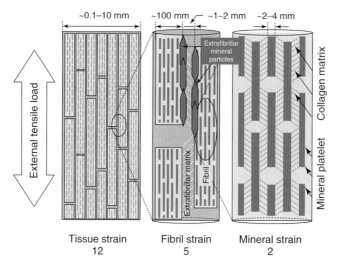

Figure 13.3 Schematic model for bone deformation in response to external tensile load at three levels in the structural hierarchy: at the tissue level (left), fibril array level (center), and mineralized collagen fibrils (right). (Center) The stiff mineralized fibrils deform in tension and transfer the stress between adjacent fibrils by shearing in the thin layers of extrafibrillar matrix (white dotted lines show direction of shear in the extrafibrillar matrix). The fibrils are covered with extrafibrillar mineral particles, shown only over a selected part of the fibrils (red hexagons) so as not to obscure the internal structure of the mineralized fibril. (Right) Within each mineralized fibril, the stiff mineral platelets deform in tension and transfer the stress between adjacent platelets by shearing in the interparticle collagen matrix (red dashed lines indicate shearing qualitatively and do not imply homogeneous deformation). (Adapted from [39].)

the mineralized fibrils and a shearing deformation in the extrafibrillar matrix. The mineralized phase provides strengthening and organic phase provides toughness. The deformation in the mineralized matrix not only occurs at the collagen fibril level but also occurs at all other hierarchical levels and different length scales. The interaction of the mineralized and organic components produces a synergistic effect that enhances mechanical properties.

In nacre, aragonite tablets and organic material is organized in such a manner that if a load is applied perpendicular to the plane of tablets, the organic material arrests the uncontrolled growth of crack by deforming inelastically and dissipating energy [53–55]. In this mechanism, the tablets approximately remain linear elastic and the large deformations are provided by significant shearing of the organic–inorganic interface. The presence of organic material between nanograins in a single tablet might also provide some resistance to tablet sliding. However, this viscoelastic mechanism is active only in hydrated state of nacre. A comparison of measurement of mechanical properties of nacre [1, 45, 53, 56–58] and of individual tablets of nacre using nanoindentation [50, 57, 59] has shown that both of them have properties similar to single-crystal aragonite, which is stiff but brittle. Considering that 95% of nacre is aragonite, dry nacre behaves very much similar to pure aragonite because organic material becomes brittle. Apart from the above-mentioned mechanism, a few other mechanisms and factors are also proposed, which contribute to hardening, energy dissipation, and damage distribution. Similar to protein unfolding domains in bone [7, 60], it is suggested that proteins in nacre also have folded modules and act as high-performance adhesives, binding the tablets, and possibly the nanograins together [54]. During deformation under loading, these biopolymers elongate in stepwise manner and absorbing significant amount of energy. The interlocking between mineral platelets [61] increases toughness by progressively failing and limiting the catastrophic failure [62]. The waviness found in mineral tablets can be significant [2], and progressively locks tablet sliding, leading to a more difficult separation of tablets from their interfaces.

13.2.2
Modeling of TC–HAP and Generic Polymer–Ceramic-Type Nanocomposites at Fundamental Length Scales

Several studies have been performed to understand mechanical behavior of bone and bonelike nanocomposite materials. Deformation, damage, and failure mechanisms at nanoscale seem to be as important as the mechanisms at macroscopic length scale. The mechanical behavior of bone with a view to understand the role of TC molecules and HAP mineral has been earlier analyzed using experiments, modeling, and simulations. Experimental approaches have focused on analyzing tensile failure of single collagen fibers and fibrils [40, 63, 64] and on analyzing structural features at the nanoscale and its relation with the bone tissue failure [7, 65, 66]. Eppell and coworkers used a microelectromechanical device to obtain the first stress–strain curve of an isolated collagen fibril. They reported a low-strain Young's modulus of 0.5 GPa and a high-strain Young's modulus of ∼12 GPa for the collagen fibril. Hansma and coworkers found that collagen in bone contains sacrificial bonds, which may be partially responsible for the toughness of bone. The time needed for these sacrificial bonds to reform after pulling correlated with the time needed for bone to recover its toughness as measured by the indentation testing based on atomic force microscopy. Modeling using the continuum approaches has focused on understanding the role played by the shear strength of TC molecules and the tensile strength of HAP mineral in flaw-tolerant hierarchical structural design of biomaterials [37, 67].

Explicit atomistic simulations using methods such as MD allow us to work at the lowest fundamental length scale and reveal structure–property relationships as well as mechanisms at the building block level. Previously, such MD schemes have focused on understanding the mechanical behavior and properties of TC molecules in different structural configurations [68–70], the hierarchical organization of TC molecules into collagen fibrils and its effect on mechanical properties [21, 71], the properties of hydrated TC molecules [24, 72], and the TC molecule stability with respect to changes in residue sequences [73].

13.2.2.1 Analytical Modeling

One of the interesting studies performed by Ji and Gao [36] proposes an analytical model called tension shear chain model (TSC), explaining the load transfer mechanism in generic hierarchical nanocomposite materials systems such as bone and bonelike materials [67]. Under uniaxial tension, the path of load transfer in the staggered nanostructure follows a TSC with mineral platelets under tension and the soft matrix under shear (Figure 13.4). The load transfer is largely accomplished by the high shear zones of protein between the long sides of mineral platelets. Under an applied tensile stress, the mineral platelets carry most of the tensile load, while the protein matrix transfers the load between mineral crystals. The mineral crystals have large aspect ratios and are much harder than the soft protein matrix, and the tensile zone in protein matrix near the ends of mineral crystals is assumed to carry no mechanical load. The work has addressed the prevalence and importance of nanoscale features in biological composites and also developed methodology for estimating interfacial shear strength, fracture localization width, and characteristic length of mineral crystal. A fractal-based analysis of the TSC model is also performed to gain understanding on the different hierarchical structural levels. The analysis shows that nanometer size of brittle mineral crystals is essential to make them insensitive to cracklike flaws [5].

The observation that the building block level structures in biological materials are always at the nanoscale has intrigued researchers for a long time. Gao and coworkers [5] found that the nanoscale dimension of a mineral may be the result of fracture strength optimization. The fracture strength becomes sensitive to cracklike flaws when the mineral size exceeds nanoscopic length scale and fails by flaw propagation under stress concentration at crack tips. Perhaps Nature finds this secret of optimum fracture strength and maximum flaw tolerance by evolution

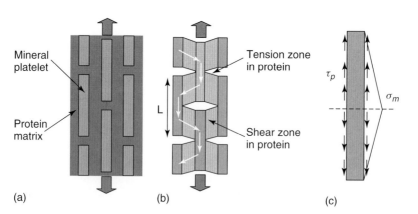

Figure 13.4 Models of biocomposites. (a) Perfectly staggered mineral inclusions embedded in protein matrix. (b) A tension–shear chain model of biocomposites in which the tensile regions of protein are eliminated to emphasize the load transfer within the composite structure. (c) The free body diagram of a mineral crystal. Images are borrowed from [36].

and hides mineral defects by designing the fundamental biomaterials structure at the nanoscale to achieve the robustness needed for survival.

Among collagenous tissues such as bone and tendon, the common structural feature is the MCF acting as a building block. Accurate modeling of such MCF is important and will require the right measurements of the dimensions, mechanical properties, and relative arrangement of the nano-length-scale constituents. Collagen molecules form ∼100–200 nm diameter fibrils, with mineral particles inside and outside the surface [74]. Most studies describe the mineral particles to be plate shaped with a wider range of geometrical dimensions. The thickness of the platelets can range from 2 to 7 nm, the length from 15 to 200 nm, and the width from 10 to 80 nm [38, 39]. TC molecule, of course, is one of the fundamental building blocks

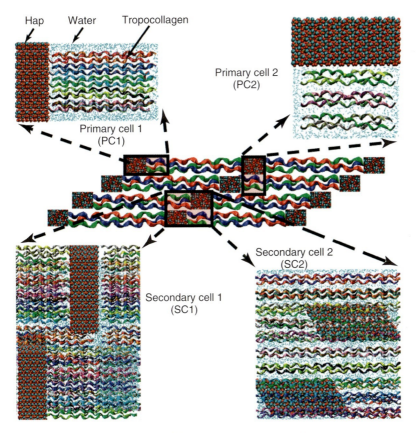

Figure 13.5 A schematic showing the derivation of PC1, PC2, SC1, and SC2 cells from the staggered and layered assembly. In PC1 and SC1 cells, tropocollagen molecules are aligned in a direction parallel to the c-axis of hydroxyapatite crystals. In PC2 and SC2 cells, tropocollagen molecules are aligned in a direction normal to the longitudinal c-axis of hydroxyapatite super cell. Dimensions of hydroxyapatite crystals are approximately the same in all cells. Tropocollagen molecules are all shown in multicolor segments, and water molecules are shown in cyan.

and has a length of ∼300 nm. Both HAP mineral and TC molecules are arranged in a staggered manner (as shown in Figure 13.1f) with a gap of ∼67 nm between two successive TC molecules creating space for mineral nucleation and growth [4]. With the current state-of-the-art methods, complete measurements for anisotropic stress-strain-strength properties for biological hydroxyapatitic mineralite and TC molecule are not available. Without these measurements, current models generally utilize elastic moduli measured for hydroxyapatitic polycrystallites. This does not lead to an accurate modeling of mineralized fibril because the hydroxyapatitic mineral in calcified tissue is usually nanocrystals.

13.2.2.2 Atomistic Modeling

In the structural studies of hard biological materials, it is observed that the mineral crystals tend to have a preferential alignment with respect to the loading direction and longitudinal axis of the polypeptide molecules [11–13, 75]. For example, in bone, both HAP crystal largest dimension (c-axis) and TC longitudinal axis are aligned along the maximum load bearing direction [76] with a specific staggered structural arrangement (Figures 13.1 and 13.5), permitting a maximum contact area. Resulting nanoscale interfacial interactions between TC and HAP phases is a strong determinant of the strength of such materials. Recent works by Dubey and Tomar [10, 20, 77, 78] present a mechanistic understanding of such interfacial interactions by examining idealized TC and HAP interfacial biomaterials. A three-dimensional atomistic modeling framework is developed, which combines both organic and inorganic cells together to form a supercell as shown in Figure 13.5. Two levels of hierarchical simulation supercells corresponding to Level n and Level $n + 1$ (Figure 13.5) were generated with two different possible types of interfaces (Figure 13.1f). Secondary supercells are formed by embedding corresponding primary supercells in TC matrix. Two different loadings, transverse and longitudinal, were applied to all supercells. For this purpose, quasi-static mechanical deformation of each supercell is performed and characteristic stress–strain curves were obtained (Figure 13.6) [10, 20]. In addition, both tensile and compressive loading cases were considered and compared (Figure 13.6). Furthermore, effect of hydration on such TC–HAP biomaterials was also studied. For failure analysis, TSC model [36, 41] was used to estimate the interfacial shear strength and fracture localization zone width.

The Young's modulus value for TC molecule was obtained to be ∼9 GPa, which is a fairly good estimate and lies in the range of modulus values obtained by other researchers [10]. The analyses confirm that relative alignment of TC molecules with respect to the HAP mineral surface such that the interfacial contact area is maximized, along with optimal direction of applied loading with respect to the TC–HAP interface orientation are important factors that contribute to making nanoscale staggered arrangement a preferred structural configuration in such biological materials. The analyses also point out that such an arrangement results in higher interfacial strength as well as higher fracture strength. In addition, such TC–HAP nanocomposite shows toughening and strain hardening behavior, which is attributed to the reconstitution of Columbic interactions between TC and HAP at the interface during sliding. The dominant tensile failure mechanism at the

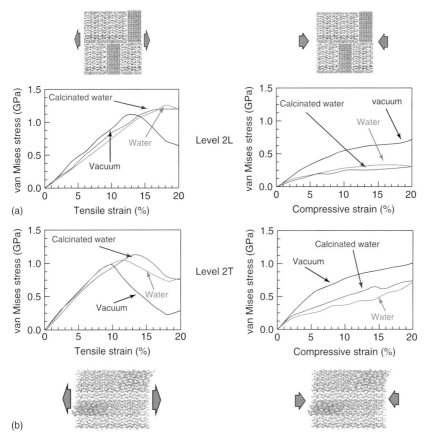

Figure 13.6 Tensile and compressive von Mises stress versus strain plots as a function of chemical environment in the case of (a) supercell layer 2L loaded in longitudinal direction and (b) supercell Level 2T loaded in transverse direction.

HAP–TC interface is simply the interfacial separation of TC and HAP without significant initial HAP deformation (Figure 13.6). The NH_3^+ and COO^- groups in TC molecules are strongly attracted to the ions in HAP surface (Ca^{2+}, PO_4^{3-}, and OH^- ions) [79]. Since TC is a flexible chainlike molecule, it elongates on applied deformation but cleaves off after the point when it is fully stretched. Such cleavage results in local nanoscale interfacial failure.

Most biological materials in physiological systems contain water as one of their constituents. It has been observed that the presence of water molecules in the TC–HAP biocomposites enhanced overall composite mechanical strength (Figure 13.6a) [10]. This is attributed to water molecule's affinity for charged surfaces, such as HAP surface [79], and NH_3^+ and COO^- groups in TC, owing to its polar nature and capability of make strong hydrogen bonds. As a result, water acts as an electrostatic bridge between HAP surface and TC molecules and strengthens the TC–HAP interface. This strengthening especially plays an important role whenever there is a relative sliding occurs between HAP surface and TC molecules at the interface. Previous studies have shown that hydration has a stabilizing effect on the collagen triple helix [22] and solvated TC molecule requires more energy to untie from the HAP surface [80]. Similar interaction behavior of water at protein–mineral interface is found in nacre as well [81, 82]. It also acts as glue between TC–TC interactions [24], and thereby, delays the failure of the overall system. Another interesting finding emerged out of a comparison between stress–strain curves for two different hierarchical levels is that the failure of such biocomposites is predominantly strain dependent and not a function of ultimate strength.

Bone diseases such as osteogenesis imperfecta are marked by extreme bone fragility and are associated with point mutations in the TC molecule. Also, there is been a debate as to whether the HAP crystals in bone tissues are plate shaped or needle shaped. Hence, a further investigation into the effect of change in mineral crystal shape and effect of change in TC residue sequences on the mechanical strength of TC–HAP biomaterials was performed. Results show that TC–HAP interface shear strength increases as the side group complexity and heterogeneity of residues increases in the TC–HAP nanocomposite, and the plate-shaped crystals are overall better in resisting load as compared to needle-shaped HAP crystals case [78]. However, the effect of change in mineral crystal morphology has a stronger effect on the mechanical strength of the TC–HAP biocomposites, as compared to change in TC residue sequences [83]. This suggests that probably mutations in TC manifest its effect by changing the mineral crystal morphology and distribution during nucleation and growth period over the life time of the animal.

13.3
Bioengineering and Biomimetics

Materials science researchers are intensively seeking to learn the engineering and design principles from Nature's hard biological materials to develop novel high-performance man-made composites. Many experiments have been carried out in this direction. Some of these efforts are discussed in this chapter. The ordered brick-and-mortar arrangement is considered to be the key strength and toughness determining structural feature of nacre. It has also been shown that the ionic cross-linking of tightly folded macromolecules is equally important. Tang and coworkers [84] have demonstrated that both structural features can be reproduced by sequential deposition of polyelectrolytes and clays (Figure 13.7). The resulting organic–inorganic hybrid reproduced both the brick-and-mortar arrangement and the crucial effect of the sacrificial bonds. The ultimate tensile strength and Young modulus of such a composite approached that of nacre and lamellar bones, respectively. This makes such materials potential candidates for bone implant applications. Further, this layer-by-layer assembly is used to develop antibacterial biocompatible version of above organic–inorganic material [85].

The architecture of nacre is such that if a stress normal to the platelet plane is applied, then the ductile organic fraction that glues the crystals will prevent

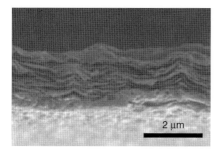

Figure 13.7 Examples of "artificial" nacre, clay-/polyelectrolyte-layered material. (Image adapted from [84].)

an uncontrolled crack growth [55]. This viscoelastic dissipative process takes place mainly in the hydrated state, and dry nacre behaves very much similar to pure aragonite because the organic material becomes brittle [2]. The transition from a brittle (glassy) state to a more elastomeric behavior was observed in a polysaccharide (chitosan) on hydration using nonconventional dynamic mechanical analysis, in which a glass transition could be detected in intermediate hydration levels [86]. Photoacoustic Fourier transform infrared spectroscopy results suggest that the water present at the nanograin interfaces also contributes significantly to the viscoelasticity of nacre [87]. It was also suggested that proteins composing the organic matrix, such as Lustrin A, exhibit a highly modular structure characterized by a multidomain architecture with folded modules and act as a high-performance adhesive, binding the platelets together [88]. On increasing stress, these biopolymers elongate in a stepwise manner as folded domains or loops are pulled open, with significant energy requirements to unfold each individual module – it was claimed that this modular elongation mechanism may contribute to the amazing toughness properties in nacre [88]. Other toughening pathways are also present such as the presence of Asperities on the surface of the aragonite tablets [89], mineral bridges [48], interlocks [90], waved surfaces [2], nanoscale nature and organization of the building blocks constituting nacre [17], and rotation and deformation of aragonite nanograins [91].

The overall function and properties of hard biological composite materials are a complicated coupled function of the structural arrangement at different hierarchical levels and individual constituent material properties. In general, it is impractical to separate structure and properties of such biocomposites to explain the material behavior. The success in mimicking even relatively simple structures, such as nacre, has been relatively modest yet. The best examples [84, 85] are far away from providing a millimeter-thick artificial nacre layer with greatly enhanced fracture toughness, a relatively simple task a mollusk has been routinely doing during its life cycle. Unlike artificial nanocomposite materials built using optimal interfacial arrangements, biological materials show remarkable strength compared to their relatively weaker constituents. In fact, sometimes artificial nanocomposites can turn out to be weaker than its constituents. Two important aspects of the biological material design arise here, which differentiate the artificial and natural nanocomposites: (i) prevalence of carefully designed organic–inorganic interfaces at almost all hierarchical levels and (ii) multilevel hierarchy with different microstructure design at different levels. First aspect points out that biological materials have extremely large amount of interfacial area. Moreover, these interfaces are formed by specific relative arrangement between the organic phases and inorganic phases for optimum load handling. Second, different microstructural design at each level provides further optimization/customization parameters for desired mechanical performance. This suggests that an in-depth understanding of chemomechanics of such interfaces, their role in load handling, and failure mechanisms at the fundamental length scales is vital to the development of both the bioengineering and biomimetics fields.

13.4
Summary

Hard biological materials such as nacre and bone exhibit remarkable mechanical performance despite the fact that they are made up of relatively very weak constituents. The two main features identified for such behavior are interfacial nanostructural design and complex hierarchy. From an engineering standpoint, they are capable of inspiring a next generation of composite materials with high strength and toughness. This requires a clear understanding of Nature's engineering design principles, fabrication pathways, and selection of appropriate materials for creating such biological composites. In terms of the underlying mechanical principles for structural design of these nanocomposites, quite a few have been suggested. For example, one principle is the alignment of mineral–protein interface along the loading directions. MD study of TC–HAP biomaterials shows that a composite is best poised to handle the load if the protein molecules are in contact with mineral crystals having their longitudinal axis parallel to the mineral surface and along the loading direction of the composite. Second principle is the staggered arrangement of hard mineral crystals in soft protein matrix, leading to a unique mechanism of load transfer where crystals bear the normal load and protein transfers the load via shear. Third principle is that the failure of such polymer–ceramic type composites is dominantly peak strain dependent instead of peak strength. Another interesting observation is that such biomaterials become flaw tolerant at nanoscale owing to special crack deflection and crack strengthening pathways. Also, the presence of moisture at the interface enhances the stability and strength of such biomaterials by supporting the cross-linking mechanism because of the polar nature of water molecule.

One common feature that strongly stands out in most hard biological materials structures is the presence of interfaces at multiple levels of hierarchy. It seems that Nature has designed these interfaces for optimum multifunctional performance during the course of evolution. Interfacial forces play key role during deformation and failure of such biomaterials. Such interfacial interaction between the soft phase and hard phase is responsible for redistribution of stresses and directly affects the toughness and strength of the biocomposite material. Furthermore, the design of the polymer–mineral interface along with the critical length of mineral constituent also contributes potentially in strengthening the biocomposite against failure and in affecting the overall mechanical performance. Unlike Nature that has a relatively restricted set of materials to choose from, engineers have more choices available for selecting materials to form composites. In order to capitalize on this advantage, a good understanding of nanoscale mechanics and structure–property function relationship of hard biological materials is vital. Therefore, characterization of such biological interfacial designs and interactions between the constituents is critical to the development of bio-inspired materials. Primarily because to swap materials in composites design one should understand and predict the effects of specific material selections and design on the overall mechanical performance of biomaterials. This is becoming feasible with increasing experimental and modeling efforts in this active research area. However, further accurate measurements and analysis is required to capture the totality of all the factors for realistic biomimicking materials development.

References

1. Currey, J.D. (1977) *Proc. R. Soc. London, Ser. B, Biol. Sci.*, **196**(1125), 443–463.
2. Barthelat, F. (2007) *Philos. Trans. R. Soc. A Math. Phys. Eng. Sci.*, **365**, 2907–2919.
3. Rho, J.Y., Kuhn-Spearing, L., and Zioupos, P. (1998) *Med. Eng. Phys.*, **20**(2), 92–102.
4. Weiner, S. and Wagner, H.D. (1998) *Ann. Rev. Mater. Sci.*, **28**, 271–298.
5. Gao, H.J., Ji, B.H., Jager, I.L., Arzt, E., and Fratzl, P. (2003) *Proc. Natl. Acad. Sci. U.S.A.*, **100**(10), 5597–5600.
6. Niebur, G. http://www.nd.edu/~gniebur.
7. Fantner, G.E., Hassenkam, T., Kindt, J.H., Weaver, J.C., Birkedal, H., Pechenik, L., Cutroni, J.A., Cidade, G.A.G., Stucky, G.D., Morse, D.E., and Hansma, P.K. (2005) *Nat. Mater.*, **4**(8), 612–616.
8. Luz, G.M. and Mano, J.F. (2009) *Philos. Trans. R. Soc. A Math. Phys. Eng. Sci.*, **367**(1893), 1587–1605.
9. Vaia, R. (2001) *MRS Bull. (USA)*, **26**(5), 394–401.
10. Dubey, D.K. and Tomar, V. (2009) *Acta Biomater.*, **5**(7), 2704–2716.
11. Landis, W.J., Hodgens, K.J., Arena, J., Song, M.J., and McEwen, B.F. (1996) *Microsc. Res. Tech.*, **33**(2), 192–202.
12. Landis, W.J., Hodgens, K.J., Song, M.J., Arena, J., Kiyonaga, S., Marko, M., Owen, C., and Mcewen, B.F. (1996) *J. Struct. Biol.*, **117**, 24–35.
13. Fratzl, P., Fratzlzelman, N., Klaushofer, K., Vogl, G., and Koller, K. (1991) *Calcif. Tissue Int.*, **48**(6), 407–413.
14. Dubey, D.K. and Tomar, V. (2009) *J. Phys.: Condens. Matter*, **21**, 205103, (13 pp).
15. Dubey, D.K. and Tomar, V. (2009) *Acta Biomater.*, doi:10.1016/j.actbio.2009.02.035
16. Dubey, D.K. and Tomar, V. (2009) *J. Mech. Phys. Solids*, **57**(10), 1702–1717.
17. Fratzl, P. and Weinkamer, R. (2007) *Prog. Mater. Sci.*, **52**, 1263–1334.
18. Meyers, M.A., Chen, P.Y., Lin, A.Y.M., and Seki, Y. (2008) *Prog. Mater. Sci.*, **53**, 1–206.
19. Launey, M.E., Munch, E., Alsem, D.H., Barth, H.B., Saiz, E., Tomsia, A.P., and Ritchie, R.O. (2009) *Acta Mater.*, **57**(10), 2919–2932.
20. Dubey, D.K. and Tomar, V. (2009) *J. Mech. Phys. Solids*, **57**(10), 1702–1717.
21. Buehler, M.J. (2008) *J. Mech. Behav. Biomed. Mater.*, **1**, 59–67.

22. Simone, A.D., Vitaglaino, L., and Berisio, R. (2008) *Biochem. Biophys. Res. Commun.*, **372**, 121–125.
23. Bhowmik, R., Katti, K.S., and Katti, D.R. (2007) *Mater. Res. Soc. Symp. Proceed.*, **978**(6), Paper #: 0978-GG14-05-FF09-05.
24. Zhang, D., Chippada, U., and Jordan, K. (2007) *Ann. Biomed. Eng.*, **35**, 1216–1230.
25. Neves, N.M. and Mano, J.F. (2005) *Mater. Sci. Eng. C Biomimet. Supramol. Syst.*, **25**(2), 113–118.
26. Sarikaya, M., Tamerler, C., Jen, A.K.Y., Schulten, K., and Baneyx, F. (2003) *Nat. Mater.*, **2**(9), 577–585.
27. Ratner, B.D. and Bryant, S.J. (2004) *Annu. Rev. Biomed. Eng.*, **6**, 41–75.
28. Sanchez, C., Arribart, H., and Guille, M.M.G. (2005) *Nat. Mater.*, **4**(4), 277–288.
29. Fratzl, P. (2007) *J. R. Soc. Interface*, **4**(15), 637–642.
30. Bhushan, B. (2009) *Philos. Trans. R. Soc. A Math. Phys. Eng. Sci.*, **367**(1893), 1445–1486.
31. Launey, M.E. and Ritchie, R.O. (2009) *Adv. Mater.*, **21**(20), 2103–2110.
32. Munch, E., Launey, M.E., Alsem, D.H., Saiz, E., Tomsia, A.P., and Ritchie, R.O. (2008) *Science*, **322**(5907), 1516–1520.
33. Palmer, L.C., Newcomb, C.J., Kaltz, S.R., Spoerke, E.D., and Stupp, S.I. (2008) *Chem. Rev.*, **108**(11), 4754–4783.
34. Leventouri, T. (2006) *Biomaterials*, **27**(18), 3339–3342.
35. Cowin, S.C. (2001) *Bone Mechanics Handbook*, CRC Press, Boca Raton, FL.
36. Ji, B. and Gao, H. (2004) *J. Mech. Phys. Solids*, **52**, 1963–2000.
37. Jager, I. and Fratzl, P. (2000) *Biophys. J.*, **79**(4), 1737–1746.
38. Fratzl, P., Gupta, H.S., Paschalis, E.P., and Roschger, P. (2004) *J. Mater. Chem.*, **14**(14), 2115–2123.
39. Gupta, H.S., Seto, J., Wagermaier, W., Zaslansky, P., Boesecke, P., and Fratzl, P. (2006) *Proc. Natl. Acad. Sci. U.S.A.*, **103**, 17741–17746.
40. Gupta, H.S., Wagermaier, W., Zickler, G.A., Aroush, D.R.-B., Funari, S.S., Roschger, P., Wagner, H.-D., and Fratzl, P. (2005) *Nano Lett.*, **5**(10), 2108–2111.
41. Ji, B.H. (2008) *J. Biomech.*, **41**(2), 259–266.
42. Wenk, H.R. and Heidelbach, F. (1999) *Bone*, **24**(4), 361–369.
43. Currey, J.D. and Taylor, J.D. (1974) *J. Zool.*, **173**(3), 395–406.
44. Sarikaya, M. and Aksay, I.A. (1995) *Biomimetic, Design and Processing of Materials. Polymers and Complex Materials*, American Institute of Physics, Woodburry, NY.
45. Wang, R.Z., Suo, Z., Evans, A.G., Yao, N., and Aksay, I.A. (2001) *J. Mater. Res.*, **16**(9), 2485–2493.
46. Sarikaya, M. (1994) *Microsc. Res. Tech.*, **27**(5), 360–375.
47. Schaffer, T.E., IonescuZanetti, C., Proksch, R., Fritz, M., Walters, D.A., Almqvist, N., Zaremba, C.M., Belcher, A.M., Smith, B.L., Stucky, G.D., Morse, D.E., and Hansma, P.K. (1997) *Chem. Mater.*, **9**(8), 1731–1740.
48. Song, F., Zhang, X.H., and Bai, Y.L. (2002) *J. Mater. Res.*, **17**(7), 1567–1570.
49. Song, F., Zhang, X.H., and Bai, Y.L. (2002) *J. Mater. Sci. Lett.*, **21**(8), 639–641.
50. Li, X.D., Chang, W.C., Chao, Y.J., Wang, R.Z., and Chang, M. (2004) *Nano Lett.*, **4**(4), 613–617.
51. Rousseau, M., Lopez, E., Stempfle, P., Brendle, M., Franke, L., Guette, A., Naslain, R., and Bourrat, X. (2005) *Biomaterials*, **26**(31), 6254–6262.
52. Peterlik, H., Roschger, P., Klaushofer, K., and Fratzl, P. (2006) *Nat. Mater.*, **5**(1), 52–55.
53. Jackson, A.P., Vincent, J.F.V., and Turner, R.M. (1988) *Proc. R. Soc. London, Ser. B Biol. Sci.*, **234**(1277), 415–440.
54. Smith, B.L., Schaffer, T.E., Viani, M., Thompson, J.B., Frederick, N.A., Kindt, J., Belcher, A., Stucky, G.D., Morse, D.E., and Hansma, P.K. (1999) *Nature*, **399**(6738), 761–763.
55. Sumitomo, T., Kakisawa, H., Owaki, Y., and Kagawa, Y. (2008) *J. Mater. Res.*, **23**(5), 1466–1471.
56. Barthelat, F., Tang, H., Zavattieri, P.D., Li, C.M., and Espinosa, H.D. (2007) *J. Mech. Phys. Solids*, **55**(2), 306–337.
57. Barthelat, F., Li, C.M., Comi, C., and Espinosa, H.D. (2006) *J. Mater. Res.*, **21**(8), 1977–1986.

58. Menig, R., Meyers, M.H., Meyers, M.A., and Vecchio, K.S. (2000) *Acta Mater.*, **48**(9), 2383–2398.
59. Bruet, B.J.F., Qi, H.J., Boyce, M.C., Panas, R., Tai, K., Frick, L., and Ortiz, C. (2005) *J. Mater. Res.*, **20**(9), 2400–2419.
60. Thompson, J.B., Kindt, J.H., Drake, B., Hansma, H.G., Morse, D.E., and Hansma, P.K. (2001) *Nature*, **414**(6865), 773–776.
61. Katti, K.S., Katti, D.R., Pradhan, S.M., and Bhosle, A. (2005) *J. Mater. Res.*, **20**(5), 1097–1100.
62. Katti, K.S. and Katti, D.R. (2006) *Mater. Sci. Eng. C Biomimet. Supramol. Syst.*, **26**(8), 1317–1324.
63. Sasaki, N. and Odajima, S. (1996) *J. Biomech.*, **29**(5), 655–658.
64. Eppell, S.J., Smith, B.N., Kahn, H., and Ballarini, R. (2005) *J. R. Soc. Interface*, **3**, 117–121.
65. Hodge, A.J. and Petruska, J.A. (1963) in *Aspects of Protein Structure. Proceedings of a Symposium* (ed. G.N. Ramachandran), London, New York: Academic Press. pp. 289–300.
66. Thurner, P.J., Erickson, B., Jungmann, R., Schriock, Z., Weaver, J.C., Fantner, G.E., Schitter, G., Morse, D.E., and Hansma, P.K. (2007) *Eng. Fract. Mech.*, **74**(12), 1928–1941.
67. Gao, H. (2006) *Int. J. Fract.*, **138**, 101–137.
68. Buehler, M.J. (2006) *J. Mater. Res.*, **21**(8), 1947–1962.
69. Buehler, M.J. (2007) *Nanotechnology*, **18**, 295102–295110.
70. Lorenzo, A.C. and Caffarena, E.R. (2005) *J. Biomech.*, **38**, 1527–1533.
71. Israelowitz, M., Rizvi, S.W.H., Kramer, J., and von Schroeder, H.P. (2005) *Protein Eng. Des. Sel.*, **18**(7), 329–335.
72. Handgraaf, J.W. and Zerbetto, F. (2006) *Proteins: Struct., Funct., Bioinformatics*, **64**(3), 711–718.
73. Radmer, R.J. and Klein, T.E. (2006) *Biophys. J.*, **90**, 578–588.
74. Hassenkam, T., Fantner, G.E., Cutroni, J.A., Weaver, J.C., Morse, D.E., and Hansma, P.K. (2004) *Bone*, **35**(1), 4–10.
75. Weiner, S., Talmon, Y., and Traub, W. (1983) *Int. J. Biol. Macromol.*, **5**(6), 325–328.
76. Landis, W.J., Song, M.J., Leith, A., McEwen, L., and McEwen, B.F. (1993) *J. Struct. Biol.*, **110**(1), 39–54.
77. Dubey, D.K. and Tomar, V. (2009) *Mater. Sci. Eng. C Mater. Biol. Appl.*, **29**(7), 2133–2140.
78. Dubey, D.K. and Tomar, V. (2010) *J. Mater. Sci. Mater. Med.*, **21**(1), 161–171.
79. Posner, A.S. and Beebe, R.A. (1975) *Semin. Arthritis Rheum.*, **4**(3), 267–291.
80. Bhowmik, R., Katti, K.S., and Katti, D.R. (2007) *Mater. Res. Soc. Symp. Proc.*, **978**, 6.
81. Barthelat, F. and Espinosa, H.D. (2007) *Exp. Mech.*, **47**(3), 311–324.
82. Ghosh, P., Katti, D.R., and Katti, K.S. (2007) *Biomacromolecules*, **8**(3), 851–856.
83. Dubey, D.K. and Tomar, V. (2010) *Appl. Phys. Lett.*, **96**(1–3), 023703.
84. Tang, Z.Y., Kotov, N.A., Magonov, S., and Ozturk, B. (2003) *Nat. Mater.*, **2**(6), 413–4U8.
85. Podsiadlo, P., Paternel, S., Rouillard, J.M., Zhang, Z.F., Lee, J., Lee, J.W., Gulari, L., and Kotov, N.A. (2005) *Langmuir*, **21**(25), 11915–11921.
86. Benesch, J., Mano, J., and Reis, R. (2008) *Tissue Eng. Part B Rev.*, **14**(4), 433–445.
87. Verma, D., Katti, K., and Katti, D. (2007) *Spectrochim. Acta, Part A: Mol. Biomol. Spectrosc.*, **67**(3–4), 784–788.
88. Wustman, B.A., Weaver, J.C., Morse, D.E., and Evans, J.S. (2003) *Connect. Tissue Res.*, **44** (Suppl. 1), 10–15.
89. Luz, G.M. and Mano, J.F. (2009) *Philos. Trans. R. Soc. London, Ser. A*, **367**(1893), 1587–1605.
90. Katti, D.R., Ghosh, P., Schmidt, S., and Katti, K.S. (2005) *Biomacromolecules*, **6**, 3267–3282.
91. Li, X., Xu, Z.-H., and Wang, R. (2006) *Nano Lett.*, **6**(10), 2301–2304.

14
Nacre-Inspired Biomaterials
Gisela M. Luz and João F. Mano

14.1
Introduction

Living organisms present uncountable examples of natural structural composites hierarchically organized to accomplish a specific function. The most outstanding feature of these structures is their mechanical properties, which are much greater than those of the individual constituents of the assembly. Minerals are one important fraction of these partnerships and have been produced by organisms belonging to all five kingdoms for more than 3500 million years. Often, these are the very brittle ones such as calcium carbonates, calcium phosphates, and amorphous silica [1].

Another important component of these natural composites is the ductile organic matrixes such as keratin, collagen, and chitin. The ceramic–organic assembly is the base of amazing structures such as antler, enamel, dentin, sea shells, egg shells, and nacre. These are all examples of mineralized structures having in common a main essential element, calcium, and an organic component that will act as glue.

Concerning minerals, one can find an impressive assortment of shapes, compositions, structures, and functions. From the nanocrystals found in bone to the large crystals found in the echinoderm spines, the variety is immense. These materials may exhibit clear anisotropic or almost isotropic properties.

Biogenic minerals, secreted within the living organisms, may be found in either crystalline or amorphous form. Sometimes, these minerals appear combined with considerable amounts of organic materials to form materials such as bone. On the other hand, some mollusk shells almost do not include any materials besides the mineral constituents.

This diversity and the properties resulting from it became a source of inspiration for the development of new biomimetic materials, constructed with the aim of achieving some of the remarkable mechanical properties found in the natural structures [2–6].

Some of the common characteristics found in structural mineralized natural composites would be of astonishing value if reproduced within a laboratory. The main features are the mild processing conditions and energy-saving processes. The bottom-up approach for the production of biominerals is carried out through the self-assembly of elementary units under mild conditions of ambient temperature, neutral, or physiological pH, always in an aqueous environment. Moreover, biology uses information and structure as the main vector for constructing materials, as compared with synthetic technologies that solve problems largely by manipulating the use of energy [7]. Another major concern is the optimization of interfaces between hard and soft materials. Even in wet conditions, the organic components that link the mineral's constituents exhibit interesting viscoelastic properties and fatigue performance in nature. Finally, structural natural materials exhibit densities that rarely exceed 3 cm^{-3} [8] and heavy elements, such as metals, are almost absent, whereas synthetic structural materials often have densities in the 4–10 g cm^{-3} range.

Part of the success of natural composites is due to their hierarchically organized structure, at the nano-, micro-, and mesolevels, normally related to the design. Natural structures exhibit a high control over the orientation of structural elements from the molecular level up to the final macroscopic structure [9]. The properties and composition along the material may vary, gradually or abruptly, adapting to the local requirements. They exhibit functionally graded properties. An example is the bilayer structure of some shells composed of nacreous aragonite and prismatic calcite [10]. In addition, this type of structure allows for multifunctionality and regeneration. For example, bone offers structural support for the body as well as it forms blood cells, maintaining its self-healing ability, through the continuous action of osteoblasts and osteoclasts.

Finally, widely variable properties are attained from apparently similar elementary units. For example, calcium minerals, especially different forms of calcium carbonates, represent ~50% of all known biogenic minerals and are produced by completely diverse living organisms [11].

Depending on the complexity of the mineral formed, its biogenesis involves confinement in a space, ion pump control, construction of nucleation sites, and control of orientation and morphology of the material. During the biomineralization process, the local environment will provide the required elements that will be taken up and incorporated by the system in order to form a new structure. The mechanisms of biomineralization involve the production of inorganic components mediated by actin present in cells that are involved in the mineralization process. The calcium phosphates will be deposited according to a very well-organized three-dimensional organic template composed of macromolecules, proteins, or polysaccharides, typically rich in carboxylate groups and, in many cases, phosphate or sulfate groups. The charged groups of these elements will interact with the mineral ions in solution or with the surface of the solid phase [10, 11].

One of the most studied mineralization models is the mollusk shell, especially those of bivalves and gastropods. These organisms present plenty of different morphological types of shell structures. For instance, in bivalve shells, it is possible to distinguish the simple prismatic, aragonite prismatic, nacreous, foliated, composite prismatic, crossed lamellar, complex crossed lamellar, and homogeneous structures [12].

Nacre is the iridescent material that composes pearls. When nacre is found in the inner layer of shells, it is called *mother of pearl*. Besides its beauty, nacre also has other peculiar characteristics that have been attracting the interest of scientists and materials researchers, namely, its notable structure and surprising excellent mechanical properties. The nacreous layer of seashells is mainly composed of $CaCO_3$ crystals conjugated with an organic matrix. The peculiar arrangement of the inorganic and organic compounds reminds one of the brick-and-mortar display and may be the reason for nacre's success. Such structure is among the reasons for the excellent mechanical properties of nacre and became the motivation for producing new layered nanocomposites through biomimetic strategies that could improve the bioactive materials that are currently being produced.

When a material is defined as bioactive, that is, if the material is in contact with body fluid, it develops a calcium phosphate layer able to bond chemically with the bone, assuring a good integration between the material and the host tissue [13]. Bioactive glasses (such as Bioglass®) and ceramics (sintered hydroxyapatite and β-tricalcium phosphate) belong to this category [13]. To obtain bioactive materials, these materials are combined with biodegradable polymers, using biomimetic methodologies [14, 15]. The mineralization of these composites can be controlled using temperature [16] or pH [17] as external stimuli.

Bioactive nanocomposites may also be produced using nanosized bioactive particles. Glass–ceramic nanoparticles are easily produced through sol–gel processes, which allow the use of these biodegradable composites for tissue engineering applications [18–20]. Owing to their size, such nanoparticles may also be applied in the production of bioactive coatings based on the nacre structure [21].

Similar to synthetic bioactive materials, nacre itself integrates well into bone tissue [22, 23] and may stimulate the differentiation of stem cells into the osteoblast lineage [24, 25]. Recurring to mild condition chemical methodologies, nacre coatings or seashells may be transformed into apatite [26, 27], assuring a variety of applications for nacre in the biomedical field, namely, in the orthopedic and dental areas.

Moreover, specific terminology emerged to describe processes related to mimicry of biomineralization [28].

- *Biomimetic materials synthesis* refers to methods very close to nature, either using living organisms or using the respective materials isolated from organisms to prepare inorganic products.
- *Bio-inspired materials synthesis* is used when concepts found in natural biomineralization are applied to the preparation of inorganic products using synthetic materials.
- *Material bionics* is related to the general imitation of structural features found in nature by any processing technique.

This chapter emphasizes nacre's most significant features, namely, the structure and the mechanical properties, and focuses on the latest and more relevant attempts to produce new nacre-inspired materials, coatings, and nanomaterials.

14.2
Structure of Nacre

Nacre presents a hierarchical organization (Figure 14.1), a common characteristic in Nature. The building blocks arrange themselves to produce an overall structure in which the nanosized details are of major importance to the final characteristics of the material. Having as example an abalone shell (Figure 14.1a), a cross section of the shell would evidence the existence of a mesostructure. Two types of layers can be seen: a prismatic calcite layer (P) and an inner nacreous aragonite layer (N) (Figure 14.1b). These layers have thickness of 0.3 mm [29]. This structural arrangement is for protection. While the outer layer avoids disruption of the shell, the inner layer is capable of dissipating mechanical energy through inelastic dissipation [30].

Moving to the next level in nacre's structural hierarchy, one can find mesolayers resulting from a periodic growth. These "growth bands" (Figure 14.1c) act as powerful crack deflectors [32], playing therefore an important role in setting the mechanical properties. A viscoelastic organic layer can be found between the mesolayers and may appear in periods of less calcification. The following level is defined by the columnar organization of the microsized aragonite platelets. These polygonal crystals will form the "bricks" of the structure (Figure 14.1d) [29].

In the abalone shell, the aragonite platelets are organized in columns, in which the intertablet boundaries form tessellated bands perpendicular to the lamellae boundaries. However, in other cases, such as in the pearl oyster, the intertablet boundaries present a random distribution. Although nacre's organization is one of the most studied, it is not the most common mineral organization of shells. The most widespread structure in mollusks would be the crossed lamellar structure in

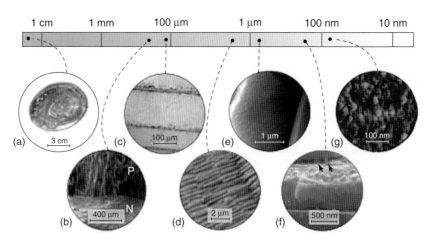

Figure 14.1 Hierarchical organization in nacre showing at least six structural levels. (Source: Images original sources: (b), (e), and (g) were reprinted with permission from Ref. [31], Copyright 2004 American Chemical Society; (c) and (d) were reprinted with permission from Ref. [32]; and (f) from Ref. [33].)

which the mineral tiles are not parallel to each other and are instead organized in a tessellated (tweed) pattern [29]. This structure may be the cause of the superior mechanical properties found in seashells owing to the enhanced ability for crack arrest at interlamellar boundaries across the different hierarchical levels [34, 35]. Knowing that all seashells of different organisms benefit from improved mechanical properties related to the biocomposite structure, it is believed that even the cross-lamellar structure, based on anisotropic basic units, will maintain an isotropic character at the macroscopic level, because of the material design organization [36].

Going down to the nanolevel, one can find 0.55 versus 10 µm aragonite tiles (Figure 14.1e) intercalated with the organic "mortar" corresponding to a volume fraction of ~5%. Some proteins composing the organic component control the shell formation process by organizing in the overall three-dimensional structure the amorphous minerals that will evolve to defined crystalline phases, and the organization of the crystals in the overall three-dimensional structure. There is a close contact between peptide chains and the inorganic fraction, and this association is mainly established through amino acids containing carboxylic acid groups [37].

Within the various models attributed to the organic fraction, it is believed that interlamellar sheets are mainly composed of highly oriented β-chitin fibrils, and next to such fibrils, discontinuous layers (or patches) of aspartic-acid-rich proteins might be detected; the silklike proteins should mainly be present within the sheets, and intracrystalline acidic glycoproteins would also be present within the mineral tablets [38].

Through the organic matrix layers, one can find some nanosized mineral columns (see arrows in Figure 14.1f) randomly distributed on the surface of the aragonite tablets. Such mineral bridges have diameters of ~46 nm and a height of 26 nm [39]. At the surface of the aragonite platelets, nanoasperities have an important role in the mechanical properties of nacre [40]. Moving forward into the nanoscale features of nacre structure, we find the last level of the hierarchical organization: nanograins with an average size of 32 nm inserted in a fine three-dimensional network of organic material [24, 31] (Figure 14.1g) compose the aragonite platelet. The nanograins are co-oriented within each platelet, which results in single crystal behavior [41]. The growth direction of such crystal is not oriented parallel to the c-axis of the individual tablet, and there are columns of co-oriented tablets, with the number of tablets per co-oriented column varying between 1 and 40. Such observations are consistent with a model for nacre growth that includes the formation of randomly distributed nucleation sites, preformed on organic matrix layers before tablet nucleation and growth [41]. In order to explain the mechanism of formation of individual nacre tablets [42], three main hypotheses are set: (i) single crystal growth, (ii) coherent aggregation of nanograins, and (iii) phase transformation from amorphous carbonate or metastable vaterite. From observations of the growth process of nacrelike tablets *in vitro*, a multistep process was suggested, involving the formation of an amorphous calcium carbonate layer, an iso-oriented growth of nanostacks, and their assembly into hexagonal tablets [42]. The growth of consecutive aragonite platelets proceeds through successive arrest

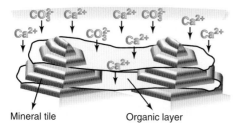

Figure 14.2 Schematic representation of growth mechanism of nacre showing intercalation of mineral tiles and organic layer arranged in a Christmas-tree pattern.

of growth by means of a protein-mediated mechanism followed by reinitiation in a Christmas-tree pattern [29, 43] (Figure 14.2). The organic fraction of nacre plays an important role in the mineralization process, as organic scaffolding was observed during the steady-state growth of tiled aragonite [33]. In vitro, precipitation rates are greatly influenced by the mixtures of water-soluble nacre proteins [44].

14.3
Why Is Nacre So Strong?

Knowing nacre's mechanical properties involved a vast diversity of extensive experiments and accurate measurements in mollusks such as bivalves, gastropods, and cephalopods [45].

Typically, stress–strain curves present an elastic region followed by a plastic behavior before failure. Values of elastic moduli of ∼70 GPa (dry) and 60 GPa (wet) are attributed to nacre from the shell of a bivalve mollusk, *Pinctada imbricata*, and tensile strength of ∼170 MPa (dry) and 130 MPa (wet). It is interesting to note that the effect of hydration is particularly important for the toughness of nacre. The work of fracture of dry nacre is ∼350–450 J m^{-2} and can go up to 1240 J m^{-2} in wet conditions [46].

The importance of the organic layer in the mechanical properties of nacre is highlighted when one compares the values obtained with nacre with the ones obtained with monolithic calcium carbonate. Indeed, the work of fracture in monolithic calcium carbonate is ∼3000 times less than the ones measured in nacre. A decrease of this parameter on drying crossed lamellar seashells can also be observed [47], proving the importance of moisture in the plasticizing effect of proteinaceous layers.

Standard tests performed on nacre determined a fracture toughness of 8 ± 3 MPa m$^{1/2}$, in four-point tests, and a fracture strength of 185 ± 20 MPa, in three-point bend tests [48]. Such values are better than those of most technical ceramics processed by conventional bulk techniques and are comparable with those of ceramic matrix (e.g., ZrO_2-Al_2O_3) and metal–matrix (WC-Co) composites. Moreover, the results in nacre represent an eightfold increase in toughness over monolithic $CaCO_3$. Quasi-static and dynamic compression and three-point bending tests on abalone shells showed that the mechanical performance exhibited strong orientation dependence on strength, as well as significant strain rate sensitivity.

The remarkable mechanical properties measured in nacre, were believed to be the outcome of its brick-and-mortar architecture. However, mimicking nacre's structure with synthetic ceramics did not lead to the same fracture toughness values of nacre [49]. Other causes must justify nacre's astonishing behavior. One of the reasons pointed out for the high toughening properties of nacre is the way the fracture occurs, by a highly tortuous fracture path occurring in the material with the exposing of the surface of ceramic platelets. However, some points fail in this theory, mainly the fact that the high amount of energy dissipation experimentally measured cannot be explained by just considering the simple brick-and-mortar architecture of nacre. Some more mechanisms able to contribute to energy dissipation in this material may be considered in order to explain precisely what happens with nacre.

The role of the organic fraction was already underlined to be very important in nacre's performance. Indeed, if a stress normal to the platelet plane is applied, a viscoelastic dissipative process will take place mainly in the hydrated state [2] and the ductile organic fraction that glues the crystals will prevent an uncontrolled crack growth [50]. Regarding the transition from a brittle (glassy) state to a more elastomeric behavior that occurs on nacre's hydration, a similar behavior was observed in chitosan using nonconventional dynamic mechanical analysis, in which glass transition could be detected in intermediate hydration levels [51, 52]. Moreover, photoacoustic Fourier transform infrared spectroscopy results suggest that the water present at the nanograin interfaces also contributes significantly to the viscoelasticity of nacre [53, 54].

Within the organic matrix, one can find proteins that can also be responsible for this material success. Indeed, Lustrin A, exhibits a highly modular structure characterized by a multidomain architecture with folded modules and acts as a high-performance adhesive, binding the platelets together [55]. When subjected to stress, a significant amount of energy will be necessary to elongate the biopolymer structural loops. This modular elongation mechanism may contribute to the amazing toughness properties in nacre [55].

The inelastic deformation attributed to nacre, both concerning shear and tension, can be due to the presence of inelastic nanosized mineral asperities present on the surface of the tablets. These asperities are believed to be an important source of shear resistance [40].

On the other hand, the nanosized grains composing the aragonite tablets (Figure 14.1g) have a ductile nature [31]. These grains when subjected to mechanical deformation have the ability to rotate and deform, assuring that the energy dissipation is mainly because of the spaces between them, filled with organic material [56].

Also, the nanosized mineral bridges (Figure 14.1f) that can be found between the aragonite tablets are believed to reinforce the weak interfaces and to allow the crack to extend according to their display [39].

Nacre's surface is convoluted, and its waviness can reach up to half of its thickness in amplitude [2]. In a stress situation, the nonflat topography will help to maintain the tablets together with their interfaces because of a progressively locked tablet sliding.

The hierarchical organization of the building blocks constituting nacre as a layered system of two different materials (inorganic and organic) with distinct elastic moduli may result in a shielding or antishielding effect at the crack tip, leading to a change of the crack driving force and energy consumed by the fracture process [57].

Also, its nanoscale nature assures that the smallest building block will tolerate in an optimal way the occurrence of flaws [58].

14.4
Strategies to Produce Nacre-Inspired Biomaterials

Several tactics of producing nanolaminates based on the use of inorganic particles have already been developed [59]. Laminate processing represents the most used approach when producing artificial nacre, although some processes involve different techniques and concepts. The most common methodologies normally used are covalent self-assembly, electrophoretic deposition, layer-by-layer (LbL) technique, template inhibition, and freeze-casting. It is worth noting the effort being made in this field. Therefore, the latest advances were analyzed and organized into six different categories.

14.4.1
Covalent Self-Assembly or Bottom-Up Approach

To produce well-defined hierarchical structures, organisms usually adopt a bottom-up approach, including complex structures that contain minerals with controlled size, shape, crystal orientation, polymorphic structure, defect texture, and particle assembly [60]. Transporting this knowledge to the laboratory, in the bottom-up methodology, an organic phase provides a template for inorganic crystals to nucleate and grow from supersaturated solution. The function of the organic phase may be to accelerate or inhibit crystal growth, and also to control molecular weight, concentration, density of functional groups on the backbone chain or side chain, and whether the polymer is adsorbed on the surface or present in solution [61]. In Nature, the formation of nacre also involves the use of organic macromolecules as templates for nucleation of the minerals and control of the final material's shape.

The great advantage of this process is the fact that there is no need of an outside source to assure the rearrangement of the molecules. The desired molecules added to the solution will assemble themselves as monolayers capable of acting as a model for the inorganic phase to precipitate [62, 63].

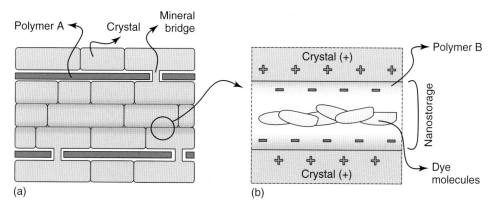

Figure 14.3 Schematic illustration of layered assembly of the units formed by the nanocrystal blocks. (a) Polymer A inhibits crystal growth. (b) Polymer B promotes growth by mineral bridges. Besides polymorphism and crystal size, polymer B is also responsible for nanostorage by incorporating organic molecules.

An organized architecture similar to a real nacreous layer can emerge from potassium sulfate (K_2SO_4) in the presence of poly(acrylic acid) [62]. The specific interactions existing between the two components during crystallization of K_2SO_4 in the presence of poly(acrylic acid) generate the nanoscopic architecture with 20 nm diameter blocks, and the switching between the modes of growth explains the formation of the nacrelike layered structure. Nacre formation requires several macromolecules as an organic template. However, poly(acrylic acid) assumes two roles in this process: (i) it mimetizes nacre's architecture through oriented assembly and (ii) owing to the composite's architecture, it also allows for nanostorage, as was proved by the incorporation of dye molecules (Figure 14.3). Following this line, $CaCO_3$ and poly(acrylic acid) were associated, and as it happens in real biominerals, the synthetic mineral was generated from bridged nanocrystals with incorporated organic polymers [63].

Figure 14.3 describes the layered assembly that emerges from the nanocrystals blocks through templating of organic polymers.

Other polymers both of natural and synthetic origin were explored with this kind of bottom-up approach, as the synthesis of chitosan and polygalacturonic acid with hydroxyapatite [64]. Composites were prepared by allowing wet precipitation of hydroxyapatite in the presence of the polymers. A significant improvement in elastic modulus, strain to failure, and compressive strength can be achieved by incorporating polycationic and polyanionic biopolymers together with hydroxyapatite. Furthermore, structural analysis of these nanocomposites revealed a multilevel organization showing resemblance with natural composites.

Simultaneous organic and inorganic polymerization reactions also lead to nanocomposite morphology through self-assembly processes. By mixing together the organic and inorganic precursors – silicates, coupling agents, surfactants, organic monomers and initiatiors in a water/ethanol solution – micellar structures will form and assemble, during dip-coating, into interfacially organized liquid crystalline mesophases. The nanocomposite morphology will be held through covalent bonds within the organic–inorganic interface [65].

Novel approaches based on self-assembly are constantly emerging. Artificial nacrelike chitosan–montmorillonite bionanocomposite films can be prepared based on the self-assembly of chitosan–montmorillonite hybrid building blocks. Montmorillonite is the most used layered silicate in polymer nanocomposites because of its swellable layered structure [66, 67]. The chitosan molecules are very easily coated onto exfoliated montmorillonite nanosheets to yield the hybrid building blocks by strong electrostatic and hydrogen-bonding interactions. These hybrid building blocks can be dispersed in distilled water and then aligned into a nacre-like lamellar microstructure by vacuum-filtration- or water-evaporation-induced self-assembly because of the role that the orientation of the nanosheets and linking of the chitosan play. The obtained bionanocomposite films have a nacre-like brick-and-mortar microstructure, which leads to their high performances in mechanical properties [68].

Bottom-up approaches may have an important role setting the design and synthesis of biomaterials for hard tissues, such as scaffolds for bone replacement or regeneration, and their main advantage is the easy feasibility of the processes associated with the self-assembly.

14.4.2
Electrophoretic Deposition

Electrophoretic deposition is a low-cost technique allowing a complex combination of different materials in order to produce thin layers, coatings, and layered materials (laminates). Through the application of an external electric field, charged nanoparticles in a suspension will migrate to an oppositely charged electrode and deposit on it [69, 70]. Moreover, the presence of an electric field will allow the manipulation of material texture by setting the particle anisotropy.

One of the materials that can be employed in this technique is poly(amic acid) and montmorillonite [71]. The association of these two materials through electrophoretic deposition results in nacrelike composite films with layered structure. The composite films produced this way have improved mechanical properties, namely, a modulus increment of 155% and strength increment of 40% when compared with those of the pure polymer film [71].

It is also possible to modify montmorillonite in order to adjust the overall mechanical characteristics. For instance, a modification of the clay with acrylamide monomers will result in layered structures produced by electrophoretic deposition, with values of hardness and modulus being 0.95 and 16.9 GPa, respectively [72].

Electrophoretic deposition can also be combined with other techniques such as hydrothermal intercalation to prepare clay-modified electrodes with a uniform and continuous polymer/clay composite film of brick-and-mortar nanolaminated structure. Acrylic anodic electrophoretic resin was chosen as the organic phase, and

Na$^+$-montmorillonite as the inorganic one. The two-step assembly process resulted in a nanocomposite mimicking nacre both in structure and composition [73].

Electrophoretic deposition was used to coat smooth gibbsite nanoplatelets with a thin layer of sol–gel silica in order to increase the surface roughness, which mimics the asperity of aragonite platelets found in nacre [74]. To avoid the severe cracking caused by the shrinkage of sol–gel silica during drying, polyelectrolyte polyethyleneimine was used to reverse the surface charge of silica-coated gibbsite nanoplatelets and increase the adherence and strength of the electrodeposited films. Polymer nanocomposite production was possible by infiltrating the interstitials of the aligned nanoplatelet multilayers with photocurable monomer followed by photopolymerization. The resulting self-standing films are highly transparent and exhibit nearly three times higher tensile strength and 1 order of magnitude higher toughness than pure polymer.

14.4.3
Layer-by-Layer and Spin-Coating Methodologies

Another suitable choice to produce nanostructured organic–inorganic multilayered films is using LbL technique. In this technique, a substrate is immersed in oppositely charged polyelectrolytes, with washing steps between the dipping stage, in order to create a strong assembly of different materials at the molecular level. One of the major advantages of this technique is that it is performed under mild conditions, which is particularly important for preserving the activity of biomolecules, since a high diversity of materials may be used [75].

Organic–inorganic films with thicknesses >5 µm can be obtained using montmorillonite clay as the mineral compound. These kinds of materials can have ultimate Young's modulus similar to that of lamellar bone [76]. LbL assembly composites inspired by nacre structure may have a promising applicability as bone implants not only because of their mechanical strength but also because other important characteristics such as antibacterial activity can be introduced in these implantable materials, for example, by alternating clay layers with starch-stabilized silver nanoparticles [77].

By playing with the chemical interactions between the chosen materials, using, for example, cross-linkers, it is possible to control the strength of the nanocomposites [78].

Osteoconductive glass ceramic nanoparticles produced by sol–gel methodologies can be combined with polymers such as poly(L-lactic acid) resulting in bioactive nanocomposites [18, 19]. Such kinds of nanocomposites can also be obtained by sequentially dipping a substrate in a suspension of bioactive glass ceramic nanoparticles (exhibiting a negatively charged surface) and in a solution of a positively charged polymer such as chitosan (Figure 14.4). Such biodegradable coatings are able to promote the deposition of apatite on immersion in simulated body fluid (SBF) (Figure 14.2d), having potential to be used in orthopedic applications [21]. As the production of such multilayers takes place in solutions, these coatings may be produced on the surface of objects with complex geometries, thus having the

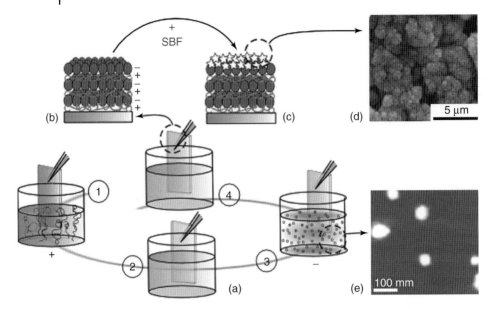

Figure 14.4 Multilayered bioactive coatings were produced by sequentially dipping a negatively charged glass substrate into chitosan solution (positive), water (for washing the excess), and nanoparticle suspension of bioactive glass (negative), represented by the AFM image (e). The previous steps were repeated for each bilayer. (a) The produced film (b) was then immersed in simulated body fluid (SBF), which resulted in the development of an apatite layer (c), as was confirmed by the SEM image (d). (Source: Data based on the results from Ref. [21].)

potential to produce osteoconductive thin coatings in nonflat implants for bone fixation or replacement and scaffolds.

Alternate spin-coating can be use as a room temperature process for fabricating organic/inorganic multilayer composites. The spin-coating LbL assembly methodology is a faster alternative to the dip-coating technique. The build up of a 100-bilayer multilayered polyelectrolyte clay film with a thickness of ∼330 nm needs <1 h, which is 20 times faster than dip-coating processes with the same characteristics [79].

Concentrations of solutions and spin speed can be changed to control layer thickness. A SiO_2/polyacrylic ester (PAE) multilayer composite without interface delamination or discontinuity of layers was successfully fabricated by this technique, using solutions of alkoxysilane and aqueous PAE emulsion. The layer thicknesses of both SiO_2 and PAE reached submicron order under optimal conditions. This approach overcomes the problem of processing temperature difference between ceramics and polymers by producing both layers from liquids. The inorganic phase is produced by a sol–gel process from a metallic alkoxide, and the organic phase is obtained from an aqueous emulsion [80].

The LbL assembly technique can easily produce laminated nanocomposites with predefined complex geometries, which is a clear advantage over the other approaches mentioned before. As a disadvantage, the LbL process requires many repeated steps to build up a film with practical thickness, and thus it is usually used to produce thin coatings or films.

14.4.4
Template Inhibition

One of the controls used in natural biomineralization is the synergistic interaction between a semirigid template and a soluble inhibitor [81]. Template inhibition is a technique where minerals can be deposited from solution onto an organic substrate that will work as growth template. A soluble inhibitor favors the lateral mineral growth over the normal growth assuring that a very thin layer develops right next to the promoting substrate. Through this strategy well-ordered two-dimensional structure of a self-assembled film on solid or liquid substrates can be obtained [3]. As the template's effect can largely influence the crystal nucleation and growth, some work has been developed in this area. For instance, template inhibition strategy can be applied in the production of thin films of $CaCO_3$ crystals, using soluble acidic polymers, such as poly(acrylic acid), as additives, which leads other polymers such as chitosan, chitin, and cellulose to act as a solid matrix in their presence. The obtained films thickness can reach values of ~ 0.8 μm [82–84]. Combination with other techniques, such as the spin-coating, allows the production of layered polymer/calcium carbonate ($CaCO_3$) composite films [84].

Within the template inhibition technique, other details of Mother Nature can be incorporated in order to improve this strategy. It is known that acidic proteins extracted from nacre and prismatic layers when preadsorbed on β-chitin and natural silk substrates induce *in vitro* aragonite and calcite formation, respectively [85].

Wu et al. [86] investigated the growth of $CaCO_3$ in a system containing insoluble chitosan matrix and soluble regenerated silk fibroin in an attempt to mimic more precisely the role that the different organic components play in the determination of polymorphism of nacre, since chitosan is a chitin derivative and regenerated silk fibroin is much more similar to silklike proteins in nacre than the conventional synthesized acidic polymers. Hybrid nanoparticles based on electrostatic interactions between $CaCO_3$ and regenerated silk fibroin lead to the formation of crystalline disks. The crystallization of $CaCO_3$ in the vicinity of the chitosan membrane was much more affected by the environment of crystallization, compared to that in bulk solution. The interaction between silk fibroin and small $CaCO_3$ particles is then very relevant in the crystallization process of certain $CaCO_3$ polymorphs.

Through template inhibition, it is possible, under mild conditions, to synthesize macroscopic and continuous calcium carbonate thin films [81]. The cooperation effect of some elements, such as poly(aspartic acid) and magnesium ions, in the $CaCO_3$ solution, allows the deposition of aragonite thin films in organic polymers, in repeating units. The resulting structure resembles nacre [87].

14.4.5
Freeze-Casting

During the formation of sea ice, the solutes present in sea water are expelled from the forming ice and entrapped within channels between the ice crystals. Such phenomenon inspired the production of composites using ceramic particles dispersed in water [88]. First, layered materials were prepared through a freeze-casting method. The porous scaffolds were then filled with a second phase, for example, an organic component, in order to produce a dense composite. This simple technique allows the production of layered composites with complex shapes and very interesting mechanical properties. The porous structures obtained by this process exhibit strong similarity to the meso- and microstructures of the inorganic component of nacre.

Using a sublimation/compression method, it is possible to prepare free-standing nacrelike composite films. Freezing a chitosan/hydroxyapatite solution allows the solidification of the mixture and the induction of solid–liquid phase separation. Through freeze-drying the solvents are removed from the structure. Consequently, flexible multilayered films are obtained. Compression transforms the obtained foamtype composites into flexible thin films. Results have shown that the initial solidification temperature and the composition ratio are important factors in determination of homogeneity and mechanical strength of organic/inorganic composites [89].

14.4.6
Other Methodologies

Despite all the efforts made by materials researchers who aim to reproduce nacre's organization, some drawbacks still need to be overcome in the common techniques used to synthesize new layered biocomposites. Therefore, imaginative approaches were proposed.

For instance, biocomposites of clay nanoplatelets as the mineral and polyimide as the organic constituent resulted in films with thickness of 10–200 µm in a very short time using a centrifugal deposition process, which resulted in an ordered nanostructure with alternating organic and inorganic layers. The mechanical properties were comparable with those of lamellar bones, with a tensile strength of 70–80 MPa, Young's modulus of 8–9 GPa, and hardness of ~1–2 GPa [90].

Although it would be much more appealing to fabricate nanolaminate structures under mild conditions, that is, aqueous solutions and environmental temperatures and pressures, Si_3N_4/BN composites with a layered microstructure resembling nacre were produced using a roll compaction technique followed by hot-pressed sintering [28].

Taking advantage of the specific behavior of Na/Ca montmorillonite in aqueous dispersion, textured films were prepared by progressive evaporation of dispersions with low concentrations of delaminated platelets. The increase in concentration of ions during evaporation changes the nature of the clay mineral platelet face interactions from repulsive to attractive. After complete drying, a dense bricklike structure is obtained when a sodium salt is used as a deflocculant. The bending strength of the textured film is strongly affected by crack formation during drying, especially when the sample thickness increases [91]. Other laminated alumina/polyimide multilayer materials were fabricated by a dry process using sputtering and vapor deposition polymerization, which, under appropriate processing conditions, could be free of cracks and delaminations [92].

In another different strategy, bio-inspired composites were designed based on microelectromechanical systems technology in order to mimic nacre structure [93]. Liu et al. [94] in an attempt to simulate nacre developed five different methods of arranging smectite clay tactoids. The rigidity of these tactoids permits the alignment of such objects, giving rise to ordered structures. The first step to prepare the nanocomposites was the dispersion of clay in water by ultrasonic agitation, followed by cleaning based on partial sedimentation. With the obtained suspension, a controlled phase separation was performed by sedimentation, centrifugation, controlled rate slip casting, filtration, and electrophoresis. The best results were obtained using the slip casting method, although all five methods led to well-aligned parallel layers of platelets. To produce the nanocomposites, polyethylene oxide was incorporated into the tactoids.

Biomimetic materials synthesis sometimes uses methods very similar to the ones used by Nature, such as the isolation of materials from the organisms to prepare inorganic products. Furthermore, biomimetic processes are complemented with other laboratory methodologies, such as the ones previously described.

Thick hybrid films were produced based on a bottom-up colloidal assembly of strong submicrometer-thick ceramic alumina platelets within a ductile chitosan polymeric matrix, exhibiting excellent stiffness and strength. Surface-modified platelets were assembled at the air–water interface to produce a highly oriented layer of platelets after ultrasonication. The 2D assembled platelets were then transferred to a flat substrate and then, covered with a polymer layer by conventional spin-coating [95].

A bottom-up approach was applied to obtain hybrid nanocomposites with alternate organic (diazo-resins and poly(acrylic acid)) and inorganic ($CaCO_3$ strata) layers, produced, respectively, by LbL and CO_2 diffusion methods [96]. Reactive diazo-resins were used for cross-linking the thin films, and poly(acrylic acid) was used for providing a surface to facilitate the biomineralization. The inorganic layers were $CaCO_3$ strata, which were prepared via the CO_2 diffusion method, which controls the thickness of the inorganic layer.

Another strategy to fabricate a nacrelike nanocomposite was by alternating $CaCO_3$ and poly(acrylic acid). The multilayered film was fabricated directly on an aragonitic nacre substrate. Aragonite growth by slow diffusion of NH_4HCO_3 vapor into a $CaCl_2$ solution alternated with poly(acrylic acid) drop coating at room temperature, leading to the construction of a multilayer inorganic/organic nanocomposite film of alternating aragonite nanostacks and porous poly(acrylic acid) [97]. The prepared multilayer is identical to natural nacre in crystallite size (30–40 nm) and single-layer thickness (~1 µm).

Similarly, block copolymer and tabular $CaCO_3$ layers were used to reproduce nacre [98]. The process relies on polymer-induced liquid precursor layer formation. After injection of $NaHCO_3$ solution in the subphase beneath a poly(styrene)-block-poly(acrylic acid) (PS-b-PAA) block copolymer monolayer, the polymer-induced liquid precursor droplets form at the air/water interface because of CO_2 evaporation, and later adsorb to the block copolymer monolayer and coalesce to form a film, which after solidification leads to an amorphous calcium carbonate layer. Subsequent LbL transfer of block copolymer/polymer-induced liquid precursor films leads to a multilayer with organic–inorganic hybrid architecture. The subsequent annealing process transforms the polymer-induced liquid precursor layers into layers of polycrystalline $CaCO_3$, which morphologically resemble those of biogenic nacre. The characteristic multilamellar architecture of the composite films is preserved through all transformation steps.

Not only nacre's structure but also its molecular glue draws the attention of materials researchers. The multifunctional macromolecules have combined binding affinities for different materials within the same molecule, thereby bridging different materials and acting as molecular glue. On the basis of this remarkable property, a multifunctional protein was designed with the purpose of self-assembly of nanofibrillar cellulose [99].

14.5
Conclusions

Nacre, with its incredible hierarchical structure and notable mechanical properties, represents a unique example of designing and producing novel materials inspired by natural structures. Besides its remarkable hierarchical organization, nacre and nacre-derived materials also show good chemical interaction with bone, making this material even more promising in the biomedical area.

Inspired by nacre's structural features, many researchers have tried to reproduce the excellent mechanical properties of this material. Biomaterial properties are being improved based on nacre's features. Still, some goals were not achieved, such as the mimicking of self-repair ability and tenacity at interfaces. Nature is an inexhaustible source of inspiration throughout the biomaterials field. Nevertheless, plenty of valuable lessons are waiting to be applied into the successful development of new biocomposites. Nacrelike laminated nanostructured composites may have great applicability in different biomedical fields. For example, structural biomaterials with improved mechanical properties and osteoconductive behavior could be used as scaffolds for bone tissue engineering and as devices for bone fixation and replacement. Nanostructured composite free-standing membranes could be very interesting in the field of guided bone regeneration or guided tissue regeneration. The coating of prostheses with multilayered nanocomposites could also improve their osteoconductive properties. Much work is still needed to demonstrate the full potential of such interesting structures in the development of products to be used in the clinical practice.

Acknowledgements

We acknowledge financial support from Portuguese Foundation for Science and Technology -FCT (Grant SFRH/BPD/45307/2008), "Fundo Social Europeu"- FSE, and "Programa Diferencial de Potencial Humano-POPH".

References

1. Mann, S. (2001) *Biomineralization: Principles and Concepts in Bioinorganic Materials Chemistry*, Oxford University Press, New York.
2. Barthelat, F. (2007) *Philos. Transact. R. Soc. A-Math. Phys. Eng. Sci.*, **365**, 2907–2919.
3. Li, C.M. and Kaplan, D.L. (2003) *Curr. Opin. Solid State Mater. Sci.*, **7**(4–5), 265–271.
4. Meldrum, F.C. (2003) *Int. Mater. Rev.*, **48**(3), 187–224.
5. Mayer, G. (2005) *Science*, **310**(5751), 1144–1147.
6. Fratzl, P. (2007) *J. R. Soc. Interface*, **4**(15), 637–642.
7. Vincent, J.F.V. et al. (2006) *J. R. Soc. Interface*, **3**(9), 471–482.
8. Wegst, U.G.K. and Ashby, M.F. (2004) *Philos. Mag.*, **84**(21), 2167–2181.
9. Fratzl, P. and Weinkamer, R. (2007) *Prog. Mater. Sci.*, **52**(8), 1263–1334.
10. Weiner, S. and Addadi, L. (1997) *J. Mater. Chem.*, **7**(5), 689–702.
11. Addadi, L. and Weiner, S. (1992) *Angew. Chem. Int. Ed. Engl.*, **31**(2), 153–169.
12. Kobayashi, I. and Samata, T. (2006) *Mater. Sci. Eng. C Biomimetic Supramolecular Syst.*, **26**(4), 692–698.
13. Kokubo, T., Kim, H.M., and Kawashita, M. (2003) *Biomaterials*, **24**(13), 2161–2175.
14. Oliveira, A.L., Mano, J.F., and Reis, R.L. (2003) *Curr. Opin. Solid State Mater. Sci.*, **7**(4–5), 309–318.
15. Rezwan, K. et al. (2006) *Biomaterials*, **27**(18), 3413–3431.
16. Shi, J., Alves, N.M., and Mano, J.F. (2007) *Adv. Funct. Mater.*, **17**, 3312–3318.
17. Dias, C.I., Mano, J.F., and Alves, N.M. (2008) *J. Mater. Chem.*, **18**(21), 2493–2499.
18. Hong, Z.K., Reis, R.L., and Mano, J.F. (2008) *Acta Biomater.*, **4**(5), 1297–1306.
19. Hong, Z., Reis, R.L., and Mano, J.F. (2009) *J. Biomed. Mater. Res. A*, **88A**(2), 304–313.
20. Luz, G.M. and Mano, J.F. (2011) *Nanotechnology*, **22**(49), 494014.
21. Couto, D.S., Alves, N.M., and Mano, J.F. (2009) *J. Nanosci. Nanotechnol.*, **9**(3), 1741–1748.
22. Atlan, G. et al. (1999) *Biomaterials*, **20**(11), 1017–1022.
23. Berland, S. et al. (2005) *Biomaterials*, **26**(15), 2767–2773.
24. Rousseau, M. et al. (2008) *J. Biomed. Mater. Res. A*, **85A**(2), 487–497.
25. Zhu, L.Q. et al. (2008) *Prog. Biochem. Biophys.*, **35**(6), 671–675.
26. Vecchio, K.S. et al. (2007) *Acta Biomater.*, **3**, 910–918.
27. Guo, Y.P. and Zhou, Y. (2008) *J. Biomed. Mater. Res. A*, **86A**(2), 510–521.
28. Wang, C.A. et al. (2000) *Mater. Sci. Eng. C Biomimetic Supramolecular Syst.*, **11**(1), 9–12.
29. Meyers, M.A. et al. (2006) *J. Manage.*, **58**(7), 35–41.
30. Sarikaya, M. and Aksay, I.A. (1995) *Biomimetic, Design and Processing of Materials. Polymers and Complex Materials*, American Institute of Physics, Woodbury.

31. Li, X.D. et al. (2004) *Nano Lett.*, **4**(4), 613–617.
32. Menig, R. et al. (2000) *Acta Mater.*, **48**(9), 2383–2398.
33. Lin, A.Y.M., Chen, P.Y., and Meyers, M.A. (2008) *Acta Biomater.*, **4**(1), 131–138.
34. Pokroy, B. and Zolotoyabko, E. (2003) *J. Mater. Chem.*, **13**(4), 682–688.
35. Kamat, S. et al. (2000) *Nature*, **405**(6790), 1036–1040.
36. Weiner, S., Addadi, L., and Wagner, H.D. (2000) *Mater. Sci. Eng. C Biomimetic Supramolecular Syst.*, **11**(1), 1–8.
37. Metzler, R.A. et al. (2008) *Langmuir*, **24**(6), 2680–2687.
38. Levi-Kalisman, Y. et al. (2001) *J. Struct. Biol.*, **135**(1), 8–17.
39. Song, F., Soh, A.K., and Bai, Y.L. (2003) *Biomaterials*, **24**(20), 3623–3631.
40. Wang, R.Z. et al. (2001) *J. Mater. Res.*, **16**(9), 2485–2493.
41. Metzler, R.A. et al. (2007) *Phys. Rev. Lett.*, **98**(26), 268102.
42. Qiao, L., Feng, Q.L., and Lu, S.S. (2008) *Cryst. Growth Des.*, **8**(5), 1509–1514.
43. Meyers, M.A. et al. (2008a) *Prog. Mater. Sci.*, **53**, 1–206.
44. Heinemann, F. et al. (2011) *Mater. Sci. Eng. C Mater. Biol. Appl.*, **31**(2), 99–105.
45. Currey, J.D. (1977) *Proc. R. Soc. Lond. Ser. B Biol. Sci.*, **196**(1125), 443–463.
46. Jackson, A.P., Vincent, J.F.V., and Turner, R.M. (1988) *Proc. R. Soc. Lond. Ser. B Biol. Sci.*, **234**(1277), 415–440.
47. Neves, N.M. and Mano, J.F. (2005) *Mater. Sci. Eng. C Biomimetic Supramolecular Syst.*, **25**(2), 113–118.
48. Sarikaya, M., Gunnison, K.E., Yasrebi, M., and Aksay, I.A. (1989) *MRS Proc.*, **174**(109), 109–116.
49. Nukala, P. and Simunovic, S. (2005) *Phys. Rev. E*, **72**(4), 041919.
50. Sumitomo, T. et al. (2008) *J. Mater. Res.*, **23**(5), 1466–1471.
51. Mano, J.F. (2008) *Macromol. Biosci.*, **8**(1), 69–76.
52. Caridade, S.G. et al. (2009) *Carbohydr. Polym.*, **75**(4), 651–659.
53. Mohanty, B. et al. (2006) *J. Mater. Res.*, **21**(8), 2045–2051.
54. Verma, D., Katti, K., and Katti, D. (2006) *Spectrochim. Acta A Mol. Biomol. Spectrosc.*, **64**(4), 1051–1057.
55. Smith, B.L. et al. (1999) *Nature*, **399**(6738), 761–763.
56. Li, X.D., Xu, Z.H., and Wang, R.Z. (2006) *Nano Lett.*, **6**(10), 2301–2304.
57. Fratzl, P. et al. (2007) *Adv. Mater.*, **19**, 2657–2661.
58. Gao, H.J. et al. (2003) *Proc. Natl. Acad. Sci. U.S.A.*, **100**(10), 5597–5600.
59. Manne, S. and Aksay, I.A. (1997) *Curr. Opin. Solid State Mater. Sci.*, **2**(3), 358–364.
60. Verma, D., Katti, K.S., and Katti, D.R. (2008) *Ann. Biomed. Eng.*, **36**(6), 1024–1032.
61. Tsortos, A. and Nancollas, G.H. (2002) *J. Colloid Interface Sci.*, **250**(1), 159–167.
62. Oaki, K. and Imai, H. (2005) *Angew. Chem. Int. Ed.*, **44**(40), 6571–6575.
63. Oaki, Y. et al. (2006) *Adv. Funct. Mater.*, **16**(12), 1633–1639.
64. Verma, D. et al. (2008) *Mater. Sci. Eng. C Biomimetic Supramolecular Syst.*, **28**(3), 399–405.
65. Sellinger, A. et al. (1998) *Nature*, **394**(6690), 256–260.
66. Brindley, G.W. and Brown, G. (1980) *Crystal Structures of Clay Minerals and their X-ray Indentification*, Mineralogical Society, London.
67. Akkapeddi, M.K. (2000) *Polym. Compos.*, **21**(4), 576–585.
68. Yao, H.B. et al. (2010) *Angew. Chem. Int. Ed.*, **49**(52), pp. 10127–10131.
69. Sarkar, P. and Nicholson, P.S. (1996) *J. Am. Ceram. Soc.*, **79**(8), 1987–2002.
70. Boccaccini, A.R. and Zhitomirsky, I. (2002) *Curr. Opin. Solid State Mater. Sci.*, **6**(3), 251–260.
71. Wang, C.A. et al. (2008) *J. Mater. Res.*, **23**(6), 1706–1712.
72. Long, B. et al. (2007) *Compos. Sci. Technol.*, **67**, 2770–2774.
73. Lin, W. et al. (2008) *Mater. Sci. Eng. C Biomimetic Supramolecular Syst.*, **28**(7), 1031–1037.
74. Lin, T.-H. et al. (2010) *J Colloid Interface Sci.*, **344**(2), 272–278.
75. Crespilho, F.N. et al. (2006) *Electrochem. Commun.*, **8**(10), 1665–1670.
76. Tang, Z.Y. et al. (2003) *Nat. Mater.*, **2**(6), 413–4U8.

77. Podsiadlo, P. et al. (2005) *Langmuir*, **21**(25), 11915–11921.
78. Podsiadlo, P. et al. (2008) *J. Phys. Chem. B*, **112**(46), 14359–14363.
79. Vertlib, V. et al. (2008) *J. Mater. Res.*, **23**(4), 1026–1035.
80. Kakisawa, H. et al. (2010) *Mater. Sci. Eng. B Adv. Funct. Solid-State Mater.*, **173**(1–3), 94–98.
81. Xu, G.F. et al. (1998) *J. Amer. Chem. Soc.*, **120**(46), 11977–11985.
82. Kato, T. and Amamiya, T. (1999) *Chem. Lett.*, (3), 199–200.
83. Kato, T. et al. (1998) *Supramol. Sci.*, **5**(3–4), 411–415.
84. Kato, T., Suzuki, T., and Irie, T. (2000) *Chem. Lett.*, (2), 186–187.
85. Falini, G. et al. (1996) *Science*, **271**(5245), 67–69.
86. Wu, Y.D. et al. (2011) *Langmuir*, **27**(6), 2804–2810.
87. Sugawara, A. and Kato, T. (2000) *Chem. Commun.*, (6), 487–488.
88. Deville, S. et al. (2006) *Science*, **311**(5760), 515–518.
89. Sun, F. et al. (2010) *Mater. Sci. Eng. C Mater. Biol. Appl.*, **30**(6), 789–794.
90. Chen, R.F. et al. (2008) *Mater. Sci. Eng. C Biomimetic Supramolecular Syst.*, **28**(2), 218–222.
91. Bennadji-Gridi, F., Smith, A., and Bonnet, J.P. (2006) *Mater. Sci. Eng. B Solid State Mater. Adv. Technol.*, **130**(1–3), 132–136.
92. Naganuma, T. et al. (2006) *J. Ceram. Soc. Jpn.*, **114**(1332), 713–715.
93. Chen, L. et al. (2007) *J. Mater. Res.*, **22**(1), 124–131.
94. Liu, T., Chen, B.Q., and Evans, J.R.G. (2008) *Bioinspiration & Biomimetics*, **3**(1), 016005.
95. Bonderer, L.J., Studart, A.R., and Gauckler, L.J. (2008) *Science*, **319**(5866), 1069–1073.
96. Wei, H. et al. (2007) *Chem. Mater.*, **19**(8), 1974–1978.
97. Hayashi, A., Nakamura, T., and Watanabe, T. (2010) *Cryst. Growth Des.*, **10**(12), 5085–5091.
98. Gong, H. et al. (2010) *Colloids Surf. A - Physicochem. Eng. Asp.*, **354**(1–3), 279–283.
99. Varjonen, S. et al. (2011) *Soft Matter*, **7**(6), 2402–2411.

15
Surfaces Inducing Biomineralization

Natália M. Alves, Isabel B. Leonor, Helena S. Azevedo, Rui. L. Reis, and João. F. Mano

15.1
Mineralized Structures in Nature: the Example of Bone

It is well known that there is a wide diversity of biological structural materials in Nature resulting from the intricate combination of inorganic minerals with organic polymers. Together, these are fashioned into a fascinating variety of shapes and forms. The study of these natural biomineralized structures has generated a growing awareness in materials science that the adaptation of biological processes may lead to significant advances in designing complex structures and controlling intricate processing routes that lead to the final shape. To date, it has not yet been possible to fully replicate these structures by nonbiological processing.

Mineralized biological tissues are produced by living organisms of all kingdoms of life for different purposes such as body support, protection of vital organs, or defense against predators. They are highly optimized materials with outstanding properties and, moreover, these materials are produced at mild temperature and pressure conditions, with relatively low energy consumption, contrary to what typically occurs in synthetic materials. The majority of the organisms have their mineralized tissues composed of calcium phosphates and carbonates or amorphous silica.

Bone is an example of a natural mineralized composite that has been extensively studied because of its unique structure and mechanical properties [1]. Bone is built according to a bottom-up approach, as typically occurs in these natural mineralized composites, in which the material is built starting at atomic and molecular scales, leading to the formation of nanostructured building blocks that in turn organize themselves in complex hierarchical structures [2]. The components at each level are related with each other, enabling the optimization of the performance for the required functions. Another characteristic of these natural composites is that the hard mineral component exhibits nanometric sizes, at least in one direction; displays an anisotropic geometry, and is immersed in a soft organic matrix. Gao [3] showed that the hierarchical organization, nanometer scale, and anisotropy of the mineral crystals are fundamental for the superior mechanical properties of these biocomposites, as evidenced in bone.

Biomimetic Approaches for Biomaterials Development, First Edition. Edited by João F. Mano.
© 2012 Wiley-VCH Verlag GmbH & Co. KGaA. Published 2012 by Wiley-VCH Verlag GmbH & Co. KGaA.

Bone is a term that in fact includes distinct bone types, and it is the major mineralized tissue of the human body. All bone types present mineralized collagen fibril as the basic building block [4], but the porosity and the structural organization of the fibrils at the meso- and microscales is different for the distinct bone types [5]. Bone contains ~60–70% w/w of calcium phosphate mineral (carbonated apatite), ~20–30% w/w of organic matrix, and 10% w/w of water. Although water is a minor component, its importance should not be minimized, as it contributes to the overall toughness of bone, acting as a plasticizer [6]. Collagen type I is the major organic component of bone, comprising about 90% of its total content. The other constituents are noncollagenous proteins (NCPs) that are highly acidic and include proteins that are rich in aspartic or glutamic acid residues or phosphorylated serine/threonine [7].

Bone is produced by osteoblasts; these cells deposit a layer of nonmineralized extracellular matrix, consisting of collagen fibrils and noncollagenous macromolecules, which eventually mineralizes a few microns away from the cells. Some osteoblasts become entrapped in the mineralizing matrix and differentiate into osteocytes [8]. The structural organization of collagen fibrillar arrays is determined by the cells during the matrix deposition stage. Another cell type, osteoclasts, is responsible for bone matrix catabolism. Bone is constantly remodeling, that is, osteocytes maintain the mineralized matrix for a certain period but eventually recruit osteoclasts to remove the old bone matrix, and thereafter, osteoblasts synthesize new bone matrix.

The structure of bone is schematically depicted in Figure 15.1. Plate-shaped crystals of carbonated hydroxyapatite (HA) aligned along their *c*-axis, with a length of 50 nm, a width of 25 nm, and a thickness between 2 and 3 nm, are embedded in the collagen type I framework. The fibrils consist of triple helix collagen chains with 1.5 nm diameter and 300 nm length. Their ends are separated by holes of about 35 nm, and the neighboring molecules are vertically offset by 68 nm. The apatite crystals are nucleated at specific regions on or within the collagen fibrils. They grow in the hole zones that exist between neighboring collagen molecules. For the particular case of lamellar bone, the fibrils are arranged in parallel arrays, with crystals aligned (sublayers). The consecutive sublayers rotate through the lamellar plane by an average of 30°, forming the so-called plywoodlike structure. As each lamella is composed of five sublayers, the total rotation is 150°, forming an asymmetric structure. Also, the collagen fibril bundles rotate around their axis within the five sublayers. The described architecture hinders crack propagation and increases toughness.

In cortical bone, the lamellae are arranged concentrically around blood vessels to form osteons, with a typical diameter of 200 µm and length of 10–20 µm. The osteons are usually oriented in the bone's longitudinal direction and present a central channel (Haversian canal) containing blood supply and bone cells. It is the orientation of the osteons that gives rise to enhanced mechanical properties in the longitudinal direction. On the other side, the structure of the so-called trabecular or cancellous bone consists of a network of interconnecting struts known as *trabeculae*

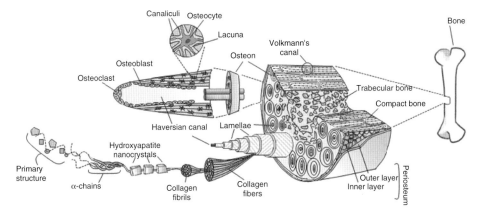

Figure 15.1 The general structure of bone, showing its hierarchical organization. (Source: Adapted from Ref. [9].)

(50–300 mm thick) [10]. The pore space of this network is filled with bone marrow. Trabecular bone is also anisotropic.

The main factor that determines the distinct mechanical properties (e.g., stiffness), depending on bone type, is the amount of mineral in the tissue [11]. Stiffness and strength values of cancellous bone vary depending on the weight- or non-weight-bearing regions. Stiffness has modulus values in the range of 1–9.7 GPa. Weight-bearing trabecular systems can sustain superior–inferior compression levels of as much as 310 MPa, and those from non-weight-bearing regions typically fail at stresses of ranging from 120 to 150 MPa [12]. Experiments on bone fracture showed that crack propagation is prevented by various complex ways and that the local criterion for fracture in human cortical bone is consistent with a strain-based criterion, rather than being stress controlled [13].

Regarding the nature of collagen–mineral interactions, they are still not well understood, although it is thought that they involve ionic bonds between side-chain carboxyls of the protein and calcium ions in the mineral particles [14]. It has also been suggested that the backbone carbonyls of proline residues form complexes with calcium ions in the mineral phase [15]. Also, although collagen has been considered the most important biopolymer in the regulation of bone structure, it is clearly not the sole component responsible for the regulation of bone mineralization, since the majority of the body is composed of collagenous tissues that never mineralize. Thus, it is thought that, besides collagen, the NCPs are also involved in distinct functions, such as cell adhesion and matrix construction, and can also affect bone mineralization in different ways, depending on their concentration, phosphorylation degree, or if they are bound to the surface or in solution, either inhibiting or promoting interactions during crystal nucleation and growth, controlling crystal shape and size, and stabilizing transient mineral phases [16–21]. Furthermore, these macromolecules are functionalized with acidic groups such as carboxylic acid, sulfonate, and phosphate groups, which allow them to be an effective metal ion chelator to interact with the inorganic matrix [16].

Studies conducted with acidic polypeptide additives, used to modify crystal growth of calcium-based minerals, have demonstrated a crystallization mechanism that proceeds through a liquid-phase mineral precursor. Various features of the crystals produced via this mechanism, such as "extruded" mineral fibers and mineralized collagen composites, have led Olszta and colleagues [22–24] to propose a new and very different view on bone mineralization. They hypothesize that an amorphous, liquid-phase precursor could play a fundamental role in the morphogenesis of calcium-based biominerals. They suggest that the charged polymer acts as a process-directing agent, by which the conventional solution crystallization is converted into a precursor process. This polymer-induced liquid-precursor (PILP) process generates an amorphous liquid-phase mineral precursor of HA, which facilitates intrafibrillar mineralization of collagen because the fluidic character of the amorphous precursor phase enables it to be drawn into the nanoscopic gaps and grooves of collagen fibrils by capillary action. Once this highly concentrated phase has infiltrated the fibers, the precursor solidifies and crystallizes on loss of hydration water into the more thermodynamically stable phase, leaving the collagen fibrils embedded with nanoscopic HA crystals.

So, by looking at the extensive research on bone mineralization, although some aspects are yet not quite understood, it is well established at this point that this process is regulated by several components of the extracellular matrix and is mediated by bone cells. The macromolecules produced by cells are key components in the overall mineralization process and, moreover, are subsequently incorporated into the biological material. The spatial regulation of the mineral nucleation and growth and of the microarchitecture development during formation of these structures is also outstanding. In the following sections, the main methodologies that have been proposed to improve calcification onto the surface of biomaterials, inspired by bone mineralization, are presented and discussed.

15.2
Learning from Nature to the Research Laboratory

The previous section evidenced that it is a true challenge for the materials scientists to try to copy or reproduce the structure, property, and performance of bone. To achieve such goal, it is necessary to build a bridge of knowledge between what we learn from Nature and what we can bring to the research laboratory, which is and will continue to be a challenging target.

Nowadays, scientists use biomineralization strategies to develop new methodologies for designing functional materials. For example, designing functionalized surfaces that are negatively charged like the macromolecules found in the mineralized structures of Nature can promote mineralization on their surfaces by establishing local ion supersaturation.

In this section, some of the research works related with surface functionalization to induce the formation of apatite layer on their surfaces is presented. This is not intended to be a detailed and exhaustive review of the subject, but works as a guide of some examples.

15.2.1
Bioactive Ceramics and Their Bone-Bonding Mechanism

Biological responses such as bone bonding of materials are very important in bone-related applications, since one important aspect of a successful regeneration is early implant stability [25]. Hench *et al.* [26, 27] showed for the first time that glass composed of SiO_2, Na_2O, CaO, and P_2O_5 was a bone-bonding material. This *bioactive glass* was defined as the one that elicits a specific biological response at the interface of the material, which results in the formation of a bond between the tissues and the material [28, 29]. This concept was considered as a window for developing new bioactive materials to repair and regenerate bone tissues [30, 31]. These materials include bioactive glasses such as Bioglass® [26, 27], bioactive glass ceramics such as Ceravital® [32], A-W glass-ceramic [33, 34], and dense calcium phosphate ceramics such as synthetic HA [35, 36], among others. These bioactive ceramics have the capacity to spontaneously form a chemically and mechanically strong bond with bone through a biologically active hydroxycarbonate apatite (HCA) layer formed on their surface that is chemically and structurally similar to the mineral phase in bone when they are implanted [26, 33, 35–37]. The presence of HCA is not observed on materials that are not bioactive, such as metals and polymers, meaning that one of the requirements for the artificial material to bond to living bone is the formation of the biologically active apatite layer on the surface [28, 32, 38, 39].

Human blood is composed of proteins and cells and, in terms of inorganic ion species, is a solution highly supersaturated compared to apatite. However, such solution is complex to reproduce *ex vivo* for designing bioactive materials [40]. To understand the mechanism of apatite formation behind these materials, Kokubo *et al.* [41, 42] proposed a protein-free and a cellular simulated body fluid (SBF) with pH 7.4 and ionic composition (Na^+ 142.0, K^+ 5.0, Ca^{2+} 2.5, Mg^{2+} 1.5, Cl^- 147.8, HCO_3^- 4.2, HPO_4^{2-} 1.0, and SO_4^{2-} 0.5 mM) nearly similar to that of human blood plasma. Thus, the capacity of the materials to induce a bone-bonding performance when implanted *in vivo* can be reproduced *in vitro* by immersing them in SBF. It is known that each surface-active ceramic has its own characteristics regarding the formation of apatite layer. For example, when Bioglass is soaked in SBF, the first reaction is ion exchange, in which the release of Ca^{2+} and Na^+ from the glass into the surrounding fluid via exchange with H_3O^+ from the solution results in an increase of the solution pH and subsequently, in the formation of a hydrated silica gel layer [27, 43]. This layer is abundant in silanol (Si–OH) groups and provides favorable sites for the calcium phosphate nucleation [27, 44, 45]. Then, the water molecules in the SBF react with the Si–O–Si bond to form additional Si–OH groups [46]. These functional groups induce apatite nucleation, and moreover, the release of Ca^{2+} and Na^+ ions accelerates apatite nucleation by increasing the ionic activity product (IAP) of apatite in the fluid. In addition, an increase of IAP in the surrounding fluid could promote calcium phosphate (CaP) nucleation and growth on the surface of bioactive ceramics [40]. So, the mineralization induced by this bioactive glass is due to the formation of specific surface functional groups such as Si–OH, which serve as effective sites for heterogeneous CaP nucleation [40]. Tanashi *et al.* [47] have also reported that Si–OH groups are effective in the induction of the apatite nucleation. Therefore, Bioglass when immersed in SBF, is likely to represent a leading *in vitro* model, giving an experimental background to the development of a detailed mechanism of apatite biomineralization on bioactive glasses [47].

15.2.2
Is a Functional Group Enough to Render Biomaterials Self-Mineralizable?

The apatite nucleation efficacy of the functional groups, such as Si–OH, is dependent not only on their composition but also on their kind [48], number [49], and arrangement [50]. Kokubo's group has demonstrated that various metal oxide gels, namely, silica (SiO_2) [51], titania (TiO_2) [52], zirconia (ZrO_2) [53], niobate (Nb_2O_5) [54], and tantalum pentoxide (Ta_2O_5) [55], prepared by sol–gel methods and then soaked in SBF, are able to form an apatite layer on their surfaces. In the case of alumina (Al_2O_3) gel [56], no apatite formation was observed. Those results indicated that Si–OH, Ti–OH, Zr–OH, Nb–OH, and Ta–OH are effective to induce the apatite nucleation, whereas the Al–OH groups are not effective [51, 52, 54–56]. Tanahashi and Matsuda [48] showed that the incorporation of bihydrogenophosphate ($-PO_4H_2$) and carboxyl (–COOH) groups on self-assembled monolayers (SAMs) was effective for the apatite nucleation but amide ($-CONH_2$), hydroxyl (–OH), amine ($-NH_2$), and methyl ($-CH_3$) groups were not effective. In the case of Kokubo's work, gels that induced apatite formation were negatively charged when soaked in SBF at pH = 7.4, with the exception of Al_2O_3 gel, which was positively charged [57].

The outcome of these studies can be an answer to our question, showing that not all functional groups are able to induce apatite layer formation and that other characteristics are needed. The functional groups able to induce apatite nucleation when immersed in water or in the blood plasma (pH $\approx$ 7.4) are expected to develop a negative charge, and in an *in vivo* environment, it is likely that these groups will lead to the apatite nucleation. Among these functional groups and in terms of acidity, the incorporation of sulfonic ($-SO_3H$) groups onto the polymer surfaces could also serve as functional group effective for the apatite nucleation [58], meaning that the high acidity of functional group can also be essential for apatite nucleation.

15.2.2.1 How the Surface Charge of Functional Group Can Be Correlated to Apatite Formation?
The functionalized surfaces are believed to be analogous to nucleation proteins in biological systems in that they provide energetically favorable interfaces for heterogeneous nucleation and growth of inorganic films from supersaturated solutions [59, 60]. Biomineralization, such as apatite formation, mainly occurs by calcium ion adsorption and complexation with a negatively charged group of the artificial material and its subsequent complexation with phosphate ions [48]. As

result, amorphous CaP with low Ca/P ratio will form, which is a metastable phase, and then will eventually transform into a stable apatite [46, 57]. It has been assumed that apatite formation is due to the electrostatic interaction between the substrate surface and specific ions in the fluid [48]. If the electrostatic interactions trigger the initial step of nucleation, it is important to know which ions are first adsorbed and conducted to the nucleation of apatite. If calcium ion adsorption triggers nucleation, such surface might be negatively charged, whereas if phosphate ion adsorption is the initial step, such surface might be positively charged [48]. Most of the functional groups show a negative charge to trigger electrostatic interaction with the calcium and phosphate ions in the fluid, and thereby to form the apatite crystals [40, 46, 57]. Tanahashi and Matsuda [48] have demonstrated that negatively charged surface groups are effective for apatite nucleation. They have found that the nonionic groups such as $-CONH_2$ and $-OH$ and the positively charged group $-NH_2$ had weaker apatite nucleating ability than $-PO_4H_2$ and $-COOH$ groups, which are negatively charged surface groups. Kim *et al.* [61, 62] have confirmed these results, by demonstrating that the surface potential of bioactive titanium is initially negative, indicating that the initialization of apatite nucleation involves an electrostatic interaction between the Ti–OH groups and calcium ions. Therefore, functional groups able to become negatively charged at blood plasma pH are assumed to be potentially effective for apatite nucleation in an *in vivo* environment.

Moreover, the process and kinetics of apatite formation on HA could be affected by other factors such as density, surface area, composition, and structure. Kim *et al.* [62] have investigated the effects of surface composition and structure on apatite formation on HA, and rationalized in terms of surface potential change. They confirmed that the process of apatite formation on HA sintered at higher temperature was slow. The higher sintered temperature leads to the initially lower negative surface charge of HA and therefore a scarcity of surface hydroxyl and phosphate groups, which are responsible for the surface negativity. Moreover, such factor might also affect not in terms of the process, but in terms of the rate of apatite formation on HA surface, through which the HA integrates with living bone.

On basis of these findings, the process of apatite formation can be interpreted in terms of electrostatic interaction between the functional groups and the ions present in the fluid, which creates a route for designing materials incorporating these groups. Therefore, a detailed insight of the apatite formation mechanism on bioactive ceramics is considered crucial for developing bioactive materials with enhanced physical, chemical, and biological functions [62].

15.2.2.2 Designing a Properly Functionalized Surface

The research work mentioned above emphasized the importance of understanding the surface chemistry of biomaterials as well as the main mechanisms associated with apatite formation. Clarifying these aspects provides us the fundamental principles and tools for the design of new bioactive materials. Some examples of how we can induce the formation of an apatite layer on materials that are not bioactive by themselves are presentedin this section.

Several design approaches have been developed to produce new bioactive materials based on the fundamental findings on bonelike apatite formation on bioactive ceramics. For example, titanium metal and its alloys are widely used as orthopedic and dental implants, although this metal is a nonbioactive material and is therefore usually coated with HA by plasma spraying, which has several disadvantages [63–65]. Kim et al. [66] have demonstrated that heterogeneous nucleation and growth of bonelike apatite layer can be induced by alkali-treated metals in body environment, that is, hydroxylation of metal oxide surfaces placed in SBF for different time periods. Thus, the formed Ti–OH groups induced apatite formation on it, through formation of amorphous calcium titanate and amorphous Ca-P. Regarding organic surfaces, their surface can be tailored to turn their surfaces more hydrophilic and capable of carrying functional groups, which is an advantage. In addition, these materials have a much higher degree of structural flexibility and may have strong surface-specific binding forces, such as the ability of the functional groups to chelate metal ions [67].

Until now, it has not been possible to develop new types of materials having not only osteoconductivity but also mechanical properties analogous to those of natural bone and, in the case of biodegradable materials, with suitable degradation kinetics. One true analog of biomineralization would be a polymer matrix, which can be placed into a metastable solution and induce precipitation within the polymer but not in the solution [67]. Money et al. [68] have reported that the process of mineral growth on biodegradable polymers can be augmented and controlled by variation of the functional groups present at the mineral nucleation site or of the ionic characteristics of the mineral growth environment. Polymer surface functionalization was achieved via hydrolysis of poly(lactide-co-glycolide) (PLGA), which results in an increase in the amount of surface carboxyl and hydroxyl groups because of scission of polyesters chains. The presence of these groups regulates the calcium binding to polymer surface followed by the heterogeneous mineral growth. Similar results were obtained by Oyane et al. [69], in which an apatite layer was formed on the surface of poly(ε-caprolactone) (PCL) when soaked in SBF, previously treated with aqueous NaOH solution. The alkaline treatment induced the formation of carboxyl groups and, then, the alternate dip in calcium and phosphate solutions induced the apatite nucleation. However, this treatment requires a long period to induce apatite nucleation in SBF, and the need to be combined with calcium ions. To reduce the nucleation time, they treated PCL with O_2 plasma followed by dipping alternately in alcoholic solutions containing calcium ions and phosphate ions. Using this treatment, an apatite layer was formed at the PCL surface within 24 h [70]. Leonor et al. [58] showed that the incorporation of $-SO_3H$ groups onto polymeric surfaces could also serve as functional groups for apatite nucleation. However, the apatite-forming ability induced by this functional group depends significantly on the incorporation of calcium ions. So, when the sulfonated samples are soaked in calcium-saturated solution, the $-SO_3H$ groups become negatively charged and interact with the positively charged Ca^{2+} ions from the solution, which increase the apatite-forming ability induced by this functional group since the release of these ions accelerates apatite formation in the SBF environment by increasing the IAP.

An apatite–polymer fiber composite with mechanical properties analogous to those of living bone, besides the osteoconductivity properties, would be considered as an ideal candidate for bone replacement. So, it has been proposed that such type of composite could be synthesized if the organic fibers are arranged in a 3D structure similar to that of collagen fibers in living bone and if they are modified to contain functional groups effective for apatite nucleation on their surface when soaking in SBF solution [71]. Oyane et al. [49, 50] successfully produced bioactive films textured on 3D template of polymers by coupling and hydrolysis of isocyanatopropyltriethoxysilane or sol–gel coupling of calcium silicate on ethylene-vinyl alcohol (EVOH) polymer. Balas et al. [72, 73] demonstrated that treating organic polymers, namely, polyethylene terephthalate (PET), EVOH, and Nylon 6, with a silane coupling agent and a titanium solution, induced the formation of bonelike apatite. In the case of polysaccharides, such as carboxymethylated chitin [71] and gellan gum gels [71], it is possible to induce apatite formation by subjecting them to a very simple alkaline treatment. Kawashita et al. [71] demonstrated that by soaking these polysaccharides, first in a saturated calcium hydroxide ($Ca(OH)_2$) solution, and then in SBF, these gels became bioactive. This was attributed to the catalytic effect of the carboxyl groups present on these polymers, which accelerated apatite nucleation. However, curdlan gels [71], which contain hydroxyl groups, did not form an apatite deposit even after the $Ca(OH)_2$ treatment. Once more, these results are in agreement with the ones obtained by Tanashi Matsuda [48]. Such results provide a fundamental condition for obtaining an apatite–polymer fiber composite with structure analogous to that of living bone using a biomimetic method [73]. Wan et al. [74] and Yamaguchi et al. [75] have reported similar research works, in which carboxymethylated chitin and chitosan were able to induce apatite formation. Kokubo et al. [76] also recently showed that carboxymethylation of chitin nonwoven fabric treated with aqueous solution of saturated $Ca(OH)_2$ induced the formation of an apatite layer within three days in SBF. This kind of composite could be useful as a flexible bioactive bone-repairing material.

In the case of starch-based polymers, it was shown that calcium hydroxide ($Ca(OH)_2$) and sodium hydroxide (NaOH) solutions could be successfully used to functionalize their surface [77]. This method is based on a wet chemistry modification, resulting in etching and/or hydrolysis that increases the amount of polar groups such as –OH and –COOH on the polymer surface. Surface oxidation methods could also be used [78]. In order to alternate nonionic starch hydroxyl groups with negatively charged –COOH groups, surface oxidation by potassium permanganate/nitric acid ($KMnO_4$)/(NHO_3) system was performed [78]. Formation of apatite layer was observed in all different types of starch blends, starch/ethylene-vinyl alcohol (SEVA-C), starch/cellulose acetate blends (SCA), and starch/polycaprolactone blends (SPCL). This $KMnO_4/NHO_3$ oxidizing system is an easily applicable method for incorporating polar groups on starch-based biomaterials.

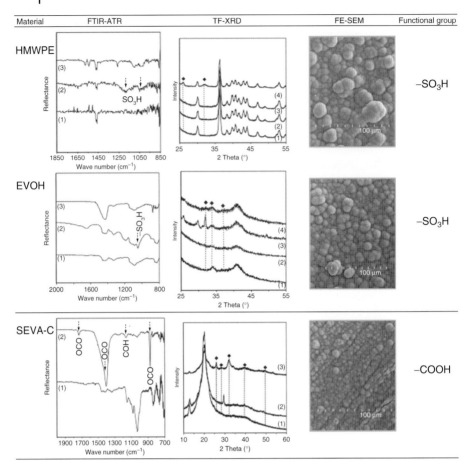

Figure 15.2 Development of bioactive polymers by the incorporation of functional groups that can efficiently induce the formation of an apatite layer onto the surface of polymeric materials that are not intrinsically bioactive. In this work, we used three different polymers, high-molecular-weight polyethylene (HMWPE), ethylene-vinyl alcohol copolymer, 32 mol% of ethylene (EVOH), and a blend of corn starch with ethylene-vinyl alcohol, 50/50 wt% (SEVA-C). These materials were subject to different treatments for the incorporation of functional groups, $-SO_3H$ and $-COOH$. For all three polymers, (1) was the control. HMWPE and EVOH were subjected to the sulfonation (2), to subsequent $Ca(OH)_2$ treatment (3), and to soaking in SBF for seven days. ATR-Attenuated Total Reflectance and FE-Field Emission (4). SEVA-C was subjected to $Ca(OH)_2$ treatment (2) and to subsequent soaking in SBF for seven days.

Figure 15.2 shows some examples of these treatments to induce the formation of an apatite layer. These results suggest that these rather simple treatments are effective for surface functionalization and subsequent mineral nucleation and growth on biodegradable polymers to be used for bone-related applications.

All studies described in this chapter share a common finding that a surface with an organized arrangement of functional groups can act as a template for the apatite growth. Such materials can be developed by simple combination of an inorganic and organic material as found in human bone. On the basis of these fundamental findings, several kinds of bone-bonding materials with different mechanical properties can be developed, and it is expected that a properly functionalized surface will enable us to program desirable and predictable cellular responses into 3D biomaterials.

15.3
Smart Mineralizing Surfaces

Responsive or smart polymers exhibit sharp and reversible conformational changes on small variations of an external signal, including temperature, pH, ionic strength, electrical field, or biological substances. Stimuli-responsive surfaces can be produced by attaching/grafting such kind of materials onto adequate substrates, allowing to switch relevant physicochemical and biological properties of materials, such as wettability [79], permeability [80], permeation of substances [81], protein binding [82], or cell adhesion [83]. Stimuli-responsive surfaces were also recently proposed to control biomineralization [84, 85]. The surface of bioactive composites of poly(L-lactic acid) reinforced with Bioglass was modified by coupling either poly(N-isopropylacrylamide) (PNIPAAm) [84], a thermo-responsive polymer, or chitosan, a pH-responsive polymer [85], using plasma activation. It was shown that surface biomimetic mineralization may be triggered by two types of stimuli: temperature or pH [84, 85].

PNIPAAm is the most studied synthetic thermo-responsive polymer and exhibits a lower critical solution temperature (LCST) at about 32 °C in aqueous solution, changing sharply from a hydrophilic to a hydrophobic state on heating [86]. It is believed that this transition involves the breakage of intermolecular hydrogen bonds between the water molecules and the amide groups in the polymeric chains, which are replaced, above the LCST, by intramolecular hydrogen bonds among the dehydrated amine groups. The thermo-responsive nature of the modified composites was easily confirmed by contact angle measurements. The water contact angle for the PNIPAAm-modified PLLA + 10% bioglass (BG) composite was $51.9 \pm 2.4°$ at 25, and at 37 °C, it changed to $58.8 \pm 2.4°$, being consistent with the increase in the surface hydrophobicity above the LCST [84]. These works showed that the conformational changes occurring at the surface affect mineral formation at the surface of these composites below and above the LCST after being immersed for two weeks in SBF. For the PNIPAAm-modified composite film with 10%BG, no mineral phase could be observed at 25 °C (Figure 15.3). However, at 37 °C, the treated film could form dense precipitates with the typical cauliflower morphology, containing needlelike nanometric structures, characteristic of apatite produced using biomimetics (Figure 15.3).

Chitosan is a pH-responsive polymer that contains groups favoring both hydrophobic ($-CH_3$) and hydrogen bonding ($-OH$, $-NH_2$, and $-C=O$). In an acidic medium, this polymer becomes positively charged because of the protonation of

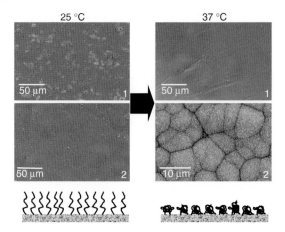

Figure 15.3 SEM images of unmodified (1) and PNIPAAM-grafted (2) films with 10% BG after immersing in SBF for two weeks at 25 and 37 °C. A scheme is also shown representing the different conformational states of the PNIPAAm chains at both temperatures. (Source: Data adapted from the results of Ref. [83].)

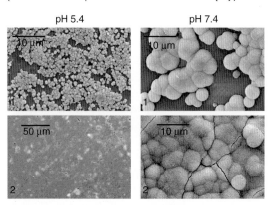

Figure 15.4 SEM images of unmodified (1) and chitosan-grafted (2) films with 30% BG after immersing in SBF for three weeks at pH values 5.4 and 7.4. (Source: Data adapted from the results of Ref. [84].)

the free amine groups (the pK_a is ∼6) [87] and polymer–polymer interactions via hydrophobic effect and/or hydrogen bonding junctions can be hindered because of electrostatic repulsion [88]. The pH-responsive behavior of the PLLA/BG composites modified with chitosan was confirmed by contact angle measurements [85]. The unmodified PLLA/BG composites revealed hydrophobic character, presenting a contact angle of 82°, independent of the pH. After modification, the contact angle changed from $88.9 \pm 4.05°$ at pH 7.4 to $67.6 \pm 2.3°$ at pH 5.4 (Figure 15.4). Apatite formation at the surface after immersion of the modified films in SBF was investigated at both pH values [85]. The results of this work showed that for the chitosan-modified composites, no apatite formation could be observed at pH 5.4 (Figure 15.4). On the other hand, a dense apatite layer was formed at pH 7.4

(Figure 15.4). For the unmodified substrates, an apatite layer was formed at both pH values (Figure 15.4).

Energy dispersive spectroscopy (EDS), thin-film X-ray diffraction (TF-XRD), and Fourier transform infrared spectroscopy (FTIR) analyses of the coatings formed at 37 °C confirmed the formation of a carbonated apatite mineral similar to the major mineral component of vertebrate bone tissue [84, 85]. Temperature and pH were chosen in the above-mentioned works; however, this concept of smart apatite formation can be extended to other sources of responsiveness and other types of mineral deposition.

Moreover, by patterning the modification of the surface, it was shown that it is also possible to combine stimuli and spatial control of biomimetic apatite formation [84]. This was achieved by just exposing some regions of the substrate surface to plasma treatment, allowing the insertion of PNIPAAm in specific chosen areas. Again, no apatite formation was observed for these modified films at 25 °C after immersion for two weeks in SBF [84]. On the other hand, apatite circular aggregates were formed at 37 °C and randomly distributed over the composite surface, being consistent with the PNIPAAm patterning generated during the plasma activation step [84]. Other apatite patterns could be produced simply by changing the mask model or using other lithographic techniques. It is recognized that apatite-coated surfaces enable the attachment, growth, and expression of osteogenic genes in osteoblastlike cells [89, 90]. Therefore, these apatite-patterned surfaces could be used in fundamental studies on differentiation, adhesion, proliferation, and cell–cell signaling of bone-related cells. These surfaces could also find application in fundamental coculture studies involving bone cells and other cell types that could be certainly useful for developing bone tissue engineering applications in the future.

15.4
In Situ Self-Assembly on Implant Surfaces to Direct Mineralization

The organization of molecules at surfaces has been widely investigated given its importance in the preparation of chemically defined and ordered surfaces. SAMs, consisting of a single layer of molecules on a substrate, are frequently used to change physical and chemical properties of surfaces [91]. SAMs can be prepared simply by adding a solution of the desired molecule onto the substrate surface and washing off the excess. Common examples are alkane thiol molecules and organosilane molecules chemisorbed onto gold and hydrated oxide surfaces, respectively. Crystallization of calcium carbonate on these organic templates has been studied extensively. Aizenberg *et al.* [92] showed that patterns of calcite crystals were formed on topographically patterned substrates supporting an acid-terminated (CO_2H) SAM. Their study suggested that these techniques may provide a simple method to pattern crystal growth that cannot be produced using other crystallization methods. HA crystallization has been studied on Langmuir–Blodgett (LB) films because of the opportunities in engineering the surface properties (functional group identity, polarity, and molecular alignment periodicity) [93]. Although these negatively charged monolayers provide two-dimensional model systems for studying biomineralization processes, most biomaterials used in bone tissue engineering strategies make use of porous three-dimensional (3D) scaffolds ideal for bone ingrowth. The ability of using self-assembling materials to present ordered structures to cells in three dimensions could be an useful strategy in tissue engineering. Hwang et al. [94] synthesized self-assembling biomaterials with molecular features designed to interact with cells and scaffolds for tissue regeneration. The molecules of these materials contain cholesteryl moieties, which have universal affinity for cell membranes, and short chains of lactic acid, a common component of biodegradable tissue engineering scaffolds. These molecular materials form layered structures and can also order into single-crystal stacks with an orthorhombic unit cell. The self-organized layered structures were found to promote improved fibroblast adhesion and spreading, although the specific mechanism for this observed response was not clarified. On the basis of this study, the Stupp and co-workers [95] proposed a triblock molecule, containing a rigid cholesteryl segment followed by a flexible oligomer of L-lactic acid and a second-generation L-lysine dendron, to modify the surface of a fibrous poly(L-lactic acid) scaffold. The two hydrophobic segments should have affinity for the hydrophobic surfaces of biodegradable polyester so that the hydrophilic dendron will be displayed on the outer surfaces. The scaffolds modified by these self-assembling molecules enhanced the adhesion of 3T3 mouse calvaria cells and produced greater population growth rates. These results demonstrated the use of self-assembly to present biologically relevant chemistry on surfaces of biomaterials. Applications of this technology include the modification of substrates for cell culture or tissue engineering scaffolds. These molecules can be further elaborated to carry bioactive molecules. For instance, the dendron part can be used to anchor proteins, peptides, sugars, and other molecules. In fact, the incorporation of bioactive ligands on the surface of biomaterials has been used as a strategy to obtain desirable biological interactions (e.g., cell proliferation and/or differentiation) [96–98]. In this type of approach, the efficacy of the molecules for stimulating specific cell responses depends on the mode by which these molecules are presented to cells. In these cases, it is important to ensure spatial arrangement, flexibility and mobility, and correct orientation of the molecules when they are presented at surfaces. Molecular self-assembly offers an effective method to modify the surface properties of common biomaterials by presenting biologically relevant chemistry in a controlled, ordered manner that may improve cell affinity for biomaterials and help to invoke a specific cellular response [95] or trigger mineralization [99]. It is expected that surface modification of polymer biomaterials using self-assembling systems will yield substantial benefits, when compared with functionalized polymer surfaces, which are disordered and complex and typically consist of a single layer. On the other hand, self-assembled multilayers offer the possibility of sustained release of bioactive molecules with potential for controlling cell behavior and function.

In the biomineralization processes, inorganic crystals precipitate onto organic matrix surfaces. As mentioned before, flat HA nanocrystals are nucleated and oriented crystallographically relative to collagen fibrils in bone. Design-similar scaffolds or templates would be difficult to create with ordinary polymers and bioceramics. An interesting approach could be the combination of 3D porous scaffolds with designed self-assembling molecules thus mimicking the varying levels of hierarchy found in natural bone (e.g., osteons). The self-assembling molecules that could be utilized include peptides, which are excellent structural units with inherent chemical functionality and ability to form a variety of supramolecular assemblies [100]. Peptide sequences, which are recognized by cells and enzymes with high efficiency and specificity, can be incorporated to form highly bioactive and organized surfaces and guide stem cells toward their differentiation into osteogenic lineages. Self-assembly offers the possibility of rational designing of functionalized biomaterials that can regulate cell behavior or promote biomineralization for applications in bone repair/regeneration. As mentioned earlier, several studies suggest that the NCPs found in the mineralizing bone matrix are modulators of bone biomineralization [101, 102]. These proteins are acidic and phosphorylated and generally enriched in glutamate (E) [103, 104]. These anionic sequences, including phosphoserine residues, are able to attract calcium ions, with subsequent formation of calcium phosphates, being ideal to trigger mineralization, and have been investigated to induce mineralization [105, 106].

While most bone tissue engineering strategies use porous three-dimensional scaffolds ideal for bone ingrowth, bioactive scaffolds for bone regeneration should also mimic, at least in part, the nanoscale architecture and function of the extracellular matrix, both to interact with cellular receptors and to act as template for the formation of apatite nanocrystals with biomimetic orientation and morphology. These scaffolds may be used to provide direct or indirect cues to cells or trigger mineralization. In an attempt to produce a novel biomimetic scaffold with mechanical, structural, and biological properties found in natural bone, Sargeant *et al.* [99] proposed a hybrid system in which nickel–titanium (NiTi) foams, with fully controllable pore architectures, were filled with a self-assembled peptide amphiphile (PA) matrix containing the cell adhesion epitope arginine, glycine, aspartic acid, serine peptide sequence (RGDS) and phosphoserine, which synergistically encourage cell adhesion and mineralization. These PA molecules exhibit the ability to self-assemble into nanofibers and can form self-supporting gels upon charge screening via changes in pH or the addition of multivalent ion or charged molecules. Both *in vitro* and *in vivo* studies were carried out to evaluate the bioactive potential of these hybrid structures. After incubating the hybrid system in osteogenic media for seven days, spherical mineral was observed using SEM, and EDS analysis revealed a Ca:P ratio of 1.71 ± 0.18, which is similar to that of HA (1.67). This spherical mineral appeared to nucleate and grow on the nanofiber. *In vitro* results demonstrated that cells (mouse calvarial preosteoblastic MC3T3-E1) could be encapsulated within nanofiber matrix of PA–Ti hybrid, and that encapsulated cells remained viable. In preliminary *in vivo* studies, cylindrical plugs 3 mm in length and 2 mm in diameter were implanted into the femur of Sprague Dawley rats. Histology showed new bone growth around the implant and inside many of the pores, indicating that the hybrid structure is osteoconductive.

Mineralization on artificial interfaces, including monolayers, metal implants, biopolymer, and synthetic polymer 3D matrices, for controlled crystal growth and emerging crystallization on patterned surfaces for the creation of patterned crystals provide additional means to control morphology, microstructure, complexity, and length scales of various inorganic nanostructured materials with two and three dimensionalities, and these materials offer significant potential in bone regeneration applications.

15.5 Conclusions

In the past years, the design of new biomaterials for tissue regeneration/substitution has been inspired by the superb integration of concepts that Nature uses to produce mineralized structures with outstanding properties and complex hierarchical organization and shape. The bone structure described in this chapter is a good example of such fascinating structures. Scientists have tried to mimic the main principles of biomineralization, as the works presented throughout this chapter illustrate, in order to control the crystallization processes of inorganic/organic hybrid materials. However, until now, a material capable of reproducing completely the structural level of bone has not been developed. Despite these limitations, the study of biomineralization and self-assembly processes can help to enhance the comprehension of bone mineralization and explore ways in which biomineralization principles can be used for the synthesis of advanced biomaterials. Mimicking the bone structure, even with very simplified synthetic systems, has already led to remarkable results concerning the generation of complex mineral morphologies and control over crystallization events. So, although Nature is, and will continue to be, the best material scientist to design complex structures and to control intricate processing routes that lead to the final shape of living creatures, we certainly can continue to learn with the study of biomineralized structures.

Acknowledgments

I. B. Leonor thanks the Portuguese Foundation for Science and Technology (FCT) for providing her a postdoctoral scholarship (SFRH/BPD/26648/2006). This work was supported by the European NoE EXPERTISSUES (NMP3-CT-2004-500283) and by the Portuguese Foundation for Science and Technology, FCT, through the projects PTDC/CTM/68804/2006, PTDC/CTM/67560/2006, and PTDC/FIS/68209/2006.

References

1. Weiner, S. and Wagner, H.D. (1998) *Annu. Rev. Mater. Sci.*, **28**, 271.
2. Currey, J.D. (2005) *Science*, **309**, 253.
3. Gao, H.J. (2006) *Int. J. Fract.*, **138**, 101.
4. Weiner, S. and Wagner, H.D. (2000) *Mater. Sci. Eng. C Biomim. Supramol. Syst.*, **11**, 1.
5. Weiner, S. and Wagner, H.D. (1998) *Annu. Rev. Mater. Sci.* **28**, 271.

6. Yan, J. (2005) Elastic–plastic fracture of compact bone. PhD Thesis Materials Science & Engineering, University of Florida, Gainesville, p. 100.
7. Benesch, J., Mano, J.F., and Reis, R.L. (2008) *Tissue Eng. B*, **14**, 433.
8. Mackie, E.J. (2003) *Int. J. Biochem. Cell Biol.*, **35**, 1301.
9. Luz, G.M. and Mano, J.F. (2010) *Compos. Sci. Technol.*, **70**, 1777.
10. Athanasiou, K.A. et al. (2000) *Tissue Eng.*, **6**, 631.
11. Currey, J.D. (1999) *J. Exp. Biol.*, **202**, 3285.
12. Brown, T.D. and Ferguson, A.B. (1980) *Acta Orthop. Scand.*, **51**, 429.
13. Nalla, R.K., Kineey, J.H., and Ritchie, R.O. (2003) *Nat. Mater.*, **2**, 164.
14. Landis, W.J. and Silver, F.H. (2009) *Cell Tissues Organs*, **189**, 20.
15. Elangovan, S., Margolis, H.C., Openheim, F.G., and Beniash, E. (2007) *Langmuir*, **23**, 11200.
16. Addadi L., Weiner, S., and Geva M. (2001) *Z. Kardiol.*, **90** (Suppl. 3), 92.
17. Weiner, S. and Addadi, L. (1997) *J. Mater. Chem.*, **7**, 689.
18. Hartgerink, J.D., Beniash, E., and Stupp, S.I. (2001) *Science*, **294**, 1684.
19. Stupp, S.I. and Braun, P.V. (1997) *Science*, **277**, 1242.
20. Teng, H.H., Dove, P.M., Orme, C.A., and De Yoreo, J.J. (1998) *Science*, **282**, 724.
21. Weiner, S. and Addadi, L. (1991) *Trends Biochem. Sci.*, **16**, 252.
22. Olszta, M.J., Cheng, X.G., Jee, S.S., Kumar, R., Kim, Y.Y., Kaufman, M.J., Douglas, E.P., and Gower, L.B. (2007) *Mater. Sci. Eng. R Rep.*, **58**, 77.
23. Olszta, M.J., Douglas, E.P., and Gower, L.B. (2003) *Calcif. Tissue Int.*, **72**, 583.
24. Olszta, M.J., Odom, D.J., Douglas, E.P., and Gower, L.B. (2003) *Connect. Tissue Res.*, **44**, 326.
25. Liu, Y., de Groot, K., and Hunziker, E.B. (2004) *Ann. Biomed. Eng.*, **32**, 398.
26. Hench, L.L. (1991) *J. Am. Ceram. Soc.*, **74**, 1487.
27. Hench, L.L. and Anderson, Ö. (1993) Bioactive Glasses, in *An Introduction to Bioceramics* (eds L.L. Hench and J. Wilson), World Scientific, London, pp. 41–62.
28. Hench, L.L. et al. (1971) *J. Biomed. Mater. Res. Symp.*, **2**, 117.
29. Hench, L.L. and Wilson, J. (1984) *Biomater. Sci.*, **226**, 630.
30. Hench, L.L. (1994) in *Bioceramics*, vol. 7 (eds Ö.H. Andersson, R.-P. Happonen, and A. Yli-Urpo), Butterworth-Heinemann Ltd, Turku, pp. 3–14.
31. Gentleman, E. and Polak, J.M. (2006) *J. Mater. Sci. Mater. Med.*, **17**, 1029.
32. Neo, M. et al. (1992) *J. Biomed. Mater. Res.*, **26**, 1419.
33. Kokubo, T. et al. (1986) *J. Mater. Sci.*, **21**, 536.
34. Yamamuro, T. (1993) in *An Introduction to Bioceramics* (eds L.L. Hench and J. Wilson), World Scientific, Singapore, pp. 89–104.
35. Aoki, H. (1991) Science and medical applications of hydroxyapatite, *Japanese Association of Apatite Science*, Takayama Press System Centre Co., Inc., Tokyo.
36. LeGeros, R.Z. and LeGeros, J.P. (1993) in *An introduction to bioceramics*, (eds L.L. Hench and J. Wilson), World Scientific, London, pp. 139–180.
37. Ducheyne, P. and Qiu, Q. (1999) *Biomaterials*, **20**, 2287.
38. Neo, M. et al. (1992) *J. Biomed. Mater. Res.*, **26**, 255.
39. Neo, M. et al. (1993) *J. Biomed. Mater. Res.*, **27**, 999.
40. Kim, H.M. (2003) *Curr. Opin. Solid State Mater. Sci.*, **7**, 289.
41. Kokubo T. et al. (1990) *J. Biomed. Mater. Res.*, **24** (6), 721.
42. Kokubo, T. and Takadama, H. (2006) *Biomaterials*, **27**, 2907.
43. Filgueiras, M.R., La, T.G., and Hench, L.L. (1993) *J. Biomed. Mater. Res.*, **27**, 445.
44. Li, P. et al. (1993) *J. Appl. Biomater.*, **4**, 221.
45. Wen, H.B. et al. (2000) *J. Biomed. Mater. Res.*, **52**, 762.
46. Kokubo, T., Kim, H.M., and Kawashita, M. (2003) *Biomaterials*, **24**, 2161.
47. Tanahashi, M. et al. (1994) *J. Appl. Biomater.*, **5**, 339.
48. Tanahashi, M. and Matsuda, T. (1997) *J. Biomed. Mater. Res.*, **34**, 305.
49. Oyane, A. et al. (1999) *J. Biomed. Mater. Res.*, **47**, 367.

50. Oyane, A. et al. (2002) *J. Ceram. Soc. Jpn.*, **110**, 248.
51. Li, P.J. et al. (1992) *J. Am. Ceram. Soc.*, **75** (8), 2094.
52. Li, P. (1994) *J. Biomed. Mater. Res.*, **28**, 7.
53. Uchida, M. et al. (2001) *J. Am. Ceram. Soc.*, **84**, 2041.
54. Miyazaki, T. et al. (2001) *J. Ceram. Soc. Jpn.*, **109**, 929.
55. Miyazaki, T. et al. (2001) *J. Sol-Gel Sci. Technol.*, **21**, 83.
56. Uchida, M. et al. (2002) *J. Ceram. Soc. Jpn.*, **110**, 710.
57. Kokubo, T. et al. (2004) *J. Mater. Sci. Mater. Med.*, **15**, 99.
58. Leonor, I.B. et al. (2007) *J. Mater. Chem.*, **17**, 4057.
59. Baskaran, S. et al. (1998) *J. Am. Ceram. Soc.*, **81**, 401.
60. Mann, S. et al. (1997) *Chem. Mater.*, **9**, 2300.
61. Kim, H.M. et al. (2003) *J. Biomed. Mater. Res. A*, **67**, 1305.
62. Kim, H.M. et al. (2005) *Biomaterials*, **26**, 4366.
63. de Groot, K. (1998) in *Calcium Phosphate Coatings: Alternatives to Plasma Spray in Bioceramics*, vol. 11 (eds R.Z. Le Geros and J.P. Le Geros), World Scientific, New York, pp. 41–43.
64. Leeuwenburgh, S. et al. (2001) *J. Biomed. Mater. Res.*, **56**, 208.
65. Shirkhanzadeh, M. (1991) *J. Mater. Sci. Lett.*, **10**, 1415.
66. Kim, H.M. et al. (1996) *J. Biomed. Mater. Res.*, **32**, 409.
67. Calvert, P. and Rieke, P. (1996) *Chem. Mater.*, **8**, 1715.
68. Murphy, W.L. and Mooney, D.J. (2002) *J. Am. Chem. Soc.*, **124**, 1910.
69. Oyane, A. et al. (2005) *Biomaterials*, **26**, 2407.
70. Oyane, A. et al. (2005) *J. Biomed. Mater. Res.*, **75A**, 138.
71. Kawashita, M. et al. (2003) *Biomaterials*, **24**, 2477.
72. Balas, F. et al. (2006) *Biomaterials*, **27**, 1704.
73. Balas, F. et al. (2007) *J. Mater. Sci. Mater. Med.*, **18** (6), 1167–1174.
74. Wan, A.C., Khor, E., and Hastings, G.W. (1998) *J. Biomed. Mater. Res.*, **41**, 541.
75. Yamaguchi, I. et al. (2001) *J. Biomed. Mater. Res.*, **55**, 20.
76. Kokubo, T. et al. (2004) *Biomaterials*, **25**, 4485.
77. Leonor, I.B., Kim, H.M., Balas, F., Kawashita, M., Reis, R.L., Kokubo, T., and Nakamura, T. (2007) *J. Tissue Eng. Regen. Med.*, **1**, 425.
78. Pashkuleva, I. et al. (2005) *J. Mater. Sci. Mater. Med.*, **16**, 81.
79. Xia, F., Feng, L., Wang, S., Sun, T., Song, W., Jiang, W., and Jiang, L. (2006) *Adv. Mater.*, **18**, 432.
80. Qu, J.-B., Chu, L.-Y., Yang, M., Xie, R., Hu, L., and Chen, W.M. (2006) *Adv. Funct. Mater.*, **16**, 1865.
81. Peppas, N.A., Keys, K.B., Torres-Lugo, M., and Lowman, A.M. (1999) *J. Controlled Release*, **62**, 81.
82. Huber, D.L., Manginell, R.P., Samara, M.A., Kim, B.-I., and Bunker, B.C. (2003) *Science*, **301**, 352.
83. Hatakeyama, H., Kukuchi, A., Yamato, M., and Okano, T. (2006) *Biomaterials*, **27**, 5069.
84. Shi, J., Alves, N.M., and Mano, J.F. (2007) *Adv. Funct. Mater.*, **17**, 3312.
85. Dias, C.I., Alves, N.M., and Mano, J.F. (2008) *J. Mater. Chem.*, **18**, 2493.
86. Schild, H.G. (1992) *Prog. Polym. Sci.*, **17**, 163.
87. Denuziere, A., Ferriera, D., and Domard, A. (1996) *Carbohydr. Polym.*, **29**, 317.
88. Cho, J., Heuzey, M.-C., Bégin, A., and Carreau, P.J. (2006) *Carbohydr. Polym.*, **63**, 507.
89. Chou, Y.-F., Huang, W., Dunn, J.C.Y., Miller, T.A., and Wu, B.M. (2005) *Biomaterials*, **26**, 285.
90. Richard, D., Dumelie, N., Benhayoune, H., Bouthors, S., Guillaume, C., Lalun, N., Balossier, G., and Laurent-Maquin, D. (2006) *J. Biomed. Mater. Res.*, **79B**, 108.
91. Whitesides, G.M. and Laibinis, P.E. (1990) *Langmuir*, **6**, 87.
92. Aizenberg, J., Black, A.J., and Whitesides, G.M. (1998) *Nature*, **394**, 868.
93. Sato, K. (2007) *Crystal. Self-Organiz. Process*, **270**, 127.
94. Hwang, J.J., Iyer, S.N., Li, L.S., Claussen, R., Harrington, D.A., and

Stupp, S.I. (2002) *Proc. Natl. Acad. Sci. U.S.A.*, **99**, 9662.

95. Stendahl, J.C., Li, L.M., Claussen, R.C., and Stupp, S.I. (2004) *Biomaterials*, **25**, 5847.
96. Zhang, H.N., Lin, C.Y., and Hollister, S.J. (2009) *Biomaterials*, **30**, 4063.
97. Ferreira, L.S., Gerecht, S., Fuller, J., Shieh, H.F., Vunjak-Novakovic, G., and Langer, R. (2007) *Biomaterials*, **28**, 2706.
98. Hsiong, S.X., Boontheekul, T., Huebsch, N., and Mooney, D.J. (2009) *Tissue Eng. A*, **15**, 263.
99. Sargeant, T.D., Guler, M.O., Oppenheimer, S.M., Mata, A., Satcher, R.L., Dunand, D.C., and Stupp, S.I. (2008) *Biomaterials*, **29**, 161.
100. Gazit, E. (2010) *Nat. Chem.*, **2**, 1010.
101. Rosenthal, A.K. (2007) *Curr. Opin. Orthop.*, **18**, 449.
102. Hunter, G.K., Hauschka, P.V., Poole, A.R., Rosenberg, L.C., and Goldberg, H.A. (1996) *Biochem. J.*, **317**, 59.
103. Baht, G.S., Hunter, G.K., and Goldberg, H.A. (2008) *Matrix Biol.*, **27**, 600.
104. Fujisawa, R., Wada, Y., Nodasaka, Y., and Kuboki, Y. (1996) *Biochim. Biophys. Acta Protein Struct. Mol. Enzymol.*, **1292**, 53.
105. Hartgerink, J.D., Beniash, E., and Stupp, S.I. (2001) *Science*, **294**, 1684.
106. Spoerke, E.D., Anthony, S.G., and Stupp, S.I. (2009) *Adv. Mater.*, **21**, 425.

16
Bioactive Nanocomposites Containing Silicate Phases for Bone Replacement and Regeneration

Melek Erol, Jasmin Hum, and Aldo R. Boccaccini

16.1
Introduction

Tissue engineering (TE) has emerged as a promising interdisciplinary research field encompassing cell biology and biomaterials science, with expanding potential for the regeneration of damaged human tissue [1]. TE approaches consider the application of engineered tissue substitutes based on biomaterial constructs, called *scaffolds*, which can sustain functionality during new tissue regeneration and eventually integrate with host tissue. Bone is one of the tissues with the highest demand for regeneration, replacement, or reconstruction [2]. Typical materials investigated for bone TE are hydroxyapatite (HA) and other calcium phosphate (bioactive) ceramics, calcium carbonate, bioactive silicate glasses, and biodegradable polymers, such as poly(lactic acid) (PLA) and poly(glycolic acid) (PGA). However, none of these materials fulfill all the requirements for creating an optimal bone scaffold, including mechanical strength, toughness, osteoinductivity, osteoconductivity, angiogenic effect, and controlled degradation.

Therefore, selected combinations of materials, for example, biopolymers and bioactive ceramics, forming composites with tailored microstructure and properties are attracting substantial research efforts for bone scaffold developments [3, 4]. Numerous composite scaffolds made from natural or artificial biomaterials have been designed to fulfill many requirements, including biocompatibility, biodegradability, and/or bioresorbability. To enhance cell–material and cell–cell interactions, composites should attain suitable mechanical properties and physicochemical behavior to induce integration of the scaffold with the surrounding environment *in vivo* and to facilitate the regaining of the original shape of the damaged tissue [1, 5]. Although many of these structures are similar to that of bone, there are still considerable challenges to develop composites that replicate the hierarchical structure and the mechanical and functional characteristic of bone tissue [6, 7].

Human bone is a fascinating natural nanocomposite (NC) material that consists of organic components, mainly collagen (COL) and other proteins and an inorganic mineral phase. Natural bone, being a natural example of inorganic–organic composites, consists in composition of ~70 wt% inorganic crystals (mainly hydroxyapatite: $Ca_{10}(PO_4)_6(OH)_2$) and 30 wt% of organic matrix (mainly Type I COL) [8–15]. The two components provide toughness and strength, based on the hierarchical combination of COL and inorganic phase, respectively. The superior mechanical properties of bone are thus related to its complex hierarchical composite structure that starts from the nanoscale and spanning over several orders of length scales up to macroscopic tissue [12].

It is known that the organization of cells and the corresponding tissue properties are dependent on the structure of the extracellular matrix (ECM). The ECM has a complex hierarchical structure with spatial arrangement at different levels, for example, covering several orders of magnitude (nanometer to centimeter). Therefore, cells in our body are predisposed to interact with nanostructured surfaces [16]. Nanomaterials have an increased number of atoms and crystals at their surfaces and possess a higher surface area to volume ratio than conventional microscale biomaterials. Numerous studies indicate that the unique properties of nanomaterials provide advantageous interactions with the proteins that control cellular function [16, 17]. For example, osteoblast (bone-forming cell) function has been shown to be greater on nanostructured compared to micron-scale materials, and thus precise nanoscale engineering of implant and scaffold surfaces is desired to provide greater control over cell behavior achieving improved tissue-implant interfaces [18, 19].

Polymer nanocomposites, particularly those incorporating a nanoscale silicate phase in a polymer matrix, are a relative new class of biocomposites developed for bone scaffolds [20]. Nanocomposite systems comprising a biodegradable polymer matrix and a nanoscaled silicate phase, for example, based on bioactive glasses (BGs), silica, or nanoclay, are the topics of this chapter, which cover the available literature on production and characterization of these relatively new class of materials and their application in bone TE. Section 16.2 discusses briefly the key characteristics of bone as a nanocomposite. In Section 16.3, a comprehensive review of composite systems incorporating BG nanoparticles and nanofibers, nanoscaled silica, and nanoclay is presented with their different synthesis methods. Finally, in Section 16.4, a summary discussion on the topic is presented and areas of future research are highlighted.

16.2
Nanostructure and Nanofeatures of the Bone

16.2.1
The Structure of Bone as a Nanocomposite

The design strategy of an ideal bone tissue scaffold necessitates understanding the fundamentals of bone composition and architecture. As mentioned above, bone is a nanocomposite based on calcium phosphate crystals, primarily hydroxyapatite (HA), and COL [10]. COL acts as a structural framework in which platelike tiny crystals of HA are embedded to strengthen the structure [21]. COL

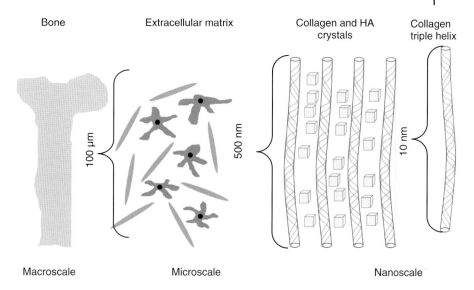

Figure 16.1 The structure of bone from macroscale to nanoscale. (Source: Adapted from Ref. [24].)

molecules assemble into triple helices that bundle into fibrils (1.5–3.5 nm diameter), which subsequently bundle into COL fibers (50–70 nm) [10]. Similarly, HA forms nanocrystals with dimensions of about 2–3 nm in thickness and tens of nanometers in length [22, 23] (Figure 16.1). Both the size and the orientation of the HA crystals are dictated specifically by the COL template, and the precise structural arrangement between the COL and hydroxyapatite crystals is critical to bone's toughness and strength [24].

For bone formation and growth, the organism uses the principles of (biologically controlled) self-assembly. This provides a control over the structure of bone at all levels of hierarchy, for example, at the macroscale (>1 mm), microscale (1 μm to 1 mm), and nanoscale (<1 μm). At the macrostructural level, bone can be classified as spongy bone and compact bone, which are organized with multilevel porosity for the establishment of multiple functions, including transportation of nutrients, oxygen, and body fluids. At the microstructural level, the repeated structural unit of compact bone consists mostly of osteon or Harversian system, which acts as load-bearing pillars while the spongy bone is formed by an interconnecting framework of trabeculae. These trabeculae have three types of cellular structures: plate/plate-like, plate/bar-like, and bar/bar-like. At the nanostructural level, as mentioned above, bone is comprised mainly of COL fibers and nanocrystals of bone mineral, particularly HA.

The hierarchical structuring of bone allows the adaptation of the material at each scale level to yield outstanding performance. For example, the extraordinary toughness of bone is due to the combined advantages of structural elements at the nanometer [25, 26] and the micrometer levels [27]. On the basis of the knowledge of the structure and properties of bone, it is recognized that the hierarchical structuring provides a major opportunity for bioinspired materials synthesis and achievement of properties for specific functions [28]. It should be noted that although several structural levels of bone tissue have been identified and bone properties have been investigated, opportunities for further research remain for a complete understanding of the mineral–matrix interactions and to explain cell–matrix and cell–cell interactions at all these levels, especially for nanoscale interactions.

16.2.2
Cell Behavior at the Nanoscale

Cells in their natural environment are surrounded by nanostructures in contact with other cells and with the ECM [29, 30]. Protein interactions at the nanoscale are crucial to control cell functions such as proliferation, migration, and ECM production [31]. It is also well known that protein adsorption characteristics on biomaterials depend on the surface features of the material [32]. At the nanometer scale, key properties determining the cell–biomaterial interactions (such as surface area, surface roughness, hydrophilicity, and wettability) are different from those in the conventional (micrometer) scale, and induce specific cell adhesion behavior [33, 34]. In general, nanoscaled biomaterials have a high ratio of surface area to volume [35], which results in greater surface energy and surface bioreactivity compared to microscale biomaterials. The larger specific surface area of nanoparticles allows not only a faster release of ions but also a higher protein adsorption, which improves bioactivity [35, 36]. In addition, the surface bioreactivity of nanoparticles is higher than that of micron-sized particles [36] and nanoparticles, and nanofibers provide nanostructured features on scaffold surfaces, which are likely to improve osteoblast cell attachment and subsequent cell behavior [34]. As mentioned above, although cells have micrometric dimensions, they evolve *in vivo* in close contact with the ECM, which is formed by structural features of nanometer size [37, 38].

16.3
Nanocomposites-Containing Silicate Nanophases

16.3.1
Nanoscale Bioactive Glasses

Silicate BGs were developed for the first time by Hench and coworkers more than 40 years ago [39]. These highly surface reactive inorganic materials consist of a silicate network incorporating sodium, calcium, and phosphorus in different proportions. The traditional 45S5 BG (45S5 Bioglass®) of composition (in weight percentage): 45 SiO_2, 24.5 Na_2O, 24.5 CaO, and 6 P_2O_5 has received approval of the US Food and Drug Administration (FDA) for clinical use in the treatment of periodontal diseases as bone filler and also in middle ear surgery [40]. Numerous BG compositions have been developed over the years incorporating selected elements such as fluorine, magnesium, strontium, iron, silver, boron, potassium, or zinc [41–47]. In the field of bone TE, bioactive silicate glasses of several compositions are being investigated [4, 48–51]. Both micron-sized and nanoscale particles are considered for fabrication of composite scaffolds, for example, combining biodegradable polymers and BG inclusions [4, 52]. In this context, it has been shown that BG dissolution products upregulate the expression of genes controlling osteogenesis [48]. There is also emerging evidence about the potential angiogenic effects of BGs, that is, increased secretion of vascular endothelial growth factor (VEGF) *in vitro* and enhancement of vascularization *in vivo* [53–55].

Nanoscale bioactive glasses (NBGs) (nanoparticles and nanofibers) are expected to enable a faster solubility of the material (higher ion release rate) and also higher protein adsorption [52, 56]. Faster deposition or mineralization of tissues such as bone or teeth is possible when the tissues are in contact with nanoscale particles [57]. This effect is related to the bone structure described above. Mimicking the nanofeatures of bone on the surface of synthetic bone implants, for example, has been shown to increase bone-forming cell adhesion and proliferation [58].

For bone TE purposes, where scaffolds made of polymer/BG composites are applicable [4, 52, 53], the use of NBGs should lead to improved scaffolds' mechanical and biological properties. For example, the introduction of nanoscale fillers with desired morphology usually increases the mechanical strength and stiffness of composites compared to the properties of the neat polymer and of composites with micron-size reinforcement [59]. The larger specific surface area of the nanoparticles also leads to an increased interface effects, and it should contribute to improved bioactivity, compared to standard composites incorporating micrometer-sized particles. A recent review has comprehensively discussed the application of NBGs in biodegradable composites [52]. The following sections briefly summarize the latest developments in the field.

16.3.1.1 Synthetic Polymer/Nanoparticulate Bioactive Glass Composites

There are increasing research efforts devoted to develop nanocomposites by combining synthetic biopolymers and NBGs, In an early study in this field poly(3hydroxybutyrate) (P(3HB))/nanoparticulate BG (Bioglass® composition) composites with different filler concentrations were fabricated by solvent casting [60]. The addition of nanoparticles had a significant effect on the elastic constant of the composite compared with the micrometer-sized counterparts. Moreover, the protein adsorption on the surface of nanocomposites was considerably higher compared to the unfilled polymer and to composites containing micron-sized BG particles [60]. More recently, P(3HB)-based 3D composite scaffolds were also produced using NBG particles to enhance antimicrobial properties (Figure 16.2) [61]. Bactericidal studies revealed that NBG particles ensure both biocompatibility and enhanced antimicrobial properties. It was also concluded that the higher specific surface area of NBG particles played a vital role in imparting the desired antimicrobial activity [61].

Zheng *et al.* [62] developed composites using poly(hydroxybutyrate-2-*co*-2-hydroxyvalerate) (PHBV) containing biomimetically synthesized nanosized

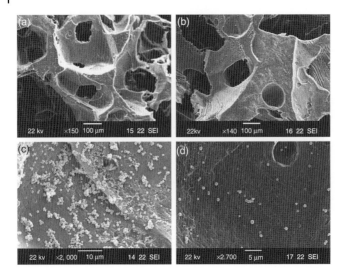

Figure 16.2 SEM micrographs of *Staphylococcus aureus* attachment after 48 h on (a,c) P(3HB) foams and (b,d) P(3HB)/NBG composite foams. A decrease in *S. aureus* cells on the surface of the P(3HB)/10 wt% NBG foams is evident (d) in comparison to P(3HB) foams (c), respectively [61]. (Source: Reproduced with permission of Elsevier.)

bioactive glass (BMBG) (CaO–P_2O_5–SiO_2). The porous composites were shown to be bioactive, and cell attachment studies indicated that the material has attractive biomineralization and cell biocompatibility [62]. Composites combining poly(L-lactic acid) (PLLA) and sol–gel derived bioactive glass ceramic (BGC) nanoparticles were fabricated by Hong et al. [56] using thermally induced phase separation. It was shown that composites containing BGC nanoparticles with lower phosphorus and higher silicon content led to higher bioactivity compared to composites incorporating BGC with lower silicon and higher phosphorus content [56]. The effect of nanoparticle content on the properties of nanocomposite scaffolds has also been investigated [63], and an improvement in the mechanical properties was measured. El-Kady et al. [64] developed a similar system using sol-gel-derived BG nanoparticles and PLA. In this system, it was shown that the scaffold's pore size decreases with the increase in glass nanoparticles content.

Nanoscaled coatings on different materials have been developed using sol-gel-derived BG nanoparticles aiming at improving mechanical and biological properties. Roohani-Esfahani et al. [65] coated the struts of biphasic calcium phosphate (BCP) scaffold with a nanocomposite layer consists of NBG and poly(ε-caprolactone) (PCL). A flat fracture surface formed for the BCP scaffold with the struts was completely disconnected, indicating the brittle nature of the structure, as seen in Figure 16.3 [65]. However, the BCP/PCL–NBG scaffolds exhibited thin fibrils before strut rupture leading to enhanced (less brittle) mechanical behavior. The effect of various NBG concentrations on the mechanical properties and *in vitro* behavior of the scaffolds was comprehensively examined and compared with that of a BCP scaffold coated with PCL and hydroxyapatite

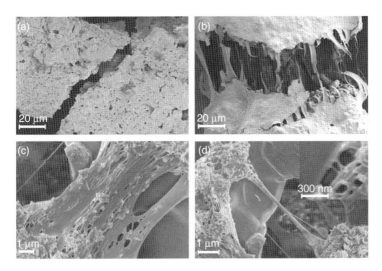

Figure 16.3 Fracture surfaces of (a) a BCP scaffold and (b–d) a BCP/PCL–NBG scaffold, according to Ref. [65]. (Source: Reproduced with permission of Elsevier.)

nanoparticles (nHA) (BCP/PCL–nHA) and a BCP scaffold coated with only a PCL layer (BCP/PCL). The scaffolds coated with PCL and NBG particles showed high compressive strengths (increased 14 times), elastic moduli (increased 3 times), interconnectivity, and porosity, as well as improved bioactivity. It was also reported that BCP/PCL–NBG scaffolds induced the differentiation of primary human bone-derived cells (HOBs), with significant upregulation of osteogenic gene expression for Runx2, osteopontin, and bone sialoprotein, compared with the other groups [65].

An approach to improve the mechanical properties of nanoparticulate BG/PLLA composites by using solvent evaporation has been reported by Liu et al. [66, 67]. It was shown that surface modification of nanosized BG particles by grafting organic molecules or polymers is a convenient solution to improve the mechanical properties of the composites. Surface-modified BG/PLLA composites exhibited much better cell proliferation ability than nonmodified BG/PLLA composites and pure PLLA [66, 67]. A related study was conducted by Kim et al. [68, 69] who developed PLA composites filled with sol-gel-derived bioactive glass nanofiber (BGNF) fabricated by electrospinning (ES). The nanocomposites were shown to consist of uniformly dispersed nanofibers within a PLA matrix (Figure 16.4). The in vitro bioactivity and osteoblast response to the developed nanocomposites were studied [69]. The nanocomposites showed excellent bioactivity, inducing CaP precipitation within 24 h of immersion in SBF. Also, the results have been confirmed by Noh et al. [70] in a similar study. In related research, Lee et al. [71] have produced PCL/sol-gel-derived BGNF nanocomposites. The glass nanofibers were homogeneously distributed in the PCL matrix, showing a much rougher surface than the pure PCL. The precipitated apatite covered the surface of the NC membrane almost completely after immersion in SBF for 14 days. Osteoblastic

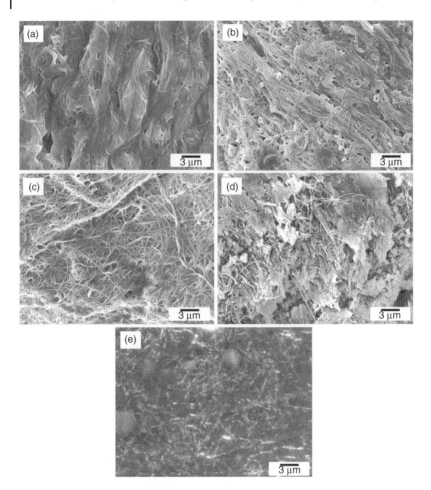

Figure 16.4 SEM morphologies of the PLA nanofiber nanocomposites: (a–d) as-dried samples containing different nanofiber concentrations: (a) 5% nano, (b) 15% nano, (c) 25% nano, and (d) 35% nano, (e) thermal-pressed surface of 25% nano [69]. (Source: Reproduced with permission of John Wiley and Sons.)

cells (MC3T3-E1) were seen to spread better and to grow with many cytoplasmic extensions, showing improved proliferation behavior compared to those on the pure PCL membrane [71]. In a similar development, Jo et al. [72] have fabricated (PCL)/sol-gel-derived BGNF composites exhibiting a highly homogeneous BGNF distribution. This microstructure resulted in a significant improvement of the biological and mechanical properties of the PCL/BGNF composites, compared to that of the micron-sized ones.

16.3.1.2 Natural Polymer/Bioactive Glass Nanocomposites

Natural polymers, such as polysaccharides (starch, chitin, and chitosan) and proteins (silk and COL) are the candidates for the preparation of nanocomposites for biomedical applications. Peter *et al.* [73, 74] have synthesized α-chitin/sol-gel-derived BGC nanoparticle and chitosan/sol-gel-derived BGC nanoparticle composite scaffolds by using lyophilization technique. Macroporous composite scaffolds with pore size in the range 150–300 µm were fabricated [74]. *In vitro* studies showed the deposition of apatite and the attachment of osteoblastlike cells (MG-63) on the surface of the composite scaffolds [73, 74]. Chitosan–gelatin/nanosized BGC nanocomposite scaffolds have also been fabricated by freezing and lyophilization technique [75]. It was reported that high surface area of BGC nanoparticles resulted in increasing protein adsorption, especially adhesive proteins.

Porous bioactive nanocomposites combining sol-gel-derived BG nanoparticles, COL, hyaluronic acid (HYA) and phosphatidylserine (PS) have been fabricated by a combination of sol–gel and freeze-drying methods [76]. A bioactive nanocomposite was also synthesized by crosslinking COL and HYA by using 1-ethyl-3-(3-dimethylaminopropyl) carbodiimide (EDC) and N-hydroxysuccinimide (NHS). Biomineralization, degradation in SBF, and mechanical strength of the EDC/NHS-cross-linked BG-COL-HYA-PS composite scaffolds were higher than those of the scaffolds without HYA, PS, and cross-linking process. PS and HYA can contribute to regulate the biomineralization process, inducing HA to precipitate on the surface of the composites. The *in vivo* bone regeneration ability of the EDC/NHS-cross-linked BG-COL-HYA-PS composite scaffolds was investigated by Xie *et al.* [77] using a rabbit radius defect model. X-ray and histological studies showed the ability of bone regeneration for both nanocomposites and nanocomposites combined with growth factors (bone morphogenetic protein (BMP)). Moreover, nanocomposites containing BMP showed a better ability of ectopic bone formation [77]. In related studies, Luz and Mano [78] have proposed a new composite membrane combining chitosan with sol-gel-derived BG nanoparticles based on both ternary (SiO_2–CaO–P_2O_5) and binary (SiO_2–CaO) systems [78, 79]. Mozafari *et al.* [80] developed macroporous bioactive nanocomposite scaffolds using cross-linked gelatin and NBG through layer solvent casting combined with freeze-drying and lamination techniques. It was observed that NBG particles are dispersed evenly among cross-linked gelatin matrices. The formation of chemical bonds between NBG particles and gelatin was also reported. The gelatin matrix structure after cross-linking was highly porous with parallelly aligned and interconnected pores, as seen in Figure 16.5 [80]. The gelatin solution was able to act as a binder for nanocomposite layers. It was found that the values of elastic modulus (E) increased with increasing concentration of NBG nanoparticles, and the Young's modulus of the scaffolds was comparable to that of natural spongy bone. The overall results demonstrated that gelatin/NBG nanocomposite scaffolds are attractive for bone engineering from different points of view; that is, including improved cell culture response, biocompatibility, bioactivity, and high reinforcing quality of BG nanoparticles [80]. Alginate, a natural copolymer derived from various species of kelp, has useful properties, such as the ability to form gels in combination with certain metal ions, which induce cross-linking of the guluronic residues of the alginate polymer without damaging cells during the gelling process. Srinivasan *et al.* [81]

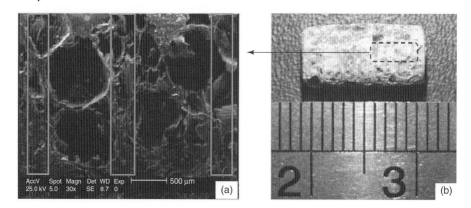

Figure 16.5 Macroporous bioactive nanocomposite scaffolds using cross-linked gelatin and NBG through layer solvent casting combined with freeze-drying and lamination techniques: (a) overall view of the final nanocomposite scaffolds after lamination technique and (b) SEM micrograph of final nanocomposite (lateral surface) [80]. (Source: Reproduced with permission of Elsevier.)

successfully fabricated alginate cross-linked with calcium ion/nBGC composite scaffolds by using lyophilization technique. Composite scaffolds exhibiting pore size of about 100–300 μm, controlled porosity, and swelling ability were fabricated. The scaffolds showed enhanced biomineralization because of the presence of nBGC in the alginate matrix. Incorporation of nBGC did not alter the viability of MG-63 cells and human periodontal ligament fibroblast (HPDLF) cells. It was also noted that nBGC helped to attain high protein adsorption, cell attachment, and cell proliferation on the scaffolds. The presence of nBGC also enhanced the alkaline phosphatase (ALP) activity of the HPDLF cells cultured on the composite scaffolds. It was suggested that the biocompatible composite scaffolds can serve as an appropriate bioactive matrix for periodontal tissue regeneration [81]. In a recent study, the physicochemical, biological, and drug-release properties of gallium-cross-linked alginate/nanoparticulate BG composite films were investigated [82]. The aim of the work was to develop biodegradable and bioactive materials with both sufficient structural integrity and prophylaxis effect against infection. As expected, incorporation of NBG nanoparticles into Ga-cross-linked alginate films significantly improved their mechanical properties when compared with films fabricated with micron-sized BG particles. Biomineralization studies in simulated body fluid (SBF) suggested the deposition of hydroxyapatite on the surface of the films indicating their bioactive nature.

The development of COL/NBG composites has been presented by Marelli *et al.* [83]. They developed plastically compressed dense collagen (DC) gels mimicking the microstructural, mechanical, and biological properties of native osteoid. This study investigated the effect of combining DC with nanosized BG particles to enhance the mineralization and cell-seeding ability of scaffolds for boneTE. It was also observed that with increasing immersion time in SBF, there was a significant increase in the scaffold compressive modulus.

It was also shown that, in the absence of osteogenic supplements, the metabolic activity and ALP production of MC3T3-E1 cells were affected by the presence of NBG, indicating accelerated osteogenic differentiation.

Kim et al. [84] have developed BGNF-COL nanocomposite in the form of both a thin membrane and as macroporous scaffold. BGNF-COL nanocomposites were seen to exhibit high bioactivity, which was assessed by the rapid formation of bonelike apatite minerals on their surfaces after immersion in SBF. The nanocomposites were shown to enhance the adhesion and growth of human osteoblastlike cells [84].

Couto et al. [85] developed injectable biodegradable materials with BG nanoparticles in order to produce thermo-responsive hydrogels for orthopedic reconstructive and regenerative medicine applications. Chitosan-β-glycerophosphate salt formulation was combined with sol-gel-derived BG nanoparticles to synthesize novel thermo-responsive hydrogels. The inner structure of the hydrogels was characterized by using cryogenic scanning electron microscopy (cryoSEM), and it was found that BG nanoparticles were well dispersed in the organic matrix [85]. *In vitro* bioactivity tests showed that the BG nanoparticles incorporated in the chitosan-based thermo-responsive system induced the formation of bonelike apatite clusters that are well integrated in the hydrogel organic structure.

16.3.2
Nanoscaled Silica

Silica is often the material of choice to enable the biological use of a wide range of inorganic particles [86–88]. For example, silica-coated semiconductor quantum dots, such as cadmium sulfide and selenide, have been demonstrated to possess high stability, chemical versatility, and biocompatibility that are crucial for many biomedical applications [89, 90]. On the basis of its high stability and biocompatibility, silica particles themselves are attractive for many biomedical applications, for example, owing to their osteogenic properties [20]. Among many structurally stable inorganic materials being investigated for drug delivery, silica materials with defined structures and surface properties represent relevant candidates. For controlled release applications, it has been shown that silica is able to store and gradually release therapeutically relevant drugs [91–93]. Drug molecules have been loaded into silica nanoparticles (SNPs), and surface modification of the SNPs with biorecognition entities can allow specific cells or receptors in the body to be located [94]. On target recognition, SNPs can release their drug payload at a rate that can be precisely controlled by tailoring the internal structure of the particles for a desired diffusion/release profile [94]. Moreover, the bioactive behavior of silica-based ordered mesoporous materials represents an added value when considering these materials for bone tissue regeneration technologies. The combination of both properties, that is, bioactivity and controlled delivery capability, is a remarkable synergy that makes mesoporous silica nanoparticle (MSNP) excellent candidates to be used as starting materials for the manufacture of 3D scaffolds for bone TE [95]. The enhanced cell response may result from the suitable protein adsorption ability, which is also influenced by the nanoscale dimensions [96]. In addition, bone-forming cells can also be activated by soluble silica. This may explain why some of the silica-releasing materials are osteoinductive [48, 97–99].

SNP have also been used as nanofiller for the preparation of polymer/SiO_2 nanocomposites, some of these composite systems are described below; however, the section is only a brief summary given the availability of recent review articles in the field [20, 100–102]. Polymer/silica nanocomposites can be prepared by physically mixing SNPs with polymers [103, 104] or by copolymerization of the organic polymers with surface-functionalized SNPs [105, 106]. Hybrids of polymer and silica have also been synthesized to serve as a support, to reinforce, and to guide new tissue ingrowth and regeneration, for example, using COL as the natural polymer phase [107]. In the last years, several studies have been devoted to the study of organic/inorganic hybrids, with the aim of combining the properties of both phases at the nanoscale level [108–111]. However, polymer/SNP hybrids are not the subject of this Chapter.

16.3.2.1 Composites Containing Silica Nanoparticles

Well-ordered spherical SNPs with stable pore mesostructures of hexagonal and cubic topology, as well as layered (vesicular) structures can be synthesized by using aerosol-based processes. Lu et al. [112] developed various metal and polymer/silica nanocomposite particles by aerosol-assisted self-assembly technique. Silica/gold nanocomposites showed an ordered hexagonal mesostructure, while silica/P123 (triblock copolymer (Pluronic-P123)) nanocomposites exhibited reticulated foam structure. It was also noted that this process can be modified for the formation of ordered mesostructured thin films [112].

Chrissafis et al. [113] used layered silicate clays (montmorillonite MMT) (see also Section 16.3.3) and SNPs to enhance the mechanical and thermal properties of PLLA ligaments. They prepared nanocomposites of PLLA containing SNPs and organically modified montmorillonite (OMMT) by solved evaporation method. It was observed that both nanoparticles were well dispersed in the PLLA matrix. All nanocomposites exhibited higher mechanical properties compared to neat PLLA. In addition, it was reported that nanoparticles also affected the thermal properties of PLLA and especially the crystallization rate, which in all nanocomposites was faster than that of neat PLLA [113].

Controlled mineralized silk films with different silica morphologies and distributions were successfully generated as three-dimensional (3D) porous networks, clustered SNPs or single SNPs by Mieszawska et al. [114]. Different relative concentrations of silica in silk films produced two different morphologies of silica deposits: a 3D network and aggregates of SNPs distributed on the silk surface. The average diameter of SNPs was fairly uniform (90 nm). Silk was proposed as the organic scaffolding to control the material stability and multiprocessing makes silk/silica biomaterials suitable for different tissue regenerative applications. Preliminary studies with human mesenchymal stem cells (hMSCs) indicated support of silk/silica films toward cell attachment and upregulation of osteogenic gene markers. These results also indicated that the silk/silica system enhances osteogenesis. Evidence for early bone formation as calcium-containing deposits was observed on silk films with silica. These results indicated the potential utility of these new silk/silica systems toward bone regeneration [114].

Madhumathia et al. [115] have developed chitin/nanosilica composite scaffolds by lyophilization method. These scaffolds showed a porous structure with smooth morphology. The obtained scaffolds were found to be bioactive in SBF and biocompatible when tested with MG 63 Cell. It was considered that silanol groups (Si–OH) played a major role in the apatite formation and chitin/nanosilica composite scaffolds were suggested for bone TE applications [115]. Hierarchical mesoporous polymer/silica nanocomposites based on chitosan were developed by Rana et al. [116]. The hierarchical structure of the composites with bimodal mesopore size distribution was investigated by transmission electron microscopy (TEM), and it was claimed that the presence of voids between SNPs would allow the composite to efficiently encapsulate an increased amount of drug molecules.

16.3.3
Nanoclays

Clays are naturally occurring minerals with variability in their constitution depending on their groups and sources. Clays used for the preparation of nanoclays belong to the smectite group clays, the most common of which are MMT and hectorite where the octahedral site is isomorphically substituted. The crystal lattice of smectite group clay consists of two-dimensional 1 nm thick layers that are made up of two tetrahedral sheets of silica (SiO_2) fused to an edge-shaped octahedral sheet of alumina. The lateral dimensions of these layers vary from 30nm to several microns depending on the particular silicate [117].

Nanoclay is one of the most affordable silicate materials that have shown promising results as reinforcement in polymers [20]. Nanoclay is made from MMT mineral deposits and is known to have "platelet" structure. Its structure consists of a one-edge-shared octahedral sheet of aluminum hydroxide fused in between two silica tetrahedrals [118]. MMT is a well-characterized layered silicate that can be made hydrophobic through either ionic exchange or modification with organic surfactant molecules to aid in dispersion [119, 120]. Relevant medical properties of MMT, such as its ability to adsorb various types of toxins and to cross the gastrointestinal barrier [121–124] along with its drug-carrying and drug-delivery abilities [125, 126] encourage its use for TE applications.

It has been reported that these clays are excreted from the body, as they are unable to be absorbed by the intestinal tract and they can also be dissolved by the acids present in the stomach or the bowel [127]. In addition to this, clays have also been used as oral laxative and as antidiarrhoetics [127]. These findings imply that clay can be suitable for TE applications, for example, to develop scaffolds, considering that the degradation products obtained from the scaffold materials can be excreted without having any adverse effect on the normal body functions [128].

16.3.3.1 Composites Containing Clay Nanoparticles

Layered smectite nanoclays, particularly of the MMT type, have interesting structural characteristics making them suitable for converting planar surfaces from hydrophilic to hydrophobic, thereby leading to more suitable clays for incorporation into organic polymer matrices. Because the surface area of these clays is very large, a small percentage of the clays, when fully dispersed and exfoliated, can saturate the host polymer (or monomer) system. Nanoscopic phase distribution can impart enhanced stiffness and strength with substantially less inorganic content than conventional mineral fillers [129]. The field of polymer nanocomposites containing clays received an impetus after breakthrough developments at Toyota in 1987 [130]. They discovered the possibility of synthesizing polymer nanocomposites based on nylon-6/organophilic MMT clay that showed dramatic improvements in mechanical and physical properties and heat distortion temperature at very low content of layered silicate [117, 130]. Polymer/clay nanocomposites have shown improvements in mechanical, thermal, and optical properties compared to pure polymers or conventional composites (composed of micron-sized particles) [20, 119]. These materials are nanosized in one dimension, and thus they act as nanoplates that sandwich polymer chains in composites. The following three methods have been developed to produce polymer/layered silicate nanocomposites:

1) **In situ polymerization**: in which a polymer precursor or monomer is inserted in between clay layers followed by expanding the layered silicate platelets into the matrix by polymerization. This method has the advantage of producing well-exfoliated nanocomposites;
2) **Solution-induced intercalation method**: it involves solvents to swell and disperse clays into a polymer solution. This approach has some difficulties for the commercial production of nanocomposites owing to the high costs of the solvents required and the required phase separation of the synthesized products from the solvents;
3) **Melt processing method**: it applies intercalation and exfoliation of layered silicates in polymeric matrices during melting. The efficiency of intercalation using this method may not be as high as that of *in situ* polymerization, and often composites produced contain a partially exfoliated layered structure [131].

Polymer layered silicate composites are ideally divided into three general types: conventional composites, intercalated nanocomposites, and exfoliated nanocomposites. The increasing demand of polymer/layered silicate nanocomposites should contribute to the expansion of applications, including in the biomedical field [20, 117].

For example, MMT was incorporated with potassium persulfate (KPS) in aqueous solution through cationic exchange first and then mixed with the acidified aqueous solution of chitosan to prepare chitosan/MMT nanocomposites [132]. It was observed that the partially exfoliated MMT exhibited a structure composed of stacked layers, where each layer had a regular shape and translucent appearance, whereas the exfoliated MMT exhibited an undefined shape with size in the range 15–25 nm. The exfoliated MMT layers were found to flatten out in parallel with the

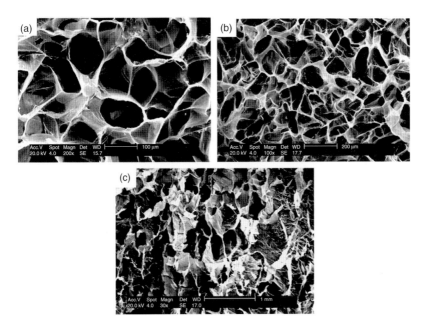

Figure 16.6 Cross-section SEM morphology of Gel/MMT–CS scaffolds (Gel/CS = 1 : 1) with different MMT contents: (a) 0%, (b) 11 wt%, and (c) 20 wt%, according to Ref. [133]. (Source: Reproduced with permission of Elsevier.)

surface, which not only increased the tensile strength of the chitosan film but also hindered degradation during *in vitro* tests [132].

Zheng *et al.* [133] developed a gelatin/montmorillonite–chitosan (Gel/MMT–CS) nanocomposite scaffold via the intercalation process and the freeze-drying technique, using ice particulates as porogen. It was reported that, as the MMT content increased to 20 wt%, the pore shape became irregular with a lower degree of interconnection (Figure 16.6) [133]. It was shown that high MMT content induces the ice particulates to nucleate and grow. The irregular growing of ice particulates causes the pores to be of less regularity. It was demonstrated that the introduced intercalation structure led to Gel/MMT–CS scaffolds with improved mechanical properties and controllable degradation rate. The mechanical properties were increased obviously by the incorporation of a small amount of MMT nanosheets, which are suitable to tailor the stiffness of the composites for specific requirements in a broad range of hard and soft tissue scaffold applications. MTT assay confirmed the good cell affinity and biocompatibility of Gel/MMT–CS nancomposites, indicating the possible application as matrix for TE. It was also noted that the *in vitro* degradation rate was greatly affected by the incorporation of MMT [133].

Katti *et al.* [134] produced a new biopolymer-based novel chitosan (chi)/MMT/hydroxyapatite (HA) nanocomposite for biomedical applications. X-ray diffraction (XRD), results indicated that an intercalated structure was formed with an increase in d-spacing of MMT, and FTIR studies revealed the evidence of molecular interaction among the three different constituents of the composite.

Atomic force microscopy (AFM) images showed well-distributed nanoparticles in the chitosan matrix. In the chi/MMT composite, clay particles were elongated in shape having a length of 500 nm and a width of 200 nm. In the case of chi/HA, globular HA particles were homogeneously dispersed in the chitosan matrix. The diameter of the HA particles was in the range 50–100 nm. The smaller particulate size in the chi/MMT/HA composite compared to chi/MMT and chi/HA indicated a better dispersion of MMT and HA in the polymer matrix (Figure 16.7) [134]. There was a significant improvement in the mechanical properties of the composites, even compared to chi/MMT and chi/HA composites. This notable improvement in mechanical property was attributed to the better dispersion of nanoparticles and the interaction between nanoparticles and the organic matrix. The results of cell culture experiments showed that the composite was biocompatible and had a better cell proliferation rate compared to chi/HA composites [134].

Ambre et al. [128] developed scaffolds based on chitosan/polygalacturonic acid (chi/PgA) complex containing MMT clay modified with 5-aminovaleric acid by

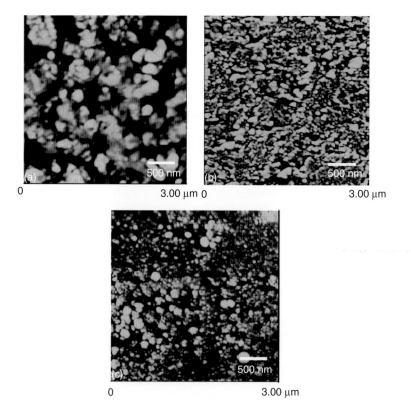

Figure 16.7 Atomic force microscope phase images of (a) chi/MMT, (b) chi/HAP, and (c) chi/MMT/HAP composites, according to Ref. [134]. (Source: Reproduced with permission of IoP Publishing.)

using freeze-drying technique. The obtained results were also compared with data from previous studies related to chi/PgA/HA composite scaffolds [135, 136]. The MTT assay revealed that the number of osteoblast cells in chi/PgA scaffolds containing the modified clay was comparable to chi/PgA scaffolds containing HA, known for its osteoconductive properties. Compressive mechanical tests showed that values of the compressive elastic modulus of chi/PgA/HA, chi/PgA/MMT, and chi/PgA/HA/MMT scaffolds were in the 4–6 MPa range and higher than those of chi/PgA scaffolds. It was concluded that the chi/PgA/HA/MMT system is a reliable material system exhibiting good biocompatibility and osteoconductive properties [136]. Zuber *et al.* [137] prepared chitin-based polyurethane bionanocomposites (PUBNCs) by emulsion polymerization. A mixture of polymer and bentonite clay enriched in MMT was formed in emulsion polymerization, in which MMT dispersed differently depending on the interaction of MMT with polymer chains. The results revealed the well-dispersed, ordered, intercalated assembled layers of bentonite formed in the polyurethane (PU) matrix. In a similar study, the same authors fabricated PUBNC by using chitin, Delite® HPS bentonite nanoclay enriched in MMT, 4,4-diphenylmethane diisocyanate (MDI), and polycaprolactone polyol (CAPA 231) [138]. It was found that the mechanical properties of the synthesized materials were improved with increased Delite® HPS bentonite nanoclay content. It was observed that a significant increase in the mechanical properties occurs with increasing content of Delite® HPS bentonite nanoclay. It was also reported that the addition of bentonite nanoclay improved the tensile strength by about three times and a linear reduction in the elongation at break took place. However, the cytotoxicity of the synthesized PU bioNC was also affected by varying the Delite HPS bentonite nanoclay content in the chemical composition of the final synthesized materials. It was revealed that cytotoxicity level increased with increased bentonite nanoclay content (Figure 16.8) [138]. The staining results also showed that increased concentration of bentonite nanoclay has adverse effects on the biocompatibility of the samples [138].

Besides natural polymers (discussed above), synthetic polymers have been used as matrices to prepare polymer/clay nanocomposites. Ozkoc *et al.* [139] synthesized PLA/clay nanocomposite (NC) by using microcompounding/injection molding and subsequent the polymer/particle leaching method. PLA was combined with NaCl (used as the particulate porogen), poly(vinyl alcohol) (PVA, used as the immiscible polymeric porogen), and organically modified layered silicate (OMLS, used as nanofiller) to fabricate nanocomposite scaffolds and investigated the effects of clay addition and clay loading level on the morphology, porosity, wettability, mechanical, and thermal properties of the scaffolds. The results showed that the incorporation of PVA in the PLA matrix as a polymeric porogen enhanced the state of dispersion of clay particles resulting in an exfoliated structure. When the pore morphology of the PLA scaffolds and NC were compared, there was no significant difference observed in terms of pore wall-structure and overall morphology. Increasing the clay content resulted in an improvement in compressive modulus of the scaffolds. It was also found that the thermal properties of the scaffolds were significantly affected by the addition of the clay. The glass transition temperature (T_g) increased, while the crystallization temperature (T_c) decreased. However, melting enthalpies

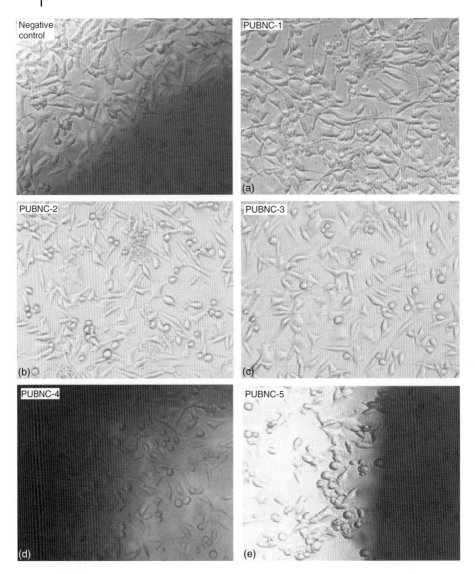

Figure 16.8 Photographs of L-929 cells interaction with sample films of chitin-based PU bionanocomposites varying nanoclay contents, (a) PUBNC1: without nanoclay, (b) PUBNC2: 1.0% nanoclay, (c) PUBNC3: 2.0% nanoclay, (d) PUBNC4: 4.0% nanoclay, and (e) PUBNC5: 8.0% nanoclay [138]. (Source: Reproduced with permission of Elsevier.)

and the degree of crystallinity values remained constant within experimental errors. Water contact angle measurements indicated that the wettability of the PLA-based nanocomposites was improved by the incorporation of clay [139].

Chen et al. [140] investigated the dispersion and nanofillers' interaction of rodlike silicates (attapulgite, ATT) in polymethylmethacrylate (PMMA) matrices. ATT modification was implemented by a novel *in situ* reaction using toluene-2,4-di-isocyanate (TDI) and mechanical mixing which resulted in homogeneous dispersion and rodlike texture of ATT nanorods. Organo-modified ATT/PMMA nanocomposites showed prominent improvements in strength, toughness, and thermal stability. It was also confirmed that mechanical properties of ATT-clay-reinforced PMMA nanocomposites strongly depend on the surface treatment of the nanorods. During the *in situ* modification, most surface hydroxyl groups of ATT reacted with TDI. This resulted in hydrophobic surfaces, which not only benefited the ATT dispersion but also improved the interfacial bonding. The improved dispersion and interfacial bonding caused enhanced mechanical and thermal properties of the nanocomposites. The uniform dispersion of *in situ* TDI-modified ATT nanorods in the PMMA, clearly visible in the TEM micrographs (Figure 16.9), was relevant in influencing the mechanical and thermal properties of the nanocomposites [140].

In a related investigation, Chang *et al.* [141] used laponite clay in the production of biocompatible poly(ethylene glycol) diacrylate (PEGDA)/laponite nanocomposite (NC) hydrogels that can support both two- and three-dimensional (2D and 3D) cell cultures. Laponite clays are disk-shaped platelets having a diameter of 30 nm and a thickness of 1 nm and are hydrophilic in nature, enabling them to disperse well in aqueous solutions. PEGDA/laponite NC hydrogels with enhanced mechanical properties were developed by harnessing the ability of PEGDA oligomers to simultaneously form chemically cross-linked networks while interacting with laponite nanoparticles through secondary interactions. The composite hydrogels supported adhesion and spreading of hMSCs. In addition to supporting 2D cell culture, the hydrogels-supported 3D cell culture of encapsulated stem cells and the laponite nanoparticles did not exhibit any detrimental effect on cells. Incorporation of laponite nanoparticles significantly enhanced both the compressive and tensile properties of PEGDA hydrogels, which were dependent on both the molecular weight of PEG and concentrations of laponite nanoparticles. It was concluded that hydrogels with enhanced mechanical properties could have potential applications as 3D scaffolds for TE. Furthermore, PEGDA/NC hydrogels can be used as injectable system for *in vivo* applications owing to their supporting 3D culture of encapsulated cells [141].

Marras *et al.* [142] prepared polymer nanocomposites, based on PCL and OMMT, by the solution intercalation technique. Fibrous membranes of neat PCL and nanocomposite were fabricated by ES. The nanocomposites exhibited improved stiffness without sacrificing polymer ductility. It was reported that increasing the clay content of the electrospun nanocomposite material resulted in the production of more fine fibrous structures. This observation could be attributed to the increase in electric conductivity and viscosity of the solution caused by the addition of inorganic clay. Furthermore, the ES process significantly affected the structure of the nanocomposite material by increasing the interlayer spacing of the inorganic mineral [142].

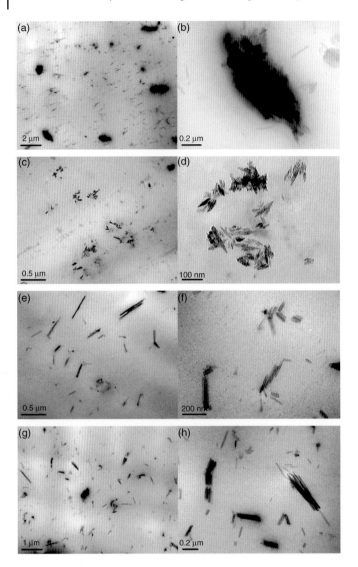

Figure 16.9 TEM images of PMMA nanocomposites: (a) and (b) containing 2 wt% original ATT clay; (c) and (d) containing 2 wt% pretreated ATT clay; (e) and (f) containing 2 wt% *in situ* modified ATT clay; and (g) and (h) containing 6 wt% *in situ* modified ATT clay, according to Ref. [140]. (Source: Reproduced with permission of John Wiley and Sons.)

16.4
Final Considerations

Table 16.1 summarizes recent investigations on the applications of silicate phase (BGs, silica, and nanoclays) containing nanocomposites in bone TE, which complements the discussions above. It is clear that the novel properties of these

Table 16.1 Selection of bioactive nanocomposites containing silicate phases for bone tissue engineering applications.

Materials	Materials system investigated	Method of fabrication	Key results achieved	References
Nanoscale bioactive glasses (NBG)	Chitosan/NBG	Combining sol-gel-derived NBG with chitosan-β-glycerophosphate salt formulation	Promoting a positive contact with surrounding tissue on injection in a bone defect	[85]
	Chitin/NBG and chitosan/NBG	Freezing/lyophilization	Improvement in cell adhesion and proliferation and increasing protein adsorption	[73–75]
	NBG-COL-HYA-PS	Freeze-drying	Better biomineralization, mechanical strength, cell attachment, and proliferation ability	[76]
	EDC/NHS-cross-linked NBG-COL-HA-PS	Freeze-drying	Ability of bone regeneration	[77]
	P(3HB)/NBG	Solvent casting/particulate leaching technique	Enhanced biocompatibility and antimicrobial properties	[61]
	P(3HB)/NBG	Solvent casting	Enhanced bioactivity, cell adhesion, and growth	[60, 61]
	PLA/NBG	Thermally induced phase separation	Improved bioactivity and mechanical properties	[56, 63]
	PLLA/surface-modified NBG	Solvent evaporation	Less nanoparticle aggregation, improved mechanical properties, bioactivity, cell adhesion, and growth	[66, 67]
	Gelatin/NBG	Solvent casting combined with lamination technique	Improved cell culture response, biocompatibility, bioactivity, and higher reinforcing quality of NBG	[80]
	PLA/BGNF	Sol–gel/ES	Excellent bioactivity and good osteoblast response	[68, 69]
	PCL/BGNF	Sol–gel/ES	Rough surface improved proliferation behavior	[71]
	PCL/BGNF	Sol–gel/ES	Significant improvement of the biological and mechanical properties and in vivo animal test results showed bone-forming ability	[72]
	Collagen/BGNF	Sol–gel/ES	High bioactivity and good cell adhesion and growth	[84]
	BCP/PCL–NBG	Foam replication and dip coating	High compressive strength, elastic modulus, and improved bioactivity	[65]

(continued overleaf)

Table 16.1 (continued)

Materials	Materials system investigated	Method of fabrication	Key results achieved	References
Silicate	P123/silica	Aerosol-assisted self-assembly technique	Reticulated foam structure	[112]
	SNP/PLLA/OMMT	Solved evaporation method	Higher mechanical properties	[113]
	SNP/silk	Solvent casting and deposition	Enhanced osteogenesis	[114]
	Chitin/SNP	Lyophilization method	Good cell adhesion and growth	[115]
Clay	Chitosan/MMT	Mixing aqueous solutions	Increased the tensile strength	[132]
	Gel/MMT–CS	Intercalation process and the freeze-drying technique	Good cell affinity and biocompatibility	[133]
	Chi/MMT/HA	Mixing aqueous solutions	Improved nanomechanical properties	[134]
	ChiPgAHAPMMT	Freeze-drying technique	Good biocompatibility and osteoconductive properties	[128]
	PLA/OMLS	Microcompounding/injection molding and subsequent polymer/particle leaching method	Improvement in compressive modulus	[137]
	PEGDA/Laponite	Mixing aqueous solutions	Enhanced mechanical properties	[139]
	PCL/OMMT	Electrospinning	More fine fibrous structures with improved stiffness	[140]

highly bioactive and resorbable nanocomposite materials are making them candidate materials for bone TE applications. In general terms, nanostructures are relevant for designing and fabricating biopolymer/silica phase (nano)composite scaffolds that mimic the nanofeatures of mineralized tissues such as bone. Thus silicate-containing bioactive nanocomposites reviewed in this chapter have a great potential for applications in the bone TE field. Scaffolds for these applications do not only need to promote bone regeneration and angiogenesis (vascularization) but also need to actively provide extended functions as required by the specific application, which may include anti-inflammatory, antibacterial, and anticancer effects, creating a new class of smart biomaterials. The clinical effectiveness of TE approaches based on this type of nanocomposites, however, still needs to be tested and fully validated in *in vivo* experiments, which should extensively consider potential toxic effects of the nanoparticles, especially in cases where highly resorbable polymer matrices are used and the silicate nanoparticles are free to migrate establishing direct contact with cells and tissues.

References

1. Khan, Y., Yaszemski, M.J., Mikos, A.G., and Laurencin, C.T. (2008) *J. Bone Joint Surg. Am.*, **90**, 36–42.
2. Stylios, G., Wan, T., and Giannoudis, P. (2007) *Inj. Int. J. Care Injured*, **3851**, 563–574.
3. Stevens, B. *et al.* (2008) *J. Biomed. Mater. Res. Part B Appl. Biomat.*, **85B**(2), 573–582.
4. Rezwan, K., Chen, Q.Z., Blaker, J.J., and Boccaccini, A.R. (2006) *Biomaterials*, **27**(18), 3413–3431.
5. Bueno, E.M. and Glowacki, J. (2009) *Nat. Rev. Rheumatol.*, **5**, 685–697.
6. Murugan, R. and Ramakrishna, S. (2005) *Compos. Sci. Technol.*, **65**, 2385–2406.
7. Burdick, J.A. and Anseth, K.S. (2002) *Biomaterials*, **23**, 4315–4323.
8. Glimcher, M.J. (1959) *Rev. Mod. Phys.*, **31**(2), 359–393.
9. Weiner, S. and Traub, W. (1986) *FEBS Lett.*, **206**(2), 262–266.
10. Rho, J.-Y., Kuhn-Spearing, L., and Zioupos, P. (1998) *Med. Eng. Phys.*, **20**(2), 92–102.
11. Landis, W.J., Song, M.J., Leith, A., McEwen, L., and McEwen, B.F. (1993) *J. Struct. Biol.*, **110**(1), 39–54.
12. Weiner, S. and Wagner, H.D. (1998) *Ann. Rev. Mater. Res.*, **28**, 271–298.
13. Fung, Y.C. (1993) *Biomechanics: Mechanical Properties of Living Tissue*, Springer-Verlag, New York.
14. Ji, B. and Gao, H. (2006) *Compos. Sci. Technol.*, **66**, 1212–1218.
15. Hollister, S.J. (2005) *Nat. Mater.*, **4**, 518–524.
16. Kaplan, F.S., Hayes, W.C., Keaveny, T.M. *et al.* (1994) in *Orthopedic Basic Science* (ed. S.R. Simon), American Academy of Orthopedic Surgeons, Columbus, OH, pp. 127–185.
17. Webster, T.J., Schadler, L.S., Siegel, R.W. *et al.* (2001) *Tissue Eng.*, **7**, 291–301.
18. Yao, C., Perla, V., McKenzie, J. *et al.* (2005) *J. Biomed. Nanotechnol.*, **1**, 68–77.
19. Cheng, G., Youssef, B.B., Markenscoff, P., and Zygourakis, K. (2006) *Biophys. J.*, **90**, 713–724.
20. Wu, C.-J., Gaharwar, A.K., Schexnailder, P.J., and Schmidt, G. (2010) *Materials*, **3**, 2986–3005. doi: 10.3390/ma3052986
21. Currey, J.D. (2002) *Bones: Structure and Mechanics*, Princeton University Press, New Jersey.
22. Lowenstam, H.A. and Weiner, S. (1989) *On Biomineralization*, Oxford University Press, New York.
23. McConnell, D. (1962) *Clin. Orthop.*, **23**, 253–268.
24. Andrew Taton, T. (2001) *Nature*, **412**, 491–492.
25. Gao, H., Ji, B., Jager, I.L., Arzt, E., and Fratzl, P. (2003) *Proc. Natl Acad. Sci. U.S.A.*, **100**, 5597–5600.
26. Gupta, H.S., Seto, J., Wagermaier, W., Zaslansky, P., Boesecke, P., and Fratzl, P. (2006) *Proc. Natl. Acad. Sci. U.S.A*, **103**, 17741–17746.
27. Peterlik, H., Roschger, P., Klaushofer, K., and Fratzl, P. (2006) *Nat. Mater.*, **5**, 52–55.
28. Tirrell, D.A. (1994) *Hierarchical Structures in Biology as a Guide for New Materials Technology*, National Academy Press, Washington, DC.
29. Clark, P. (1993) Cell behaviour on micropatterned surfaces, *2nd CEC Workshop on Bioelectronics – Interfacing Biology with Electronics*, Elsevier Advanced Technology, Frankfurt.
30. Norman, J.J. and Desai, T.A. (2006) *Ann. Biomed. Eng.*, **34**(1), 89–101.
31. Benoit, D.S.W. and Anseth, K.S. (2005) *Biomaterials*, **26**, 5209–5220.
32. Wilson, C.J., Clegg, R.E., Leavesley, D.I. *et al.* (2005) *Tissue Eng.*, **11**, 1–18.
33. Webster, T.J., Ergun, C., Doremus, R.H., Siegel, R.W., and Bizios, R. (2000) *Biomaterials*, **21**, 1803–1810.
34. Wei, G. and Ma, P.X. (2008) *Adv. Funct. Mater.*, **18**, 3568–3582.
35. Palin, E., Liu, H.N., and Webster, T.J. (2005) *Nanotechnology*, **16**(9), 1828–1835.
36. Loher, S., Reboul, V., Brunner, T.J., Simonet, M., Dora, C., Neuenschwander, P., and Stark, W.J. (2006) *Nanotechnology*, **17**(8), 2054–2061.

37. Engel, E. et al. (2008) *Trends Biotechnol.*, **26**(1), 39–47.
38. Curtis, A. (2009) in *Cellular Response to Biomaterials* (ed. D.L.D. Silvio), Woodhead Publishing Limited, London, p. 429.
39. Hench, L.L., Splinter, R.J., Allen, W.C., and Greenlee, T.K. (1971) *J. Biomed. Mater. Res.*, **5**(6), 117–141.
40. Hench, L.L. (1998) *J. Am. Ceram. Soc.*, **81**(7), 1705–1728.
41. Sepulveda, P., Jones, J.R., and Hench, L.L. (2001) *J. Biomed. Mater. Res.*, **58**(6), 734–740.
42. Saravanapavan, P., Jones, J.R., Pryce, R.S., and Hench, L.L. (2003) *J. Biomed. Mater. Res. A*, **66A**(1), 110–119.
43. Oki, A., Parveen, B., Hossain, S., Adeniji, S., and Donahue, H. (2004) *J. Biomed. Mater. Res. A*, 69A (2), 216–221.
44. Vallet-Regi, M., Salinas, A.J., Roman, J., and Gil, M. (1999) *J. Mater. Chem.*, **9**(2), 515–518.
45. Liu, X., Huang, W., Fu, H., Yao, A., Wang, D., Pan, H., Lu, W.W., Jiang, X., and Zhang, X. (2009) *J. Mater. Sci. Mater. Med.*, **20**, 1237–1243.
46. Lao, J., Nedelec, J.M., and Jallot, E. (2009) *J. Mater. Chem.*, **19**, 2940–2949.
47. Varanasi, V.G., Saiz, E., Loomer, P.M., Ancheta, B., Uritani, N., Ho, S.P., Tomsia, A.P., Marshall, S.J., and Marshall, G.W. (2009) *Acta Biomater.*, **5**, 3536–3547.
48. Xynos, I.D., Hukkanen, M.V.J., Batten, J.J., Buttery, L.D., Hench, L.L., and Polak, J.M. (2000) *Calcif. Tissue Int.*, **67**(4), 321–329.
49. Yao, J., Radin, S., Leboy, S., and Ducheyne, P. (2005) *Biomaterials*, **26**(14), 1935–1945.
50. Hench, L.L. and Polak, J.M. (2002) *Science*, **295**(5557), 1014–1017.
51. Chen, Q.Z.Z., Thompson, I.D., and Boccaccini, A.R. (2006) *Biomaterials*, **27**(11), 2414–2425.
52. Boccaccini, A.R., Erol, M., Stark, W.J., Mohn, D., Hong, Z., and Mano, J. (2010) *Compos. Sci. Technol.*, **70**, 1764–1776.
53. Day, R.M., Boccaccini, A.R., Shurey, S., Roether, J.A., Forbes, A., Hench, L.L., and Gabe, S.M. (2004) *Biomaterials*, **25**, 5857–5866.
54. Gorustovich, A.A., Roether, J.A., Boccaccini, A.R. (2010) *Tissue Eng. Part B Rev.*, **16** (2), 199–207. http://www.ncbi.nlm.nih.gov/pubmed/19831556.
55. Leu, A. and Leach, J.K. (2008) *Pharm. Res.*, **25**, 1222–1229.
56. Hong, Z., Reis, R.L., and Mano, J.F. (2009) *J. Biomed. Mater. Res. A*, **88**(2), 304–313.
57. Alves, N.M., Leonor, I.B., Azevedo, H.S., Reis, R.L., and Mano, J.F. (2010) *J. Mater. Chem.*, **20**, 2911–2921.
58. Kay, S., Thapa, A., Haberstroh, K.M., and Webster, T.J. (2002) *Tissue Eng.*, **8**, 753–761.
59. Koo, J.H. (2006) *Polymer Nanocomposites: Processing, Characterization, and Applications*, McGraw-Hill, New York.
60. Misra, S.K., Mohn, D., Brunner, T.J., Stark, W.J., Philip, S.E., Roy, I., Salih, V., Knowles, J.C., and Boccaccini, A.R. (2008) *Biomaterials*, **29**(12), 1750–1761.
61. Misra, S.K., Ansari, T.I., Valappil, S.P., Mohn, D., Philip, S.E., Stark, W.J., Roy, I., Knowles, J.C., Salih, V., and Boccaccini, A.R. (2010) *Biomaterials*, **31**, 2806–2815.
62. Zheng, H., Yingjun, W., Chunrong, Y., Xiaofeng, C., and Naru, Z. (2007) *Key Eng. Mater.*, **336–338**, 1534–1537.
63. Hong, Z., Reis, R.L., and Mano, J.F. (2008) *Acta Biomater.*, **4**, 1297–1306.
64. El-Kady, A.M., Ali, A.F., and Farag, M.M. (2010) *Mater. Sci. Eng. C*, **30**(1), 120–131.
65. Roohani-Esfahani, S.I., Nouri-Khorasani, S., Lu, Z.F., Appleyard, R.C., and Zreiqat, H. (2011) *Acta Biomater.*, **7**, 1307–1318.
66. Liu, A., Hong, Z., Zhuang, X., Chen, X., Cui, Y., Liu, Y., and Jing, X. (2008) *Acta Biomater.*, **4**(4), 1005–1015.
67. Liu, A., Wei, J., Chen, X., Jing, X., Cui, Y., and Liu, Y. (2009) *Chin. J. Polym. Sci.*, **27**(3), 415–426.
68. Kim, H.W., Kim, H.E., and Knowles, J.C. (2006) *Adv. Funct. Mater.*, **16**(12), 1529–1535.
69. Kim, H.W., Lee, H.H., and Chun, G.S. (2008) *J. Biomed. Mater. Res. A*, **85A**(3), 651–663.

70. Noh, K.T., Lee, H.Y., Shin, U.S., and Kim, H.W. (2010) *Mater. Lett.*, **64**, 802–805.
71. Lee, H.H., Yu, H.S., Jang, J.H., and Kim, H.W. (2008) *Acta Biomater.*, **4**, 622–629.
72. Jo, J.H., Lee, E.J., Shin, D.S., Kim, H.E., Kim, H.W., Koh, Y.H., and Jang, J.H. (2009) *J. Biomed. Mater. Res. B Appl. Biomater.*, **91B**(1), 213–220.
73. Peter, M., Kumar, P.T.S., Binulal, N.S., Nair, S.V., Tamura, H., and Jayakumar, R. (2009) *Carbohydr. Polym.*, **78**(4), 926–931.
74. Peter, M., Binulal, N.S., Soumya, S., Nair, S.V., Furuike, T., Tamura, H., and Jayakumar, R. (2010) *Carbohydr. Polym.*, **79**(2), 284–289.
75. Peter, M., Binulal, N.S., Nair, S.V., Selvamurugan, N., Tamura, H., and Jayakumar, R. (2010) *Chem. Eng. J.*, **158**, 353–361.
76. Wang, Y., Yang, C., Chen, X., and Zhao, N. (2006) *Macromol. Mater. Eng.*, **291**, 254–262.
77. Xie, E., Hu, Y., Chen, X., Bai, J., Ren, L., and Zhang, Z. (2008) A novel nanocomposite and its application in repairing bone defects. Proceedings of the 3rd IEEE International Conference on Nano/Micro Engineered and Molecular Systems, Sanya, China, pp. 943–946.
78. Luz, G.M. and Mano, J.F. (2010) *Mater. Sci. Forum*, **636–637**, 31–35.
79. Luz, G.M. and Mano, J.F. (2011) *Nanotechnology*, **22**, 494014 (11 pp).
80. Mozafari, M., Moztarzadeh, F., Rabiee, M., Azami, M., Maleknia, S., Tahriri, M., Moztarzadeh, Z., and Nezafati, N. (2010) *Ceram. Int.*, **36**, 2431–2439.
81. Srinivasan, S., Jayasree, R., Chennazhi, K.P., Nair, S.V., and Jayakumar, R. (2012) *Carbohydr. Polym.*, **87**, 274–283.
82. Mourino, V., Newby, P., Pishbin, F., Cattalini, J.P., Lucangioli, S., and Boccaccini, A.R. (2011) *Soft Matter*, **7**, 6705–6712.
83. Marelli, B., Ghezzi, C.E., Mohn, D., Stark, W.J., Barralet, J.E., Boccaccini, A.R., and Nazhat, S.N. (2011) *Biomaterials*, **32**, 8915–8926.
84. Kim, H.W., Song, J.H., and Kim, H.E. (2006) *J. Biomed. Mater. Res. A*, **79**(3), 698–705.
85. Couto, D.S., Hong, Z., and Mano, J.F. (2009) *Acta Biomater.*, **5**(1), 115–123.
86. Bottini, M., D'Annibale, F., Magrini, A., Cerignoli, F., Arimura, Y., Dawson, M.I., Bergamaschi, E., Rosato, N., Bergamaschi, A., and Mustelin, T. (2007) *Int. J. Nanomed.*, **2**, 227–233.
87. Gerion, D., Herberg, J., Bok, R., Gjersing, E., Ramon, E., Maxwell, R., Kurhanewicz, J., Budinger, T.F., Gray, J.W., Shuman, M.A., and Chen, F.F. (2007) *J. Phys. Chem. C*, **111**, 12542–12551.
88. Graf, C., Dembski, S., Hofmann, A., and Ruehl, E. (2006) *Langmuir*, **22**, 5604–5610.
89. Tushar, R.S., Amit, A., and Sming, N. (2006) *Anal. Chem.*, **78**, 5627–5632.
90. Dalia, H.L., Peter, P.G., Jaclyn, F., Geng, Z., Michael, J.T., Mark, I.G., Daniel, A.H., and Ramachandran, M. (2008) *Methods*, **46**, 25–32.
91. Meseguer-Olmo, L., Ros-Nicolas, M.J, Vicente-Ortega, V, Alcaraz-Banos, M, Clavel-Sainz, M, Arcos, D, Ragel, C.V, Vallet-Regi, M, and Meseguer-Ortiz, C (2006) *J. Orthop. Res.*, **24**, 454–460.
92. Radin, S., El-Bassyouni, G., Vresilovic, E.J., Schepers, E., and Ducheyne, P. (2004) *Biomaterials*, **26**, 1043–1052.
93. Kortesuo, P., Ahola, M., Karlsson, S., Kangasniemi, I., and Yli-Urpo, A. (1999) *J. Biomater*, **21**, 193–198.
94. Huh, S., Wiench, J.W., Yoo, J.-C., Pruski, M., and Lin, V.S.Y. (2003) *Chem. Mater.*, **15**, 4247–4256.
95. Vallet-Reg, M., Ruiz-Gonz'alez, L., Izquierdo-Barba, I., and Gonz'alez-Calbet, J.M. (2006) *J. Mater. Chem.*, **16**(1), 26–31.
96. Webster, T.J., Ergun, C., Doremus, R.H., Siegel, R.W., and Bizios, R. (2000) *J. Biomed. Mater. Res.*, **51**, 475–483.
97. Xynos, I., Edgar, A., Buttery, L., Hench, L., and Polak, J. (2001) *J. Biomed. Mater. Res.*, **55**, 151–157.
98. Hench, L.L. (2001) *Key Eng. Mater.*, **192–195**, 575–580.
99. Yamaguchi, T., Chattopadhyay, N., Kifor, O., Butters, R.R. Jr.,

Sugimoto, T., and Brown, E.M. (1998) *J. Bone Miner. Res.*, **13**, 1530–1538.
100. Gaharwar, A.K., Rivera, C.P., Wu, C.-J., and Schmidt, G. (2011) *Acta Biomater.*, **7**, 4139–4148.
101. Wei, L., Hu, N., and Zhang, Y. (2010) *Materials*, **3**(7), 4066–4079. doi: 10.3390/ma3074066
102. Arcos, D. and Vallet-Regí, M. (2010) *Acta Biomater.*, **6**(8), 2874–2888.
103. Huang, S.L., Chin, W.K., and Yang, W.P. (2005) *Polymer*, **46**, 1865–1877.
104. Constantini, A., Luciani, G., Annunziata, G., Silvestri, B., and Branda, F. (2006) *J. Mater. Sci.: Mater. Med.*, **17**, 319–325.
105. Hajji, P., David, L., Gerard, J.F., Pascault, J.P., and Vigier, G. (1999) *J. Polym. Sci., Part B: Polym. Phys.*, **37**, 3172–3187.
106. Liu, Y.L., Hsu, C.Y., and Hsu, K.Y. (2005) *Polymer*, **46**, 1851–1856.
107. Alt, V., Koegelmaier, D.V., Lips, K.S., Witt, V., Pacholke, S., Heiss, C., Kampschulte, M., Heinemann, S., Hanke, T., Schnettler, R., and Langheinrich, A.C. (2011) *Acta Biomater.*, **7**, 3773–3779.
108. Vallés-Lluch, A., Costa, E., Gallego Ferrer, G., Monleón Pradas, M., and Salmerón-Sánchez, M. (2010) *Compos. Sci. Technol.*, **70**, 1789–1795.
109. Lee, K.-Y., Lee, Y.-H., Kim, H.-M., Koh, M.-Y., Ahn, S.-H., and Lee, H.-K. (2005) *Curr. Appl. Phys.*, **5**(5), 453–457.
110. Rhee, S.-H. (2003) *Biomaterials*, **24**(10), 1721–1727.
111. Lluch, A.V., Ferrer, G.G., and Pradas, M.M. (2009) *Colloids Surf., B Biointerfaces*, **70**(2), 218–225.
112. Lu, Y., Fan, H., Stump, A., Ward, T.L., Rieker, T., and Brinker, C.J. (1999) *Nature*, **398**, 223–226.
113. Chrissafis, K., Pavlidou, E., Paraskevopoulos, K.M., Beslikas, T., Nianias, N., and Bikiaris, D. (2010) *J. Therm. Anal. Caalorim.*, **105**(1), 313–323. doi: 10.1007/s10973-010-1168-z
114. Mieszawska, A.J., Nadkarni, L.D., Perry, C.C., and Kaplan, D.L. (2010) *Chem. Mater.*, **22**, 5780–5785.
115. Madhumathia, K., Sudheesh Kumara, P.T., Kavyaa, K.C., Furuikeb, T., Tamurab, H., Nair a, S.V., and Jayakumar, R. (2009) *Int. J. Biol. Macromol.*, **45**, 289–292.
116. Rana, V.K., Park, S.S., Parambadath, S., Kim, M.J., Kim, S.H., Mishra, S., Singh, R.P., and Ha, C.S. (2011) *Med. Chem. Commun.*, **2**, 1162–1166.
117. Patel, H.A., Somani, R.S., Bajaj, H.C., and Jasra, R.V. (2006) *Bull. Mater. Sci.*, **29**(2), 133–145.
118. Bhat, G., Hegde, R.R., Kamath, M.G., and Deshpande, B. (2008) *J. Eng. Fiber. Fabric.*, **3**(3), 22–34.
119. Alexandre, M. and Dubois, P. (2000) *Mater. Sci. Eng. R Rep.*, **28**(1–2), 1–63.
120. Zerda, A.S. and Lesser, A.J. (2001) *J. Polym. Sci., Part B: Polym. Phys.*, **39**, 1137–1146.
121. Lee, W.-F. and Fu, Y.-T. (2003) *J. Appl. Polym. Sci.*, **89**(13), 3652–3660.
122. Dong, Y. and Feng, S.-S. (2005) *Biomaterials*, **26**(30), 6068–6076.
123. Lin, F.-H., Chen, C.-H., Cheng, W.T.K., and Kuo, T.-F. (2006) *Biomaterials*, **27**(17), 3333–3338.
124. Viseras, C., Aguzzi, C., Cerezo, P., and Lopez-Galindo, A. (2007) *Appl. Clay Sci.*, **36**(1–3), 37–50.
125. des Rieux, A., Fievez, V., Garinot, M., Schneider, Y.-J., and Préat, V. (2006) *J. Controlled Release*, **116**(1), 1–27.
126. Sun, B., Ranganathan, B., and Feng, S.-S. (2008) *Biomaterials*, **29**(4), 475–486.
127. Carretero, M.I. (2002) *Appl. Clay Sci.*, **21**(3–4), 155–163.
128. Ambre, A.H., Katti, K.S., and Katti, D.R. (2010) *J. Nanotechnol. Eng. Med.*, **1**, 031013–0310139.
129. Kamena, K. (2005) Nanoclays: multi-dimensional new nano-tools in the polymer development toolbox. 5th SPE Automotive Composites Conference, Society of Plastic Engineers, September 12–14, 2005.
130. Fukushima, Y. and Inagaki, S. (1987) *J. Inclusion Phenom.*, **5**, 473–482.
131. Gao, F. (2004) *Mater. Today*, **11**, 50–55.
132. Lin, K.-F., Hsu, C.-Y., Huang, T.-S., Chiu, W.-Y., Lee, Y.-H., and Young, T.-H. (2005) *J. Appl. Polym. Sci.*, **98**, 2042–2047.

133. Zheng, J.P., Wang, C.Z., Wang, X.X., Wang, H.Y., Zhuang, H., and Yao, K.D. (2007) *React. Funct. Polym.*, **67**, 780–788.
134. Katti, K.S., Katti, D.R., and Dash, R. (2008) *Biomed. Mater.*, **3**, 034122 (12 pp).
135. Verma, D., Katti, K.S., and Katti, D.R. (2009) *Mater. Sci. Eng., C*, **29**(7), 2079–2084.
136. Verma, D., Katti, K.S., and Katti, D.R. (2010) *Philos. Trans. R. Soc. London, Ser. A*, **368**(1917), 2083–2097.
137. Zuber, M., Zia, K.M., Mahboob, S., Hassan, M., and Bhatti, I.A. (2010) *Int. J. Biol. Macromol.*, **47**, 196–200.
138. Zia, K.M., Zuber, M., Barikani, M., Hussain, R., Jamild, T., and Anjum, S. (2011) *Int. J. Biol. Macromol.*, **49**, 1131–1136.
139. Ozkoc, G., Kemaloglu, S., and Quaedflieg, M. (2010) *Polym. Compos.*, **31**, 674–683.
140. Chen, F., Lou, D., Yang, J., and Zhong, M. (2011) *Polym. Adv. Technol.*, **22**, 1912–1918.
141. Chang, C.-W., Spreeuwel, A., Zhang, C., and Varghese, S. (2010) *Soft Matter*, **6**, 5157–5164.
142. Marras, S.I., Kladi, K.P., Tsivintzelis, I., Zuburtikudis, I., and Panayiotou, C. (2008) *Acta Biomater.*, **4**, 756–765.

Part IV
Systems for the Delivery of Bioactive Agents

17
Biomimetic Nanostructured Apatitic Matrices for Drug Delivery

Norberto Roveri and Michele Iafisco

17.1
Introduction

Among biomimetic inorganic nanostructured biomaterials, synthetic hydroxyapatites (HAs) probably have received the highest attention and interest as matrices for drug delivery. Also silica xerogels represent very interesting inorganic matrices for drug delivery, but the bioactive molecules are only made to physically interact with the inorganic surface and the drug release kinetics appears to be controlled more difficultly [1, 2]. On the contrary, nanostructured biomimetic apatite surfaces chemically link the bioactive molecules that can be released with a kinetics strictly related to the disordered distribution of charges on the inorganic surface. Nanocrystalline apatite matrices easily bind bioactive molecules on their surface due their high surface area and presence of available unbalanced ionic sites (Ca^{2+} and PO_4^{3-}). The Ca/P molar ratio of 1.7 can be obtained only in the apatite crystal bulk, while the Ca/P ratio can decrease to 1.4 on the apatite crystal surface not only for calcium deficiencies but also for phosphate and hydroxyl group replacement by carbonate anions. The lack of neutralization of surface charges induces on the apatite surface the binding of biomolecules and drugs, which can be released spontaneously in physiological liquids, but this also occurs as a consequence of changes in pH, ionic strength, or temperature. These variations can be observed in biological environment under physiological and pathological stimuli. The physicochemical characteristics of biomimetic apatites allow consideration of this inorganic matrix as being particularly suitable as drug delivery stimuli. Nanostructured HA matrices can be synthesized not only as nanosized crystals with morphology, structure, and surface area closely mimicking those of bone apatite crystals but also as porous scaffolds mimicking bone sponginess. HAs are biocompatible and bioresorbable and can be synthesized by biomimetic chemical methods. In fact, they can be synthesized in water without organic solvents, in mild and soft conditions, and mimicking biological mechanisms such as template-mediated synthesis.

17.2
Biomimetic Apatite Nanocrystals

Biomimetism of synthetic calcium phosphates can be carried out at different levels, such as chemical composition, structure, crystal dimension, morphology, bulk, and surface physicochemical properties. Biomaterials can be turned into biomimetic imprinting of all these characteristics to not only optimize their interaction with biological tissues but also mimic biogenic materials in their functionalities. Detailed information on calcium phosphates, their synthesis, structure, chemistry, other properties, and biomedical application have been comprehensively reviewed recently [3, 4]. This section describes the physicochemical characteristics and some methods to synthesize some different biomimetic nanostructured apatites that can successfully act as drug deliverers with controlled kinetics.

17.2.1
Properties

In biological systems, apatites are the principal inorganic constituents of normal (bones, teeth, fish enameloid, deer antlers, and some species of shells) and pathological (dental and urinary calculi and stones, atherosclerotic lesions) calcifications [5]. Except for small portions of the inner ear, all hard tissues of the human body are formed of calcium phosphates. Structurally, they occur mainly in the form of poorly crystallized nonstoichiometric F-, Na-, Mg-, and carbonate-substituted hydroxyapatite $Ca_{10}(PO_4)_6(OH)_2$ (HA) [6]. The study of nanocrystalline calcium phosphate properties and, subsequently, the possibility of reproducing bone mineral for the development of new advanced biomaterials is constantly growing [7]. The possibility of synthesizing in the laboratory biomimetic apatites makes it possible to use these systems as a bone mineral "model," enabling to investigate the interaction between bonelike apatite nanocrystals and the components of surrounding fluids (ions, proteins, antibodies, ...) and to follow the surface functionalization with drugs to be delivered *in vivo*.

Among the characteristics of these compounds, their irregular platelike morphology (elongated toward the *c*-axis) is one of the noticeable features in comparison to regular HA. A comparison of the well-crystallized HA with the biomimetic one shows another principal difference: the mean crystallite size, with length of the order of 15–30 nm and width about 6–9 nm [4]. Moreover, from a chemical point of view, the composition of nanocrystalline apatites strongly differs from that of HA. Although it has been the object of much controversy during the past decades, the global chemical composition of biological apatites (or their synthetic analogs) can generally be described as

$$Ca_{10-x}(PO_4)_{6-x}(HPO_4 \text{ or } CO_3)_x(OH \text{ or } 1/2\, CO_3)_{2-x} \text{ with } 0 \leq x \leq 2$$

(except maybe for very immature nanocrystals that may depart from this generic formula). In particular, this expression underlines the presence of vacancies in both Ca and OH sites. For example, Legeros *et al.* [8] analyzed various cortical bone specimens, suggesting the following relatively homogeneous composition, unveiling high vacancy contents

$$Ca_{8.3} (PO_4)_{4.3} (HPO_4 \text{ or } CO_3)_{1.7} (OH \text{ or } 1/2\, CO_3)\, 0.3$$

Minor substitutions are also found in biological apatites involving, for example, monovalent cations (especially Na^+). In this case, charge compensation mechanisms have to be taken into account.

It is worth to highlight that in this particular material the chemical formulas only allow a "global" insight on the nature and amount of ions present in the compound, but it does not reflect possible local variations that may be observed in the nanocrystals or *in vivo* within the osteons [9], linked to the local apatite crystal formation, and in vivo, linked to continuous bone remodeling.

Therefore, the characterization of nanocrystalline apatites is a relatively arduous task because of their poor crystallinity and metastability. But despite the complexity of these systems, recent investigations have succeeded in revealing their particular surface features (and related reactivity in wet media) that may be exploited for the setup of new biomaterials or for a better understanding of natural or pathological biomineralization phenomena. Recent advances in the characterization of apatite nanocrystals were made thanks to the use of spectroscopic techniques and, in particular, Fourier transform infrared (FT-IR) spectroscopy. The FT-IR method is useful for drawing conclusions on the local chemical environment of phosphate, carbonate, and hydroxide ions as well as water molecules in such systems. Detailed analyses of the phosphate groups by FT-IR spectroscopy have enabled to distinguish in nanocrystalline apatites the presence of additional bands that cannot be attributed to phosphate groups in a regular apatitic environment [10, 11]. These chemical environments have been referred to by Rey as *"nonapatitic"* environments.

Taking into account all the above data, nanocrystalline apatites (whether biological or their synthetic analogs prepared under close-to-physiological conditions) may most probably be described as the association of an apatitic core (often nonstoichiometric) and a structured fragile hydrated surface layer containing water molecules and rather labile ions (e.g., Ca^{2+}, HPO_4^{2-}, CO_3^{2-}, ...) occupying nonapatitic crystallographic sites (although in the case of biomimetic apatites, the layer is directly exposed on the surface and not included in a "sandwichlike" structure between two "apatitic" layers). A schematic model for such nanocrystalline apatites is given in Figure 17.1.

The presence of this hydrated surface layer is thought to be responsible for most of the properties of biomimetic apatites and, in particular, their high surface reactivity in relation to the surrounding fluids (which is probably directly linked to a high mobility of ionic species contained within this layer), and this reactivity may explain, from a physicochemical viewpoint, the role of bone mineral in homeostasis *in vivo* [12]. This layer indeed contains labile ions that can potentially be exchanged for other ions from the surrounding solution, or by small molecules, which may be exploited for couplings with proteins or drugs. It is interesting to remark that the typical nonapatitic features mentioned above tend to progressively disappear during the aging of the nanocrystals in solution. This process is referred to as *"maturation"*

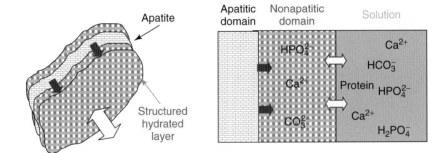

Figure 17.1 Schematic model of a biomimetic nanocrystal – apatite nanocrystal (3D view) (a) and its interaction with surrounding fluids – apatite nanocrystal in solution (profile) (b).

and has been related to the progressive growth of apatite domains at the expense of the surface hydrated layer [13, 14]. This maturation process is thought to be linked to the metastability of such poorly crystallized nonstoichiometric apatites, which steadily evolve in solution toward stoichiometry and better crystallinity. This evolution can be, for example, witnessed by the decrease in the amount of nonapatitic HPO_4^{2-} ions on aging, or else by the decreased potentialities to undergo ion exchanges [14]. One example of such effects can be given, for example, by the decreased exchangeability of HPO_4^{2-} by CO_3^{2-} observed on carbonated apatites matured for increasing amounts of time. In addition, beside this compositional evolution, some structural and microstructural features also tend to evolve such as the mean crystallite size, which increases on maturation. The control of synthesis parameters such as pH, temperature, or maturation time can thus enable one to tailor the physicochemical properties of biomimetic apatites [13] and, in particular, their surface reactivity, so to mimic, for example, mature bone mineral or else newly formed bone apatite.

Sakhno *et al.* [12] have compared two types of nano-HA synthesized at two different temperatures (40 and 95 °C) [12]. The apatite synthesized at low temperature displays plateletlike morphology and is composed of a crystalline core coated by an amorphous surface layer 1–2 nm thick. By increasing the preparation temperature, the platelet morphology was retained but apatite nanoparticles exhibited a higher degree of crystallinity (evaluated by X-ray diffraction analysis) (Figure 17.2).

17.2.2
Synthesis

The biomimetism of materials can refer not only to their chemical composition, structure, morphology, and physicochemical properties but also their synthetic process. In fact, it is possible to synthesize biomimetic biomaterials in water without organic solvents, in mild and soft conditions, and mimicking biological mechanisms such as in template-mediated synthesis.

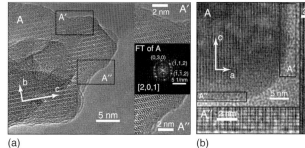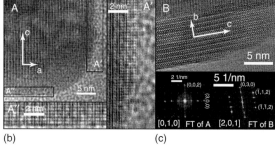

Figure 17.2 (a) High-resolution TEM image of a portion of apatite synthesized at 40 °C (A) and related FT. Original magnification: 800k × (a). (b,c) High-resolution TEM image of a portion of apatite synthesized at 95 °C and related FT (right, bottom). Panels A′, A″: zoomed view of two enframed border regions in A. Original magnification: 800k × (b, c). (Source: From Ref. [12] with permission.)

Many different methodologies have been proposed to prepare nanosized and/or nanocrystalline HAs [15–17]. These are wet chemical precipitation [18–20], sol–gel synthesis [21–23], coprecipitation [24, 25], hydrothermal synthesis [26, 27], mechanochemical synthesis [28], microwave processing [29–31], vapor diffusion [32, 33], silica gel template [34], emulsion-based syntheses [35], and several other methods by which nanocrystals of various shapes and sizes can be obtained [36, 37]. In general, shape, size, and specific surface area of the apatite nanocrystals appear to be very sensitive to both the reaction temperature and the reactant addition rate. Apatites with different stoichiometry and morphology have been prepared, and the effects of varying powder synthesis conditions on stoichiometry, crystallinity, and morphology have been analyzed. The effects of varying the concentration of the reagents, the reaction temperature and time, the initial pH, the aging time, and the atmosphere within the reaction vessel have also been studied [38]. In order to optimize the specific biomedical applications of synthetic biomimetic HA, especially drug delivery function, the physicochemical features that should be tailored in them are dimensions, porosity, morphology, and surface properties [39].

Recently, Iafisco *et al.* [32, 40] reported a new methodology, based on the vapor diffusion sitting drop micromethod, to precipitate carbonate-substituted HA nanoparticles. The method was developed by using an innovative device called the *"crystallization mushroom"* [41, 42]. The advantages that the "crystallization mushroom" offers compared to other crystallization devices are the reduced consumption of reagents during the crystallization process, since the volume of microdroplets is around 40 µl, and the high reproducibility because of the possibility of running 12 batches of crystals for each experiment. Therefore, this setup may be suitable to evaluate interactions and/or the cocrystallization of HA with small amounts of proteins, polymers, or drugs for studies in the fields of biomineralization and biomaterials. By using this methodology, it has been found that mixtures containing 50 mM $Ca(CH_3COO)_2$ and 30 mM $(NH_4)_2HPO_4$ in microdroplets and 3 ml of a

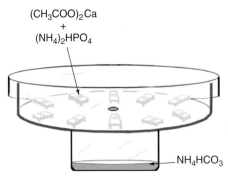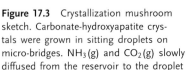

Figure 17.3 Crystallization mushroom sketch. Carbonate-hydroxyapatite crystals were grown in sitting droplets on micro-bridges. $NH_3(g)$ and $CO_2(g)$ slowly diffused from the reservoir to the droplet through the small opening in the plate at the bottom (a). TEM image of carbonate apatite crystallized by vapor diffusion after one week. Scale bar is 200 nm (b).

40 mM NH_4HCO_3 solution in the gas generation chamber were ideal to precipitate carbonate–HA nanocrystals after seven days of reaction. The nanocrystals were produced by solvent-mediated phase transformation of octacalcium phosphate (OCP) to HA, with OCP most probably acting as a temporal template for the heterogeneous nucleation of apatite nuclei. The obtained crystals displayed nanometric dimensions, carbonate ions in the crystal lattice, platelike morphology, and low degree of crystallinity, closely resembling the inorganic phase of bones (Figure 17.3). Nassif et al. [33] also precipitated carbonate-HA by the vapor diffusion method. They used $CaCl_2$–NaH_2PO_4 mixed solutions in the volume range of milliliters (macromethod) and either an NH_4OH and $NaHCO_3$ solution or solid $(NH_4)_2CO_3$ to generate the gas phase, which led to the precipitation of B- or A-type carbonate–apatite phases, respectively. They concluded that the best similarity between synthetic and natural HA was obtained by using an aqueous carbonate precursor, which is in agreement with the results obtained in the work of Iafisco et al. [32].

Porous apatitic biomimetic scaffolds in simulating spongy bone morphology has been prepared using various technologies to control pore dimension, shape, distribution, and interconnections. However, HA ceramics processed by high-temperature treatment present a significant reduction of bioreactivity and growth kinetics of new bone because of the lack of resorbability [43].

Porous bioceramics with a low degree of crystallinity and an appreciable bioresorbability have been obtained by synthetic methods at lower temperatures [44]. Colloidal processing [45], starch consolidation [46], gel casting, and foam out [47] have allowed the production of bioceramics with a bimodal distribution of the pore size that can be tailored as a function of the sintering conditions. However, the low resorbability of sintered HA ceramics appears useful when they have to be implanted with a defined 3D form. In these cases, the HA ceramic high porosity induces bone formation inside the implant, increasing HA degradation, but without reaching its complete resorption and allowing the implant to conserve its crystalline architecture. Porous HA can be synthesized by a hydrothermal method directly from natural sea corals [48] and cuttlefish bones [49], where the HA phase replaces aragonite while preserving its natural porous structure. The interconnected network of pores not only promotes bone ingrowth but also allows bioceramics to be utilized as drug delivery agents, by inserting different bioactive molecules or by filling the macro- and micropores with gelatin, which can act as delivery agent of bioactive molecules [50].

Following the biomimetic approach, inspired by Nature, natural wood templates have been selected as a starting point to obtain open-pore geometries with high surface area and microstructure allowing cell ingrowth and reorganization and providing the necessary space for vascularization [51, 52]. In fact, the alternation of fiber bundles and channel-like porous areas makes the wood a material to be used as template in starting the development of new bone substitute biomaterials by an ideal biomimetic hierarchical structure [53]. Biomimetic apatite bone scaffolds characterized by highly organized hierarchical structures have been recently obtained by chemically transforming native wood through a sequence of thermal and hydrothermal processes. The whole chemical conversion has been carried out in five chemical steps from native wood to porous HA: (i) pyrolysis of ligneous raw materials to produce carbon templates characterized by the natural complex anisotropic pore structure, (ii) carburization process by vapor or liquid calcium permeation to yield calcium carbide, (iii) oxidation process to transform calcium carbide into calcium oxide, (iv) carbonation by hydrothermal process under CO_2 pressure for the further conversion into calcium carbonate, and (v) phosphating process through hydrothermal treatment to achieve the final HA phase [54]. The SEM image reported in Figure 17.4a demonstrates that the structured anisotropy typical of the native wood was preserved on the macroscale, exhibiting pore sizes in the range 100–300 µm, revealing an ordered fastening of parallel microtubes 100–150 µm long and 15–30 µm wide with a hollow core of about 10–25 µm in

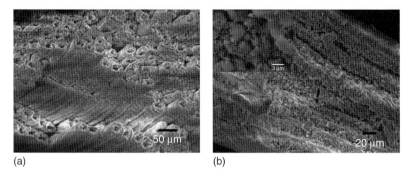

(a)   (b)

Figure 17.4 Detailed SEM images of pinewood-derived hydroxyapatite: (a) microstructure of wood-derived parallel fastened hydroxyapatite microtubes and (b) typical needlelike hydroxyapatite nuclei grown on the microtube surface; inset in (b) shows higher magnification of picture (b). (Source: From Ref. [54] with permission.)

diameter, and organized similar to the cell morphology of the natural wood used as the starting template for its synthesis. The SEM image reported in Figure 17.4b of the newly formed HA surface morphology shows typical needlelike nuclei grown on the surface, proving the concomitant occurrence of a dissolution/precipitation process at submicron level, in agreement with the hypotheses previously proposed. This surface nanostructure morphology of the unidirectional fastened hollow HA microtubules allows biological systems such as cells to utilize biomorphic scaffolds on the micrometer level, which are also biomimetic for composition and structure at the nanometer scale.

17.3
Biomedical Applications of Biomimetic Nanostructured Apatites

The main driving force behind the use of apatites as materials for health care is their chemical similarity to the mineral component of mammalian bones and teeth, as previously explained [6]. As a result, in addition to being nontoxic, they are biocompatible, not recognized as foreign materials in the body, and, most importantly, exhibit bioactive behavior and integrate into living tissue by the same processes active in remodeling healthy bone. This leads to an intimate physicochemical bond between the implants and bone, termed *osteointegration*. More to the point, calcium phosphates are also known to support osteoblast adhesion and proliferation. Even so, the major limitations to use apatites as load-bearing biomaterials are their mechanical properties, namely, their brittleness and poor fatigue resistance [55]. The poor mechanical behavior is even more evident for highly porous ceramics and scaffolds because porosity >100 µm is considered a requirement for proper vascularization and bone cell colonization [56]. That is why, in biomedical applications, calcium phosphates are used primarily as fillers and coatings [57].

Commercially, pure titanium and Ti–6Al–4V alloys are the most commonly used metallic implant materials, as they are highly biocompatible materials with good mechanical properties and corrosion resistance [58, 59]. The biocompatibility of titanium implants is limited to the stable oxide layer (with a thickness of 3–10 nm) that spontaneously forms when titanium is exposed to oxygen [60]. This reaction prevents the formation of fibrous tissue around the implant and creates direct contact to osseous tissue. However, when applying Ti(O_2) as implant material, a nonphysiological surface is exposed to a physiological environment. Nevertheless, by generating a coating onto a titanium surface that mimics the organic and inorganic components of living bone tissue, a physiological transition between the nonphysiological titanium surface and the surrounding bone tissue can be established. In this way, the coated titanium implant functions as scaffold for improved bone cell attachment, proliferation, and differentiation. Such a coating is supposed to further enhance early and strong fixation of a bone-substituting implant by stimulating bone formation starting from the implant surface. As such, a continuous transition from tissue to implant surface can be induced. Consequently, research efforts have focused on modifying the surface properties of titanium to control the interaction between the implant and its biological surrounding. As mentioned previously, apart from the living cells and the collagenous extracellular matrix the main constituent of bones and teeth is HA, and the ideal surgical implant would be made from the same materials. However, HA is weak and brittle, making it unsuitable for replacing parts of the body. To eliminate this problem, it was suggested that the titanium implants could be improved by coating them at the surface with HA. This approach would combine the mechanical strength of titanium with the bioactive properties of HA in the bone–prosthesis interface.

Calcium phosphate coatings for orthopedic and dental implants were introduced by de Groot and Geesink [61, 62]. Thereafter, numerous reports have been published about the osteoconductive properties of calcium-phosphate-coated implants; in fact, osteoconduction refers to the ability of a biomaterial to support the growth of bone over its surface. Different HA coatings are described to induce an increased bone-to-implant contact [63, 64] to improve the implant fixation [65] and to facilitate the bridging of small gaps between the implant and the surrounding bone [66]. The calcium phosphate layer guides bone growth along the implant surface and, as a result, bone formation now occurs from both the surrounding tissue and the implant surface, in which calcium phosphates function as physiological transition between the nonphysiological titanium surface and the surrounding bone.

Actually, the most diffused method to apply calcium phosphate coatings to implants has been the plasma spraying technique, owing to its high deposition rate and the ability to coat large areas [67, 68]. The plasma spray is a technique in which a so-called plasma gun creates an electric arc current of high energy between a cathode and an anode. An inert gas is directed through the space between these electrodes. Subsequently, the arc current ionizes the gas and a plasma is formed. The electrons and ions in this plasma are separated form each other and are accelerated toward the cathode and anode, respectively. These rapidly moving particles then collide with other atoms or molecules in the gas, which results in expansion owing to the temperature increase. Then, a plasma flame is formed emerging from the gun toward the substrate at velocities approaching or exceeding the speed of sound. Afterward, HA ceramic powder particles are fed into the plasma flame. The particles melt and are deposited on the substrate at which the gun is aimed. The quality of plasma spray coatings can be influenced by several parameters, such as the temperature of the plasma, the nature of the plasma gas, the particle size of the powder, and the chemical nature of the ceramic powder. In this way, the HA coating deposited is quite different form bone mineral apatite. In fact, during this procedure, the overheating and melting can change the synthetic HA powder [69]. The osteoconductive and bone-bonding behavior of plasma-sprayed coatings was assessed by numerous studies. However, it should be noted that despite recent advances, especially with low-energy plasmas, the chemical composition of plasma coatings generally involve a mixture of several phases, including amorphous calcium phosphate, and not always in a reproducible manner. Therefore, researchers have been continuously inspired in the past two decades to explore alternative or complementary techniques for deposition of coatings onto an implant surface.

Various deposition methods have been proposed, including magnetron sputtering [70, 71], electrophoretic deposition [72], hot isostatic pressing, sol–gel deposition [73], pulsed laser deposition [74], biomimetic deposition [75], ion beam dynamic mixing deposition [76], electrospray deposition (ESD) [64], electrolytic deposition, and induction heating deposition from solution [67]. Generally, the properties of the produced coatings differ considerably in terms of chemical composition, structure, thickness, mechanical properties, and so on. The electrochemically assisted surface deposition on the metallic prosthesis of a biomimetic coating appears as the most interesting resolution. This method allows both to overcome the difficulty of depositing protein and bioactive organic molecules by plasma spray or physical vapor deposition and to control the coating process easily.

The electrodeposition of an adherent biomimetic coating of nanostructured HA on a metallic surface can be performed by an electrolytic cell, which can be made of three electrodes (work, reference, and control electrodes) or two electrodes (work and reference electrodes) [77]. On applying a DC potential between work and reference electrodes, we have a growth of pH value, from an initial value of about 3.5–4.0) to a final pH of about 8.0–11.0 around the work electrode. In fact, during electrodeposition we have the following reactions.

Cathodic reactions that induce H_2 development.

$$2H_2PO_4^- + 2e^- \rightarrow 2HPO_4^{2-} + H_2$$
$$2HPO_4^{2-} + 2e^- \rightarrow 2PO_4^{3-} + H_2$$
$$2H^+ + 2e^- \rightarrow H_2$$
$$2H_2O + 2e^- \rightarrow H_2 + 2OH^-$$

Anodic reaction that induces O_2 development.

$$4OH^- \rightarrow O_2 + 2H_2O + 4e^-$$

To have an apatitic cathodic electrodeposition, the negatively charged work electrode is immersed in an electrolytic 0.5 and 0.001 M salt solution commonly containing: calcium ions ($Ca(NO_3)_2 \cdot 4H_2O$, $Ca(NO_3)_2 \cdot 2H_2O$, $Ca(CH_3COO)_2$); phosphate ions ($NH_4H_2PO_4$, K_2HPO_4, KH_2PO_4, H_3PO_4); and carbonate ions ($CaCO_3$, CO_2, Na_2CO_3, K_2CO_3). Applying a DC current ranging between 10 and 300 mA and a voltage ranging between 1 and 100 V at a temperature varying between 2 and 70 °C for time ranging between 5 s and 3 h, we have migration of calcium ions (at the beginning) and phosphate group (in the second phase) on the same work electrode, producing an apatite biomimetic coating [77]. The increase of the cell temperature increases the degree of crystallinity of the apatite crystals. A microbiological investigation provided evidence for the high bioreactivity of electrodeposited nanostructured HA coatings, which induces a quick cell proliferation on the coated titanium surface that is enormously reduced on uncoated titanium plate. In order to enhance the biomimetism of the above-described electrodeposited apatite coating, the same electrochemically assisted deposition process has been successfully utilized to produce HA/collagen nanostructured hybrid coatings that closely mimic bone tissue composition. In preparing the electrolyte, proper amount of Ca(NO$_3$)$_2$·2H$_2$O and NH$_4$H$_2$PO$_4$ are dissolved separately in distilled water to make two different solutions: the first one with 42 mM of [Ca^{2+}] and the second with 25 mM of [PO$_4^{3-}$]. These two solutions have to be mixed in the same measure. In order to obtain a biomimetic hybrid coating containing apatite and collagen, a molecular suspension of type I collagen (10% wt) in an amount of about 5–15% respect the global volume of the solution has to be added. During the electrochemical process, a basic environment is formed near the work electrode favoring both the HA nanocrystal growth and the nanocollagen fibril reconstitution on the same electrode.

During the past decade, BioSiC prepared by Si vapor or by Si-melted reactive infiltration of carbon templates, previously obtained by pyrolysis of different kinds of wood, received great attention [78, 79]. BioSiC morphology composed of open-pore geometries with wide surface area and microstructure allowing cell ingrowth and reorganization and providing the necessary space for vascularization allows one to consider this porous material to resemble a bone filler and substitute in orthopedics, odontology, and dental and maxillofacial implantations. Nanostructured HA/collagen biomimetic coating has been recently obtained by the electrochemically assisted deposition on BioSiC [80]. This is an innovative hybrid material to prepare innovative bone substitutes and bone tissue engineering scaffolds, with the possibility of synergistically joining the porous bio-inspired morphology, the mechanical property of biomorphic silicon carbide, and the good biocompatibility and biological response of the surface HA/collagen biomimetic hybrid coating (Figure 17.5).

In addition to pure calcium phosphates, many different biologically functional molecules can be immobilized onto titanium surfaces to enhance bone regeneration at the interface of implant devices. However, most techniques used to prepare inorganic HA coatings are performed either at extremely high temperatures or under nonphysiological conditions, which preclude the incorporation of biomolecules. The recent investigation trend has attempted to avoid this difficulty by adsorbing biological agents onto the surfaces of preformed inorganic layers [81, 82]. However, these superficially adsorbed molecules will be rapidly released in

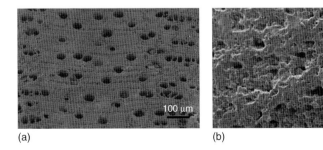

Figure 17.5 SEM image of wood silicon carbide. Transverse cross section of the sample (a) before and (b) after electrochemically assisted biomimetic hydroxyapatite/collagen surface deposition. (Source: From Ref. [80] with permission.)

an uncontrollable single burst on implantation [83]. Hence, coating procedures that incorporate biomolecules into the HA coating create a more sustained-release profile and are therefore of high interest [84]. In this way, the molecules can both sustain their biological activity for a considerable period of time and support the mechanical properties of the coating in case of structural extracellular matrix (ECM) components such as collagen [85]. Both the biomimetic and ESD processes are among the most promising techniques for generating organic–inorganic composite coatings on implant materials because of their physiological process conditions [86, 87]. ESD involves atomization of a precursor solution by applying a high voltage to the liquid surface, which then disperses into an aerosol spray of microsized charged droplets (<10 µm). This is accomplished by pumping the solution through a nozzle. Usually, a spherical droplet is formed at the tip of the nozzle, but when a high voltage is applied between the nozzle and substrate, this droplet converts into a conical shape and fans out form a spray of highly charged droplets. These charged droplets are attracted toward a grounded substrate, where a thin biofilm is formed after solvent evaporation. ESD technique has several advantages: very simple and cheap setup; high deposition efficiency, since the electric field directs charged droplets to the substrate; control over the coating composition, and the possibility of tailoring the morphology of the deposited coatings [88–90].

It is worth to notice that caution should always be taken when directly comparing the success rates of these coating techniques without a proper understanding of the physicochemical nature of the specific calcium phosphate coatings. Generally, the conclusions about the biological/clinical performance of coatings cannot be made without a complete set of characterizations that enable correlation of material properties to biological response [91]. Clinically, each application demands specific requirements, and in that respect, the wide range of available coating techniques offers the possibility to select the most appropriate deposition method for each specific implant application.

17.4
Biomimetic Nanostructured Apatite as Drug Delivery System

Over the past few decades, the rise of modern pharmaceutical technology and the amazing growth of the biotechnology industry have revolutionized the approach to the development of drug delivery systems [92]. For most of the industry's existence, pharmaceuticals have primarily consisted of relatively simple, fast-acting chemical compounds that are dispensed orally (as solid pills and liquids) or injected. During the past three decades, however, complex formulations that control the rate and period of drug delivery (i.e., time-release medications) and that target specific areas of the body for treatment have become increasingly common. A controlled-release drug delivery system should be able to achieve the following benefits: (i) maintenance of optimum therapeutic drug concentration in the blood with minimum fluctuation, (ii) predictable and reproducible release rates for extended duration, (iii) enhancement of activity duration for short half-life drugs, (iv) elimination of side effects, frequent dosing, and wastage of drug, and (v) optimized therapy and improved patient compliance [93–95].

To optimize the benefits, the design of a controlled-release system requires simultaneous consideration of several factors [96], such as the chemical and physical properties of the drug, the route of administration [97], the nature of the delivery vehicle, the mechanism of drug release, the potential for targeting, and biocompatibility. Drug delivery technologies are classified according to the route through which a drug is administered into the body. In the oral route, a drug is taken by mouth into the gastrointestinal (GI) tract to be absorbed. Other important routes include intravenous injection; intramuscular injection; subcutaneous injection; pulmonary, ocular, buccal (through the internal wall of the cheeks), sublingual (under the tongue), nasal, vaginal, rectal, transdermal routes; and implantation (inside a body cavity). Highly sophisticated drug delivery systems are being developed to exert their effects after they are administered into the body, in particular, into the systemic circulation. They usually consist of particles in the size range of nanometers that have unique combinations of capacities such as controlled release and biological site-specific targeting.

Furthermore, one of the most interesting thing is the implantable drug delivery system (IDDS). These techniques have the advantage of maintaining a steady release of drug to the specific action site so that they are safer and more reliable. IDDS can be classified into three major categories: biodegradable or nonbiodegradable implants, implantable pump systems, and the newest atypical class of implants. Biodegradable and nonbiodegradable implants are available as monolithic systems or reservoir systems. The release kinetics of drugs from such systems depend both on the solubility and diffusion coefficient of the drug in the polymer, the drug load, as well as the *in vivo* degradation rate of the polymer in the case of the biodegradable systems. The third IDDS system, the atypical class, includes recently developed systems such as ceramic HA antibiotic systems used in the treatment of bone infections, intraocular implants for the treatment of glaucoma, and transurethral implants utilized in the treatment of impotence. The major advantages of these systems include targeted local delivery of drugs at a constant rate, requirement of lesser quantities of the drug to treat the disease state, minimization of possible side effects, and enhanced efficacy of treatment [98]. Besides, these forms of delivery systems are capable of protecting drugs that are unstable *in vivo* and that would normally require frequent dosing intervals. Owing to the development of such sustained-release formulations, it is now possible to administer unstable drugs once a week to once a year, which in the past required frequent daily dosing.

Preliminary studies using these systems have shown superior effectiveness over conventional treatment methods. However, some of the most recently discovered implants are in the early developmental stages and more rigorous clinical testing is required before their use in standard practice [99].

Apatites and other calcium phosphates may find several applications in implant and injectable drug device. In fact, as drug carriers, calcium phosphate nanoparticles have some advantageous properties. Apatites particles with tunable phase composition and thus with tunable resorption rate offer an advantage of increasing the efficiency and local character of drug delivery as well as a controlled release over time. A reduced dosage and frequency of administration, and consequently fewer side effects, would result thereby [100]. By means of calcium phosphate particles that are stable before injection and then comparatively slowly resorbed in the body, the active compound is protected until the particles are injected and delivered to the targeted site, after which a controlled release, tunable to anywhere between a few days to a few weeks, may be set to take place. Calcium phosphate particles are also able to permeate the cell membrane and dissolve in the cell, which makes them an attractive candidate for gene delivery agents too [101, 102]. Furthermore, nano-calcium phosphates hold larger load amounts of drugs because of their larger specific surface areas than coarse particles. The encapsulation of DNA into calcium phosphate nanoparticles protects the nucleic acid from cytoplasmic environment and enables its efficient delivery into the cell nucleus. Nonviral gene delivery systems have been, in fact, intensively investigated as possible alternatives to viral vectors for the transfection purposes. Although nonviral gene delivery vectors typically possess lower transfection efficiencies compared to viral ones, their excellent safety profiles are appealing for gene therapy [103]. Several nanomaterial-based vectors have been developed using functionalized multiwalled and single-wall carbon nanotubes; metallic, bimetallic, TiO_2, and silica nanoparticles; magnesium phosphate; manganous phosphate; and various dendrimers [92]. However, unlike many of these options, including poly-L-lactic acid (PLLA) and a few other polymeric particles proposed as drug/gene delivery carriers (PLLA, e.g., is typified by a very slow dissolution rate in the body, which disables a prompt release of an encapsulated active ingredient, whereas in the later stages of dissolution, it undergoes a self-catalytic degradation reaction owing to an increased release of acidic products harmful for the surrounding tissues [104]), the degradation of calcium phosphates does not produce any toxic chemicals apart from the release of Ca^{2+} (which may in the worst-case scenario initiate protein aggregation), and these particles are also scalable in size and imply low manufacturing costs. In that sense, nanosized calcium phosphate particles present one of the most viable options for gene delivery systems, and for this reason, they have been extensively used as an *in vitro* gene delivery agent for over 30 years and are currently being investigated for *in vivo* purposes as well [105, 106]. Moreover, structural and chemical features of calcium phosphate particles may provide all the essential requirements for their efficient use as genetic transfection agents, including the binding affinity of DNA molecules onto HA crystals [107], the composite particle stability in extracellular space, as well as an efficient cellular uptake and resorbability of the particles, followed by a gradual release of the DNA and its escape from the endosomal network into the cytoplasmic intracellular space, and eventual cytosolic transport and nuclear localization for transcription [108].

Compared with other drug delivery carriers, calcium phosphates possess the following advantages: (i) favorable biodegradability and biocompatibility properties in general; (ii) more soluble and less toxic than silica, quantum dots, carbon nanotubes, or magnetic particles [109]; (iii) more stable/robust than liposomes, which predisposes them to a more controlled and reliable drug delivery – contrary to liposomes and other micelle-based carriers, which are subject to dissipation below specific critical concentrations (which presents a clear obstacle on injecting them into the bloodstream), calcium-phosphate-based systems and particularly those with Ca/P molar ratio close to the one of HA are negligibly soluble in blood, which is by itself supersaturated with respect to HA; and (iv) higher biocompatibility and pH-dependent dissolution compared to polymers [110]. The dissolution of calcium phosphates is accelerated in low-pH media, which are typically found in endolysosomes and in the vicinity of tumors, providing an advantage delivering the drugs into malign zones or cell organelles. Also, unlike most metallic and oxide nanoparticles, including Au, Ag, Co, Cr, Cd, Se, Te, TiO_2, CuO, and ZnO, which have all been shown to induce damage to DNA, produce oxidative lesions, increase mutation frequency, and decrease cell viability, nanosized calcium phosphates belong to the class of safest nanomaterials evaluated so far (together with SiO_2 and most likely Fe_2O_3 and C60) [111]. Another important advantage of calcium phosphates is the low production costs and excellent storage abilities (not easily subjected to microbial degradation). As previously explained, the preparation of nanosized calcium phosphates is not a complicated task. Unlike most other ceramics, nanosized calcium phosphates can be prepared *in situ*, under ambient conditions, and in a wide array of morphologies, from spheres to platelets to rods to fibers. Calcium phosphates can also be prepared with a variety of phase compositions, thereby enabling fine-tuning of the dissolution properties *in vivo* at the structural level as well.

17.4.1
Adsorption and Release of Drugs

Numerous bioactive molecules can associated with calcium phosphates, including antibiotics, growth factors, and anticancer and antiosteoporosis drugs [39]. However, only a very few studies have been carried out to understand adsorption behavior of therapeutic molecules on apatitic calcium phosphates, especially nanocrystalline apatites. Yet these observations can have several consequences on the regulatory role of bone, determining drug uptake and release. Adsorption on biomimetic biomaterials appears as a powerful way to control the release and the activity range of therapeutic molecules.

Bioactive molecules often exhibit high binding affinity for calcium phosphate surfaces. Generally, the adsorption equilibrium of such drugs strongly interacting with an apatitic surface is well described by a Langmuir isotherm, a behavior similar to that presented by proteins and amino acids [112–115], characterized by two adsorption parameters: the affinity constant (K) and the maximum coverage (N). Historically, two types of adsorption sites have been recognized on apatite, calcium, and phosphate surface sites. Negatively charged molecules were considered to bind to calcium sites, and positively charged ones to phosphate sites [116].

Antibiotic interaction on calcium phosphate surface has been studied in several works. The nature of the charged molecules plays a significant role in the adsorption process. It appeared from several studies that antibiotics containing carboxylic groups in their chemical structure were better incorporated into calcium phosphate biomaterials than others [117]. For example, it has been published that antibiotics containing carboxylic groups, such as cephalothin, have better interaction with calcium at the surface of the apatite coating applied onto titanium implants, resulting in better binding and greater incorporation into the calcium phosphate coating [117]. Other studies have shown that carboxyl-containing molecules are good binders of calcium [118].

Concerning bone morphogenetic protein (BMP) growth factors, only a very few studies have been carried out on the adsorption of these molecules on apatitic calcium phosphates. The study of interaction between such proteins and apatitic surfaces is of great biological and medical interest as there have been attempts to apply BMPs for the reconstruction of bone defects resulting from trauma, surgical resection of tumors, and congenital anomalies in orthopedic and maxillofacial surgery. Few works show a strong affinity of growth factors for apatite surfaces, enhanced in the simultaneous presence of calcium ions in the solution, as in the case of the adsorption of other proteins [119, 120]. It was published for the adsorption of BMP-2 growth factor onto HA that the isotherm is Langmuirian in shape [121] and that adsorbate–adsorbent interactions can take place during protein adsorption, depending on the protein structure and the nature of the surface of the adsorbent [121]. A computed theoretical modeling based on the interactions of the (001) HA surface with functional groups of BMP-2 has recently shown three types of interactions involved in the adsorption process with this type of protein: (i) an electrostatic interaction of charged COO^- groups from the protein with Ca^{2+} ions on the apatite surface (ii) a water-bridged H-bonds between OH and NH_2 residues of BMP-2 and (iii) with PO_4^{3-} ions on the surface [122].

Finally, the adsorption of bioactive molecules onto apatitic surfaces appears generally irreversible with respect to dilution, as observed for proteins [123]. Washing an apatitic sample with adsorbed drugs accompanied by an ion exchange process does not create any possibility of reverse exchange. No desorption should be expected by simple dilution of the adsorption solution or even on washing of the samples, except for faint release effects because of the dissolution of the apatite. This behavior could be useful in view of the controlled release of such agents from apatitic biomaterials *in vivo*.

To better understand the adsorption behavior of bioactive molecules *in vivo*, it is important to consider nanocrystalline apatites as a better model of bone material than HA, and to consider bone tissue as a dynamic system. The adsorption behavior can be affected not only by the chemical properties of drugs but also by the physicochemical characteristics of the apatitic support. Very few studies have been performed on the adsorption of organic molecules on nanocrystalline apatites as model of bone material. For the adsorption of the antineoplastic drug cisplatin, greater activity of the poorly crystalline samples in the cisplatin adsorption process was observed, compared to well-crystallized apatitic HA [124]. The adsorption process does not differ strongly from that of well-crystallized apatites; however, an interesting aspect is the variation of the adsorption parameters with the characteristics of the nanocrystals. Adsorption on nanocrystalline apatite of bioactive agents indicates an increase of the affinity constant and a decrease of the amount adsorbed as maturation time increases [125]. The increase of the affinity constant may be correlated with an increase of interfacial energy of nanocrystals during maturation in aqueous media [126]. The hydrated layer probably reduces the surface energy of the nanocrystals in aqueous media. Another explanation could be that ions in the hydrated layer become less exchangeable as the maturation time increases, which is effectively observed [127]. It showed that the maturation of these compounds was directly related to the decrease in the concentration of nonapatitic phosphate groups. Moreover, the morphology and size of the crystals also account for the higher reactivity of these materials.

The decrease of the number of adsorption sites with maturation time can be interpreted as a decrease of potentially exchangeable ions in the hydrated layer or a change in the equilibrium constants. It can be seen that the ability of a poorly crystalline apatite to adsorb is related to its number of nonapatitic sites and its low crystallinity. As the adsorption process can be explained by ion exchange process at the surface of the apatitic support, the amount of bioactive molecules adsorbed could be related to the amount of labile mineral ions available in the hydrated layer and then to the extension of the hydrated layer. As the maturation proceeds, the amount of such exchangeable species diminishes and the adsorption capacity of the solid decreases [128].

As previously reported, apatites and other calcium phosphates are widely used as biomaterials and may find several applications in implant drug devices. In fact, as drug carriers, calcium phosphate nanoparticles have some advantageous properties. The nanoparticles can be made to fluoresce by the incorporation of lanthanide ions, and they can also act as carriers for different drugs. For example, one of the most efficient ways to improve the bone-forming ability of biomaterials is their association with BMPs [122]. Thus the choice of the process used to prepare implantable bone biomaterials has an influence on the release of the associated bioactive molecules. Various techniques associating drugs with a calcium phosphate biomaterial have been reported: adsorption and impregnation allow the therapeutic agent to be incorporated at the surface of the biomaterial, whereas centrifugation and vacuum-based-techniques enable it to enter into pores of biomaterials [129, 130].

Generally, bioactive molecules, such as growth factors, are incorporated into biomaterials by simple impregnation followed by drying, and the type of bonding with the substrate and the release rate are often undetermined [131]. It is suspected that such associations do not allow the chemical bonding of the growth factor to the biomaterial, and thus the release rate is often difficult to control. For example, precipitation and clustering of the growth factor molecules may occur and the release is only determined by local dissolution and diffusion rules [122]. The uncontrolled release of growth factors has, in some instances, been related to an accelerated resorption of bone tissue and of the implant [122]. Since growth factors can stimulate the degradation as well as the formation of bone (depending on their local concentrations), they could impair the osteoconductivity of the coated implant surface [132]. Similarly, bisphosphonate (BP) molecules, by affecting bone remodeling, could also block the bone repair process: the drug at very high concentration could have detrimental effects on the fixation of the implant over longer periods of time. Zoledronate grafted to HA coating on titanium implants shows a dose-dependent effect on the inhibition of resorption activity according to the amount of zoledronate loaded [133]. Local and slow administration of antineoplasic drugs such as methotrexate (MTX) is also useful to avoid systemic side effects and because its time effect (the sensitivity of cells to this drug increases with time) is greater than the dose effect [134].

On the contrary, adsorption leads to stable association and control of the amount of bioactive molecules contained in the solid implant and thus of the dose released. Generally, the release is rather slow because most of the bioactive molecules adsorbed are irreversibly bound and are not spontaneously released in a cell culture media. They can only be displaced by mineral ions and/or soluble proteins with a stronger affinity for apatite surfaces but in a predicable manner, or by cell activity. This characteristic has been observed for various growth factors such as BMP-2 or VEGF [122, 134], antiosteoporisis agents [135, 136], and anticancer drugs such as MTX and cisplatin [124, 137]. It has been reported that slow release of MTX from calcium phosphate is due to the porosity like in most of the cases but mainly due to the adsorption of MTX [137]. Figure 17.6 illustrates release of antibiotics incorporated into apatitic calcium phosphate coatings on titanium implant using biomimetic method [117]. A local release system of antibiotics could be used to prevent postsurgical infections, favoring early osteointegration of prosthesis. Since some of the antibiotics containing carboxylic groups in their chemical structure,

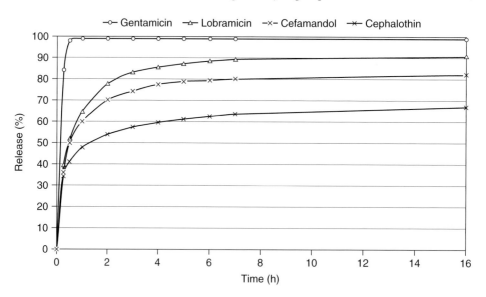

Figure 17.6 Antibiotic release from carbonated apatitic calcium phosphate coatings on titanium implants into phosphate buffered saline at pH 7.3. (Source: From Ref. [117] with permission.)

such as cephalothin, are better adsorbed than others into calcium phosphate biomaterials, these molecules are slowly released from the carrier. Moreover, the high binding capability of apatitic support for a wide range of therapeutic agents allows its surface functionalization with linking agents, such as BPs, to anchor biologically active molecules that can be released breaking the linkage as a consequence of external stimuli or internal chemical factors, such as pH and ionic force variation due to physiological or pathological biological processes [138]. For example, some works have investigated the adsorption and "smart release" of antitumoral platinum complexes containing BPs onto apatitic nanocrystals as bone-specific drug delivery devices to be used for the treatment of bone tumors on local implantation [138].

It is important to notice that, as the adsorption properties of the bioactive molecules are strongly dependent on the nature of the calcium phosphate support and maturation stage, it is probable, although this has not been systematically investigated, that the release rate is related to the physicochemical properties of the apatite samples. For example, the cisplatin release rate can be controlled by use of a more crystalline apatite for a faster release rate or a less crystalline material for a slower rate [124, 139]. Similarly, some authors show that the coating of sintered biphasic calcium phosphate granules (HA and β-tricalcium phosphate (β-TCP)) with nanocrystalline apatite with nanocrystalline apatite improves BMP-2 growth factor's adsorption (the rate of adsorption was increased). Only a low quantity of protein was spontaneously released in cell culture medium during the first weeks.

The adsorption and release of bioactive molecules are strongly affected not only by the chemical properties of the drug molecule but also by the chemical and structural characteristics of the HA substrates. The adsorption and release of cisplatin, alendronate, and di(ethylenediamineplatinum)medronate have been investigated using two biomimetic synthetic HA nanocrystal materials with either plate-shaped or needle-shaped morphologies and with different physicochemical surface properties by Palazzo *et al.* [139]. These bioactive molecules were chosen in order to compare the behavior of metal-based drugs to that of a classical organic drug (alendronate), evaluating the effect of the overall charge of a drug molecule on drug affinity for apatite nanocrystals with variable structural and chemical properties. The HA surface area and surface charge (Ca/P ratio), as well as the charge on the adsorbed molecules and their mode of interaction with the HA surface, influence the adsorption and release kinetics of the three drugs investigated. The results demonstrated that HA nanocrystals and antitumor drugs can be selected in such a way that the bioactivity of the drug–HA conjugate could be tailored for specific therapeutic applications.

The adsorption of two different platinum complexes with cytotoxic activity, bis-{ethylenediamineplatinum(II)}-2-amino-1-hydroxyethane-1,1-diyl-bisphosphonate (hereafter complex A) and the bis-{ethylenediamineplatinum(II)} medronate (hereafter complex B) on the synthesized biomimetic nanocrystals complex A having a greater affinity for calcium phosphate nanocrystals. The release profiles of the platinum complexes from the HA nanoparticles follow an inverted trend (complex B > complex A) when compared with the adsorption process.

Most probably, the less effective desorption in the case of complex A could be due to the amino group present on the BP that remains anchored to the HA matrix, coordinating and holding some of the Pt(en) residues. Unmodified and HA-adsorbed Pt complexes were tested for their cytotoxicity toward human cervical carcinoma cells (HeLa). The HA-loaded Pt complexes were more cytotoxic than the unmodified compounds A and B, and their cytotoxicity was comparable to that of [PtCl$_2$(en)], thus indicating a common active species. The above results demonstrate that HA nanocrystals and antitumor drugs can be conjugated such that a smart bone filler delivery system is obtained, acting both as bone substitute and as platinum drug releasing agent, with the final goal of locally inhibiting the tumor regrowth and reducing the systemic toxicity. The system described in this chapter can not only ensure a prolonged release of active species but also improve the performance of the unmodified drug. Moreover, these results suggest the possibility of using the physicochemical differences of HA nanocrystals, above all degree of crystallinity, crystal size, and surface area, in order to strongly tailor the Pt complex release kinetics. Considering the biomimetic apatite nanocrystal's functionalization effects, an attractive goal could be to obtain a drug delivery process characterized by a stimuli-responsive kinetics. This aim induces to surface-functionalize HA nanocrystals with different linking agents, such as BPs, to anchor biologically active molecules, which can be released breaking the linkage as a consequence of external stimuli or internal chemical factors, such as pH and ionic force variation due to physiological or pathological process.

17.5
Adsorption and Release of Proteins

Despite the many and deep investigations, the mechanisms involved in protein and peptide adsorption on inorganic substrate is still far from being fully understood. In fact, this is a complex process consisting of many events, including protein conformational changes and coadsorption of different ions. The most important forces involved in molecular adsorption are expected to be of electrostatic nature, associated with the charged groups exposed at the protein/biomaterial interface [140]. Also protein/protein binding and protein unfolding are expected, and the protein conformational changes that result in entropic gain contribute to the protein surface adhesion with hydrophobic/hydrophilic forces. However, the effective immobilization of proteins on the surface of inorganic nanocrystals remains a challenge to the life science industry, because of the requirement of (i) a sufficiently strong and specific affinity of proteins to the surface, (ii) site-directed immobilization, and (iii) preservation of a native conformation and biological function of proteins in their adsorbed stages [141]. Most immobilization methods developed until now involve modification or coating of the surface with appropriate substances in order to immobilize the proteins by physical adsorption via van der Waals, hydrophobic, or electrostatic forces or via chemical bonding [142].

In addition, to test and develop a biomaterial device, the interaction of protein with HA is important in the biomineralization science. In fact, Nature has evolved sophisticated strategies for engineering hard tissues through the interaction of proteins and, ultimately, cells with inorganic mineral phases [143, 144]. The remarkable material properties of bones and teeth thus result from the activities of proteins that function at the organic–inorganic interface. The underlying molecular mechanisms that control biomineralization are of significant interest to both medicine and dentistry, since disruption of biomineralization processes can lead to bone and tooth hypomineralization or hypermineralization, artherosclerotic plaque formation, artificial heart valve calcification, kidney and gall stone formation, dental calculus formation, and arthritis, among other pathological calcification processes. A better understanding of the biomolecular mechanisms used to promote or retard crystal growth could provide important design principles for the development of calcification inhibitors and promoters in orthopedics, cardiology, urology, and dentistry. At the level of fundamental science, it is important to note the paucity of molecular structure information available for biomineralization of proteins in general and for mammalian proteins that directly control calcification processes in hard tissue in particular. Even the most fundamental questions about how the proteins interact at the biomineral surface, such as their general structure and orientation on the calcium phosphate surfaces, or whether the acidic residues are truly interacting directly with the crystal surface remain largely uncharacterized at the experimental level.

Interactions of cytokines or proteins from blood with calcium phosphates have been studied to improve knowledge on osteconduction mechanism, trombogenesis, or phagocytosis and have provided useful information to develop successful surgical devices or more efficient carriers for drugs. Recently, in evaluating the interaction of HA with serum and BMPs, the interaction of HA with recombinant human mature bone morphogenetic protein 2 (rhBMP-2m) [121], bovine serum albumin (BSA), human serum albumin (HSA) [112], lactoferrin [114], and myoglobin (Mb) [113] has been clearly elucidated.

RhBMP-2m has been expressed to study its adsorption onto HA, and the influence of different parameters on the adsorption process such as pH and concentration of calcium and phosphate has been investigated [121]. The adsorption proceeds rapidly at the initial stages, and the maximum adsorbed amount is reached after 4 h. The adsorption data fit well into a Langmuir adsorption isotherm. The process is notably influenced by adding calcium or phosphate to the system, but while calcium ions increase the adsorption of rhBMP-2m onto HA, phosphate ions inhibit it. The influence of pH and NaCl concentration are notable but less important than those of calcium and phosphate. To explain this finding, the HA surface has been modeled and widely studied. HA surface contains Ca and P sites, related to emerging calcium and phosphate ions in the surface structure. In aqueous solution, hydroxylated species form on these sites, with the corresponding surface equilibrium given by Eqs. (17.1) and (17.2), where S(HA) represents the HA surface.

$$S(HA)\text{-}CaOH + H^+ \leftrightarrow S(HA)\text{-}Ca(OH_2) + \quad (17.1)$$
$$S(HA)\text{-}POH \leftrightarrow S(HA)\text{-}PO\text{-} + H^+ \quad (17.2)$$

Since the ionic species in solution such as H^+, OH^-, Na^+, Cl^-, Ca^{2+}, and HPO_4^{2-}/PO_4^{3-} influence surface equilibrium as well as surface complexation, the net surface charge of the HA will basically depend on the surface structure, pH, and the concentration of the mentioned in solution. There is a wide range of zero point charge (zpc) values reported in the literature [145, 146]. Nevertheless, most of them are situated within the acidic and nearly neutral pH range. Yin et al. [147] found that, for HA the ζ-potential, at any explored pH, always exhibited negative values, except when the authors added calcium ions to the solution. Kawasaki et al. [148] also reported a negative ζ-potential value for HA, even in the presence of adsorbed proteins. These results can be interpreted by the predominance of phosphate or −PO− species at the surface. To better explain the HA surface behavior as a function of pH, Boix et al. drew the surface structure as a mixture of the model of surface complexation and the model of surface hydroxylation (Figure 17.7).

Calculations using the values reported by Wu et al. [149] for the surface equilibrium equations (17.1) and (17.2) indicate that at the pHs used in the work of Boix et al., there is a small predominance of negative phosphate ions at the surface over positive calcium species . Then, a small net negative surface charge is expected. From the point of view of the adsorbate, the rhBMP-2m presents acidic and basic residues, whose ionization depends on the solution pH. The isoelectric point (IEP) of rhBMP-2m is around pH 7.9. The fact that the working pH values are close to the IEP of the protein leads to small net charge. Phosphate anions in solution compete with carboxylate groups of the protein for adsorption on Ca sites. When phosphate is added, it adsorbs on Ca sites replacing hydroxylated species and the surface charge becomes more negative. Although the global charge of the protein is slightly positive, the adsorption is inhibited. Smith et al. have observed by chemical force

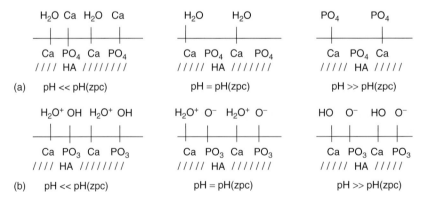

Figure 17.7 Models for HA surface proposed by Boix et al. (a) Model of surface complexation and (b) model of surface hydroxylation. (Source: From Ref. [121] with permission.)

microscopy (CFM) that in the presence of phosphate ions the adsorption of BSA onto HA is inhibited [150]. However, they observed that some molecules remain adsorbed, although not detectable macroscopically, in specific sites. On the other hand, when Ca^{2+} ions are added in solution, they adsorb on phosphate sites of the HA surface, allowing a more favorable protein adsorption through the carboxylate groups. Although the HA surface charge becomes positive and the global charge of the protein is slightly positive, the global electrostatic interaction is unfavorable and an increase in adsorption is observed. The adsorption results indicate that rhBMP-2m adsorbs or binds preferably on the Ca sites and through the carboxylate groups of the acidic residues. This type of interaction has also been proposed by other authors studying protein adsorption on HA [128].

The interaction mechanism of BSA and (HSA) with two kinds of biomimetic HA nanocrystals with different dimensions, surface areas, crystallinity degree, and surface properties has been elucidated by Iafisco et al. [112], pointing out some important discriminating factors. First, the two proteins reach a different maximum coverage especially toward the HA nanocrystals with higher surface area, smaller size, lower crystallinity degree, and plate shape morphology. Second, the adsorption isotherms could be described by both the Langmuir and Freundlich models because the matrix behavior is a mixture of an energetically homogeneous and an energetically heterogeneous support surface, with nonidentical adsorption sites, in agreement with the typical surface of the biomimetic HA nanocrystals with low degree of crystallinity and a nonstoichiometric Ca/P molar ratio. Although HSA adsorption on needle-shaped (HAns) and plate-shaped (HAps) HA nanocrystals appears quite similar, on the basis of both models adopted, Iafisco et al. concluded that the HAps surface has a greater affinity toward both BSA and HSA. The different cross-sectional surface areas of BSA on HA led to the hypothesis that BSA interacts with HAps preferentially oriented in an end-on way, while it interacts with HAns with a side-on orientation. On the contrary, no conclusion can be drawn on the orientation of the adsorbed HSA molecules on the HA nanoparticles because of the comparable values of cross-sectional surface areas. The protein conformations were differently affected by the inorganic substrates, showing a quite different structural stability. In particular, by using CD and FT-IR spectroscopies, it was found that BSA and HSA behaviors after interaction with the surface of both the synthetic HAns and HAps nanocrystals were different because of the adsorption effects on both the adducts and the detached proteins, depending on the nature of the sorbent surface and, possibly, on the mode of interaction. From FT-IR spectroscopy, it can be generally appreciated that the protein adhesion on HA surface resulted in a significant reduction in helicity for both proteins on the two substrates, as a function of the coating extent as well as of the β-structure, which appears significantly modified. From CD spectroscopy on the protein solutions exchanged after adsorption, it can be concluded that the protein helical structure is partly regained, but not up to the extent of the native protein. This is particularly evident on desorption from HAns for both BSA and HSA. In the tertiary structure region, a distinct behavior is shown by HSA, in particular after desorption from

HAps, as far as the environment at the Phe and Tyr residues, and possibly disulfide bridges, is involved.

Mb can be considered as a model protein because of its well-known structure and properties, commercial availability, and relatively small size. A new trend of research is to compare the protein interaction not only with the "clean" substrate but also with the functionalized materials. In particular, the Mb affinity versus the alendronate-surface-functionalized biomimetic HA with respect to unfunctionalized HA has been tested by Iafisco et al. [113]. The recognition and assembly of Mb on drug-loaded apatite crystals not only allows the better understanding of the protein adhesion mechanism but also aids in the development of surface coatings to improve the biocompatibility of bone implantable biomaterials and for hard tissue engineering and regeneration technologies. The HA–Mb interaction mechanism has been elucidated, and, in particular, using UV–Vis and surface-enhanced Raman spectroscopy, it has been found that the spin state of Mb heme moiety changes from the six-coordination high-spin native state to six-coordination low-spin state as a consequence of the interaction with biomimetic HA nanocrystals. The spin state of Mb is relevant in relation to the catalytic activity of the protein. The surface electrostatic potential map of protein allows us to hypothesize a preferential interactive mechanism through one defined region of protein surface, in spite of random adhesion mechanisms toward other inorganic supports. The heme moiety is attracted toward the apatitic surface as a consequence of surface disorder of the nanocrystals, which is connected to a negative surface charge with respect to the crystalline core. These results have highlighted that the immobilization of BPs onto the HA is an important strategy to set up a bone-specific drug delivery device. Considering that the interaction of protein with HA/biomolecules conjugates tailored for specific therapeutic applications plays a key role as a biological probe, the Mb affinity toward the HA/alendronate bioconjugates has been tested. The alendronate avoids both the adsorption and conformational changes of Mb heme moiety. Mb behavior toward alendronate-grafted HA crystals shows that this functionalization imprints surface selectivity to HA and drives the response of biological environment toward it.

17.5.1
Adsorption and Release of Bisphosphonates

It is noteworthy to discuss the merits of BP in this chapter. At present, BPs are well established as successful antiresorptive agents for the prevention and treatment of postmenopausal osteoporosis [151, 152]. BPs were first studied over 30 years ago as synthetic analogs of inorganic pyrophosphate, characterized by a nonhydrolyzable P-C-P structure with two phosphonic acid groups bonded to the same carbon (Figure 17.8).

Their mechanisms of action are physicochemical effects and biological activity: like pyrophosphate, BPs present a high affinity to adsorb onto apatite crystals because of the high attraction of phosphonate groups for calcium and thereby prevent mineral dissolution and bone resorption, inhibiting osteoclast activity

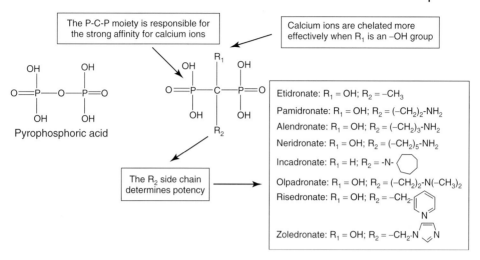

Figure 17.8 Structures of the endogenous pyrophosphonate and its synthetic analog, bisphosphonate (BP), which exhibit strong bone affinity. The geminal carbon in BP typically contains two separate substituents, R_1 and R_2, which may significantly affect both the mineral affinity and the pharmacological activity. Examples of BP class of compounds currently used in the clinical setting are listed. The compounds have been categorized into two classes based on the presence of an amino group in the R_2 side chain. The amino BPs typically exhibit a higher potency of antiresorptive effects, the primary clinical utility of BPs. Most of the BPs contain a geminal –OH group that enhances the mineral affinity of the compound.

[152, 153]. The P–C–P backbone is responsible for the strong affinity for bone mineral and inhibitory effects on bone dissolution [151]. The ability of the BPs to bind to bone mineral is increased when the R_1 functional group is able to coordinate to calcium, as hydroxyl group (OH) [151]. Varying the R_2 functional group can influence antiresorptive potency; those containing a nitrogen atom within a heterocyclic ring (such as risedronate, zoledronate) have the most potent antiresorptive effect [154, 155].

Preclinical studies with different BPs show that these antiresorptive agents not only inhibit bone loss but also increase bone mass and resistance to fracture [156]. The classical BP treatment is the systemic way, which is by oral administration or intravenous injection. However, systemic use of BPs can result in undesirable side effects such as fever [157], GI disturbances such as ulcers [158], or osteonecrosis of the jaw, especially caused by intravenous preparations [159, 160]. Moreover, low bioavailability is commonly observed for oral administration [161]. On the basis of this, the development of strategies for local administration of BPs within osteoporotic sites becomes even more interesting.

BP molecules present a high affinity for apatite and were found to prevent mineral dissolution and bone resorption. BPs are thus well established as successful antiresorptive agents for the prevention and treatment of bone diseases such as osteoporosis [151]. However, although it is commonly admitted that BPs attach to bone mineral crystals, a very few studies have been carried out on these aspects.

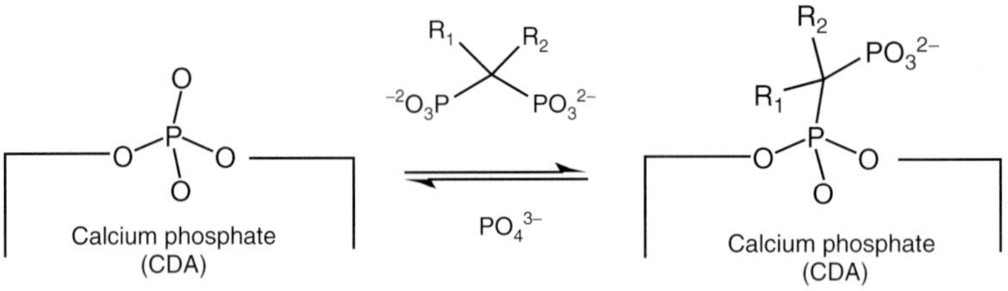

Figure 17.9 Schematic representation of chemical association of BPs onto the surface of calcium-deficient apatite (CDA) particles, via BP/phosphate exchange process. (Source: From Ref. [163] with permission.)

In the case of risedronate adsorbed onto HA, the isotherm is Langmuirian in shape. Interestingly, the follow-up of the variation in content of mineral ions of the solution indicates that the adsorption of risedronate is associated with a release of phosphate ions, whereas the calcium concentration remains almost unchanged. Similar adsorption characteristics were reported for the interaction of other BPs with calcium phosphates [162]. Considering that the dissolution of HA is congruent (i.e., the Ca/P ratio is the same for the solid and the solution), the analysis of the suspensions obtained after adsorption revealed that the binding of one risedronate ion onto the HA surface was accompanied by the release of one phosphate ion in the solution. The adsorption reaction can thus be described as an ion exchange process with a one-to-one substitution in the case of the risedronate [163] (Figure 17.9):

It is interesting to notice that the amount of BP molecules adsorbed at saturation can vary considerably depending on different parameters such as the temperature, pH, or initial phosphate and calcium content in the solution [163]. The maximum amount adsorbed is expected to correspond to the saturation of the surface adsorption sites, and such variations suppose that the maximum number of sites that can be occupied (N) varies according to these various factors. Considering the proposed ion exchange reactions, the chemical equilibrium involved cannot explain the variation of the adsorption limit: when the concentration of absorbate increases to large values, all sites should be occupied. Steric considerations, strength of attachment, and surface mobility of the adsorbed bioactive molecules have then to be considered and could be involved in the reorganization of such agents on the apatitic surface.

In all these mechanisms of drug adsorption onto apatitic surface, several processes can disturb the adsorption equilibrium and have to be taken into account including the possible precipitation of calcium (or phosphate) salts, the dissolution equilibrium of the apatite, and the surface hydrolysis reactions of apatite. For example, with zoledronate, a BP molecule, formation of a crystalline zoledronate complex onto calcium phosphate surface was observed [163]. In the case of risedronate adsorbed onto HA (accompanied by the release of phosphate ions in the solution), it should be noted that although the solubility equilibrium of HA would lead to a decrease of the free calcium ions in the solution, this is not observed and the relative stability (even the slight increase) of the calcium content with the increase of risedronate in solution can be interpreted as an indication of the probable formation of ion pairs or complexes between risedronate molecules and calcium ions in solution. Concerning the interactions of the antineoplasic drug cisplatin with bone substitutes, it has been reported that phosphate species are able to displace chloride ions from cisplatin molecules to form phosphate complexes, which may be more reactive toward the apatite surface.

Few studies have tried to evaluate the effect of local BP delivery from implants. Josse and coworkers [162] studied associations of zoledronate onto different types of calcium phosphate matrices such as calcium-deficient apatites (CDA) and β-tricalcium phosphate (β-TCP). When zoledronate was grafted onto calcium phosphate powders, the adsorption process was associated with the release of phosphate ions in the solution from the apatitic structure. At high BP concentrations, the formation of a crystalline zoledronate complex on the surface of the calcium phosphate support was evidenced by SEM and solid-state NMR. The biological activity of zoledronate-loaded calcium phosphate materials, resulting in the inhibition of osteoclastic activity, was evaluated using a specific *in vitro* bone resorption model.

Peter *et al.* [133] chemically grafted zoledronate onto an HA coating of titanium implants. The implants were immersed in zoledronate solution of variable concentration to favor BP adsorption. *In vivo* experiments established that when the implants were inserted in rats, the mechanical fixation of the orthopedic implants measured using a tensile testing machine was increased by the local presence of zoledronate. Moreover, the increase in peri-implant density appears to be zoledronate concentration dependent: at higher zoledronate concentrations, the mechanical stability decreases and a zolendronate concentration leading to an optimal implant fixation can thus be evidenced. Several studies have reported that BPs at high concentration could have negative effects on the fixation of implants over longer periods of time.

All these studies demonstrate the interest in preparation of new biomaterials for hard tissue repair with local action of these antiresorptive agents. Local and slow BP release from implants seems to be able to influence bone cells and promote bone repair. Moreover, calcium-phosphate-based biomaterials appear to be good candidates for releasing bioactive BP molecules.

17.6
Conclusions and Perspectives

Biomimetic HA nano- and microcrystals can be synthesized to exhibit excellent properties such as biocompatibility, bioactivity, osteoconductivity, and direct bonding to bone and exciting new applications in the fields of bone tissue engineering and orthopedic and odontoiatric therapies. Biogenic materials are nucleated in defined nano-/microdimensional sites inside the biological environments in which chemistry can be spatially controlled. The spatial delimitation is essential to biological mechanisms for controlling the size, shape, and structural organization of biomaterials. This strategy employing natural material genesis has attracted a lot of attention in designing bio-inspired materials. In the past, industrial HA has been subjected to milling to improve HA surface area and reactivity. Recently, the development of nanotechnology has opened new opportunities in obtaining HA micro/nanoparticles by the "bottom-up" methods. These HAs are nanostructured and have higher surface area and consequently higher reactivity.

Biomedical applications, especially drug delivery function, are optimized by the physical and chemical features that could be tailored in synthetic biomimetic HA. The surface functionalization of HA nanocrystals with bioactive molecules enables them to transfer information and to act selectively on the biological environment, and cell activity represents the main challenge for innovative biomaterials, opening interesting future developments.

In order to improve their functionality when used as implant devices, the above-mentioned biomaterials can be functionalized release drugs, growth factors, enzymes, nanoparticles, and general bioactive agents into the surrounding environment, lowering systemic toxicity. In the drug delivery field, nanostructured biomaterials offer a highly improved performance over their larger particle size counterparts, because of their large surface-to-volume ratio and unusual chemical/electronic synergistic effects. The role of some of these features (porosity and surface area) has been clarified in the literature, while some parameters linked to other surface characteristics still have to be evaluated (surface electrification, hydrophobicity, hydrophilicity). Biomimetic structural surface features of the nanostructured matrices are strongly connected with the loading and release of biologically active agents. In the case of nanodimension matrices, some of the biomimetic characteristics of such biomaterials, such as surface area due to the nanometric scale and structural and compositional surface disorder, need to be modulated in order to tailor the release of bioactive molecules with the aim of obtaining a controlled kinetics. Moreover, in the case of nanostructured porous materials, the nanoporosity allows the absorption of a consistent amount of bioactive molecules, while macroporosity can be utilized for cell growth, transforming these materials into a real cellular scaffolds for tissue engineering.

An ideal biomaterial with drug delivery function should be able to target these biological agents toward specific organism sites without compromising their efficacy. Furthermore, it should be able to release the biologically active agents with kinetics that can be controlled by the modification of structural characteristics of matrices or by means of either a physiological or a chemical trigger. In this way, researchers could obtain a biomimetic release, where the kinetic release of bioactive molecules is controlled by pathophysiological factors, such as variation of pH, ionic force, and enzyme and protein activities. Therefore, we could obtain a "stimuli-responsive" nanodevice system, in which the stimulus is represented by a change induced by a biological, pathological process (inflammation, infection, metastasis, etc.). Nowadays, many "smart materials" result from the assembly of different components into a single system: one material acts as a sensor, another one as an actuator, and the third one as the processor. However, it is possible to design a unique material integrating all these functions.

In consideration of the above, nanotechnology coupled with a biomimetic approach offers a unique, innovative strategy to overcome the shortcomings of many conventional drug delivery materials. From nanomedicine to nanofabrics, this promising technology has encompassed almost all disciplines of human life. In the case of biomaterial science, it offers the double advantage of mimicking Nature and improving material performance.

Acknowledgments

We thank the Alma Mater Studiorum, Università di Bologna (funds for selected research topics) and the Consorzio Interuniversitario di Ricerca in Chimica dei Metalli nei Sistemi Biologici (C.I.R.C.M.S.B).

References

1. Roveri, N., Morpurgo, M., Palazzo, B., Parma, B., and Vivi, L. (2005) *Anal. Bioanal. Chem.*, **381**(3), 601–606.
2. Margiotta, N., Ostuni, R., Teoli, D., Morpurgo, M., Realdon, N., Palazzo, B. *et al.* (2007) *Dalton Trans.*, **29**, 3131–3139.
3. Uskokovic, V. and Uskokovic, D.P. (2011) *J. Biomed. Mater. Res. B Appl. Biomater.*, **96**(1), 152–191.
4. Dorozhkin, S.V. (2010) *Acta Biomater.*, **6**(3), 715–734.
5. Palmer, L.C., Newcomb, C.J., Kaltz, S.R., Spoerke, E.D., and Stupp, S.I. (2008) *Chem. Rev.*, **108**(11), 4754–4783.
6. Roveri, N. and Palazzo, B. (2007) *Hydroxyapatite Nanocrystals as Bone Tissue Substitute*, Wiley-VCH Verlag GmbH & Co. KGaA.
7. Bongio, M., Van Den Beucken, J.J.J.P., Leeuwenburgh, S.C.G., and Jansen, J.A. (2010) *J. Mater. Chem.*, **20**(40), 8747–8759.
8. Legros, R., Balmain, N., and Bonel, G. (1987) *Calcif. Tissue Int.*, **41**(3), 137–144.
9. Paschalis, E.P., DiCarlo, E., Betts, F., Sherman, P., Mendelsohn, R., and Boskey, A.L. (1996) *Calcif. Tissue Int.*, **59**(6), 480–487.
10. Rey, C., Collins, B., Goehl, T., Dickson, I.R., and Glimcher, M.J. (1989) *Calcif. Tissue Int.*, **45**(3), 157–164.
11. Rey, C. (1990) *Biomaterials*, **11**, 13–15.
12. Sakhno, Y., Bertinetti, L., Iafisco, M., Tampieri, A., Roveri, N., and Martra, G. (2010) *J. Phys. Chem. C*, **114**(39), 16640–16648.
13. Drouet, C., Bosc, F., Banu, M., Largeot, C., Combes, C., Dechambre, G. *et al.* (2009) *Powder Technol.*, **190**(1–2), 118–122.
14. Cazalbou, S., Eichert, D., Ranz, X., Drouet, C., Combes, C., Harmand, M.F. *et al.* (2005) *J. Mater. Sci.: Mater. Med.*, **16**(5), 405–409.
15. Schmidt, H.K. (2000) *Mol. Cryst. Liq. Cryst.*, **353**, 165–179.
16. Cushing, B.L., Kolesnichenko, V.L., and O'Connor, C.J. (2004) *Chem. Rev.*, **104**(9), 3893–3946.
17. Mao, Y., Park, T.J., Zhang, F., Zhou, H., and Wong, S.S. (2007) *Small*, **3**(7), 1122–1139.
18. Wang, J.W. and Shaw, L.L. (2007) *Adv. Mater.*, **19**(17), 2364–2369.
19. Ganesan, K. and Epple, M. (2008) *New J. Chem.*, **32**(8), 1326–1330.
20. Zhang, Y.J. and Lu, J.J. (2007) *J. Nanopart Res.*, **9**(4), 589–594.
21. Ben-Nissan, B. and Choi, A.H. (2006) *Nanomedicine*, **1**(3), 311–319.
22. Choi, A.H. and Ben-Nissan, B. (2007) *Nanomedicine*, **2**(1), 51–61.
23. Chai, C.S. and Ben-Nissan, B. (1999) *J. Mater. Sci.: Mater. Med.*, **10**(8), 465–469.
24. Tas, A.C. (2000) *Biomaterials*, **21**(14), 1429–1438.
25. Lopez-Macipe, A., Gomez-Morales, J., and Rodriguez-Clemente, R. (1998) *Adv. Mater.*, **10**(1), 49–53.

26. Guo, X., Gough, J.E., Xiao, P., Liu, J., and Shen, Z. (2007) *J. Biomed. Mater. Res. A*, **82**(4), 1022–1032.
27. Chaudhry, A.A., Haque, S., Kellici, S., Boldrin, P., Rehman, I., Fazal, A.K. et al. (2006) *Chem. Commun.*, **21**, 2286–2288.
28. Yeon, K.C., Wang, J., and Ng, S.C. (2001) *Biomaterials*, **22**(20), 2705–2712.
29. Siva Rama Krishna, D., Siddharthan, A., Seshadri, S.K., and Sampath Kumar, T.S. (2007) *J. Mater. Sci.: Mater. Med.*, **18**(9), 1735–1743.
30. Liu, J.B., Li, K.W., Wang, H., Zhu, M.K., Xu, H.Y., and Yan, H. (2005) *Nanotechnology*, **16**(1), 82–87.
31. Rameshbabu, N., Rao, K.P., and Kumar, T.S.S. (2005) *J. Mater. Sci.*, **40**(23), 6319–6323.
32. Iafisco, M., Morales, J.G., Hernandez-Hernandez, M.A., Garcia-Ruiz, J.M., and Roveri, N. (2010) *Adv. Eng. Mater.*, **12**(7), B218–BB23.
33. Nassif, N., Martineau, F., Syzgantseva, O., Gobeaux, F., Willinger, M., Coradin, T. et al. (2010) *Chem. Mater.*, **22**(12), 3653–3663.
34. Iafisco, M., Marchetti, M., Morales, J.G., Hernandez-Hernandez, M.A., Ruiz, J.M.G., and Roveri, N. (2009) *Cryst. Growth Des.*, **9**(11), 4912–4921.
35. Phillips, M.J., Darr, J.A., Luklinska, Z.B., and Rehman, I. (2003) *J. Mater. Sci.: Mater. Med.*, **14**(10), 875–882.
36. Ye, F., Guo, H.F., and Zhang, H.J. (2008) *Nanotechnology*, **19**(24).
37. Layrolle, P. and Lebugle, A. (1994) *Chem. Mater.*, **6**(11), 1996–2004.
38. Koutsopoulos, S. (2002) *J. Biomed. Mater. Res.*, **62**(4), 600–612.
39. Roveri, N., Palazzo, B., and Iafisco, M. (2008) *Expert Opin. Drug Deliv.*, **5**(8), 861–877.
40. Iafisco, M., Delgado-López, J.M., Gómez-Morales, J., Hernández-Hernández, M.A., Rodríguez-Ruiz, I., and Roveri, N. (2011) *Crys t. Res. Technol.* **46**(8), 841–846.
41. Gómez-Morales, J., Hernández-Hernández, A., Sazaki, G., and García-Ruiz, J.M. (2009) *Cryst. Growth Des.*, **10**(2), 963–969.
42. Hernández-Hernández, A., Rodríguez-Navarro, A.B., Gómez-Morales, J., Jiménez-Lopez, C., Nys, Y., and García-Ruiz, J.M. (2008) *Cryst. Growth Des.*, **8**(5), 1495–1502.
43. Rodriguez-Lorenzo, L.M., Vallet-Regi, M., and Ferreira, J.M. (2001) *Biomaterials*, **22**(6), 583–588.
44. Tampieri, A., Celotti, G., Sprio, S., Delcogliano, A., and Franzese, S. (2001) *Biomaterials*, **22**(11), 1365–1370.
45. Tadic, D., Beckmann, F., Schwarz, K., and Epple, M. (2004) *Biomaterials*, **25**(16), 3335–3340.
46. Rodríguez-Lorenzo, L.M., Vallet-Regí, M., Ferreira, J.M.F., Ginebra, M.P., Aparicio, C., and Planell, J.A. (2002) *J. Biomed. Mater. Res.*, **60**(1), 159–166.
47. Padilla, S., Román, J., and Vallet-Regí, M. (2002) *J. Mater. Sci.: Mater. Med.*, **13**(12), 1193–1197.
48. Roy, D.M. and Linnehan, S.K. (1974) *Nature*, **247**(5438), 220–222.
49. Ivankovic, H., Gallego Ferrer, G., Tkalcec, E., Orlic, S., and Ivankovic, M. (2009) *J. Mater. Sci.: Mater. Med.*, **20**(5), 1039–1046.
50. Tampieri, A., Celotti, G., Landi, E., Montevecchi, M., Roveri, N., Bigi, A. et al. (2003) *J. Mater. Sci.: Mater. Med.*, **14**(7), 623–627.
51. Zimmerman, A.F., Palumbo, G., Aust, K.T., and Erb, U. (2002) *Mater. Sci. Eng. A*, **328**(1–2), 137–146.
52. Li, X.F., Fan, T.X., Liu, Z.T., Ding, J., Guo, Q.X., and Zhang, D. (2006) *J. Eur. Ceram. Soc.*, **26**(16), 3657–3664.
53. Singh, M., Martínez-Fernández, J., and de Arellano-López, A.R. (2003) *Curr. Opin. Solid State Mater. Sci.*, **7**(3), 247–254.
54. Tampieri, A., Sprio, S., Ruffini, A., Celotti, G., Lesci, I.G., and Roveri, N. (2009) *J. Mater. Chem.*, **19**(28), 4973–4980.
55. Dorozhkin, S.V. (2010) *Biomaterials*, **31**(7), 1465–1485.
56. Palazzo, B., Sidoti, M.C., Roveri, N., Tampieri, A., Sandri, M., Bertolazzi, L. et al. (2005) *Mater. Sci. Eng., C: Biomimetic Supramol. Syst.*, **25**(2), 207–213.
57. Iafisco, M., Varoni, E., Battistella, E., Pietronave, S., Prat, M., Roveri, N.

et al. (2010) *Int. J. Artif. Organs*, **33**(11), 765–774.
58. Balazic, M., Kopac, J., Jackson, M.J., and Ahmed, W. (2007) *Int. J. Nano Biomater.*, **1**(1), 3–34.
59. Liu, X., Chu, P.K., and Ding, C. (2004) *Mater. Sci. Eng. R*, **47**(3–4), 49–121.
60. Sul, Y.T., Johansson, C.B., Petronis, S., Krozer, A., Jeong, Y., Wennerberg, A. et al. (2002) *Biomaterials*, **23**(2), 491–501.
61. Geesink, R.G.T., De Groot, K., and Klein, C.P.A.T. (1987) *Clin. Orthop. Relat. Res.*, **225**, 147–170.
62. De Grootl, K., Wolke, J.G.C., and Jansen, J.A. (1998) *Proc. Inst. Mech. Eng. [H]*, **212**(2), 137–147.
63. Thomas, K.A., Cook, S.D., Haddad, R.J. Jr., Kay, J.F., and Jarcho, M. (1989) *J. Arthroplasty*, **4**(1), 43–53.
64. Leeuwenburgh, S.C., Wolke, J.G., Siebers, M.C., Schoonman, J., and Jansen, J.A. (2006) *Biomaterials*, **27**(18), 3368–3378.
65. Soballe, K., Hansen, E.S., Brockstedt-Rasmussen, H., and Bunger, C. (1993) *J. Bone Joint Surg. Ser. B*, **75**(2), 270–278.
66. Soballe, K., Hansen, E.S., Brockstedt-Rasmussen, H., Hjortdal, V.E., Juhl, G.I., Pedersen, C.M. et al. (1991) *Clin. Orthop. Relat. Res.*, **272**, 300–307.
67. Gómez Morales, J., Rodríguez Clemente, R., Armas, B., Combescure, C., Berjoan, R., Cubo, J. et al. (2004) *Langmuir*, **20**(13), 5174–5178.
68. MacDonald, D.E., Betts, F., Stranick, M., Doty, S., and Boskey, A.L. (2001) *J. Biomed. Mater. Res.*, **54**(4), 480–490.
69. Zyman, Z., Weng, J., Liu, X., Li, X., and Zhang, X. (1994) *Biomaterials*, **15**(2), 151–155.
70. Lusquinos, F., De Carlos, A., Pou, J., Arias, J.L., Boutinguiza, M., Leon, B. et al. (2003) *J. Biomed. Mater. Res. A*, **64A**(4), 630–637.
71. Yang, Y.Z., Kim, K.H., and Ong, J.L. (2005) *Biomaterials*, **26**(3), 327–337.
72. Wei, M., Ruys, A.J., Swain, M.V., Kim, S.H., Milthorpe, B.K., and Sorrell, C.C. (1999) *J. Mater. Sci.: Mater. Med.*, **10**(7), 401–409.
73. Wang, D.G., Chen, C.Z., Ting, H., and Lei, T.Q. (2008) *J. Mater. Sci.: Mater. Med.*, **19**(6), 2281–2286.
74. Hashimoto, Y., Kawashima, M., Hatanaka, R., Kusunoki, M., Nishikawa, H., Hontsu, S. et al. (2008) *J. Mater. Sci.: Mater. Med.*, **19**(1), 327–333.
75. Muller, L., Conforto, E., Caillard, D., and Muller, F.A. (2007) *Biomol. Eng.*, **24**(5), 462–466.
76. Choi, J.M., Kim, H.E., and Lee, I.S. (2000) *Biomaterials*, **21**(5), 469–473.
77. Manara, S., Paolucci, F., Palazzo, B., Marcaccio, M., Foresti, E., Tosi, G. et al. (2008) *Inorg. Chim. Acta*, **361**(6), 1634–1645.
78. López-Álvarez, M., de Carlos, A., González, P., Serra, J., and León, B. (2010) *J. Biomed. Mater. Res. Part B: Appl. Biomater.*, **95B**(1), 177–183.
79. de Arellano-López, A.R., Martínez-Fernández, J., González, P., Domínguez, C., Fernández-Quero, V., and Singh, M. (2004) *Int. J. Appl. Ceram. Technol.*, **1**(1), 56–67.
80. Lelli, M., Foltran, I., Foresti, E., Martinez-Fernandez, J., Torres-Raya, C., Varela-Feria, F.M. et al. (2010) *Adv. Eng. Mater.*, **12**(8), B348–BB55.
81. Lind, M., Overgaard, S., Glerup, H., Søballe, K., and Bünger, C. (2001) *Biomaterials*, **22**(3), 189–193.
82. Ono, I., Gunji, H., Kaneko, F., Saito, T., and Kuboki, Y. (1995) *J. Craniofac. Surg.*, **6**(3), 238–244.
83. Wozney, J.M. and Rosen, V. (1998) *Clin. Orthop. Relat. Res.*, (346), 26–37.
84. De Jonge, L.T., Ju, J., Leeuwenburgh, S.C.G., Yamagata, Y., Higuchi, T., Wolke, J.G.C. et al. (2010) *Thin Solid Films*, **518**(19), 5615–5621.
85. de Jonge, L.T., Leeuwenburgh, S.C.G., van den Beucken, J.J.J.P., teRiet, J., Daamen, W.F., Wolke, J.G.C. et al. (2010) *Biomaterials*, **31**(9), 2461–2469.
86. Liu, Y., Hunziker, E.B., Layrolle, P., De Bruijn, J.D., and De Groot, K. (2004) *Tissue Eng.*, **10**(1–2), 101–108.
87. Liu, Y., Hunziker, E.B., Randall, N.X., De Groot, K., and Layrolle, P. (2003) *Biomaterials*, **24**(1), 65–70.

88. Leeuwenburgh, S.C.G., Heine, M.C., Wolke, J.G.C., Pratsinis, S.E., Schoonman, J., and Jansen, J.A. (2006) *Thin Solid Films*, **503**(1–2), 69–78.
89. Uematsu, I., Matsumoto, H., Morota, K., Minagawa, M., Tanioka, A., Yamagata, Y. et al. (2004) *J. Colloid Interface Sci.*, **269**(2), 336–340.
90. Leeuwenburgh, S., Wolke, J., Schoonman, J., and Jansen, J. (2003) *J. Biomed. Mater. Res. A*, **66**(2), 330–334.
91. Shepperd, J.A.N. and Apthorp, H. (2005) *J. Bone Joint Surg. Br.*, **87B**(8), 1046–1049.
92. Pietronave, S., Iafisco, M., Locarno, D., Rimondini, L., and Maria Prat, M. (2009) *J. Appl. Biomater. Biomech.*, **7**(2), 77–89. Aug;
93. Tigani, D., Zolezzi, C., Trentani, F., Ragaini, A., Iafisco, M., Manara, S. et al. (2008) *J. Mater. Sci.: Mater. Med.*, **19**(3), 1325–1334.
94. Brayden, D.J. (2003) *Drug Discov. Today*, **8**(21), 976–978.
95. Hughes, G.A. (2005) *Nanomed.: Nanotechnol. Biol. Med.*, **1**(1), 22–30.
96. Koo, O.M., Rubinstein, I., and Onyuksel, H. (2005) *Nanomed.: Nanotechnol. Biol. Med.*, **1**(1), 77–84.
97. Lin, C.-C. and Metters, A.T. (2006) *Adv. Drug Delivery Rev.*, **58**(12–13), 1379–1408.
98. Szymura-Oleksiak, J., Slósarczyk, A., Cios, A., Mycek, B., Paszkiewicz, Z., Szklarczyk, S. et al. (2001) *Ceram. Int.*, **27**(7), 767–772.
99. Dash, A.K. and CudworthIi, G.C. (1998) *J. Pharmacol. Toxicol. Methods*, **40**(1), 1–12.
100. Epple, M., Ganesan, K., Heumann, R., Klesing, J., Kovtun, A., Neumann, S. et al. (2010) *J. Mater. Chem.*, **20**(1), 18–23.
101. Zhang, M. and Kataoka, K. (2009) *Nano Today*, **4**(6), 508–517.
102. Bisht, S., Bhakta, G., Mitra, S., and Maitra, A. (2005) *Int. J. Pharm.*, **288**(1), 157–168.
103. Wiethoff, C.M. and Middaugh, C.R. (2003) *J. Pharm. Sci.*, **92**(2), 203–217.
104. Zhang, Q., Zhao, D., Zhang, X.-Z., Cheng, S.-X., and Zhuo, R.-X. (2009) *J. Biomed. Mater. Res. Part B: Appl. Biomater.*, **91B**(1), 172–180.
105. Watson, A. and Latchman, D. (1996) *Methods*, **10**(3), 289–291.
106. Chowdhury, E.H. and Akaike, T. (2006) *J. Controlled Release*, **116**(2), e68–ee9.
107. Okazaki, M., Yoshida, Y., Yamaguchi, S., Kaneno, M., and Elliott, J.C. (2001) *Biomaterials*, **22**(18), 2459–2464.
108. Olton, D., Li, J., Wilson, M.E., Rogers, T., Close, J., Huang, L. et al. (2007) *Biomaterials*, **28**(6), 1267–1279.
109. Faraji, A.H. and Wipf, P. (2009) *Bioorg. Med. Chem.*, **17**(8), 2950–2962.
110. Kester, M., Heakal, Y., Fox, T., Sharma, A., Robertson, G.P., Morgan, T.T. et al. (2008) *Nano Lett.*, **8**(12), 4116–4121.
111. Singh, N., Manshian, B., Jenkins, G.J.S., Griffiths, S.M., Williams, P.M., Maffeis, T.G.G. et al. (2009) *Biomaterials*, **30**(23–24), 3891–3914.
112. Iafisco, M., Sabatino, P., Lesci, I.G., Prat, M., Rimondini, L., and Roveri, N. (2010) *Colloids Surf., B Biointerfaces*, **81**(1), 274–284.
113. Iafisco, M., Palazzo, B., Falini, G., Foggia, M.D., Bonora, S., Nicolis, S. et al. (2008) *Langmuir*, **24**(9), 4924–4930.
114. Iafisco, M., Foggia, M.D., Bonora, S., Prat, M., and Roveri, N. (2011) *Dalton Trans.*, **40**(4), 820–827.
115. Ouizat, S., Barroug, A., Legrouri, A., and Rey, C. (1999) *Mater. Res. Bull.*, **34**(14–15), 2279–2289.
116. Palazzo, B., Walsh, D., Iafisco, M., Foresti, E., Bertinetti, L., Martra, G. et al. (2009) *Acta Biomater.*, **5**(4), 1241–1252.
117. Stigter, M., Bezemer, J., de Groot, K., and Layrolle, P. (2004) *J. Controlled Release*, **99**(1), 127–137.
118. Misra, D.N. (1998) *Colloids Surf., A: Physicochem. Eng. Aspects*, **141**(2), 173–179.
119. Barroug, A., Fastrez, J., Lemaitre, J., and Rouxhet, P. (1997) *J. Colloid Interface Sci.*, **189**(1), 37–42.
120. Barroug, A., Lernoux, E., Lemaitre, J., and Rouxhet, P.G. (1998) *J. Colloid Interface Sci.*, **208**(1), 147–152.

121. Boix, T., Gomez-Morales, J., Torrent-Burgues, J., Monfort, A., Puigdomenech, P., and Rodriguez-Clemente, R. (2005) *J. Inorg. Biochem.*, **99**(5), 1043–1050.
122. Autefage, H., Briand-Mesange, F., Cazalbou, S., Drouet, C., Fourmy, D., Goncalves, S. et al. (2009) *J. Biomed. Mater. Res. B*, **91B**(2), 706–715.
123. Dong, X.L., Wang, Q., Wu, T., and Pan, H.H. (2007) *Biophys. J.*, **93**(3), 750–759.
124. Barroug, A., Kuhn, L.T., Gerstenfeld, L.C., and Glimcher, M.J. (2004) *J. Orthop. Res.*, **22**(4), 703–708.
125. Rey, C., Combes, C., Drouet, C., Sfihi, H., and Barroug, A. (2007) *Mater. Sci. Eng., C: Biomimetic Supramol. Syst.*, **27**(2), 198–205.
126. Errassif, F., Menbaoui, A., Autefage, H., Benaziz, L., Ouizat, S., Santran, V. et al. (2010) Adsorption on apatitic calcium phosphates: applications to drug delivery, in *Advances in Bioceramics and Biotechnologies* (eds R. Narayan, M. Singh, and J. McKittrick), Wiley-VCH Verlag GmbH & Co. KGaA.
127. Cazalbou, S., Combes, C., Eichert, D., Rey, C., and Glimcher, M.J. (2004) *J. Bone Miner. Metab.*, **22**(4), 310–317.
128. Quizat, S., Barroug, A., Legrouri, A., and Rey, C. (1999) *Mater. Res. Bull.*, **34**(14–15), 2279–2289.
129. Gautier, H., Daculsi, G., and Merle, C. (2001) *Biomaterials*, **22**(18), 2481–2487.
130. Gautier, H., Merle, C., Auget, J.L., and Daculsi, G. (2000) *Biomaterials*, **21**(3), 243–249.
131. Alam, M.I., Asahina, I., Ohmamiuda, K., Takahashi, K., Yokota, S., and Enomoto, S. (2001) *Biomaterials*, **22**(12), 1643–1651.
132. Liu, Y.L., Enggist, L., Kuffer, A.F., Buser, D., and Hunziker, E.B. (2007) *Biomaterials*, **28**(35), 5399–5399.
133. Peter, B., Pioletti, D.P., Laib, S., Bujoli, B., Pilet, P., Janvier, P. et al. (2005) *Bone*, **36**(1), 52–60.
134. Midy, V., Hollande, E., Rey, C., Dard, M., and Plouet, J. (2001) *J. Mater. Sci.: Mater. Med.*, **12**(4), 293–298.
135. Yoshinari, M., Oda, Y., Ueki, H., and Yokose, S. (2001) *Biomaterials*, **22**(7), 709–715.
136. McLeod, K., Kumar, S., Smart, R.S.C., Dutta, N., Voelcker, N.H., Anderson, G.I. et al. (2006) *Appl. Surf. Sci.*, **253**(5), 2644–2651.
137. Lebugle, A., Rodrigues, A., Bonnevialle, P., Voigt, J.J., Canal, P., and Rodriguez, F. (2002) *Biomaterials*, **23**(16), 3517–3522.
138. Iafisco, M., Palazzo, B., Marchetti, M., Margiotta, N., Ostuni, R., Natile, G. et al. (2009) *J. Mater. Chem.*, **19**(44), 8385–8392.
139. Palazzo, B., Iafisco, M., Laforgia, M., Margiotta, N., Natile, G., Bianchi, C.L. et al. (2007) *Adv. Funct. Mater.*, **17**(13), 2180–2188.
140. Kandori, K., Masunari, A., and Ishikawa, T. (2005) *Calcif. Tissue Int.*, **76**(3), 194–206.
141. Makrodimitris, K., Masica, D.L., Kim, E.T., and Gray, J.J. (2007) *J. Am. Chem. Soc.*, **129**(44), 13713–13722.
142. Baeza, A., Izquierdo-Barba, I., and Vallet-Regí, M. (2010) *Acta Biomater.*, **6**(3), 743–749.
143. Zhu, L., Uskokovic, V., Le, T., Denbesten, P., Huang, Y., Habelitz, S. et al. (2010) *Arch. Oral Biol.* **56**(4), 331–336.
144. Fan, Y.W., Sun, Z., Wang, R., Abbott, C., and Moradian-Oldak, J. (2007) *Biomaterials*, **28**(19), 3034–3042.
145. El Shafei, G.M.S. and Moussa, N.A. (2001) *J. Colloid Interface Sci.*, **238**(1), 160–166.
146. Wu, L., Forsling, W., and Schindler, P.W. (1991) *J. Colloid Interface Sci.*, **147**(1), 178–185.
147. Yin, G., Liu, Z., Zhan, J., Ding, F.X., and Yuan, N.J. (2002) *Chem. Eng. J.*, **87**(2), 181–186.
148. Kawasaki, K., Kambara, M., Matsumura, H., and Norde, W. (2003) *Colloids Surf., B: Biointerfaces*, **32**(4), 321–334.
149. Wu, L.M., Forsling, W., and Schindler, P.W. (1991) *J. Colloid Interface Sci.*, **147**(1), 178–185.
150. Smith, D.A., Connell, S.D., Robinson, C., and Kirkham, J. (2003) *Anal. Chim. Acta*, **479**(1), 39–57.
151. Rodan, G.A. and Fleisch, H.A. (1996) *J. Clin. Invest.*, **97**(12), 2692–2696.

152. Russell, R.G. and Rogers, M.J. (1999) *Bone*, **25**(1), 97–106.
153. Neves, M., Gano, L., Pereira, N., Costa, M.C., Costa, M.R., Chandia, M. et al. (2002) *Nucl. Med. Biol.*, **29**(3), 329–338.
154. Coxon, F.P., Thompson, K., and Rogers, M.J. (2006) *Curr. Opin. Pharmacol.*, **6**(3), 307–312.
155. Nancollas, G.H., Tang, R., Phipps, R.J., Henneman, Z., Gulde, S., Wu, W. et al. (2006) *Bone*, **38**(5), 617–627.
156. Lalla, S., Hothorn, L., Haag, N., Bader, R., and Bauss, F. (1998) *Osteoporos. Int.*, **8**(2), 97–103.
157. Monkkonen, J., Simila, J., and Rogers, M.J. (1998) *Life Sci.*, **62**(8), Pl95–Pl102.
158. Elliott, S.N., McKnight, W., Davies, N.M., MacNaughton, W.K., and Wallace, J.L. (1997) *Life Sci.*, **62**(1), 77–91.
159. Kos, M. and Luczak, K. (2009) *Biosci. Hypotheses*, **2**(1), 34–36.
160. Kos, M., Kuebler, J.F., Luczak, K., and Engelke, W. (2010) *J. Cranio-Maxillofac. Surg.*, **38**(4), 255–259.
161. Hoffman, A., Stepensky, D., Ezra, A., Van Gelder, J.M., and Golomb, G. (2001) *Int. J. Pharm.*, **220**(1–2), 1–11.
162. Josse, S., Faucheux, C., Soueidan, A., Grimandi, G., Massiot, D., Alonso, B. et al. (2005) *Biomaterials*, **26**(14), 2073–2080.
163. Roussière, H., Fayon, F., Alonso, B., Rouillon, T., Schnitzler, V., Verron, E. et al. (2007) *Chem. Mater.*, **20**(1), 182–191.

18
Nanostructures and Nanostructured Networks for Smart Drug Delivery

Carmen Alvarez-Lorenzo, Ana M. Puga, and Angel Concheiro

18.1
Introduction

Natural materials and living systems do not follow homogeneous monotonic assembly strategies, but they are formed by multiple building blocks that arrange according to hierarchical orders. Highly specific patterning from the molecular level to the nanoscale, microscale, and macroscale has been developed along the species evolution to create materials with amazing performances and to construct biological machinery with spectacular diverse functionality [1]. Nature uses a few common materials but arranges them according to a complex interplay between morphology and properties. The recipes contained in the genetic code of the living beings interact with the environmental conditions providing flexibility. Adaptive changes and better use of the resources are possible because of the hierarchical self-organization of the materials [2]. The understanding of how the natural objects and processes work is being already used to guide the design of nanomaterials, nanodevices, and procedures with outstanding performances, hardly to forecast few years ago. The improvement in the knowledge of the natural hierarchical structures and the advances in polymer science and nanotechnology are driven interdisciplinary approaches for the synthesis of biomimetic materials [3–5]. Bio-inspired networks, surfaces, and technologies are often the result of biomimetic investigations whereby the natural mechanism is modified or altered to meet specific requirements [6, 7]. Excellent general reviews of objects and processes found in nature and the applications of biomimetic materials or systems under development or already commercialized have been recently published [1, 8].

Bio-inspired materials for being used in biomedicine can be divided into two categories: structurally inspired and functionally inspired [9]. Liposomes (bilayers resembling cell membrane) can be considered the first structurally inspired drug carriers, followed by virus-inspired vesicles such as virosomes and polymersomes [9]. Some examples of functionally inspired systems are those bearing surfaces that are bioadhesive even in the presence of water, mimicking mussel adhesive proteins or bacterium flagella [10, 11]. Different from other biomimetic materials that are

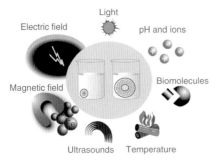

Figure 18.1 Scheme of the volume phase transition of a nanostructured system triggered by internal or external stimuli.

intrinsically static, the stimuli-responsive networks can act as sensors of certain physical or chemical variables of the environment and as actuators undergoing specific changes (as the living systems do) [12]. In the biomedical field, such changes should be preferably predictable and repeatable [13, 14]. For other applications, defined changes in key properties over time may be even more desirable [15]. Sensitiveness to internal (e.g., pH, temperature, and biomolecules) or external (e.g., light, irradiation, electric, or magnetic field) signals of the body is typically communicated to the material through functional groups that modify their properties proportionally to the intensity of the signal and that enable the transduction into changes in the material features [16–18]. Such changes may result in irreversible structure disintegration (e.g., disassembly of micelle unimers or gel polymer chains) or may lead to reversible changes in conformation, which are commonly shown as phase-volume transitions, or in affinity toward other chemical groups or molecular entities. Strictly speaking, only the stimuli-sensitive materials that selectively recognize a signal and respond to it reversibly, being activated when the stimulus is applied/appears and deactivated when it stops/disappears, can be considered as intelligent or smart (Figure 18.1) [19, 20]. To be able to do that, the polymer network has to undergo a first-order phase transition, accompanied by a change in the specific volume of the polymer, when the stimulus appears [21, 22]. The theoretical basis of the critical phenomena in cross-linked networks has been detailed reported elsewhere [12, 23]. Collapse of swollen networks mainly occurs when the solvent becomes poor, which causes that the attraction forces between the polymer chains exceed the repulsion forces associated with the excluded volume [22, 23].

In the particular case of smart drug delivery systems, most polymer-based nanostructures and nanostructured networks benefit simultaneously from being structurally and functionally bio-inspired. The combination of both issues has been graphically depicted considering drug nanocarriers as cell-like entities with a membrane able to sense the external environment, with specific ligands for active targeting and switchable gates through which certain substances can enter and the drug molecules can be pulsately released [17, 24]. The inner network can be endowed with several compartments, each of them responsible for specific functions that feedback regulate drug release or perform other therapeutic functions, such as enzymelike activity for producing beneficial substances or degrading adverse ones. Although still far from such sophistication, clinical trials mainly carried out with anticancer agents formulated in advanced lipidic and polymeric particles have demonstrated the benefits of such an approach and some formulations are already available in the market [25]. Other areas of interest include (i) site-specific release of drugs unstable in physiological fluids (e.g., peptides, proteins) that should exert their effect only in certain tissues (e.g., inflamed, infarcted, infected), or that should enter into cells or cellular structures that are not easily accessible from the general circulation (for example, in gene therapy) and (ii) temporal or biorhythm-specific release of hormones, gastric acid inhibitors, β-blockers, or drugs for heart rhythm disorders or asthma [26–28]. The following sections deal with polymer-based nanostructures and nanostructured networks that apply the stimuli responsiveness to achieve an improved and precisely tuned ability for loading drugs and controlling their release. First, a general overview of the basis of each responsive approach is given. Then, synthesis routes to prepare polymer-based nanostructures and nanostructured networks are briefly explained together with relevant examples of their applications in the drug delivery field.

18.2
Stimuli-Sensitive Materials

18.2.1
pH

Several pH gradients can be found in the human body under normal and physiopathological conditions. The gastrointestinal tract is characterized by regions of widely different pH, varying in the range of 1.00–3.00 in stomach, 4.80–8.20 in upper gut, and 7.00–7.50 in colon [29]. Inside cells, the differences of pH among cytosol (7.4), Golgi apparatus (6.40), endosome (5.5–6.0), and lysosome (4.5–5.0) are also considerable [30]. The extracellular pH of tumor tissues (6.5–7.0) is slightly lower than that of the blood and the healthy tissues (7.4) [31]. Similarly, a drop in pH up to 6.5 is observed after 60 h of the onset of an inflammatory process [32]. By contrast, the wounds commonly show an increase in pH from 4 to 6 (normal skin) up to values that can be as high as 8.9 [33].

Molecules that behave as weak acids or bases are suitable components of pH-responsive systems. Carboxylic, sulfonate, and primary and tertiary amino groups change from neutral to ionized state because of pH modifications in the physiological range. The changes in the degree of ionization can dramatically alter the conformation and the affinity of the chains for the solvent as well as the interactions among them. The neutralization of the charges makes water to become a poor solvent; networks bearing acid groups swell at alkaline pH but collapse at low pH, while those bearing bases swell in acid medium and shrink when pH raises. Polyampholyte hydrogels containing both types of monomers show the maximum swelling at neutral pH, that is, when both acids and bases are partially ionized

[34]. The conformational changes may be translated in the disassembly of weakly bonded components or the swelling/shrinking of covalent networks. To be useful from a biomedical point of view, the polymer has to be designed to have a pK_a that enables sharp changes in the ionization state at the pH of interest. Although critical phenomena of most polyelectrolytes takes places at extreme values of pH, copolymerization with hydrophobic monomers bearing n-alkyl groups enables the modulation of the critical pH and also of the magnitude of the changes in the volume of the network [35–37]. The content in ionic monomer and the cross-linking degree of the network also determine the pH responsiveness [38].

18.2.2
Glutathione

Glutathione (GSH) is an ubiquitous intracellular substance suitable as stimulus for triggering drug release inside cells. The intracellular compartments (cytosol, mitochondria, and nucleus) contain 2–3 orders higher level of GSH tripeptide (2–10 nM) than the extracellular fluids (2–20 µM) [39]. Thus, GSH/glutathione disulfide is the major redox couple in animal cells. GSH is kept reduced by NADPH and glutathione reductase, although the intracellular levels of GSH are also dependent on other redox couples [40]. Block copolymers or polymer networks bearing disulfide (-S-S-) bonds are suitable to undergo reduction reactions in the presence of GSH, leading to the rupture of the bond to form −SH end groups [41]. As a consequence of the redox process, the nanostructure swells or disassemblies and the drug is released. The bond rupture is a priori reversible, although this is not a foreseeable situation inside the cells.

18.2.3
Molecule-Responsive and Imprinted Systems

Molecular recognition plays a decisive role in the biological functions of living organisms. Membrane receptors, enzymes, and antibodies perform their activities because they can recognize specific molecules. Similarly, recognition of specific biomarkers may enable a precise control and feedback regulation of drug release [42]. Networks can be endowed with ability to recognize signaling molecules by conjugation of polymers with biological molecules such as carbohydrates [43], DNA [44], enzymes [42, 45], antibodies [46], or other proteins [47]. For example, the trigger stimulus can be an enzyme−substrate reaction that results in the modification of the inner pH of the network. To do that, the enzyme is attached to the pH-responsive network. In the absence of the substrate, the network does not modify its conformation, but when the substrate concentration reaches a certain level, it reacts with the enzyme resulting in a product that modifies the local pH and, consequently, induces the network to change the degree of swelling. Once the substrate is consumed, the pH restores to the original value and the network to the original degree of swelling. A proof of the interest of this type of hydrogels for drug

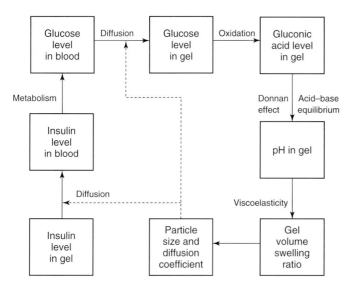

Figure 18.2 Block diagram of a closed-loop control process. Glucose diffuses into the hydrogel along a concentration gradient. Within the gel, glucose is oxidized to form gluconic acid, which causes the pH to decrease within the gel. The pH change results in swelling of the gel network, which leads to an increase in both the particle size and the diffusion coefficients of solutes through the gel. This causes an increase in insulin infusion, which promotes glucose uptake. As the glucose production rate decreases, diffusion will slow down, causing less oxidation, an increase in pH, a decrease in volume swelling and diffusion, and a decrease in insulin release. (Source: Reprinted from [48] with permission of American Chemical Society.)

delivery was shown for glucose-responsive insulin release from polymethacrylic acid networks containing glucose oxidase (Figure 18.2) [48].

A different approach consists of designing synthetic networks with domains capable of mimicking the recognition features of the biological macromolecules. Various semisynthetic ringtype substances, such as crown ethers, cryptates, cyclophanes, or cyclodextrins, can host in their cavity certain molecules forming inclusion complexes [49–53]. The selectivity of the interactions with these materials is conditioned by the size of the cavities and the arrangement of their chemical groups. The same ring can guest a bunch of structurally related substances with similar affinity, and in consequence, the specificity of the recognition is quite limited [54]. A more biomimetic and selective approach applies the molecular imprinting technology to reproduce the small, but critical, part of the biomacromolecules responsible for the interaction with the target molecule. This approach pursues the creation of cavities (artificial receptors) that are sterically and chemically complementary to the target molecule, recognizing it with high selectively. To do that, the substance of interest is used as a template during polymerization in order to induce an adequate arrangement of the monomers, particularly of those suitable to interact with it that are called *functional monomers* (Figure 18.3) [55]. The functional monomer/template association can be achieved through (i) covalent bonds

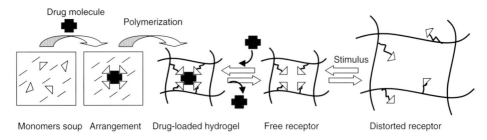

Figure 18.3 Diagram of the synthesis of an imprinted hydrogel and the washing out/release and reloading processes. The effect of stimuli on the conformation of the drug receptors is also depicted.

(*preorganized approach*) or (ii) noncovalent interactions, such as hydrogen bonds and ionic, hydrophobic, or charge-transfer interactions (*self-assembly approach*). Comprehensive reviews of the molecular imprinting technology have been published elsewhere [56–59]. The noncovalent imprinting protocol, which resembles the recognition pattern observed in Nature, allows more versatile combinations of templates and monomers and provides faster bond association and dissociation kinetics than the covalent imprinting approach [60, 61]. Conventional imprinted networks are highly rigid in order to maintain the physical stability of the imprinted cavities. Preparation of stimuli-responsive imprinted networks is more challenging since memorization of the cavities structure is required in order to maintain the recognition ability after several swelling/collapse cycles [62]. Such an advanced feature can be achieved also by mimicking how biomacromolecules work. Proteins find their desired conformation out of an infinite number of conformations because that desired conformation corresponds to a global minimum of energy. Thus, to design a polymer network that can always fold back into a given conformation after being stretched, that is, to thermodynamically memorize a conformation, it should be synthesized in the presence of the template in a conformation that corresponds to the global minimum of energy [62, 63]. The imprinted cavities develop affinity for the template molecules when the functional monomers come into proximity, but when they are separated, the affinity diminishes. The proximity is controlled by the reversible phase transition of the gel that consequently controls the adsorption/release of the template (Figure 18.3). The theoretical basis of the dependence of the affinity of an imprinted hydrogel for a certain molecule on the functional monomer concentration, the cross-linker proportion, and of the ionic strength of the medium has been summarized in the Tanaka equation [62].

18.2.4
Temperature

The temperature of the body is homeostatically maintained in a short range of values under healthy conditions. Pyrogens associated to infections lead to fever reactions increasing the overall temperature of the body. There are also many pathological conditions (inflammation, infarction, tumor, etc.) that evolve with local increases in temperature [64]. On the other hand, local increase in temperature can be achieved applying external sources on the skin or can be remotely induced through various types of radiation to drug delivery systems that incorporate elements that transform the energy into heat.

Most temperature-responsive materials are based on components that modify the affinity for the solvent as a function of temperature [65]. Swelling or shrinking of the network when temperature increases depends on the predominant interactions that determine the three-dimensional structure of the macromolecules, namely van der Waals forces, hydrophobic and electrostatic interactions, and hydrogen bonds [22, 66]. If van der Waals forces or hydrogen bonds can be established among the polymer chains, an increase in temperature causes the swelling, while a decrease in temperature induces the shrinking. Oppositely, if the collapse is caused by hydrophobic or ionic interactions, the network is swollen at low temperature and shrinks as temperature raises [22, 67]. This is the case of polymers bearing hydrophobic groups, such as alkyl chains. Changes in the hydrophilicity/hydrophobicity of the chemical groups may alter the intra- and intermolecular interaction leading to coil-helix or loose globule-compact coil phase-volume transitions [16]. Detailed lists of polymers having a critical solubility temperature (CST) have been published elsewhere [68, 69]. Poly-N-alkylacrylamides, particularly poly-N-isopropylacrylamide (PNIPAAm) and poly-N,N-diethylacrylamide, poly(methyl vinyl ether) (PMVE), poly-N-vinylcaprolactam (PVCL), and poly(ethylene oxide)-poly(propylene oxide) (PEO-PPO) block copolymers have been widely explored as components of temperature-responsive drug delivery systems since their CST is near to the body temperature [70]. PNIPAAm hydrogel is swollen at temperatures lower than 33 °C and collapses at higher temperatures [71]. The CST of PNIPAAm networks is quite sensitive to the copolymerization with hydrophobic or ionizable monomers, which, respectively, decrease and raise the critical temperature because of an overall change in the polymer hydrophilicity [72]. Adsorption of components such as surfactants and interpenetration of an ionic polymer also modify the temperature of collapse [73]. Furthermore, the time required for the phase transition is highly dependent on the overall size of the network [74] and its porosity [75]. The incidence of such variables on the CST should be taken into account when designing dually responsive networks, namely pH- and temperature-sensitive systems [76, 77].

Natural polysaccharides and linear polypeptides can also provide temperature responsiveness [69]. For example, homopolypeptides made of a single amino acid type have a well-defined collapse temperature: 24 °C for valine, 40 °C for proline, 45 °C for alanine, and 55 °C for glycine [78]. Combination of various amino acids renders "elastinlike polypeptides" with tunable CST (Figure 18.4) [79].

18.2.5
Light

Human beings are exposed to a wide range of electromagnetic radiations mainly due to solar irradiation. In addition, there have been developed numerous equipments that can apply innocuous electromagnetic radiation of very specific wavelengths in

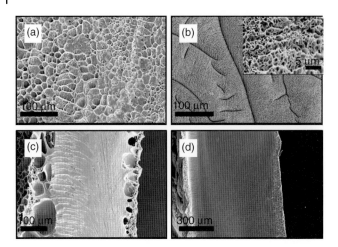

Figure 18.4 Swelling behavior and scanning electron micrographs of elastinlike hydrogels made of block peptides containing 20 Lys, 10 Arg, 41 Glx, 20 Asp, and 171 Val cross-linked with hexamethylene diisocyanate. Surface views: (a) 4 °C and (b) 37 °C; cross sections: (c) 4 °C and (d) 37 °C. (Source: Reprinted from [79] with permission of the American Chemical Society.)

the range from 2500 to 380 nm. Ultraviolet (UV), visible, and near-infrared (NIR) radiation sources can switch drug release on and off at an specific site of the body, offering a very precise control of the release and reducing the effect of radiation on the adjacent tissues to a minimum [80, 81]. UV light can only trigger the release from formulations placed on the skin or the mucosa, since radiation of wavelength below 700 nm does not penetrate more than 1 cm deep into the body [82–84]. A deeper penetration can be achieved using NIR light within the range of wavelengths from 650 to 900 nm. This is because hemoglobin (the principal absorber of visible light) and water and lipids (the main absorbers of infrared light) have their lowest absorption coefficient in the NIR region [85].

Biomimetic visible-light-responsive hydrogels have been prepared by adding chlorophyll (the natural light absorbent of plants) into a PNIPAAm network [86]. However, light responsiveness is most commonly provided by photoactive groups (azobenzene, cinnamoyl, spirobenzopyran, or triphenylmethane) than undergo reversible structural changes under UV–vis light (Figure 18.5) [87–89]. For example, the trans–cis isomerization of the azobenzene chromophore that occurs on exposure to UV light is accompanied by a change in the dipole moment from 0.5 D (trans isomer) to 3.1 D (cis isomer) [68]. This results in an increase in the hydrophilicity, which can lead to the disassembly of polymeric micelles or to the swelling of networks also containing temperature-sensitive components [90, 91]. Cinnamoyl groups isomerize to more hydrophilic species because of the electric charge generation or dimerization, while spirobenzopyran groups are transformed in zwitterion species (Figure 18.5).

On the other hand, metals can be used to absorb visible or NIR light and to efficiently transform the radiant energy into local heating, triggering the release from

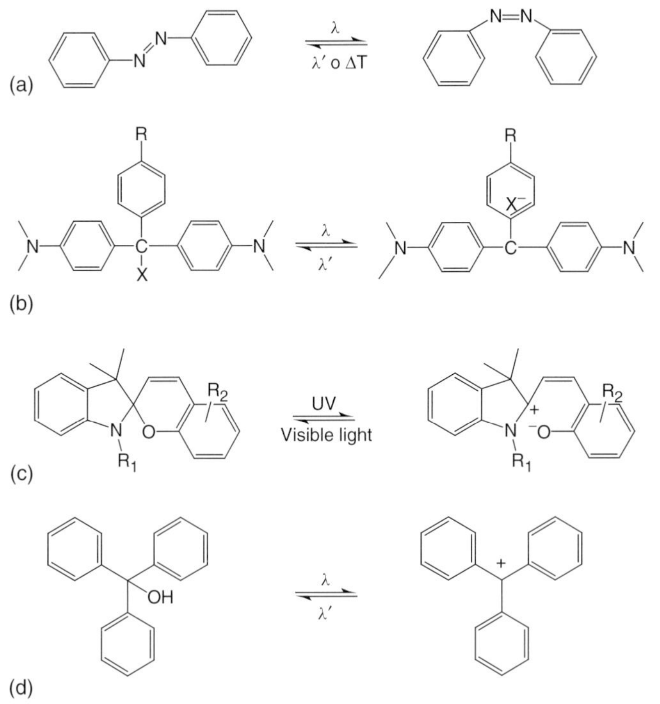

Figure 18.5 Effect of light irradiation on the structure of some photosensitive groups: (a) trans-to-cis isomerization of azobenzene, (b) ionization of cinnamoyl, (c) zwitter ionization of spirobenzopyran, and (d) ionization of triphenylmethane.

temperature-sensitive networks [92, 93]. Under NIR irradiation, gold nanoparticles cause a local increase of several degrees above body temperature. Gold-containing drug delivery systems may combine thermal ablation and chemotherapy for the management of tumors [94–96].

18.2.6
Electrical Field

Electrical stimuli can be generated using commercially available equipment for transdermal delivery, which enables a precise control of the intensity, the amount of current, the duration of the pulses, and the intervals between successive pulses [97]. Electrically sensitive networks can be made of polyelectrolytes with a high density in ionizable groups, similar to those used for preparing pH-responsive systems. Biocompatible polyanions from natural sources or of synthetic origin, which have a long history as pharmaceutical excipients, have been the most tested. Polycations and amphoteric polyelectrolytes may also be used. Electrically responsive networks can be used as components of injectable drug-loaded microparticles or implants for subcutaneous insertion. An electric field can be applied through an electroconducting patch placed on the skin over the polyelectrolyte network [97]. The changes in pH that occur near the electrodes because of the movement of the protons to the cathode makes the network to shrink and the drug is released as water is being squeezed [98]. The intensity of the electrical field and the time of application regulate drug release rate and duration. When the electrical field is switched off, the hydrogel swells again. Therefore, alternant shrinking (release on) and swelling (release off) may be achieved by applying pulses of electricity.

An alternative to the polyelectrolytes is the use of intrinsically conducting polymers (ICPs), also called *synthetic metals*, which are polymers that possess the electrical, electronic, magnetic, and optical properties of a metal [99]. These materials are different from conducting polymers, which are merely a physical mixture of a nonconductive polymer with a conducting metal or carbon powder. ICP properties are intrinsic to a doped form of the polymer; the electrical conductivity is due to the uninterrupted and ordered π-conjugated backbone [100]. ICPs are electrochemically formed as a continuous film on the surface of a working electrode through oxidative polymerization of monomer units. When a current is applied, the ICP undergoes reversible redox reactions that cause changes in polymer charge, conductivity, and swelling. The drug entrapped in the ICP layer (during or after synthesis) is released when an electrical stimulus is applied, using step potential or cyclic voltametry [101–104].

18.2.7
Magnetic Field

Strong and small magnetos can act at a distance on superparamagnetic materials inside the body enabling delicate maneuvers [105]. Although various metal combinations have been tested, magnetite (Fe_3O_4) and maghemite (γ-Fe_2O_3) are currently the only nontoxic paramagnetic materials acceptable for biomedical applications, including magnetic resonance imaging (MRI) contrast enhancement, targeted drug delivery, hyperthermia, biosensors, and diagnostic medical devices [106, 107]. Superparamagnetic iron oxide nanoparticles (SIONPs) are commonly obtained by coprecipitation from aqueous ferric or ferrous salt solutions by the addition of a concentrated base solution. The resultant colloidal suspensions are known as *ferrofluids* [105]. However, since the particles tend to aggregate reducing the paramagnetic features, they should be covered with polymers that can confer steric stability [108] or incorporated into micelles or cross-linked polymer networks [109]. Magnetic drug carriers containing temperature-responsive polymers possess three unique features: (i) visualization of the drug carrier into the body (MRI); (ii) tissue distribution controlled through an external magnetic field, which may be helped if decorated with cell ligands; and (iii) triggering of drug release due to a local increase in temperature when an alternating magnetic field is applied [27, 110]. Different from other responsive systems that do not allow by themselves tissue guidance, drug carriers bearing magnetic particles can be concentrated into a specific region by applying high-gradient magnetic fields. This enables a high local

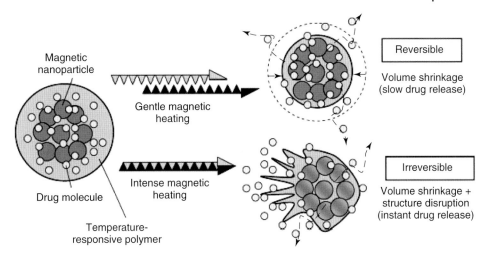

Figure 18.6 Two drug release mechanisms under magnetic heating. Gentle magnetic heating causes temperature-responsive polymer to shrink, squeezing drug out from the nanoparticle. Intense magnetic heating additionally ruptures the nanoparticle, triggering a burstlike drug release. (Source: Reproduced from [69] with permission of Elsevier.)

concentration even though the total injected dose is low. The drug can be released while the magnetic field is on, leading to site-specific treatment [111]. When the magnetic field is off, the drug carriers will disperse in the whole body but, since the dose administered was low, no untoward systemic effects are expected [112]. The increase in temperature can be modulated by the frequency of oscillation and the time of application of the magnetic field; a small increase triggers reversible pulsate drug squeezing as the temperature-responsive network shrinks, while a strong increase may lead to the rupture of the carrier followed by a burst drug release and a simultaneous thermal ablation of the surrounding tissues (Figure 18.6) [69]. Hyperthermia also increases the permeability of the tumor vasculature facilitating the extravasation of the particles into the tumor region. Interestingly, cancer cells are destroyed at temperatures close to 43 °C, while normal cells can stand such temperature [113]. Magnetic drug carriers can permit real-time tumor tracking by MRI on controlled release of anticancer drugs in cancer cells [114].

18.2.8
Ultrasounds

Ultrasounds can be applied to the body, using common physiotherapeutic equipment, to facilitate the penetration of nanostructures into specific regions and to trigger drug release [115]. Disassembly of polymeric micelles and polymersomes has been shown to occur when ultrasounds are applied [116, 117]. Networks containing gas-filled microbubbles are also suitable systems to be disrupted applying ultrasounds [118]. Furthermore, ultrasound imaging enables the monitoring of the therapy [119].

18.2.9
Autonomous Responsiveness

Chemical and biomechanical oscillators are being explored to render autonomous volume transitions or changes in permeability useful for a rhythmic release of drugs [120]. These systems are inspired in the fact that reactions in the cells do not occur in a homogeneous, well-stirred medium, but inside compartments delimited by membranes that regulate the transfer of reactants and products. Such a heterogeneous condition enables biological processes that would not be possible in a homogeneous medium [28]. Different from the above-described responsive polymers that switch on/off the release when a stimulus external to the network triggers the process, the autonomously responsive networks combine in a unique structure the polymer and the stimulus, undergoing rhythmic reversible phase transitions. Such a behavior enables to mimic the pulsate circadian levels of certain hormones, which is hardly to achieve with the current drug delivery systems(DDSs).

Chemical oscillating reactions in which large changes in pH occur are mainly based on the Belousoz–Zhabotinsky reaction that involves the oxidation and reduction of salts, such as permanganates, iodates, sulfates, chlorates, or bromates [121]. Poly(2-acrylamido-2-methyl-1-propane-sulfonic acid) (PAMPS), as well as other sulfonic acid polymers, can also undergo chemical oscillating reactions [122]. Similar reactions have also been shown to be useful for inducing the self-oscillation of temperature-responsive networks [123]. To be useful in drug delivery, the network has to be designed in such a way that the time required for the drug to diffuse out is shorter than the oscillation period [124].

Networks made of polyelectrolytes of different charge, bearing a polyelectrolyte and a temperature-responsive polymer, or with a grafted enzyme have also been shown to undergo autonomous oscillations because of positive and negative feedbacks that forbid the system from reaching a stationary state [28]. Siegel et al. [125, 126] have developed oscillators that work under physiological conditions. A pH oscillator that modulates the ionization state of a model drug, benzoic acid, has been shown to regulate the permeation of the drug through a lipophilic membrane. Coupling of swelling of a PNIPAAm-*co*-methacrylic acid (MAA) hydrogel membrane to an enzymatic reaction (involving glucose oxidase, catalase, and gluconolactonase), with the hydrogel controlling access of substrate to the enzyme, and the enzyme's product controlling the hydrogel's swelling state, was shown to be useful for delivery of gonadotropin-releasing hormone (GnRH) in rhythmic pulses, with periodicity of the same order as observed in sexually mature adult humans [28].

18.3
Stimuli-Responsive Nanostructures and Nanostructured Networks

On the basis of materials described above, nano- and microparticles as well as macroscopic networks can be created according to the "top-down" and the "bottom-up" approaches [127]. The top-down method consists of obtaining particles from a greater size piece using physical, chemical, or mechanical means, which commonly lead to defective materials [128]. The bottom-up method, also known as *self-assembly*, creates the structure starting from building blocks (small monomers, presynthesized polymers, clusters) that are subsequently joint by physical or chemical interactions [17, 129]. This approach enables a precise control of the structure and morphology of the networks. In Nature, the "bottom-up" strategy is the predominant one, and several synthetic routes have been developed to assemble monomers or polymers. Although ecofriendly synthesis methods based on biofabrication by microorganisms or using plant extracts are being implemented [127], totally synthetic routes are still preferred for biomedical purposes, since the final structure and performance can be precisely predicted, the risk of adverse reaction or immunogenicity is lower, and the large-scale production is still more economic [130]. In the following sections, the most common bottom-up methods to prepare polymer-based stimuli-responsive nanostructures are summarized.

18.3.1
Self-Assembled Polymers: Micelles and Polymersomes

Amphiphilic copolymers spontaneously aggregate in water-forming nanometric structures with a hydrophobic core surrounded by a hydrophilic shell (polymeric micelles) or forming vesicles similar to liposomes but with alternating layers of water and amphiphilic copolymers organized as a palisade (polymersomes) [131]. Polymersomes are inherently more stable than liposomes since the polymeric layers are thicker, stronger, and tougher than the lipidic ones [132].

Polymersomes and polymeric micelles can host nonpolar drugs in the hydrophobic domains and relatively polar substances in the hydrophilic domains [133, 134]. Polymeric micelles are more kinetically stable than regular micelles. Hence, on dilution, polymeric micelles disassemble very slowly even when the final concentration is below the critical micellar concentration, enabling longer residence times in the biological environment. In addition, micelles with hydrophilic blocks composed of PEO are sterically stabilized in the aqueous medium; opsonization and further uptake by macrophages being less feasible [135, 136]. When parenterally administered, polymeric micelles can passively accumulate in tissues exhibiting enhanced permeability and retention (EPR) of macromolecules, such as those suffering inflammatory or tumoral processes [134, 137]. If the shell is decorated with ligands or antibodies, active targeting to very specific regions can be achieved [138]. The size of polymeric micelles, in the range of that of virus, lipoproteins, and other biological systems of transport, enables the entrance into the cells [134]. Intelligent micelles can be designed to not release the drug until a change in the physiological conditions or an external stimulus alters the hydrophilicity or the conformation of the unimers [88, 139]. The number of micelles that disintegrate or destabilize and, consequently, the drug release profile, depends on the intensity of the stimulus. As soon as the stimulus stops, the micelles are reformed and the release is interrupted. Polymersomes can also be designed to behave in a similar way, starting from copolymers with blocks that have stimulus-dependent

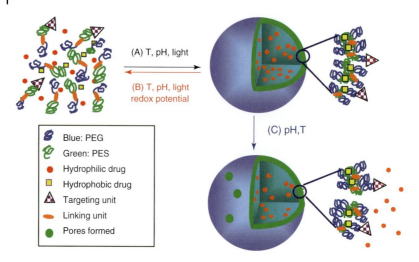

Figure 18.7 Scheme of the reversible formation/dissociation of stimuli-sensitive polymersomes and creation of membrane pores for enabling switchable drug release. (Source: Reproduced from [132] with permission of the American Chemical Society.)

solubility and that can lead to the disintegration of the polymersome or to the creation of pores in the layers when the stimulus appears (Figure 18.7) [132, 140]. Furthermore, the morphology of the micelles and polymersomes could be tailored to obtain novel nanostructures, including cylinders, "onions," "knitting ball," "golf ball," or "virus"-like particles, not explored yet as drug carriers [141].

pH-responsive micelles for tumor-targeted release have been prepared using amphiphilic copolymers with amino groups in one of the blocks, such as poly(2-vinylpyridine)-*b*-poly(ethylene oxide), P2VP-*b*-PEO; poly(*N*-vinylpyrrolidone)-*b*-poly(2-acrylamido-2-methyl-1-propanesulfonic acid), PVP-PAMPS; poly(*N*-vinylpyrrolidone)-*b*-poly(*N*,*N*-dimethylaminoethyl methacrylate), PVP-PDMAEMA; poly(2-(dimethylamine)ethyl methacrylate)-*b*-poly(ethylene oxide), PEO-*b*-DMAEMA; poly(ethylene glycol)-*b*-poly(L-histidine), PEG-*b*-PLH; and methyl ether poly(ethylene glycol)-*b*-poly(β-amino ester), MPEG-PAE [142–145]. At blood pH 7.4, the nonionized amino blocks self-aggregate forming the micellar core. The micelles retain the drug while in the bloodstream and passively accumulate in the tumor tissue. Once they enter into the tumor cells forming endocytic vesicles that fuse with lysosome (pH 5–6), drug release occurs because of the ionization of the hydrophobic blocks and the disassembly of the micelles [30, 139, 146]. However, tumor cells and normal cells have both the same endosomal acid pH. To achieve a more tumor-specific release, MPEG-PAE micelles were designed to disassembly at the weakly acidic tumor microenvironment, undergoing a brusque demicellization at pH 6.4–6.8 [145]. MPEG-PAE micelles loaded with camptothecin released *in vitro* <20% dose at pH 7.4 but more than 80% dose at pH 6.4 in 12 h. Furthermore, the micelles protected the lactone ring from chemical degradation [145], which is critical for the antitumor effect of camptothecin. Protective effects against hydrolysis have been reported for other micellar systems [147–149]. *In vivo* test with MDA-MB231 human breast tumor-bearing mice demonstrated that intravenously injected MPEG-PAE micelles loaded with a fluorescent dye preferentially accumulated in the tumor, with minimal distribution to the healthy tissues. This behavior clearly contrasts with the nonselective distribution observed for nonresponsive polymeric micelles [145]. In a later study, it was shown that MPEG-PAE micelles can encapsulate Fe_3O_4 nanoparticles in the hydrophobic cores for a selective release in cerebral ischemic area, taking profit of the acidic environment caused by this pathological condition. The Fe_3O_4 nanoparticles enabled the MRI of the ischemic region [150].

pH-sensitive polymeric micelles have also a great potential as carriers of DNA or siRNA [134, 151]. DNA and siRNA interact with the amino groups of the copolymer, forming a complex inside the micelle (micelleplex), which protects nucleic acids from the enzymatic degradation [152–156]. Despite the use of pH-responsive micelles is still quite incipient, promising results have been already obtained. Micelles of poly(ethylene glycol)-*b*-poly[(3-morpholinopropyl) aspartamide]-*b*-poly(L-lysine) (PEG-*b*-PMPA-*b*-PLL) combine the buffering capacity of PMPA with the excellent aptitude to condense DNA of PLL, resulting in a high transfectional efficiency [151]. Micelles of DMAEMA, acrylic acid, and butyl methacrylate that were loaded with siRNA exhibited enhanced transfection efficiency and lower toxicity than common polyplexes [157].

Cross-linking of micelles with agents bearing disulfide groups that can be reduced in the presence of glutathione has been explored as a way to prevent extracellular drug leakage [158, 159]. Rapid release in the presence of GSH has been achieved by designing shell-detachable micelles [160], while a very precise intracellular release has been obtained using micellar systems that respond to the low pH and the reductive conditions of the cancer cells [161].

Temperature-sensitive micelles have been prepared with copolymers possessing PNIPAAm either at the hydrophilic block or at the hydrophobic one [162]. If PNIPAAm is at the shell, the micelles are formed at temperatures below LCST, while above this temperature, the micelles destabilize and the drug is released [163]. For example, the LCST of block copolymers of poly(lactic acid) and of PNIPAAm copolymerized with dimethylacrylamide, PLA-*b*-(PNIPAAm-*co*-DMAAm), is close to 40 °C. Micelles loaded with doxorubicin slowly release the drug at 37 °C, but the release becomes faster when the temperature rises to 42 °C [164]. Similarly, polymersomes consisting of poly(ε-caprolactone) and PNIPAAm have shown reversible disintegration in response to tiny changes of temperature [165].

Copolymers with azobenzene groups that, when exposed to the light, undergo reversible alterations of the hydrophilic–lipophilic balance (HLB) have been assayed as components of light-sensitive micelles [88]. The trans to cis isomerization alters the hydrophilicity of the copolymer, being hydrophobic at trans (forming micelles) and hydrophilic at cis (staying as unimers). The wavelength that triggers the isomerization depends on the nature of the substituent groups and, thus, can easily be tuned [166]. The cis form is unstable at the corporal temperature so that, in darkness or if exposed to higher wavelength radiation, it reverts to the trans form. Therefore, cycles of micellar destabilization/reconstitution can be obtained applying light pulses [90]. In addition, the conformation of the azobenzene groups determines the interactions of the copolymers with hydrophobic regions of other macromolecules, enabling a precise control of the release of proteins [167].

Polymeric micelles with a dual response to temperature and radiation have been prepared from copolymers of NIPAAm and azobenzene monomers [168]. Under UV light, the hydrophilicity of the copolymer rises, the CST increases above body temperature, and the micelles disintegrate. Similarly, combination of temperature-sensitive Pluronics and copolymers with azobenzene groups rendered liquid dispersions of low viscosity in the dark that undergo a sol to gel transition when irradiated at 365 nm [91]. These changes in the viscosity have great potential in modulating drug release rate.

Temperature- and pH-dependent micellization was demonstrated for the pentablock copolymer, consisting of poly(2-diethylaminoethyl-methyl methacrylate)-poly(ethylene oxide)-poly(propylene oxide)-poly(ethylene oxide)-poly(2-diethylaminoethyl-methyl methacrylate) (PDEAEM25-PEO100- PPO65-PEO100-PDEAEM25) [169]. Such responsiveness enables the preparation of *in situ* gelling solutions that, once at body temperature, can regulate drug release precisely as a function of small changes in pH. Micellization/demicellization cycles due to changes in temperature and pH have also been reported for copolymers of NIPAAm, N,N-dimethylacrylamide, and N-acryloylvaline [170].

Ultrasound-triggered release from intravenously injected polymeric micelles has been tested for a pulsatile drug delivery in tumors [171]. The ultrasounds have to be applied when maximum micelle accumulation in the tumor is reached; namely between 4 and 8 h for doxorubicin-loaded Pluronic P-105 polymeric micelles or PEO-diacylphospholipid mixed micelles. The amount of drug released can be modulated tuning the frequency, the power density, the pulse length, and the interpulse intervals [172–174]. Drug release from the micelles is reversible, namely, during interpulse intervals exceeding 0.5 s, the drug can be completely reencapsulated into the restored micelles. In a recent study carried out with doxorubicin encapsulated in PEG-*co*-polycaprolactone micelles or perfluoro-15-crown-5-ether (PFCE) nanoemulsions, an increase in cell uptake and penetration into the nuclei was observed when ultrasounds were applied (Figure 18.8) [175].

Recently, polymeric micelles with shells that mimic the lipidic structure and protein channels of cell membranes are being explored as a way to achieve reversible stimuli-triggered drug release. The protein channels can be inserted in the micelle shell [176] or can be artificially created using copolymers with the same hydrophobic core but possessing different hydrophilic blocks [177]. In the second case, the copolymers render nanostructured shells that underwent phase separation phenomena when the stimulus is applied. The phase separation of the components of the shell results in the opening of pores through which the drug can be released (Figure 18.9) [178]. Temperature-, pH-, or ionic strength-responsive channels are under evaluation for obtaining micelles with switchable drug release [177].

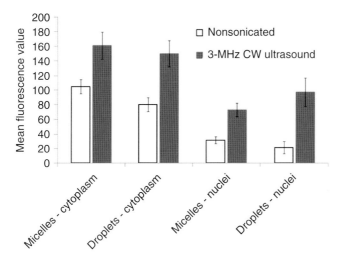

Figure 18.8 Mean fluorescence values and 95% confidence intervals for the cytoplasm and nuclei of the cells incubated (white bars) or sonicated (black bars) with doxorubicin encapsulated in poly(ethylene glycol)-*co*-polycaprolactone micelles or in perfluoro-15-crown-5-ether nanodroplets. (Source: Reproduced from [175] with permission of the American Chemical Society.)

18.3.2
Treelike Polymers: Dendrimers

Dendrimers are ramified polymers with a central core from which repeating branches are built up via a stepwise manner [179]. Their size (in nanometers range) is determined by the nature and the number of branches, which are named generations. Dendrimer architecture leads to three well-defined regions: the initiator core, the interior (building branches), and the exterior (surface terminal groups). The inner regions are suitable for hosting drugs, while the various functional terminal groups enable the grafting of stimulus-responsive components [180, 181]. Poly(amidoamine) (PAMAM) dendrimers have shown ability to act as carriers of acidic drugs, increasing drug solubility, stability, and circulation time and enabling controlled release and even drug targeting [182, 183]. Functionalization with temperature-responsive chains, such as PNIPAAm, together with PEGylation renders temperature-triggered dendritic drug delivery systems having a high cytocompatibility (Figure 18.10) [184].

Dendrimers can be endowed with pH responsiveness by incorporating labile groups that hydrolyze at certain pH and lead to the disintegration of the nanostructure [185] or by introduction of protonizable amino groups at the external surface of the dendrimers, which can enable the release of the entrapped drug within a narrow pH region [186, 187]. Recently, pH-triggered supramolecular self-assembly of hyperbranched polymers to form hierarchical structures has been achieved, which may open novel possibilities for regulating drug release [188].

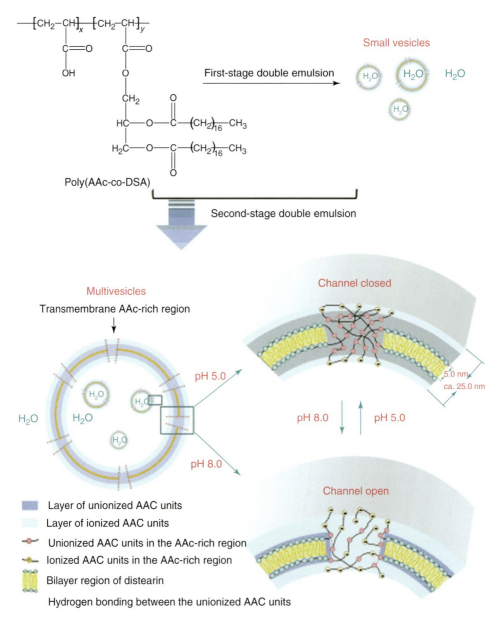

Figure 18.9 Scheme of multivesicle assemblies equipped with pH-responsive transmembrane channels from two-stage double emulsion of copolymers comprising acrylic acid and acrylate of 1,2-distearoyl-rac-glycerol, named poly(AAc-co-DSA). The AAc-rich regions and the bilayer islets within the vesicle membrane are not drawn to scale. (Source: Reproduced from [178] with permission of Wiley-VCH Verlag GmbH & Co. KGaA.)

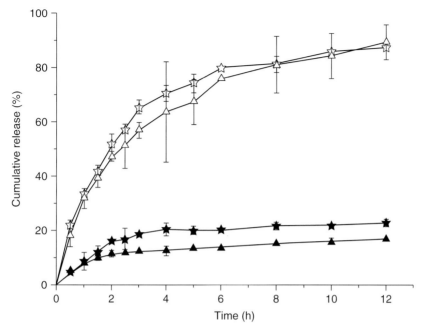

Figure 18.10 Indomethacin release profiles from PAMAM-g-PNIPAAm dendrimer at 30 °C (open triangle) and 37 °C (filled triangle) and from PAMAM-g-PNIPAAm-co-PEG dendrimer at 30 °C (open star) and 37 °C (filled star). PAMAM, PNIPAAM, and PEG represent poly(amidoamine), poly(N-isopropylacrylamide), and poly(ethylene glycol), respectively. (Source: Reproduced from [184] with permission of Elsevier.)

Functionalization with azobenzenes at any of the three regions of dendrimer architecture renders light-responsive systems. The process requires starting from azobenzenes that have a single reactive group if intended to be used as terminal groups, two identical reactive groups if azobenzene is placed at the core, or two different reactive groups if azobenzene is used as branch in order to allow the growing of the structure [189]. The location of the azobenzene group(s) determines the effect of the light irradiation. Minor responsiveness occurs when azobenzene is placed at the external part of the dendrimer (Figure 18.11). The trans to cis isomerization of azobenzene usually causes a decrease in the hydrodynamic volume of the dendrimer, although the opposite effect has also been reported [190, 191].

Azobenzene dendrimers can self-assembly as multilayer (onionlike) vesicles that can host drugs between the layers or inside each layer. The azobenzene moieties behave as valves with "off–on" switchability. On irradiation, the azobenzene groups in cis conformation increase the permeability and the drug can diffuse out. Such structures have been shown to be useful for regulating the release of hydrophilic (calcein) or hydrophobic (nil red) molecules [192].

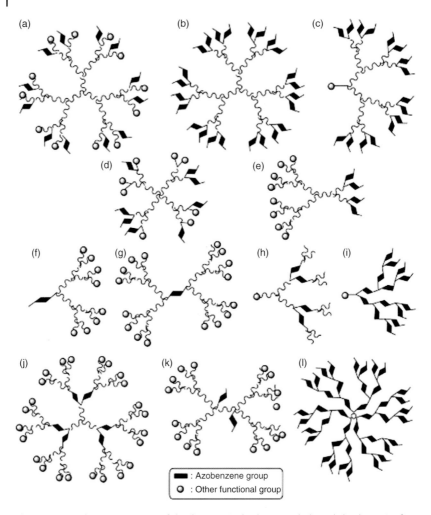

Figure 18.11 The various types of dendrimers (a, b, d, e, g, j, k, l) and dendrons (c, f, h, i) having azobenzene group(s) in their structure. (Source: Reproduced from [189] with permission of Elsevier.)

18.3.3
Layer-by-Layer Assembly of Preformed Polymers

The layer-by-layer (LbL) techniques have been applied to prepare stimuli-responsive films and hollow capsules; the drug being loaded during the layers deposition or in a latter step [193–195]. The layer assembly is usually driven by electrostatic interactions or hydrogen bonding, although other driven forces, such as avidin–biotin recognition, are also possible [196]. Despite the technique is <20 years old, biomaterial prototypes with responsiveness to pH, ionic strength, temperature, light, magnetic field, or specific biological molecules have been already developed

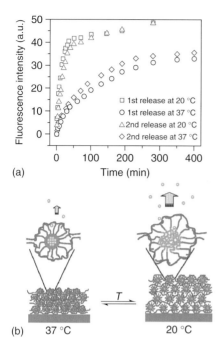

Figure 18.12 (a) Release kinetics of pyrene from a [BCM/PMAA]$_{10}$ film in 30 ml of pH 5.0 buffer solution at 20 and 37 °C. Pyrene release was monitored by measuring fluorescence intensity of pyrene accumulated in solution ($\lambda_{ex} = 338$ nm, $\lambda_{em} = 373$ nm). (b) Schematic representation of reversible temperature-triggered swelling of BCM/PMAA (block copolymer micelles/poly(methacrylic acid)) films. (Source: Reproduced from [201] with permission of the American Chemical Society.)

[197, 198]. For example, incorporation of magnetic nanoparticles to the surface of temperature-responsive microgels was possible by including the nanoparticles into layers deposited on the microgels. The temperature responsiveness was not altered by the coating, while the magnetic nanoparticles provided heat-triggered drug release [199]. On the other hand, the LbL technique may enable the codelivery of multiple drugs with specific release profiles, being the release of each drug or of different doses of the same drug triggered by a different stimulus [200].

Combination of poly(methacrylic acid) and temperature-responsive micelles of poly(N-vinylpyrrolidone)-b-poly(N-isopropylacrylamide) (PVPON-b-PNIPAM) rendered stable hydrogen-bonded multilayers that reversibly swelled as a function of temperature and salt concentration. Such responsiveness enabled stimuli-regulated release of a fluorescent probe (Figure 18.12) [201]. Temperature-responsive multilayered films have also been obtained combining elastinlike polymer and chitosan [202].

Starting from negatively charged polystyrene cores, successive layers of poly-(4-vinylpyridine hydrochloride) and poly(sodium styrene sulfonate) were deposited on the cores, which were then dissolved in order to obtain hollow capsules [203]. These capsules have a pH-dependent permeability that makes the loading of proteins as well as the control of the release possible. Successive layers of cross-linked PVP and chitosan have also been shown to be useful in preparing hollow microcapsules for pH-dependent release of doxorubicin [204]. Thin films that retain the drug at the pH of the stomach but release it at the weakly acid/neutral pH of the gut were obtained by alternate deposition of insulin and poly(vinyl sulfate), poly(acrylic acid), or dextran sulfate [205].

18.3.4
Polymeric Particles from Preformed Polymers

Coacervation/precipitation, followed or not of chemical cross-linking, is a common approach used to prepare particlelike structures starting from preformed polymers [17]. Microscopic polymer-rich droplets are formed in a polymer dispersion during a phase separation (coacervation) caused by a change in pH or temperature, addition of a nonsolvent, or incorporation of an interacting substance (Figure 18.13) [206, 207]. Polymer aggregation is reversible, that is, the non-cross-linked particles can be resolubilized when enter into contact with the physiological fluids or when

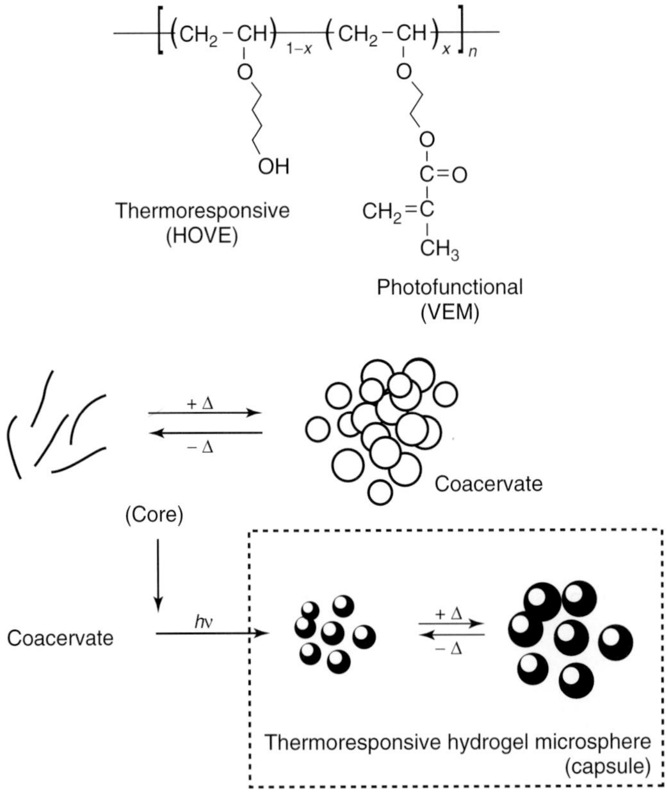

Figure 18.13 Synthesis of thermoresponsive hydrogel microspheres (capsules). (Source: Reproduced from [206] with permission of the American Chemical Society.)

the conditions that induced the phase separation are reversed. This is the case of colloidal particles obtained from oppositely charged polymers, which have an intrinsically pH- and ionic strength-responsive solubility, and can regulate drug release rate as a function of these variables [208]. If the coacervate particles are photo or chemically cross-linked, stable smart microgel particles are obtained [209, 210]. Temperature-responsive nanoparticles and microspheres can be prepared from PNIPAAm [206, 207, 211], and pH-sensitive microgels can be obtained from poly(N-methacryloyl-L-valine) and poly(N-methacryloyl-L-phenylalanine) [212]. Dual pH- and temperature-responsive particles were prepared by copolymerization of acrylic acid and NIPAAm monomers in the presence of chitosan [213], but the replacing NIPAAm for Fe_3O_4 cores and the subsequent cross-linking of chitosan rendered pH- and magnetic-responsive particles [214]. Complex coacervation has also been applied to obtain core–shell or micellelike particles [215].

Water-in-oil (W/O) heterogeneous gelation and aqueous homogeneous gelation are other two gentle ways of obtaining micro/nanogels from natural polysaccharides [216]. A high content of functional groups, including hydroxyl, amino, and carboxylic acid groups, makes polysaccharides and other saccharide derivatives (such as cyclodextrins) very suitable for being cross-linked under mild conditions [217]. Interpenetrated networks of chemically cross-linked cyclodextrins and poly(acrylic acid) have shown a high drug loading ability and pH-responsive release [218]. Cyclodextrins are also useful for regulating the nanostructure of coalescent polymers [219] or to create novel nano- and microstructures for drug delivery [220, 221]. Recently, nanocapsules of chitosan and Pluronic F127 were prepared by an emulsification/solvent evaporation method to form a cross-linked structure with the chitosan placed not only at the surface but also in the shell of Pluronic micelles [222]. These nanocapsules exhibited 200-times increase in volume when the temperature dropped from 37 to 4 °C allowing drug diffusion. Owing to their small size (37 nm) at 37 °C, the nanocapsules can be efficiently internalized into both cancerous and noncancerous mammalian cells. Application of cryotherapy (i.e., subzero temperatures) leads to a fast drug release into the cytosol.

18.3.5
Polymeric Particles from Monomers

In general, the level of cross-linker and the volume of solvent determine the physical nature of a polymer network prepared under given polymerization conditions [223]. According to the pseudophase diagram depicted in Figure 18.14, geltype polymers are obtained at relatively low cross-linker ratios (<5%) and, if a solvent compatible with the polymer network is used, also at high cross-linker ratios. Such materials have very low specific surface areas in the dry state but swell significantly in thermodynamically good solvents. Macroporous monoliths are obtained when high cross-linking ratios and poor solvent media are used. Under these conditions, the growing polymer phase separates from the liquid phase producing a network of aggregated polymers with empty spaces (pores) more or less interconnected among themselves and with the outer surface of the network. The pores are permanent,

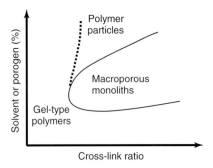

Figure 18.14 Polymer pseudophase diagram showing three distinct regions: geltype polymers, macroporous monoliths, and polymer particles. (Source: Adapted from [223] with permission of Elsevier.)

even in the dry state. If the high cross-linking ratio is maintained, but the volume of the solvent is further increased, discrete polymer particles can be formed and recovered [224].

Commonly used approaches to obtain micro- and nanoparticles are summarized in Figure 18.15. The choice of the polymerization process and the nature of the steric stabilizer, which mainly depend on the nature of the monomers, greatly impact the internal structure and size of the particles and even their performance for a given purpose [225–227]. The selection of the most adequate preparation procedure should be made on the basis of the knowledge of the relationships between the properties of the particles and their performance in each specific application and on the simplicity with which the method can be scaled up [224, 228].

The multistep swelling method consists of several subsequent steps, which start with the preparation of a latex (usually polystyrene particles of about 1 μm diameter) by emulsifier-free emulsion polymerization (Figure 18.15a). The particles are then swollen in a microemulsion formed by a free radical initiator and a low-molecular-weight "activating solvent" (such as dibutyl phthalate) dispersed in water, previously obtained by sonication (sodium dodecyl sulfate may be used as the stabilizer). Once the particles have absorbed the droplets of the internal phase of the microemulsion, they are transferred to a second oil-in-water (O/W) emulsion in which the monomers and cross-linker are in the internal phase. Polymerization is initiated when the monomers and the cross-linker diffuse into the particles. This method leads to monodisperse particles in the micrometer range (2–50 μm) [229]. Although not widely applied in the pharmaceutical field, this technique has been shown to be useful to obtain imprinted polymers for the active enantiomers R-propranolol, S-ibuprofen, and S-ketoprofen. The functional monomers were MAA for propranolol and 4-vinylpyridine (VPD) in the case of ibuprofen and ketoprofen. Granules prepared with imprinted polymers and racemic drugs released the active enantiomer (i.e., the one used as a template) more slowly than the other enantiomer, being promising for achieving enantioselective release [230].

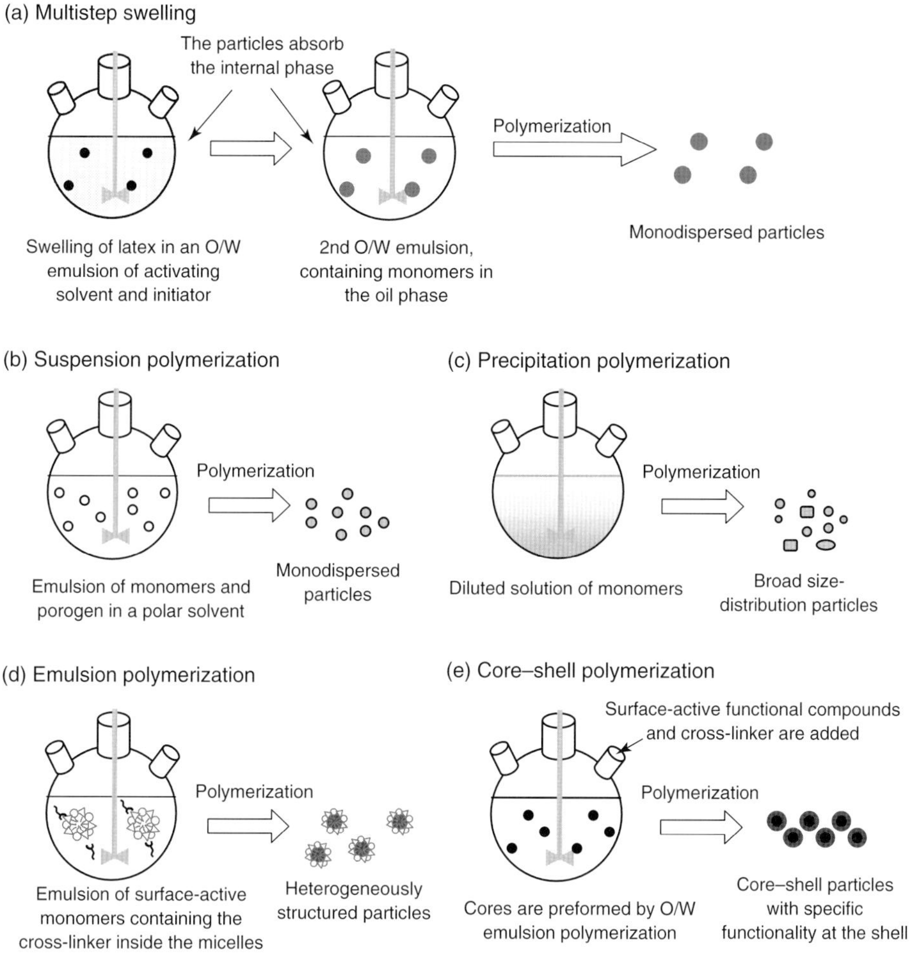

18.3 Stimuli-Responsive Nanostructures and Nanostructured Networks | 441

(a) Multistep swelling

Swelling of latex in an O/W emulsion of activating solvent and initiator → The particles absorb the internal phase → 2nd O/W emulsion, containing monomers in the oil phase → Polymerization → Monodispersed particles

(b) Suspension polymerization

Emulsion of monomers and porogen in a polar solvent → Polymerization → Monodispersed particles

(c) Precipitation polymerization

Diluted solution of monomers → Polymerization → Broad size-distribution particles

(d) Emulsion polymerization

Emulsion of surface-active monomers containing the cross-linker inside the micelles → Polymerization → Heterogeneously structured particles

(e) Core–shell polymerization

Cores are preformed by O/W emulsion polymerization → Surface-active functional compounds and cross-linker are added → Polymerization → Core–shell particles with specific functionality at the shell

Figure 18.15 (a–e) Schematic view of the preparation methods of nano and microparticles based on different polymerization procedures. The polymerized part of the particles is shown in a gray color.

Suspension polymerization in aqueous or highly polar solvents is the most traditional method for making polymer beads to be used in any field [231]. In a typical procedure, the monomers are dissolved in an organic solvent. The resultant solution is mixed with water (4–10 volumes) containing a suitable suspension stabilizer, which is commonly poly(vinylalcohol) or poly(vinylpyrrolidone), and a dispersion of small organic droplets is formed (Figure 18.15b). After polymerization, a suspension of polymer particles with a size similar to that of the starting droplets is obtained [232, 233]. Omeprazole-imprinted nanoparticles-on-microspheres were prepared using methacryloyl quinine and methacryloyl quinidine as functional monomers and ethylene glycol dimethacrylate as a cross-linker. The latex

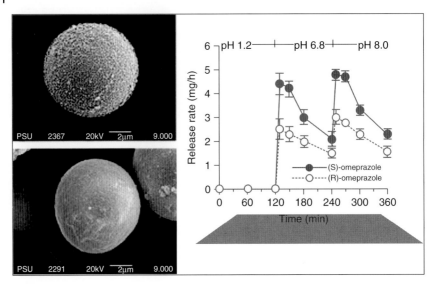

Figure 18.16 Molecularly imprinted nanoparticle-on-microsphere cinchona polymer can be utilized in the formulation of enantioselective-controlled delivery of omeprazole in response to changes in pH. (Source: Reproduced from [234] with permission of Elsevier.)

obtained was incorporated into a pH-responsive network of poly(hydroxyethyl methacrylate) and polycaprolactone-triol enabling the triggering of drug release at pH > 6.8; the MIP (molecularly imprinted polymer) regulating the release rate of each enantiomer (Figure 18.16) [234].

Precipitation polymerization is one of the simplest approaches to synthesize uniform microspheres [235, 236]. The method is based on the use of diluted monomer solutions (e.g., 2%) in a poor solvent for the polymer. The phase separation that takes places during polymerization leads to small (<1 μm), spherical and with a narrow size distribution particles (Figure 18.15c). A comprehensive review on precipitation polymerization synthesis of temperature- and pH-responsive microgels based on PNIPAAm and PVCL and the possibilities of tuning size distribution, surface charge, and microstructure has been recently published [237]. Mismatched reactivity ratios for monomer and cross-linker during precipitation polymerization have been shown to cause heterogeneities in the microgel structure, which may be tuned to modulate stimuli responsiveness and eventually microgel degradation [238]. Applying a precipitation polymerization approach, microgels of poly(acrylic acid) and Pluronic with pH- and temperature-responsive drug loading and release features have been obtained [239]. The amount of drugs and proteins loaded at specific environmental conditions was shown to be tuned by the degree of cross-linking of the microgels, which also regulates the degree of swelling when the change in pH or temperature occurs. Temperature-responsive nanohydrogels prepared with poly(N-isopropylacrylamide-co-acrylamide) have shown to load hydrophobic drugs, such as 5-fluorouracil or docetaxel, and to release them at constant rate

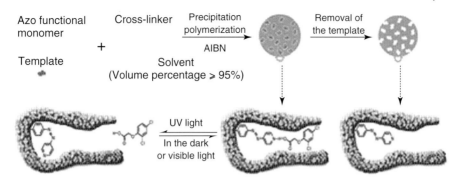

Figure 18.17 Photoresponsive binding affinity of the imprinted sites in azo-containing MIP microspheres. The affinity for the template decreased on UV light irradiation, whereas it could be recovered during the subsequent thermal (or visible light-induced) back-isomerization. (Source: Reproduced from [243] with permission of The Royal Society of Chemistry.)

at 43 °C. Nanohydrogels loaded with docetaxel were less toxic than the free drug and showed an antitumor efficacy in Kunming mice-bearing S180 sarcoma, which was enhanced by hyperthermia [240]. Imprinted particles prepared via precipitation polymerization have also a great potential in the controlled drug delivery field. Sulfasalazine-imprinted particles can sustain drug release more efficiently than the corresponding nonimprinted particles both at pH 1 and 6.8 [241]. Theophylline-imprinted systems with modulated loading capacities and release rates were obtained using different proportions of MAA and methylmethacrylate (MMA) functional monomers [242]. Recently, light-responsive-imprinted microparticles using a methacrylate azo functional monomer have been prepared applying precipitation polymerization under dark conditions. The template 2,4-dichlorophenoxyacetic acid only interacts with the functional monomer in the trans conformation (Figure 18.17). Thus, the affinity and the number of specific binding points notably decrease on UV light irradiation, whereas they can be recovered after thermal or visible light-induced back-isomerization [243]. The uptake/release of the target molecules was shown to be highly repeatable under UV light on/off cycles.

Emulsion and core–shell polymerization have also been applied to obtain responsive systems (Figure 18.15d,e). Nanosized particles can be prepared by polymerization of monomers with surface-active features that are placed at the interface of two liquid phases, such as those of an O/W or W/O emulsion. Commonly, the monomers and the cross-linker are dispersed in an organic solvent that is then emulsified in a large volume of aqueous phase [244]. The nanoparticles resultant from the emulsion polymerization or other above-described method can be further modified to obtain core–shell particles by dispersion in a solvent-containing monomers, cross-linker, and initiator. Once the polymerization is induced, a shell is formed around the starting nanoparticles that act as nuclei [245]. A comprehensive review on preparation of core–shell nanoparticles has been recently published [246]. Polystyrene cores with pH and temperature shells made of poly[2-(dimethylamino)ethyl methacrylate]-*b*-poly[methyl methacrylate] (PDMA-*b*-PMMA) rendered stimuli-sensitive colloidosomes [247]. "Snowman"-like particles with PMMA core and an ultrathin shell of polyaniline have been prepared to combine the electrical responsiveness of polyaniline with the transparent, nonspherical cores that make the response to the electrical field faster [248]. On the other hand, nanogels of poly(*N*-isopropylacrylamide-*co*-acrylic acid) bearing doxorubicin conjugated through a pH-labile bond exhibited dual temperature/pH-dependent cellular uptake and cytotoxicity. The reduction in size of the nanogels when temperature raised from 37 to 43 °C favored their cellular internalization, while the pH responsiveness enabled a fine control of drug release rate [249]. Poly(acrylonitrile-*co*-4-vinylpyridine) core–shell nanoparticles responsive to pH and magnetic field have been shown to take up naproxen and to sustain its release for several hours at pH 7.4 [250].

18.3.6
Chemically Cross-Linked Hydrogels

As explained above, continuous covalently cross-linked hydrogel layers can be obtained under specific polymerization conditions. Other feasible methods to prepare membranous hydrogel systems, mainly from preformed polymers, involve phase separation, electrospinning, foaming, particle leaching, emulsion freeze drying, and sintering [251, 252].

The macroscopic size of these hydrogels may be a handicap for the achievement of a fast responsiveness to the stimuli, unless the hydrogels are sufficiently thinner. However, for some applications, a more sustained actuation of the network after being triggered by the stimulus may be beneficial, particularly to avoid a burst release of certain drugs, which may be toxic if the whole dose is rapidly delivered. Since stimulus-sensitive hydrogels for drug delivery were among the first responsive devices to be explored, the number of papers and reviews on this topic is outstanding [253, 254]. Nevertheless, information about switchable and repeatable performance under *in vivo* mimicking conditions is much less abundant [20, 76]. Here, we review those systems responsive to various stimuli and last advances in the field. Very precise spatiotemporal release can be achieved with nanostructures responsive to a combination of more than one stimulus, for example, pH and temperature, or the simultaneous presence of two biomarkers. If only one signal is present, the drug will not be released [17].

Dually, pH- and temperature-responsive hydrogels have been shown to be useful for adjusting the release of thrombolytic agents, such as heparin or streptokinase, to the small changes of pH and temperature that accompany the formation of the thrombi [255]. Interpenetrated chitosan-NIPAAm provided pH and temperature-dependent release of diclofenac [73]. Similarly, hydrogels of NIPAAm, butylmethacrylate, and acrylic acid have been designed to coat the vaginal tissue and to release an anti-HIV microbicide in response to semen-induced pH changes [256]. On the other hand, molecule-responsive materials are particularly attractive for the design of self-regulated DDSs able to control the delivery of the drug as a function of the substance concentration that serves as an index of the evolution of a pathological state [42]. Antigen-sensitive hydrogels have been prepared by copolymerizing monomers of an antigen and its antibody using the antigen–antibody interactions as cross-linking points [46]. If free antigen appears in the medium, it can compete for the binding to the antibodies and the hydrogel swells; the swelling enabling the release of an entrapped drug. Detailed descriptions of key factors to have in mind when designing stimulus-responsive drug-imprinted network can be found elsewhere [56, 57, 257–259]. Temperature-responsive hydrogels based on PNIPAAm and imprinted for 4-aminopyridine and l-pyroglutamic acid had significantly larger saturation and affinity constants than the nonimprinted ones and were also highly selective. The hydrogels were able to sorb and release similar amounts of drug after several shrunken–swollen cycles [260, 261]. Poly(N-tertbutylacrylamide-co-acrylamide/maleic acid) hydrogels synthesized in the presence of serum albumin (BSA) exhibited both pH- and temperature-switchable affinity for the protein [262]. At a low temperature (swollen state), BSA interacts with hydrogel through hydrogen bonds and the adsorption is maximal. By contrast, as the temperature rises, the gel collapses and the imprinted cavities and the nature of the interactions are altered. A similar behavior has been observed with ibuprofen-imprinted thermoresponsive cryogels synthesized in a frozen aqueous medium, which showed drug binding constants of 119 and 5 M^{-1} in the collapsed and swollen states, respectively [263]. The abrupt change in affinity during the gel volume phase transition allows drug release to be switched on and off. Differently, light-responsive networks have been designed to regulate paracetamol uptake and release through subtle changes in the conformation of the imprinted cavities [264]. The hydrogels were prepared from acrylamide, 4-[(4-methacryloyloxy)phenylazo]benzenesulfonic acid, and N,N'-hexylenebismethacrylamide, and applying the imprinting technology. The azobenzene chromophores in the hydrogel undergo reversible photoisomerization under alternating irradiation at 353 and 440 nm, changing their binding affinity for paracetamol (Figure 18.18). In the dark, the hydrogels sorbed paracetamol from aqueous media. Irradiation at 353 nm triggered the release, which was almost complete after 120 min of irradiation. Subsequent irradiation at 440 nm caused the imprinted receptors to recover their initial conformation, being able to capture the paracetamol that had been previously released into the buffer solution. After several light cycles, a progressive decrease in the amount of paracetamol that the hydrogels can rebind is observed, probably because a gradual deformation of the imprinted receptors. Paracetamol-imprinted hydrogels exhibited a notable selectivity for the template drug, compared to other structural analogs [264].

A prototype of multiresponsive drug delivery system has been designed to detect and analyze subtle multiple signals generated in the biological medium and to adjust drug release rate [265]. The device consists of an electroresponsive hydrogel (made of sodium acrylate), sensors to detect biological or external signals (concentration of glucose, light, temperature, pH, electrical field, or their binary

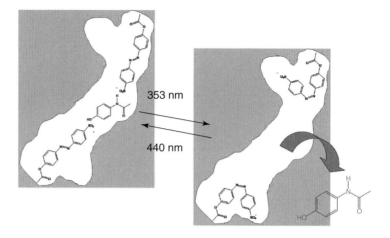

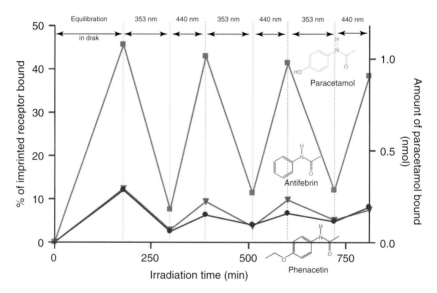

Figure 18.18 Photoregulated release and uptake of paracetamol, antifebrin, or phenacetin by paracetamol-imprinted 4-[(4-methacryloyloxy)phenylazo]benzene-sulfonic acid-containing polyacrylamide hydrogel in aqueous HEPES buffer at pH 7.16. Nonspecific binding of substrate to nonimprinted control hydrogels has already been subtracted from the binding data. (Source: Reproduced from [264] with permission of the American Chemical Society.)

combinations), and a computer analysis system. The microcomputer processes the signals and emits the necessary orders to the hydrogel, as specific electric stimuli through a number of electrodes, to switch drug release on/off [266]. A simultaneous and independent control of the release of several drugs registering different stimuli can be achieved using such a device; for example, the release of a certain drug

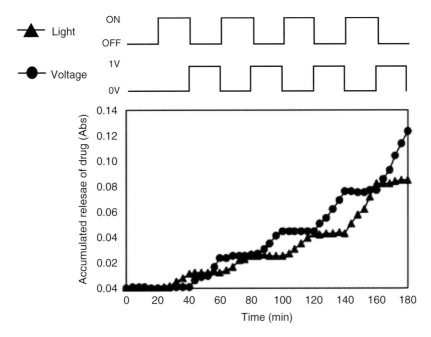

Figure 18.19 Accumulated released amounts of two different drugs from binary inputs–binary outputs intelligent DDS. (Source: Reprinted from [265] with permission from Elsevier.)

can be externally regulated by applying light pulses, whereas the release of another drug can be adjusted by voltage pulses (Figure 18.19) [265].

18.3.7
Grafting onto Medical Devices

Concerns about inflammation, cell proliferation, or even microorganism biofilm formation after implantation of certain medical devices, such as stents, prosthesis, or catheters, have prompted the search of approaches to elute drugs for prophylactic or therapeutic aims [267–269]. Direct compounding of the drug with the device components may alter the bulk properties of the medical device. Therefore, surface modification with polymers that can interact with the drug and to control the release seems to be more advantageous [270, 271]. Grafting of stimuli-responsive polymers may endow the medical device surface with improved biocompatibility and controlled drug release features, while the responsive network may benefit from the mechanical stability offered by the substrate in order to stand better the *in vivo* stress conditions [272–275].

The modification by γ-irradiation is one of the preferable methods for surface functionalization of polymeric materials because of the uniform and rapid creation of active radical sites, rendering high values of grafting in a clean and rapid way [274, 276, 277]. Detailed information about preirradiation, preirradiation oxidative,

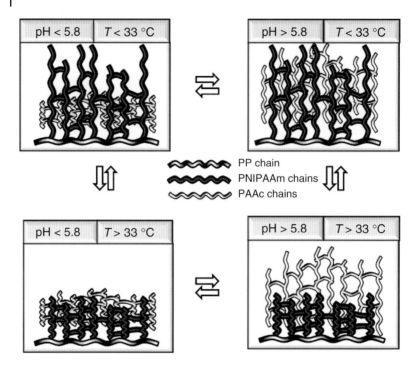

Figure 18.20 Scheme of the dual temperature and pH responsiveness of net-PP-g-PNIPAAm-inter-net-PAAc. (Source: Reproduced from [284] with permission of Elsevier.)

or direct grafting methods applying γ-ray sources can be found elsewhere [278, 279]. Grafting of MAA to medical grade silk suture and twisted yarn has been shown to be useful for the loading of microbicide doses of 8-hydroxy quinoline [280]. Similarly, the graft polymerization of 1-vinylimidazole and of acrylonitrile onto polypropylene monofilaments rendered sutures that can elute ciprofloxacin or tetracycline hydrochloride in a sustained way under *in vivo* conditions [281, 282].

Monomers suitable for endowing polypropylene and polyethylene substrates with affinity for vancomycin were screened using isothermal titration calorimetry in order to establish a rational surface functionalization protocol [283]. Acrylic acid sodium salt was identified as suitable for interacting with the drug. Then, interpenetrated networks of poly(acrylic acid) and PNIPAAm were prepared by grafting to the substrate one of the networks followed by polymerization/cross-linking of the second network [284]. The resultant functionalized surfaces were dually pH and temperature responsive (Figure 18.20), exhibited a synergistic loading mechanism, and were able to release vancomycin at a rate suitable to prevent *in vitro* the formation of methicillin-resistant *Staphylococcus aureus* (MRSA) biofilms [283, 285].

Polypropylene surfaces bearing grafted networks of NIPAAm and N-(3-aminopropyl) methacrylamide hydrochloride (APMA) have been shown to host/elute nalidixic acid to prevent the growth of *Escherichia coli* [286] and nonsteroidal anti-inflammatory drugs (NSAIDs) to avoid eliciting inflammatory reactions [287]. Furthermore, such a surface modification resulted in improved blood and cell compatibility and decreased the friction coefficient [286, 287].

18.4
Concluding Remarks

Improved knowledge about Nature's hierarchical materials and biological complex mechanisms is an unvaluable tool for the design of advanced drug carriers. Optimized therapeutic effect can be achieved by reproducing the recognition and the delivery behavior of biological systems, particularly when the drugs require a very rigorous control of their concentration in specific areas of the body or must be delivered according to rhythms that are hardly predictable. Stimuli-responsive drug delivery systems may benefit from being both structurally bio-inspired and functionally biomimetic. Novel architectures and assemblies of polymeric materials, inspired by natural vesicles and cell compartments, and even molecular arrangement of the chemical groups mimicking the receptor site of proteins, can provide an adequate environment (both regarding affinity and protection) to host drugs, being able to prevent premature leakage when distributed along the body toward the place the drug should act. Fine spatiotemporal regulation of drug release can be achieved through remotely triggered or illness-regulated changes in the polymer conformation. Careful selection of polymers able to perform as sensors and actuators during hundreds or even thousands of cycles is thus required. The changes should be reversible and proportional to the intensity of the stimulus in order to enable truly switchable drug release. There is much work done in this field, but extra efforts are being made to transform the stimuli-responsive polymer-based systems into suitable medicines. Multidisciplinary approaches to the design of nanostructures and nanostructured networks for smart drug delivery may enable rapid advances, particularly due to better identification of appropriate internal stimulus (peculiar features of the diverse regions of the body, from organs to cells compartments, and changes during malfunctions or disorders), development of suitable external triggering sources (innocuous and easy-to-use under clinical conditions), and improved synthesis procedures for obtaining novel formats of truly stimuli-reversible systems with enough reproducibility and purity.

Acknowledgments

Work supported by MICINN(SAF2008-01679; SAF2011-22771; PRI-AIBPT-2011-1211), FEDER and Xunta de Galicia (10CSA203013PR), Spain. A.M. Puga is grateful to MICINN for a FPI grant (BES-2009-024735).

References

1. Bushan, B. (2009) *Philos. Trans. R. Soc. London, Ser. A*, **367**, 1445–1486.
2. Fratzl, P. and Weinkamer, R. (2007) *Prog. Mater. Sci.*, **52**, 1263–1334.
3. Vissokov, G. and Tzvetkoff, T. (2003) *Eurasian Chem. Tech.*, **5**, 185–191.
4. Sahoo, S.K. and Labhasetwar, V. (2003) *Drug Discov. Today*, **8**, 1112–1120.
5. Ruiz-Hitzky, E., Darder, M., Aranda, P., and Ariga, K. (2010) *Adv. Mater.*, **22**, 323–336.
6. Scardino, A.J. and de Nys, R. (2011) *Biofouling*, **27**, 73–86.
7. Bhushan, B. and Jung, Y.C. (2011) *Prog. Mater. Sci.*, **56**, 1–108.
8. Roeder, R.K. (2010) *JOM-US*, **62**, 49–55.
9. Lee, H. and Messersmith, P.B. (2007) Bio-inspired nano-materials for a new generation of medicine, in *Nanotechnology in Biology and Medicine*, Chapter 3 (ed. T. Vo-Dinh), CRC Press, Boca Raton, FL.
10. Salman, H.H., Gamazo, C., Campanero, M.A., and Irache, J.M. (2005) *J. Controlled Release*, **106**, 1–13.
11. Cha, H.J., Hwang, D.S., and Lim, S. (2008) *Biotechnol. J.*, **3**, 631–638.
12. Liu, F. and Urban, M.W. (2010) *Prog. Polym. Sci.*, **35**, 3–23.
13. Kost, J. and Langer, R. (2001) *Adv. Drug Delivery Rev.*, **46**, 125–148.
14. Sershen, S. and West, J. (2002) *Adv. Drug Delivery Rev.*, **54**, 1225–1235.
15. Yoshida, M. and Lahann, J. (2008) *ACS Nano*, **2**, 1101–1107.
16. Schmaljohann, D. (2006) *Adv. Drug Delivery Rev.*, **58**, 1655–1670.
17. Motornov, M., Roiter, Y., Tokarev, I., and Minko, S. (2010) *Prog. Polym. Sci.*, **35**, 174–211.
18. Pasparakis, G. and Vamvakaki, M. (2011) *Polym. Chem.*, **2**, 1234–1248.
19. Alexander, C. and Shakesheff, K.M. (2006) *Adv. Matter.*, **18**, 3321–3328.
20. Alvarez-Lorenzo, C. and Concheiro, A. (2008) *Mini Rev. Med. Chem.*, **8**, 1065–1074.
21. Hirokawa, Y. and Tanaka, T. (1984) *J. Chem. Phys.*, **81**, 6379–6380.
22. Galaev, I.Y. (1995) *Russ. Chem. Rev.*, **84**, 471–489.
23. Shibayama, M. and Tanaka, T. (1993) in *Responsive Gels: Volume Transitions I* (ed. K. Dusek) Springer, Berlin, pp. 1–62.
24. Sanvicens, N. and Marco, M.P. (2008) *Trends Biotechnol.*, **26**, 425–433.
25. Patel, S., Bhirde, A.A., Rusling, J.F., Chen, X., Gutkind, J.S., and Patel, V. (2011) *Pharmaceutics*, **3**, 34–52.
26. Youan, B.B.C. (2004) *J. Controlled Release*, **98**, 337–353.
27. You, J.O., Almeda, D., Ye, G.J.C., and Auguste, D.T. (2010) *J. Biol. Eng.*, **4**(15).
28. Siegel, R. (2010) in *Nonlinear Dynamics with Polymers* (eds J.A. Pojman and Q. Tran-Cong-Miyata), Wiley-VCH Verlag GmbH, Weinheim, pp. 189–217.
29. Washington, N., Washington, C., and Wilson, C.G. (2001) *Physiological Pharmaceutics: Barriers to Drug Absorption*, 2nd edn, Taylor & Francis, London.
30. Nishiyama, N., Bae, Y., Miyata, K., Fukushima, S., and Kataoka, K. (2005) *Drug Discov. Today: Technol.*, **2**, 21–26.
31. Ojugo, A.S.E., Mesheedy, P.M.J., McIntyre, D.J.O., McCoy, C., Stubbs, M., Leach, M.O., Judson, I.R., and Griffiths, J.R. (1999) *NMR Biomed.*, **12**, 495–504.
32. Ganta, S., Iyer, A., and Amiji, M. (2010) in *Targeted Delivery of Small and Macromolecular Drugs* (eds R.I. Mahato and A.S. Narang), Taylor & Francis, CRC Press, Boca Raton, FL, pp. 555–585.
33. Schneider, L.A., Korber, A., Grabbe, S., and Dissemond, J. (2007) *Arch. Dermatol. Res.*, **298**, 413–420.
34. Alvarez-Lorenzo, C., Hiratani, H., Tanaka, K., Stancil, K., Grosberg, A.Yu., and Tanaka, T. (2001) *Langmuir*, **17**, 3616–3622.
35. Philippova, O.E., Hourdet, D., Audebert, R., and Khokhlov, A.R. (1997) *Macromolecules*, **30**, 8278–8285.
36. Siegel, R. and Firestone, B.A. (1988) *Macromolecules*, **21**, 3254–3259.
37. Mayo-Pedrosa, M., Cachafeiro-Andrade, N., Alvarez-Lorenzo, C.,

Martinez-Pacheco, R., and Concheiro, A. (2008) *Eur. Polym. J.*, **44**, 2629–2638.
38. Yoo, M.K., Seok, W.K., and Sung, Y.K. (2004) *Macromol. Symp.*, **207**, 173–186.
39. Schafer, F.Q. and Buettner, G.R. (2001) *Free Radical Biol. Med.*, **30**, 1191–1212.
40. Go, Y.M. and Jones, D.P. (2008) *Biochim. Biophys. Acta*, **1780**, 1273–1290.
41. Cheng, R., Feng, F., Meng, F., Deng, C., Feijen, J., and Zhong, Z. (2011) *J. Controlled Release*, **152**, 2–12.
42. Miyata, T., Uragami, T., and Nakamae, K. (2002) *Adv. Drug Delivery Rev.*, **54**, 79–98.
43. Benoit, D.S.W., Collins, S.D., and Anseth, K.S. (2007) *Adv. Funct. Mater.*, **17**, 2085–2093.
44. Li, Z., Zhang, Y., Fullhart, P., and Mirkin, C.A. (2004) *Nano Lett.*, **4**, 1055–1058.
45. Ulijn, R.V. (2006) *J. Mater. Chem.*, **16**, 2217–2225.
46. Miyata, T., Asami, N., and Uragami, T. (1999) *Nature*, **399**, 766–769.
47. Hoffman, A.S. (2000) *Clin. Chem.*, **46**, 1478–1486.
48. Farmer, T.G., Edgar, T.F., and Peppas, N.A. (2008) *Ind. Eng. Chem. Res.*, **47**, 10053–10063.
49. Uekama, K. (2004) *Chem. Pharm. Bull.*, **52**, 900–915.
50. Rodríguez-Tenreiro, C., Rodríguez-Perez, A., Alvarez-Lorenzo, C., Concheiro, A., and Torres-Labandeira, J.J. (2006) *Pharm. Res.*, **23**, 121–130.
51. Rodriguez-Tenreiro, C., Alvarez-Lorenzo, C., Rodriguez-Perez, A., Concheiro, A., and Torres-Labandeira, J.J. (2007) *Eur. J. Pharm. Biopharm.*, **66**, 55–62.
52. Rosa dos Santos, J.F., Alvarez-Lorenzo, C., Silva, M., Balsa, L., Couceiro, J., Torres-Labandeira, J.J., and Concheiro, A. (2009) *Biomaterials*, **30**, 1348–1355.
53. Otero-Espinar, F., Torres-Labandeira, J.J., Alvarez-Lorenzo, C., and Blanco-Méndez, J. (2010) *J. Drug Deliv. Sci. Technol.*, **20**, 289–301.
54. Klein, C. Th., Polheim, D., Viernstein, H., and Wolschann, P. (2000) *Pharm. Res.*, **17**, 358–365.
55. Maeda, M. and Bartsch, R.A. (1998) in *Molecular and Ionic Recognition with Imprinted Polymers*, ACS Symposium Series, Vol. **703** (eds R.A. Barstch and M. Maeda), American Chemical Society, Washington, DC, pp. 1–8.
56. Alvarez-Lorenzo, C. and Concheiro, A. (2004) *J. Chromatogr., B*, **804**, 231–245.
57. Alvarez-Lorenzo, C. and Concheiro, A. (2006) in *Biotechnology Annual Review*, vol. **12** (ed. M.R. El-Gewely), Elsevier, Amsterdam, pp. 225–268.
58. Ye, L. and Mosbach, K (2008) *Chem. Mater.*, **20**, 859–868.
59. Whitcombe, M.J., Chianella, I., Larcombe, L., Piletsky, S.A., Noble, J., Porter, R., and Horgan, A. (2011) *Chem. Soc. Rev.*, **40**, 1547–1571.
60. Mayes, A.G. and Whitcombe, M.J. (2005) *Adv. Drug Delivery Rev.*, **57**, 1742–1778.
61. Ribeiro, A., Veiga, F., Santos, D., Torres-Labandeira, J.J., Concheiro, A., and Alvarez-Lorenzo, C. (2011) *Biomacromolecules*, **12**, 701–709.
62. Ito, K., Chuang, J., Alvarez-Lorenzo, C., Watanabe, T., Ando, N., and Grosberg, A. Yu. (2004) *Prog. Polym. Sci.*, **28**, 1489–1515.
63. Alvarez-Lorenzo, C., Guney, O., Oya, T., Sakai, Y., Kobayashi, M., Enoki, T., Takeoka, Y., Ishibashi, T., Kuroda, K., Tanaka, K., Wang, G., Grosberg, A. Yu., Masamune, S., and Tanaka, T. (2000) *Macromolecules*, **33**, 8693–8697.
64. Sutton, D., Nasongkla, N., Blanco, E., and Gao, J.M. (2007) *Pharm. Res.*, **24**, 1029–1046.
65. Gil, E.S. and Hudson, S.M. (2004) *Prog. Polym. Sci.*, **29**, 1173–1222.
66. Ilmain, F., Tanaka, T., and Kokufuta, E. (1991) *Nature*, **349**, 400–401.
67. Grinberg, N.V., Dubovik, A.S., Grinberg, V.Y., Kuznetsov, D.V., Makhaeva, E.E., Grosberg, A.Y., and Tanaka, T. (1999) *Macromolecules*, **32**, 1471–1475.
68. Galaev, I.Y. and Mattiasson, B. (1993) *Enzyme Microb. Technol.*, **15**, 354–366.
69. Liu, T.Y., Hu, S.H., Liu, D.M., Chen, S.Y., and Chen. I.W. (2009) *Nano Today*, **4**, 52–65.
70. Bikram, M. and West, J.L. (2008) *Expert Opin. Drug Del.*, **5**, 1077–1091.

71. Schild, H.G. (1992) *Prog. Polym. Sci.*, **17**, 163–249.
72. Feil, H., Bae, Y.H., Feijen, J., and Kim, S.W. (1993) *Macromolecules*, **26**, 2496–2500.
73. Alvarez-Lorenzo, C., Concheiro, A., Dubovik, A.S., Grinberg, N.V., Burova, T.V., and Grinberg, V.Y. (2005) *J. Controlled Release*, **102**, 629–641.
74. Masuo, E.S. and Tanaka, T. (1988) *J. Chem. Phys.*, **89**, 1695–1703.
75. Yu, H. and Grainger, D.W. (1994) *Macromolecules*, **27**, 4554–4560.
76. Alvarez-Lorenzo, C. and Concheiro, A. (2002) *J. Controlled Release*, **80**, 247–257.
77. Salgado-Rodriguez, R., Claverie, A.L., and Arndt, K.F. (2004) *Eur. Polym. J.*, **40**, 1931–1946.
78. Urry, D.W. (1997) *J. Phys. Chem. B*, **101**, 11007–11028.
79. Martin, L., Alonso, M., Girotti, A., Arias, F.J., and Rodriguez-Cabello, J.C. (2009) *Biomacromolecules*, **10**, 3015–3022.
80. Lohmann, D. and Petrak, K. (1989) *Crit. Rev. Ther. Drug Carrier Syst.*, **5**, 263–320.
81. Jiang, J., Tong, X., Morris, D., and Zhao, Y. (2006) *Macromolecules*, **39**, 4633–4640.
82. Normand, N., Valamanesh, F., Savoldelli, M., Mascarelli, F., BenEzra, D., Courtois, Y., and Behar-Cohen, F. (2005) *Mol. Vis.*, **11**, 184–191.
83. Donnelly, R. F., Juzenas, P., McCarron, P.A., Woolfson, A.D., and Moan, J. (2006) *Trends Cancer Res.*, **2**, 1–20.
84. McCoy, C. P., Rooney, C., Edwards, C.R., Jones, D.S., and Gorman, S.P. (2007) *J. Am. Chem. Soc.*, **129**, 9572–9573.
85. Weissleder, R. and Ntziachristos, V. (2003) *Nat. Med.*, **9**, 123–128.
86. Suzuki, A. and Tanaka, T. (1990) *Nature*, **346**, 345–347.
87. Shum, P., Kim, J.M., and Thompson, D.H. (2001) *Adv. Drug Delivery Rev.*, **53**, 273–284.
88. Rijcken, C.J.F., Soga, O., Hennink, W.E., and van Nostrum, C.F. (2007) *J. Controlled Release*, **120**, 131–148.
89. Alvarez-Lorenzo, C., Bromberg, L., and Concheiro, A. (2009) *Photochem. Photobiol.*, **85**, 848–860.
90. Tong, X., Wang, G., Soldera, A., and Zhao, Y. (2005) *J. Phys. Chem. B*, **109**, 20281–20287.
91. Alvarez-Lorenzo, C., Deshmukh, S., Bromberg, L., Hatton, T.A., Sandez, I., and Concheiro, A. (2007) *Langmuir*, **23**, 11475–11481.
92. Sershen, S.R., Westcott, S.L., Hallas, N.J., and West, J.L. (2000) *J. Biomed. Mater. Res.*, **51**, 293–298.
93. Gorelikov, I., Field, L.M., and Kumacheva, E. (2004) *J. Am. Chem. Soc.*, **126**, 15938–15939.
94. Hirsch, L.R., Stafford, R.J., Bankson, J.A., Sershen, S.R., Rivera, B., Price, R.E., Hazle, J.D., Halas, N.J., and West, J.L. (2003) *Proc. Natl. Acad. Sci. U.S.A.*, **100**, 13549–13554.
95. Angelatos, A.S., Radt, B., and Caruso, F. (2005) *J. Phys. Chem. B*, **109**, 3071–3076.
96. Wu, W.T., Shen, J., Banerjee, P., and Zhou, S.Q. (2011) *Biomaterials*, **32**, 598–609.
97. Murdan, S. (2003) *J. Controlled Release*, **92**, 1–17.
98. Kishi, R. and Osada, Y. (1989) *J. Chem. Soc., Faraday Trans. 1*, **85**, 655.
99. MacDiarmid, A.G. (2001) *Angew. Chem. Int. Ed.*, **40**, 2581–2590.
100. Svirskis, D., Travas-Sejdic, J., Rodgers, A., and Garg, S. (2010) *J. Controlled Release*, **146**, 6–15.
101. Wadhwa, R., Lagenaur, C.F., and Cui, X.T. (2006) *J. Controlled Release*, **110**, 531–541.
102. Okner, R., Oron, M., Tal, N., Mandler, D., and Domb, A.J. (2007) *Mater. Sci. Eng. C*, **27**, 510–513.
103. Abidian, M.R. and Martin, D.C. (2009) *Adv. Funct. Mater.*, **19**, 573–585.
104. Richardson, R.T., Wise, A.K., Thompson, B.C., Flynn, B.O., Atkinson, P.J., Fretwell, N.J., Fallon, J.B., Wallace, G.G., Shepherd, R.K., Clark, G.M., and O'leary, S.J. (2009) *Biomaterials*, **30**, 2614–2624.
105. Medeiros, S.F., Santos, A.M., Fessi, H., and Elaissari, A. (2011) *Int. J. Pharm.*, **403**, 139–161.

106. Müller, R., Steinmetz, H., Hiergeist, R., and Gawalek, W. (2004) *J. Magn. Magn. Mater.*, **272–276**, 1539–1541.
107. Oh, J.W. and Park, J.M. (2011) *Progr. Polym. Sci.*, **36**, 168–189.
108. Bromberg, L., Chang, E.P., Hatton, T.A., Concheiro, A., Magariños, B., and Alvarez-Lorenzo, C. (2011) *Langmuir*, **27**, 420–429.
109. Huang, C., Zhou, Y., Jin, Y., Zhou, X., Tang, Z., Guo, X., and Zhou, S. (2011) *J. Mater. Chem.*, **21**, 5660-5670.
110. Kumar, C.S.S.R. and Mohammad, F. (2011) *Adv. Drug Deliv. Rev.*, **63**, 789-808.
111. Alexiou, C., Schmidt, R.J., Jourgons, R., Kremer, M., Wanner, G., Bergemann, C., Huenges, E., Nawroth, T., Arnold, W., and Parak, F.P. (2006) *Eur. Biophys. J.*, **35**, 446–450.
112. Scherer, C. and Neto, A.M.F. (2005) *Braz. J. Phys.*, **35**, 718–727.
113. Hafeli, U.O. (2004) *Int. J. Pharm.*, **277**, 19–24.
114. Manthe, R.L., Foy, S.P., Krishnamurthy, N., Sharma, B., and Labhasetwar, V. (2010) *Mol. Pharm.*, **7**, 1880–1898.
115. Mason, T.J. (2011) *Ultrason. Sonochem.*, **18**, SI 847–SI 852.
116. Husseini, G.A., de la Rosa, M.A.D., AlAqqad, E.O., Al Mamary, S., Kadimati, Y., Al Baik, A., and Pitt, W.G. (2011) *J. Franklin 1*, **348**, SI 125–SI 133.
117. Pangu, G.D., Davis, K.P., Bates, F.S., and Hammer, D.A. (2010) *Macromol. Biosci.*, **10**, 546–554.
118. Hernot, S. and Klibanov, A.L. (2008) *Adv. Drug Delivery Rev.*, **60**, 1153–1166.
119. Deckers, R. and Moonen, C.T.W. (2010) *J. Controlled Release*, **148**, 25–33.
120. Siegel, R.A. (2009) in *Chemomechanical Instabilities in Responsive Materials*, NATO Science for Peace and Security Series A (eds P. Borckmans, P. De Kepper, A.R. Khokhlov, and S. Métens), Chemistry and Biology, Springer, Berlin, pp. 175–201.
121. Giannos, S.A. and Dinh, S.M. (1999) in *Intelligent Materials for Controlled Release*, ACS Symposium Series, Vol. 728 (eds S.M. Ding, J.D. DeNuzzio, and A.R. Comfort), American Chemical Society, Washington, DC, pp. 87–95.
122. Topham, P.D., Howse, J.R., Crook, C.J., Gleeson, A.J., Bras, W., Armes, S.P., Jones, R.A.L., and Ryan, A.J. (2007) *Macromol. Symp.*, **256**, 95–104.
123. Sakai, T. and Yoshida, R. (2004) *Langmuir*, **20**, 1036–1038.
124. Yoshida, R., Yamaguchi, T., and Ichijo, H. (1996) *Mater. Sci. Eng. C*, **4**, 107–113.
125. Misra, G.P. and Siegel, R.A. (2002) *J. Controlled Release*, **81**, 1–6.
126. Dhanarajan, A. and Siegel, R.A. (2005) *Macromol. Symp.*, **227**, 105–114.
127. Kumar, S.A. and Khan, M.I. (2010) *J. Nanosci. Nanotech.*, **10**, 4124–4134.
128. Lyuksyutov, S.F., Vaia, R.A., Paramonov, P.B., Juhl, S., Waterhouse, L., Ralich, R.M., Sigalov, G., and Sancaktar, E. (2003) *Nat. Mat.*, **2**, 468–472.
129. Stupp, S.I., LeBonheur, V., Walker, K., Li, L.S., Huggins, K.E., Keser, M., and Amstutz, A. (2007) *Science*, **276**, 384–389.
130. Roy, I. and Gupta, M.N. (2003) *Chem. Biol.*, **10**, 1161–1171.
131. Letchford, K. and Burt, H. (2007) *Eur. J. Pharm. Biopharm.*, **65**, 259–269.
132. Meng, F., Zhong, Z., and Feijen, J. (2009) *Biomacromolecules*, **10**, 197–209.
133. Alvarez-Lorenzo, C., Concheiro, A., and Sosnik, A. (2009) in *Handbook of Hydrogels* (ed. D.B. Stein), Novapublishers, New York, pp. 449–484.
134. Miyata, K., Christie, R.J., and Kataoka, K. (2011) *React. Funct. Polym.*, **71**, 227–234.
135. Croy, S.R. and Kwon, G.S. (2006) *Curr. Pharm. Design*, **12**, 4669–4684.
136. Alvarez-Lorenzo, C., Rey-Rico, A., Sosnik, A., Taboada, P., and Concheiro, A. (2010) *Front. Biosci.*, **E2**, 424–440.
137. Haag, R. (2004) *Angew. Chem. Ind. Ed.*, **43**, 278–282.
138. You, J., Li, X., Cui, F., Du, Y.Z., Yuan, H., and Hu, F.Q. (2008) *Nanotechnology*, **19**, 045102.
139. Rapoport, N. (2007) *Prog. Polym. Sci.*, **32**, 962–990.

140. Massignani, M., Lomas, H., and Battaglia, G. (2010) in *Modern Techniques for Nano- and Microreactors/-Reactions* (ed. F. Caruso), Springer, Berlin, pp. 115–154.
141. Pinna, M., Hiltl, S., Guo, X., Böker, A., and Zvelindovsky, A.V. (2010) *ACS Nano*, **4**, 2845–2855.
142. Martin, T.J., Prochazka, K., Munk, P., and Webber, S.E. (1996) *Macromolecules*, **29**, 6071–6073.
143. Luo, Y.L., Yuan, J.F., Liu, X.J., Xie, H., and Gao, Q.Y. (2010) *J. Bioact. Comp. Polym.*, **25**, 292–304.
144. Lui, S., Weaver, J.V.M., Tang, Y., Billingham, N.C., Armes, S.P., and Tribe, K. (2002) *Macromolecules*, **35**, 6121–6131.
145. Min, K.H., Kim, J.H., Bae, S.M., Shin, H., Kim, M.S., Park, S. et al. (2010) *J. Controlled Release*, **144**, 259–266.
146. Bae, Y., Nishiyama, N., Fukushima, S., Koyama, H., Yasuhiro, M., and Kataoka, K. (2005) *Bioconjugate Chem.*, **16**, 122–130.
147. Barreiro-Iglesias, R., Bromberg, L., Temchenko, M., Hatton, A.T., Alvarez-Lorenzo, C., and Concheiro, A. (2004) *J. Controlled Release*, **97**, 537–549.
148. Gonzalez-López, J., Sández, I., Concheiro, A., and Alvarez-Lorenzo, C. (2010) *J. Phys. Chem. C*, **114**, 1181–1189.
149. Alvarez-Lorenzo, C. and Concheiro, A. (2010) *J. Drug Deliv. Sci. Technol.*, **20**, 249–257.
150. Gao, G.H., Lee, J.W., Nguyen, M.K., Im, G.H., Yang, J., Heo, H. et al. (2011) *J. Controlled Release*, **155**, 11-17.
151. Fukushima, S., Miyata, K., Nishiyama, N., Kanayama, N., Yamasaki, Y., and Kataoka, K. (2010) *J. Am. Chem. Soc.*, **127**, 2810–2811.
152. Alvarez-Lorenzo, C., Barreiro-Iglesias, R., Concheiro, A., Iourtchenko, L., Alakhov, V., Bromberg, L., Temchenko, M., Deshmukh, S., and Hatton, T.A. (2005) *Langmuir*, **21**, 5142–5148.
153. Bromberg, L., Deshmukh, S., Temchenko, M., Iourtchenko, L., Alakhov, V., Alvarez-Lorenzo, C., Barreiro-Iglesias, R., Concheiro, A., and Hatton, T.A. (2005) *Bioconjugate Chem.*, **16**, 626–633.
154. Bromberg, L., Raduyk, S., Hatton, T.A., Concheiro, A., Rodriguez-Valencia, C., Silva, M., and Alvarez-Lorenzo, C. (2009) *Bioconjugate Chem.*, **20**, 1044–1053.
155. Sun, T.M., Du, J.Z., Yao, Y.D., Mao, C.Q., Dou, S., Huang, S.Y. et al. (2011) *ACS Nano*, **5**, 1483–1494.
156. Gary, D.J., Lee, H., Sharma, R., Lee, J.S., Kim, Y., Cui, Z.Y. et al. (2011) *ACS Nano*, **5**, 3493-3505.
157. Convertine, A.J., Diab, C., Prieve, M., Paschal, A., Hoffman, A.S., Johnson, P.H. et al. (2010) *Biomacromolecules*, **11**, 2904–2911.
158. Koo, A.N., Lee, H.J., Kim, S.E., Chang, J.H., Park, C. et al. (2008) *Chem. Commun.*, 6570–6572.
159. Lv, L.P., Xu, J.P., Liu, X.S., Liu, G.Y., Yang, X., and Ji, J. (2010) *Macromol. Chem. Phys.*, **211**, 2292–2300.
160. Tang, L.Y., Wang, Y.C., Li, Y., Du, J.Z., and Wang, J. (2009) *Bioconjugate Chem.*, **20**, 1095–1099.
161. Wei, C., Guo, J., and Wang, C. (2011) *Macromol. Rapid Commun.*, **32**, 451–455.
162. Alexander, C. (2006) *Expert Opin. Drug Deliv.*, **3**, 573–581.
163. Chung, J.E., Yokoyama, M., and Okano, T. (2000) *J. Controlled Release*, **65**, 93–103.
164. Kohori, F., Sakai, K., Aoyagi, T., Yokoyama, M., Yamato, M., Sakurai, Y. et al. (1999) *Colloids Surf., B Biointerface*, **16**, 195–205.
165. Zhang, Y., Juang, M., Zhao, J., Ren, X., Chen, D., and Zhang, G. (2005) *Adv. Funct. Mater.*, **15**, 695–699.
166. El Halabieh, R.H., Mermut, O., and Barrett, C.J. (2004) *Pure Appl. Chem.*, **76**, 1445–1465.
167. Pouliquen, G. and Tribet, C. (2006) *Macromolecules*, **39**, 373–383.
168. Sugiyama, K. and Sono, K. (2001) *J. Appl. Polym. Sci.*, **81**, 3056–3063.
169. Determan, M.D., Cox, J.P., and Mallapragada, S.K. (2007) *J. Biomed. Mater. Res.*, **81A**, 326–333.
170. Lokitz, B.S., York, A.W., Stempka, J.E., Treat, N.D., Li, Y., Jarrett, W.L.,

and McCormick, C.L. (2007) *Macromolecules*, **40**, 6473–6480.
171. Rapoport, N. (2006) in *Smart Nanoparticles in Nanomedicine, MML Series*, vol. **8** (eds R. Arshady and K. Kono), Kentus Books, London, pp. 305–362.
172. Husseini, G.A., Myrup, G.D., Pitt, W.G., Christensen, D.A., and Rapoport, N.Y. (2006) *J. Controlled Release*, **69**, 43–52.
173. Ghaleb, A.H., Stevenson-Abouelnasr, D., Pitt, W.G., Assaleh, K.T., Farahat, L.O., and Fahadi, J. (2010) *Colloids Surf., A Physicochem. Eng. Aspects*, **359**, 18–24.
174. Husseini, G.A. and Pitt, W.G. (2008) *Adv. Drug Delivery Rev.*, **60**, 1137–1152.
175. Mohan, P. and Rapoport, N. (2010) *Mol. Pharm.*, **7**, 1959–1973.
176. Broz, P., Driamov, S., Ziegler, J., Ben-Haim, N., Marsch, S., Meier, W., and Hunziker, P. (2006) *Nano Lett.*, **6**, 2349–2353.
177. Ma, R. and Shi, L. (2010) *Macromol. Biosci.*, **10**, 1397–1405.
178. Chiu, H.C., Lin, Y.W., Huang, Y.F., Chuang, C.K., and Chern, C.S. (2008) *Angew. Chem. Int. Ed.*, **47**, 1875–1878.
179. Astruc, D., Boisselier, E., and Ornelas, C. (2010) *Chem. Rev.*, **110**, 1857–1959.
180. Lee, C.C., Mackay, J.A., Frechet, J.M.J., and Szoka, F.C. (2005) *Nat. Biotechnol.*, **23**, 1517–1526.
181. Kojima, C. (2010) *Expert Opin. Drug Deliv.*, **7**, 307–319.
182. Tekade, R.K., Dutta, T., Gajbhiye, V., and Jain, N.K. (2009) *J. Microencapsul.*, **26**, 287–296.
183. Svenson, S. (2009) *Eur. J. Pharm. Biopharm.*, **71**, 445–462.
184. Zhao, Y., Fan, X., Liu, D., and Wang, Z. (2011) *Int. J. Pharm.*, **409**, 229–236.
185. Xu, S., Luo, Y., and Haag, R. (2007) *Macromol. Biosci.*, **7**, 968–974.
186. Sideratou, Z., Tsiourvas, D., and Paleos, C.M. (2000) *Langmuir*, **16**, 1766–1769.
187. Hui, H., Xiao-dong, F., and Zhong-lin, C. (2005) *Polymer*, **46**, 9514–9522.
188. Jin, H., Zhou, Y., Huang, W., and Yan, D. (2010) *Langmuir*, **26**, 14512–14519.
189. Deloncle, R. and Caminade, A.M. (2010) *J. Photochem. Photobiol. C*, **11**, 25–45.
190. Gabriel, C.J. and Parquette, J.R. (2006) *J. Am. Chem. Soc.*, **128**, 13708–13709.
191. Wang, S., Wang, X., Li, L., and Advincula, R.C. (2004) *J. Org. Chem.*, **69**, 9073–9084.
192. Park, C., Lim, J., Yun, M., and Kim, C. (2008) *Angew. Chem. Int. Ed.*, **47**, 2959–2963.
193. Ochs, C.J., Such, G.K., Yan, Y., van Koeverden, M.P., and Caruso, F. (2010) *ACS Nano*, **4**, 1653–1663.
194. Wang, Y., Hosta-Rigau, L., Lomas, H., and Caruso, F. (2011) *Phys. Chem. Chem. Phys.*, **13**, 4782–4801.
195. Such, G.K., Johnston, A.P.R., and Caruso, F. (2011) *Chem. Soc. Rev.*, **40**, 19–29.
196. Inoue, H. and Anzai, J. (2005) *Langmuir*, **21**, 8354–8359.
197. Sukhisvili, S.A. (2005) *Curr. Opin. Colloid Interface Sci.*, **10**, 37–44.
198. Wong, J.E. and Rictering, W. (2008) *Curr. Opin. Colloid Interface Sci.*, **13**, 403–412.
199. Wong, J.E., Gaharwar, A.K., Müller-Schulte, D., Bahadur, D., and Richtering, W. (2008) *J. Colloid Interface Sci.*, **324**, 47–54.
200. Zelikin, A.N. (2010) *ACS Nano*, **4**, 2494–2509.
201. Zhu, Z. and Sukhishvili, S.A. (2009) *ACS Nano*, **3**, 3595–3605.
202. Barbosa, J.S., Costa, R.R., Testera, A.M., Alonso, M., Rodríguez-Cabello, J.C., and Mano, J.F. (2009) *Nanoscale Res. Lett.*, **4**, 1247–1253.
203. Basset, C., Harder, C., Vidaud, C., and Déjugnat, C. (2010) *Biomacromolecules*, **11**, 806–814.
204. Manna, U. and Patil, S. (2009) *J. Phys. Chem. B*, **113**, 9137–9142.
205. Yoshida, K., Sato, K., and Anzai, J. (2010) *J. Mater. Chem.*, **20**, 1546–1552.
206. Sugihara, S., Ohashi, M., and Ikeda, I. (2007) *Macromolecules*, **40**, 3394–3401.
207. Maeda, T., Akasaki, Y., Yamamoto, K., and Aoyagi, T. (2009) *Langmuir*, **25**, 9510–9517.
208. Sarmento, B., Ribeiro, A., Veiga, F., Ferreira, D., and Neufeld, R. (2007) *Biomacromolecules*, **8**, 3054–3060.

209. Barreiro-Iglesias, R., Coronilla, R., Concheiro, A., and Alvarez-Lorenzo, C. (2005) *Eur. J. Pharm. Sci.*, **24**, 77–84.
210. Bodnar, M., Hartman, J.F., and Borbely, J. (2005) *Biomacromolecules*, **6**, 2521–2527.
211. Konak, C., Panek, J., and Hruby, M. (2007) *Colloid Polym. Sci.*, **285**, 1433–1439.
212. Filippov, S., Hruby, M., Konak, C., Mackova, H., Spirkova, M., and Stepanek, P. (2008) *Langmuir*, **24**, 9295–9301.
213. Chuang, C.Y., Don, T.M., and Chiu, W.Y. (2009) *J. Polym. Sci., A: Polym. Chem.*, **47**, 2798–2810.
214. Wu, Y., Guo, J., Yang, W.L., Wang, C.C., and Fu, S.K. (2006) *Polymer*, **47**, 5287–5294.
215. Voets, I.K., Keizer, A., and Stuart, M.A.C. (2009) *Adv. Colloid Interface Sci.*, **147–148**, 300–318.
216. Oh, J.K., Lee, D.I., and Park, J.M. (2009) *Prog. Polym. Sci.*, **34**, 1261–1282.
217. Antoniette, M. and Landfester, K. (2002) *Prog. Polym. Sci.*, **27**, 689–757.
218. Rodriguez-Tenreiro, C., Diez-Bueno, L., Concheiro, A., Torres-Labandeira, J.J., and Alvarez-Lorenzo, C. (2007) *J. Controlled Release*, **123**, 56–66.
219. Tonelli, A.E. (2008) *J. Inclusion Phenom. Macrocycl. Chem.*, **60**, 197–202.
220. Johnson, J.A., Turro, N.J., Koberstein, J.T., and Mark, J.E. (2010) *Prog. Polym. Sci.*, **35**, 332–337.
221. Marui, Y., Kida, T., and Akashi, M. (2010) *Chem. Mater.*, **22**, 282–284.
222. Zhang, W., Gilstrap, K., Wu, L., Bahadur, R., Moss, M.A., Wang, Q., Lu, X., and He, X. (2010) *ACS Nano*, **4**, 6747–6759.
223. Cormack, P.A.G. and Elorza, A.Z. (2004) *J. Chromatogr., B*, **804**, 173–182.
224. Alvarez-Lorenzo, C. and Concheiro, A. (2006) in *Smart Nano- and Microparticles* (eds R. Arshady and K. Kono), Kentus Books, London, pp. 279–336.
225. Pérez-Moral, N. and Mayes, A.G. (2004) *Anal. Chim. Acta*, **504**, 15–21.
226. Sanson, N. and Rieger, J. (2010) *Polym. Chem.*, **1**, 965–977.
227. Lee, A., Tsai, H.Y., and Yates, M.Z. (2010) *Langmuir*, **26**, 18055–18060.
228. Fairhurst, R.E., Chassaing, C., Venn, R.F., and Mayes, A.G. (2004) *Biosens. Bioelectron.*, **20**, 1098–1105.
229. Nakamura, N., Ono, M., Nakajima, T., Ito, Y., Aketo, T., and Haginaka, J. (2005) *J. Pharm. Biomed. Anal.*, **37**, 213–237.
230. Suedee, R., Srichana, T., and Rattananont, T. (2002) *Drug Deliv.*, **9**, 19–30.
231. Horak, D., Lednicky, F., Rehak, V., and Svec, F. (1993) *Polym. Sci.*, **49**, 2041–2050.
232. Ansell, R.J. and Mosbach, K. (1997) *J. Chromatogr., A*, **787**, 55–66.
233. Kim, K. and Kim, D. (2005) *J. Appl. Polym. Sci.*, **96**, 200–212.
234. Suedee, R., Jantarat, C., Lindner, W., Viernstein, H., Songkro, S., and Srichana, T. (2010) *J. Controlled Release*, **142**, 122–131.
235. Ye, L., Weiss, R., and Mosbach, K. (2000) *Macromolecules*, **33**, 8239–8245.
236. Wang, D., Hong, S.P., Yang, G., and Row, H.H. (2003) *Korean J. Chem. Eng.*, **20**, 1073–1076.
237. Pich, A. and Richtering, W. (2010) in *Chemical Design of Responsive Microgels* (eds A. Pich and W. Richtering), Springer, Berlin, pp. 1–37.
238. Smith, M.H., Herman, E.S., and Lyon, L.A. (2011) *J. Phys. Chem. B*, **115**, 3761–3764.
239. Bromberg, L., Temchenko, M., and Hatton, T.A. (2003) *Langmuir*, **19**, 8675–8684.
240. Zhang, J., Qian, Z.Y., and Gu, Y.Q. (2009) *Nanotechnology*, **20**, 325102.
241. Puoci, F., Iemma, F., Muzzalupo, R., Spizzirri, U.G., Trombino, S., Cassano, R., and Picci, N. (2004) *Macromol. Biosci.*, **4**, 22–26.
242. Ciardelli, G., Cioni, B., Cristallini, C., Barbani, N., Silvestri, D., and Giusti, P. (2004) *Biosens. Bioelectron.*, **20**, 1083–1090.
243. Fang, L., Chen, S., Zhang, Y., and Zhang, H. (2011) *J. Mater. Chem.*, **21**, 2320–2329.
244. Vaihinger, D., Landfester, K., Kräuter, I., Brunner, H., and Tovar, G.E.M. (2002) *Macromol. Chem. Phys.*, **203**, 1965–1973.

245. Carter, S.R. and Rimmer, S. (2004) *Adv. Funct. Mater.*, **14**, 553–561.
246. Schartl, W. (2010) *Nanoscale*, **2**, 829–843.
247. Yuan, Q., Cayre, O.J., Fujii, S., Armes, S.P., Williams, R.A., and Biggs, S. (2010) *Langmuir*, **26**, 18408–18414.
248. Liu, Y.D., Fang, F.F., and Choi, H.J. (2010) *Langmuir*, **26**, 12849–12854.
249. Xionga, W., Wang, W., Wang, Y., Zhao, Y., Chena, H., Xu, H., and Yang, X. (2011) *Colloid Surf., B Biointerfaces*, **84**, 447–453.
250. Sahiner, N. and Ilgin, P. (2010) *Polymer*, **5**, 3156–3163.
251. Shaikh, R.P., Pillay, V., Choonara, Y.E., du Toit, L.C., Ndesendo, V.M.K., Bawa, P., and Cooppan, S. (2010) *AAPS PharmSciTech.*, **11**, 441–459.
252. Huang, C., Soenen, S.J., Rejman, J., Lucas, B., Braeckmans, K., Demeester, J., and De Smedt, S.C. (2011) *Chem. Soc. Rev.*, **40**, 2417–2434.
253. Prabaharan, M. and Mano, J.F. (2006) *Macromol. Biosci.*, **6**, 991–1008.
254. Bajpai, AK., Shukla, S.K., Bhanu, S., and Kankane, S. (2008) *Prog. Polym. Sci.*, **33**, 1088–1118.
255. Vakkalanka, S.K., Brazel, C.S., and Peppas, N.A. (1996) *J. Biomater. Sci. Polym. Ed.*, **8**, 119–129.
256. Gupta, K.M., Barnes, S.R., Tangaro, R.A., Roberts, M.C., Owen, D.H., Katz, D.F., and Kiser, P.F. (2007) *J. Pharm. Sci.*, **96**, 670–681.
257. Kanekiyo, Y., Naganawa, R., and Tao, H. (2002) *Chem. Commun.*, 2698–2699.
258. Kanekiyo, Y., Naganawa, R., and Tao, H. (2003) *Angew. Chem.*, **42**, 3014–3016.
259. Byrne, M.E., Park, K., and Peppas, N.A. (2002) *Adv. Drug Delivery Rev.*, **54**, 149–161.
260. Liu, X.Y., Ding, X.B., Guan, Y., Peng, Y.X., Long, X.P., Wang, X.C. et al. (2004) *Macromol. Biosci.*, **4**, 412–415.
261. Liu, X.Y., Guan, Y., Ding, X.B., Peng, Y.X., Long, X.P., Wang, X.C. et al. (2004) *Macromol. Biosci.*, **4**, 680–684.
262. Demirel, G., Ozcetin, G., Turan, E., and Caykara, T. (2005) *Macromol. Biosci.*, **5**, 1032–1037.
263. Burova, T.V., Grinberg, N.V., Kalinina, E.V., Ivanov, R.V., Lozinsky, V.I., Alvarez-Lorenzo, C. et al. (2011) *Macromol. Chem. Phys.*, **212**, 72–80.
264. Gong, C., Wong, K.L., and Lan, M.H.W. (2008) *Chem. Mater.*, **20**, 1353–1358.
265. Sakata, S., Uchida, K., Kaetsu, I., and Kita, Y. (2007) *Radiat. Phys. Chem.*, **76**, 733–737.
266. Sakata, S., Uchida, K., Kaetsu, I., Kita, Y., and Tsuji, D. (2007) *Radiat. Phys. Chem.*, **76**, 738–740.
267. Ikada, Y. (1994) *Biomaterials*, **15**, 725–736.
268. Venkatraman, S. and Boey, F. (2007) *J. Controlled Release*, **120**, 149–160.
269. Pavithra, D. and Doble, M. (2008) *Biomed. Mater.*, **3**, 034003.
270. Dwyer, A. (2008) *Semin. Dial.*, **21**, 542–546.
271. Goddard, J.M. and Hotchkiss, J.H. (2007) *Prog. Polym. Sci.*, **32**, 698–725.
272. Pekala, W., Rosiak, J., Rucinska-Rybus, A., Burczaka, K., Galanta, S., and Czollczyñska, T. (1986) *Radiat. Phys. Chem.*, **27**, 275–285.
273. Cole, M.A., Voelcker, N.H., Thissen, H., and Griesser, H.J. (2009) *Biomaterials*, **30**, 1827–1850.
274. Alvarez-Lorenzo, C., Bucio, E., Burillo, G., and Concheiro, A. (2010) *Expert Opin. Drug Deliv.*, **7**, 173–185.
275. Tokarev, I. and Minko, S. (2010) *Adv. Mater.*, **22**, 3446–3462.
276. Shim, J.K., Na, H.S., Lee, Y.M., Huh, H., and Nho, Y.C. (2001) *J. Membr. Sci.*, **190**, 215–226.
277. Alves, P., Coelho, J.F.J., Haack, J., Rota, A., Bruinink, A., and Gil, M.H. (2009) *Eur. Polym. J.*, **45**, 1412–1419.
278. Bucio, E. and Burillo, G. (2009) *J. Radioanal. Nucl. Chem.*, **280**, 239–243.
279. Burillo, G. and Bucio, E. (2009) in *Gamma Radiation Effects on Polymeric Materials and its Applications* (eds C. Barrera-Díaz and G. Martínez-Barrera), Research Signpost, Trivandrum, Kerala, India, pp. 45–62.
280. Singh, H. and Tyagi, P.K. (1989) *Angew. Makromol. Chem.*, **172**, 87–102.
281. Gupta, B., Anjum, N., Gulrez, S.K.H., and Singh, H. (2007) *J. Appl. Polym. Sci.*, **103**, 3534–3538.

282. Gupta, B., Jain, R., and Singh, H. (2008) *Polym. Adv. Technol.*, **19**, 1698–1703.
283. Ruiz, J.C., Alvarez-Lorenzo, C., Taboada, P., Burillo, G., Bucio, E., De Prijck, K *et al.* (2008) *Eur. J. Pharm. Biopharm.*, **70**, 467–477.
284. Muñoz-Muñoz, F., Ruiz, J.C., Alvarez-Lorenzo, C., Concheiro, A., and Bucio, E. (2009) *Eur. Polym. J.*, **45**, 1859–1867.
285. Ruiz, J.C., Burillo, G., and Bucio, E. (2007) *Macromol. Mater. Eng.*, **292**, 1176–1188.
286. Contreras-García, A., Bucio, E., Concheiro, A., and Alvarez-Lorenzo, C. (2010) *React. Funct. Polym.*, **70**, 836–842.
287. Contreras-García, A., Alvarez-Lorenzo, C., Taboada, C., Concheiro, A., and Bucio, E. (2011) *Acta Biomater.*, **7**, 996–1008.

19
Progress in Dendrimer-Based Nanocarriers
Joaquim M. Oliveira, João F. Mano, and Rui L. Reis

19.1
Fundamentals

Dendrimers are classified on the basis of their architectural structure as belonging to dendritic polymers [1]. The first report on dendrimers synthesis is attributed to Vögtle *et al.* [2], but Tomalia *et al.* [3] research works in 1980s potentiate its exploitation in a variety of biomedical applications. Figure 19.1 illustrates the hierarchical organization of the typical dendrimer and dendrons structures, where R is a reactive functionality [4, 5]. Dendrimers are composed of a core, branched and repeated units, and terminal functional groups. The core is covalently linked to the highly regular branching units that are organized in layers called *generations (G)* [1]. The functional groups at the periphery may perform different functions such as covalent bonding to antibodies, drugs, natural-based polymers or macromolecules, and fluorescent probes.

Dendrimers can be synthesized by divergent [6, 7] and convergent methods [8–10]. Using these methods, dendrimers are obtained in a precise and controlled fashion, while both molecular weight ($M_w/M_n = 1.0000$–1.05) and external functional groups may be fine-tuned [1, 11]. The type and number of the capping groups dictate its final physico-chemical, pharmacokinetic behavior, and biological properties [12].

The commercially available dendrimers, such as the polyamidoamine (PAMAM) dendrimers and poly(propyleneimine) (PPI) dendrimers are obtained by the divergent method. To produce core-shell structures, branch cell construction and a series of stepwise and iterative stages are required *in situ* around a desired core. By its turn, in the convergent method or Fréchet approach [13], dendrons are synthesized according to the divergent approach and then are anchored to a multifunctional core. This route can afford the so-called "Christmas tree"-like compounds, which are characterized by the best structural control and purity [14, 15]. Fréchet-type dendrimers are also advantageous as they can be designed to obtain various (un)symmetrical dendrimers [16], such as amphiphilic dendrimers.

Dendrimers are typically limited in size, which is in the order of ~10–20 nm [17, 18]. Despite further surface engineering and decoration with different polymers

Biomimetic Approaches for Biomaterials Development, First Edition. Edited by João F. Mano.
© 2012 Wiley-VCH Verlag GmbH & Co. KGaA. Published 2012 by Wiley-VCH Verlag GmbH & Co. KGaA.

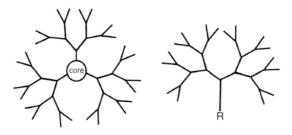

Dendrimers (dendron-like)   Dendrons (tree-like)

Figure 19.1 Hierarchical organization of dendrimers and dendrons illustrating its dendron-like or treelike architecture, respectively.

and bioactive agents it is also possible, which can allow obtaining macromolecules of higher dimensions. Dendrimeric macromolecules may be designed to: (i) be stimuli-responsive, (ii) include fluorescent probes or tags, (iii) possess high payload efficiency, (iv) decrease drug dosages needs and redosage frequency, (v) target delivery by covalent bonding to antibodies, and (vi) surface modification with polymers for improving biocompatibility and bioavailability.

More details on chemistry of dendrimers are reviewed elsewhere [19–22].

19.2
Applications of Dendrimer-Based Polymers

In this section, the enzymelike action and nanomedicine applications of dendrimers are addressed.

19.2.1
Biomimetic/Bioinspired Materials

Biomimetic materials have long been developed using inspiration from nature. The concept of biomimesis in materials science is mainly applied in the context of compounds with enzymatic activity. Dendrimers have been designed and synthesized [1, 6, 23] with precise M_w even at high generation, in order to behave as catalysts and mimic the activity of different enzymes (Figure 19.2).

It has been shown [6] that the dendrimers possessing a size equal to or greater than G4 can assume a densely packed globular shape. In fact, partial-core-shell-filled tecto(dendrimers) can be synthesized by covalently assembling PAMAM dendrimers (nucleophilic or electrophilic) around other electrophilic or nucleophilic core dendrimers [24]. Alternatively, functionalization of the dendritic interior also possibly obtains a catalytically active core [25]. In order to further demonstrate the versatility of dendrimers as biomimetic macromolecules, Wei et al. [26] showed that dendrimers in solution can have their remote catalytic groups folded back into its core. In their work, it was demonstrated that dendrimers share the flexible folding

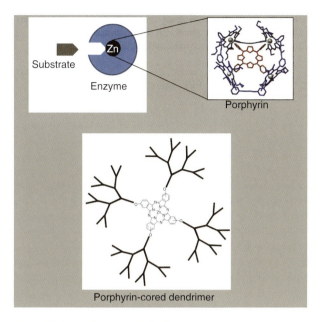

Figure 19.2 Porphyrin-cored dendrimers as synthetic enzyme mimics.

of proteins and thus support the idea that dendrimers can act as multifunctional catalysts and reactants. As conceptually advanced by Oliveira et al. [4], dendrimers may also be used as an alternative to serum proteins in culture media and thus provide an adequate substratum for superior cell culturing. Other reports on the dendrimers such as synthetic enzyme mimics provide both a comprehensive survey and state-of-the-art examples that borrow concepts from enzymes [27, 28].

19.2.2
Drug Delivery Systems

The development of nanocarriers for the efficient delivery of bioactive agents and proteins or growth factors has been attracting great deal of attention. Dendrimers are the ideal candidates for drug delivery applications as they possess precise M_w and thus can provide a reproducible pharmacokinetic behavior. Moreover, they can accommodate different drugs in its interior while may be linked to different molecules for both traceability and targeting specific organs, tissues, cells, or even subcellular compartments. The interior cavity of dendrimers needs to be hydrophobic so that a drug can be loaded, but the outer shell of the dendrimers, however, should be preferably more hydrophilic (e.g., surface modified with polyethylene glycol, PEG) to improve the bioavailability in the body [29]. Dendrimers can be modified with carbohydrates in order to improve affinity toward their receptors (e.g., lectin type) [30]. In order to refine the biological activities and drug delivery of dendritic polymers, efficient carbohydrate ligands of the glycodendrimer-type are emerging as potent ligands for carbohydrate-binding proteins. Thus surface engineering using a carbohydrate "coating" on the dendrimeric system (also known as *dendronized polymers*) constitutes a new class of biopolymers with many interesting applications. Such types of dendrimers have been proposed by Roy *et al.* [31] and have been synthesized by means of functionalization of the dendrimers with lactogroups (also called *lactodendrimers*). Thus dendronized polymers can be conveniently produced with a good yield using both divergent and convergent approaches or by using chemoenzymatic approaches as well [32].

It has been shown, however, that dendrimers of higher generations (G7) and with amine-capping groups cause hemolysis [33] and often reveal toxicity *in vitro* and *in vivo* [34, 35]. Thus the surface engineering of dendrimers with carbohydrates should also make it possible to avoid the cytotoxic effects of cationic and high-generation dendrimers and reduce the hemolytic toxicity by reduction/shielding of the positive charge at the surface. The pioneering work of Sashiwa *et al.* [36–39] allowed to produce different dendronized polymers, based on chitosan linked to a dendrimeric core, such as the commercially available PAMAM and poly(ethyleneimine) (PEI) dendrimers. Our group [40] has also proposed to surface engineer PAMAM (low generation) with the water-soluble carboxymethylchitosan (CMCht). The authors demonstrated that the fluorescent-labeled CMCht/PAMAM dendrimer nanoparticles (NPs) are internalized by different cell types such as rat bone marrow stromal cells, RBMSCs, and central nervous system cells (Figure 19.3), while dexamethasone-loaded NPs were found to be noncytotoxic and promote osteogenesis (2D system).

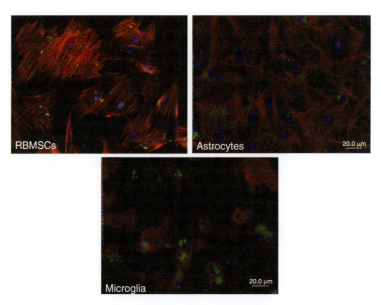

Figure 19.3 Fluorescence microscopy images of RBMSCs, astrocytes, and microglia cells cultured in the presence of FITC-labeled CMCht/PAMAM dendrimer nanoparticles (green). Nuclear DNA and cytoskeleton were labeled with Hoechst 33258 (blue) and Texas Red phalloidin (red), respectively.

Complementarily, *in vitro* and *in vivo* studies have shown that combination of scaffolds, bone marrow stromal cells, and Dex-loaded CMCht/PAMAM dendrimer NPs enhanced osteogenesis *in vitro* (3D systems) and promoted superior *de novo* bone formation, *in vivo* [41–43].

Dendrimers can also possibly enhance the solubility of hydrophobic drugs such as dexamethasone, nifedipine, camphotecin, quinolones, and indomethacin [7, 44–46]. Others showed that PAMAM dendrimers improved the solubility of nonsteroidal anti-inflammatory drugs (NSAIDs), namely, ketoprofen, ibuprofen, and diflunisal [47, 48].

The attractiveness of dendronized polymers as drug delivery systems is closely related to the possibility of tailoring drug delivery profile. Dendrimers allow a temporally controlled delivery of single or multiple compounds, and delivery can be triggered by means of applying an external stimuli such as temperature, pH, light, electric field, chemicals, and ionic strength [49–51]. By its turn, photochemical-internalization (PCI) and site-specific delivery, that is, escape of the macromolecules from lysosomes to the cytosol, is also possible and is seen as most advantageous in gene delivery [52, 53].

19.2.3
Gene Delivery Systems

The ability of dendrimers to be taken up by cells, and escape the lysosomes may allow us to envision its use in the delivery of genetic material. As dendrimers possess high charge density and tunable surface functional groups, it is also possible to easily tune condensation with DNA and formation of the dendriplexes. For example, the amine-terminated PAMAM dendrimers, which are polycationic, can interact with negatively charged DNA (Figure 19.4).

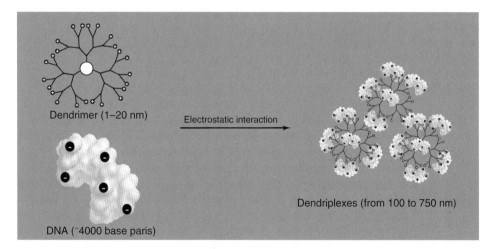

Figure 19.4 Schematic representation of the dendriplexes obtained through the electrostatic interaction of amine-terminated PAMAM dendrimers and negatively charged DNA.

Dendriplexes with size comprised between 100 and 750 nm are thus characterized by high stability and provide more efficient transport of DNA into the nucleus as compared to that of viruses or liposomes. It has been shown that the high transfection efficiency of dendrimers is due to both well-defined shape and low pK of the amines (3.9 and 6.9) external groups. The low pK permits the dendrimer to buffer the pH change in the endosomal–lysosomal compartment. This can be beneficial as it can direct toward polymeric swelling, thus allowing disrupting the membrane of the organelle and promoting the complex release, that is,, they have an intrinsic endosomolytic escape capacity, or can mediate their escape by degradable spacers. Another advantage is that these vectors can also prevent fast degradation of DNA by endo nucleases and exonucleases. PAMAM and other cationic dendrimers such as PPI thus fulfill these requirements. There are other factors affecting the efficiency of nonviral gene delivery systems. For successful gene therapy, the genetic material needs to reach nucleus and be permanently integrated into the DNA and then expressed by cells. Loup et al. [54] demonstrated that cationic phosphorus-containing dendrimers can be used as *in vitro* DNA-transfecting agents. The researchers showed that the generation of these dendrimers has an important influence on both efficiency of transfection and cytotoxicity.

It has been attempting to clarify the pathway(s) of transfection, in order to develop efficient gene delivery systems. Interestingly, Manunta et al. [55] demonstrated that internalization is dependent on the cell type, and it is a successful example of dendrimer application as a nonviral vector for gene delivery. This work demonstrated that transfection of the dendriplexes may occur via different pathways, but there are evidences that gene delivery might be occurring by a caveolin-dependent pathway, at least in cells expressing caveolae.

The protection of DNA from *in vivo* degradation by the vectors is another important issue that should be considered in transfection strategies. Diaz-Mochon et al. [56] showed that a hybrid combination of PAMAM and peptide dendrimers (known as *peptoid dendrimers*) were able to transfect cells with higher efficiency than the PAMAM counterpart and were also nontoxic. This work demonstrated that combination of primary and secondary amines generates a "proton sponge" effect that can facilitate the DNA transfection. Thus by means of facilitating the release of the plasmid from the cytoplasmic lysosome, it is possible to enhance transfection efficiency.

The major disadvantage of cationic dendrimers (high generation) is their reported cytotoxicity. Anionic dendrimers, on the other hand, have shown no cytotoxic effect on cells over a broad range of concentrations. A different strategy has been proposed based on PCI gene delivery [57]. The authors developed ternary complexes that were enveloped in the aryl ether dendrimers (carboxyl-terminated groups, G2) with a phtalocyanine core (photosensitizer). *In vitro* transfection studies using human cervical epithelioid carcinoma cells demonstrated that this ternary system significantly enhance transgene expression as compared to conventional reagents such as PEI and Lipofectamine. Moreover, it was shown that PCI-mediated gene delivery reduced phototoxicity of the ternary complexes, for the *in vivo* studies revealed that the ternary complex positively induced gene delivery by PCI. That work demonstrated that polyplex-polycations significantly affects both transfection efficiency and toxicity. Therefore, further optimization of this type of systems should involve research on the use of other polycations, but PCI offers the opportunity to develop efficient light-inducible gene delivery systems.

19.2.4
Biosensors

The versatility of dendrimers may explain their potential as materials for optic, electronics, and magnetic applications [58–60]. Actually, PAMAM dendrimers have adsorbed on native oxide surfaces of silicon wafers as a strategy to direct the integration of metal electrodeposition with silicon microfabrication processes and selective deposition by dendrimer patterning [61].

Wei-Jie et al. [62] reported on the development of an DNA electrochemical biosensor. This biosensor was fabricated by immobilizing PAMAM dendrimers (G1) on the glassy carbon electrode (GCE) surface by coupled activation agent, which was treated via electrochemical oxidation. The study showed that the complementary ssDNA segment in the solution was recognized and detected by means of measuring the oxidation–deoxidation peaks current. The DNA biosensor possessed high sensitivity and allowed to distinguish complementary and mismatching sequences, and to detect the target DNA as well. Similarly, Ningning et al. [63] modified Au electrodes with submonolayers of mercaptoacetic acid (RSH) and produced thin films with PAMAM dendrimers (G4). A DNA probe was then immobilized on the films in order to produce stable recognition layers. The biosensor exhibited a high selectivity, sensitivity, and stability for the measurement of DNA hybridization.

The development of an amperometric enzyme electrode for monitoring small molecules such as glucose has also been proposed [64]. The glucose electrode was developed by electrostatic immobilization of the enzyme glucose oxidase (GOX) on C and Pt electrodes modified with mixed ferrocene-cobaltocenium dendrimers. Another study [65], PPI dendrimer cores functionalized with octamethylferrocenyl units were deposited on a Pt electrode. The amperometric biosensor for determining the concentration of glucose was prepared by immobilization of GOX on the electrodes. Enzyme field effect transistor (ENFET) has also been developed for glucose [66]. This type of biosensor was fabricated using dendrimer-encapsulated Pt NPs and GOX via a layer-by-layer (LbL) self-assembly method. The developed biosensor possessed good stability, enhanced sensitivity, and extended lifetime as compared to others.

By its turn, Perinotto et al. [67] developed a biosensor for ethanol by means of immobilizing the enzyme alcohol dehydrogenase (ADH) on Au-interdigitated electrodes (IDEs), in conjunction with layers of PAMAM dendrimers using the LbL technique. The biosensor was able to detect and distinguish between ethanol solutions with concentrations as low as 1 ppm by volume.

19.2.5
Theranostics

During cancer treatment, anticancer agents can damage both malignant and normal cells in a similar way. One-package nanocarriers that can target the diseased tissues, while releasing one or multiple therapeutic agents are being investigated for treating cancer. Moreover, the combination of the drug delivery ability and imaging can allow the production of nanotools to be used as bioimaging probes for theranostics, that is, diagnosis in personalized medicine for delivering drugs, following their distribution, and monitoring therapy.

To achieve this goal, the nanocarriers should: (i) possess stealth properties that prevent them from being opsonized or cleared before reaching the target cells (biostable), (ii) be able to be internalized by cells, cross different biological barriers, and diffuse within tissues, (iii) comprise a functionality to target specific subcellular compartments, cells, tissues, and organs, (iv) be biodegradable, (v) be compatible with external activation by magnetic field, ultrasound, X-ray, or optics to trigger the release of the bioactive compound, and (vi) ensure excellent signal-to-noise ratio for detection.

There is a big gap between design and application, and it is easier to develop new functionalities for desired effects as compared to obtain new properties by multiplying known functions [68]. Also, it can be found in literature that some works show great developments in dendrimers technology for theranostics. Kobayashi et al. [69] showed that a dendrimer-based magnetic resonance contrast agent may be useful for *in vivo* detection of renal tubular damage. Rietveld et al. [70] developed dendrimers with tetrabenzoporphyrin cores for *in vivo* oxygen imaging. Other authors [71, 72] have linked dendrimers to different antibodies for receptor specificity, as the conjugates bound to specific antigen expressing cells. Another interesting work by Baek and Roy [73] reported on T-antigen-linked-glycoPAMAM dendrimers aimed at finding applications in the detection and immunotherapy of carcinomas such as breast cancer.

Several ligands are known to be associated with tumor. Ligand-based dendrimeric systems are gaining interest for targeting cancer cells [74] and as nanotools in cancer treatments [75]. The work of Citro et al. [76] demonstrated that PEGylated poly(cyanoacrylate) NPs conjugated with transferrin were able to deliver paclitaxel (PTX), an antitumor drug.

Another good work on multifunctional devices based on PAMAM dendrimer (target the desire cells, releasing the desired drug and monitoring their internalization-fluorescent probe) has been reported by Islam et al. [77]. Partially acetylated PAMAM dendrimers (G5) were then conjugated with fluorescein isothiocyanate (FITC), focal adhesion (FA), and methotrexate (MTX), for targeting tumor cells through the folate receptor, while releasing intracellularly an antitumor drug. Yang et al. [78] have also synthesized the FITC-labeled and biotin-linked PAMAM dendrimer (G5) conjugates and further studied its ability to target cancer cells.

Thus the investigations herein highlighted clearly demonstrate that the dendrimeric nanopackages may comprise certain functionalities, such as targetability, traceability, drug delivery capacity, or reactivity, which offer new possibilities for the diagnosis and management of cancer. Details of other available dendrimeric systems for these applications may be found in recent original works [79–82].

19.3
Final Remarks

Dendrimers are one of the most useful drug/gene delivery systems. Their ability to be internalized and transfect cells without inducing toxicity and be tuned for stimuli-induced delivery confers a great advantage over other carriers namely viral vectors.

Despite most of preclinical reports dealing with dendrimeric drug delivery systems have shown promising data, there is the need, however, to intensify the comprehensive research (*in vitro* and *in vivo*) for elucidating their stimuli-responsiveness, biodegradability, biocompatibility, biodistribution, and bioelimination. Future looks bright for dendrimer-based technologies in theranostics especially in cancer. Dendrimers as synthetic enzyme mimics are also most promising in tissue engineering applications and catalysis. Other interesting applications include their use as photoluminescent, photoelectric, and nonlinear optic materials.

References

1. Tomalia, D.A. (2005) *Prog. Polym. Sci.*, **30**(3–4), 294–324.
2. Buhleier, E., Wehner, W., and Vögtle, F. (1978) *Synthesis*, **2** 155–158.
3. Tomalia, D.A., Baker, H., Dewald, J.R., Hall, M., Kallos, G., Martin, S. et al. (1985) *Polym. J.*, **17**, 117–132.
4. Oliveira, J.M., Salgado, A.J., Sousa, N., Mano, J.F., and Reis, R.L. (2010) *Prog. Polym. Sci.*, **35**(9), 1163–1194.
5. D'Emanuele, A., Jevprasesphant, R., Penny, J., and Attwood, D. (2004) *J. Controlled Release*, **95**, 5447–5453.
6. Esfand, R. and Tomalia, D.A. (2001) *Drug Discov. Today*, **6**(8), 427.
7. Devarakonda, B., Hill, R.A., and de Villiers, M.M. (2004) *Int. J. Pharm.*, **284**(1–2), 133–140.
8. Fréchet, J.M.J., Jiang, Y., and Hawker, C.J. (eds) (1989) A. E. Proceedings of IUPAC International Symposium, Macromolecules, Seoul, Korea.
9. Leu, C.-M., Shu, C.F., Teng, C.-F., and Shiea, J. (2001) *Polymer*, **42**, 2339–2348.
10. Leon, J.W. and Fréchet, J.M.J. (2008) *Adv. Drug Delivery Rev.*, **60**(9), 1037–1055.
11. Wolinsky, J.B. and Grinstaff, M.W. (2008) *Adv. Drug Delivery Rev.*, **60**(9), 1037–1055.
12. Domanski, D.M., Klajnert, B., and Bryszewska, M. (2004) *Bioelectrochemistry*, **63**(1–2), 193–197.
13. Hawker, C.J. and Frechet, J.M.J. (1990) *J. Am. Chem. Soc.*, **112**(21), 7638–7647.
14. Angurell, I., Turrin, C.-O., Laurent, R., Maraval, V., Servin, P., Rossell, O. et al. (2007) *J. Organomet. Chem.*, **692**(10), 1928–1939.
15. Séverac, M., Leclaire, J., Sutra, P., Caminade, A.-M., and Majoral, J.-P. (2004) *Tetrahedron Lett.*, **45**(15), 3019–3022.
16. Lee, J.W., Kim, J.H., Kim, B.-K., Shin, W.S., and Jin, S.-H. (2006) *Tetrahedron*, **62**(5), 894–900.
17. Tomalia, D.A. (2005) *Mater. Today*, **8**(3), 34.

18. Lee, C.C., MacKay, J.A., Fréchet, J.M., and Szoka, F.C. (2005) *Nat. Biotechnol.*, **23**, 1517–1526.
19. DNT Inc. (2005) *Focus Surfactants*, **2005**(8), 4.
20. DNT Inc. (2006) *Focus Surfactants*, **2006**(9), 3.
21. Tomalia, D.A. (2010) *Soft Matter*, **6**(3), 456–474.
22. Gorman, C., Buschow, K.H.J., Robert, W.C., Merton, C.F., Bernard, I., Edward, J.K. et al. (2001) Dendrimers: Polymerization and Properties. Encyclopedia of Materials: Science and Technology, Elsevier, Oxford, p. 2042.
23. Tomalia, D.A., Huang, B., Swanson, D.R., Brothers, I.I.H.M., and Klimash, J.W. (2003) *Tetrahedron*, **59**, 3799–3813.
24. Tomalia, D.A., Brothers, H.M., Piehler, L.T., Durst, H.D., and Swanson, D.R. (2002) *Proc. Natl. Acad. Sci. U.S.A.*, **99**(8), 5081–5087.
25. Thayumanavan, S., Bharathi, P., Sivanandan, K., and Rao Vutukuri, D. (2003) *C. R. Chim.*, **6**(8–10), 767–778.
26. Wei, S., Wang, J., Venhuizen, S., Skouta, R., and Breslow, R. (2009) *Bioorg. Med. Chem. Lett.*, **19**(19), 5543–5546.
27. Liang, C. and Fréchet, J.M.J. (2005) *Prog. Polym. Sci.*, **30**(3–4), 385–402.
28. Smith, D.K., Hirst, A.R., Love, C.S., Hardy, J.G., Brignell, S.V., and Huang, B. (2005) *Prog. Polym. Sci.*, **30**(3–4), 220–293.
29. Yellepeddi, V.K., Kumar, A., and Palakurthi, S. (2009) *Expert Opin. Drug Deliv.*, **6**(8), 835–850.
30. Bezouska, K. (2002) *Rev. Mol. Biotechnol.*, **90**, 269–290.
31. Pavlov, G.M., Errington, N., Harding, S.E., Korneeva, E.V., and Roy, R. (2001) *Polymer*, **42**(8), 3671–3678.
32. Kobayashi, K., Akaike, T., and Usui, T. (1994) *Methods Enzymol.*, **242**, 226–235.
33. Malik, N., Wiwattanapatapee, R., Klopsch, R., Lorenz, K., Frey, H., Weener, J.W. et al. (2000) *J. Controlled Release*, **65**, 133–148.
34. Domanski, D.M., Klajnert, B., and Bryszewska, M. (2004) *Bioelectrochemistry*, **63**(1–2), 189–191.
35. Dutta, T., Agashe, H.B., Garg, M., Balakrishnan, P., Kabra, M., and Jain, N.K. (2007) *J. Drug Targeting*, **15**, 89–98.
36. Sashiwa, H., Shigemasa, Y., and Roy, R. (2002) *Carbohydr. Polym.*, **49**(2), 195–205.
37. Sashiwa, H., Shigemasa, Y., and Roy, R. (2002) *Carbohydr. Polym.*, **47**(2), 201–208.
38. Sashiwa, H., Shigemasa, Y., and Roy, R. (2002) *Carbohydr. Polym.*, **47**(2), 191–199.
39. Sashiwa, H. and Aiba, S.-I. (2004) *Prog. Polym. Sci.*, **29**(9), 887–908.
40. Oliveira, J.M., Kotobuki, N., Marques, A.P., Pirraco, R.P., Benesch, J., Hirose, M. et al. (2008) *Adv. Funct. Mater.*, **18**, 1840–1853.
41. Oliveira, J.M., Sousa, R.A., Malafaya, P.B., Silva, S.S., Kotobuki, N., Hirose, M. et al. (2011) *Nanomed. Nanotechnol., Biol. Med.*, **7**(6), 914–924.
42. Oliveira, J.M., Sousa, R.A., Kotobuki, N., Tadokoro, M., Hirose, M., Mano, J.F. et al. (2009) *Biomaterials*, **30**(5), 804–813.
43. Oliveira, J.M., Kotobuki, N., Tadokoro, M., Hirose, M., Mano, J.F., and Reis, R.L. et al. (2010) *Bone*, **46**(5), 1424–1435.
44. Chauhan, A.S., Sridevi, S., Chalasani, K.B., Jain, A.K., Jain, S.K., Jain, N.K. et al. (2003) *J. Controlled Release*, **90**(3), 335–343.
45. Cheng, Y., Li, M., and Xu, T. (2008) *Eur. J. Med. Chem.*, **43**(8), 1791–1795.
46. Cheng, Y., Qu, H., Ma, M., Xu, Z., Xu, P., Fang, Y. et al. (2007) *Eur. J. Med. Chem.*, **42**(7), 1032–1038.
47. Yiyun, C. and Tongwen, X. (2005) *Eur. J. Med. Chem.*, **40**(11), 1188–1192.
48. Yiyun, C., Tongwen, X., and Rongqiang, F. (2005) *Eur. J. Med. Chem.*, **40**(12), 1390–1393.
49. Mano, J.F. (2008) *Adv. Eng. Mater.*, **10**(6), 515–527.
50. Hui, H., Xiao-dong, F., and Zhong-lin, C. (2005) *Polymer*, **46**(22), 9514–9522.
51. Pistolis, G., Malliaris, A., Tsiourvas, D., and Paleos, C.M. (1999) *Chem. Eur. J.*, **5**, 1440–1444.
52. Lai, P.-S., Lou, P.-J., Peng, C.-L., Pai, C.-L., Yen, W.-N., Huang, M.-Y. et al. (2007) *J. Controlled Release*, **122**(1), 39–46.

53. Perumal, O.P., Inapagolla, R., Kannan, S., and Kannan, R.M. (2008) *Biomaterials*, **29**(24–25), 3469.
54. Loup, C., Zanta, M.-A., Caminade, A.-M., Majoral, J.-P., and Meunier, B. (1999) *Chem. Eur. J.*, **5**(12), 3644–3650.
55. Manunta, M., Nichols, B., Hong Tan, P., Sagoo, P., Harper, J., and George, A.J.T. (2006) *J. Immunol. Methods*, **314**, 134–146.
56. Diaz-Mochon, J.J., Fara, M.A., Sanchez-Martin, R.M., and Bradley, M. (2008) *Tetrahedron Lett.*, **49**(5), 923–926.
57. Nishiyama, N., Iriyama, A., Jang, W.-D., Miyata, K., Itaka, K., Inoue, Y. *et al.* (2005) *Nat. Mater.*, **4**, 934–941.
58. Alvarez-Venicio, V., Jiménez-Nava, B., Carreón-Castro, Md.P., Rivera, E., Méndez, I.A., Huerta, A.A. *et al.* (2008) *Polymer*, **49**(18), 3911–3922.
59. Bao, C., Jin, M., Lu, R., Zhang, T., and Zhao, Y.Y. (2003) *Mater. Chem. Phys.*, **81**(1), 160–165.
60. Cho, M.Y., Kang, H.S., Kim, K., Kim, S.J., Joo, J., Kim, K.H. *et al.* (2008) *Colloids Surf. A Physicochem. Eng. Asp.*, **313–314**, 431–434.
61. Arrington, D., Curry, M., Street, S., Pattanaik, G., and Zangari, G. (2008) *Electrochim. Acta*, **53**(5), 2644–2649.
62. Wei-Jie, S., Shi-Yun, A., Jin-Huan, L., and Lu-Sheng, Z. (2008) *Chin. J. Anal. Chem.*, **36**(3), 335–338.
63. Ningning, Z., Yunfeng, G., Zhu, C., Pingang, H., and Yuzhi, F. (2006) *Electroanalysis*, **18**(21), 2107–2114.
64. Alonso, B., Armada, P.G., Losada, J., Cuadrado, I., González, B., and Casado, C.M. (2004) *Biosens. Bioelectron.*, **19**(12), 1617–1625.
65. Armada, M.P.G., Losada, J., Zamora, M., Alonso, B., Cuadrado, I., and Casado, C.M. (2006) *Bioelectrochemistry*, **69**(1), 65–73.
66. Yao, K., Zhu, Y., Yang, X., and Li, C. (2008) *Mater. Sci. Eng. C*, **28**(8), 1236–1241.
67. Perinotto, A.C., Caseli, L., Hayasaka, C.O., Riul, A. Jr., Oliveira, O.N. Jr., and Zucolotto, V. (2008) *Thin Solid Films*, **516**(24), 9002–9005.
68. Marco Fischer, F.V. (1999) *Angew. Chem. Int. Ed.*, **38**(7), 884–905.
69. Kobayashi, H., Kawamoto, S., Jo, S.-K., Sato, N., Saga, T., Hiraga, A. *et al.* (2002) *Kidney Int.*, **61**, 1980–1985.
70. Rietveld, I.B., Kim, E., and Vinogradov, S.A. (2003) *Tetrahedron*, **59**(22), 3821.
71. Thomas, T.P., Patri, A.K., Myc, A., Myaing, M.T., Ye, J.Y., Norris, T.B. *et al.* (2004) *Biomacromolecules*, **5**, 2269–2274.
72. Shukla, R., Thomas, T.P., Peters, J.L., Desai, A.M., Kukowska-Latallo, J., Patri, A.K. *et al.* (2006) *Bioconjug. Chem.*, **17**, 1109–1115.
73. Baek, M.-G. and Roy, R. (2002) *Bioorg. Med. Chem.*, **10**(1), 11–17.
74. Choi, Y., Thomas, T., Kotlyar, A., Islam, M.T., and Baker, J.J.R. (2005) *Chem. Biol.*, **12**(1), 35–43.
75. Agarwal, A., Saraf, S., Asthana, A., Gupta, U., Gajbhiye, V., and Jain, N.K. (2008) *Int. J. Pharm.*, **350**(1–2), 3–13.
76. Citro, G., Perrotti, D., Cucco, C., D'Agnano, I., Sacchi, A., Zupi, G. *et al.* (1992) *Proc. Natl. Acad. Sci. U.S.A.*, **89**, 7031–7035.
77. Islam, M.T., Majoros, I.J., and Baker, J.J.R. (2005) *J. Chromatogr. B*, **822**(1–2), 21–26.
78. Yang, W., Cheng, Y., Xu, T., Wang, X., and Wen, L.-P. (2009) *Eur. J. Med. Chem.*, **44**(2), 862–868.
79. Johansson, E.M.V., Dubois, J., Darbre, T., and Reymond, J.-L. (2010) *Bioorg. Med. Chem.*, **18**(17), 6589–6597.
80. Lu, H.-L., Syu, W.-J., Nishiyama, N., Kataoka, K., and Lai, P.-S. (2011) *J. Controlled Release*, **155**(3), 458–464.
81. Mahato, R., Tai, W., and Cheng, K. (2011) *Adv. Drug Delivery Rev.*, **63**(8), 659–670.
82. Xu, X., Zhang, Y., Wang, X., Guo, X., Zhang, X., Qi, Y. *et al.* (2011) *Bioorg. Med. Chem.*, **19**(5), 1643–1648.

Part V
Lessons from Nature in Regenerative Medicine

Biomimetic Approaches for Biomaterials Development, First Edition. Edited by João F. Mano.
© 2012 Wiley-VCH Verlag GmbH & Co. KGaA. Published 2012 by Wiley-VCH Verlag GmbH & Co. KGaA.

20
Tissue Analogs by the Assembly of Engineered Hydrogel Blocks

Shilpa Sant, Daniela F. Coutinho, Nasser Sadr, Rui L. Reis, and Ali Khademhosseini

20.1
Introduction

The broad goal of tissue engineering and regenerative medicine is to create functional human tissue equivalents for organ repair and replacement using cells combined with the biomaterial scaffolds. Engineered tissues also hold a significant promise as *in vitro* model systems to study human physiology, pathophysiology, and drug safety/efficacy before clinical trials [1, 2]. However, current tissue-engineered scaffolds fail to mimic the structural and functional complexity because of the inability to recreate heterogeneous and spatiotemporal aspects of cellular, biochemical, and mechanical properties *in vivo*. For functional tissue engineering, it is important that engineered tissues closely mimic unique cellular microenvironment found in native tissues to reestablish the complex cell–matrix and cell–cell interactions that regulate tissue morphogenesis, function, and regeneration [3, 4].

Cellular microenvironments in living tissues consist of the extracellular matrix (ECM), neighboring cells, and surrounding soluble factors. The ECM is a highly hydrated, viscoelastic three-dimensional (3D) network containing a rich variety of proteoglycans, proteins, and glycoproteins. Molecular ECM components organize into hierarchical insoluble suprastructures (fibrils, microfibrils) giving them unique tissue-specific structural properties [5]. The ECM not only provides structural support but also determines critical cellular functions through cell–matrix interactions such as biochemical and mechanical cues [5]. The ECM contains chemical concentration gradients of soluble growth factors, chemokines, and cytokines that play an important role in biological phenomena such as chemotaxis, morphogenesis, and wound healing [6, 7]. The mechanical properties of the ECM are equally important to control cellular functions through mechanotransduction by structural rearrangements of the cytoskeleton and immobilized proteins as reviewed in detail elsewhere [8–10]. Cells, in turn, can interact and remodel the surrounding ECM. Thus, the bidirectional crosstalk between the cells and ECM plays an important role in tissue development, homeostasis, and disease progression [11].

Given the importance of ECM in tissue morphogenesis and homeostasis, it is important to engineer scaffolds that mimic ECM structure and function. Among

Biomimetic Approaches for Biomaterials Development, First Edition. Edited by João F. Mano.
© 2012 Wiley-VCH Verlag GmbH & Co. KGaA. Published 2012 by Wiley-VCH Verlag GmbH & Co. KGaA.

biomaterial scaffolds, hydrogel scaffolds provide biomimetic microenvironments because of their ECM-like viscoelastic and diffusive transport characteristics [12, 13]. Since hydrogel chemistry, cross-linking density, and response to environmental stimuli (e.g., heat, light, electrical potential, chemicals, and biological agents) can be manipulated, they are ideal for producing tailored 3D cellular microenvironments. Moreover, advances in materials chemistry, microscale technologies, and microfluidics have enabled fabrication of hydrogels with spatiotemporal control over their physicochemical and mechanical properties [4, 14–16]. For instance, "gradient hydrogels", which exhibit continuous spatial change in their physicochemical properties, are emerging as promising tools to create spatially patterned scaffolds for tissue engineering applications [14].

While considering the tissue/organ heterogeneity and complexity, it is of critical importance to generate scaffolds that have specific geometries, contain spatially organized cells, and present a differential set of biochemical and mechanical cues to the cells. Traditional tissue engineering strategies rely on "top-down" approaches where it is hoped that the cells seeded on the scaffolds will reorganize, creating their own ECM and microarchitecture with the help of biochemical and mechanical cues engineered within the scaffolds. On the other hand, "bottom-up" approaches aim to create biomimetic structures at the microscale that can be used as building blocks to generate larger tissues [17]. For example, microscale hydrogel blocks can be first engineered with specific geometries, spatially organized cells, and biochemical/mechanical components separately and then can be assembled together using various physicochemical or mechanical processes to more closely mimic the *in vivo* tissues as depicted in Figure 20.1.

In this chapter, we discuss biologically relevant examples to highlight the importance of tissue architecture and mechanical, biochemical, and cell–cell contact cues during tissue morphogenesis and regeneration. We further discuss current efforts in the field to engineer hydrogel blocks with the specific cues either alone or in combination. Finally, we review various techniques to assemble the engineered hydrogel blocks to create biomimetic tissue equivalents *in vitro*.

20.2
Tissue/Organ Heterogeneity *In Vivo*

Each organ contains a multitude of cell populations that are located in a spatially organized manner. Organs such as liver, kidney, pancreas, and bones are made up of functional units (liver lobules, nephrons, pancreatic islets, and osteons, respectively), where cells along with the surrounding ECM are organized spatially to perform tissue-specific functions. Similarly, the structural, biochemical, and mechanical cues defining the cell physiological milieu emulate precise spatial organization with distribution of each constituent being accurately defined by the tissue architecture resulting in dramatically heterogeneous microenvironments.

Structurally, the unique architecture and characteristics of tissues and organs are determined by the ECM and the cells that produce it. ECM molecules are generally

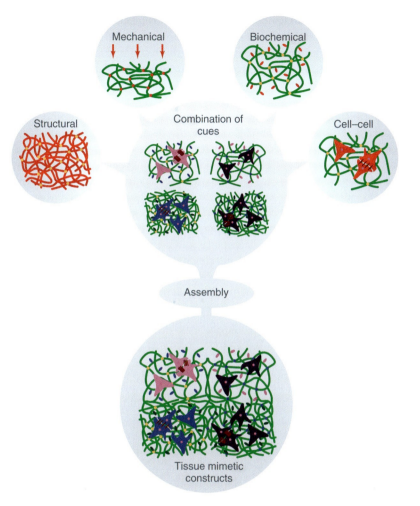

Figure 20.1 Schematics of structural, mechanical, biochemical, and cell–cell contact cues necessary to recapitulate unique microenvironments *in vivo*. To generate functional tissue equivalents *in vitro*, engineered scaffolds should mimic the natural tissue/organ complexity by recapitulating each of the single cues and by combining them within single constructs. The construct obtained by the combination of specific set of these cues could then be assembled by means of several techniques into 3D structures with higher level of organization and functionality, to generate tissue mimetic constructs *in vitro*.

classified into collagens, elastins, structural glycoproteins, and proteoglycans. Collagens and elastins form hierarchical structures such as collagen fibrils, elastin fibers, which in turn interact with proteoglycan and structural glycoproteins to form tissue-specific connective tissues (for a detailed discussion, see reviews [5, 18]). For instance, collagen fibers in tendons and cornea share collagen I as the predominant molecular species. However, collagen fibers in tendons are organized hierarchically into densely packed thick fiber bundles to provide high tensile strength [18]. On the other hand, corneal collagen fibers are organized as the orthogonal lamellae formed by thin parallel small-diameter collagen fibrils with constant spacing. The peculiar organization of ECM components is functional to the specific need of minimal light scattering, imparting transparency to the cornea. At the same time, it provides high tensional strength in two dimensions (2D) [19, 20]. Indeed, the ECM composition and organization also determine tissue rigidity. For instance, bone is stiff because of the extensive, highly cross-linked collagen fibrils and matrix mineralization, whereas the aorta, lung, and skin can withstand repeated stretching because of the presence of elastic fibers. Such mechanical control of tissue development and morphogenesis has been discussed in various reviews [9, 10, 21].

ECM also interacts with various growth factors and signaling molecules to regulate cellular behaviors such as adhesion, proliferation, migration, and differentiation. Differential binding affinities of morphogens to ECM components generate spatial concentration gradients of morphogens. For instance, strong affinity of fibroblast growth factor 10 (FGF10) to heparin sulfate (HS) keeps it near its source (tip of the cell) leading to cell elongation, whereas lower affinity of FGF7 for HS allows its free diffusion creating long-range gradients and promoting branching of salivary glands [22]. Another example of biochemical signaling provided by ECM is that of bone. In addition to collagen, fibronectin, laminin, and mineral crystals, the ECM of natural bone houses a conspicuous number of specific noncollagenous proteins, especially bone sialoprotein (BSP), osteopontin (OPN), osteonectin (ON), and osteocalcin (OC), which are involved in regulating the mineralization of the growth plate and bone [23]. Biochemical signals also result from secretion of diffusible factors secreted by the cells. A reciprocal interaction between epithelial and mesenchymal cells plays an important role in formation of various organs such as tooth, heart, pancreas, and liver [24]. The two compartments interact with each other through growth factors such as bone morphogenic protein (BMP), FGF and through signaling pathways such as Wingless Int (Wnt) and sonic hedgehog (Shh) [24]. Cellular interactions are not restricted only to biochemical signals, but also involve cell-cell contact. Cells can communicate by direct interaction between membrane molecules of the two adjacent cells (adherens and tight junctions). For example, direct cell–cell contact between endothelial cells (ECs) and osteoblasts (OBs) is required to drive vasculogenesis in bone [25–27].

It is important to highlight that heterogeneous distribution of structural, mechanical, biochemical, and cellular components is not only limited to differences between organs and tissues but also relevant within a particular tissue/organ. Mature heart valves consist of highly organized ECM that is present in three layers, namely, fibrosa, spongiosa, and either ventricularis (in case of semilunar valves) or atrialis (atrioventricular valves). The fibrosa is predominantly composed of fibrillar collagens (types I and III) oriented circumferentially and provide tensile stiffness. The spongiosa is composed of proteoglycan with interspersed collagen fibers and provided with tissue compressibility and integrity. The ventricularis/atrialis layers mainly consist of radially oriented filamentous elastin fibers that facilitate tissue motion by extension and recoil of the valve during cardiac cycle [28, 29]. The functional cellular components of valve include valve interstitial cells (VICs) responsible for matrix remodeling and repair and valve endothelial cells (VECs) lining the blood-contacting surfaces of the valve [29]. Despite the common functional requirements of all the heart valves (mitral, atrioventricular, aortic, and pulmonary), each valve is structurally different. Moreover, the valve thickness varies by the valve type and valve region [28].

Given such structural and spatiotemporal complexity of native tissues and organs, regenerative medicine approaches must focus on strategies that will allow for engineering artificial tissues with controlled spatial and temporal heterogeneity observed *in vivo* as a necessary step toward successful clinical translation. Over the years, several techniques have been developed to individually replicate the components of the microenvironment to ultimately develop a synthetic bio-microenvironment. The following sections are devoted to understanding how researchers are individually engineering the structural, mechanical, biochemical, and cellular crosstalk cues within hydrogel systems.

20.3
Hydrogel Engineering for Obtaining Biologically Inspired Structures

20.3.1
Structural Cues

For an effective regulation of cellular microenvironments, one has to consider the structural heterogeneity of tissues [30–32]. As tissues have multiple levels of structural organization, a multiscale approach could be applied to develop hierarchical structures that provide a physically confined environment to the cells. In fact, engineered hydrogels with structural cues (defined here as specific geometries, topographies, and porosities) have been shown to regulate cell spreading [33–35] and function [36–39].

Microengineering geometries at a length scale relevant to cells (10–100 μm) allows for replicating these structural cues *in vitro* and presenting them to the cells locally [15]. For example, multilayer 3D hepatic tissue structures were fabricated by photopatterning single-layer poly(ethylene glycol) (PEG) hydrogels. Hepatocytes were encapsulated at a high density into honeycomb microarchitectured hydrogels. Three-layer constructs were perfused in a continuous flow reactor and encapsulated cells were stained for viability. This geometry improved the viability of hepatocytes by minimizing the barriers of nutrient transport to these highly metabolic cells [40].

Besides substrate geometry, cells can be greatly affected by the substrate topography. Topographical cues can be engineered on hydrogels, providing an *in vitro* system with grooves and posts pertinent to cellular and signaling processes, namely, cell attachment [35, 41], alignment [35, 41], and function [36]. For instance, a mouse fibroblast cell line was used to demonstrate the importance of surface topography on cell adhesion [35]. The fibroblasts remained rounded and formed clusters on the unpatterned star-polyethylene glycol (starPEG) surfaces (Figure 20.2a (2, 3)) and thin actin protrusions with no pronounced stress fiber network could be detected (Figure 20.2a (1)). On the other hand, when the same hydrogel was patterned with micrometer posts, the cells showed flat lamellipodia, contacting and wrapping around the posts, instead of contacting the top surface of the post (Figure 20.2a (4, 5)) [35]. In a recent approach, an *in vitro* model of the myocardium was engineered with anisotropic nanotopographic features using UV-assisted capillary molding technique. Results suggested that the level of expression of the major gap junction protein (Connexin 43) increased with increasing ridge/groove width, demonstrating that the mechanical stretching experienced by cells varies according to the substrate structure and dimension [36]. Recently, we have photopatterned cell-laden rectangular hydrogel constructs using methacrylated gelatin and demonstrated that the cells with the intrinsic potential to form aligned tissues *in vivo* self-organize into aligned tissues *in vitro* when provided the appropriate microarchitecture [42].

Apart from local presentation of structural cues through nano/microscale geometry/topography, hydrogel engineering efforts have also focused on global presentation of such cues to the cells. Recently, 3D hydrogels that provide independent control over porous structure and macroscopic structure hav been developed by combining sphere templating technique and photolithography [46]. Myoblasts are aligned circumferentially along macroporous walls and appeared to form a network of fibrillar structures throughout the void space of the scaffolds [46]. Similarly, porogen leaching was applied to poly(ethylene glycol) diacrylate (PEGDA) hydrogels, creating a macroporous environment that enhanced BMP-2 expression by encapsulated human Mesenchymal stem cells (hMSCs), suggesting their differentiation into osteogenic lineage [38]. Similarly, we have used uncross-linked gelatin microspheres at 4 °C as sacrificial component to generate cell-laden alginate hydrogels with controlled porosity and homogeneous cell distribution [47]. Porous hydrogels showed enhanced proliferation and albumin production by hepatocarcinoma (HepG2) cells.

Together, these examples demonstrate the importance of engineering structural cues in the hydrogels to decrease the gap between natural and synthetic cellular microenvironments.

20.3.2
Mechanical Cues

To serve their intended purposes, biological tissues often rely on specific mechanical characteristics of a specialized ECM. In the human body, indeed, ECM compositions and microscopic organizations are tuned in order to optimally match specialized structural functions (e.g., in bone, cartilage, or heart valves) [48]. Besides providing macroscopic load bearing capacity, matrix stiffness is known to affect cell–microenvironment interactions [49–51]. It is therefore clear that a rational design of the appropriate mechanical milieu (e.g., elasticity, compressibility,

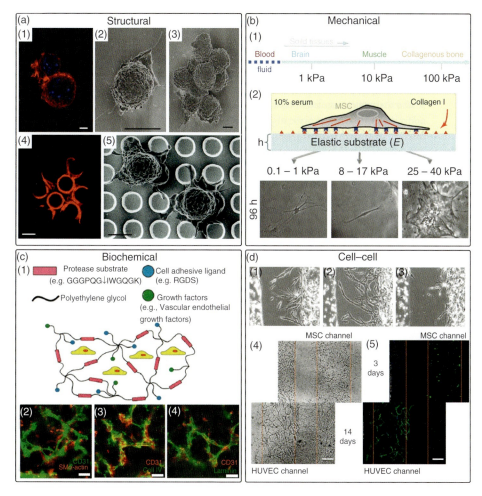

Figure 20.2 Engineering hydrogels with structural, mechanical, biochemical, and cell–cell contact cues. (a) Structural cues, such as surface topography, have been shown to affect cell adhesion. Cells poorly adhered on unpatterned starPEG and maintain a round shape (1,2) with a tendency of cluster formation, (3) whereas cells showed flattened adherent body with protrusions around the posts when cultured on the same polymer patterned with posts after 4 h (4,5). Scale bar, 5 μm [35]. (b) Mechanical cues engineered in hydrogels direct MSC differentiation toward neural, myoblastic, or osteogenic differentiation. Scale bar, 20 μm [43]. (c) Biochemical cues such as RGD adhesion peptide or VEGF have been engineered within MMP-sensitive PEG (1) used as a substrate for coculture of 10T1/2 and HUVECs, demonstrating cellular organization and tubule-like formation after six days (2–4). Scale bar, 50 μm [44]. (d) Cell–cell contact is essential in many biological phenomena. For instance, human ECs-MSCs when cocultured together by means of micropatterned fibrin channels demonstrated the ability of HUVEC to drive MSC sprouting (1,2 and 3; scale, 100 μm), with upregulated α-SMA (green) expression only after 14 days when both the cell types contact each other (4 and 5; scale bar, 250 μm) [45].

viscoelastic behavior, tensile strength, and failure strain) is critical to biomimetic tissue engineering [4, 52–54].

In fact, hydrogel biomechanics has been shown to regulate cell apoptosis [55], migration [56], proliferation [57], and lineage specification [58]. Indeed, cells sense the specific mechanical properties of the tissues *in vivo* (Figure 20.2b (1)), and the design of hydrogels with controlled elastic moduli have been shown to direct MSCs differentiation toward neural, myoblastic, or osteogenic differentiation (Figure 20.2b (2)) [43].

Despite their similarity to the highly hydrated molecular-based composition, hydrogels are often too soft as compared to the most natural ECMs. To overcome this limitation, the primary approach has been to screen various biocompatible polymers with an engineering focus on controlling the monomer/polymer molecular rigidity [59], the types of cross-linking molecules [60], the polymer concentration, and the cross-linking density [61, 62]. However, single polymer approaches often fall short of satisfying the complex set of tissue-specific biological and physical requirements. A promising alternative is the multipolymer approach, adopted, for instance, to match the mechanical and surface charge properties for neural repair [63]. Similarly, synthetic and natural polymers have been used to design interpenetrating/semi-interpenetrating networks, which have shown improved mechanical properties while maintaining proper cell–substrate interactions [64, 65].

Polymer fibers have been added to hydrogels also in the form of electrospun poly(L-lactide) (PLLA) fibers, obtaining inhomogeneous structures where the interactions of fibrous and gel compartments allowed recapitulation of mechanical and viscoelastic properties of bone and ligament more closely [66, 67]. Similarly, addition of inorganic components such as nanocrystalline hydroxyapatite to a blend of Poly(2-hydroxyethyl methacrylate) (pHEMA) and self-assembled DNA hydrogels was used to further improve mechanical and cytocompatibility properties for bone tissue engineering [68].

Dynamic external control of local mechanical and chemical properties has been demonstrated using photodegradable hydrogels that were then used to modulate the hMSCs differentiation to chondrocytes [69–71]. Finally, secondary photocross-linking has been proposed to create mechanical patterns in hydrogels, widening the possible complexity of the mechanical properties design in space [39, 72].

20.3.3
Biochemical Cues

In vivo, biochemical cues can be provided by the ECM or cell-secreted signaling molecules in a controlled manner either as soluble factors or ECM-sequestered molecules. A strategy to coordinate a number of signaling pathways that lead to cell migration [73], proliferation [74], and differentiation [75] is the inclusion of bioactive cues on the hydrogel matrices. Various chemicals have been integrated within the hydrogel matrices with different affinities: (i) solubilized within the polymer solution [76], conferring a controlled release of these molecules to the surrounding microenvironment, or (ii) immobilized on the polymer chain [77].

Single biomolecules have been incorporated in order to enhance the tissue function [78–80]. Cartilage regeneration was stimulated *in vitro* in 3D hydrogel systems by suspending transforming growth factor-β1 (TGF-β1) in photocross-linkable gelatin. The glycosaminoglycan secreted by the encapsulated chondrocytes increased with time and was higher in the hydrogels engineered with TGF-β1 [78]. The physical entrapment of soluble factors in hydrogel matrices allowed the development of dynamic systems for cell adhesion and controlled release of drugs. Matrix metalloproteinase (MMP)-responsive PEG hydrogels with entrapped thymosin β4 (Tβ4) supported human umbilical vein endothelial cell (HUVEC) adhesion, survival, migration, and formation of a vascular network [79].

Hydrogels that exhibit dynamic properties in response to the cell microenvironments can be engineered by incorporating proteases as degradable cross-linkers in the backbone of the polymer, combined with RGD motifs for improved cell adhesion and migration [44, 73]. For example, through Michael-type addition chemistry, MMP cross-linkers were introduced in hyaluronic acid (HA) hydrogels. The degradation of the hydrogel by the MMPs was required for MSCs spreading and migration [73]. In another example, proteolytically degradable and cell-adhesive PEG hydrogels were used as angiogenic matrices both, *in vitro* and *in vivo* (Figure 20.2c). RGDS motifs and vascular endothelial growth factor (VEGF$_{165}$) were conjugated to acrylated PEG and combined with MMP-sensitive PEG precursor at different ratios. Coculture of MSCs and ECs within these bioactive hydrogels promoted infiltration of host blood vessel when transplanted into mouse corneas [44].

Multiple soluble growth factors have been included in hydrogel networks. Microfluidic networks have been used to create cross-gradients of soluble factors in hydrogels. Gradients of molecules allow spatial distribution of the solutes, mimicking the molecule gradients observed in natural processes such as angiogenesis [74] or simulating tissue interfaces [81–83]. The biological relevance of these techniques lies in the fact that many of the signaling pathways, for instance, for stem cell differentiation, require the action of growth factors at distinct stages. Angiogenesis was simulated by covalently binding VEGF$_{165}$ to agarose-sulfide hydrogels. In this study, it was demonstrated that ECs can be guided to form a tubule-like structure by a gradient of VEGF in 3D engineered hydrogels [74]. Similarly, a gradient of the osteogenic transcription factor Runx2/Cbfa1 was achieved by controlled deposition of poly(L-lysine) (PLL) densities. A tissue-interface-like structure was achieved by spatially regulating the genetic modification of fibroblasts into OBs [75].

As many cellular processes involve a complex network of multiple signaling pathways at various time points, schemes for sequential delivery of multiple growth factors are crucial. Therefore, it is envisioned that combined soluble and immobilized growth factors can be beneficial for enhanced biochemical signaling in tissue regeneration.

20.3.4
Cell–Cell Contact

In the body, cells are in contact, or in close proximity, with neighboring cells and ECM in a highly organized manner [9, 84–86]. Within this organization, interaction between cells governs important biological processes such as cell proliferation and differentiation [87, 88]. Some cell–cell interactions are temporary, such as the contact of cells in an inflammatory response [89], whereas others require stable connections and are crucial for cell function and therefore tissue regeneration [90, 91]. The selective adhesion between specific cells is mediated by transmembrane proteins such as integrins, cadherins, or connexins [87]. Therefore, tissue engineering constructs that aim to mimic the molecular ECM of native tissues will benefit from methods of controlling cell–cell interactions (for a detailed discussion, see review [92]).

Engineered systems have been mostly focusing on the stable connections, essential for tissue regeneration. Some of the early work in this area has been done by developing micropatterned substrates with adhesive islands to control the degree of cell–cell contacts and cell shape [93, 94]. In one example, photolithography was used to coculture hepatocytes and NIH-3T3 fibroblasts on micropatterned substrates, revealing many critical interactions for maintaining hepatocyte phenotype in culture [94]. In another example, patterned coculture was developed using the cell-resistant properties of HA. A multistep adsorption of fibronectin and deposition of PLL onto HA yielded hydrogel substrates with distinctly localized murine embryonic stem cells (mESCs) and NIH-3T3 cells [95].

A promising alternative to these 2D approaches is the engineering of cell cocultures within hydrogels by mixing the cells with the polymer solution. These 3D systems help mediating cell–cell interactions, increasing the potential use of cocultures *in vivo*. For example, HUVECs and hMSCs were cocultured by injecting each cell type, previously suspended in fibrin hydrogels in distinct microchannels. Two days after coculture, the cells from different MSCs sources started sprouting toward the patterned HUVECs channel (Figure 20.2d (1–3)). After 14 days, green staining of smooth muscle actin (α-SMA), an early marker of smooth muscle cell (SMC) differentiation, illustrated the alignment of MSCs with endothelial tubular structures. α-SMA$^+$ cells were not present in monocultures of MSC or HUVECs, demonstrating the importance of cell–cell contact (Figure 20.2d (4, 5)) [45]. When HUVECs and human umbilical vein smooth muscle cells (hUVSMCs) were printed in close contact using a biological laser printing system, they seemed to establish cell–cell junctions around lumen-like structures, demonstrating a symbiotic relationship between cells, essential for their development [96]. Regulation of engineered coculture systems has been also achieved *in vivo* [97, 98]. Bone-marrow-derived MSCs were cocultured with OBs, chondrocytes, or ECs within a hydrogel system. The coculture of stem cells with primary cells resulted in a superior differentiation of MSCs when transplanted [97].

Overall, the studies reported demonstrate that patterned cocultures can be applied to control the degree of homotypic or heterotypic interactions between cells, ultimately improving tissue function.

20.3.5
Combination of Multiple Cues

In living tissues, structural, mechanical, biochemical, and cellular cues interact dynamically and contribute synergistically to tissue formation. Cell–cell interactions and mechanical cues can generate biochemical signals, which further guide cell behaviors. During organogenesis, individual cells sense changes in physical forces and transduce them into intracellular signals that drive the alterations in cell shape, polarity, growth, migration, and differentiation through distinct molecular pathways such as Rho-associated kinases (reviewed in [99]). On the other hand, mechanical signals can instruct the cells to secrete proteolytic enzymes to modulate matrix degradation and remodeling during morphogenesis, inflammation, and wound healing. Thus, tissue engineering approaches that combine multiple cues in a single hydrogel may aid in developing *in vitro* tissues that would better model human diseases or be useful as tissue constructs for clinical applications. Advances in micro-/nanoscale technologies and materials science have enabled the integration of multiple cues in a single material. For instance, hydrogels have been engineered with combined chemical and mechanical cues [69, 100–103], as well as combination of structural and biochemical cues [100, 104–106]. In one example, Hahn *et al.* [100] used single-photon absorption photolithography to spatially control the mechanical and transport properties of PEGDA hydrogels at the microscale. The authors further generated internal, spatially immobilized patterns of varying feature sizes, geometries, and RGDS concentrations using two-photon absorption photolithographic technique. It was further shown that cells were confined to, invaded, and migrated into only RGDS-containing regions that were patterned inside collagenase-degradable hydrogels [100]. Recently, Bott *et al.* engineered PEG-based MMP-sensitive hydrogels with varying degradability, integrin binding sites, and stiffness (Figure 20.3a (1)) [101]. They found that despite the presence of integrin binding sites and the proteolytic degradability at higher stiffness, the matrix acts as a physical barrier for cells in 3D gels, impeding their proliferation and migration (Figure 20.3a (2)).

Gradient hydrogels are another important area of research that is relevant to the field of tissue engineering as physicochemical/mechanical cues are presented to the cells in the gradient manner in living tissues; this is more prominent especially at the tissue interfaces (for a detailed review, see [14]). Recently, polyacrylamide hydrogel was designed with the chemical (collagen type I) and mechanical (stiffness) gradient at an interfacial region in opposing directions [102]. One side of the interface was stiff (high Young's modulus) with a low collagen concentration, whereas the other side of the interface was compliant (low Young's modulus) with a high collagen concentration. When fibroblasts were seeded on the gels, they preferentially migrated toward the high-collagen-compliant

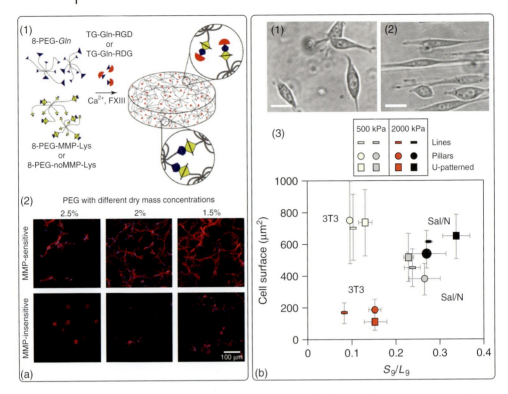

Figure 20.3 Current engineering efforts aim to combine specific structural, mechanical, biochemical, and cell–cell contact cues within single hydrogels. (a) MMP-sensitive hydrogels presenting bioactive molecules (the cell-integrin binding peptides, TG-Gln-RGD) and controlled mechanical properties (the percentage of dry mass) have been used to evaluate the effect of biochemical/mechanical properties on fibroblast proliferation and morphology (1 and 2). Scale bar, 100 μm [101]. (b) Mechanical cues were combined with structural cues (surface topography) demonstrating that cells seeded on unpatterned (1) or patterned (2) surfaces polarize depending on the cell source as well as specific topography and rigidity of the substrate hydrogels (3) [106].

side of the interfacial region showing dominant effect of chemical gradients [102].

Various cell behaviors including cell adhesion and migration are influenced by ECM structure and rigidity. When studied in combination, topography (unpatterned, pillars, lines) and mechanical properties (500–2000 kPa) of microfabricated polydimethylsiloxane (PDMS) substrates showed synergistic influence on normal and cancer cell morphology (Figure 20.3b (1)) and surface area (Figure 20.3b (2)) [106]. Such studies on hydrogels engineered with multiple cues will help decouple and understand effects of various cues in combination on cell behaviors.

20.4
Assembly of Engineered Hydrogel Blocks

Although the ability to engineer multiple cues in single hydrogel blocks has improved our understanding about cell–materials interactions *in vitro*, such approaches often fall short of recreating the intricate microstructural features. Approaches such as modular tissue engineering aim to generate biologically relevant complex structures by bringing together small hydrogel blocks engineered with specific set of properties. This field holds great biological relevance, as many tissues are composed of repeating units, such as osteons (bones) and lobules (liver), as described in Section 20.2. Thus, this bottom-up approach has a great potential to generate structures that can mimic functional units of tissues/organs. The basis of modular tissue engineering is to design microscale features (structural, mechanical, biochemical, and cell–cell contact) and then assemble these modules to create larger clinically relevant tissues. The instructive building blocks can be engineered in a number of ways, described in Section 20.3. Once engineered, these building blocks can be assembled into more complex 3D structures by a number of methods. Here, we highlight some of the most promising techniques for assembling engineered modules into structures with higher level of organization and functionality.

The simplest method for assembling engineered building blocks employs a reduction of space and the removal of the liquid where the subunits are suspended. According to a simple *packing-based mechanism*, MSC-laden alginate microspheres were compacted and glued together with collagen [107]. Combining this methodology with layer-by-layer technology, these packed osteogenic microspheres were glued to chondrogenic microspheres, creating a layered hydrogel with distinct biological properties [107]. However, the simplicity of this method can compromise either the mechanical integrity of the structure or the viability of encapsulated cells, depending greatly on the type and amount of gluing material. Another disadvantage of this technique is the lack of control over the assembly dynamics. Therefore, *microfluidic-based assembly* can be combined with these techniques to physically confine the assembly of the building blocks, dictating the shape of the larger tissue [108, 109]. In one example, a layer-by-layer microfluidic system was employed to sequentially flow different cell types within distinct polymer solutions in microchannels forming a multilayer tissue-like structure [108]. One benefit of this combined approach is the ability to assemble different types of cells within different hydrogel matrices in a precise and controlled manner. A different approach to microfluidic-based assembly consists of loosely assembling the microblocks in microfluidic channels. Collagen cylinders were assembled and perfused through a channel, creating interconnected structures that resemble capillary networks [109]. While promising for the development of complex vascularized structures, they might lack mechanical stability if the final structures are not stabilized by additional cross-linking mechanisms. Although this physical template-based assembly allows control over the final geometry based on the channel, it is difficult to determine the order in which the blocks are assembled. A promising alternative consists of guiding and assembling the microblocks inside the microfluidic

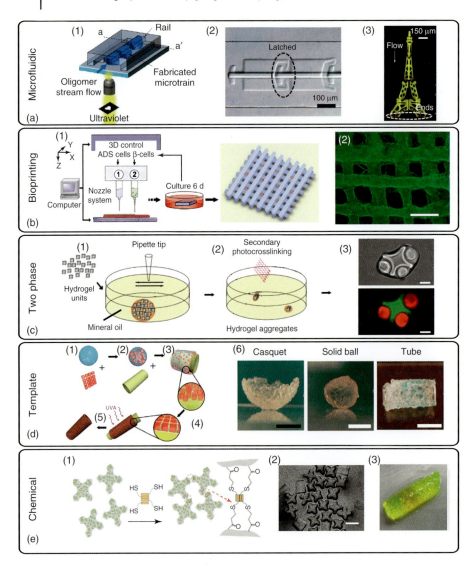

channels (Figure 20.4a) [110]. "Railed microfluidics" approach was used where grooves (rails) fabricated on top of the PDMS channels served as guidance for the complementary microgels (microtrains). This microfluidic-based assembly enables the development of complex structures composed of more than 50 micrometer-size units. The combination of this system with cross-solution movement allows increasing the complexity of the system and directs the assembly of blocks with different cell types using microfluidics [110].

With a similar concept as the layer-by-layer technology, *tissue printing* (Figure 20.4b) offers a high potential for controlling cell and ECM positioning and thus to tightly control the assembly of the instructive microcues [111, 115]. With

Figure 20.4 Overview of the techniques used to assemble engineered microgels into 3D structures with higher level of organization and functionality. (a) Railed *microfluidic* channels (1) were used to generate hydrogel blocks and control their assembly that was stabilized either by secondary photocross-linking or by geometrical constrains such as latches (2). Complex structures composed of more than 50 microstructures were created (3) [110]. (b) *Bioprinting*-based assembly; 3D control system directed by a computer was used (1) to print adipose-derived stromal (ADS) cells and pancreatic islets in a multicellular construct (2) to capture pathological features of the disease, serving also as a platform for drug testing [111]. (c) Microgel units synthesized by photolithography (1) were assembled by *two-phase* approaches (scale bar, 1mm), in which hydrophobic/hydrophilic forces drive controlled assembly before secondary photocross-linking step (2). Rheological characteristics of the solution, process parameters, with geometrical constrains can control the degree of specificity of the assembly (3; scale bar, 200 μm) [112]. (d) *Template*-based assembly, Microgels of desired shapes in prepolymer (1, 2) were placed in contact with a high affinity PDMS template and tightly packed (3, 4) before being further photocross-linked and the template removed (5). This technique was successfully adopted to create casquet, solid sphere, and hollow tubes (6; scale bars, 12, 3, and 5 mm, respectively) [113]. (e) *Chemical* addition reaction between the acrylate groups on the surface of the shape-specific (star, circle, and square) microgels and the thiol groups on the cross-linker (1) was used to assemble the final constructs with controlled porosity, permeability, and pore interconnectivity (2,3; scale bar, 500 μm) [114].

---

this technique, a specific pattern of cells can be designed and printed, as a regular printer injects ink. After stabilization, subsequent layers can be printed on top of each other, developing complex but highly controlled 3D structures. Also, the use of different nozzle systems helps to simultaneously print different solutions with different biological cues, increasing the complexity of the system [111]. However, one disadvantage of this technique lies on the impact on cell viability, since the printing process can be long.

Recent research from our laboratory has combined microgels fabricated with photocross-linkable materials with directed assembly techniques to create systems with increasing complexity. *Two-phase systems* (Figure 20.4c) have been applied for the assembly of microgels. When the microgels were transferred into a two-phase oil-aqueous solution, the surface tension created on the hydrophilic microgels caused their aggregation in order to reduce the total surface area in contact with the hydrophobic solution. Stabilization of the assembled structure was achieved by secondary cross-linking of the construct. By varying the size and aspect ratio of the building blocks, the ultimate dimension and shape of the engineered tissue can be changed, demonstrating the high potential of this technique [112]. We have also assembled the microgels at the interface between air and hydrophobic solution (perfluorodecalin). The hydrophilic microgels were randomly placed on the surface of the high-density hydrophobic solution. The surface tension between the air and the solution led to the aggregation of the microgels at the interface [116]. However, these approaches may lead to the development of uncontrolled 3D structures. Therefore, the two-phase-based approach was taken one step further and combined with *template-based methodologies* (Figure 20.4d) to improve the control over the developed structures [113, 117]. 2D templates patterned with hydrophilic and hydrophobic regions were used to direct the assembly of microgels. Once placed in contact with the template, hydrophilic gels assembled within the hydrophilic patterns and stabilized after secondary cross-linking [117]. Confined to the assembly on flat surfaces, this approach is limited to 2D applications. To move this concept to 3D, an oxidized PDMS surface was used as the hydrophilic template. Hydrogel blocks suspended in polymer solution were deposited on the oxidized PDMS surface. The affinity between the hydrophilic structures attracted the microgels to the PDMS, and on removal of the excess solution, the blocks assembled on the template as a result of the capillary forces of the remaining polymer [113]. Secondary cross-linking was used to stabilize the assembled structure and allowed the removal of the template. Various shapes can be developed with this technique, by molding the PDMS template into the desired complementary shapes [113].

The assembly of hydrogel building blocks can also be *chemically directed*. One approach consists of including within the hydrogel blocks specific components that will interact with the ones in the neighboring blocks. Specifically, the nucleation and growth of collagen fibers present at the interface between different modules led to the assembly of these collagen fibers, thus anchoring together different hydrogel matrices [118]. In another approach, Michael-type addition reaction between the acrylate groups on the surface of the microgels and the thiol groups on the cross-linker was used to assemble engineered hydrogel blocks (Figure 20.4e) [114]. Physical properties (porosity, permeability, and pore interconnectivity) of the assembled constructs were modulated by controlling the initial shape of individual microgels, star-shaped microgels exhibiting favorable properties than circle and square microgels. However, this chemically directed assembly is dependent on the engineered chemistry of the bulk and/or surface of the building blocks, potentially decreasing their application.

A relative recent approach to direct the assembly of individually engineered hydrogel units is through *magnetic forces*. Magnetic microbeads were suspended in the polymer solution before microgel production [119]. The magnetically labeled microgels were manipulated individually with a magnet, directing their assembly. Although simple and efficient to operate, the major drawback of this technique is the deleterious effect of magnetic field on the cell growth and function [120]. However, Bong *et al.* used stop flow lithography to control spatial distribution of the magnetic beads inside the microgels. The "magnetic barcoded particles" have a defined 90 μm region with the magnetic particles, through which the assembly was performed. This spatial distribution allows encapsulation of cells in the region free of magnetic particles, thus improving the biological success of the construct [121].

20.5
Conclusions

Micro-/nanofabrication techniques coupled with advances in materials chemistry have allowed for synthesis of hydrogels with spatially controlled local microenvironments. Emergence of dynamic hydrogels that allow real-time manipulation of their mechanical properties (e.g., photocleavable hydrogels) offers exciting possibilities for engineering temporal component into such hydrogels. However, it is still unclear what minimum level of engineering is required in order to achieve fully functional cell-seeded hydrogel constructs. Although various techniques have been developed for assembling hydrogel blocks, there is an unmet need for the development of quantitative analytical methods to demonstrate the functionality of engineered tissue analogs beyond cell viability to shift the current focus from engineering to biology. Top-down approaches have been successful in combining multiple components (structural, biochemical, mechanical, and cellular cues) of tissue microenvironments into single hydrogels; however, bottom-up approaches are mostly focused on fine-tuning techniques for assembling hydrogel blocks engineered with single components. Future efforts that integrate top-down and bottom-up approaches can bring us closer to the ultimate goal of engineering more complex clinically relevant tissues for therapeutic applications.

Acknowledgments

AK acknowledges funding from the National Institutes of Health (**EB009196; DE019024; EB007249; HL099073; AR057837**), the National Science Foundation CAREER Award (**DMR0847287**), and the Office of Naval Research Young Investigator Award. SS is grateful for the System-based Consortium for Organ Design and Engineering (SysCODE) postdoctoral training fellowship. DFC acknowledges the Foundation for Science and Technology (FCT), Portugal, and the MIT-Portugal Program for the personal grant **SFRH/BD/37156/2007**. NS acknowledges the Fondazione Fratelli Agostino and Enrico Rocca for sustaining his research through the Progetto Rocca Postdoctoral Fellowship.

References

1. Khademhosseini, A., Vacanti, J.P., and Langer, R. (2009) *Sci. Am.*, **300**(5), 64–71.
2. Perez-Castillejos, R. (2010) *Mater. Today*, **13**(1–2), 32–41.
3. Freytes, D.O., Wan, L.Q., and Vunjak-Novakovic, G. (2009) *J. Cell. Biochem.*, **108**(5), 1047–1058.
4. Lutolf, M.P. and Hubbell, J.A. (2005) *Nat. Biotechnol.*, **23**(1), 47–55.
5. Bruckner, P. (2010) *Cell Tissue Res.*, **339**(1), 7–18.
6. Wang, F. (2009) *Cold Spring Harb. Perspect. Biol.*, **1**(4), a002980.
7. Swartz, M.A. (2003) *Curr. Opin. Biotechnol.*, **14**(5), 547–550.
8. Vogel, V. and Sheetz, M. (2006) *Nat. Rev. Mol. Cell Biol.*, **7**(4), 265–275.
9. Ingber, D.E. (2005) *Proc. Natl. Acad. Sci. U.S.A.*, **102**(33), 11571–11572.
10. Mammoto, T. and Ingber, D.E. (2010) *Development*, **137**(9), 1407–1420.
11. Gjorevski, N. and Nelson, C.M. (2009) *Cytokine Growth Factor Rev.*, **20**(5–6), 459–465.
12. Lutolf, M.P. (2009) *Integr. Biol.*, **1**(3), 235–241.
13. Slaughter, B.V., Khurshid, S.S., Fisher, O.Z., Khademhosseini, A., and Peppas, N.A. (2009) *Adv. Mater.*, **21**(32–33), 3307–3329.

14. Sant, S., Hancock, M.J., Donnelly, J.P., Iyer, D., and Khademhosseini, A. (2010) *Can. J. Chem. Eng.*, **88**(6), 899–911.
15. Khademhosseini, A., Langer, R., Borenstein, J., and Vacanti, J.P. (2006) *Proc. Natl. Acad. Sci. U.S.A.*, **103**(8), 2480–2487.
16. Jia, X.Q. and Kiick, K.L. (2009) *Macromol. Biosci.*, **9**(2), 140–156.
17. Nichol, J.W. and Khademhosseini, A. (2009) *Soft Matter*, **5**(7), 1312–1319.
18. Tsang, K.Y., Cheung, M.C.H., Chan, D., and Cheah, K.S.E. (2010) *Cell Tissue Res.*, **339**(1), 93–110.
19. Hassell, J.R. and Birk, D.E. (2010) *Exp. Eye Res.*, **91**(3), 326–335.
20. Quantock, A.J. and Young, R.D. (2008) *Dev. Dyn.*, **237**(10), 2607–2621.
21. Patwari, P. and Lee, R.T. (2008) *Circ. Res.*, **103**(3), 234–243.
22. Makarenkova, H.P., Hoffman, M.P., Beenken, A., Eliseenkova, A.V., Meech, R., Tsau, C., Patel, V.N., Lang, R.A., and Mohammadi, M. (2009) *Sci. Signal.*, **2**(88), 10.
23. Murshed, M. and McKee, M.D. (2010) *Curr. Opin. Nephrol. Hypertens.*, **19**(4), 359–365.
24. Thiery, J.P., Acloque, H., Huang, R.Y.J., and Nieto, M.A. (2009) *Cell*, **139**(5), 871–890.
25. Fuchs, S., Hofmann, A., and Kirkpatrick, C.J. (2007) *Tissue Eng.*, **13**, 2577–2588.
26. Grellier, M., Bordenave, L., and Amedee, J. (2009) *Trends Biotechnol.*, **27**(10), 562–571.
27. Grellier, M., Ferreira-Tojais, N., Bourget, C., Bareille, R., Guillemot, F., and Amedee, J. (2009) *J. Cell. Biochem.*, **106**(3), 390–398.
28. Hinton, R.B. and Yutzey, K.E. (2011) *Ann. Rev. Physiol.*, **73**(1), 29–36.
29. Schoen, F.J. (2008) *Circulation*, **118**(18), 1864–1880.
30. Decking, U.K.M. (2002) *News Physiol. Sci.*, **17**(6), 246–250.
31. Sansalone, V., Naili, S., Bousson, V., Bergot, C., Peyrin, F., Zarka, J., Laredo, J.D., and Haïat, G. (2010) *J. Biomech.*, **43**(10), 1857–1863.
32. Federico, S., Grillo, A., La Rosa, G., Giaquinta, G., and Herzog, W. (2005) *J. Biomech.*, **38**(10), 2008–2018.
33. Mazzoccoli, J.P., Feke, D.L., Baskaran, H., and Pintauro, P.N. (2010) *Biotechnol. Prog.*, **26**(2), 600–605.
34. Doraiswamy, A. and Narayan, R.J. (2010) *Philos. Trans. R. Soc. A Math. Phys. Eng. Sci.*, **368**(1917), 1891–1912.
35. Schulte, V.A., Diez, M., Moller, M., and Lensen, M.C. (2009) *Biomacromolecules*, **10**(10), 2795–2801.
36. Kim, D.H., Lipke, E.A., Kim, P., Cheong, R., Thompson, S., Delannoy, M., Suh, K.Y., Tung, L., and Levchenko, A. (2010) *Proc. Natl. Acad. Sci. U.S.A.*, **107**(2), 565–570.
37. Guvendiren, M. and Burdick, J.A. (2010) *Biomaterials*, **31**(25), 6511–6518.
38. Betz, M.W., Yeatts, A.B., Richbourg, W.J., Caccamese, J.F., Coletti, D.P., Falco, E.E., and Fisher, J.P. (2010) *Biomacromolecules*, **11**(5), 1160–1168.
39. Khetan, S. and Burdick, J.A. (2010) *Biomaterials*, **31**(32), 8228–8234.
40. Tsang, V.L., Chen, A.A., Cho, L.M., Jadin, K.D., Sah, R.L., DeLong, S., West, J.L., and Bhatia, S.N. (2007) *FASEB J.*, **21**(3), 790–801.
41. Bian, W.N., Liau, B., Badie, N., and Bursac, N. (2009) *Nat. Protoc.*, **4**(10), 1522–1534.
42. Aubin, H., Nichol, J.W., Hutson, C.B., Bae, H., Sieminski, A.L., Cropek, D.M., Akhyari, P., and Khademhosseini, A. (2010) *Biomaterials*, **31**(27), 6941–6951.
43. Engler, A.J., Sen, S., Sweeney, H.L., and Discher, D.E. (2006) *Cell*, **126**(4), 677–689.
44. Moon, J.J., Saik, J.E., Poche, R.A., Leslie-Barbick, J.E., Lee, S.H., Smith, A.A., Dickinson, M.E., and West, J.L. (2010) *Biomaterials*, **31**(14), 3840–3847.
45. Trkov, S., Eng, G., Di Liddo, R., Parnigotto, P.P., and Vunjak-Novakovic, G. (2010) *J. Tissue Eng. Regen. Med.*, **4**(3), 205–215.
46. Bryant, S.J., Cuy, J.L., Hauch, K.D., and Ratner, B.D. (2007) *Biomaterials*, **28**(19), 2978–2986.
47. Hwang, C.M., Sant, S., Masaeli, M., Kachouie, N.N., Zamanian, B., Lee, S.H., and Khademhosseini, A. (2010) *Biofabrication*, **2**(3), 1–12.

48. Fung, Y. (1993) *Biomechanics: Mechanical Properties of Living Tissues*, 2nd edn, Springer.
49. Griffith, L.G. and Swartz, M.A. (2006) *Nat. Rev. Mol. Cell Biol.*, **7**(3), 211–224.
50. Butler, D.L., Goldstein, S.A., and Guilak, F. (2000) *J. Biomech. Eng.*, **122**(6), 570–575.
51. Guilak, F., Cohen, D.M., Estes, B.T., Gimble, J.M., Liedtke, W., and Chen, C.S. (2009) *Cell. Stem Cell*, **5**(1), 17–26.
52. Drury, J.L. and Mooney, D.J. (2003) *Biomaterials*, **24**(24), 4337–4351.
53. Lee, K.Y. and Mooney, D.J. (2001) *Chem. Rev.*, **101**(7), 1869–1879.
54. Brandl, F., Sommer, F., and Goepferich, A. (2007) *Biomaterials*, **28**(2), 134–146.
55. Wang, H.B., Dembo, M., and Wang, Y.L. (2000) *Am. J. Physiol. Cell Physiol.*, **279**(5), C1345–C1350.
56. Peyton, S.R. and Putnam, A.J. (2005) *J. Cell. Physiol.*, **204**(1), 198–209.
57. Hadjipanayi, E., Mudera, V., and Brown, R.A. (2009) *J. Tissue Eng. Regen. Med.*, **3**(2), 77–84.
58. Pek, Y.S., Wan, A.C., and Ying, J.Y. (2010) *Biomaterials*, **31**(3), 385–391.
59. Carr, L., Cheng, G., Xue, H., and Jiang, S. (2010) *Langmuir*, **26**(18), 14793–14798.
60. Lee, K.Y., Rowley, J.A., Eiselt, P., Moy, E.M., Bouhadir, K.H., and Mooney, D.J. (2000) *Macromolecules*, **33**(11), 4291–4294.
61. Coutinho, D.F., Sant, S.V., Shin, H., Oliveira, J.T., Gomes, M.E., Neves, N.M., Khademhosseini, A., and Reis, R.L. (2010) *Biomaterials*, **31**(29), 7494–7502.
62. Bakota, E.L., Aulisa, L., Galler, K.M., and Hartgerink, J.D. (2011) *Biomacromolecules*, **12**(1), 82–87.
63. Zuidema, J.M., Pap, M.M., Jaroch, D.B., Morrison, F.A., and Gilbert, R.J. (2011) *Acta Biomater.*, **7**(4), 1634–1643.
64. Liu, Y. and Chan-Park, M.B. (2009) *Biomaterials*, **30**(2), 196–207.
65. Brigham, M.D., Bick, A., Lo, E., Bendali, A., Burdick, J.A., and Khademhosseini, A. (2009) *Tissue Eng. A*, **15**(7), 1645–1653.
66. Freeman, J.W., Woods, M.D., Cromer, D.A., Ekwueme, E.C., Andric, T., Atiemo, E.A., Bijoux, C.H., and Laurencin, C.T. (2011) *J. Biomech.*, **44**(4), 694–699.
67. Xu, W., Ma, J., and Jabbari, E. (2010) *Acta Biomater.*, **6**(6), 1992–2002.
68. Zhang, L., Rodriguez, J., Raez, J., Myles, A.J., Fenniri, H., and Webster, T.J. (2009) *Nanotechnology*, **20**(17), 175101.
69. Kloxin, A.M., Kasko, A.M., Salinas, C.N., and Anseth, K.S. (2009) *Science*, **324**(5923), 59–63.
70. Kloxin, A.M., Kloxin, C.J., Bowman, C.N., and Anseth, K.S. (2010) *Adv. Mater.*, **22**(31), 3484–3494.
71. Tibbitt, M.W., Kloxin, A.M., Dyamenahalli, K.U., and Anseth, K.S. (2010) *Soft Matter*, **6**(20), 5100–5108.
72. Marklein, R.A. and Burdick, J.A. (2010) *Soft Matter*, **6**(1), 136–143.
73. Lei, Y.G., Gojgini, S., Lam, J., and Segura, T. (2011) *Biomaterials*, **32**(1), 39–47.
74. Aizawa, Y., Wylie, R., and Shoichet, M. (2010) *Adv. Mater.*, **22**(43), 4831–4835.
75. Phillips, J.E., Burns, K.L., Le Doux, J.M., Guldberg, R.E., and Garcia, A.J. (2008) *Proc. Natl. Acad. Sci. U.S.A.*, **105**(34), 12170–12175.
76. Brynda, E., Houska, M., Kysilka, J., Praadny, M., Lesny, P., Jendelovaa, P., Michaalek, J., and Sykovaa, E. (2009) *J. Mater. Sci. Mater. Med.*, **20**(4), 909–915.
77. Porter, A.M., Klinge, C.M., and Gobin, A.S. (2010) *Biomacromolecules*, **12**(1), 242–246.
78. Hu, X.H., Ma, L., Wang, C.C., and Gao, C.Y. (2009) *Macromol. Biosci.*, **9**(12), 1194–1201.
79. Kraehenbuehl, T.P., Ferreira, L.S., Zammaretti, P., Hubbell, J.A., and Langer, R. (2009) *Biomaterials*, **30**(26), 4318–4324.
80. Rahman, N., Purpura, K.A., Wylie, R.G., Zandstra, P.W., and Shoichet, M.S. (2010) *Biomaterials*, **31**(32), 8262–8270.
81. Choi, N.W., Cabodi, M., Held, B., Gleghorn, J.P., Bonassar, L.J., and Stroock, A.D. (2007) *Nat. Mater.*, **6**(11), 908–915.
82. Wang, X.Q., Wenk, E., Zhang, X.H., Meinel, L., Vunjak-Novakovic, G.,

and Kaplan, D.L. (2009) *J. Controlled Release*, **134**(2), 81–90.

83. Du, Y., Hancock, M.J., He, J., Villa-Uribe, J.L., Wang, B., Cropek, D.M., and Khademhosseini, A. (2010) *Biomaterials*, **31**(9), 2686–2694.

84. Bissell, M.J., Rizki, A., and Mian, I.S. (2003) *Curr. Opin. Cell Biol.*, **15**(6), 753–762.

85. Engler, A.J., Humbert, P.O., Wehrle-Haller, B., and Weaver, V.M. (2009) *Science*, **324**(5924), 208–212.

86. Rivron, N.C., Rouwkema, J., Truckenmuller, R., Karperien, M., De Boer, J., and Van Blitterswijk, C.A. (2009) *Biomaterials*, **30**(28), 4851–4858.

87. Alberts, B., Johnson, A., Lewis, J., Raff, M., Roberts, K., and Walter, P. (2007) *Molecular Biology of the Cell*, 5th edn, Garland Science.

88. Murtuza, B., Nichol, J.W., and Khademhosseini, A. (2009) *Tissue Eng. B Rev.*, **15**(4), 443–454.

89. Dustin, M.L. (2008) *Immunol. Rev.*, **221**(1), 77–89.

90. Giehl, K. and Menke, A. (2008) *Front. Biosci.*, **13**, 3975–3985.

91. Matsuo, K. and Irie, N. (2008) *Arch. Biochem. Biophys.*, **473**(2), 201–209.

92. Kaji, H., Camci-Unal, G., Langer, R., and Khademhosseini, A. (2010) *Biochim. Biophys. Acta Gen. Subj.*, **1810**(3), 239–250.

93. Chen, C.S., Mrksich, M., Huang, S., Whitesides, G.M., and Ingber, D.E. (1997) *Science*, **276**(5317), 1425–1428.

94. Bhatia, S.N., Balis, U.J., Yarmush, M.L., and Toner, M. (1998) *Biotechnol. Prog.*, **14**(3), 378–387.

95. Khademhosseini, A., Suh, K.Y., Yang, J.M., Eng, G., Yeh, J., Levenberg, S., and Langer, R. (2004) *Biomaterials*, **25**(17), 3583–3592.

96. Wu, P.K. and Ringeisen, B.R. (2010) *Biofabrication*, **2**(1), 014111.

97. Park, J.S., Yang, H.N., Woo, D.G., Kim, H., Na, K., and Park, K.H. (2010) *Biomaterials*, **31**(28), 7275–7287.

98. Yang, H.N., Park, J.S., Na, K., Woo, D.G., Kwon, Y.D., and Park, K.H. (2009) *Biomaterials*, **30**(31), 6374–6385.

99. Ingber, D.E. (2006) *FASEB J.*, **20**(7), 811–827.

100. Hahn, M.S., Miller, J.S., and West, J.L. (2006) *Adv. Mater.*, **18**(20), 2679–2684.

101. Bott, K., Upton, Z., Schrobback, K., Ehrbar, M., Hubbell, J.A., Lutolf, M.P., and Rizzi, S.C. (2010) *Biomaterials*, **31**(32), 8454–8464.

102. Hale, N.A., Yang, Y., and Rajagopalan, P. (2010) *ACS Appl. Mater. Interfaces*, **2**(8), 2317–2324.

103. Sharma, R.I. and Snedeker, J.G. (2010) *Biomaterials*, **31**(30), 7695–7704.

104. Luo, Y. and Shoichet, M.S. (2004) *Nat. Mater.*, **3**(4), 249–253.

105. Leslie-Barbick, J.E., Shen, C., Chen, C., and West, J.L. (2011) *Tissue Eng. A*, **17**(1–2), 221–229.

106. Tzvetkova-Chevolleau, T., Stephanou, A., Fuard, D., Ohayon, J., Schiavone, P., and Tracqui, P. (2008) *Biomaterials*, **29**(10), 1541–1551.

107. Cheng, H.-w., Luk, K.D.K., Cheung, K.M.C., and Chan, B.P. (2011) *Biomaterials*, **32**(6), 1526–1535.

108. Tan, W. and Desai, T.A. (2004) *Biomaterials*, **25**(7–8), 1355–1364.

109. McGuigan, A.P. and Sefton, M.V. (2006) *Proc. Natl. Acad. Sci. U.S.A.*, **103**(31), 11461–11466.

110. Chung, S.E., Park, W., Shin, S., Lee, S.A., and Kwon, S. (2008) *Nat. Mater.*, **7**(7), 581–587.

111. Xu, M.G., Wang, X.H., Yan, Y.N., Yao, R., and Ge, Y.K. (2010) *Biomaterials*, **31**(14), 3868–3877.

112. Du, Y., Lo, E., Shamsher, A., and Khademhosseini, A. (2008) *Proc. Natl. Acad. Sci. U.S.A.*, **105**(28), 9522–9527.

113. Fernandez, J. and Khademhosseini, A. (2010) *Adv. Mater.*, **22**(23), 2538–2541.

114. Liu, B., Liu, Y., Lewis, A.K., and Shen, W. (2010) *Biomaterials*, **31**(18), 4918–4925.

115. Norotte, C., Marga, F.S., Niklason, L.E., and Forgacs, G. (2009) *Biomaterials*, **30**(30), 5910–5917.

116. Zamanian, B., Masaeli, M., Nichol, J.W., Khabiry, M., Hancock, M.J., Bae, H., and Khademhosseini, A. (2010) *Small*, **6**(8), 937–944.

117. Du, Y.A., Ghodousi, M., Lo, E., Vidula, M.K., Emiroglu, O., and Khademhosseini, A. (2010) *Biotechnol. Bioeng.*, **105**(3), 655–662.

118. Gillette, B.M., Jensen, J.A., Tang, B.X., Yang, G.J., Bazargan-Lari, A., Zhong, M., and Sia, S.K. (2008) *Nat. Mater.*, **7**(8), 636–640.
119. Ekici, S., Ilgin, P., Yilmaz, S., Aktas, N., and Sahiner, N. (2011) *Appl. Surf. Sci.*, **257**(7), 2669–2676.
120. Hsieh, C.H., Lee, M.C., Tsai-Wu, J.J., Chen, M.H., Lee, H.S., Chiang, H., Wu, C.H.H., and Jiang, C.C. (2008) *Osteoarthr. Cartil.*, **16**(3), 343–351.
121. Bong, K.W., Chapin, S.C., and Doyle, P.S. (2010) *Langmuir*, **26**(11), 8008–8014.

21
Injectable *In-Situ*-Forming Scaffolds for Tissue Engineering

Da Yeon Kim, Jae Ho Kim, Byoung Hyun Min, and Moon Suk Kim

21.1
Introduction

Tissue engineering and regenerative medicine is "an interdisciplinary field that applies the principles of engineering and life sciences toward the repair/regeneration of damaged human organs" [1]. Tissue engineering strategies use a suitable scaffold matrix to repair the damaged human tissue, to provide structural support, and to deliver cells and/or growth factors that have the ability to form tissues within the body on transplantation. Conventionally, scaffolds have been prepared *ex vivo* as three-dimensional scaffolds fabricated from a synthetic or natural material [2]. However, it is prohibitively difficult to perform *ex vivo* fabrication of certain complicated scaffold geometries.

As a possible alternative, injectable *in-situ*-forming scaffolds seems to be one of the key research points for the fabrication of complicated scaffold geometries [3]. The injectable *in-situ*-forming scaffold is based on the idea that if a certain biomaterial undergoes a simple liquid-to-solid phase transition under physiological conditions, this biomaterial could be injected as a liquid, and then form a solid gel *in situ* that acts as scaffold. Injectable *in-situ*-forming scaffolds have the potential benefit of being able to trap the various cells and/or growth factors by simple mixing and offer the advantage of minimally invasive nonsurgical procedures for the repair and regeneration of damaged tissues.

Injectable *in-situ*-forming scaffolds form spontaneously or in response to certain biological stimulus. The most focusing stimulus is body temperature, which can trigger responses that fall into two main categories: self-assembly by reversible either electrostatic or hydrophobic interactions and nonreversible chemical reactions after introducing nonreversible covalent bonds [4]. Among these injectable *in-situ*-forming scaffolds, the aim of the current chapter provides an overview of injectable *in-situ*-forming scaffolds that can be formed by physical interactions in response to body temperature.

21.2
Injectable *In-Situ*-Forming Scaffolds Formed by Electrostatic Interactions

The electrostatic interaction can occur between cationic and anionic polyelectrolytes in aqueous solutions. The intrinsic strength of polyelectrolytes is very important point to create certain self-assembly. There are various anionic and cationic polymers. If we make a mixture of anionic and cationic polymers in aqueous solutions, the mixture is capable of forming *in-situ*-forming scaffolds via electrostatic interactions in response to temperature (Figure 21.1).

Recent studies have shown that a mixture of various anionic and cationic polymers is capable of forming a gel *in situ* at body temperatures [5, 6]. The formation of a gel can be attributed to the dominance of electrostatic association between electric charges such as cationic and anionic group inside the polymers. Typical cationic polyelectrolyte compounds include spermine, spermidine, polyethylenimine, chitosan, polylysine, and so on, and typical anionic polyelectrolyte compounds are sodium carboxymethyl cellulose, hydroxypropyl methyl cellulose, alginate, tripolyphosphate, polyacrylic acid derivatives, and so on. The structure, mechanical properties, and gelation of such scaffolds depend on the concentration and native properties of polyelectrolytes. Among several anionic and cationic polyelectrolytes, introduction of injectable *in-situ*-forming scaffolds based on alginate and chitosan as well-known example of electrostatic interactions are described below.

Alginate, derived from seaweed, an important industrial product that is now commercially available in different grades, has become one of the most studied anionic linear polysaccharide polymers with mannuronic and glucoronic acid as repeating units. Several reports demonstrated *in-situ*-forming alginate scaffolds as carriers of various cells for cartilage, bone, and nerve repair [4]. Gelation time and

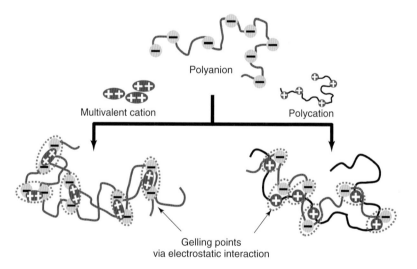

Figure 21.1 Schematic representation of injectable *in-situ*-forming gels formed by electrostatic interactions.

mechanical properties of *in-situ*-forming alginate scaffolds have a dependence on the anionic alginate concentration and the type of cation. Aqueous solutions of alginate form three-dimensional *in-situ*-forming scaffolds formed by electrostatic interactions when mixed with divalent cation such as $CaCl_2$, $CaCO_3$, and $CaSO_4$ or multivalent cation. Calcium cross-linking has been shown to yield gels with good mechanical properties. The use of alginate as an *in-situ*-forming alginate scaffolds exhibits for enhanced immunogenicity and poor bioresorbability, which may lead to adverse tissue interactions.

Chitosan, an amino-polysaccharide obtained by alkaline deacetylation of chitin, is a natural and abundant polymer. Their degradation occurs primarily by enzyme action and hydrolysis, and gives saccharides and glucosamines, which are constituents of normal metabolites in mammals. Chitosan is also suggested to have minimal foreign body reaction, antithrombogenic properties, and stimulating effects on the immunosystem against viral and bacterial infections.

Chitosan have the primary ammonium cations on the chitosan chains. It has been suggested that the cationic nature of chitosan provides a suitable point to act with anionic materials. Electrostatic interaction can occur between the polycationic chitosan and anionic compounds such as anionic glycerol phosphate disodium salt, acyllic acid, sodium deoxycholate, polylactide, hyaluronic acid, thioglycolic acid, calcium phosphate cement, fucoidan or heparin. The chitosan solutions with and without anionic compounds has shown temperature-dependent viscosity changes to give *in-situ*-forming chitosan scaffolds with the desired mechanical properties [7].

The *in-situ*-forming chitosan scaffolds possess a pore structure that ensures nutrient diffusion through the scaffold and thus foster cell growth after injection into the body at specific sites. With the eventual goal of applying *in-situ*-forming chitosan scaffolds for injectable therapeutic tissue engineering, the *in-situ*-forming chitosan scaffolds can be used in tissue-engineered skin, cartilage, cardiac, bone, and nerve regeneration.

21.3
Injectable *In-Situ*-Forming Scaffolds Formed by Hydrophobic Interactions

By far, the most studied type of physical interactions useful for *in-situ*-forming scaffolds are hydrophobic interactions. If certain biomaterial contain amphiphilic block segment, this biomaterial solution has strong hydrophobic interactions of the hydrophobic block segments, inducing liquid-to-solid phase transition (Figure 21.2). The phase transition indicates the formation of a structured network of amphiphilic block polymer solution as a function of temperature.

Some examples of various amphiphilic block polymers reported so far are shown in Figure 21.3. Poly(ethylene glycol) (PEG) is a nontoxic and biocompatible material with outstanding physicochemical and biological properties, especially as an ingredient for scaffolds. This makes for a longer acting medicinal effect and reduces toxicity, and it allows longer dosing intervals. For this reason, PEG may provide a useful starting point to make amphiphilic block polymers.

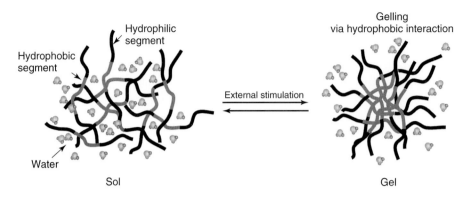

Figure 21.2 Schematic representation of injectable *in-situ*-forming gels formed by hydrophobic interactions.

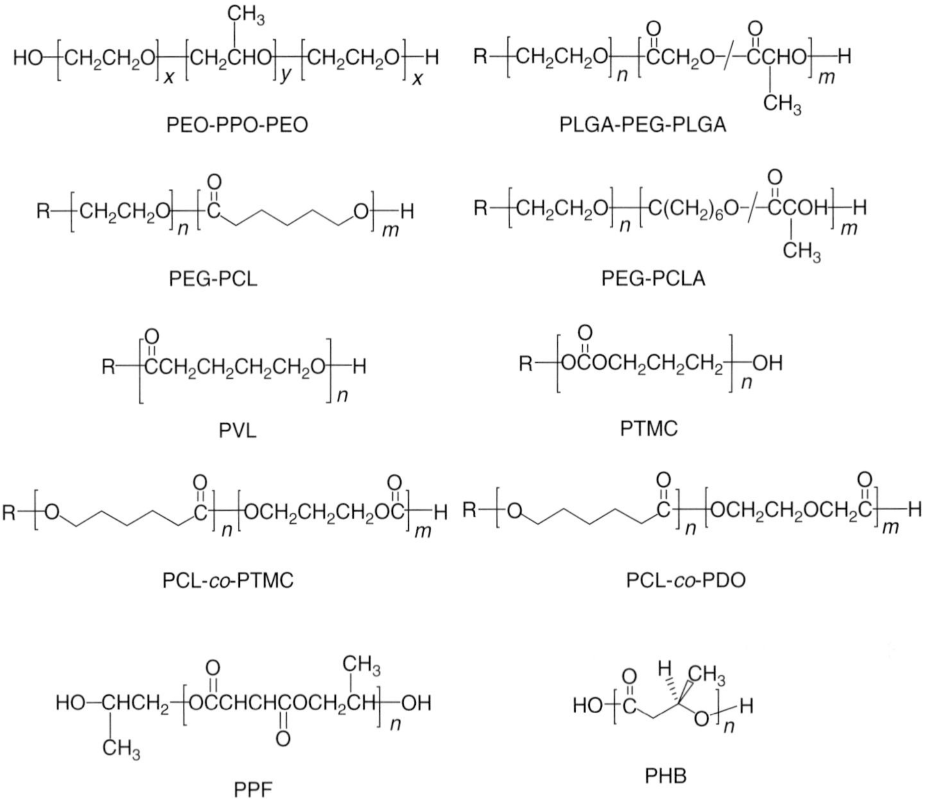

PEO-PPO-PEO

PLGA-PEG-PLGA

PEG-PCL

PEG-PCLA

PVL

PTMC

PCL-*co*-PTMC

PCL-*co*-PDO

PPF

PHB

Figure 21.3 Typical amphiphilic copolymers with biodegradable polyester segments.

21.3 Injectable In-Situ-Forming Scaffolds Formed by Hydrophobic Interactions

By far, the most studied amphiphilic block polymers are poly(ethylene oxide) (PEO) and poly(propylene oxide) (PPO) block copolymers, well known as *Poloxamers*™ and *Pluronics*™. Aqueous solutions of commercial Poloxamers and Pluronics series exhibit a temperature-induced sol–gel phase transition. The phase transition of these block copolymer solutions is rapid, but the gels formed are mechanically weak and display limited stability, short residence times, high permeabilities, and nonbiodegradability. Despite these disadvantages, Poloxamers and Pluronics series have been studied for possible use as *in-situ*-forming scaffolds in tissue engineering applications.

For successful application in the tissue engineering, further important properties of amphiphilic block polymer are adjustable degradation, which should correspond to the production rate of new extracellular matrix by the incorporated cells or biological factors to maintain a balance of assembly of new tissue and degradation of the scaffold material.

An important advance can be achieved by incorporating biodegradable polyester segments into PEG block with the aim of creating biodegradable amphiphilic block copolymers. The reported polyester blocks are polycaprolactone (PCL), poly(ε-caprolactone-*co*-D,L-lactic acid) (PCLA), poly(δ-valerolactone) (PVL), poly(trimethylene carbonate) (PTMC), poly(ε-caprolactone-*co*-trimethylene carbonate) (PCL-*co*-PTMC), poly(ε-caprolactone-*co*-1,4-dioxan-2-one) (PCL-*co*-PDO), poly(propylene fumarate) (PPF), poly-[(R)-3-hydroxybutyrate] (PHB), and so on (Figure 21.3).

By far, the most studied type is the PLGA-PEG-PLGA triblock copolymer, which exhibits a temperature-dependent reversible sol–gel transition in water [8]. The sol–gel transition temperature is influenced by the copolymer composition and specifically by the PEG molecular weight and the lactic acid/glycolic acid ratio of the poly(lactic-*co*-glycolic acid) (PLGA) blocks.

Recently, a diblock copolymer formed from PEG and PCL exhibit sol–gel transitions at body temperature [9]. Although PCL undergoes hydrolytic degradation, the rate of degradation *in vivo* is rather slow (two to three years) compared to that of poly(lactic acid) (PLA), poly(glycolic acid) (PGA), or PLGA under physiological conditions. Both *in vivo*-formed methoxy poly(ethyleneglycol)-b-polycaprolactone (MPEG-b-PCL) gels as *in-situ*-forming scaffolds maintain their structure for at least 10 months without degradation. Because of this slow-degradation property, such injectable *in-situ*-forming scaffold may be promising for long-term scaffold. By incorporation of poly(L-lactic acid) (PLLA) into the PCL segment, PEG-b-(PCL-ran-PLLA) diblock copolymers reported as *in-situ*-forming scaffold [10]. The degradation of PEG-b-(PCL-ran-PLLA) should depend on chain scissions within the PLLA segment. Degradation was observed to depend on chemical composition and was accelerated in proportion to the amount of PLLA segment present in the PCL-ran-PLLA structure. The degradation period can be controlled from a few weeks to a few months by changing the segments of block copolymers.

These amphiphilic block polymers, as injectable *in-situ*-forming scaffolds, must possess specific bulk and surface properties, easy handling at room temperature, amenability to sterilization, and structural integrity during *in vivo* application.

21.4
Immune Response of Injectable *In-Situ*-Forming Scaffolds

Scaffolds are in direct and sustained contact with tissues, and some degrade *in situ*. Such an intimate and/or prolonged contact between the scaffolds and biological tissues can induce a severe immune response [11]. This response to host tissues has limited use of the scaffolds for tissue engineering applications. One of the major goals in tissue engineering is an approach to control the interplay between scaffolds and biological tissues to create environments with the suppressed immune response. Considered from this viewpoint, the development of scaffolds having biocompatible specific structure in which pore structure, surface area to volume ratio, texture, and surface topography are manipulated to control cell shape, alignment and organization, and nontoxic scaffolds with degradation products that are equally devoid of toxicity is important.

Thus an understanding of the mechanisms that induce immune responses to *in-situ*-forming scaffolds is a key to developing scaffolds that are biocompatible with the targeted tissue. In considering immune responses to *in-situ*-forming scaffolds, ideally, the selected *in-situ*-forming scaffolds will specifically induce transplant tolerance or at least minimize rejection. Most of the implanted *in-situ*-forming scaffolds contained macrophages, neutrophils, and/or lymphocytes. From the results of ED1 staining, which recognizes the macrophage marker CD68, considered a unique *in vivo* indicator of an inflammatory response, the extent of host cell infiltration, and inflammatory cell accumulation within and near the transplanted *in-situ*-forming scaffolds depend on the characteristic properties of the used biomaterials. This inflammation response of natural biomaterials was less pronounced than that to the FDA-approved biomaterial, PLGA, or polyester as synthetic biomaterials [5–7, 9, 10]. Over time, however, there was a marked decrease in macrophages and neutrophils in most of the *in-situ*-forming scaffolds.

Although these previous works evaluated the scaffolds-mediated immune responses, it still remains unknown why and how immune responses execute and what mediators are involved in the process. Thus the final goals are minimizing the immune response to the scaffolds through thoughtful biomaterial selection and three-dimensional design of *in-situ*-forming scaffolds and a sufficient understanding of the immune response to enable effective antirejection strategies with minimal side effects.

21.5
Injectable *In-Situ*-Forming Scaffolds for Preclinical Regenerative Medicine

The clinical use of the injectable *in-situ*-forming scaffolds is currently the embryonic stage, because most research of injectable *in-situ*-forming scaffolds is conducted in animal models. Several animal studies based on the use of various injectable *in-situ*-forming scaffolds using alginate, chitosan, synthetic polymers, and so on (Table 21.1). Preclinical studies describing the use of injectable

Table 21.1 The use of injectable *in-situ*-forming scaffolds for organ regeneration.

Target organ	Biomaterial	Seeded cell source	References
Cartilage	Chitosan	Chondrocyte/Rabbit, Sheep	[12, 13]
	Oligo(poly(ethylene glycol) fumarate) (OPF)	Mesenchymal stem cells	[14]
	PEG-PCL	Chondrocyte	[15]
	Hyaluronic acid	Mesenchymal stem cells/rat	[16]
Bone	Chitosan-alginate	Mesenchymal stem cells/mouse	[17]
	PLA-PGA	Rabbit	[18]
	Carboxymethyl cellulose (CMC)	Rat	[19]
Heart	HEMA-polyTMC Poly(NIPAAm-*co*-AAc-*co*-HEMAPTMC)	Rat	[20]
Urethra	Gelatin	Rat	[21]
Ocular	Poloxamer	Rabbit	[22]
Periodontal	Chitosan	Dog	[23]
Muscle	MPEG-PCL	Adipose tissue derived stem cells/mouse	[24]
	PLGA	Adipose tissue derived stem cells/mouse	[25]

in-situ-forming scaffolds for cartilage and bone regeneration as well as vascular autografts are included. The basic and preclinical applications of injectable *in-situ*-forming scaffolds have been used for the treatment of damaged or diseased tissue using various cell sources such as embryonic stem cells, adult stem cells. Even though a number of injectable *in-situ*-forming scaffolds are available, research must continue; so that we can understand more details of how tissues develop in the injectable *in-situ*-forming scaffolds and which cell type should be applied to clinical application.

21.6
Conclusions and Outlook

Research in the area of development of injectable *in-situ*-forming scaffolds for tissue engineering has been well established in the past years. With the rapid progress in regenerative medicine, the larger demand for injectable *in-situ*-forming scaffolds is increasing. Although injectable *in-situ*-forming scaffolds may demonstrate success in animal model, new challenges are faced when a strategy is translated in humans. Further research for human application must address immunological issues, integration of host, and the variability of tissue development in human environment, depending on surrounding disease processes, age, or physical activity. It is likely that *in-situ*-forming hydrogel scaffolds can undoubtedly fulfill the requirements for specific applications in human in the near future.

References

1. Jenkins, D.D., Yang, G.P., Lorenz, H.P., Longaker, M.T., and Sylvester, K.G. (2003) *Clin. Plast. Surg.*, **30**, 581–588.
2. Barrilleaux, B., Phinney, D.G., Prockop, D.J., and O'Connor, K.C. (2006) *Tissue Eng.*, **12**, 3007–3019.
3. Mano, J.F., Silva, G.A., Azevedo, H.S., Malafaya, P.B., Sousa, R.A., Silva, S.S., Boesel, L.F., Oliveira, J.M., Santos, T.C., Marques, A.P., Neves, N.M., and Reis, R.L. (2007) *J. R. Soc. Interface*, **4**, 999–1030.
4. Gutowska, A., Jeong, B., and Jasionowski, M. (2001) *Anat. Rec.*, **263**, 342–349.
5. Kim, K.S., Lee, J.Y., Kang, Y.M., Kim, E.S., Lee, B., Chun, H.J., Kim, J.H., Min, B.H., Lee, H.B., and Kim, M.S. (2009) *Tissue Eng. Part A*, **15**, 3201–3209.
6. Lee, J.Y., Kang, Y.M., Kim, E.S., Kang, M.L., Lee, B., Kim, J.H., Min, B.H., Park, K., and Kim, M.S. (2010) *J. Mater. Chem.*, **20**, 3265–3271.
7. Kim, K.S., Ahn, H.H., Lee, J.H., Lee, J.Y., Lee, B., Lee, H.B., and Kim, M.S. (2008) *Biomaterials*, **29**, 4420–4428.
8. Jeong, B., Bae, Y.H., Lee, D.S., and Kim, S.W. (1997) *Nature*, **388**, 860–862.
9. Ahn, H.H., Kim, K.S., Lee, J.H., Lee, J.Y., Lee, I.W., Chun, H.J., Kim, J.H., Lee, H.B., and Kim, M.S. (2009) *Tissue Eng. Part A*, **15**, 1821–1832.
10. Kang, Y.M., Lee, S.H., Lee, J.Y., Son, J.S., Kim, B.S., Lee, B., Chun, H.J., Min, B.H., Kim, J.H., and Kim, M.S. (2010) *Biomaterials*, **31**, 2453–2460.
11. Badylak, S.F. and Gilbert, T.W. (2008) *Semin. Immunol.*, **20**, 109–116.
12. Hoemann, C.D., Sun, J., Légaré, A., McKee, M.D., and Buschmann, M.D. (2005) *Osteoarthr. Cartil.*, **13**, 318–329.
13. Hao, T., Wen, N., Cao, J.K., Wang, H.B., Lü, S.H., Liu, T., Lin, Q.X., Duan, C.X., Wang, C.Y. (2010) *Osteoarthr. Cartil.*, **18**, 257–265.
14. Park, H., Temenoff, J.S., Tabata, Y., Caplan, A.I., and Mikos, A.G. (2007) *Biomaterials*, **28**, 3217–3227.
15. Park, J.S., Woo, D.G., Sun, B.K., Chung, H.M., Im, S.J., Choi, Y.M., Park, K., Huh, M.K., and Park, K.H. (2007) *J. Controlled Release*, **124**, 51–59.
16. Kim, J., Kim, I.S., Cho, T.H., Lee, K.B., Hwang, S.J., Tae, G., Noh, I., Lee, S.H., Park, Y., and Sun, K. (2007) *Biomaterials*, **28**, 1830–1837.
17. Park, D.J., Choi, B.H., Zhu, S.J., Huh, J.Y., Kim, B.Y., and Lee, S.H. (2005) *J. Craniomaxillofacial Surg.*, **33**, 50–54.
18. Rimondini, L., Nicoli-Aldini, N., Fini, M., Guzzardella, G., Tschon, M., and Giardino, R. (2005) *Oral Surg. Oral Med. Oral Pathol. Oral Radiol. Endod.*, **99**, 148–154.
19. Reynolds, M.A., Aichelmann-Reidy, M.E., Kassolis, J.D., Prasad, H.S., and Rohrer, M.D. (2007) *J. Biomed. Mater. Res. Part B Appl. Biomater.*, **83**, 451–458.
20. Fujimoto, K.L., Ma, Z., Nelson, D.M., Hashizume, R., Guan, J., Tobita, K., and Wagner, W.R. (2009) *Biomaterials*, **30**, 4357–4368.
21. Takahashi, S., Chen, Q., Ogushi, T., Fujimura, T., Kumagai, J., Matsumoto, S., Hijikata, S., Tabata, Y., and Kitamura, T. (2006) *J. Urol.*, **176**, 819–823.
22. Kwon, J.W., Han, Y.K., Lee, W.J., Cho, C.S., Paik, S.J., Cho, D.I., Lee, J.H., and Wee, W.R. (2005) *J. Cataract Refract. Surg.*, **31**, 607–613.
23. Ji, Q.X., Deng, J., Xing, X.M., Yuan, C.Q., Yu, X.B., Xu, Q.C., and Yue, J. (2010) *Carbohydr. Polym.*, **82**, 1153–1116.
24. Kim, M.H., Hong, H.N., Hong, J.P., Park, C.J., Kwon, S.W., Kim, S.H., Kang, G., and Kim, M. (2010) *Biomaterials*, **31**, 1213–1218.
25. Kim, M.H., Hong, H.N., Hong, J.P., Park, C.J., Kwon, S.W., Kim, S.H., Kang, G., and Kim, M. (2006) *Biochem. Biophys. Res. Commun.*, **348**, 386–392.

22
Biomimetic Hydrogels for Regenerative Medicine

Iris Mironi-Harpaz, Olga Kossover, Eran Ivanir, and Dror Seliktar

22.1
Introduction

With many tissues in the body comprising 90% or more water, it is of little surprise that hydrogels with their high water content possess tissue and cell compatibility unmatched by any other man-made material. Initial research efforts on biomedical hydrogels were focused on applying new polymer formulations for producing better, more tissue-compatible materials [1]. From these efforts, the first generation of hydrogel implants were designed to be biologically inert, mechanically and chemically stable, and physically functional devices [2]. At the same time, biologically and pharmacologically interactive hydrogels were also used for applications in tissue regeneration and drug delivery. A substantial amount of progress has been made in developing novel polymers with unique features and reconstituting them in suitable ways to contribute to these advancing applications in medicine [3].

Recently, there has been a renaissance of research on hydrogels, wherein scientists are now focused on studying how different hydrogels can be engineered and designed with application-specific form and function [4]. Such a strategy allows scientists to develop customized biomedical hydrogels based on a set of desired properties, rather than solely relying on empirical formulations and iterative trial-and-error experimentation [5]. This shift to a more systematic hydrogel design strategy not only will impact hydrogel development but also will enable researchers in mainstream biomedicine to more easily employ sophisticated biomimetic hydrogels in their work. In this chapter, we cover some of the basic principles of how biomimetic hydrogels are developed for biomedical research and applications.

22.2
Natural and Synthetic Hydrogels

Some of the earliest works on hydrogels were performed with naturally occurring materials where the polymer chains are made of biological constituents found in

Biomimetic Approaches for Biomaterials Development, First Edition. Edited by João F. Mano.
© 2012 Wiley-VCH Verlag GmbH & Co. KGaA. Published 2012 by Wiley-VCH Verlag GmbH & Co. KGaA.

Figure 22.1 Synthetic hydrogels can be cast into many shapes and sizes for biomedical applications.

organic specimens. Biological hydrogels are very abundant in nature and are most often composed of large water-soluble proteins or carbohydrate macromolecules cross-linked together into a network that can imbibe a significant quantity of water. These materials serve as the extracellular matrix (ECM) of plants and animals and are found in a variety of sea plants, small organisms, and higher species, including proteins such as collagen, fibrin, and silk or polysaccharides such as alginate and hyaluronic acid (HA) [6]. Although these natural polymers have been characterized extensively, their properties as reconstituted hydrogels are often unpredictable and poorly reproducible, making them less suitable for biomedical use where precision engineering is a requisite design feature. Nonetheless, natural polymers are inherently bioactive and mildly cross-linked so as to be favored in applications requiring excellent biological compatibility with cells and tissues.

As an alternative, man-made analogs to biological hydrophilic polymers have been developed using synthetic polymer chemistry techniques [2]. Over the past 60 years, numerous synthetic hydrophilic polymers – which serve as the structural backbone of polymeric networks – have been conceived and have thus added a great deal of versatility to the development of more precisely controllable biomedical hydrogels (Figure 22.1). New hydrophilic polymers are now routinely synthesized using either chain transfer polymerization (condensation) or free radical polymerization techniques. These techniques have indirectly contributed to the establishment of better control over both intramolecular and intermolecular arrangements of polymer chains in a hydrogel network, altering the desired properties of the material [7].

Synthetic polymer hydrogels are not always able to provide a useful alternative to naturally derived materials. For example, it may not always be possible to artificially synthesize a protein-like molecule using straightforward hydrocarbon chemistry and still retain the complex structure and function of the surrogate. This is particularly true for certain proteins that display desired biological recognition sites with distinct structural features [8]. For these purposes, smaller portions of the protein, or synthetic protein domains, have been artificially reconstituted using either solid-state synthesis of individual amino acids or recombinant biology of engineered amino acid domains inspired from a natural protein [4]. Hydrogels can be assembled from these artificial protein building blocks alone, or in combination with additional synthetic or natural polymers, to enhance desired biomimetic features of the network [9]. This approach to polymer hybridization is used effectively to create semisynthetic materials, where key desired physical properties as well as reproducibility are not compromised by the enhancement of biological activity.

22.3
Hydrogel Properties

By definition, a hydrogel is a continuous network of hydrophilic polymer chains that is formed when the soluble macromolecular precursors phase transition into a hydrated insoluble (or semisoluble) phase through specific or nonspecific associations with each other or other polymer chains, or with other smaller molecules in solution [2]. Reconstituted hydrogels are typically formed from a liquid precursor solution cross-linked by a distinct chemical reaction or molecular associations, that is, self-assembly. The properties of the hydrogel network that is formed depend first and foremost on the nature of polymer chains and also on the degree of the interactions between the chains, their molecular arrangement, and the amount of water they absorb.

The associations between polymer chains are so important for the stability and physical properties of a hydrogel network that cross-linking is routinely categorized by the nature of the distinct type of intermolecular and/or intramolecular interactions between the polymer chains, as well as their relationship with the surrounding water molecules. Consequently, much of the water molecules in a hydrogel polymer network are either bound to the polymer chains (bound water) or fill the space between the network chains, in the center of larger pores or voids (free water or bulk water) that make up the physical structure of the cross-linked network [2]. At the final swollen state of a hydrogel, the relative void volume, or mesh size, is used to characterize important spatial features that affect the interface with cells and the mass transport of small and large molecules within the network by diffusion-mediated mechanisms. Depending on the relative mesh size, soluble factors may become entrapped in the network, or else diffuse freely between voids [10]. Changing the mesh size can readily be accomplished by altering the cross-linking density and composition of polymer or with postprocessing methods including simple freeze-thawing techniques. Freezing hydrogels results in a local rearrangement of the polymer chains as water molecules crystallize; these rearrangements sometimes lead to the formation of irreversibly insoluble macromolecular network structures [7].

Numerous chemical interactions can take place within a cross-linked hydrogel network. When reversible molecular entanglements or secondary bonds that are formed between polymer chains through ionic interactions, H–H interactions, or hydrophobic interactions, dominate connections between the polymers chains, this is referred to as a *physically cross-linked hydrogel network* (i.e., physical hydrogel). Many of the reconstituted biological protein and polysaccharide hydrogels are assembled in this way, including reconstituted collagen gels, Matrigel™, alginate, gelatin, and HA. Chemically cross-linked hydrogel networks (i.e., chemical hydrogels) are formed when polymer chains are held together by covalent bonds either with or without molecular linkers. For example, polymer chains can be linked directly to each other by covalent bonding during a chemical cross-linking reaction, provided the functional and reactive groups are accessible to each other, or they can be connected through a molecular cross-linker.

In addition to some well-established reaction schemes for cross-linking chemical hydrogels, several cross-linking chemistries have recently been revisited to enable a wider range of gelation options for forming chemical hydrogels specifically for biomedical exploitation [11]. *Ex situ* gelation via free radicals initiated by peroxides or Fenton's reagent has been employed routinely to cross-link hydrogels by bridging between pendant vinyl groups on the polymer chains. Because radicals can also attack cellular proteins and polysaccharides, *in situ* gelation can be performed using a more tightly controlled light-activated free radical polymerization, or photopolymerization. Photopolymerization provides an additional advantage for controlling spatial patterns and gradients of cross-linking reactions based on traditional photolithographic techniques or more recently, multiphotonic patterning of hydrogel cross-linking [12–15]. Schiff base formation reactions have also be used for *in situ* cross-linking of hydrogels, most notably for protein hydrogels using aldehydes. Another option for *in situ* gelation is cross-linking through Click chemistry such as with a Michael-type addition reaction [16]. Usually, Click chemistry requires a metal atom for catalysis; however, in the case of Michael-type addition, the reaction is metal free, and thereby more compatible with cells and tissues [17]. Disulfide bonding can also be applied for *in situ* gelation if the polymer chains contain free thiols. Enzymatic cross-linking is another way to achieve a controlled covalent bond formation between hydrophilic polymer chains toward assembly of a hydrogel network [18]. Covalently cross-linking polymers can also be accomplished using a combination of different chemistries. Moreover, hydrogels can be assembled by combining both physical and chemical cross-linking, as exemplified with fibrin hydrogels that are first physically cross-linked with the proteolytic removal of fibrinopeptides leading to fibrin self-assembly, and then chemically cross-linked with factor XIII transglutaminase-mediated (TG) reaction.

22.4
Engineering Strategies for Hydrogel Development

First-generation hydrogels introduced as early as the 1960s were simple polymeric systems, which were mostly inert, designed to be physically stable, degradable or nondegradable, and biologically compatible with tissues in the body [19]. For example, dermal fillers introduced in the 1980s were made using cross-linked HA; they absorbed water very well and displayed prolonged skin-lifting capacity, but after time, the polymer chains broke down and the hydrogel was eventually resorbed, leaving little evidence of its presence [20]. However, as more discoveries about hydrogels were made, it was clear that a next generation of hydrogels would require a more precise interaction with the body and its cells. In these next-generation hydrogels, it is not sufficient for the material to be inert, biologically compatible, or resorbable; the hydrogels must now be bioactive, with precise means of controlling the cell–material interactions [21, 22]. A next-generation dermal filler, for example, will not merely fill and occupy tissue volume, but as it is resorbed, it will be able to sustain better tissue composition by stimulating the cells to produce more collagen and elastin, thereby ensuring better long-term results. The notion that one can develop a hydrogel to display a specific set of desired biomimetic properties has been the focus of intense research in the biomedical hydrogel field and has recently led to the emergence of many new hydrogels that actively control cellular and tissue interactions.

The desirable characteristic of a more sophisticated hydrogel – either physical or biological attributes – can be incorporated into the network through the structural and compositional arrangements of the polymeric macromolecules as described earlier in this chapter. Theoretically, by taking the building blocks (polymer backbone), molecular additives, cross-linking methods, and other processing procedures, one could combine in a sequential fashion to end up with a final desired hydrogel product. However, this process is not straightforward because specialized bioactive hydrogels are very complex and may require several desired features, some of which may conflict with others. To address this challenge, two practical approaches have been applied for constructing tailor-made biomimetic hydrogels.

In the first approach, naturally occurring biological hydrogels are broken down to gain insight into their compositional subdomains. As each subdomain of the hydrogel is understood in greater detail, the mechanisms of elementary structure and function in that arrangement can be reproduced into synthetic biomimetic analogs or modified in the natural reconstituting macromolecules using genetic engineering tools. For example, in hydrogels designed to undergo reversible gelation based on environmental changes (e.g., temperature or pH), coiled-coil engineered proteins were cross-linked into a polymer network [23]. Others developed similar stimuli-responsive hydrogels using coiled-coil or β-sheet protein domains grafted to synthetic polymers [24].

The alternative approach is to design biomedical hydrogel to give rise to a system possessing a set of desired features [25]. The individual chemical units are first specified according to the desired features, and then these elements are linked together to form a hydrogel, which can be modified further with additional components [14]. Soft contact lens hydrogels, for example, were originally designed to exhibit tissue compatibility, good mechanical strength, and desired optical properties using monomers of hydroxyethyl methacrylate (HEMA) cross-linked with ethylenegylcol dimethacrylate (EDMA). The polyHEMA hydrogels, which were inherently nonadherent to proteins, exhibited excellent optical and mechanical properties but unfortunately had limited gas permeability, which in turn affected their tissue compatibility after extended wear. Eventually, these polymers were copolymerized with gas-permeable silicone motifs (siloxane) to design an elastomeric hydrogel with excellent permeability to oxygen, and of course, the other desired properties for a soft contact lens. In contrast to the polymer chemistry used in synthetic polymers such as polyHEMA, natural proteins and polysaccharides are more challenging to modify because of their limited solubility in the organic solvents in which most chemical modifications are best performed. Still, natural polymers can also be successfully ameliorated with chemical motifs for additional man-made control features, such as when naturally derived HA is modified with functional methacrylate groups on its backbone in order to make it photochemically cross-linkable [21].

Although the development of hydrogels using a methodical approach has resulted in remarkably versatile biomaterials, the method has its drawbacks. For example, not all desired features can be systematically incorporated into the hydrogel. In these cases, certain desired characteristics of the hydrogel determine the choice of the features to be included in the material design. These characteristics are usually categorized by groupings such as the mode of degradation, mechanical properties, optical properties, biological recognition, and structural anisotropy. For example, if one chooses anisotropic biodegradation as a primary characteristic of the hydrogel, the design of the material will first and foremost be centered on providing precise control over that feature through chemical composition, additives, choice of cross-linking method, and postgelation processing. This is nicely exemplified in the design of photolytically or enzymatically degradable hydrogel [14, 17]. More often than not, hydrogels need to be designed based on two or more desired properties such as with synthetic, encapsulating cell-responsive 3D culture materials [26, 27]. In this case, it is helpful to focus on proteolytic degradability and cell adhesive bioactivity, and not on the other features such as mechanical properties or porosity.

22.5
Applications in Biomedicine

With so many possibilities available for creating customized hydrogels, choosing the desired properties should be primarily motivated by the required specification of the indented application. In first-generation hydrogels, this strategy was used with success to provide different biomedical products. This design paradigm is now leading to new materials that extend well beyond clinical therapeutics, to applications that may ultimately alter the way in which scientists approach biomedical problems in areas such as stem cell research [8], developmental biology [28], drug delivery [29], cancer research [30], biotechnology [31], biosensors [7], and pharmacological screening [10].

In some biomedical applications, hydrogels must exhibit biodegradation. Controlling the rate of biodegradation in synthetic gels is achieved by clearing the implanted hydrogel based on a predetermined rate of hydrolytic breakdown of ester bonds in the polymer network. On the other hand, naturally reconstituted

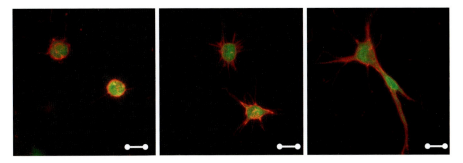

Figure 22.2 Semisynthetic PEG-Fibrinogen hydrogels are used as a scaffold for 3D culture of mesenchymal cells in biomedical research (scale bar = 25 micrometers).

protein gels are more difficult to control in terms of their proteolytic susceptibility and resorption, often leading to premature dissolution. To address this problem, a number of synthetic polymer conjugation schemes have been proposed. Most recently, protein PEGylation has been applied to natural fibrinogen and fibrin hydrogels in order to slow down degradation *in situ* while maintaining chemotactic and other biological activity of the PEGylated protein network [32, 33]. The relative composition and molecular weight of the PEG provide an accurate means of controlling both the cross-linking density of the hydrogel and also the proteolytic degradation half-life of the materials (Figure 22.2).

Similar to proteolytic degradation of natural polymers, synthetic hydrogels can also be designed to degrade by proteolytic enzymes; this property can give the synthetic hydrogel a biomimetic advantage when cells try to mediate a timely liberation of bound growth factor (GF) or to locally disassemble the hydrogel during invasion associated with tissue regeneration. This can be achieved by cross-linking synthetic polymers using protease sensitive cross-linker molecule made from oligopeptides whose amino acid sequences are copied from proteolytic substrates on natural proteins [27]. The reactive groups on these difunctional oligopeptides can rapidly form chemical hydrogels *in situ*, depending on the type of chemical reaction used. For example, a Michael-type addition reaction was used to cross-link di-thiol oligopeptides with vinyl-sulfone-functionalized branched PEG macromeres to form a collagenase-sensitive hydrogel network [34]. In this case, the oligopeptides were made from a sequence that was a substrate for matrix-metalloproteinases (MMPs); alterations to the amino acid sequence affected the rate of hydrogel degradation, which potentially provides a means of controlling the *in vivo* resorption kinetics. Importantly, the oligopeptide additives and chemistries used to cross-link them are mild and enable *in situ* gelation of these materials in the presence of cells and tissues. Consequently, different *in situ* chemistries have been reported for cross-linking polymers and oligopeptides in this way, most notable is using a TG-mediated cross-linking reaction between glutamine-containing oligopeptides and lysine-containing oligopeptides [35]. In principle, any synthetic polymer backbone can be combined with a proteolytically sensitive or otherwise degradable oligopeptide cross-linker molecule under the proper reaction conditions to form a highly specialized hydrogel, endowed with specific resorption characteristics according to the oligopeptide's attributes [17].

In addition to biodegradation, hydrogels may need to adhere *in situ* for specific applications. For example, resorbable poly(ethylene glycol) hydrogels were clinically applied as a postsurgical barrier to prevent abdominal adhesions using an acrylate-functionalized primer that localized the hydrogel (Focalseal, Genzyeme). The primer was used to form covalent attachments between the acrylate-functionalized PEG polymer chains and the tissue-bound vinyl groups during free radical light-activated photopolymerization [36]. Other primer systems were designed for synthetic cartilage regeneration implants using the naturally occurring chondroitin sulfate (CS) polysaccharide; the CS was modified with both methacrylate and aldehyde functionalities on the backbone in order to covalently link acrylate-functionalized hydrogels with tissue proteins during radical photopolymerization [37]. With greater use of biomimetic additives to localize hydrogels *in situ*, including concepts inspired from nature [38], rapid immobilization anywhere in the body is possible, including for uses such as oral and intestinal drug delivery [10].

Many applications in biomedicine require the localized activation of biological events in cells and tissues. The use of hydrogels bearing GFs to activate cells is now well established, and stable GF immobilization schemes have been recently proposed [39–41]. For example, Hubbell and Sakiyama-Elbert [42] devised an immobilization scheme using a TG substrate domain engineered into either bidomain fusion peptides or directly onto recombinant GFs. Using this approach, they were able to enhance the regenerative bioactivity of the reconstituted fibrin gels during fibrinogen coagulation and factor XIIIa cross-linking, which also binds the engineered TG domains to the fibrin. Similar peptides designed with both protease cleavage sites and affinity sites to specific GFs may be used as simple molecular additives that can be combined with soluble factors and reconstituted hydrogels to provide cells with on-demand bioactivity to enhance tissue repair in bone, skin, and other clinical indications.

Cell encapsulation in hydrogels can be very useful for cell therapy and tissue engineering [25]. In traditional cell therapy, where cells are injected to the tissue site by bolus of saline, there are typically poor *in situ* retention and cell survival rates. Therefore, the encapsulation of cells in a hydrogel – which simultaneously protects the cells *in situ* and facilitates a better integration of the graft with the host tissue – holds promise for improving clinical outcomes. Unfortunately, the initial attempts at this approach failed to provide tangible clinical outcomes, as exemplified by earlier work done on an artificial pancreas using insulin-producing xenografted β-cells [31]. Here, the synthetic encapsulating hydrogel – which was supposed to provide a protective, semipermeable barrier that could facilitate the diffusion of glucose, insulin, nutrients, and waste products but block considerably larger inflammatory cytokines – integrated poorly with the host because of interface problems [43]. However, with the advent of new synthetic hydrogels that employ mild *in situ* gelation conditions and display a multitude of control features that are available to encapsulate and then mediate the release or containment of these cellular constituents, the field of cell therapeutics may be advanced. For example, cardiologists are augmenting their clinical applications of cardiac cell therapy to use encapsulating hydrogels formed *in situ* for the localized immobilization and eventual release of their cellular payload in a timely manner [44].

Some of the new sophisticated hydrogels have combined several desired features to accommodate the more stringent requirements of a specific application. For example, with regards to substrate materials for advanced cell culture studies, most scientists would prefer to conduct *in vitro* experiments with cells on a representative three-dimensional milieu rather than using two-dimensional culture plastic substrates [45, 46]. However, few suitable hydrogel materials are currently available because cell culture matrices must possess biodegradation, bioactivity, and mechanical stability. In cancer research, for instance, most *in vitro* tumor invasion studies are currently performed with Matrigel™ on multi-wall cell inserts. Unfortunately, Matrigel™, similar to the other cell compatible bioactive hydrogels, is limited in terms of mechanical stability, biodegradation kinetics, reproducibility, and immunological reliability. Thus, a hydrogel designed with multiple properties – such as one that displays controlled proteolytic sensitivity, reliable mechanical properties, and specific bioactivity – can easily accommodate these research needs while providing more reliability, control, and consistency to the experimental system [5, 13, 15, 18, 47–52]. Designing hydrogels to be compatible enough to work with different cell types and *in vitro* experimental systems still remains a challenge to be taken up by the field; however, if synthetic hydrogels such as those described can prove more effective in these applications, the use of tissue culture plastic substrates can eventually be replaced altogether.

Indeed, multiproperty hydrogels have already made an impact in other fields such as stem cell research, most notably for examining how material properties such as stiffness, degradation, or ligand density can alter cell fate determination [5, 8, 48]. As these sophisticated multifeatured materials become more readily available to the mainstream scientific community, providing better alternatives to staple commodities such as Matrigel™, more and more scientists will employ these materials for their rudimentary explorations of cell biology, drug screening, cancer, and developmental research. Thus, the future of biomedical hydrogels will provide the ability to design materials for many more research and clinical applications.

References

1. Wichterle, O. and Lim, D. (1960) *Nature*, **185**, 117–118.
2. Hoffman, A.S. (2002) *Adv. Drug Delivery Rev.*, **54**, 3–12.
3. Patterson, J., Martino, M.M., and Hubbell, J.A. (2010) *Mater Today*, **13**, 14–22.
4. Lutolf, M.P. and Hubbell, J.A. (2005) *Nat. Biotechnol.*, **23**, 47–55.
5. Discher, D.E., Mooney, D.J., and Zandstra, P.W. (2009) *Science*, **324**, 1673–1677.
6. Lee, K.Y. and Mooney, D.J. (2001) *Chem. Rev.*, **101**, 1869–1879.
7. Peppas, N.A., Hilt, J.Z., Khademhosseini, A., and Langer, R. (2006) *Adv. Mater.*, **18**, 1345–1360.
8. Lutolf, M.P., Gilbert, P.M., and Blau, H.M. (2009) *Nature*, **462**, 433–441.

9. Krishna, O.D. and Kiick, K.L. (2010) *Biopolymers*, **94**, 32–48.
10. Hoare, T.R. and Kohane, D.S. (2008) *Polymer*, **49**, 1993–2007.
11. Hennink, W.E. and van Nostrum, C.F. (2002) *Adv. Drug Delivery Rev.*, **54**, 13–36.
12. Burdick, J.A. and Khetan, S. (2010) *Biomaterials*, **31**, 8228–8234.
13. Burdick, J.A. and Khetan, S. (2011) *Soft Matter*, **7**, 830–838.
14. Kloxin, A.M., Kasko, A.M., Salinas, C.N., and Anseth, K.S. (2009) *Science*, **324**, 59–63.
15. Kloxin, A.M., Kloxin, C.J., Bowman, C.N., and Anseth, K.S. (2010) *Adv. Mater.*, **22**, 3484–3494.
16. van Dijk, M., Rijkers, D.T., Liskamp, R.M., van Nostrum, C.F., and Hennink, W.E. (2009) *Bioconjugate Chem.*, **20**, 2001–2016.
17. Anseth, K.S., DeForest, C.A., and Polizzotti, B.D. (2009) *Nat. Mater.*, **8**, 659–664.
18. Jo, Y.S., Rizzi, S.C., Ehrbar, M., Weber, F.E., Hubbell, J.A., and Lutolf, M.P. (2010) *J. Biomed. Mater. Res. A*, **93**, 870–877.
19. Burdick, J.A. and Prestwich, G.D. (2011) *Adv. Mater.*, **23**, H41–H56.
20. Brandt, F.S. and Cazzaniga, A. (2008) *Clin. Interv. Aging*, **3**, 153–159.
21. Baier Leach, J., Bivens, K.A., Patrick, C.W. Jr., and Schmidt, C.E. (2003) *Biotechnol. Bioeng.*, **82**, 578–589.
22. Young, J.L. and Engler, A.J. (2011) *Biomaterials*, **32**, 1002–1009.
23. Petka, W.A., Harden, J.L., McGrath, K.P., Wirtz, D., and Tirrell, D.A. (1998) *Science*, **281**, 389–392.
24. Wang, C., Stewart, R.J., and Kopecek, J. (1999) *Nature*, **397**, 417–420.
25. Tibbitt, M.W. and Anseth, K.S. (2009) *Biotechnol. Bioeng.*, **103**, 655–663.
26. Lee, S.H., Moon, J.J., and West, J.L. (2008) *Biomaterials*, **29**, 2962–2968.
27. Lutolf, M.P., Lauer-Fields, J.L., Schmoekel, H.G., Metters, A.T., Weber, F.E., Fields, G.B. *et al.* (2003) *Proc. Natl. Acad. Sci. U.S.A.*, **100**, 5413–5418.
28. Nelson, C.M., Vanduijn, M.M., Inman, J.L., Fletcher, D.A., and Bissell, M.J. (2006) *Science*, **314**, 298–300.
29. Bhattarai, N., Gunn, J., and Zhang, M. (2010) *Adv. Drug Delivery Rev.*, **62**, 83–99.
30. Loessner, D., Stok, K.S., Lutolf, M.P., Hutmacher, D.W., Clements, J.A., and Rizzi, S.C. (2010) *Biomaterials*, **31**, 8494–8506.
31. Hildgen, P., Rabanel, J.M., Banquy, X., Zouaoui, H., and Mokhtar, M. (2009) *Biotechnol. Prog.*, **25**, 946–963.
32. Almany, L. and Seliktar, D. (2005) *Biomaterials*, **26**, 2467–2477.
33. Liu, H., Collins, S.F., and Suggs, L.J. (2006) *Biomaterials*, **27**, 6004–6014.
34. Lutolf, M.P., Weber, F.E., Schmoekel, H.G., Schense, J.C., Kohler, T., and Muller, R. *et al.* (2003) *Nat. Biotechnol.*, **21**, 513–518.
35. Messersmith, P.B., Brubaker, C.E., Kissler, H., Wang, L.J., and Kaufman, D.B. (2010) *Biomaterials*, **31**, 420–427.
36. Lauder, C.I., Garcea, G., Strickland, A., and Maddern, G.J. (2010) *Dig. Surg.*, **27**, 347–358.
37. Wang, D.A., Varghese, S., Sharma, B., Strehin, I., Fermanian, S., Gorham, J. *et al.* (2007) *Nat. Mater.*, **6**, 385–392.
38. Lee, H., Lee, B.P., and Messersmith, P.B. (2007) *Nature*, **448**, 338–341.
39. Wylie, R.G., Ahsan, S., Aizawa, Y., Maxwell, K.L., Morshead, C.M., and Shoichet, M.S. (2011) *Nat. Mater.*, **10**, 799–806.
40. Hudalla, G.A. and Murphy, W.L. (2011) *Adv. Funct. Mater.*, **21**, 1754–1768.
41. Murphy, W.L. and Hudalla, G.A. (2010) *Langmuir*, **26**, 6449–6456.
42. Sakiyama-Elbert, S.E. and Hubbell, J.A. (2000) *J. Controlled Release*, **65**, 389–402.
43. Orive, G., Hernandez, R.M., Gascon, A.R., Calafiore, R., Chang, T.M., De Vos, P. *et al.* (2003) *Nat. Med.*, **9**, 104–107.
44. Davis, M.E., Hsieh, P.C., Grodzinsky, A.J., and Lee, R.T. (2005) *Circ. Res.*, **97**, 8–15.
45. Discher, D.E., Engler, A.J., Sen, S., and Sweeney, H.L. (2006) *Cell*, **126**, 677–689.
46. Discher, D.E., Janmey, P., and Wang, Y.L. (2005) *Science*, **310**, 1139–1143.
47. Cushing, M.C. and Anseth, K.S. (2007) *Science*, **316**, 1133–1134.

48. Gilbert, P.M., Havenstrite, K.L., Magnusson, K.E., Sacco, A., Leonardi, N.A., Kraft, P. et al. (2010) *Science*, **329**, 1078–1081.
49. Huebsch, N., Arany, P.R., Mao, A.S., Shvartsman, D., Ali, O.A., Bencherif, S.A. et al. (2010) *Nat. Mater.*, **9**, 518–526.
50. Mann, B.K. and West, J.L. (2002) *J. Biomed. Mater. Res.*, **60**, 86–93.
51. Martino, M.M., Tortelli, F., Mochizuki, M., Traub, S., Ben-David, D., Kuhn, G.A. et al. (2011) *Sci. Transl. Med.*, **3**, 100ra89.
52. Seliktar, D., Zisch, A.H., Lutolf, M.P., Wrana, J.L., Hubbell, J.A., and Res, A. (2004) *J. Biomed. Mater.*, **68**, 704–716.

23
Bio-inspired 3D Environments for Cartilage Engineering
José Luis Gómez Ribelles

to Concha

23.1
Articular Cartilage Histology

Hyaline cartilage is a very specialized tissue in which a small number of cells, chondrocytes, are distributed in a quite particular organization. The extracellular matrix (ECM) is composed mainly of collagen type II fibers and proteoglycan aggregates, which are mainly formed by the association of aggrecan with a large number of glycosaminoglycan chains. The combination of the collagen fiber stiffness and the high water sorption capacity of glycosaminoglycans (GAGs) produces a hard tissue that is able to sustain the high compression loading to which articular cartilage is subjected [1–3] and is very permeable to water-soluble substances allowing the diffusion of nutrients and waste products of cell metabolism, which is crucial in the avascular tissue. Healthy articular cartilage contains between 70% and 75% of water, the GAGs content can be up to 20% depending on the joints and cartilage sites [4, 5], and collagen type II can represent between 11% and 20% of the wet weight [6].

Furthermore, ECM composition, and consequently its mechanical properties, cell shape, and cell distribution are not homogeneous. Different tissue layers, going from articular surface to subchondral bone, with quite different characteristics can be distinguished [7–11]. The articular surface layer contains flat chondrocytes dispersed in a network of collagen type II fibers preferentially aligned parallel to the surface, which gives the tissue a special ability to sustain the shear stress to which the articular surface is subjected. The transition zone contains dispersed rounded chondrocytes in a disordered network of collagen fibers. The deep zone contains large rounded chondrocytes ordered in columns perpendicular to articular surface and collagen fibers aligned in the same direction. Finally, the calcified cartilage is the interface with subchondral bone. It is characterized by hypertrophic chondrocytes, a high content of collagen type X in the ECM and by tissue mineralization. The level of GAGs is lower in the surface layer, increases in the transition zone, and is the highest in the deep zone. The elastic modulus of hyaline cartilage continuously increases with depth from the articular surface. Schinagl *et al.* [10] measured values

Biomimetic Approaches for Biomaterials Development, First Edition. Edited by João F. Mano.
© 2012 Wiley-VCH Verlag GmbH & Co. KGaA. Published 2012 by Wiley-VCH Verlag GmbH & Co. KGaA.

between 0.08 at the surface and 2.1 MPa at the deep zone of bovine cartilage, while the apparent modulus of the whole cartilage piece was 0.38 MPa.

The regenerative capacity of articular cartilage tissue engineering techniques comes from the plasticity of the mesenchymal lineage to which connective tissue pertains. Cells of fibrous tissue, fibroblasts, chondrocytes from fibrocartilage, and chondrocytes from hyaline cartilage have essentially different phenotypic characteristics that allow them to produce specialized tissues with very different properties. Fibrous tissue contains mainly collagen type I arranged in such a way that is able to sustain high tension loading. On the contrary, as explained above, chondrocytes of hyaline cartilage produce an ECM consisting in collagen type II with a large amount of GAGs optimized to resist dynamic compression loading. Fibrocartilage is something in between, its chondrocytes produce both collagen type I and collagen type II and may sustain in some degree both compression and tension loading [12]. The phenotype of these cells *in vivo* is maintained by complex signaling pathways. Growth factors produced by cells in surrounding bone, sinovial, and different zones of cartilage itself must regulate gene transcription [13], as shown by *in vitro* coculture [9, 14], but mechanotransduction is thought to play an important role as well [15]. Cell phenotype is thought to be influenced by tissue loading through cell to matrix contact that, in the case of hyaline cartilage, is mediated by integrins connected to ligands found in the internal surfaces of the lacunae, which contain, in particular, collagen type VI, connections to hyaluronic acid ligands seems to be relevant as well, regulating important cell functions [16–19]. When cells are separated from their biological niche, their phenotype changes. Interestingly, monolayer culture of fibroblasts and chondrocytes of hyaline cartilage or fibrocartilage yield proliferative cells that become quite similar to each other after two or three passages [12, 20]. Furthermore, and more importantly, they acquire the potentiality to differentiate into other fibroblastic or chondrocytic cell types, as well as other mesenchymal lineages (as osteoblasts), provided the differentiation medium is adequate [12]. On the other hand, cells residing in adult tissues such as bone marrow mesenchymal stem cells (bMSCs) or adipose-derived stem cells (ADSCs) also have chondrogenic potential. Although they are not differentiating actively *in vivo* to cartilage cells, they can be induced to chondrogenic differentiation *in vivo* or *in vitro* by different procedures that are analyzed in the following.

Characteristic markers of the phenotype of hyaline cartilage chondrocytes are the expression of collagen type II and aggrecan in messenger RNA, while collagen type I and versican are signs of dedifferentiation. Martin *et al.* [6] measured gene expression by real-time PCR in healthy tissue and determined that the absolute values of these markers were quite variable, while the ratio of collagen type II to collagen type I or aggrecan to versican present less variability between different individuals. Nevertheless, quantitative determination of these major markers provides only a rough characterization of the cells, and a number of other markers of minor components of the ECM are important because if they are absent, the cartilage function or durability is compromised (as proved by knockout animal models) [21]. In fact, major markers alone do not account for the differences among the chondrocytes in the different zones of articular cartilage, which have different morphology and biosynthetic behavior.

To assess chondrogenic differentiation, quantification of GAGs, and collagen type II production in the ECM by histology or other quantitative methods is frequently used both in *in vitro* and *in vivo* studies, in addition to phenotypic markers. For instance, different biosynthetic production of GAGs is characteristic of the behavior of chondrocytes of the different cartilage zones in 3D culture [9, 14, 22–24].

23.2
Spontaneous and Forced Regeneration in Articular Cartilage

Osteoarthritis produces defects in the articular surface with loss of cartilaginous tissue. The size and depth of the defect can enlarge with loss of cartilage volume and degeneration of the cartilage tissue adjacent to the defect wall. The presence of the unbalanced compression stress in the tissue adjacent to the defect can induce degeneration, cell death, and loss of tissue [25].

The healing potential of articular cartilage in partial thickness defects is very limited. Several reasons can explain the lack of reparative potential: the inability of chondrocytes to migrate into the site of injury, the avascular nature of cartilage that hinders the arrival of pluripotent cells, hypoxia, and others [26, 27].

Spontaneous regeneration of articular cartilage in osteoarthritic knees can take place when the enlargement of cartilage defect injures subchondral bone producing bleeding in the zone of the defect. The formation of a fibrin clot allows migration of bMSCs, coming from subchondral bone, with chondrogenic potential, that are able to initiate new tissue formation. Timing of spontaneous new cartilage formation in the defect site has been established. Five days after bleeding, bMSCs invade the fibrin clot. Fibrin is bioresorbed in around seven days. After 14 days, bMSCs have differentiated to chondrocytes that are dispersed in a GAG rich matrix. By eight weeks, the tissue resembles cartilage but most of the times with the characteristics of fibrocartilage and rapidly degenerates [27, 28].

Various therapeutic interventions in the treatment of osteoarthritis aim to induce what could be called *forced regeneration*: abrasion chondroplasty, drilling [29], spongialization, and microfracture [30] are different techniques that injure in different ways the subchondral bone, producing bleeding in the zone of osteoarthritic lesion and the formation of clot that initiate the regenerative process. But, as it happens in spontaneous regeneration, regenerated tissue is in practice fibrocartilage, or degenerate to fibrocartilage with time [27, 31].

23.3
What Can Tissue Engineering Do for Articular Cartilage Regeneration?

Probably the first step in the design of a tissue engineering approach to regenerate articular cartilage should be to identify as much as possible the reasons why spontaneous regeneration fails to produce functional hyaline cartilage in the site of the osteoarthritic defect, in order to be able to propose strategies to overcome the shortcuts found. What is wrong in spontaneous cartilage regeneration? According to the previous sections, it is possible to identify different reasons why hyaline cartilage is not properly regenerated: (i) maybe the cell source is not adequate – although extensive *in vitro* experimentation has shown the chondrogenicity of bMSCs, this is only in particular laboratory conditions; (ii) maybe the growth factors that reach the cells in the cartilage environment are not inducing the right chondrogenic differentiation of bMSCs, in fact this is not a natural process in normal adult cartilage; (iii) maybe the biomechanical environment is not the correct one because of the initial lack of ECM what probably implies an incorrect stress transmission to the cells; and (iv) maybe hypoxia compromise bMSC that migrate from the highly vascularized bone marrow. Some or all these factors and probably others can contribute to make spontaneous regeneration noneffective.

Tissue engineering approach can contribute to modify the course of regeneration in three different lines.

- On the one hand, the cell source can be changed. The work by Brittberg *et al.* [32] introduced the idea of implanting autologous expanded chondrocytes to the site of the cartilage defect that has been broadly applied in clinics. Other cell source proposed are mesenchymal stem cells, MSCs, obtained from different tissues, directly implanted in the defect in order to differentiate *in vivo* or differentiated to well-characterized chondrocytes that will be then implanted. Embryonic stem cells [33] or reprogrammed pluripotent cells, IPS, or directly from dermal fibroblasts [34–37] have proven also the ability to differentiate into chondrocytes, but for different reasons, they have been less studied. Nevertheless, the use of bMSC migrating from subchondral bone cannot be ruled out since it is the easiest means to bring chondrogenic cells to the site of the defect.
- The use of a vehicle for cell transplantation, acting as scaffolding system to avoid cell dispersion and to control biomechanical stimulation of the transplanted cells, seems to be crucial for the maintenance of chondrocyte phenotype independently of the type of cells transplanted. The literature shows the suitability of a variety of different materials including *in situ* cross-linked hydrogels, injectable systems consisting of microparticles that act as three-dimensional scaffold, or chondral or osteochondral macroporous supports. How different scaffold architectures can be produced with natural or synthetic materials has also been shown.
- Growth factor supply has been extensively used for chondrogenesis in *in vitro* studies, but not so much work has been done with respect to *in vivo* delivery of these growth factors during regeneration.

In the following sections, these three ingredients of tissue engineering strategy are briefly analyzed. This paper does not pretend to give an exhaustive revision of the state of the art, but to express an opinion on the lines of thought that could direct the design of a cartilage engineering strategy. The reader can find very detailed and comprehensive reviews of cartilage cartilage regeneration and cartilage engineering techniques in Refs [26, 27, 38–42].

23.4
Cell Sources for Cartilage Engineering

23.4.1
Bone Marrow Mesenchymal Cells Reaching the Cartilage Defect from Subchondral Bone

The strategy based on bMSCs colonization of an artificial scaffold is straightforward; this is the simplest way to drive chondrogenic cells to the cartilage defect since cells are autologous and a single surgical act that can be performed by arthroscopy is required. Histological analysis of the tissue formed after microfracture shows that cell density in the regenerated cartilage is high enough, at least in the order of that of natural hyaline cartilage, thus it seems that there is no need of additional cell implantation to obtain the required cell density. As mentioned above, several aspects of this cell source are challenging. On the one hand, a quite heterogeneous population of cells arrives to the site of the defect from subchondral bone and perhaps the ability of bMSCs to differentiate into a variety of cells of mesenchymal lineage could make that only a fraction of them yield functional chondrocytes with the phenotype of hyaline cartilage cells [21, 27, 43, 44]. Furthermore, growth factor signaling *in vivo* perhaps do not contain all the components needed for chondrogenic differentiation of bMSCs. It is worth to explore the possibilities of acting on the other two ingredients of tissue engineering strategy: the scaffold characteristic and artificial delivery of growth factors to address differentiation in the hyaline cartilage chondrocyte phenotype and improve the organization of regenerated tissue.

The course of spontaneous regeneration is in some way similar to artificial tissue engineering: bMSCs invade a 3D fibrin scaffold used as provisional ECM for migration, attachment, and differentiation. Interesting enough, fibrin clot is resorbed around one week after bleeding, quite soon compared with new tissue formation that needs at least several weeks [45]. This is one of the factors that must be considered when analyzing regeneration from the point of view of tissue engineering. Perhaps, a major contribution to cartilage regeneration is simply to provide bone marrow MSCs with a scaffold that might transmit compression loading to regenerative cells in a controlled way during a longer period of time. It has been shown by *in vitro* culture of bMSCs that early mechanical stimulation of bMSCs is not favorable to chondrogenic differentiation [21, 46, 47], while compression loading transmission to mature chondrocytes is essential for tissue organization [48–50]. The adequate design of the scaffold compliance can protect bMSCs from excessive loading while differentiation takes place (around 14 days after bleeding in the cartilage defect [45]).

The results obtained in animal models are encouraging, showing that the combination of injuring subchondral bone and scaffold implant in the defect zone can produce a well-organized hyaline cartilage [51–55]. Unfortunately, the research effort related with this approach has not been compared much to the use of other cell sources. More research is needed on artificial delivery of growth factors *in vivo* that could take place from the scaffold itself (Section 23.6); the role of the scaffold properties, in particular, scaffold stiffness and the degradation time; and on the histological and mechanical properties of the regenerated tissue.

23.4.2
Autologous Mesenchymal Stem Cells from Different Sources

Nevertheless, it is possible that limitations inherent to the heterogeneity of the cell population invading the scaffold or difficulties in the *in vivo* course of differentiation make preferable to implant in the site of the defect cells with a well-characterized and homogeneous phenotype. Much research activity has been addressed to find the optimal conditions to obtain functional chondrocytes *in vitro* from autologous pluripotent cells of mesenchymal lineage obtained from different sources such as bMSCs [56–59], ADSCs [60–65], muscle-derived stem cells (MDSCs) [66, 67], or stem cells isolated from synovium [67–71] or periosteum [68, 72]. These cells can be expanded in monolayer or on microcarriers in bioreactor culture to the high cell numbers needed for cartilage regeneration while preserving their chondrogenic potential [73]. Chondrogenic differentiation can be induced when they are cultured in pellets, in which a high concentration of cells is obtained by centrifugation [74–76], or micromasses, obtained from droplets of a highly concentrated cell suspension [76, 77] in chondrogenic media containing the adequate growth factors. These 3D environments allow cell-to-cell contacts similar to those encountered during embryonic development of cartilage. Nevertheless, important limitations are that nutrient supply does not reach the core of the pellet where cell necrosis takes place [74–76], a problem that can be reduced by producing micropellets, the size of which would lead nutrients to reach the core of the cell aggregate [75]. On the other hand, the phenotype of the differentiated cells and the characteristics of the ECM are close to that of fibrocartilage, with collagen type I expression and also presence of degenerative collagen type X.

Interestingly, chondrogenic differentiation of MSCs can also be induced when pluripotent cells are seeded in gels, thus isolated from each other but fully surrounded by the gel polymeric chains such as collagen [74, 78, 79], agarose [60, 80], fibrin [81], gelatin or alginate [60], and chitosan [82]. Chondrogenic differentiation of MSCs has also been performed in nanofibrous scaffolds [83] or in 3D environments containing or formed by microspheres [84, 85].

Transforming growth factor (TGFβ) family and bone morphogenetic proteins (BMPs) [73, 86] are the most studied growth factor for *in vitro* differentiation, in particular, TGFβ1 [78, 85], TGFβ2 [65, 87], TGFβ3 [88, 89], BMP2 [69, 78], BMP7 [87], also basic fibroblast growth factor (βFGF) [90], and insulin-like growth factor I (IGF-I) [83].

It has been shown that the chondrogenic markers expressed by cells produced *in vitro* by MSCs differentiation highly depend on culture conditions: growth factors, two-dimensional or three-dimensional culture support materials and surface characteristics, hypoxia, mechanical stimulation, culture in bioreactor or in static conditions, and others. The objective of determining the most appropriate conditions to address pluripotent cells to the phenotype of the hyaline cartilage chondrocyte is to overcome the difficulty of the limited characterization of the chondrocytes obtained by differentiation *in vitro*. Chondrogenic positive (such as collagen type II, sox 9, or aggrecan) and negative (such as collagen type I) markers [46, 91] or the quantification of the production of ECM components: collagen and GAGs can be useful for comparing the performance of different differentiation protocols or in the identification of cell populations of the highest chondrogenic potential, but probably the *in vivo* performance can hardly be predicted from currently available *in vitro* characterization. In fact, mechanical properties of the produced cartilage not always correlate with histological or phenotypic characterization that shows the importance of mechanical characterization of the tissue synthesized "*in vitro* or *in vivo*" [21, 55, 92] and its comparison with those of natural cartilage [93].

On the other hand, the fact that the cells implanted in the zone of the defect have the right phenotype is not a guarantee of success in the regenerative process. Owing to the plasticity of these cells, one can expect that they change their phenotype if they found an unfavorable environment when transplanted *in vivo*. In particular, it seems that biomechanical conditions can be crucial for the maintenance of chondrocyte phenotype. Mechanotransduction is an important signaling pathway that regulates cell functions that are crucial in chondrogenesis [15, 94]. Articular cartilage is subjected to complex loading reaching stresses up to 20 MPa and strains in the order of 5–10% [1, 3]. It has been shown by *in vitro* studies that loading of chondrocytes seeded in a scaffolding material affect their morphology, proliferation, metabolism, and phenotype [1, 94, 95]. Although dynamic compression can stimulate ECM production to a higher or lower extent depending on the characteristics of the scaffold and the sequence of loading application [5, 48–50, 94, 96, 97], low-frequency loading or static stress can also suppress ECM production [94]. Unbalanced compression loading in the borders of the osteoarthritic defect is one of the factors for the loss of tissue [3, 25], so one can expect that improper transmission of stresses to the cells lodged in the scaffold can result in their dedifferentiation. Thus one comes back to the importance of determining the required mechanical behavior of the three-dimensional support for the maintenance of chondrocyte phenotype *in vivo*.

23.4.3
Mature Autologous Chondrocytes

Much research and most of the clinical applications of cartilage engineering are related to the substitution of subchondral bone bMSCs by mature chondrocytes as cell source for tissue regeneration in the zone of the osteoarthritic defect. In the autologous chondrocyte implant (ACI) technique, a low number of chondrocytes are obtained by enzymatic digestion of a small portion of healthy hyaline cartilage obtained by a biopsy. These chondrocytes are expanded *in vitro* to obtain a large number of cells, in the order of 30 million, that are implanted in the zone of the defect and covered by a periosteum flap. The results of this technique are encouraging but tissue seems to degenerate to fibrocartilage with time [98–100].

The performance of chondrocytes obtained from other cartilage sources has also been studied [101, 102].

Chondrocytes can be isolated from the articular cartilage tissue by enzymatic digestion to obtain a cell suspension and plated on a flat substrate [103]. Interaction between cells and the substrate material is mediated by the proteins adsorbed on the material. With the same mechanisms that cells interact with natural ECM, transmembrane proteins of the cell, mainly integrin pairs, are able to recognize specific ligands in the proteins adsorbed on the substrate. Once integrins adhere to the ligands outside the cell, they cluster, forming focal adhesions, while in the cytoplasmatic domains, by joining a number of proteins, they form adhesion complexes that initiate actin cytoskeleton polymerization. At the end, a complex signaling cross talk between the nucleus and the ECM is established, which will condition cell proliferation and cell phenotype [104].

Adhesion of the chondrocyte to the substrate induces fast proliferation but cells are quite different from *in vivo* chondrocytes in what respects to cell morphology, due to spreading on the substrate, the development of actin cytoskeleton, and protein expression. Although further changes in the cell environment could make the monolayer-expanded cells to recover the expression of characteristic markers of hyaline cartilage chondrocytes, it can be pointed that dedifferentiation could compromise irreversibly the cells making them to produce a defective tissue when implanted in a cartilage defect. Expansion of mature chondrocytes has also been performed in microcarriers in bioreactor culture [105, 106]; it has been reported that doubling rates in gelatin microcarriers can be twice as much as on flat substrates. Nevertheless, the behavior of chondrocytes in monolayer culture can vary significantly depending on the nature of the substrate. Cell attachment, spreading, proliferation, and viability depend on the surface properties. Since, as mentioned above, cell–substrate interaction is mediated by proteins, the surface properties of the substrate material can influence biological response by selecting the type and amount of proteins adsorbed from culture medium and controlling their conformation [107–110]. Thus, it has been shown that laminin adsorbed on hydrophilic poly(hydroxyethyl acrylate) substrates adopt a globular conformation that do not favor the exposition of the ligands that integrins are able to recognize. Globular forms or at least not fully extended conformation were also found in hydrophobous poly(hydroxyethyl acrylate). Nevertheless, laminin adopt fully extended conformation on copolymers in which hydroxyethyl acrylate and ethyl acrylate alternate in domains of nanometric dimensions [111]. This fact correlates with the finding that chondrocytes cultured on these copolymer substrates attach and proliferate more than in the homopolymers [112]. Fibronectin adsorption has also been extensively studied on this kind of substrates [113–119]. Not only chemical characteristics of the surface but also its topology strongly influence cell attachment and morphology. Thus, chondrocytes cultured on rough polylactide (PLA) surfaces adopted elongated shape, while in smoother surfaces of the same material, they adopt the typical spread morphology. This behavior could also been related to the conformation of fibronectin adsorbed on the surface [120, 121]. Chondrocyte phenotype is closely related to cell morphology and cytoskeleton development. The characteristics of the substrate surface also influence the formation of actin stress fibers or at least the time in which they are formed. Thus, in chondrocytes cultured on PLLA substrates, the actin cytoskeleton is formed much slower than in glass or polystyrene controls [120]. The adhesive behavior of chondrocytes to different proteins depends on the degree of dedifferentiation. Thus, the initial adhesive ability to collagen type II is gradually lost when the cells are dedifferentiated by monolayer culture, while adhesion to other proteins such as fibronectin is enhanced [122]. Related to this is the fact that cell behavior highly depends on its capacity to remodel the extracellular protein mesh and substitutes the proteins they encounter initially by those produced by the cells themselves [123, 124]. The composition of the culture medium is also crucial for chondrocyte dedifferentiation, while fetal bovine serum FBS, enhance cell proliferation and dedifferentiation, media without serum-containing insulin transferrin selenium (ITS) and/or growth factors, such as TGFβ, BMP, IGF, favor rounded morphology of the cultured cells. The possibility of the expansion of chondrocytes in monolayer while maintaining their phenotype has been reported in collagen-type-II-grafted substrates [125]. In the same line, proliferation of chondrocytes in three-dimensional substrates such as gels in which the cells are encapsulated or in scaffolds has also been pursued as a way of reaching a high cell number without changes in phenotype [106].

The success in the expansion protocol, whatever it is, must be proven by the ability of the obtained chondrocytes to redifferentiate in a three-dimensional environment in chondrogenic medium or in *in vivo* models as performed in ACI practice. With respect to that, the problem is not very different to that exposed above with respect to chondrogenic differentiation of pluripotent cells. Expanded chondrocytes can be redifferentiated in high-density pellet [126–128] or micromass culture [129–131], or in gels such as alginate [23, 132, 133], poly(ethylene glycol) (PEG) [134–136], hyaluronic acid [137], chondroitin sulfate [137], fibrin glue [138–140], self-assembling polypeptides [141], collagen [142–144], or poly(N-isopropylacrylamide)-based gels [145]. Other three-dimensional scaffolding systems have been tested, such as microparticles [90, 146] or macroporous 3D supports (see below).

Redifferentiation *in vitro* takes place in specific chondrogenic culture media containing growth factors such as TGFβ1 [146–148], TGFβ3 [46, 149], βFGF [150], IGF-1 [151], BMP7 [152], or insulin [126, 153], and it is strongly influenced by other factors, in particular, mechanical stimulation, that are shown to enhance the production of extracellular GAG in the matrix by chondrocytes seeded in both gels and macroporous scaffolds [5, 48, 49, 97, 141, 154]. Proliferation of chondrocytes seems to be affected by compression loading, static compression applied on day 4 after seeding of chondrocytes in a cartilage-support-suppressed proliferation [95], inhibition of proliferation by dynamic compression was also shown in cells seeded in PEG hydrogels [96]. Nevertheless, in other studies, cell numbers were increased with dynamic loading [94, 155]. It seemed that the effect of mechanical loading can be very different depending on the supporting material itself, growth factors added to the culture medium, maturation of the tissue before loading application [156], and sequence of loading application. Thus, Nicodemus [157] found no significant effect of mechanical stimulation on chondrocytes seeded in PEG hydrogels of varying cross-linking density. Bryant *et al.* [96] showed increased GAG production when chondrocytes were cultured in highly cross-linked hydrogels but no effect in loosely cross-linked systems. On the other hand, it has been pointed that the effect of mechanical stimulation is inhibited by the presence of FBS in culture medium [158].

23.5
The Role and Requirements of the Scaffolding Material

23.5.1
Gels Encapsulating Cells as Vehicles for Cell Transplant

One of the problems to solve in cell transplant is the retention of the cells in the site of the implant. In the first generation of ACI technique, the cell suspension is injected in a pocket created by suturing a thin layer of periosteum membrane to the articular surface covering the defect [159]. Cell transplant to the site of the defect can be performed using *in situ* forming gels. Cells are suspended in a solution that gelifies when injected in the site of the cartilage defect. Thus, cells start the regeneration process in a chondrogenic environment as has been well characterized *in vitro*. Different cross-linking reactions are non-unfavorable for cell viability and yield cross-linked networks whose mechanical properties and degradation rate can be modulated by cross-linking density. A broad set of different materials have been used to produce gel-encapsulating cells [160]: proteins such as collagen type I or collagen type II [78, 144, 161], fibrin [139, 140, 162], elastin-like polypeptides [163]; polysaccharides [144] such as hyaluronic acid [5, 164], chitosan [82, 165–168], chondroitin sulfate [5]; agar gel [101], gellan gum [169, 170]; and synthetic hydrogels such as cross-linked PEG [1, 5], or oligo(poly(ethylene glycol) fumarate) hydrogels [85, 146].

23.5.2
Macroporous Scaffolds: Pore Architecture

Macroporous three-dimensional supports have been produced and tested in *in vitro* or *in vivo* models for cartilage engineering (the reader can find comprehensive reviews in Refs [171–173]). As in the case of hydrogels, many different materials have been used as components of macroporous scaffolds, some examples are proteins such as collagen type I or collagen type II [144, 147, 174], gelatin [175], elastin-like [176, 177]; and other polypeptides; polysaccharides such as hyaluronic acid [178], chitosan [147, 178, 179], chondroitin sulfate [174]; synthetic hydrogels such as cross-linked poly(ethylene oxide); hydrophobous biodegradable polyesters as the series of polymers and copolymers based on biodegradable polyesters such as polylactide, polyglicolide (PGA), polycaprolactone (PCL) [112, 180–183]; biodegradable poly(ether ester) multiblock copolymers [184–186], poly(3-hydroxybutyrate-*co*-3-hydroxyhexanoate) [187], and biostable polymers [53, 188–191].

The pore architecture can be important with respect to the mechanical properties of the scaffold, the ability for tissue ingrowth or cell seeding, and permeability to oxygen, nutrients, and waste products of cell metabolism. The pore geometry and interconnectivity is many times related to the material selected, according to the possibilities of processing it into complex three-dimensional forms.

It is very difficult to reach conclusions from the existing literature not only on the materials that allow the best performance but also on the desirable properties of the scaffold to obtain *in vivo* functional hyaline cartilage. The question about what the presence of the scaffold in the site of the osteoarthritic defect can add to cartilage regeneration is still pertinent. Some of the general requirements usually stated for a scaffold in tissue engineering literature must be adapted to the case of articular cartilage engineering.

23.5.3
Cell Adhesion Properties of the Scaffold Surfaces

It is commonly said that the scaffold must provide cells with large surface for adhesion in order to proliferate and start ECM production. Probably, this is not the case of chondrocytes or MSCs transplanted to an osteoarthritic defect into a scaffold. Cell adhesion to the pore walls induces dedifferentiation of the chondrocytes that behave on the pore walls as in a monolayer culture at least in *in vitro* static culture. It is not clear in what extent this behavior changes when implanted *in vivo* or under mechanical or flow stimulation in bioreactor culture. On the contrary, as mentioned above, if the scaffold is colonized by bMSCs coming from subchondral bone they must migrate through a fibrin clot previously formed inside the pores. In this way, little contact between cells and pore walls must be expected, and from this point of view, the chemical structure and other characteristics of the scaffold surfaces are not so important. Perhaps, more crucial is how easily fluids are able to invade the macropore structure, both for cell seeding and for blood invasion. An interesting approach is to fill pores with a gel encapsulating the cells what simulates the environment of bMSCs when migrate into the cartilage defect through the fibrin clot [81, 101]. On the other hand, coating of pore walls with a polysaccharide or protein layer improves cell seeding [180, 192, 193].

23.5.4
Mechanical Properties

One of the important missions of the scaffold is to provide cells with the adequate biomechanical environment. It has been proved that chondrocyte phenotype is strongly influenced by shear and compression loading. Let us assume a macroporous scaffold in which cells have been seeded and implanted in a full depth defect of the articular cartilage. When this construct is subjected to high dynamic compression loading, stress transmission to the cells is controlled by scaffold compliance. Scaffold stiffness is a result of the material elastic modulus, porosity, and pore architecture [194]. A highly porous material (say with 80% volume fraction of pores) can hardly attain the elastic modulus of native articular cartilage. But construct compliance is also restricted by fluid dynamics inside the scaffold produced by the sponge effect. Water absorbed in the construct is distributed in the pores, in the newly formed tissue, and in the scaffold material itself. Compression loading forces water to flow toward less loaded zones. Fluid dynamics inside the scaffold is determined by water diffusion through the scaffold walls and by pore interconnectivity. Some materials, for instance, hydrophobous polyesters, are expected to impose strong restrictions to water permeation, and thus, fluid flow must be ensured by pore interconnectivity, taking place through the volume occupied by cells and ECM. In case of implanting cells encapsulated in a hydrogel of natural or synthetic origin, permeability across the material should allow water motion. If water permeation across the construct is impeded, construct stiffness increases since water is highly incompressible.

It could be firsthypothesized that a design of the cell/scaffold construct yielding global compressive modulus in the order of that of native tissue could be favorable for chondrogenic differentiation. The deformation suffered by a too stiff macroporous material under compression in physiological conditions could be insufficient to transmit compressive stress to the cells lodged in its pores, producing an effect analogous to the stress shielding described in hip prosthetic implants [195, 196]. On the other hand, a too compliant scaffold could collapse in the site of the implant.

The fact that the mechanical behavior of the cells/scaffold construct is a complex combination of the scaffold stiffness, water permeability, and amount and properties of regenerated tissue can explain that adequate mechanical properties of the construct can be achieved with the variety of different materials mentioned in the Section 23.5.2 whose mechanical properties are very different from each other, from very soft to quite stiff ones. Bioresorption of the scaffold materials and ECM production along time makes the time dependence of biomechanical neighborhood of the cells quite difficult to predict.

This discussion highlights the importance of mechanical testing under conditions that simulates as much as possible the *in vivo* environment and the use of model simulation that could predict time evolution of construct properties.

23.5.5
Can Scaffold Architecture Direct Tissue Organization?

Articular cartilage structure is very complex; morphologically different cells can be observed in different positions between articular surface and subchondral bone. A particular spatial organization of cells lacunae is also clear in histological observation. A series of layers of tissue with varying composition, in particular, varying GAG content and collagen fiber organization, and consequently different elastic moduli can be distinguished at different depths from articular surface, as mentioned earlier. This complex architecture can hardly be induced artificially.

Some studies have aimed to induce *in vitro* differentiation toward chondrocytes with the characteristics of those of the different articular cartilage layers identified by the different GAGs content of the ECM they produce [5, 8, 9, 14, 23, 24, 197]. It could be thought that a complex scaffold with layered architecture could be seeded with different cells at different levels with the hope that they produce adequate ECM once implanted. A scaffold with pore architecture able to organize the characteristic columns of cells found in the central part of the cartilage can be designed. A surface layer with different pore structure could be produced as well trying and induce the observed organization of hyaline cartilage in the regenerated tissue. Nevertheless, the evolution of tissue remodeling in the site of the implant makes that this a priori design can hardly be maintained during time.

Probably, cell phenotype in the different cartilage layers, tissue organization, and mechanical stress distribution are highly cross-correlated. It seems far from current possibilities to artificially design highly complex interrelated structures. Nevertheless, as encountered sometimes in tissue engineering, it might suffice to help the organism in some aspects to trigger the natural regeneration of perfectly organized tissue. In the case of articular cartilage, it has been observed the natural chondrogenic differentiation of bMSCs in the site of the defect, perhaps, the modification of stress distribution due to the presence of the scaffold could induce the right chondrocyte phenotype and tissue organization. Interesting results has been found when an empty scaffold is implanted in the site of a full thickness defect, injuring subchondral bone and allowing bleeding and filling of the pore structure with the animal blood. In rabbit models, it was found that tissue is formed inside the pores of the scaffold that are certainly filled by cartilaginous tissue, and a layer of well-organized hyaline cartilage is formed in the articular surface on the scaffold surface [53, 55]. The thickness of this layer with well-organized hyaline structure grows with implant time while it pushes continuously the scaffold toward subchondral bone. Bone remodeling has been found to start immediately after implantation allowing the scaffold to penetrate into bone [191]. This behavior must be compared with that found after microfracture without scaffold, when a tissue with the organization of fibrocartilage is formed. It seems that the presence of the scaffold has been able, in the experiments of references, to modify the sequence of facts of the regenerative process, probably simply by changing the biomechanical environment to which the cells arriving at the upper surface of the scaffold, in the articular surface, are subjected.

23.5.6
Scaffold Biodegradation Rate

When tissue regeneration takes place inside the pores of a scaffold or comes from cells encapsulated in a hydrogel, hyaline tissue organization is disturbed by the presence of the scaffold walls. Bioresorption of the scaffold material and further tissue remodelation is required to form a well-structured tissue. Thus the question of the appropriate rate of scaffold degradation arises. Again, materials with very different degradation times have been tested *in vitro* in static and in bioreactor culture and in animal models as well. It has been shown that the degradation kinetics of the scaffolding material influences chondrocyte phenotype [134, 198, 199], but it is difficult to reach a general conclusion about the stage of tissue regeneration at which the scaffolding material should disappear.

On the other hand, mechanical compressive testing of the region in which a PCL scaffold was implanted in a rabbit knee joint, at an implant time at which few polymer degradation is expected (PCL is a quite slow degrading polymer), shows that a well-integrated construct consisting of a scaffold with the pores filled by regenerated tissue is fully functional [55]. Long-term bioresorbable materials or even biostable ones, such as a biointegrable cartilaginous prosthesis, must not be completely discarded provided their long-term stability could be demonstrated [53, 191].

23.6
Growth Factor Delivery *In Vivo*

The effect of growth factor supply during *in vitro* cell culture of mature chondrocytes or MSCs has been extensively studied and a series of growth factors that positively influences chondrogenic differentiation have been identified.

Growth factor delivery *in vivo* can be controlled by diffusion from the scaffold itself, for instance, growth factors can be encapsulated in biodegradable polyester liberated when the matrix degrades [145, 200]. Hydrogels of natural or synthetic origin can be loaded with the growth factor and deliver it by diffusion [88, 201]. Loaded microspheres can be adhered to a greater or lesser extent to the scaffold walls or encapsulated in a gel at the same time than the cells and deliver the proteins by diffusion or microsphere degradation at a rate that is controlled independently from scaffold degradation [200]. Scaffold wall coatings able to deliver a growth factor have been proposed as well [202, 203].

The interest of this approach could be to supplement the growth factors that diffuse to the site of the defect from neighboring tissues. If the strategy consists of injuring the subchondral bone and implanting an empty scaffold, growth factor delivery could address the differentiation of bMSCs to the chondrogenic phenotype. If differentiated chondrocytes are implanted within an scaffolding material then growth factor delivery should contribute to the maintenance of this phenotype until newly formed tissue is developed.

23.7
Conclusions

Cartilage regeneration that follows bleeding and the formation of a clot in the site of the cartilage defect that is invaded by bone marrow MSCs coming from subchondral bone can be improved by the implant of a scaffold that regulates the biomechanical environment in which the cells are exposed and could artificially deliver growth factors that have been identified as favorable for the induction and maintenance of the phenotype of the hyaline cartilage chondrocyte. This is probably the simplest way to induce tissue regeneration, but perhaps, this strategy could be unsatisfactory in the sense that the heterogeneous population of cells reaching the site of the defect and the signaling factors encountered there could determine a cell fate that makes it unable to produce well-organized hyaline cartilage, even with all the artificial interventions that can be performed. Then the conclusion that a homogeneous and well-characterized chondrocyte population that is preferable would be reached. If so, a number of chondrogenic cells high enough to regenerate an osteoarthritic defect can be obtained from the expansion of bMSCs or mature chondrocytes. These cells can be transplanted to the site of the cartilage defect, expecting that differentiation and tissue regeneration take place in the *in vivo* environment. The problem of this approach is that the *in vivo* niche could be unsuitable to direct proper differentiation toward hyaline cartilage chondrocyte phenotype, similar to what was mentioned for bMSCs invading the site of the defect after injuring the subchondral bone. The alternative is to induce differentiation *in vitro* in a chondrogenic culture medium and under adequate conditions that include hypoxia, mechanical stimulation, the right cell density, a 3D scaffolding vehicle in which cells are seeded, and others. Many works can be found in the literature that identify the role of a particular parameter or a few of these parameters in chondrogenesis. Nevertheless, there is still much work to be done to find the right differentiation conditions that assure that a pull of cells with the phenotype of hyaline cartilage are available to be transplanted to the cartilage defect. On the other hand, the plasticity of the mesenchymal lineage obliges to expect that the phenotype of the chondrocytes implanted evolves with time once they are implanted and to find a new environment different from that of *in vitro* culture. Little is known about this issue, and it seems crucial for the success of any tissue engineering therapy for the regeneration of articular cartilage to identify what the scaffold characteristics, including degradation time and long-term drug supply, can perform to stabilize chondrocyte phenotype until the new tissue is completely functional avoiding the tissue degeneration shown by clinical reports.

Acknowledgment

Support of the Spanish Ministry of Education through project No. MAT2010-21611-C03-01(including the FEDER financial support) is acknowledged.

References

1. Bryant, S.J., Anseth, K.S., Lee, D.A., and Bader, D.L. (2004) *J. Orthop. Res.*, **22**(5), 1143–1149.
2. Wong, B.L., Kim, S.H.C., Antonacci, J.M., McIlwraith, C.W., and Sah, R.L. (2010) *Osteoarthr. Cartil.*, **18**(3), 464–471.
3. Wong, B.L. and Sah, R.L. (2010) *J. Orthop. Res.*, **28**(12), 1554–1561.
4. Martin, I., Suetterlin, R., Baschong, W., Heberer, M., Vunjak-Novakovic, G., and Freed, L.E. (2001) *J. Cell. Biochem.*, **83**(1), 121–128.

5. Nguyen, L.H., Kudva, A.K., Guckert, N.L., Linse, K.D., and Roy, K. (2011) *Biomaterials*, **32**(5), 1327–1338.
6. Martin, I., Jakob, M., Schafer, D., Dick, W., Spagnoli, G., and Heberer, M. (2001) *Osteoarthr. Cartil.*, **9**(2), 112–118.
7. Hunziker, E.B., Quinn, T.M., and Hauselmann, H.J. (2002) *Osteoarthr. Cartil.*, **10**(7), 564–572.
8. Klein, T.J., Malda, J., Sah, R.L., and Hutmacher, D.W. (2009) *Tissue Eng. Part B: Rev.*, **15**(2), 143–157.
9. Klein, T.J., Rizzi, S.C., Reichert, J.C., Georgi, N., Malda, J., Schuurman, W., Crawford, R.W., and Hutmacher, D.W. (2009) *Macromol. Biosci.*, **9**(11), 1049–1058.
10. Schinagl, R.M., Gurskis, D., Chen, A.C., and Sah, R.L. (1997) *J. Orthop. Res.*, **15**(4), 499–506.
11. Yang, P.J. and Temenoff, J.S. (2009) *Tissue Eng. Part B: Rev.*, **15**(2), 127–141.
12. Freemont, A.J. and Hoyland, J. (2006) *Eur. J. Radiol.*, **57**(1), 32–36.
13. Hunziker, E.B., Driesang, I.M.K., and Morris, E.A. (2001) *Clin. Orthop. Relat. Res.*, (391), S171–S181.
14. Sharma, B., Williams, C.G., Kim, T.K., Sun, D.N., Malik, A., Khan, M., Leong, K., and Elisseeff, J.H. (2007) *Tissue Eng.*, **13**(2), 405–414.
15. Palomares, K.T.S., Gerstenfeld, L.C., Wigner, N.A., Lenburg, M.E., Einhorn, T.A., and Morgan, E.F. (2010) *Arthritis Rheum.*, **62**(4), 1108–1118.
16. Grigolo, B., De Franceschi, L., Roseti, L., Cattini, L., and Facchini, A. (2005) *Biomaterials*, **26**(28), 5668–5676.
17. Knudson, W., Aguiar, D.J., Hua, Q., and Knudson, C.B. (1996) *Exp. Cell Res.*, **228**(2), 216–228.
18. Patti, A.M., Gabriele, A., Vulcano, A., Ramieri, M.T., and Della Rocca, C. (2001) *Tissue Cell*, **33**(3), 294–300.
19. Yoo, H.S., Lee, E.A., Yoon, J.J., and Park, T.G. (2005) *Biomaterials*, **26**(14), 1925–1933.
20. Brodkin, K.R., Garcia, A.J., and Levenston, M.E. (2004) *Biomaterials*, **25**(28), 5929–5938.
21. Huang, A.H., Farrell, M.J., and Mauck, R.L. (2010) *J. Biomech.*, **43**(1), 128–136.
22. Jadin, K.D., Wong, B.L., Bae, W.C., Li, K.W., Williamson, A.K., Schumacher, B.L., Price, J.H., and Sah, R.L. (2005) *J. Histochem. Cytochem.*, **53**(9), 1109–1119.
23. Lee, C.S.D., Gleghorn, J.P., Won Choi, N., Cabodi, M., Stroock, A.D., and Bonassar, L.J. (2007) *Biomaterials*, **28**(19), 2987–2993.
24. Ng, K.W., Ateshian, G.A., and Hung, C.T. (2009) *Tissue Eng. Part A*, **15**(9), 2315–2324.
25. Creaby, M.W., Wang, Y., Bennell, K.L., Hinman, R.S., Metcalf, B.R., Bowles, K.A., and Cicuttini, F.M. (2010) *Osteoarthr. Cartil.*, **18**(11), 1380–1385.
26. Beris, A.E., Lykissas, M.G., Papageorgiou, C.D., and Georgoulis, A.D. (2005) *Inj.-Int. J. Care Inj.*, **36**, S14–S23.
27. Hunziker, E.B. (2001) *Osteoarthr. Cartil.*, **10**(6), 432–463.
28. Shapiro, F., Koide, S., and Glimcher, M.J. (1993) *J. Bone Joint Surg.-Am.*, **75A**(4), 532–553.
29. Insall, J. (1974) *Clin. Orthop. Relat. Res.*, (101), 61–67.
30. Steadman, J.R., Rodkey, W.G., Briggs, K.K., and Rodrigo, J.J. (1999) *Orthopade*, **28**(1), 26–32.
31. Pelttari, K., Steck, E., and Richter, W. (2008) *Inj.-Int. J. Care Inj.*, **39**, S58–S65.
32. Brittberg, M., Lindahl, A., Nilsson, A., Ohlsson, C., Isaksson, O., and Peterson, L. (1994) *N. Engl. J. Med.*, **331**(14), 889–895.
33. Nakagawa, T., Lee, S.Y., and Reddi, A.H. (2009) *Arthritis Rheum.*, **60**(12), 3686–3692.
34. Hiramatsu, K., Sasagawa, S., Outani, H., Nakagawa, K., Yoshikawa, H., and Tsumaki, N. (2011) *J. Clin. Invest.*, **121**(2), 640–657.
35. Medvedev, S.P., Grigor'eva, E.V., Shevchenko, A.I., Malakhova, A.A., Dementyeva, E.V., Shilov, A.A., Pokushalov, E.A., Zaidman, A.M., Aleksandrova, M.A., Plotnikov, E.Y., Sukhikh, G.T., and Zakian, S.M. (2011) *Stem Cells Dev.*, **20**(6), 1099–1112.

36. Outani, H., Okada, M., Hiramatsu, K., Yoshikawa, H., and Tsumaki, N. (2011) *Biochem. Biophys. Res. Commun.*, **411**(3), 607–612.
37. Yin, S., Cen, L., Wang, C., Zhao, G.Q., Sun, J., Liu, W., Cao, Y.L., and Cui, L. (2011) *Tissue Eng. Part A*, **16**(5), 1633–1643.
38. Chung, C. and Burdick, J.A. (2008) *Adv. Drug Deliv. Rev.*, **60**(2), 243–262.
39. Darling, E.M. and Athanasiou, K.A. (2003) *Tissue Eng.*, **9**(1), 9–26.
40. Mano, J.F. and Reis, R.L. (2007) *J. Tissue Eng. Regen. Med.*, **1**(4), 261–273.
41. Martin, I., Miot, S., Barbero, A., Jakob, M., and Wendt, D. (2007) *J. Biomech.*, **40**(4), 750–765.
42. Temenoff, J.S. and Mikos, A.G. (2000) *Biomaterials*, **21**(5), 431–440.
43. Mareddy, S., Crawford, R., Brooke, G., and Xiao, Y. (2007) *Tissue Eng.*, **13**(4), 819–829.
44. Vogel, W., Grunebach, F., Messam, C.A., Kanz, L., Brugger, W., and Buhring, H.J. (2003) *Haematologica*, **88**(2), 126–133.
45. Hunziker, E.B. (1999) *Osteoarthr. Cartil.*, **7**(1), 15–28.
46. Lima, E.G., Bian, L., Ng, K.W., Mauck, R.L., Byers, B.A., Tuan, R.S., Ateshian, G.A., and Hung, C.T. (2007) *Osteoarthr. Cartil.*, **15**(9), 1025–1033.
47. Thorpe, S.D., Buckley, C.T., Vinardell, T., O'Brien, F.J., Campbell, V.A., and Kelly, D.J. (2008) *Biochem. Biophys. Res. Commun.*, **377**(2), 458–462.
48. Appelman, T.P., Mizrahi, J., Elisseeff, J.H., and Seliktar, D. (2009) *Biomaterials*, **30**(4), 518–525.
49. Appelman, T.P., Mizrahi, J., Elisseeff, J.H., and Seliktar, D. (2011) *Biomaterials*, **32**(6), 1508–1516.
50. Wang, Y., de Isla, N., Huselstein, C., Wang, B.H., Netter, P., Stoltz, J.F., and Muller, S. (2008) *Bio-Med. Mater. Eng.*, **18**, S47–S54.
51. Dorotka, R., Windberger, U., Macfelda, K., Bindreiter, U., Toma, C., and Nehrer, S. (2005) *Biomaterials*, **26**(17), 3617–3629.
52. Gille, J., Schuseil, E., Wimmer, J., Gellissen, J., Schulz, A.P., and Behrens, P. (2010) *Knee Surg. Sports Traumatol. Arthrosc.*, **18**(11), 1456–1464.
53. Gómez Ribelles, J., Monleón Pradas, M., García Gómez, R., Forriol, F., Sancho Tello, M., and Carda, C. (2010) in *Biodevices 2010* (eds A.L.N. Fred, J. Filipe, and H. Gamboa), INSTICC, Portugal Valencia, pp. 229–234.
54. Khubutiya, M.S., Kliukvin, I.Y., Istranov, L.P., Khvatov, V.B., Shekhter, A.B., Vaza, A.Y., Kanakov, I.V., and Bocharova, V.S. (2008) *Bull. Exp. Biol. Med.*, **146**(5), 658–661.
55. Martinez-Diaz, S., Garcia-Giralt, N., Lebourg, M., Gomez-Tejedor, J.A., Vila, G., Caceres, E., Benito, P., Pradas, M.M., Nogues, X., Ribelles, J.L.G., and Monllau, J.C. (2010) *Am. J. Sports Med.*, **38**(3), 509–519.
56. Huang, A.H., Stein, A., and Mauck, R.L. (2010) *Tissue Eng. A*, **16**(9), 2699–2708.
57. Johnstone, B., Hering, T.M., Caplan, A.I., Goldberg, V.M., and Yoo, J.U. (1998) *Exp. Cell Res.*, **238**(1), 265–272.
58. Pittenger, M.F., Mackay, A.M., Beck, S.C., Jaiswal, R.K., Douglas, R., Mosca, J.D., Moorman, M.A., Simonetti, D.W., Craig, S., and Marshak, D.R. (1999) *Science*, **284**(5411), 143–147.
59. Prockop, D.J. (1997) *Science*, **276**(5309), 71–74.
60. Awad, H.A., Wickham, M.Q., Leddy, H.A., Gimble, J.M., and Guilak, F. (2004) *Biomaterials*, **25**(16), 3211–3222.
61. Betre, H., Ong, S.R., Guilak, F., Chilkoti, A., Fermor, B., and Setton, L.A. (2006) *Biomaterials*, **27**(1), 91–99.
62. Dragoo, J.L., Carlson, G., McCormick, F., Khan-Farooqi, H., Zhu, M., Zuk, P.A., and Benhaim, P. (2007) *Tissue Eng.*, **13**(7), 1615–1621.
63. Estes, B.T., Diekman, B.O., and Guilak, F. (2008) *Biotechnol. Bioeng.*, **99**(4), 986–995.
64. Jin, X.B., Sun, Y.S., Zhang, K., Wang, J., Shi, T.P., Ju, X.D., and Lou, S.Q. (2007) *Biomaterials*, **28**(19), 2994–3003.
65. Jin, X.B., Sun, Y.S., Zhang, K., Wang, J., Shi, T.P., Ju, X.D., and Lou, S.Q. (2008) *J. Biomed. Mater. Res. A*, **86A**(4), 1077–1087.
66. Adachi, N., Sato, K., Usas, A., Fu, F.H., Ochi, M., Han, C.W., Niyibizi, C., and

Huard, J. (2002) *J. Rheumatol.*, **29**(9), 1920–1930.
67. Nawata, M., Wakitani, S., Nakaya, H., Tanigami, A., Seki, T., Nakamura, Y., Saito, N., Sano, K., Hidaka, E., and Takaoka, K. (2005) *Arthritis Rheum.*, **52**(1), 155–163.
68. Emans, P.J., Pieper, J., Hulsbosch, M.M., Koenders, M., Kreijveld, E., Surtel, D.A.M., van Blitterswijk, C.A., Bulstra, S.K., Kuijer, R., and Riesle, J. (2006) *Tissue Eng.*, **12**(6), 1699–1709.
69. Park, Y., Sugimoto, M., Watrin, A., Chiquet, M., and Hunziker, E.B. (2005) *Osteoarthr. Cartil.*, **13**(6), 527–536.
70. Sakaguchi, Y., Sekiya, I., Yagishita, K., and Muneta, T. (2005) *Arthritis Rheum.*, **52**(8), 2521–2529.
71. Yokoyama, A., Sekiya, I., Miyazaki, K., Ichinose, S., Hata, Y., and Muneta, T. (2005) *Cell Tissue Res.*, **322**(2), 289–298.
72. Stevens, M.M., Qanadilo, H.F., Langer, R., and Shastri, V.P. (2004) *Biomaterials*, **25**(5), 887–894.
73. Jorgensen, C., Gordeladze, J., and Noel, D. (2004) *Curr. Opin. Biotechnol.*, **15**(5), 406–410.
74. Chang, C.H., Lin, H.Y., Fang, H.W., Loo, S.T., Hung, S.C., Ho, Y.C., Chen, C.C., Lin, F.H., and Liu, H.C. (2008) *Artif. Organs*, **32**(7), 561–566.
75. Markway, B.D., Tan, G.K., Brooke, G., Hudson, J.E., Cooper-White, J.J., and Doran, M.R. (2010) *Cell Transplant.*, **19**(1), 29–42.
76. Zhang, L.M., Su, P.Q., Xu, C.X., Yang, J.L., Yu, W.H., and Huang, D.S. (2010) *Biotechnol. Lett.*, **32**(9), 1339–1346.
77. Scharstuhl, A., Schewe, B., Benz, K., Gaissmaier, C., Buhring, H.J., and Stoop, R. (2007) *Stem Cells*, **25**(12), 3244–3251.
78. Noth, U., Rackwitz, L., Heymer, A., Weber, M., Baumann, B., Steinert, A., Schutze, N., Jakob, F., and Eulert, J. (2007) *J. Biomed. Mater. Res. A*, **83A**(3), 626–635.
79. Pound, J.C., Green, D.W., Roach, H.I., Mann, S., and Oreffo, R.O.C. (2007) *Biomaterials*, **28**(18), 2839–2849.
80. Mauck, R.L., Yuan, X., and Tuan, R.S. (2006) *Osteoarthr. Cartil.*, **14**(2), 179–189.
81. Ho, S.T.B., Cool, S.M., Hui, J.H., and Hutmacher, D.W. (2010) *Biomaterials*, **31**(1), 38–47.
82. Sa-Lima, H., Caridade, S.G., Mano, J.F., and Reis, R.L. (2010) *Soft Matter*, **6**(20), 5184–5195.
83. Janjanin, S., Li, W.J., Morgan, M.T., Shanti, R.A., and Tuan, R.S. (2008) *J. Surg. Res.*, **149**(1), 47–56.
84. Hui, T.Y., Cheung, K.M.C., Cheung, W.L., Chan, D., and Chan, B.P. (2008) *Biomaterials*, **29**(22), 3201–3212.
85. Park, H., Temenoff, J.S., Tabata, Y., Caplan, A.I., and Mikos, A.G. (2007) *Biomaterials*, **28**(21), 3217–3227.
86. Vinatier, C., Mrugala, D., Jorgensen, C., Guicheux, J., and Noel, D. (2009) *Trends Biotechnol.*, **27**(5), 307–314.
87. Lim, S.M., Oh, S.H., Lee, H.H., Yuk, S.H., Im, G.I., and Lee, J.H. (2010) *J. Mater. Sci. Mater. Med.*, **21**(9), 2593–2600.
88. Park, J.S., Woo, D.G., Yang, H.N., Lim, H.J., Park, K.M., Na, K., and Park, K.H. (2010) *J. Biomed. Mater. Res. A*, **92A**(3), 988–996.
89. Ravindran, S., Roam, J.L., Nguyen, P.K., Hering, T.M., Elbert, D.L., and McAlinden, A. (2011) *Biomaterials*, **32**(33), 8436–8445.
90. Huang, S., Wang, Y.J., Liang, T., Jin, F., Liu, S.X., and Jin, Y. (2009) *Mater. Sci. Eng. C: Biomimetic Supramol. Syst.*, **29**(4), 1351–1356.
91. Grogan, S.P., Barbero, A., Diaz-Romero, J., Cleton-Jansen, A.M., Soeder, S., Whiteside, R., Hogendoorn, P.C.W., Farhadi, J., Aigner, T., Martin, I., and Mainil-Varlet, P. (2007) *Arthritis Rheum.*, **56**(2), 586–595.
92. Han, S.H., Kim, Y.H., Park, M.S., Kim, I.A., Shin, J.W., Yangs, W.I., Jee, K.S., Park, K.D., Ryu, G.H., and Lee, J.W. (2008) *J. Biomed. Mater. Res. A*, **87A**(4), 850–861.
93. Grellmann, W., Berghaus, A., Haberland, E.J., Jamali, Y., Holweg, K., Reincke, K., and Bierogel, C. (2006) *J. Biomed. Mater. Res. A*, **78A**(1), 168–174.
94. Lee, D.A. and Bader, D.L. (1997) *J. Orthop. Res.*, **15**(2), 181–188.
95. Li, K.W., Falcovitz, Y.H., Nagrampa, J.P., Chen, A.C., Lottman, L.M., Shyy,

J.Y.J., and Sah, R.L. (2000) *J. Orthop. Res.*, **18**(3), 374–382.
96. Bryant, S.J., Chowdhury, T.T., Lee, D.A., Bader, D.L., and Anseth, K.S. (2004) *Ann. Biomed. Eng.*, **32**(3), 407–417.
97. Mauck, R.L., Seyhan, S.L., Ateshian, G.A., and Hung, C.T. (2002) *Ann. Biomed. Eng.*, **30**(8), 1046–1056.
98. Brittberg, M., Peterson, L., Sjogren-Jansson, E., Tallheden, T., and Lindahl, A. (2003) *J. Bone Joint Surg.-Am.*, **85A**, 109–115.
99. Nehrer, S., Domayer, S., Dorotka, R., Schatz, K., Bindreiter, U., and Kotz, R. (2006) *Eur. J. Radiol.*, **57**(1), 3–8.
100. Smith, G.D., Knutsen, G., and Richardson, J.B. (2005) *J. Bone Joint Surg.-Br.*, **87B**(4), 445–449.
101. Gong, Y.H., He, L.J., Li, J., Zhou, Q.L., Ma, Z.W., Gao, C.Y., and Shen, J.C. (2007) *J. Biomed. Mater. Res. B: Appl. Biomater.*, **82B**(1), 192–204.
102. Yamaoka, H., Asato, H., Ogasawara, T., Nishizawa, S., Takahashi, T., Nakatsuka, T., Koshima, I., Nakamura, K., Kawaguchi, H., Chung, U.I., Takato, T., and Hoshi, K. (2006) *J. Biomed. Mater. Res. A*, **78A**(1), 1–11.
103. Kuettner, K.E., Pauli, B.U., Gall, G., Memoli, V.A., and Schenk, R.K. (1982) *J. Cell Biol.*, **93**(3), 743–750.
104. Roskelley, C.D., Srebrow, A., and Bissell, M.J. (1995) *Curr. Opin. Cell Biol.*, **7**(5), 736–747.
105. Malda, J., Van den Brink, P., Meeuwse, P., Grojec, M., Martens, D.E., Tramper, J., Riesle, J., and van Blitterswijk, C.A. (2004) *Tissue Eng.*, **10**(7–8), 987–994.
106. Melero-Martin, J.M., Dowling, M.A., Smith, M., and Al-Rubeai, M. (2006) *Biomaterials*, **27**(15), 2970–2979.
107. Boyan, B.D., Hummert, T.W., Dean, D.D., and Schwartz, Z. (1996) *Biomaterials*, **17**(2), 137–146.
108. Cutler, S.M. and Garcia, A.J. (2003) *Biomaterials*, **24**(10), 1759–1770.
109. Garcia, A.J. (2005) *Biomaterials*, **26**(36), 7525–7529.
110. Keselowsky, B.G., Collard, D.M., and Garcia, A.J. (2003) *J. Biomed. Mater. Res. Part A*, **66A**(2), 247–259.
111. Hernandez, J.C.R., Sanchez, M.S., Soria, J.M., Ribelles, J.L.G., and Pradas, M.M. (2007) *Biophys. J.*, **93**(1), 202–207.
112. Olmedilla, M.P., Garcia-Giralt, N., Pradas, M.M., Ruiz, P.B., Ribelles, J.L.G., Palou, E.C., and Garcia, J.C.M. (2006) *Biomaterials*, **27**(7), 1003–1012.
113. Ballester-Beltran, J., Rico, P., Moratal, D., Song, W.L., Mano, J.F., and Salmeron-Sanchez, M. (2011) *Soft Matter*, **7**(22), 10803–10811.
114. Gonzalez-Garcia, C., Sousa, S.R., Moratal, D., Rico, P., and Salmeron-Sanchez, M. (2010) *Colloids Surf. B: Biointerfaces*, **77**(2), 181–190.
115. Gugutkov, D., Altankov, G., Hernandez, J.C.R., Pradas, M.M., and Sanchez, M.S. (2010) *J. Biomed. Mater. Res. A*, **92A**(1), 322–331.
116. Llopis-Hernandez, V., Rico, P., Ballester-Beltran, J., Moratal, D., and Salmeron-Sanchez, M. (2011) *PLoS ONE*, **6**(5), e19610.
117. Perez-Garnes, M., Gonzalez-Garcia, C., Moratal, D., Rico, P., and Salmeron-Sanchez, M. (2011) *Int. J. Artif. Organs*, **34**(1), 54–63.
118. Rico, P., Hernandez, J.C.R., Moratal, D., Altankov, G., Pradas, M.M., and Salmeron-Sanchez, M. (2009) *Tissue Eng. A*, **15**(11), 3271–3281.
119. Salmeron-Sanchez, M., Rico, P., Moratal, D., Lee, T.T., Schwarzbauer, J.E., and Garcia, A.J. (2011) *Biomaterials*, **32**(8), 2099–2105.
120. Costa Martínez, E., Rodríguez Hernández J.C., Machado, M., Mano, J.F., Gómez Ribelles, J.L., Monleón Pradas, M., and Salmerón Sánchez, M. (2008) *Tissue Eng. A*, **14**(10), 1751–1762.
121. Costa Martínez, E., Escobar Ivirico, J.L., Muñoz Criado, I., Gómez Ribelles, J.L., Monleón Pradas, M., and Salmerón Sánchez, M. (2007) *J. Mater. Sci. Mater. Med.*, **18**(8), 1627–1632.
122. Schmal, H., Mehlhorn, A.T., Fehrenbach, M., Muller, C.A., Finkenzeller, G., and Sudkamp, N.P. (2006) *Tissue Eng.*, **12**(4), 741–750.
123. Altankov, G., Grinnell, F., and Groth, T. (1996) *J. Biomed. Mater. Res.*, **30**(3), 385–391.

124. Altankov, G. and Groth, T. (1994) *J. Mater. Sci. Mater. Med.*, **5**(9–10), 732–737.
125. Barbero, A., Grogan, S.P., Mainil-Varlet, P., and Martin, I. (2006) *J. Cell. Biochem.*, **98**(5), 1140–1149.
126. Andreas, K., Zehbe, R., Kazubek, M., Grzeschik, K., Sternberg, N., Baumler, H., Schubert, H., Sittinger, M., and Ringe, J. (2011) *Acta Biomater.*, **7**(4), 1485–1495.
127. Cheuk, Y.C., Wong, M.W.N., Lee, K.M., and Fu, S.C. (2011) *J. Orthop. Res.*, **29**(9), 1343–1350.
128. Gohring, A.R., Lubke, C., Andreas, K., Haupl, T., Sittinger, M., Ringe, J., Kaps, C., Pruss, A., and Perka, C. (2010) *Biotechnol. Prog.*, **26**(4), 1116–1125.
129. Anderer, U. and Libera, J. (2002) *J. Bone Miner. Res.*, **17**(8), 1420–1429.
130. Dehne, T., Schenk, R., Perka, C., Morawietz, L., Pruss, A., Sittinger, M., Kaps, C., and Ringe, J. (2010) *Gene*, **462**(1–2), 8–17.
131. Schrobback, K., Klein, T.J., Schuetz, M., Upton, Z., Leavesley, D.I., and Malda, J. (2011) *J. Orthop. Res.*, **29**(4), 539–546.
132. Coleman, R.M., Case, N.D., and Guldberg, R.E. (2007) *Biomaterials*, **28**(12), 2077–2086.
133. Choi, B.H., Woo, J.I., Min, B.H., and Park, S.R. (2006) *J. Biomed. Mater. Res. Part A*, **79A**(4), 858–864.
134. Bryant, S.J., Bender, R.J., Durand, K.L., and Anseth, K.S. (2004) *Biotechnol. Bioeng.*, **86**(7), 747–755.
135. Bryant, S.J., Davis-Arehart, K.A., Luo, N., Shoemaker, R.K., Arthur, J.A., and Anseth, K.S. (2004) *Macromolecules*, **37**(18), 6726–6733.
136. Nicodemus, G.D. and Bryant, S.J. (2008) *Tissue Eng. B: Rev.*, **14**(2), 149–165.
137. Banu, N. and Tsuchiya, T. (2007) *J. Biomed. Mater. Res. A*, **80A**(2), 257–267.
138. Chou, C.H., Cheng, W.T.K., Kuo, T.F., Sun, J.S., Lin, F.H., and Tsai, J.C. (2007) *J. Biomed. Mater. Res. A*, **82A**(3), 757–767.
139. Eyrich, D., Brandl, F., Appel, B., Wiese, H., Maier, G., Wenzel, M., Staudenmaier, R., Goepferich, A., and Blunk, T. (2007) *Biomaterials*, **28**(1), 55–65.
140. Peretti, G.M., Xu, J.W., Bonassar, L.J., Kirchhoff, C.H., Yaremchuk, M.J., and Randolph, M.A. (2006) *Tissue Eng.*, **12**(5), 1151–1168.
141. Kisiday, J.D., Jin, M.S., DiMicco, M.A., Kurz, B., and Grodzinsky, A.J. (2004) *J. Biomech.*, **37**(5), 595–604.
142. Dorotka, R., Bindreiter, U., Macfelda, K., Windberger, U., and Nehrer, S. (2005) *Osteoarthr. Cartil.*, **13**(8), 655–664.
143. Dorotka, R., Bindreiter, U., Vavken, P., and Nehrer, S. (2005) *Tissue Eng.*, **11**(5–6), 877–886.
144. Vickers, S.M., Squitieri, L.S., and Spector, M. (2006) *Tissue Eng.*, **12**(5), 1345–1355.
145. Park, J.S., Woo, D.G., Yang, H.N., Na, K., and Park, K.H. (2009) *J. Biomed. Mater. Res. A*, **91A**(2), 408–415.
146. Park, H., Temenoff, J.S., Holland, T.A., Tabata, Y., and Mikos, A.G. (2005) *Biomaterials*, **26**(34), 7095–7103.
147. Lee, J.E., Kim, K.E., Kwon, I.C., Ahn, H.J., Lee, S.H., Cho, H.C., Kim, H.J., Seong, S.C., and Lee, M.C. (2004) *Biomaterials*, **25**(18), 4163–4173.
148. Lee, J.E., Kim, S.E., Kwon, I.C., Ahn, H.J., Cho, H., Lee, S.H., Kim, H.J., Seong, S.C., and Lee, M.C. (2004) *Artif. Organs*, **28**(9), 829–839.
149. Byers, B.A., Mauck, R.L., Chiang, I.E., and Tuan, R.S. (2008) *Tissue Eng. A*, **14**(11), 1821–1834.
150. Huang, X., Yang, D.S., Yan, W.Q., Shi, Z.L., Feng, J., Gao, Y.B., Weng, W.J., and Yan, S.G. (2007) *Biomaterials*, **28**(20), 3091–3100.
151. Bonassar, L.J., Grodzinsky, A.J., Srinivasan, A., Davila, S.G., and Trippel, S.B. (2000) *Arch. Biochem. Biophys.*, **379**(1), 57–63.
152. Gavenis, K., Klee, D., Pereira-Paz, R.M., von Walter, M., Mollenhauer, J., Schneider, U., and Schmidt-Rohlfing, B. (2007) *J. Biomed. Mater. Res. B: Appl. Biomater.*, **82B**(2), 275–283.
153. Malafaya, P.B., Oliveira, J.T., and Reis, R.L. (2010) *Tissue Eng. A*, **16**(2), 735–747.

154. Wang, Y., de Isla, N., Decot, V., Marchal, L., Cauchois, G., Huselstein, C., Muller, S., Wang, B.H., Netter, P., and Stoltz, J.F. (2008) *Biorheology*, **45**(3–4), 527–538.
155. Dewitt, M.T., Handley, C.J., Oakes, B.W., and Lowther, D.A. (1984) *Connect. Tissue Res.*, **12**(2), 97–109.
156. Demarteau, O., Wendt, D., Braccini, A., Jakob, M., Schafer, D., Heberer, M., and Martin, I. (2003) *Biochem. Biophys. Res. Commun.*, **310**(2), 580–588.
157. Nicodemus, G.D., and Bryant, S.J. (2008) *J. Biomech.*, **41**(7), 1528–1536.
158. Bilgen, B., Orsini, E., Aaron, R.K., and Ciombor, D.M. (2007) *J. Tissue Eng. Regen. Med.*, **1**(6), 436–442.
159. Marlovits, S., Zeller, P., Singer, P., Resinger, C., and Vecsei, V. (2006) *Eur. J. Radiol.*, **57**(1), 24–31.
160. Kim, I.L., Mauck, R.L., and Burdick, J.A. (2011) *Biomaterials*, **32**(34), 8771–8782.
161. Zscharnack, M., Poesel, C., Galle, J., and Bader, A. (2009) *Cells Tissues Organs*, **190**(2), 81–93.
162. Ahmed, T.A.E., Dare, E.V., and Hincke, M. (2008) *Tissue Eng. B: Rev.*, **14**(2), 199–215.
163. Hrabchak, C., Rouleau, J., Moss, I., Woodhouse, K., Akens, M., Bellingham, C., Keeley, F., Dennis, M., and Yee, A. (2010) *Acta Biomater.*, **6**(6), 2108–2115.
164. Chung, C., Mesa, J., Randolph, M.A., Yaremchuk, M., and Burdick, J.A. (2006) *J. Biomed. Mater. Res. A*, **77A**(3), 518–525.
165. Cho, J.H., Kim, S.H., Park, K.D., Jung, M.C., Yang, W.I., Han, S.W., Noh, J.Y., and Lee, J.W. (2004) *Biomaterials*, **25**(26), 5743–5751.
166. Hu, X.H., Zhou, J., Zhang, N., Tan, H.P., and Gao, C.Y. (2008) *J. Mech. Behav. Biomed. Mater.*, **1**(4), 352–359.
167. Jin, R., Teixeira, L.S.M., Dijkstra, P.J., Karperien, M., van Blitterswijk, C.A., Zhong, Z.Y., and Feijen, J. (2009) *Biomaterials*, **30**(13), 2544–2551.
168. Tan, H.P., Chu, C.R., Payne, K.A., and Marra, K.G. (2009) *Biomaterials*, **30**(13), 2499–2506.
169. Oliveira, J.T., Martins, L., Picciochi, R., Malafaya, I.B., Sousa, R.A., Neves, N.M., Mano, J.F., and Reis, R.L. (2010) *J. Biomed. Mater. Res. A*, **93A**(3), 852–863.
170. Oliveira, J.T., Picciochi, R., Santos, T.C., Martins, L., Pinto, L.G., Malafaya, P.B., Sousa, R.A., Marques, A.P., Castro, A.G., Mano, J.F., Neves, N.M., and Reis, R.L. (2008) *Tissue Eng. A*, **14**(5), 748–748.
171. Hutmacher, D.W. (2000) *Biomaterials*, **21**(24), 2529–2543.
172. Lee, J., Cuddihy, M.J., and Kotov, N.A. (2008) *Tissue Eng. B: Rev.*, **14**(1), 61–86.
173. Liu, C., Xia, Z., and Czernuszka, J.T. (2007) *Chem. Eng. Res. Des.*, **85**(A7), 1051–1064.
174. van Susante, J.L.C., Pieper, J., Buma, P., van Kuppevelt, T.H., van Beuningen, H., van der Kraan, P.M., Veerkamp, J.H., van den Berg, W.B., and Veth, R.P.H. (2001) *Biomaterials*, **22**(17), 2359–2369.
175. Xia, W.Y., Liu, W., Cui, L., Liu, Y.C., Zhong, W., Liu, D.L., Wu, J.J., Chua, K.H., and Cao, Y.L. (2004) *J. Biomed. Mater. Res. B: Appl. Biomater.*, **71B**(2), 373–380.
176. Annabi, N., Fathi, A., Mithieux, S.M., Martens, P., Weiss, A.S., and Dehghani, F. (2011) *Biomaterials*, **32**(6), 1517–1525.
177. Nettles, D.L., Kitaoka, K., Hanson, N.A., Flahiff, C.M., Mata, B.A., Hsu, E.W., Chilkoti, A., and Setton, L.A. (2008) *Tissue Eng. A*, **14**(7), 1133–1140.
178. Yamane, S., Iwasaki, N., Majima, T., Funakoshi, T., Masuko, T., Harada, K., Minami, A., Monde, K., and Nishimura, S. (2005) *Biomaterials*, **26**(6), 611–619.
179. Duarte, A.R.C., Mano, J.F., and Reis, R.L. (2010) *Adv. Mater. Forum V*, **636–637** (Pt 1 and 2), 22–25.
180. Antunes, J.C., Oliveira, J.M., Reis, R.L., Soria, J.M., Gomez-Ribelles, J.L., and Mano, J.F. (2010) *J. Biomed. Mater. Res. A*, **94A**(3), 856–869.
181. Garcia-Giralt, N., Izquierdo, R., Nogues, X., Perez-Olmedilla, M., Benito, P., Gomez-Ribelles, J.L., Checa, M.A., Suay, J., Caceres, E., and Monllau, J.C. (2008) *J. Biomed. Mater. Res. A*, **85A**(4), 1082–1089.

182. Ivirico, J.L.E., Salmeron-Sanchez, M., Ribelles, J.L.G., Pradas, M.M., Soria, J.M., Gomes, M.E., Reis, R.L., and Mano, J.F. (2009) *J. Biomed. Mater. Res. B: Appl. Biomater.*, **91B**(1), 277–286.

183. Izquierdo, R., Garcia-Giralt, N., Rodriguez, M.T., Caceres, E., Garcia, S.J., Ribelles, J.L.G., Monleon, M., Monllau, J.C., and Suay, J. (2008) *J. Biomed. Mater. Res. A*, **85A**(1), 25–35.

184. Malda, J., Woodfield, T.B.F., van der Vloodt, F., Wilson, C., Martens, D.E., Tramper, J., van Blitterswijk, C.A., and Riesle, J. (2005) *Biomaterials*, **26**(1), 63–72.

185. Miot, S., Woodfield, T., Daniels, A.U., Suetterlin, R., Peterschmitt, I., Heberer, M., van Blitterswijk, C.A., Riesle, J., and Martin, I. (2005) *Biomaterials*, **26**(15), 2479–2489.

186. Woodfield, T.B.F., Malda, J., de Wijn, J., Peters, F., Riesle, J., and van Blitterswijk, C.A. (2004) *Biomaterials*, **25**(18), 4149–4161.

187. Wang, Y., Bian, Y.Z., Wu, Q., and Chen, G.Q. (2008) *Biomaterials*, **29**(19), 2858–2868.

188. Barry, J.J.A., Gidda, H.S., Scotchford, C.A., and Howdle, S.M. (2004) *Biomaterials*, **25**(17), 3559–3568.

189. Brígido Diego, R., Pérez Olbedilla, M., Serrano Aroca, A., Gómez Ribelles, J.L., Monleón Pradas, M., and Salmerón Sánchez, M. (2005) *J. Mater. Sci. Mater. Med.*, **16**(8), 693–698.

190. Brígido Diego, R., Gómez Ribelles, J.L., and Salmerón Sánchez, M. (2007) *J. Appl. Polym. Sci.*, **104**(3), 1475–1481.

191. Maher, S.A., Doty, S.B., Torzilli, P.A., Thornton, S., Lowman, A.M., Thomas, J.D., Warren, R., Wright, T.M., and Myers, E. (2007) *J. Biomed. Mater. Res. A*, **83A**(1), 145–155.

192. Gamboa Martínez, T., Gómez Ribelles, J.L., and Gallego Ferrer, G. (2011) *J. Bioact. Compat. Polym.*, **26**(5), 464–477.

193. Mano, J.F., Hungerford, G., and Gómez Ribelles, J.L. (2008) *Mater. Sci. Eng. C-Biomimetic Supramol. Syst.*, **28**(8), 1356–1365.

194. Gibson, L.J. and Ashby, M.F. (1999) *Cellular Solids – Structure and Properties*, Cambridge University Press, Cambridge.

195. Hallab, N.J., Jacobs, J.J., and Katz, A.J.L. (2004) in *Biomaterials Science* (eds B.D. Ratner, A.S. Hoffman, F.J. Schoen, and J.E. Lemons), Elsevier and Academic Press, Amsterdam, pp. 527–555.

196. Ronca, D. and Guida, G. (2002) in *Integrated Biomaterials Science* (ed. R. Barbucci), Kluwer Academic and Plenum Publishers, New York, pp. 527–550.

197. Hwang, N.S., Varghese, S., Lee, H.J., Theprungsirikul, P., Canver, A., Sharma, B., and Elisseeff, J. (2007) *FEBS Lett.*, **581**(22), 4172–4178.

198. Bryant, S.J. and Anseth, K.S. (2003) *J. Biomed. Mater. Res. A*, **64A**(1), 70–79.

199. Rice, M.A. and Anseth, K.S. (2004) *J. Biomed. Mater. Res. A*, **70A**(4), 560–568.

200. Solorio, L.D., Fu, A.S., Hernandez-Irizarry, R., and Alsberg, E. (2010) *J. Biomed. Mater. Res. A*, **92A**(3), 1139–1144.

201. Hunziker, E.B. (2001) *Osteoarthr. Cartil.*, **9**(1), 22–32.

202. Chung, H.J. and Park, T.G. (2007) *Adv. Drug Deliv. Rev.*, **59**(4–5), 249–262.

203. Prabaharan, M., Rodriguez-Perez, M.A., de Saja, J.A., and Mano, J.F. (2007) *J. Biomed. Mater. Res. B: Appl. Biomater.*, **81B**(2), 427–434.

24
Soft Constructs for Skin Tissue Engineering

Simone S. Silva, João F. Mano, and Rui L. Reis

24.1
Introduction

Inspired by nature's models, processes, and elements, numerous investigations have led to the development of new technologies and advanced skin tissue engineering (TE) construct materials that can enhance skin regeneration [1–4]. Some of those strategies are based on the crossroads between biomaterials, cells, and the mechanism of skin healing [1, 5–7]. In fact, skin TE refers to skin products made by matrix materials or cells or combination of both, which are required in the healing process to proceed [8, 9]. Moreover, considering the complexity of the human skin, the success of the certain skin products will depend on the possibilities to recreate the native skin environment. By its turn, nature can provide a powerful resource of biopolymers with a wide range of structures and functions that are capable of modifying and adapting to a range of different environments. Some examples of natural polymers applied for skin TE are hyaluronic acid, alginate, chitin/chitosan, and fibrin [10, 11]. Furthermore, synthetic polymers such as poly(lactic-*co*-glycolic acid) (PLGA) [12], poly(ε-caprolactone) (PCL) [13], and polyurethane (PU) [14] have been extensively studied for the same purpose. Particularly for skin TE, the properties of such materials can be designed from the functional requirements and application of engineering principles to the further generation of a determined soft construct. Many works [2, 15–18] involving *in vitro* and *in vivo* investigations on the mentioned macromolecules suggest their potential use in the wound healing process as well as the creation of novel skin substitutes, dermal replacements, and epidermal sheets.

24.2
Structure of Skin

The skin is a multilayered material with well-defined anatomical regions, namely, the epidermis, the dermis, and the hypodermis [5, 19–21] (Figure 24.1). The epidermis is an avascular tissue that has keratinocytes as the major cell type,

Biomimetic Approaches for Biomaterials Development, First Edition. Edited by João F. Mano.
© 2012 Wiley-VCH Verlag GmbH & Co. KGaA. Published 2012 by Wiley-VCH Verlag GmbH & Co. KGaA.

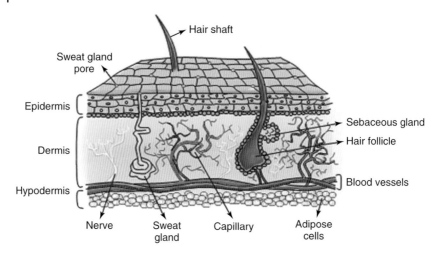

Figure 24.1 Schematic representation of the human skin. (Source: Adapted from Refs [5, 19, 22].)

representing approximately 95% of the total cell population [20, 21]. These cells produce a large variety of polypeptide growth factors and cytokines, which act as signals between cells and help to regulate skin function [19]. The function of epidermis is to prevent moisture and heat loss from the skin and bacterial infiltration from the environment. Situated directly below the epidermis is the dermis, a vascular layer with a thickness between 1 and 4 mm [21]. It is composed of glycosaminoglycans – GAGs and collagen with elastin [5]. The dermis plays an active role in wound healing as fibroblasts, the major cell type present in the dermis, synthesize new collagen and also produce proteolytic enzymes necessary for remodeling [5]. The hypodermis is located underneath the dermis. It consists of well-vascularized adipose tissue that also contributes to the mechanical properties of the skin, as well as to its ability to regulate body temperature [5, 19].

24.2.1
Wound Healing

When the skin is wounded, a complex series of cellular and chemical events are initiated that will act on the damaged tissues – blood vessels, dermis, and epidermis. By definition, a wound is a disruption of normal anatomical structure and function that involves complex processes, which results in the restoration of anatomical continuity and function [23]. Basically, the wound healing process is the tissue response to injury and the process of regeneration. There are many published reports in the literature [24–26] describing the various biological and physiological stages of the wound healing process. These stages can be summarized into the following phases: hemostasis and inflammation, reepithelialization and granulation tissue formation, and finally, tissue remodeling [27, 28]. Each phase in itself is complex, requiring interactions between biochemical mediators, cell types, matrix components, and environmental factors. This sequential process requires the interaction of cells in the dermis and epidermis, as well as the activity of chemical mediators released from inflammatory cells, fibroblasts, and keratinocytes [28].

On the basis of wound healing processes, there are two types of wounds: acute and chronic wounds [29]. Acute wounds can be caused by mechanical damage induced by sheer, blunting, or exposure to extreme heat, irradiation, and corrosive chemicals [29]. These wounds may take 8–12 weeks to heal [29]. Clinically, chronic wounds are produced as a result of specific diseases such as venous leg ulcers, diabetic foot ulcers, tumors, and severe physiological contaminations [30, 31]. These wounds could take more than 12 weeks to heal, and recurrence of the wounds is not uncommon [31, 32]. In addition to the above wound types, one can classify wounds according to the layers involved: (i) superficial wounds, which involve only the loss of epidermis; (ii) partial thickness wounds, which involve the epidermis and dermis; and (iii) full-thickness wounds, including the total structure of skin (epidermis, dermis, and hypodermis) [1].

24.3
Current Biomaterials in Wound Healing

Nowadays, the increasing number of skin products composed by natural polymers continues to expand, aiming to improve the wound care [1–3, 33], which has also been followed by the increase of intellectual property. Natural polymers such as cellulose, alginate, and chitin/chitosan have been proposed for skin TE because of their well-known physicochemical and biological properties [10, 11, 34]. These polymers have also a wide variety of structures, which confer those interesting properties such as easy gelation, solubility in water, antimicrobial activity, high hydrophilic character, self-assembly ability, and wound healing properties [11, 35–40]. Despite these advantages and depending on how they are processed, the resulting materials can present some weaknesses related to their structural stability, inappropriate mechanical properties, and contraction (shrinkage). Cross-linking reactions, combinations with other biopolymer or synthetic materials, or even modification either at surface or bulk level are some of the solutions for the mentioned problems [10, 11, 34]. Particularly for skin TE, efficient wound management requires an understanding of the tissue repair process and knowledge of the properties of dressing materials available. Properties and characteristics of selected polymers currently used in skin TE are described in the following sections.

24.3.1
Alginate

Alginate, a polymer derived from sea algae [34], has been widely used in the wound management industry for the production of absorbent products such as gels, foams, and fibrous dressings used to cover wounds [36]. On contact with wound exudates, the alginate changes from a soft fibrous texture into a gel that helps to keep a moist interface between the dressing and the wound surface, which can assist in the healing process [41, 42]. Alginate-based dressings available in the market of medical devices include Sorbsan® (Maersk), Kaltostat® (Conva Tec), and Algisite® M (Smith&Nephew) [43]. Furthermore, conjugation of alginate with silk fibroin [44], chitosan [45], and collagen [46] has lead to formation of biomatrices, which have also been proposed as wound dressings. By comparing the wound healing effect of silk fibroin/alginate sponges on rat full-thickness wound model, the authors observed a significant increase in the size of reepithelialization via rapid proliferation of epithelial cells [44]. Clinical studies [46] indicated the efficacy and safety of a combination of collagen (90%) and alginate (10%), commercialized as FIBRACOL PLUS Dressing (Johnson & Johnson Gateway®) in the treatment of diabetic foot ulcers.

24.3.2
Cellulose

Cellulose is an important skeletal component in plants and the most abundant organic polymer in the world [38]. Numerous cellulose derivatives such as oxidized regenerated cellulose (ORC), hydroxypropyl cellulose (HPC), and sodium carboxymethylcellulose (SCMC) have emerged in the health care sector as wound dressings and artificial skin products [47–49]. ORC, for example, is a very useful material as a wound dressing since it includes antimicrobial activity because of its ability to reduce pH [48]. Commercially, Promogran®, a spongy matrix containing ORC (45%) and collagen (55%), has been introduced to both the USand EU markets. In the presence of chronic wound exudates, this material physically binds and inactivates matrix mettalloproteases (MMPs), which have a detrimental effect on wound healing when present in excessive quantities [50].

Pure SCMC, a sodium salt of carboxymethyl ether cellulose, has been used alone or in combination with drug and coexcipients as a wound dressing for the treatment of partial thickness wound and deep diabetic foot ulcer and as a dermal filler [49]. Bacterial cellulose (BC) has also been proposed to be used as an effective wound material because of its unique properties such as versatility, moldability *in situ*, biocompatibility, high-water-holding ability, cost effect production, and high mechanical strength in the wet state [51, 52].

Recently, cellulose has been dissolved and modified or functionalized using a new class of solvents called ionic liquids (ILs) [53, 54], creating novel possibilities for its application. In a particular example, Park *et al.* [54] produced cellulose/chitosan composite nanofibers using 1-ethyl-3-methylimidazolium acetate, which could be useful as an antibacterial reagent to treat skin ulcers.

24.3.3
Chitin/Chitosan

Chitin is the second most abundant natural polymer found in the shell of crustacean, cuticles of insects, and cell walls of fungi [55–57]. Chitin and its deaceatylated derivative, chitosan, have been widely used as a base material in the production of matrices for wound management because of its hemostatic properties, stimulation of healing, biocompatibility, biodegradability, antimicrobial, and hydrating properties [55, 58–60]. Previous studies [61, 62] have also shown that chitin derivatives, for example, carboxymethyl chitin, water-soluble chitin, and dibutylchitin (DBC) with good solubility are particularly effective as wound healing accelerator and as wound dressings. Commercial wound dressings based on chitin and chitosan are described in Table 24.1. Most of them are manufactured only in a few countries (the United States, Japan, and some European countries) [60].

Typically, chitosan solutions have been used to produce membranes, nanofibers, fibers, sponges, and hydrogels, which are proposed as wound dressings and dermal equivalents [63–67]. Chitosan-based materials can be used to prevent or treat wound and burn infections not only because of its intrinsic antimicrobial properties but also by virtue of its ability to deliver extrinsic antimicrobial agents to wounds and burns [68]. Other studies [58] also indicated that chitosan has hemostatic properties, which help in natural blood clotting and block nerve endings, hence reducing pain.

Among the distinct chitosan matrices, chitosan membranes have been widely investigated for the purpose of wound covering because of their easy production and long shelf life [64, 65, 67, 69]. Nevertheless, the development of chitosan membranes with desirable properties sometimes requires modification at surface or bulk level by chemical and physical means [64, 67, 70]. The application of plasma surface modification on chitosan membranes, for instance, is an interesting strategy to improve cellular adhesion, as reported in the literature [64, 67, 70]. Promising results on wound healing have also been achieved using polyelectrolyte complexes (PECs) formed by chitosan and alginate [45] or gelatin [71, 72]. These PECs can be processed into sponges [71] as well as bilayer scaffolds for use in human skin fibroblast and keratinocyte transplantation [72, 73]. Other studies have indicated that collagen/chitosan bilayer (scaffold/membrane) has the ability to regenerate a damaged dermis and to support the angiogenesis [74]. Besides collagen and gelatin, chitosan has also been widely associated to other proteins such as silk fibroin [75], soy protein [18, 76, 77], keratin [78], and heparin [79]. For example, Silva *et al.* [18, 76], proposed the use of chitosan/soy protein (CSS) blended membranes as wound dressings. Moreover, *in vivo* studies [18] demonstrated that the CSS membranes accelerated skin wound healing in rats after two weeks of dressing. Also, chitosan derivatives with good mechanical, water solubility, and biological properties have been proposed for skin applications [10, 49, 56, 60]. Furthermore, the use of ILs as green solvents for chitin/chitosan suggested an improvement in their processability for widespread application of these polymers [10, 15]. For instance, *in vitro* assays on chitosan/silk fibroin (CSF) hydrogel-based constructs, produced using ILs, demonstrated that CSF hydrogels supported the adhesion and

Table 24.1 Commercial wound dressings based on chitin and chitosan derivatives [60, 68, 81, 82].

Trade name	Company	Composition	Properties/applications
Beschitin®	Unitika	Nonwoven manufactured using chitin	Accelerates granulation phase and prevents scar formation.
Choriochit®	Tissue Bank of the Blood Donor Centre in Katowice and the Silesian Medical Academy in Katowice	Lyophilized human placenta blended with microcrystalline chitosan	Good handling, good wound isolation, and ability to reduce pathogens growth.
ChitiPack® S	Eisai Co.	Spongelike chitin from squid pen	Favors early granulation, no scar formation; applied in traumatic wounds, surgical tissue defects.
ChitiPack® P	Eisai Co.	Chitin suspension on poly(ethylene terephthalate) nonwoven fabric	Applied in the treatment of large skin defects, namely, the defects difficult to suture; favors early granulation.
ChitiPack® C	Eisai Co.	Fibrous wound dressing made by spinning of chitosan acetate solution coagulated in bath, containing mixture of ethylene glycol, cold water, and sodium or potassium hydroxide	Regeneration and reconstruction of skin.
Chitidine®	IMS	Powdered chitosan containing elementary iodine	Disinfection and cleaning of wounded skin; primary wound dressing.
HemCon® Bandage ChitoFleX® Hemostatic dressing	HemCon Medical Technologies	Freeze-dried chitosan acetate salt	Positively charged chitosan salt has strong affinity to bond with red blood cells, activates the platelets, and forms a clot that stops massive blood bleeding

Table 24.1 (continued).

Trade name	Company	Composition	Properties/applications
Tromboguard®	Tricomed SA	Two-layer dressings consisting of hydrophilic polyurethane sponges affixed to the biologically active layer containing chitosan, sodium alginate/calcium, and silver salts	Hemostatic activity, antibacterial behavior, and acceleration of wound healing.
Tegarsob®	3M	Chitosan particles that swell absorbing exudates and producing a soft gel	Applied in leg ulcers, sacral wounds, and chronic wounds

growth of primary human dermal fibroblasts, which could be useful for future skin healing studies [15].

Overall, the applications of chitosan-based matrices for skin TE are promising, but occasionally, contradictory findings can be found, which are attributed in part to the source of chitin (fungal or arthropod), implant site, types of cells (cell lines, primary, and stem cells), surface properties, and also chitosan material characteristics such as molecular weight, residual ash content, and manufacturing process [64, 80]. As described above, strategies involving the blending of chitosan as well as different surface and bulk modification level seem to solve some of those issues.

24.3.4
Hyaluronic Acid

Hyaluronic acid or *hyaluronan* (HA) is a water-soluble polysaccharide widely distributed throughout the extracellular matrix (ECM) of all connective tissues in humans and other animals [83]. There are some limitations on the direct use of HA in wound management because of its solubility, rapid resorption, and short tissue residence time [84]. However, by controlling the degree of esterification, HA derivatives can be processed as membranes by solvent casting, or as sponges or microspheres by freeze-drying, spray-drying, or extrusion [83, 84]. Some HA-based matrices have been commercialized as wound dressings and dermal substitutes [85–87]. For example, Hyaff®, a benzyl-esterified derivative of HA, demonstrated its efficacy in healing extensive burns [6, 88]. HA-based products such as Hyalofill® PA, Hyalograf3D™, and Hyalomatrix® can be found in the market [89]. Besides esterification, HA-based matrices have also been constructed by its conjugation with chitosan [87, 90], gelatin [90], collagen [41, 86], and chondroitin sulfate [91]. These resulting matrices showed potential for wound healing applications, in skin TE, as artificial skin and wound dressings.

24.3.5
Collagen and Other Proteins

Collagen is the most abundant protein in animals and is also the main component of ECM [42, 92]. Collagen is used as a building material for the production of fibers, films, hydrogels, and scaffolds, some of which could be applied as wound dressings, as bilayered skin equivalents, or as component of engineered acellular dermis [16, 40, 41, 46]. In fact, collagen-containing wound dressings should have a beneficial effect on wound healing, as collagen possesses a high binding capacity for different inflammatory mediators, such as proteases and cytokines and antioxidant potential *in vitro* [93, 94]. A number of collagen-based dressings composed of a variety of carriers/combining agents in the form of gels, sheet, and sponges are commercially available, for example, ColActive™ (Smith & Nephew), Promogram® (Johnson & Johnson), and Puraply® (Royce Medical) [95]. In the pioneer work of Yannas and Burke [96], a porous composite of bovine collagen chondroitin sulfate with an outer silicone covering was used to produce a bilayer artificial skin, which is actually called *Integra*® (Integra Life Sciences Holding Corporation, New Jersey, USA). Although improvements in collagen matrices have been achieved, concerns about their variability in terms of physicochemical and degradation properties, high cost of pure collagen, and wound contraction and scarring, still need to be solved [6].

Silk is a diverse family of proteins with extraordinary mechanical properties and biological compatibility [39, 97, 98]. Silk can be processed by various techniques into gels, sponges, scaffolds, nanofibers, and films [99–101]. Silk can also be chemically modified [75, 99, 101] by conjugation with polysaccharides [15, 44, 102], whose properties can be designed to specific skin TE applications. Some studies demonstrated the use of silk matrices as skin supports [101]. Furthermore, fibroin films [75] and sponges [44] have been found to enhance skin wound healing *in vivo*, compared to clinically used materials.

Gelatin is produced by thermal denaturation of collagen [10, 11]. It is a promising protein for developing matrices for skin repair [73, 90, 91, 103]. For example, a gelatin hydrogel was shown to have potential to mimic the ECM, which may lead to the needed wound healing for tissue regeneration [104]. But the high solubility of gelatin can be problematic when it is used as wound dressing. Therefore, the gelatin must be cross-linked or conjugated with other biopolymers to maintain its stability and also to modify the biological and mechanical behavior of the produced gelatin-based matrices [73, 105].

Soy protein, the major component of the soybean, has the advantages of being economically competitive and presents good water resistance, as well as storage stability [106]. The combination of these properties associated with its reduced susceptibility to thermal degradation makes soy an ideal template to be used as a biomaterial for skin TE. Some researchers [107] proposed the use of soy protein films with gentamicin incorporated as wound dressings, which have the desired mechanical and physical properties, as well as drug release behavior to protect against bacterial infection.

Fibrin, a protein formed in the human body, has also been explored for skin applications [108, 109]. Fibrin, associated with fibronectin, has been shown to support keratinocyte and fibroblast growth both *in vitro* and *in vivo* and may enhance the cellular motility in the wound. When used as a delivery system for cultured keratinocytes and fibroblasts, fibrin glue may provide similar advantages to those obtained with conventional skin grafts [109].

24.3.6
Synthetic Polymers

Current synthetic polymers such as poly(L-lactic acid) (PLLA), poly(ethylene glycol) (PEG), PLGA, PCL, and PU, have been proposed as wound dressings and dermal substitutes or full-thickness skin equivalent [13, 110–112]. PU wound dressings have been extensively studied since they are impermeable to bacteria and water, but permeable to gas in moist environment. Commercially, PU-based membranes (Tegarderm™, 3M Health Care, and Opsite) are cost-effective for covering small-sized split-thickness skin graft, but show limited adherence to the wound bed [6]. Despite their versatility, reproducibility process, and, in some cases, low cost, synthetic polymers, for example, PCL, present some drawbacks, such as high hydrophobicity, limited cell recognition sites, neutral charge distribution, and slow degradation rate [113]. To improve synthetic matrices properties toward skin TE, biomimetic protein sequences, such as the RGD (R, arginine; G, glycine; D, aspartic acid) sequences, can be incorporated into these matrices [114, 115]. The inclusion of these RGD sequences into self-assembling hydrogels facilitated the migration and persistence of human adult dermal fibroblasts and resulted in natural cell morphology but also in increased cell–matrix interactions such as contraction [116]. In addition, the combination of synthetic polymers with natural polymers can be necessary to balance the advantages and disadvantages of each material in order to produce suitable scaffolds for skin regeneration [105]. Chen *et al.* [117] showed that the hybridization of knitted polyglycolic acid (PGA) mesh with collagen increased the efficiency of cell seeding, improved cell distribution, and therefore facilitated rapid formation of dermal tissue after two weeks.

24.4
Wound Dressings and Their Properties

Even though the majority of skin wounds can heal naturally, a proper, immediate, and permanent or temporary coverage of the wound surface may be needed to accelerate wound healing. Many types of wound dressings and devices have targeted different aspects of the wound healing process. Owing to their diversity, dressings can be classified as inactive, interactive, and active [24]. When compared to inactive dressings (e.g., gauze), interactive wound dressings (e.g., polymeric films, hydrogels) present several advantages such as (i) reduction in pain on removal, (ii) flexibility, (iii) less dressing changes, and (iv) sometimes antimicrobial properties. Furthermore, bioactive wound dressing materials are produced from a variety of biopolymers such as collagen, HA, alginate, among others [24, 31]. Besides, active compounds such as antimicrobials (as e.g., gentamicin), antibiotics (oflaxicin), and growth factors (e.g., epidermal growth factor – EGF, vascular endothelial growth factor – VEGF) have been incorporated into these materials, promoting their release to prevent contaminations and infections, and then helping in the treatment of chronic wounds and burns [118]. In fact, growth factors can be released in all phases of wound healing, contributing to it by controlling the cell proliferation and migration that modulate epithelialization, angiogenesis, and collagen metabolism [119].

24.5
Biomimetic Approaches in Skin Tissue Engineering

Biomimetic approaches involving hierarchical organization, hybridization, and adaptability of biopolymers and processes could have considerable potential in the development of new skin substitutes. Skin substitutes are a heterogeneous group of products aimed at replacing, either temporarily or permanently, the form and function of lost skin [33, 120]. The temporary skin substitutes are used to help healing the partial thickness burns, closing the clean excised wound until skin is available for grafting, and typically there are no living cells present [121]. By its turn, permanent skin substitutes are used to (i) replace lost skin, providing either epidermis or dermis, or both and (ii) to provide a higher quality of skin than a thin skin graft [121]. The successful application of tissue-engineered skin substitutes requires that the morphological and ultrastructural organization of the epidermis, dermis, and dermal–epidermal junction mimic the normal skin structure as closely as possible [122]. On the basis of these considerations, skin substitute production has relied on creating organized three-dimensional (3D) structures such as hydrogels, porous sponge, woven fiber, and honeycomb mesh. These matrices should provide support for both dermal fibroblasts and the overlying keratinocytes needed for skin TE [22, 123]. To fulfil these requirements, an ideal scaffold for skin TE should have, for instance, a suitable microstructure, porosity, controllable biodegradability, good biocompatibility, and suitable mechanical properties [124]. Most 3D engineered skin products, for example, dermal products, have been developed using different strategies (Figure 24.2 and Table 24.2). In a skin construct, the 3D culture can use a biopolymer or blended systems with fibroblasts, keratinocytes, and other cell types such as myofibroblasts and endothelial cells that will enrich the dermal environment [125]. Then, the resulting products may possess biological

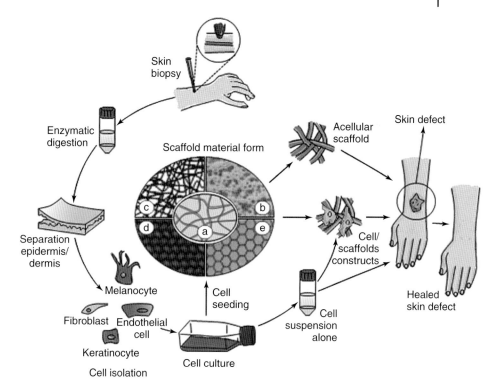

Figure 24.2 Current TE strategies for dermal substitute production. Initially, a patient's skin biopsy is treated enzymatically, where epidermis and dermis are separated, followed by isolation of different cells (epidermal, keratinocytes, melanocytes, dermal fibroblasts, and vascular endothelial cells). These cells can be cultured in structures produced in different forms: (a) hydrogels, (b) porous sponge, (c) nonwoven fiber (nanofiber), (d) woven fiber, and (e) honeycomb mesh. Furthermore, cell suspensions can be directly applied to skin defects as a liquid or spray. Finally, matrices with or without cells are implanted into the defect promoting its healing. (Source: Adapted from Refs [22, 123].)

and pharmacological properties that resemble the human skin, thus allowing the new tissue growth.

Regarding processing, new techniques such as bioprinting, solid freeform fabrication, prototyping, and electrospinning have allowed the production of structures with increasing complexity [17, 126, 127], as well as the production of 3D microenvironment that actively control cell behavior. In an interesting work [126], a full-skin equivalent was built up using a robotic printing platform with two types of cells (fibroblasts and keratinocytes) in a collagen hydrogel. Further development of this approach would allow for the deposition of substitute skin *in situ*, perfectly following the shape of the afflicted body part. On the other hand, both freeze-drying and electrospinning are very popular techniques to produce macromaterials and nanomaterials. In fact, many studies suggested that the materials obtained from those techniques can be used as culture skin substitutes. Besides, biomimetic

Table 24.2 Examples of developed matrices (scaffolds, nanofibers, and hydrogels) processed from biopolymer alone or blended systems intended for skin tissue engineering.

Composition	Processing methodology	Cell type	Animal model	References
Rutin-conjugated chitosan	Cross-linking (EDC), NPC activation, hydrogel formation	Fibroblasts (L929)	Sprague Dawley rats	[130]
Chitosan-alginate-fucoidan	—	Human dermal fibroblast/dermal microvascular endothelial	Sprague Dawley rats	[63]
Chitosan-collagen-silicone	Trimethylation, freeze-drying, and VEGF	HUVEC	Porcine model	[131]
Chitosan-gelatin	Freeze-drying, cross-linking (GA)	Cocultures of keratinocytes and fibroblasts	—	[124]
	Electrospinning	Human fibroblasts	—	[132]
Chitosan-gelatin-chondroitin sulfate	Freeze-drying, cross-linking (GA)	Fibroblasts (L929)	—	[133]
Collagen-alginate	Freeze-drying, cross-linking (EDC; CaCl$_2$)	Human keratinocytes; human fibroblasts	Adult male mice	[134]
Collagen-elastin	Cross-linking (EDC), freeze-drying	Fibroblasts	Wistar rats	[16]
Collagen-glycosaminoglycan	Freeze-drying, cross-linking (DHT)	Human dermal fibroblasts	—	[135]
Collagen-HA	Freeze-drying	Dermal fibroblasts	Guinea pig model	[41, 86]
Hyaluronate–collagen	Electrospinning, cross-linking (EDC)	Human fibroblasts	—	[136]
PCL-collagen	Immobilization of EGF, electrospinning	Human dermal keratinocyte cell line	—	[137]
PLGA/collagen	Electrospinning	Human fibroblasts	Sprague Dawley rats	[12]
PLACL/gelatin	Electrospinning	Human dermal fibroblasts	—	[110]
RGD-g-PLLA	Gas-foaming method, cross-linking (EDC)	Endothelial progenitor cells	Athymic nude mice	[112]
Silk/alginate	Freeze-drying	—	Rats	[44]

BMSCs, bone marrow mesenchymal stem cells; bFGF, bFGF-basic fibroblast growth factor; CaCl$_2$, calcium chloride; DHT, dehydrothermal treatment; EDC, 1-ethyl-3-(3-(dimethylaminopropyl))carbodiimide; EGF, epidermal growth factor; GA, glutaraldehyde; NHS, N-hydroxysuccinimide; NPC, p-nitrophenyl chlroformate; PEG, poly(ethylene glycol); PLGA, poly(lactic acid-co-glycolic acid); PLACL, poly(lactic acid)-co-poly(caprolactone); RGD, Arginine-Glycine-Aspartic acid; PLLA, poly (L-lactic acid); PCL, polycaprolactone; HUVECs, human umbilical vein endothelial cells.

strategies based on techniques such as electrospinning could be useful in mimicking some aspects of skin hierarchy. However, very limited *in vivo* skin applications on nanofibrous materials have been reported. Examples of the developed matrices that were proposed for skin TE are showed in Table 24.2. Apparently, the complexity of the compositions has been one of the alternatives found by different researchers to achieve an ideal skin matrix. For example, sponges composed by freeze-dried collagen-GAG sponges populated with autologous keratinocytes and fibroblasts have been shown to effectively close full-thickness burn wounds [128]. Pan *et al.* [129] demonstrated that the use of PLGA/dextran nanofibers favored the interaction with fibroblasts and resembled the dermal architecture.

24.5.1
Commercially Available Skin Products

There are several commercial skin products that can be found in market [2, 3, 121]. A representative list of engineered skin substitutes that are commercially available is shown in Table 24.3. Unfortunately, currently there are no engineered skin substitutes that can completely simulate the complexity of human skin, either in form or function. However, with the advances in TE and biotechnology, there are many skin substitutes that can be used for replacement of one or both skin layers [138]. Furthermore, skin substitutes have also been used in basic research to elucidate fundamental processes in the skin, and also as model systems to identify irritative, toxic, or corrosive properties of chemical agents that come in contact with human skin [2, 139].

The majority of currently available skin substitutes use acellular collagen scaffolds or sheets of biopolymers (e.g., collagen, hyaluronic acid, and PLGA) containing fibroblasts and/or keratinocytes. Despite the established role of skin substitutes in the management of a variety of wounds, particularly burns, there is a range of problems that have not been well resolved, such as high price, wound contraction, scar formation, poor integration with host tissue, and delayed vascularization [22]. For example, the formation of new blood capillaries to supply essential nutrients and oxygen can be difficult because of delayed vascularization. Some solutions for mentioned problems involve the many strategies such as (i) use of biomaterials (Section 24.3) to enhance angiogenesis, (ii) optimization of the scaffold design to the release of growth factors (e.g., bFGF) from 3D structures, (iii) addition of angiogenic factors, (iv) use of endothelial cells, and (v) *in vivo* prevascularization [7, 140, 141].

24.6
Final Remarks

Successful use of materials derived from natural sources and synthetic ones for the development of soft constructs for skin TE and wound healing has been encouraging. The developed strategies combining appropriate biomaterials with or without

Table 24.3 Examples of skin substitutes commercially available, containing biomaterials in their composition.

Trade name	Skin construct type	Company	Layers	Prominent cell type	Uses
Apligraf™ (earlier name: Graftskin™)	Dermoepidermal	Organogenesis Incorporation	Bovine collagen	Allogeneic keratinocytes and fibroblasts	Venous ulcers, diabetic foot ulcers
Biobrane™	Dermal	UDL Laboratories	Porcine dermal collagen, silicone film, nylon mesh	—	Temporary covering of full-thickness burn wounds
Dermagraft	Dermal	Advanced BioHealing	PGA/PLA, ECM	Neonatal allogeneic fibroblasts	Temporary
Hyalograft3D™	Dermal	Fidia Advanced Biopolymers	Microperforated hyaluronic acid membrane	Autologous fibroblasts	Diabetic foot ulcers, venous leg ulcers
Integra™	Dermal	Integra NeuroSciences	Bovine collagen, chondroitin sulfate, and silicone sheet	—	Chronic and traumatic wounds
Laserskin™ or Vivoderm	Epidermal	Fidia Biopolymers	Microperforated hyaluronic acid membrane	Autologous keratinocytes	Superficial and partial thickness burns

Matriderm®	Dermal	Dr Suwelack Skin and HealthCare AG	Bovine dermal collagen and elastin	—	Burns and reconstruction, nonhealing wounds that require skin grafts, such as diabetic foot ulcers
Orcel™	Dermoepidermal	Ortec International	Bovine collagen sponge	Allogeneic keratinocytes and fibroblasts	Burns
TransCyte®	Dermal	Advanced BioHealing Inc.	Nylon coated with porcine dermal collagen, bonded to a silicone membrane	Neonatal fibroblasts	Excised full-thickness burns and nonexcised partial thickness burns
TISSUEtech Autograft System	Dermoepidermal	Fidia Advanced Biopolymers	Esterified hyaluronic acid matrix	Keratinocytes and fibroblasts	Chronic skin lesions

ECM, extracellular matrix; HAM, hyaluronic acid membrane; ORC, oxidized regenerated cellulose; PGA, polyglycolic acid; PLA, polylactic acid; PEO/PBT, polyethylene oxide terephthalate/polybutylene terephthalate.
Source: From Refs [2, 22, 25, 121].

in-vitro-cultured cells associated to interesting techniques (e.g., rapid prototyping) had demonstrated promising findings. It is also suggested that some current problems associated with the produced scaffolds and artificial skin substitutes may be overcome using these soft constructs. Interestingly, an impressive amount of works proposed the use of nanofibers for wound healing, where the findings suggested new opportunities in the processing of soft constructs for skin TE.

However, a significant portion of the work cited in this chapter has been dedicated to laboratory, namely, *in vitro* and *in vivo* assays, and then forward movement to clinical performance of the constructs has been slow. Besides, the lack of specific protocols to achieve the certification of the developed skin products has been an obstacle to their expansion and commercialization.

Therefore, biomimetic strategies that could stimulate skin regeneration and healing should clearly involve better knowledge of cellular mechanisms and interactions of cells with instructive biomaterials and surfaces toward the development of products based on the own biological tools taking place in the native regeneration process.

In the future, it is expected that other strategies such as the use of cell sheet engineering technology and stem cells or the association of synthetic or biological matrices together with molecular biology and genetic engineering would lead to the production of tissue-engineered human-skin-based products resembling natural human skin.

Acknowledgments

We are especially grateful to Dra. Marta Silva (3B's Research Group) and Gisela Luz (3B's Research Group) for the critical reading and illustrations of the manuscript, respectively.

We acknowledge financial support from Portuguese Foundation for Science and Technology -FCT (Grant SFRH/BPD/45307/2008), "Fundo Social Europeu" - FSE, and "Programa Diferencial de Potencial Humano-POPH".

List of Abbreviations

Ag	silver
BC	bacterial cellulose
CSF	chitosan/silk fibroin
CSS	chitosan/soy protein
DBC	dibutylchitin
ECM	extracellular matrix
EDC	1-ethyl-3-(3-dimethylaminopropyl)-carboiimide
EGF	epidermal growth factor
bFGF	basic fibroblast growth factor
GAGs	glycosaminoglycans
HA	hyaluronic acid
HAM	hyaluronic acid membrane
HPC	hydroxypropyl cellulose

ILs	ionic liquids
MMPs	matrix mettalloproteases
NHS	N-hydroxysuccinimide
NPC	p-nitrophenyl chloroformate
ORC	oxidized regenerated cellulose
PBT	polybutylene terephthalate
PCL	poly(ε-caprolactone)
PDGF	platelet-derived growth factor
PEC	polyelectrolyte complexes
PEO	polyethylene oxide terephthalate
PLGA	poly(lactic-co-glycolic acid)
PLLA	poly(lactic acid)
PGA	polyglycolic acid
PVA	poly(vinyl alcohol)
PU	polyurethane
RGD	R, arginine; G, glycine; D, aspartic acid
SCMC	sodium carboxymethylcellulose
TE	tissue engineering
TGF	transforming growth factor
VEGF	vascular endothelial growth factor
3D	three-dimensional

References

1. Shevchenko, R.V., James, S.L., and James, S.E. (2010) *J. Roy. Soc. Interface*, **7** (43), 229–258.
2. Groeber, F., Holeiter, M., Hampel, M., Hinderer, S., and Schenke-Layland, K. (2011) *Adv. Drug Deliv. Rev.*, **63** (4–5), 352–366.
3. Huang, S. and Fu, X.B. (2010) *J. Control. Release*, **142**(2), 149–159.
4. Macneil, S. (2007) *Nature*, **445**, 874–880.
5. Metcalfe, A.D. and Ferguson, M.W.J. (2007) *J. Roy. Soc. Interface*, **4**(14), 413–437.
6. Zhong, S.P., Zhang, Y.Z., and Lim, C.T. (2010) *Wiley Interdiscipl. Rev.: Nanomed. Nanobiotechnol.*, **2**(5), 510–525.
7. Sheila, M. (2008) *Mater. Today*, **11**(5), 26–35.
8. Ponec, M. (2002) *Adv. Drug Deliv. Rev.*, **54** (Suppl. 1), S19–S30.
9. Orgill, D. and Blanco, C. (2009) in *Biomaterials for Treating Skin Loss* (eds D. Orgill and C. Blanco), Woodhead Publishing Limited, Cambridge, pp. 3–8.
10. Silva, S.S., Mano, J.F., and Reis, R.L. (2010) *Crit. Rev. Biotechnol.*, **30**(3), 200–221.
11. Gomes, M.E., Azevedo, H.S., Malafaya, P.B., Silva, S.S., Oliveira, J.M., Sousa, R.A., Mano, J.F., and Reis, R.L. (2008) in *Tissue Engineering* (eds C. Blitterswijk *et al.*), Elsevier, New York, pp. 145–192.
12. Liu, S.-J., Kau, Y.-C., Chou, C.-Y., Chen, J.-K., Wu, R.-C., and Yeh, W.-L. (2010) *J. Membr. Sci.*, **355**(1–2), 53–59.
13. Chong, E.J., Phan, T.T., Lim, I.J., Zhang, Y.Z., Bay, B.H., Ramakrishna, S., and Lim, C.T. (2007) *Acta Biomater.*, **3**(3), 321–330.
14. Kim, H.Y., Khil, M.S., Cha, D.I., Kim, I.S., and Bhattarai, N. (2003) *J. Biomed. Mater. Res. B: Appl. Biomater.*, **67B**(2), 675–679.
15. Silva, S.S., Santos, T.C., Cerqueira, M.T., Marques, A.P., Reys, L.L., Silva, T.H., Caridade, S.G., Mano, J.F., and

Reis, R.L. (2012) *Green Chem.*, **14**, 1463–1470.

16. Matsumoto, Y., Ikeda, K., Yamaya, Y., Yamashita, K., Saito, T., Hoshino, Y., Koga, T., Enari, H., Suto, S., and Yotsuyanagi, T. (2011) *Biomed. Res. (Tokyo)*, **32**(1), 29–36.

17. van der Veen, V.C., Boekema, B.K.H.L., Ulrich, M.M.W., and Middelkoop, E. (2011) *Wound Repair Regen.*, **19**, S59–S65.

18. Santos, T.C., Marques, A.P., Silva, S.S., Oliveira, J.M., Mano, J.F., and Reis, R.L. (2011) *Tissue Eng. A*, under revision.

19. Metcalfe, A.D. and Ferguson, M.W.J. (2007) *Biomaterials*, **28**(34), 5100–5113.

20. McGrath, J.A., Eady, R.A.J., and Pope, F.M. (2008) *Rook's Textbook of Dermatology*, Blackwell Publishing, Inc., pp. 45–128.

21. Saxena, V. (2009) in *Biomaterials for Treating Skin Loss* (eds D. Orgill and C. Blanco), Woodhead Publishing Limited, Cambridge, pp. 18–24.

22. Bottcher-Haberzeth, S., Biedermann, T., and Reichmann, E. (2010) *Burns*, **36**, 450–460.

23. Cochrane, C., Rippon, M.G., Rogers, A., Walmsley, R., Knottenbelt, D., and Bowler, P. (1999) *Biomaterials*, **20**(13), 1237–1244.

24. Boateng, J.S., Matthews, K.H., Stevens, H.N.E., and Eccleston, G.M. (2008) *J. Pharm. Sci.*, **97**(8), 2892–2923.

25. Li, J., Chen, J., and Kirsner, R. (2007) *Clin. Dermatol.*, **25**(1), 9–18.

26. Velnar, T., Bailey, T., and Smrkoli, V. (2009) *J. Int. Med. Res.*, **37**(5), 1528–1542.

27. Eming, S., Krieg, T., and Davidson, J. (2007) *J. Invest. Dermatol.*, **127**, 517–525.

28. Shakespeare, P. (2001) *Burns*, **27**(5), 517–522.

29. Whitney, J.D. (2005) *Nurs. Clin. N. Am.*, **40**(2), 191–205.

30. Moore, K., McCallion, R., Searle, R.J., Stacey, M.C., and Harding, K.G. (2006) *Int. Wound J.*, **3**(2), 89–98.

31. Zahedi, P., Rezaeian, I., Ranaei-Siadat, S.O., Jafari, S.H., and Supaphol, P. (2010) *Polym. Adv. Technol.*, **21**(2), 77–95.

32. Ferreira, M., Tuma Júnior, P., Carvalho, V., and Kamamoto, F. (2006) *Clinics*, **61**, 571–578.

33. Shores, J., Gabriel, A., and Gupta, S. (2007) *Adv. Skin Wound Care*, **20**, 493–508.

34. Mano, J.F., Silva, G.A., Azevedo, H.S., Malafaya, P.B., Sousa, R.A., Silva, S.S., Boesel, L.F., Oliveira, J.M., Santos, T.C., Marques, A.P., Neves, N.M., and Reis, R.L. (2007) *J. Roy. Soc. Interface*, **4**(17), 999–1030.

35. Rinaudo, M. (2006) *Prog. Polym. Sci.*, **31**(7), 603–632.

36. Qin, Y. (2008) *Polym. Int.*, **57**, 171–180.

37. Dechert, T.A., Ducale, A.E., Ward, S.I., and Yager, D.R. (2006) *Wound Repair Regen.*, **14**(3), 252–258.

38. Klemm, D., Heublein, B., Fink, H.-P., and Bohn, A. (2005) *Angew. Chem. Int. Ed.*, **44**(22), 3358–3393.

39. Altman, G.H., Diaz, F., Jakuba, C., Calabro, T., Horan, R.L., Chen, J.S., Lu, H., Richmond, J., and Kaplan, D.L. (2003) *Biomaterials*, **24**, 401–416.

40. Sai, K.P. and Babu, M. (2000) *Burns*, **26**(1), 54–62.

41. Park, S.N., Kim, J.K., and Suh, H. (2004) *Biomaterials*, **25**, 3689–3698.

42. Wiegand, C. and Hipler, U.-C. (2010) *Macromol. Symp.*, **294**(2), 1–13.

43. Paul, W. and Sharma, C. (2004) *Trends Biomater. Artif. Organs*, **18**, 18–23.

44. Roh, D.H., Kang, S.Y., Kim, J.Y., Kwon, Y.B., Kweon, H.Y., Lee, K.G., Park, Y.H., Baek, R.M., Heo, C.Y., Choe, J., and Lee, J.H. (2006) *J. Mater. Sci. Mater. Med.*, **17**(6), 547–552.

45. Wang, L.H., Khor, E., Wee, A., and Lim, L.Y. (2002) *J. Biomed. Mater. Res.*, **63**(5), 610–618.

46. Donaghue, V.M., Chrzan, J.S., Rosenblum, B.I., Giurini, J.M., Habershaw, G.M., and Veves, A. (1998) *Adv. Wound Care*, **11**, 114–119.

47. Hart, J., Silcock, D., Gunnigle, S., Cullen, B., Light, N.D., and Watt, P.W. (2002) *Int. J. Biochem. Cell Biol.*, **34**(12), 1557–1570.

48. Spangler, D., Rothenburger, S., Nguyen, K., Jampani, H., Weiss, S., and Shubhangi Bhende, S. (2003) *Surg. Infect.*, **4**, 255–262.

49. Ramli, N.A. and Wong, T.W. (2011) *Int. J. Pharm.*, **403**(1–2), 73–82.
50. Wysocki, A.B., Staiano-Coico, L., and Grinnell, F. (1993) *J. Invest. Dermatol.*, **101**(1), 64–68.
51. Brown, R.M., Czaja, W.K., Young, D.J., and Kawecki, M. (2007) *Biomacromolecules*, **8**(1), 1–12.
52. Svensson, A., Nicklasson, E., Harrah, T., Panilaitis, B., Kaplan, D.L., Brittberg, M., and Gatenholm, P. (2005) *Biomaterials*, **26**(4), 419–431.
53. Takegawa, A., Murakami, M., Kaneko, Y., and Kadokawa, J. (2009) *Polym. Compos.*, **30**(12), 1837–1841.
54. Park, T.-J., Jung, Y., Choi, S.-W., Park, H., Kim, H.J., Kim, E., Lee, S., and Kim, J. (2011) *Macromol. Res.*, **19**, 213–215.
55. Kumar, M.N.V.R. (2000) *Reactive Funct. Polym.*, **46**(1), 1–27.
56. Kumar, M.N.V.R., Muzzarelli, R.A.A., Muzzarelli, C., Sashiwa, H., and Domb, A.J. (2004) *Chem. Rev.*, **104**, 6017–6084.
57. Kurita, K. (2001) *Prog. Polym. Sci.*, **26**, 1921–1971.
58. Hamblin, M.R., Dai, T.H., Tanaka, M., and Huang, Y.Y. (2011) *Expert Rev. Anti Infect. Ther.*, **9**(7), 857–879.
59. Jayakumar, R., Prabaharan, M., Kumar, P.T.S., Nair, S.V., and Tamura, H. (2011) *Biotechnol. Adv.*, **29**(3), 322–337.
60. Muzzarelli, R. and Muzzarelli, C. (2005) in *Polysaccharides I* (ed. T. Heinze), Springer, Berlin and Heidelberg, pp. 151–209.
61. Pielka, S., Paluch, D., Kus, J.S., Zywicka, B., Solski, L., Szosland, L., and Zaczynska, E. (2003) *Fibres Textiles East. Eur.*, **11**, 79–84.
62. Muzzarelli, R.A.A., Guerrieri, M., Goteri, G., Muzzarelli, C., Armeni, T., Ghiselli, R., and Cornelissen, M. (2005) *Biomaterials*, **26**, 5844–5854.
63. Murakami, K., Aoki, H., Nakamura, S., Nakamura, S.-i., Takikawa, M., Hanzawa, M., Kishimoto, S., Hattori, H., Tanaka, Y., Kiyosawa, T., Sato, Y., and Ishihara, M. (2010) *Biomaterials*, **31**(1), 83–90.
64. Silva, S.S., Luna, S.M., Gomes, M.E., Benesch, J., Paskuleva, I., Mano, J.F., and Reis, R.L. (2007) *Macromol. Biosci.*, **8**, 568–576.
65. Azad, A.K., Sermsintham, N., Chandrkrachang, S., and Stevens, W.F. (2004) *J. Biomed. Mater. Res. B: Appl. Biomater.*, **69B**(2), 216–222.
66. Lu, S., Gao, W., and Gu, H.Y. (2008) *Burns*, **34**(5), 623–628.
67. Luna, S.M., Silva, S.S., Gomes, M.E., Mano, J.F., and Reis, R.L. (2011) *J. Biomater. Appl.*, **26**(1), 101–116.
68. Muzzarelli, R.A.A. (2009) *Carbohydr. Polym.*, **76**(2), 167–182.
69. Altiok, D., Altiok, E., and Tihminlioglu, F. (2010) *J. Mater. Sci. Mater. Med.*, **21**(7), 2227–2236.
70. Zhu, X., Chian, K.S., Chan-Park, M.B.E., and Lee, S.T. (2005) *J. Biomed. Mater. Res. A*, **73A**, 264–274.
71. Deng, C.M., He, L.Z., Zhao, M., Yang, D., and Liu, Y. (2007) *Carbohydr. Polym.*, **69**, 583–589.
72. Liu, H.F., Yin, Y.J., and Yao, K.D. (2007) *J. Biomater. Appl.*, **21**(4), 413–430.
73. Mao, J., Zhao, L., De Yao, K., Shang, Q., Yang, G., and Cao, Y. (2003) *J. Biomed. Mater. Res. A*, **64**, 301–308.
74. Ma, L., Shi, Y.C., Chen, Y.X., Zhao, H.H., Gao, C.Y., and Han, C.M. (2007) *J. Mater. Sci. Mater. Med.*, **18**(11), 2185–2191.
75. Kweon, H., Ha, H.C., Um, I.C., and Park, Y.H. (2001) *J. Appl. Polym. Sci.*, **80**(7), 928–934.
76. Silva, S., Santos, M., Coutinho, O., Mano, J., and Reis, R. (2005) *J. Mater. Sci. Mater. Med.*, **16**, 575–579.
77. Silva, S.S., Goodfellow, B.J., Benesch, J., Rocha, J., Mano, J.F., and Reis, R.L. (2007) *Carbohydr. Polym.*, **70**, 25–31.
78. Tanabe, T., Okitsu, N., Tachibana, A., and Yamauchi, K. (2002) *Biomaterials*, **23**(3), 817–825.
79. Kratz, G., Amander, C., Swedenborg, J., Back, M., Falk, C., Gouda, I., and Larm, O. (1997) *J. Plast. Reconstr. Hand Surg.*, **31**, 119–123.
80. Hamilton, V., Yuan, Y., Rigney, D., Puckett, A., Ong, J., Yang, Y., Elder, S., and Bumgardner, J. (2006) *J. Mater. Sci. Mater. Med.*, **17**(12), 1373–1381.
81. Niekraszewicz, A. (2005) *Fibres Textiles East. Eur.*, **13**, 16–18.

82. Burkatovskaya, M., Castano, A.P., Demidova-Rice, T.N., Tegos, G.P., and Hamblin, M.R. (2008) *Wound Repair Regen.*, **16**(3), 425–431.
83. Kogan, G., Soltes, L., Stern, T., and Gemeiner, P. (2007) *Biotechnol. Lett.*, **29**, 17–25.
84. Kennedy, J. and Knill, C. (2006) in *Medical Textiles and Biomaterials for Healthcare* (eds S. Anand et al.), CRC Press, Cambridge, pp. 3–72.
85. Kubo, K. and Kuroyanagi, Y. (2003) *J. Artif. Organs*, **6**, 64–70.
86. Park, S.N., Lee, H.J., Lee, K.H., and Suh, H. (2003) *Biomaterials*, **24**, 1631–1641.
87. Xu, H., Ma, L., Shi, H., Gao, C., and Han, C. (2007) *Polym. Adv. Technol.*, **18**(11), 869–875.
88. Ramakrishna, S., Fujihara, K., Teo, W.-E., Lim, T.-C., and Ma, Z. (2005) *An Introduction to Electrospinning and Nanofibers*, World Scientific.
89. Price, R.D., Berry, M.G., and Navsaria, H.A. (2007) *J. Plast. Reconstr. Aesthetic Surg.*, **60**(10), 1110–1119.
90. Liu, H., Mao, J., Yao, K., Yang, G., Cui, L., and Cao, Y. (2004) *J. Biomater. Sci. Polym. Ed.*, **15**, 25–40.
91. Wang, T.W., Wu, H.C., Huang, Y.C., Sun, J.S., and Lin, F.H. (2006) *Artif. Organs*, **30**, 141–149.
92. Shoulders, M. and Raines, R. (2009) *Annu. Rev. Biochem.*, **78**, 929–958.
93. Wiegand, C., Abel, M., Ruth, P., Wilhelms, T., Schulze, D., Norgauer, J., and Hipler, U.-C. (2009) *J. Biomed. Mater. Res. Part B: Appl. Biomater.*, **90B**(2), 710–719.
94. Schönfelder, U., Abel, M., Wiegand, C., Klemm, D., Elsner, P., and Hipler, U.-C. (2005) *Biomaterials*, **26**(33), 6664–6673.
95. Rangaraj, A., Harding, K., and Leaper, D. (2011) *Wound UK*, **7**(2), 54–63.
96. Yannas, I.V. and Burke, J.F. (1980) *J. Biomed. Mater. Res.*, **14**, 65–81.
97. Vepari, C. and Kaplan, D.L. (2007) *Prog. Polym. Sci.*, **32**, 991–1007.
98. MacIntosh, A.C., Kearns, V.R., Crawford, A., and Hatton, P.V. (2008) *J. Tissue Eng. Regen. Med.*, **2**, 71–80.
99. Silva, S.S., Maniglio, D., Motta, A., Mano, J.F., Reis, R.L., and Migliaresi, C. (2008) *Macromol. Biosci.*, **8**(8), 766–774.
100. Motta, A., Fambri, L., and Migliaresi, C. (2002) *Macromol. Chem. Phys.*, **203**(10–11), 1658–1665.
101. Guan, G.P., Bai, L., Zuo, B.Q., Li, M.Z., Wu, Z.Y., Li, Y.L., and Wang, L. (2010) *Bio-Med. Mater. Eng.*, **20**(5), 295–308.
102. Silva, S.S., Motta, A., Rodrigues, M.T., Pinheiro, A.F.M., Gomes, M.E., Mano, J.F., Reis, R.L., and Migliaresi, C. (2008) *Biomacromolecules*, **9**(10), 2764–2774.
103. Powell, H.M. and Boyce, S.T. (2008) *J. Biomed. Mater. Res. A*, **84A**(4), 1078–1086.
104. Balakrishnan, B., Mohanty, M., Umashankar, P.R., and Jayakrishnan, A. (2005) *Biomaterials*, **26**, 6335–6342.
105. Kwon, I.K., Kim, S.E., Heo, D.N., Lee, J.B., Kim, J.R., Park, S.H., and Jeon, S. (2009) *Biomed. Mater.*, **4**(4), 044106. http://dx.doi.org/10.1088/1748-6041/4/4/044106
106. Seal, R. (1980) in *Applied Protein Chemistry* (ed. R. Grant), Applied Science Publishers, London, pp. 87–112.
107. Peles, Z. and Zilberman, M. (2012) *Acta Biomater.*, **8**(1), 209–2177.
108. Krasna, M., Planinsek, F., Knezevic, M., Arnez, Z.M., and Jeras, M. (2005) *Int. J. Pharm.*, **291**(1–2), 31–37.
109. Currie, L.J., Sharpe, J.R., and Martin, R. (2001) *Plast. Reconstr. Surg.*, **108**(6), 1713–1726.
110. Chandrasekaran, A.R., Venugopal, J., Sundarrajan, S., and Ramakrishna, S. (2011) *Biomed. Mater.*, **6**(1), doi: 10.1088/1748-6041/6/1/015001
111. Min, B.-M., You, Y., Kim, J.-M., Lee, S.J., and Park, W.H. (2004) *Carbohydr. Polym.*, **57**(3), 285–292.
112. Kim, K.L., Han, D.K., Park, K., Song, S.-H., Kim, J.Y., Kim, J.-M., Ki, H.Y., Yie, S.W., Roh, C.-R., Jeon, E.-S., Kim, D.-K., and Suh, W. (2009) *Biomaterials*, **30**(22), 3742–3748.
113. Ray, F., Pashley, D., Williams, M., RAina, R., Loushine, R., Weller, R., Kimbrogh, W., and King, N. (2005) *J. Endod.*, **31**, 593–598.

114. Hersel, U., Dahmen, C., and Kessler, H. (2003) *Biomaterials*, **24**(24), 4385–4415.
115. Shin, H., Jo, S., and Mikos, A.G. (2003) *Biomaterials*, **24**(24), 4353–4364.
116. Zhou, M., Smith, A.M., Das, A.K., Hodson, N.W., Collins, R.F., Ulijn, R.V., and Gough, J.E. (2009) *Biomaterials*, **30**(13), 2523–2530.
117. Chen, G., Sato, T., Ohgushi, H., Ushida, T., Tateishi, T., and Tanaka, J. (2005) *Biomaterials*, **26**(15), 2559–2566.
118. Kim, M.S., Kim, G.H., Kang, Y.M., Kang, K.N., Kim, D.Y., Kim, H.J., Min, B.H., and Kim, J.H. (2011) *Tissue Eng. Regen. Med.*, **8**(1), 1–7.
119. Grazul-Bilska, A.T., Johnson, M.L., Bilski, J.J., Redmer, D.A., Reynolds, L.P., Abdullah, A., and Abdullah, K.M. (2003) *Drugs Today*, **39**, 787.
120. Damanhuri, M., Boyle, J., and Enoch, S. (2007) *Wounds Int.*, **2**, 1–11.
121. Horch, R.E., Kopp, J., Kneser, U., Beier, J., and Bach, A.D. (2005) *J. Cell. Mol. Med.*, **9**(3), 592–608.
122. Damour, O., Black, A.F., Bouez, C., Perrier, E., Schlotmann, K., and Chapuis, F. (2005) *Tissue Eng.*, **11**(5–6), 723–733.
123. Hodgkinson, T. and Bayat, A. (2011) *Arch. Dermatol. Res.*, **303**(5), 301–315.
124. Mao, J.S., Liu, H.F., Yin, Y.J., and Yao, K.D. (2003) *Biomaterials*, **24**(9), 1621–1629.
125. Brohem, C.A., da Silva Cardeal, L.B., Tiago, M., Soengas, M.S., deMoraes Barros, S.B., and Maria-Engler, S.S. (2011) *Pigm. Cell Melanoma Res.*, **24**(1), 35–50.
126. Lee, W., Debasitis, J.C., Lee, V.K., Lee, J.-H., Fischer, K., Edminster, K., Park, J.-K., and Yoo, S.-S. (2009) *Biomaterials*, **30**(8), 1587–1595.
127. Martins, A., Reis, R., and Neves, N. (2008) *Int. Mater. Rev.*, **53**, 257–274.
128. Boyce, S., Kagan, R., Yakuboff, K., Meyer, N., Rieman, M., Greenhalgh, D., and Warden, G. (2002) *Ann. Surg.*, **235**, 269–279.
129. Pan, H., Jiang, H., and Chen, W. (2006) *Biomaterials*, **27**(17), 3209–3220.
130. Tran, N., Joung, Y., Lih, E., and Park, K. (2011) *Biomacromolecules*, **12**, 2872–2880.
131. Guo, R., Xu, S., Ma, L., Huang, A., and Gao, C. (2011) *Biomaterials*, **32**(4), 1019–1031.
132. Jafari, J., Emami, S.H., Samadikuchaksaraei, A., Bahar, M.A., and Gorjipour, F. (2011) *Bio-Med. Mater. Eng.*, **21**(2), 99–112.
133. Emami, S.H., Abad, A.M.A., Bonakdar, S., Tahriri, M.R., Samadikuchaksaraei, A., and Bahar, M.A. (2010) *Int. J. Mater. Res.*, **101**(10), 1281–1285.
134. Kim, G., Ahn, S., Kim, Y., Cho, Y., and Chun, W. (2011) *J. Mater. Chem.*, **21**(17), 6165–6172.
135. Varkey, M., Ding, J., and Tredget, E.E. (2011) *Biomaterials*, **32**(30), 7581–7591.
136. Payne, W., Posnett, J., Alvarez, O., Jameson, G., Brown-Etris, M., Wolcott, R., Dharma, H., Hartwell, S., and Ochs, D. (2009) *Ostomy Wound Manage.*, **55**(2), 50–55.
137. Gumusderelioglu, M., Dalkiranoglu, S., Aydin, R.S.T., and Cakmak, S. (2011) *J. Biomed. Mater. Res. A*, **98A**(3), 461–472.
138. Balasubramani, M., Kumar, T.R., and Babu, M. (2001) *Burns*, **27**(5), 534–544.
139. Robinson, M.K., Osborne, R., and Perkins, M.A. (1999) *J. Pharmacol. Toxicol. Methods*, **42**(1), 1–9.
140. Ma, H., Perng, C.K., Wang, Y.J., and Tsi, C.H. (2011) *J. Surg. Res.*, **168**(1), 9–15.
141. Hosseinkhani, H., Hosseinkhani, M., Khademhosseini, A., Kobayashi, H., and Tabata, Y. (2006) *Biomaterials*, **27**(34), 5836–5844.

Index

a

A. diadematus 80, 82, 85
abalone shell 316
acellular double-layered skin construct 98
acidic polypeptide additives 336
aciniform silk 77
acrylamide 286
actin–myosin interactions 194
actomyosin 168
actomyosin contractility 168
Adenocystis utricularis 12
adhesives suitable for medical applications 273
– barnacles 279–280
– bio-inspired synthetic adhesive polymers 284–288
– brown algae 280–281
– mussels 274–278
– recombinant forms of 281–284
– tube worms 278–279
adsorptions on PS surfaces 249–250
alcohol dehydrogenase (ADH) 465
alginate 8–9, 496
alginate hydrogels 14
aligned collagen hydrogels 68
alkaline phosphatase (ALP) 4
Alloderm® 55
Al_2O_3 gel 338
amino acid 3,4-dihydroxyphenylalanine (DOPA) 5
amino acids 160
ampullate silk 77, 79
– ADF-3 and ADF-4 80
– regeneration of silk fibers from 83
– toughness of major 82

angiogenesis 46
anion–pi interactions 125
anterior cruciate ligament fibroblasts (ACLFs) adhesion 85
(antibody-Trx)–(Trx-ELP) complex 47
antigens 118
antimicrobial-functionalized protein-based biomaterials 81
apatite formation at surfaces 344–345
apatite nucleation 338
– surface charge and 338–339
apatite–polymer fiber composite 341
Apligraf® 67, 98
APMA/acrylamide copolymer 287
aqueous homogeneous gelation 439
Arabidopsis thaliana 80
aragonite tiles 299
arginine-glycine-aspartic acid (RGD) tripeptide 174
articular cartilage
– histology 515–517
– spontaneous and forced regeneration in 517
– tissue engineering approach to regeneration 517–518
artificial protein, biosynthesis of 102
aryl ether dendrimers 464
Asperities 307
atelocollagen solutions 65
ATT-clay-reinforced PMMA nanocomposites 371
autologous bone material, harvesting of 69
autologous chondrocyte transplantation (ACT) 70
avidity 128
A-W glass-ceramic 338
azobenzene dendrimers 435

Biomimetic Approaches for Biomaterials Development, First Edition. Edited by João F. Mano.
© 2012 Wiley-VCH Verlag GmbH & Co. KGaA. Published 2012 by Wiley-VCH Verlag GmbH & Co. KGaA.

b

bacterial protein expression 101
barnacles 279–280
base pairing 140
Belousoz–Zhabotinsky reaction 428
BGNF-COL nanocomposite 363
bihydrogenophosphate (–PO$_4$H$_2$) group 338
binding event 123
– avidity and cooperativity among receptors 128
bioactive ceramics 337–338
bioactive glasses 315, 338
– nanoscale 356–363
– natural polymer 360–363
– synthetic polymer/nanoparticulate 357–360
bioactive glass nanoparticles 4
bioactive nanocomposites 315
bioactive silicate glasses 353
bioadhesive peptides 174–175
biodegradable amphiphilic block copolymers 499
biodegradable hydrogels for cell invasion 41–43
biodegradable polymers 353
Bioglass® 16, 315–338, 356
bioinspired adhesives 259
– future of 270
– for wet conditions 268–269
bio-inspired materials 417
bio-inspired materials synthesis 315
bio-inspired synthetic adhesive polymers 284–288
– based on polysaccharides 285–286
– DOPA based 286–287
biologicalmarine adhesives 273
biologicalmolecular recognition 118
biomimetic adhesives 274
biomimetic apatite nanocrystals
– biomedical application of 390–394
– bisphosphonates, adsorption and release of 406–409
– in drug delivery systems 394–397
– drugs, adsorption and release of 397–402
– properties of 384–386
– proteins, adsorption and release of 402–406
– synthesis of 386–390
biomimetic materials synthesis 315
biomimetic protein-engineered hydrogels
– biodegradable hydrogels for cell invasion 41–43
– cell–cell binding domains 45–46
– cell–ECM binding domains 43–45

– delivery of soluble cell signaling molecules 46–48
– immobilization of bioactive factors into and onto hydrogels 47
– incorporating bioactive ligands in amino acid sequence 43–44
– mechanical properties for cellular response 40–41
– in regulating cell behavior 43–48
– release of VEGF 47
biomimetic receptive materials 140–141
biomimetics 3
biomimetism 94
biomineralization process 314
biomineralization strategies 336–337
biosensing surfaces 108
biotin 180
biotin–avidin linkages 226
bisphosphonates, adsorption and release using nanocrystals 399, 406–409
block copolymer micelle lithography (BCML) 217–219
blood platelet adhesion and surface wettability 251
Bombyx mori silk 75–77, 85
– disulfide linkage between light- and heavy-fibroin chains 76
– formation of P25/heavy-chain/light-chain complex 77
– sericins 77
– Takahashi's model 76
bone 3
– carbonated hydroxyapatite (HA) alignment of 334
– cell behavior at nanoscale 356
– deformation in response to external tensile load 300
– human 353
– mineralization of 335
– multilevel hierarchical structural design of 295–296
– as a nanocomposite 354–356
– nanometer-sized crystals of carbonate apatite 295
– natural mineralized composite of 333
– from osteoblasts 334
– repair using collagen 69
– tissue engineering 16
bone bonding of materials 337–338
bonelike apatite formation on bioactive ceramics 340
bone marrow stem cells (BMSCs) 85
bone morphogenetic protein (BMP) 361

bone morphogenetic protein-2 (BMP-2) 47, 86
bone morphogenetic protein (BMP) growth factors 398
bone morphogenic protein (BMP) 476
bone sialoprotein (BSP) 4, 86
bottom-up approach 217, 320–322, 327–328
bovine capillary endothelial (BCE) cells 176
bovine spongiform encephalopathy (BSE) 9
Broussignac process 7
brown algae 280–281

c
cadherins 46
calcium alginate 17
calcium alginate/chitosan 17
calcium carbonate 4, 16, 353
calcium phosphate (bioactive) ceramic 353
calcium phosphate coatings for orthopedic and dental implants 391
calcium phosphates (CaP) 16, 314
calixarenes 134, 137–138
Ca(OH)$_2$ treatment 341
carbohydrate-binding proteins 461
carbohydrates 142–143
carbonate–HA nanocrystals 384, 387–388
carbon nanotube (CNT) 261
carboxyl (–COOH) group 338
carboxylic acid 335
carboxymethylchitosan (CMCht) 462
carrageenan hydrogels 14
carrageenans 9, 17
cartilage 10
cartilage engineering
– autologous chondrocyte implant (ACI) technique 521
– effect of growth factor 528
– role of scaffolding materials 524–528
– tissue engineering approach to regeneration 517–518
– using autologous chondrocytes 521–524
– using autologous mesenchymal stem cells 520–521
– using bone marrow mesenchymal cells 519–520
cartilage repair using collagen 70
Cassie–Baxter model of surface wettability 240, 251
cationic dendrimers 464
cationic polyelectrolyte compounds 496
cation–pi interactions 124–125
cell adhesion 166–167
cell-adhesion resistant bovine serum albumin (BSA) 178

cell–cell binding domains 45–46
– by cadherins 46
– by Ig superfamily proteins 46
– by selectins 45–46
cell–ECM binding domains 43–45
– interactions with biomaterials 215
cell interfaces, biomimetics of
– cell responses to nanostructured materials 227–233
– dimensionality aspects 223–224
– interligand spacing on a surface 221–222
– ligand density 222–223
– linking systems 226–227
– nanofeatures, importance of 220–221
– substrate mechanical properties 223
– surface patterning technique 216–219
– variation of surface physical parameters at the nanoscale 219–224
cell-mediated FN fibrillogenesis 189–190
– by adsorption of FN 201
– by applying mechanical tension to FN 198
– by applying shear forces to an FN solution 200
– cell-free routes to induce 195–202
– high-resolution SEM images of 199
– importance of molecular unfolding 197
– surface-induced FN fibrillogenesis 201–202
cell-responsive hydrogels 41–42
cell-sheet-based tissue engineering 183
Cell-Tak™ 274, 282–283
cellular communication
– host responses 170
– integrin-mediated signaling 166–167
– intercellular 165–166
– intracellular 164–165
cellular fibronectin (cFN) 190
cement 279–280
Ceravital® 338
chelation 135
chemically cross-linked hydrogels 444–447
chemical sensing, using biomimetic receptive material 118
chemomechanics of hard biomaterials 297
chi/MMT/HAP composite 368
Chinese hamster ovarian (CHO) cells 44
chi/PgA scaffolds 369
chitin 296
– difference between chitosan and 6
– isolation method of 7
– scaffolds 15
β-chitin 8
chitin–chitosan scaffolds 16
β-chitin fibrils 317

chitin/nanosilica composite scaffolds   365
α-chitin/sol-gel-derived BGC nanoparticle   361
chitosan   4, 6–8, 321, 323, 343, 497
– commercial production methods   7–8
– deacetylation degree of   6
– difference between chitin and   6
– in drug delivery systems   17
– hydrogels   13
– scaffolds   15
chitosan-based membranes   13
chitosan (chi)/ MMT/hydroxyapatite (HAP) nanocomposite   367
chitosan-β-glycerophosphate salt formulation   363
chitosan—hydroxyapatite system   16
chitosan-NIPAAm   444
chitosan/sol-gel-derived BGC nanoparticle composite scaffolds   361
2-chloro-DOPA residues   278
chondrocytes   70
chondroitine-6 sulfate   65
chondroitin sulfate   11, 15
chondroitin sulfate (CS) polysaccharide   510
Chondrosia reniformis   10
Chondrus   9
CH-pi stacking interactions   138
Christmas tree-like compounds   459
α-chymotrypsin   250
cisplatin   409
clathrates   134
CMCht/PAMAM dendrimer nanoparticles (NPs)   462
coacervates   279, 284
collagenase target–PEG hydrogel   42
collagen/chitosan composite hydrogels   14
collagen hydrogels
– aligned   64
– from atelocollagen   63
– category II   63–65
– chemical cross-linking   62
– classical   59–61
– concentrated   61
– cross-linked   62–63
– dense matrices of   61–62
– enzymatic cross-linking   62–63
– extruded collagen fibers   65
– from plastic compression   61
collagen–mineral interactions, nature of   335
collagens   9–10, 190
– bone repair using   69
– cartilage repair using   70
– conduits   68
– fibrillogenesis   56–57

– fibrils   5
– from fish skins   10
– interactions of cells with   57–58
– intervertebral disk degeneration and   69–70
– multichannel conduits   66
– multi-channeled collagen–calcium phosphate scaffolds   66
– nerve repair using   68
– networks in connective tissues   57
– porcine products from   10
– skin repair using   66–67
– sponges   65–66
– structure   56
– tendon repair using   68
– tissues mimicked by   66–70
– *in vitro*   59
COL/NBG composites   362
complex coacervation   279
concatemerization   102
contact angle on a surface   162
contact splitting principle   261
cooperativity among receptors   128
core–shell polymerization   443
covalent-based imprinting   132
covalently cross-linked protein hydrogels   38–39
covalent self-assembly model biomaterials   320–322
cp-19k protein   283
cp-20k protein   283
cross-linked collagen sponges   67
crown ethers   134
cryo-high-resolution scanning electron microscopy (cryo-HRSEM)   196
cryptophanes   134
crystallization mushroom   387
CS5 domain   44
cyclotriveratrylenes   134
cylindriform silk   77

d

D-banded fibrils   65
deacetylation methods   7
Delite®HPS bentonite nanoclay   369
dendrimeric macromolecules   460
dendrimers   459–460
– biomimetic/bioinspired materials   460–461
– as biosensors   465
– drug delivery systems   461–463
– gene delivery systems   463–465
– technology for theranostics   466–467
dendriplexes   464

dendronized polymers 462
dermis 98, 537–538
Dex-loaded CMCht/PAMAM dendrimer NPs 463
2,4-dichlorophenoxyacetic acid 443
3,4-dihydroxyphenylalanine (DOPA) 277
– containing adhesive proteins 259, 268
– DOPA-modified block copolymers 287
– DOPA-modified PMMA-PMAA-PMMA membrane 287
– DOPA-modified poly(ethylene glycol)s (PEGs-DOPAs) 5
– DOPA-quinone cross-links 287
– DOPA-quinones 277
3-(4,5,dimethoxy-2-nitrophenyl)-2-butyl ester (DMNPB) 178
4,4-diphenylmethane diisocyanate (MDI) 369
dip-pen nanolithography 216, 233
direct-write lithographic technique 216
display techniques, evolutionary 146–150
β-D-mannuronic acid 8
DNA-modified nanoparticles 48
dopamine methacrylate (DMA) 286
doxorubicin 107, 438
dragline silk 77
drug delivery systems
– stimuli-responsive nanostructures and nanostructured networks 428–449
– stimuli-sensitive materials 419–428
drugs, adsorption and release using nanocrystals 397–402
DTT-inducedmultimerization process 195–196

e
E-cadherin 46
EDC/NHS-cross-linked BG-COL-HYA-PS composite scaffolds 361
E-isomer 181
elastin-like macromolecules 94
– applications as biomaterial 97–99
– characteristics of 95–96
– content of elastin in skin and ligament 94
– elastin gene 94–95
– elastin-like recombinamers 99–110
– elastogenesis and associated disorders 97
– nature-inspired biosynthetic elastins 99–103
– skin repairs 97–98
– in vascular constructs 98–99
elastin-like polymers (ELPs) 93
elastin-like recombinamers (ELRs) 93
– biocompatibility feature of 103
– biosynthesis of 100–101
– cross-linked matrices of 100
– in drug delivery systems 106–108
– ELR-based amphiphilic block copolymers 107
– ELR-REDV sequence 104–105
– from *Escherichia coli* strains 101
– formation of secondary structures 102–103
– general properties 99–100
– lysine-containing 108
– RGD-containing 109
– self-assembly properties of 109–110
– surface enginerring 108–110
– tissue-specific properties of 104–106
– triblock–ELR copolymer 107
– use of, modified with polyethyleneimine (PEI) or polyacrylic acid (PAA) 109
elastogenesis 97
electrical potential of a material 162
electrochemical off switching 179–181
emerging technologies in biomaterials
– electrochemical desorption methods 176–178
– electrochemical off switching 179–181
– oxidative release mechanism 177–178
– photoactivation method 178–179
– photobased desorption methods 178
– reversible photoactive switching 181–182
– reversible temperature-based switching 182–184
emphysema 97
Engelbreth-Holm-Swarm tumors (EHS-LAM) 198
engineering biomolecular materials
– peptides 146–150
– periplasmic binding protein superfamily 150
– phosphate-binding protein 150
– trinitrotoluene 150
– *in vitro* selection of RNA/DNA aptamers 144–146
enzyme field effect transistor (ENFET) 465
epidermis 97, 537–538
ERK/MAPK signaling pathway 232
Escherichia coli 79–80, 281, 283, 449
– *E. coli* signal peptide (OmpASP) 282
– *E. coli* S-30 *in vitro* system 148
– ELRs from 101
– phage library expression in 147–148
1-ethyl-3-(3-dimethylaminopropyl) carbodiimide (EDC) 361
ethylene-vinyl alcohol (EVOH) polymer 341
Euprosthenops australis dragline silk 83

expanded poly(tetrafluoroethylene) (ePTFE) 98
ex situ gelation 506
extracellular matrix (ECM) 93, 354, 473, 515
– assembly of FN in 190
– cell–material interactions 160
– function of 159
– interaction between components 159
– nanoscale ECM molecular organization 214
– secreted by bone-lining cells 213
– sensitivity of cells and 213
– structural elements within 160
– structure of 159
– surface functionalization for controlled presentation of 224–227
extruded collagen fibers 65, 68

f
F-fucoidan 12
fibrillin-1 97
fibrillogenesis 56–57
fibrillogenesis *in vitro* 59
fibrils with 3D contact shapes 262–263
fibroblast growth factor7 (FGF7) 98
fibroblast growth factor 10 (FGF10) 476
fibroins 3 and 4 (ADF-3 and ADF-4) 77
fibronectin (FN)
– amyloid fibril formation and 197
– C-terminus of FN molecule 192
– domains of assembly 192–194
– EIIIA and EIIIB modules 191
– enrichment 198
– FN–FN interaction on the PEA substrate 206
– FN–FN interactions 192, 195
– formation of FN network on PEA 203–206
– globular FN molecules 198
– heparin-induced FN precipitation 196
– 70 kDa amino-terminal domain of 192, 203
– potential integrin binding sites in 193
– reduction by dithiothreitol (DTT) 195
– and regulation of matrix assembly 194–195
– repeating modules of 191
– RGD motif of 192–193
– structure 190–192
– α_v-class-produced 193
flagelliform silk fiber 77
Floridean starch 9
Flory–Huggins solution theory 131
fluorescein isothiocyanate (FITC) 466
fluorophore-labeled glutamine binding proteins (QBPs) 108

focal adhesion (FA) 466
focal adhesion kinase (FAK) 232
fp-151-RGD 282–283
Fréchet-type dendrimers 459
freeze-drying–particle leaching method 15
freeze-drying technique 7
– of a collagen solution 65–66
fucoidan 12
– biologic properties 12
Fucus serratus 280–281

g
Ga-cross-linked alginate films 362
gallium-cross-linked alginate/nanoparticulate BG composite films 362
Gecko's adhesive pads 259
Gecko's dry adhesion mechanism 259–261
gelatin 10
gelatin/montmorillonite–chitosan (Gel/MMT–CS) nanocomposite scaffold 367
gelatin/NBG nanocomposite scaffolds 361
gelling property of carrageenans 9
Gigartina 9
glass–ceramic nanoparticles 315
glassy carbon electrode (GCE) surface 465
glycosaminoglycans (GAGs) 5, 191, 515
1–4 glycosidic bonds 8
G monomers 8
gold nanoparticle patterns on hydrogels 219
G-protein-coupled odorant receptors 118–119
G-protein-coupled receptors 144
grafting-from approaches 172
grafting of stimuli-responsive polymers 447–449
grafting-to approaches 172
green algae 11
green solvents 15

h
HA-loaded Pt complexes 402
HA–Mb interaction mechanism 406
hard biomaterials
– atomistic modeling framework 304–306
– bioengineering of 306–308
– biomimetics of 306–308
– bone 298
– chemomechanics of 297
– chemomechanics of the organic–inorganic interfaces 297
– general mechanical performance 295
– nacre 299
– propagation of crack 299

- role of interfaces 299–301
- tension shear chain model (TSC) 302–304
- toughness and strain-hardening behavior of 297

heavy-chain fibroin 76
hepatocarcinoma (HepG2) cells 478
hepcidin 81
hernias 97
heterocalixaromatics 140
heterogeneity of a material 162
hierarchical structures 265–266
high-throughput analysis 252–253
honeycomb worms 278
host–guest chemistry 123, 126, 135, 140
- case of lock and key binding 119–120
- induced fit mechanism 120–121
- pioneering works 119
- preexisting equilibrium model (PEEM) 121

HSA adsorption on needle-shaped (HAns) 405
human cervical carcinoma cells (HeLa) 402
human mesenchymal stem cells (hMSCs) 364
human periodontal ligament fibroblast (HPDLF) cells 362
human plasma fibronectin (HFN) 249–250, 252
human serum albumin (HSA) 249–250, 252
human umbilical vein endothelial cells (HUVECs) 44, 482
hyaline cartilage 515
hyaluronic acid (HA) 10–11, 283, 504
hyaluronic acid (HA) hydrogels 481
hybrid atelocollagen II/GAGs hydrogels 69
hydrogel engineering
- assembly of engineered hydrogel blocks 485–488
- biochemical cues 480–481
- cell–cell interactions 482–483
- combination of multiple cues 483–484
- gradient hydrogels 483–484
- mechanical cues 478–480
- PEGDA hydrogels at themicroscale 483
- structural cues 477–478

hydrogels 13–15, 173–174
- to activate cells 510
- application in biomedicine 508–511
- cell encapsulation in 510
- chemically cross-linked 444–447
- engineering strategies for development of 506–508
- from marine-based polymers 14

- natural vs synthetic 503–505
- pH-responsive 444
- poly(ethylene glycol) diacrylate (PEGDA)/laponite nanocomposite (NC) 371
- poly(ethylene glycol) (PEG) 477
- poly(N-tertbutylacrylamide-co-acrylamide/maleic acid) 445
- properties of 505–506
- role in in situ for specific applications 510
- surface wettability and substrate preparation 254
- temperature-responsive 444

hydrogen bonding 124
hydrophilic–lipophilic balance (HLB) 431
hydrophobic domains 161
hydrophobic interactions 125
hydroxyapatite (HAP) 296–299
hydroxyapatites (HAs)
- carbonate-substituted 384, 387–388
- ceramics 353, 388
- chemical conversion from wood to porous 389–390
- cocrystallization of 387
- nanocrystalline 387
- properties 383
- well-crystallized vs biomimetic 384

hydroxycarbonate apatite (HCA) layer 338
hydroxyethyl methacrylate (HEMA) cross-linked with ethyleneglycol dimethacrylate (EDMA) 507
hypodermis 537–538

i

immobilization binding sites 48
implantable drug delivery system (IDDS) 395
implant surfaces, self-assembling biomaterials on 345–348
imprint lithography 252
inert linkers 226
in-situ-forming scaffolds
- by electrostatic interaction 496–497
- by hydrophobic interactions 497–499
- immune response of 500
- for organ regeneration 501
- for preclinical regenerative medicine 500–501

in situ gelation 506
in situ self-assembly method 345–348
Integra 67
Integra®DPT 98
integrin clustering 167

integrin-mediated cell adhesion 166–169
– focal adhesion (FA) components 167
– and host responses 170
– integrin–FN adhesion 168–169
– model systems for controlling 170–171
intervertebral disk 69
intervertebral disk degeneration, repairing of 69–70
inverse temperature transition (ITT) 100
Invista Dacron® 98
in vivo peptide screening systems 148
isothermal titration calorimetry (ITC) 127

k

70 kDa amino-terminal region of FN 192, 203
keratinocytes 65, 97
Kurita process 7

l

lactodendrimers 462
lactoferrin 403
β-lactoglobulin 250
lamellar architecture 4
Langerhans cells 97
Langmuir–Blodgett (LB) films 345
layer-by-layer (LbL) assembly methodology 323–324
– of preformed polymers 436–438
layer-by-layer (LbL) technique 4
α-L-guluronic acid 8
linking systems for biomimetics 226–227
Lipofectamine™ 283
L-lactic acid 346
L-lysine dendron 346
loaded scaffolds containing bioactive molecules, development of 15
lower critical solution temperature (LCST) 100, 182
Lustrin A 307
lysozyme 250

m

macrocycles 135
macrocyclic host systems 134–135
macroporous monoliths 439
magnetic microbeads 488
major ampullate spidroin protein 1 and 2 (MaSp1 and MaSp2) and 77
Marfan's syndrome 97
marine-based tissue engineering approaches
– hydrogels 13–15
– tridimensional porous structures 15–17

– using materials in particulate form 17–18
– using membranes 12–13
marine biomimetic approach 4
marine-origin biopolymers
– alginate 8–9
– carrageenans 9
– chitosan 6–8
– chondroitin sulfate 11
– collagen 9–10
– fucoidan 12
– hyaluronic acid 10–11
– ulvan 11–12
marine-origin polyanions 13
marine sponge collagen particles 18
material bionics 315
material driven FN fibrillogenesis 202–210
– biological activity of 206–210
– physiological organization of fibronectin at the material interface 203–206
materials in particulate form 17–18
matrix metalloproteinase (MMP)-responsive PEG hydrogels 481, 509
matrix metalloproteinases (MMPs) 41
maturation process 386
MC3T3-E1 adhesion 250
MC3T3-E1 cells 363
Megabalanus rosa 276, 280
melanocytes 97
mesoporous silica nanoparticle (MSNP) 364
metal–ligand chelates 278
metalloporphyrins 134, 138–139
methicillin-resistant *Staphylococcus aureus* (MRSA) biofilms 448
methotrexate (MTX) 400, 466
methyl ether poly(ethylene glycol)-*b*-poly(â-amino ester) 430
methylmethacrylate (MMA) functional monomers 443
MG-63 cells 362
micelles 429–432
Michael-type addition reaction 488
microelectronics 252
microengineering geometries 477
microfabrication masking techniques 252
microfluidic-based assembly 485
microfluidics 252
micropatterning 109
microscopic polymer-rich droplets 438
Millepora dichotoma 17
mineralized collagen fibrils (MCFs) 298
mineralized natural composites, characteristics of 313–314
MMP-sensitive-PEG hydrogel 42

modular tissue engineering   485
molecular-imprinting-based biomimetic sensors
– brief history   129
– inherent problem with MIP   133
– molecularly imprinted polymer (MIP)   129
– polymer matrix design   130–131
– practical use of MIPs   133
– removal of template from MIPs   132–133
– strategies for forming MIP   129–132
– use of metal complexes and covalent imprinting techniques   133
molecularly imprinted polymer (MIP)   129
– inherent problem with   133
– polymer matrix design   130–131
– polyurethane-based   132
– practical use of   133
– removal of template from   132–133
– strategies for forming   129–132
– use of metal complexes and covalent imprinting techniques   133
molecular recognition events
– biomimetic surfaces for   121–123
– covalent template–monomer complex   130
– formation of template–monomer complexes   131
– foundation of   123
– Gibbs free energy and   127
– host–guest interactions and   128
– kinetic rates of association and dissociation of binding constant   127
– merits   127–128
– noncovalent interactions   123–125
– protein–protein interactions and   128
– self-assembly approach for forming   129–131
– strengths of representative bonds and interactions in   125
– thermodynamics of   125–127
molecular self-assembly method   345–348
mollusk shell   314
mollusk shells   4
monoacryloxyethyl phosphate (MAEP)   286
montmorillonite   322
montmorillonite MMT   364
mouse calvarial preosteoblastic MC3T3-E1   347
MPEG-PAE micelles   431
mRNA display   149
MSC-laden alginate microspheres   485
multi-channeled collagen–calcium scaffolds   66
multiresponsive drug delivery system, prototype   445–447

muscle   5
mussel adhesive proteins (MAPs)   5, 275–278
mussels   274–278
mutable collagenous tissue (MCT)   5
myoglobin (Mb)   403
Mytilus   275–276

n

Na/Ca montmorillonite   326
nacre   4–5
– calcium phosphate layer of   315
– characteristics of   315
– covalent self-assembly model biomaterials   320–322
– fracture toughness of   318
– hierarchical structural design of   297
– inelastic deformation attributed to   319
– mechanical properties of   317–320
– nacreous aragonite layer (N) of   316
– nanolaminates based on   320
– organic fraction of   318
– organization of aragonite tablets and organic material   300
– prismatic calcite layer (P) of   316
– role of the organic fraction   319
– structure of   316–318
– toughness of   299, 318–320
nacre-inspired biomaterials, techniques for
– bottom-up approach   327–328
– electrophoretic deposition in nacrelike composite films   322–323
– freeze-casting   326
– layer-by-layer (LbL) assembly methodology   323–324, 328
– polymer-induced liquid precursor layer formation   328
– spin-coating LbL assembly methodology   324
– template inhibition technique   325
nacrelike chitosan–montmorillonite bionanocomposite films   322
N-(3-aminopropyl) methacrylamide hydrochloride (APMA)   286, 448
nanocarriers   466
nanoclays   365–372
nanocomposite morphology   321–322
nanocomposite systems   354
nanocontact printing   216–217
nanoimprint lithography   217
nanoindentation of collagen fibrils   213
nanoparticulate BG/PLLA composites   359
nanoscale bioactive glasses (NBGs)   356–363
nanoscaled silica   363–365

nanostructured materials, cell responses to
- cell adhesion and migration   228–230
- cell–cell communication   230–231
natural vs synthetic biomaterials   94
Nb–OH   338
Nephila clavipes   77, 81–83
nerve growth factor (NGF)   47
nerve guidance conduits   68
nerve repair using collagen   68
Neuragen®   68
NeuraGen™   66
neutrophil defensins (2 HNP-2 and 4 HNP-4)   81
N-hydroxysuccinimide (NHS)   361
nickel-nitrilotriacetic acid (Ni-NTA)   226
NIH-3T3 fibroblasts   483
1-(2-nitrophenyl)ethyl-5-trichloro silylpentanoate (NPE-TCSP)   178
nitroveratryloxycarbonyl (NVOC)-protected hydroquinone-terminated SAM   178
nonapatitic crystallographic sites   385
nonconventional nanolithography   216–217
noncovalent interactions   123, 161
nonsteroidal anti-inflammatory drugs (NSAIDs)   449, 463
NP degeneration   69

o

oligo(ethylene glycol)methacrylate (OEGMA)   172
oligoethylene-glycol (OEG)-terminated thiol SAMs   177
oligomerization   102
oligonucleotides   79, 143–144
omeprazole-imprinted nanoparticles-on-microspheres   441
OrCel®   67
organically modified montmorillonite (OMMT)   364
O-silyl hydroquinone   179
osteoblast (bone-forming cell) function   354
osteoblastic cells (MC3T3-E1)   359–360
osteoblastlike cells (MG-63)   361
osteocalcin   86
osteocalcin (OC)   4
osteoconductive glass ceramic nanoparticles   323
osteopontin   214
oxidative release mechanism   177–178

p

packing-based mechanism   485
paratope   118

partially acetylated PAMAM dendrimers (G5)   466
Pc-1 and Pc-2 proteins   279
PCL/sol-gel-derived BGNF nanocomposites   359
pectin/chitosan microcapsules   17
PEG-*co*-polycaprolactone micelles   432
PEG diacrylate   287
PEGylated poly(cyanoacrylate) NPs   466
PEGylated protein network   509
pentapeptide poly(VPGVG)   99
peptoid dendrimers   464
perfluoro-15-crown-5-ether (PFCE) nanoemulsions   432
peripheral nerve repair   68
phage display   147
P25/heavy-chain/light-chain complex of *Bombyx mori* silk   77
photoactivation method   178–179
photobased desorption methods   178
photolithography   252, 264
Phragmatopoma californica   276, 278
pH-responsive microgels   442
phtalocyanine core (photosensitizer)   464
Phymatolithon calcareum   17
physiological-like FN network   202
Pichia pastoris   10
pillar fabrication   266
Pinctada imbricata   318
Pinctada maxima   4
pi–pi stacking   124
PLA-*b*-(PNIPAAm-*co*-DMAAm)   431
PLA/clay nanocomposite (NC)   369
plasma fibronectin (pFN)   190
plasmin target–PEG hydrogel   42
plastic antibody   131
platelet-derived growth factor (PDGF)   47
plate-shaped (HAps) HA nanocrystals   405
PLGA-PEG-PLGA triblock copolymer   499
PLLA/BG composites   343–344
PLL/HA nanofilms   233
Pluronics™   499
PNIPAAm-*co*-methacrylic acid (MAA) hydrogelmembrane   428
Poloxamers™   499
poly(2-acrylamido-2-methyl-1-propane-sulfonic acid) (PAMPS)   428
poly(acrylic acid)   327
poly(amidoamine) (PAMAM) dendrimers   433, 459
polyanionic biopolymer   321
poly(APGVG)   101
polyaromatic macrocycles   137

poly(ε-caprolactone-co-1,4-dioxan-2-one) (PCL-co-PDO) 499
poly(ε-caprolactone-co-d,L-lactic acid) (PCLA) 499
poly(ε-caprolactone-co-trimethylene carbonate) (PCL-co-PTMC) 499
polycaprolactone (PCL) 499, 537, 545
polycaprolactone polyol (CAPA 231), 369
polycationic biopolymer 321
poly (D, L-lactide-co-glycolide) (PLGA) 99, 537, 545
poly(2-diethylaminoethyl-methyl methacrylate)-poly(ethylene oxide)-poly(propylene oxide)-poly(ethylene oxide)-poly(2-diethylaminoethyl-methyl methacrylate) (PDEAEM25-PEO100-PPO65-PEO100-PDEAEM25) 432
poly[(3,4-dihydroxystyrene)-co-styrene] 286
poly(2-(dimethylamine)ethyl methacrylate)-b-poly(ethylene oxide) 430
poly[2-(dimethylamino)ethyl methacrylate]-b-poly[methyl methacrylate] (PDMA-b-PMMA) 444
polydimethylsiloxane (PDMS) 202, 484
polydopamine films 287
poly(ethyl acrylate) (PEA) 203
poly(ethylene glycol)-b-poly(l-histidine) 430
poly(ethylene glycol)-b-poly[(3-morpholinopropyl) aspartamide]-b-poly(l-lysine) (PEG-b-PMPA-b-PLL) 431
poly(ethylene glycol) diacrylate (PEGDA)/laponite nanocomposite (NC) hydrogels 371
poly(ethylene glycol) (PEG)-DOPA 286
poly(ethylene glycol) (PEG) groups 171, 227, 477, 497, 545
– diacrylates 219
poly(ethylene oxide) (PEO) 499
poly(ethyleneoxide)-poly(propyleneoxide)-poly(ethyleneoxide) (PEO-PPO-PEO) block copolymers 287
polyethylene terephthalate (PET) 341
poly(glycolic acid) (PGA) 353
poly(Gly-Lys) 285
poly(Gly-Tyr) 285
polyHEMA hydrogels 507–508
poly(2-hydroxyethyl methacrylate) (pHEMA) 480
poly(lactic acid) (PLA) 353
poly(lactic-co-glycolic acid) (PLGA) 537
poly(L-lactic acid) (PLLA) 323, 545
– fibers 480
poly(L-lysine) (PLL) 283, 481

polymer brush systems 172–173
polymeric particles
– from monomers 439–444
– from preformed polymers 438–439
polymer-induced liquid-precursor (PILP) 336
polymer/layered silicate nanocomposites 366
polymer nanocomposites 354
polymer poly(N-isopropylacrylamide) (PIPAAm) 182
polymer/silica nanocomposites 364
polymer/SiO$_2$ nanocomposites 364
polymersomes 429–432
poly(meth)acrylate polymers 286
poly(methacrylic acid) (PMAA) 287
poly(methyl acrylate) (PMA) 203
poly(methyl methacrylate) (PMMA) 220, 245
poly(NIPAMM) surfaces 182
poly(N-isopropylacrylamide-co-acrylamide) 442
poly(N-isopropylacrylamide-co-acrylic acid) bearing doxorubicin 444
poly(N-isopropylacrylamide) (PIPAAm) 202, 245
poly(N-isopropylacrylamide) (PNIPAAm) 343–345
poly(N-methacryloyl-L-phenylalanine) 439
poly(N-methacryloyl-L-valine) 439
poly(N-tertbutylacrylamide-co-acrylamide/ maleic acid) hydrogels 445
poly(N-vinylpyrrolidone)-b-poly(2-acrylamido-2-methyl-1-propanesulfonic acid) 430
poly(N-vinylpyrrolidone)-b-poly(N-isopropylacrylamide) (PVPON-b-PNIPAM) 437
poly (N-vinylpyrrolidone)-b-poly(N, N-dimethylaminoethyl methacrylate) 430
polypeptides 141–142, 160
poly(propylene fumarate) (PPF) 499
poly(propyleneimine) (PPI) dendrimers 459
poly(propylene oxide) (PPO) block copolymers 499
poly-[(R)-3-hydroxybutyrate] (PHB) 499
polysaccharides 11, 13, 16, 360
poly(sodium styrene sulfonate) 437
polystyrene-block-poly(2-vinylpyridine) (PS-b-P2VP) 218
poly(tetrafluoroethylene) 287
poly(trimethylene carbonate) (PTMC) 499
poly(Tyr-Lys) 285

polyurethane-based MIPs 132
polyurethane (PU) 369, 537, 545
poly(δ-valerolactone) (PVL) 499
poly(vinyl acetate) 245
polyvinylidene difluoride (PVDF) membrane 108
poly(vinylidene fluoride) 245
poly(2-vinylpyridine)-*b*-poly(ethylene oxide) 430
poly-(4-vinylpyridine hydrochloride) 437
poly(VPGVG) concept matrix 104, 107
poly(X-Tyr-Lys) 285
porcine products 10
porous apatitic biomimetic scaffolds 388
potassium permanganate/nitric acid (KMnO$_4$)/(NHO$_3$) system 341
precipitation polymerization 442
preorganization of host structures 132, 135
proline (P) 76
propranolol 440
protein adsorption phenomenon
– basics 161–162
– kinetics of 162–163
– packing density and 163
– protein orientation *vs* activity 163
protein-engineered hydrogels
– cellular behavior for therapeutic applications 25–26
– covalently cross-linked 38–39
– design of 28–30
– development of biomimetic 41–48
– production of 30–32
– protein-graft-PEG hydrogel 42
– self-assembled 32–38
– structural diversity and applications of 32–39
– for tissue engineering applications 41–48
protein engineering, principles of 26–28
proteins, adsorption and release using nanocrystals 402–406
proteins of marine origin as scaffolds 16
protein structure
– in functionalizing nanopatterned surfaces 224–226
– primary 160
– quaternary 161
– secondary 160
– tertiary 160
PS superhydrophilic surface 250
P2VP-*b*-PEO 430
pyriform silk 77
pyrophosphate 406

r

recombinant DNA technology 101
recombinant hybrid adhesive proteins 282
recombinant mussel adhesive proteins 281–282
recursive directional ligation (RDL) 102
repetitive polypeptides, biosynthesis of 101
responsive adhesion patterns 265–268
reversible photoactive switching 181–182
reversible temperature-based switching 182–184
RGD ligands 43, 45
RGD oligopeptide 180
RhBMP-2m 403
Rho-kinase 167, 194
ribosome display 148
risedronate 408
robotic DNA spotter 252
R-propranolol 440
Runx2/Cbfa1 482

s

Saccharomyces cerevisiae 281
sandcastle worms 278
45S5 BG 356
scaffolds for tissue engineering 15–16
scanning probe block copolymer lithography 219
sea cucumbers 5
sea urchins 5
SELEX screening 146
self-assembled lipid layers 227
self-assembled monolayers (SAMs) 171–172, 227
– advantage of 171
– medical applicarions 172–173
– mixed 171
– nonfouling 171
– oligo(ethylene) glycol coating of 227
– oligoethylene-glycol (OEG)-terminated thiol 177
– silane-grafted 178
self-assembled peptide amphiphile (PA) matrix 347
self-assembled polymers 429–432
self-assembled protein hydrogels 32–38
self-cleaning effect 237
shape-memory polymer 266, 268
shape theory 119
shared-electron number (SEN) method 136
shear-dependent fibrillogenesis 200
S-ibuprofen 440
silica nanoparticles (SNPs) 363–364
silk-based biomaterials

- biomedical applications 84–87
- *Bombyx mori* silk 75–77
- in the field of vascular tissue engineering 85–86
- mechanical properties 82–84
- non-load-bearing applications 83
- recombinant silk 79–82
- silk chimeric proteins with potential application in the biomedical field 81
- silk proteins expressed in *E. coli* and yeast recombinant systems 80
- spider silk 75, 77–79
- in tissue constructs for endothelial keratoplasty 84
- in the treatments for peripheral nerve injuries 86
silk–elastinlike polypeptide (SELPs) 48
"similis simili gaudet" principle 125
Si_3N_4/BN composites 326
single-stranded DNA (ssDNA) 143
Si–OH 338
SiO_2/polyacrylic ester (PAE) multilayer composite 324
SIS® 55
S-ketoprofen 440
skin, structure of 537–538
skin repair, using collagen 66–67
skin repair, using elastin 97–98
skin tissue engineering (TE) construct materials 537
- biomimetic approaches in 546–549
- skin substitutes, currently available 549
- in wound healing 538–545
slow-degrading hydrogel 42
smart mineralizing surfaces 343–345
smart surfaces 108
soft lithography 216, 252
sol-gel-derived bioactive glass nanofiber (BGNF) 359, 361
sol–gel silica 323
solvent casting 12
solvent casting–particle leaching method 15
spider silk 75, 77–79, 245
- amino-terminal sequence 78
- carboxy-terminal region 78
- crystalline domains 78
- formation of 77
- formation of helical structures 78
- hydrophobic poly-A and GA blocks 78
- protein expression 77
- recombinant spider silk 4RepCT 84
- spinning process of 83
- toughness of major 82
spin-coating LbL assembly methodology 324

β-spiral conformation 99, 103
4S-StarPEG (four-arm polyethylene glycolsuccinimidyl glutarate pentaerythritol) 64, 70
starch/cellulose acetate blends (SCA) 341
starch/ethylene-vinyl alcohol (SEVA-C) 341
starch/polycaprolactone blends (SPCL) 341
stem-cell-based applications in tissue engineering 18
stem cell biology 232–233
stimuli-responsive nanostructures and nanostructured networks 428–449
- chemically cross-linked hydrogels 444–447
- dendrimers 433–436
- grafting of stimuli-responsive polymers 447–449
- layer-by-layer (LbL) techniques of preformed polymers 436–438
- polymeric particles from monomers 439–444
- polymeric particles from preformed polymers 438–439
- self-assembled polymers 429–432
stimuli-responsive surfaces 343
stimuli-sensitive materials
- autonomous responsiveness 428
- electrical field 425–426
- glutathione (GSH) 420
- light 423–425
- magnetic field 426–427
- molecule-responsive and imprinted systems 420–422
- pH gradients 419–420
- temperature of the body 422–423
- ultrasounds 427
streptavidin–biotin complex 126
Strombus gigas 4
sulfonic ($-SO_3H$) group 338
superfibronectin 197
superhydrophilic TiO_2 nanotube layers 251
superhydrophobic surfaces 237–238
- anisotropic superhydrophobic surfaces 244–245
- blood platelet adhesion on 251
- bovine serum albumin (BSA) adsorption on 250
- fabricated using nature 241–245
- inspired by lotus leaf 241–243
- inspired by the legs of the water strider 243–244
- silicone nanofilaments 250
- substrate preparation and surface wettability 254

superhydrophobic surfaces (*contd.*)
– superhydrophobic poly(l-lactic acid) (PLLA) 241
– superhydrophobic PTFE 251
superhydrophobic Teflon AF substrates 253
supramolecular chemistry 121
– chelate effect 135
– effective supramolecular receptors for biomimetic sensing 137–140
– macrocyclic effect 134–135
– modeling methods 136
– preorganization of host structures 135
– rational design of structural properties 136
– templating effect 136–137
supravalvular aortic stenosis (SVAS) 97
surface-active ceramic 338
surface bioreactivity of nanoparticles 356
surface charge 161
surface composition of a material 162
surface hydrophobicity 161
surface-induced FN fibrillogenesis 201–202
surface-initiated atom-transfer radical polymerization (SI-ATRP) 172
surface-modified BG/PLLA composites 359
surface nanotopography 220–221
surface wettability, application in biomedical field
– amino-ended surfaces 250
– blood interactions with surfaces 251
– carboxyl-ended surfaces 250
– cell interactions 246–249
– high-throughput screening 252–253
– protein interactions 249–251
– substrates for preparing hydrogel and polymeric particles 254
surface wettability, theory of
– Cassie–Baxter model 240, 251
– transition between the Cassie–Baxter and Wenzel Models 240–241
– Wenzel's model 240
– Young's model 239–240
suspension polymerization 441
synthetic hydroxyapatites (HAs) 383
synthetic *vs* natural biomaterials 94

t

T-antigen-linked-glyco PAMAM dendrimers 466
Ta–OH 338
TC–HAP biomaterials 304–306
template inhibition technique 325, 487
templating 252
tendon-derived fibroblasts 64
tendon repair using collagen 68
tetraethylene glycol (TEG) spacer 180
"thermodynamically reversible addressing of proteins" (TRAPs) technology 108
thermosensitive shape-memory polymers 266
thioredoxin (Trx) protein 47
three-dimensional scaffolds 16
threonine (T) 76
Ti–6Al–4V alloys 390
tilted fibrillar structures 263–264
Ti–OH 338
tissue engineered scaffolds for disk regeneration 69
tissue engineering 4
tissue engineering and regenerative medicine (TERM) 93, 106
tissue/organ heterogeneity *in vivo* 474–477
tissue plasminogen activator (tPA) 42
tissue printing 486
tissues mimicked by collagen biomaterials
– bone 69
– cartilage 70
– intervertebral disk 69–70
– nerves 68
– skin 66–67
– tendons 68
tobacco 80
toluene-2,4-di-isocyanate (TDI) 371
trabeculae 334
tree frogs' attachment pads 260, 269
tree trunk 3
triblock–ELR copolymer 107
β-tricalcium phosphate (β-TCP) 409
tridimensional porous structures 15–17
Tris–HCl buffer 10
tropocollagen (TC) 296–299
tryptophan (W) 76
tube worms 278–279
two-phase systems 487
tyrosine-isoleucine-glycine-serine-arginine (YIGSR) oligopeptide 174

u

U-fucoidan 12
Ulva 11
ulvan 11–12
urokinase plasminogen activator (uPA) 42

v

valve endothelial cells (VECs) 477
valve interstitial cells (VICs) 477
vancomycin 448
Verongida sponges 15

4-vinylpyridine (VPD)   440
Vroman effect   163–164, 170

w

water-in-oil (W/O) heterogeneous gelation   439
Watson–Crick base pairing   140
weight-bearing trabecular systems   335
Wenzel's model of surface wettability   240
Williams syndrome   97
wound dressings and their properties   545–546
wound healing biomaterials
– alginate   539–540
– cellulose   540
– chitin/chitosan   541–543
– collagen   67, 544–545
– elastin   97–98
– fibrin   545
– gelatin   544
– hyaluronic acid or hyaluronan (HA)   543–544
– polymeric membranes   12–13
– silk   544
– soy protein   544–545
– synthetic polymers   545

y

Young's model of surface wettability   239–240

z

Z-isomer   181
zoledronate   400, 408–409
ZrO_2-Al_2O_3 318
Zr–OH   338